S0-AKY-170

NEW TRENDS IN PHYSICS AND PHYSICAL CHEMISTRY OF POLYMERS

NEW TRENDS IN PHYSICS AND PHYSICAL CHEMISTRY OF POLYMERS

Edited by

Lieng-Huang Lee
Xerox Corporation
Webster, New York

PLENUM PRESS • NEW YORK AND LONDON

Library of Congress Cataloging in Publication Data

International Symposium on New Trends in Physics and Physical Chemistry of Polymers (1988: Toronto, Ont.)
New trends in physics and physical chemistry of polymers / edited by Lieng-Huang Lee.
p. cm.
"Proceedings of an International Symposium on New Trends in Physics and Physical Chemistry of Polymers, sponsored by the Polymer Division of the American Chemical Society, held June 6–10, 1988, in Toronto, Canada"—T.p. verso.
Includes bibliographical references.
ISBN 0-306-43383-4
1. Polymers—Congresses. I. Lee, Lieng-Huang, 1924- . II. American Chemical Society. Division of Polymer Chemistry. III. Title.
QD380.I582 1988 89-22965
547.7—dc20 CIP

Proceedings of an International Symposium on New Trends in Physics and Physical Chemistry of Polymers, sponsored by the Polymer Division of the American Chemical Society, held June 6–10, 1988, in Toronto, Canada

A Division of Plenum Publishing Corporation
233 Spring Street, New York, N.Y. 10013

Printed in the United States of America

PREFACE

Between June 6-10, 1988, the Third Chemical Congress of North America was held at the Toronto Convention Center. At this rare gathering, fifteen thousand scientists attended various symposia. In one of the symposia, Professor Pierre-Gilles de Gennes of Collège de France was honored as the 1988 recipient of the American Chemical Society Polymer Chemistry Award, sponsored by Mobil Chemical Corporation. For Professor de Gennes, this international setting could not be more fitting. For years, he has been a friend and a lecturer to the world scientific community. Thus, for this special occasion, his friends came to recount many of his achievements or report new research findings mostly derived from his theories or stimulated by his thoughts.

In this volume of Proceedings, titled New Trends in Physics and Physical Chemistry of Polymers, we are glad to present the revised papers for the Symposium and some contributed after the Symposium. In addition, we intend to include most of the lively discussions that took place during the conference.

This volume contains a total of thirty-six papers divided into six parts, primarily according to the nature of the subject matter:

- Adsorption of Colloids and Polymers.
- Adhesion, Fractal and Wetting of Polymers.
- Dynamics and Characterization of Polymer Solutions.
- Diffusion and Interdiffusion of Polymers.
- Entanglement and Reptation of Polymer Melts and Networks.
- Phase Transitions and Gel Electrophoresis.

Since these six parts in no way can cover the breadth of Prof. de Gennes' contributions, we chose to list his entire publications in the appendix for our readers to look up either in English or in French.

As the Chairman of the Symposium, I would like to take this opportunity to thank Professors P. Pincus, H. Yu and A. Silberberg for co-chairing several sessions, and all speakers for their cooperation, especially, those who traveled long distances to get to Toronto.

We would like to acknowledge the financial support provided by the Petroleum Research Fund of the American Chemical Society, E.I. duPont de Nemours Company, Eastman Kodak Company and 3M Company. The able assistance of the Treasurer of the Division of Polymer Chemistry, Professor R.M. Ottenbrite is sincerely appreciated.

As the editor, I wish to thank Ms. Dee Costenoble of our Webster Research Center for her patience in typing the entire volume and Mr. F.G. Belli and his group of the Technical Information Center for helping prepare the indexes. I sincerely appreciate Drs. T. Kavassalis and J. Noolandi of Xerox Research Center of Canada for providing one of their illustrations to be the cover of this book.

L.H. Lee

September, 1989

Prof. A. Silberberg, Prof. P. G. de Gennes and Dr. L. H. Lee

CONTENTS

PART ONE: ADSORPTION OF COLLOIDS AND POLYMERS

PART TWO: ADHESION, FRACTAL AND WETTING OF POLYMERS

PART THREE: DYNAMICS AND CHARACTERIZATION OF POLYMER SOLUTIONS

PART ONE:

ADSORPTION OF COLLOIDS AND POLYMERS

PIERRE-GILLES DE GENNES

PIERRE-GILLES DE GENNES–1988 RECIPIENT OF THE ACS AWARD IN POLYMER CHEMISTRY

P. Pincus

Materials Department
University of California
Santa Barbara, California 93106

This year's recipient of the ACS Award in Polymer Chemistry is Pierre-Gilles de Gennes of the college de France and the Ecole Superieure de Physique et Chimie Industrielle de Paris. We, who are here in Toronto attending this symposium in his honor, are well aware of his many seminal contributions to polymer science including: the introduction of the n = 0 analogy to magnetic critical phenomena as a tool to describe chain conformations in good solvents; the "reptation" concept (for the slithering motion of a polymer coil trapped in an entangled network) which has led to molecular theories of the constitutive relations of polymer rheology; the major impetus for the description of the static properties of polymers at interfaces and, with his own contribution to this Symposium, a tentative theory for the dynamics of adsorbed layers; the prediction of the "polyelectrolyte peak" in the scattering from semidilute charged polymers solutions. I am cognizant of the fact that de Gennes' ideas in our field are so numerous that each among you might well have chosen a rather different selection to cite. Of course, many of his predictions have been and some remain controversial. Indeed, his creative ingenuity coupled with a propensity for making bold speculations and then suggesting the crucial experimental tests is the root of his scientific stature. (It is important to note here, however, that de Gennes does not become so emotionally attached to his own suggestions that he defends them beyond reason; he willingly owns up to his own mistakes.) Thus at any given time, scores of scientists are experimentally probing and theoretically massaging even his notes that are hidden away in the Comptes Rendus.

But, much of this assembly may be rather less familiar with de Gennes' equally central contributions to other areas of condensed matter physics. Allow me to cite a few examples from disparate fields: his PhD dissertation laid the foundation for the understanding of critical scattering of neutrons by spin wave fluctuations in magnetic materials; he extended the BCS and GLAG theories of superconductivity to describe elementary excitations in gapless superconductors, surface superconductivity, proximity effects, etc.; he developed the Landau-de Gennes theory for the static and dynamic properties of liquid crystals including spatial inhomogeneities, smectic phases, and defects; his work on fluctuations of curvature energy dominated interfaces appears to be the key concept required for the description of middle phase microemulsions; he has set down the basic microscopic principles associated with the wetting of solids by fluids. A significant fraction of de Gennes' works and excellent examples of his approach to science appear in his three monographs on the physics of superconductivity, liquid crystals, and polymer solutions. Over the years, many recognitions of his accomplishments have been noted including his award of the Gold Medal of the CNRS, 1980 (France's highest award for scientific achievement) and election as a Fellow of the Royal Society.

While these awards and accomplishments are impressive indeed, de Gennes' uniqueness resides as much with his scientific style. His approach to problems is always to begin by abstraction in order to isolate the underlying physical basis for the effect under investigation. Then, rather than solving for the most general case, which is often cumbersome, he considers the various limiting situations which generally yields a complete picture throughout the relevant parameter space. An important consideration is to attack the problem with the minimum arsenal of formalism - if the results can be obtained by using Newton's Second Law, that's what is done! This is not to say the de Gennes only does "back of the envelope calculations." He is well versed in all the tools of modern theoretical physics and used them as necessary - for example, his use of critical phenomena and field theory concepts in gelation and percolation, renormalization group study of polyelectrolytes, etc. Furthermore, when his results depend upon imperfectly based assumption or subtle analyses, de Gennes insists upon arriving at his answer by substantially different channels. Only then does his level of confidence in the result become more secure.

Another essential characteristic of the de Gennes style is boldness. He is willing and eager to leap-frog over the straightforward evolutionary development of a subject to come to grips with important, often complex, issues. He searches to identify the central physics and then, with reasonable assumptions and speculations, makes contact with the existing empirical observations and, most importantly, makes specific predictions which are feasible for experimental test. Two specific examples are: the prediction of M^{-2} self-diffusion in polymer melts on the basis of the reptation concept; the prediction that the elastic modulus of a gel in the vicinity of the gelation transition is related to the electrical conductivity of a percolating resistor network.

The de Gennes boldness manifest itself in another form through his joy in confronting unusual physical situations. Let me briefly cite two examples where I had some personal involvement. About twenty years ago, there was reported an observation (later proven erroneous) that a particular molten transition metal alloy was ferromagnetic. Immediately, de Gennes remarked that a) there was no obvious fundamental principles precluding the existence of liquid ferromagnets and b) this was potentially very exciting if the liquid was a typical isotropic fluid. Such a system would then be ideal in the sense that there would be no anisotropy energy. Then, the magnetic healing length characterizing the thickness of a Bloch wall would diverge leading to a variety of unusual magnetic behaviors. The de Gennes approach was not to "pooh-pooh" the observations out of hand as many would, but think about their implications. A second example occurred in the seventies while I was on sabbatical leave in Paris. One day after a seminar in the College de France, a visitor from the US told us about rumors of the discovery at du Pont of nematic solutions of flexible polymers. With this information alone, de Gennes began to speculate about how flexible coils could exhibit orientational order--an apparent paradox. Within a few days he came up with the concept of induced rigidity associated with incipient helix-coil transitions. While this idea may not be the dominant mechanism for Kevlar-like systems, it surely plays a role for polypeptide solutions. Again, we have some realizations of the creative mind at work.

In these paragraphs I have tried to provide an impression of my own observations of de Gennes' unique approach to science. I should add that over the years he has demonstrated remarkable taste in discovering those fields for which the time is ripe for development. I cannot fail to include my own pleasure over the last thirty years of collaboration with Pierre-Gilles. Not only has he been a wonderful colleague but a marvelous friend!

Award Address

DYNAMICS OF ADSORBED POLYMERS

P.G. de Gennes

Collège de France
11, Place Marcelin Berthelot
75231 Paris Cedex 05
France

ABSTRACT

When a polymer coil approaches an adsorbed layer (made with the same polymer), two successive processes take place: a) entry, where the coil "reptates" in a hostile environment of other chains, up to the moment where it establishes a first direct contact with the adsorbing wall. b) spreading, where the chain crawls in the vicinity of the adsorbing wall, and ultimately reaches its large final number of wall contacts. The spreading process has recently been detected in pulsed experiments of Cohen-Stuart and Tamai, where the hydrodynamic thickness of the layer is monitored after chain addition. We present tentative scaling laws for both entry and spreading.

I. INTRODUCTION

Adsorbed polymers play an important role in colloid protection.[1] When two grains, each covered with a "corona" of polymer coils, are nearly in contact, the coronas repel each other (We always assume that the polymers are in good solvent conditions). Then the grains also repel each other and do not flocculate. However, if some chains manage to establish bridges between the two grains, a certain form of flocculation occurs. An important (remote) goal is to understand the dynamics of this bridging process.

The present paper has more modest aims: we want to discuss the basic kinetic processes taking place in a single adsorbed layer. The simplest one, conceptually, is the crawling motion of one labeled chain inside the layer, leading to a certain diffusion coefficient $D_{\|}$. This is reviewed in Section II, following Ref. 2. Then, in Section III we analyze the entry of a chain in the adsorbed layer (which creates a repulsive cloud): this entry process has a striking similarity with quantum mechanical tunneling through a repulsive barrier, as shown in Ref. 3.

The central result of Ref. 3 is that the barrier effects are rather weak, thus explaining why finite rates are observed in kinetic experiments where an adsorbed layer exchanges molecules with a solution.[4,5]

We define entry as the moment when a chain, coming from the bulk solution, has just achieved one first contact with the adsorbing wall (Fig. 1). After entry, there is another step, where the chain spreads along the surface, and ultimately makes all the required numbers of contacts. A recent experiment of the Wageningen group probes an unsaturated layer of polyvinyl pyrrolidone on glass. The layer is exposed for a short time to a solution of other chains. The hydrodynamic thickness e_H jumps rapidly to a higher value, and then decreases after exposure, during times of order 5 minutes. Following Ref. 6, our (tentative) interpretation of this behavior is that at the moment of entry, the ingoing chains are weakly bound and protrude outside. But during the spreading process, they are sucked inwards and e_H is reduced. A crude theoretical discussion of the spreading time is also given in Section III.

All our discussion is based on the assumption that an adsorbed layer tends to reach an equilibrium state. For fully saturated layers, this equilibrium is associated with a remarkable self-similar structure[7] (Fig. 2). Some evidence for self-similarity has accumulated during the recent years via measurements of hydrodynamic thickness, ellipsometry and neutron scattering.[8]

For a saturated layer in usual conditions (strong adsorption), the "self-similar grid" is valid for distances z to the wall which extends from the monomer size a up to the coil size $R_F \simeq N^{3/5}a$ (N = number of monomers per chain). For starved layers with a coverage (number of monomers/cm^2)

$$\Gamma = \Gamma_o(1-x) \,, \qquad (1 >> x > N^{-1/5}), \tag{1}$$

one expects that the maximum size ℓ is reduced: a detailed calculation of this process has been given recently by Pincus and Rossi.[9] The scaling relation between ℓ and x, derived in Ref. 5, is

$$\ell = a x^{-3}, \qquad (1 >> x > N^{-1/5}). \tag{2}$$

(A brief derivation of Eq.(2) will be given below).

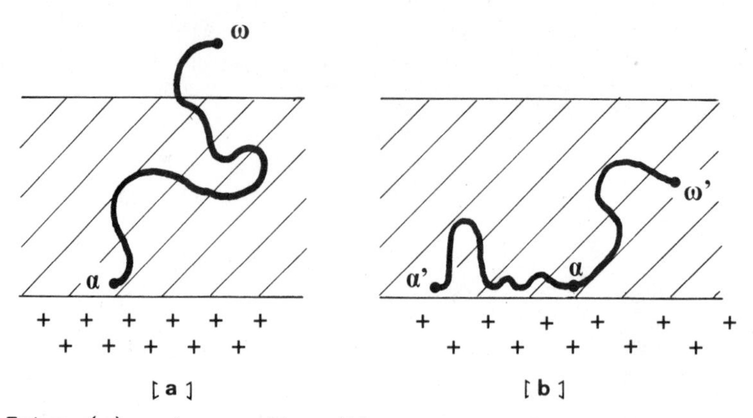

Fig. 1. Entry (a) and spreading (b) processes for a chain coming from solution and penetrating the adsorbed layer (hatched area).

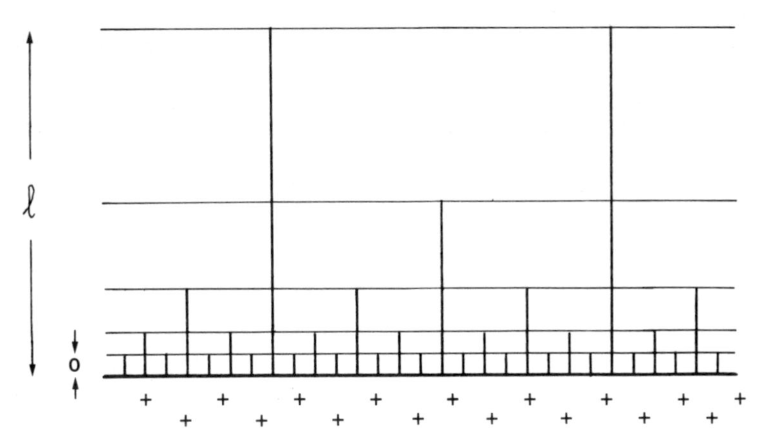

Fig. 2. The "self-similar grid": at a distance z from the absorbing wall, the adsorbed layer behaves like a solution of local mesh size equal to z. Instead of representing the polymers by random coils, they are depicted as portions of straight lines (to generate a simpler drawing).

The difference in chemical potential $\Delta\mu$ (per monomer) between the starved layer and the fully saturated layer, should scale like the difference in chemical potentials between two bulk solutions of mesh sizes ℓ and R_F(*)

$$\Delta\mu \simeq kT\left(\frac{1}{g(\ell)} - \frac{1}{N}\right) \simeq \frac{kT}{g(\ell)} \simeq kTx^{-5}, \tag{3}$$

where $g(\ell) = (\ell/a)^{5/3}$ is the number of monomers in a free coil of size ℓ. This $\Delta\mu$ will play the role of driving force for the spreading process, and will thus be of use in Section IV.

At distance $z < \ell$, the concentration profile $a^{-3}\phi(z)$ is expected to have the self-similar form[7]

$$a^{-3}\phi(z) = \frac{g(z)}{z^3} = \left(\frac{a}{z}\right)^{4/3}. \qquad (a < z < \ell) \tag{4}$$

More details on these static features can be found in Ref. 8. For instance, one can derive Eq.(2) for ℓ by writing the missing part in a starved layer in the form

$$\Gamma_o x = \int_\ell^\infty a^{-3}\phi(z)\,dz \simeq a^{-2}\left(\frac{a}{\ell}\right)^{1/3}.$$

To discuss the dynamics, one further remark is important: recent experiments[9] (on polymers adsorbed at the free surface of a liquid) suggest that the first few Ångströms of the adsorbed layer are <u>glassy</u>[10]: in such a case, we have to distinguish two types of friction:

a) for monomers which are not close to the wall, the dominant friction is against the solvent: it is described by a friction coefficient (ratio of force to velocity)

$$\zeta_s \simeq \eta_s a, \tag{5}$$

where η_s is the solvent viscosity, and a the monomer size.

b) for monomers which are very close to the wall ($z \sim a$), there is a much stronger friction coefficient due to glassy features or to special bindings at the wall.

$$\zeta_1 = f\eta_s a, \qquad (f >> 1) \tag{6}$$

where the dimensionless factor f measures the special frictions in the first layer. It will be of importance to analyse the relative importance of the factor f in the various relaxation process listed above.

II. SELF-DIFFUSION ALONG THE ADSORBING PLANE

The type of motion considered here is shown in Fig. 3. We follow one labeled chain in the layer. The chain reptates[12] on the self-similar grid. It is subdivided into subunits (i), with a spatial size z_i equal to the distance to the wall, and a number of monomers $g_i = g(z_i)$. Instead of computing directly the diffusion constant $D_{||}$, we use as an intermediate the mobility $\mu = D_{||}/kT$ of the chain (where kT is the thermal energy). When the chain reptates, the subunit (i) has a certain curvilinear velocity U_{ci}. It is important to realize that U_{ci} is different for different subunits: what is conserved is the "tube current J", i.e., the number of monomers, crossing one given point on the tube, per second.

$$J = U_{ci} g_i / z_i. \tag{7}$$

The velocity of the center of mass is

$$\mathbf{V} = \mathbf{R} N^{-1} J, \tag{8}$$

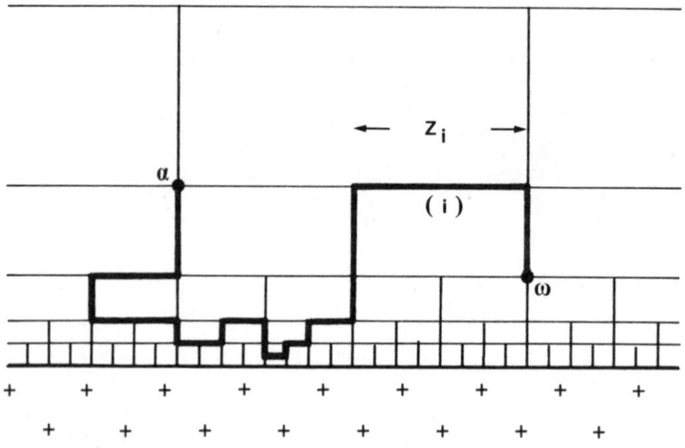

Fig. 3. One labeled chain, in the adsorbed layer, can be represented as a random walk on a self-similar grid. This is made of subunits (i) with characteristic size z_i and a number of monomers $g_i = (z_i/a)^{5/3}$. The reptation processes described in the third section assume that the chain moves along its own path.

where **R** is the end-to-end vector. The dissipation associated with these motions is

$$TS^{o} = \sum_{i} \zeta_{1} U_{ci}^{2}, \tag{9}$$

where ζ_i is a friction coefficient for subunit i. Using Eq. (7), we get an entropy source:

$$TS^{o} = \sum_{i} \zeta_{i} \left(\frac{z_{i}}{g_{i}}\right)^{2} J^{2}, \tag{10}$$

On the other hand, we may define the mobility μ via

$$TS^{o} = \mu^{-1} V^{2}. \tag{11}$$

Comparing Eqs.(10) and (11), and making use of Eq.(8), we get

$$\mu = \frac{1}{3N^{2}} \frac{\mathbf{R}_{x}^{2}}{\sum_{j} \zeta_{j} \left(\frac{\mathbf{R}_{j}}{g_{j}}\right)^{2}} . \tag{12}$$

We still have to perform an average over the chain conformations. Some years ago, we showed that the (weight averaged) distribution for loops (or tails) of n subunits in the self-similar structure is[13]

$$f_{n} \simeq n^{-6/5}. \qquad (1 < n < g(\ell)) \tag{13}$$

This transforms Eq.(12) into

$$3N^{2}\mu = \frac{\sum_{n} 1/n}{\sum_{m} \zeta_{m} m^{-3}} . \tag{14}$$

Let us now discuss the friction coefficients ζ_m. In the absence of any special friction at the wall, we would set

$$\zeta_{m} = \eta_{s} z_{m} ,$$

where z_m is the size of the subunit ($z_m \sim m^{3/5} a$). When this is done, we can check on Eq.(14) that the sum Σ/m is rapidly convergent, and dominated by its first term. This is even more true if glassy features show up. Thus, we may write

$$\mu \sim \frac{1}{N^{2}\zeta_{1}} \sum_{1}^{g(\ell)} \frac{1}{n} = \frac{1}{N^{2}\zeta_{1}} \ell n [g(\ell)] . \tag{15}$$

Consider first the case of a saturated layer ($\ell = R_F$ and $g(\ell) = N$). Then μ (and $D_{\parallel}$) is proportional to $N^{-2}\ell n N$. The friction is dominated by the small loops, and the only effect of the large loops is to give a weak (logarithmic) enhancement with respect to the reptation in a dense system.

Can we see this diffusion? Various optical techniques (forced Rayleigh scattering, photobleaching,...) using evanescent waves are being considered. The main difficulty is rooted in the factor f of Eq.(6): if f is extremely large, $D_{\parallel}$ will be too small to be detected.

Finally, let us mention briefly the differences induced if we have a starved layer: then Eq.(15) for the mobility becomes

$$\mu \simeq \frac{1}{N^2\zeta_1} \ell n \frac{1}{x}, \tag{16}$$

where we have made use of $g(\ell) \sim x^{-5}$ as explained in Eq.(3). For weakly starved layers (x small), μ is slightly increased over the standard reptation form $((N^2\zeta_1)^{-1})$: this is again an effect of the large loops and tails.

III. ENTRY AND SPREADING

The two operations have been illustrated in Fig. 1. Entry may again be considered as a reptation process, inside the adsorbed layer, provided that we incorporate two adjustments:

a) during the entry there are no direct contacts to the wall: thus this process is related to the weak friction coefficient ζ_s Eq.(5).

b) there is a tunneling barrier opposing the entry, and described[3] by a transmission coefficient ϑ

$$\vartheta = \begin{cases} N^{-0.3}; & (saturated) \\ [g(\ell)]^{-0.3} \simeq x^{3/2} & (starved) \end{cases} \tag{17}$$

Ignoring all details, logarithmic factors,...this leads to an entry time

$$\tau_e = \tau_o N^3 \vartheta^{-1}. \tag{18}$$

$$(\tau_o = \zeta_s a^2/kT)$$

The first factor ($\tau_o N^3$) in Eq.(18) describes the conventional reptation, while ϑ^{-1} describes all barrier effects. We then get explicit formulas for τ_e by insertion of Eq. (17) for ϑ. For instance, with a saturated layer, we expect $\tau_e \sim N^{3.3}$.

Let us now turn to the spreading process (Fig. 1b). Again we can describe it in terms of a reptation, but

a) the friction is now dominated by the wall ($\zeta_1 = \zeta_s f$);

b) there is a suction force: when the number of spreading monomers increases from n to $n+1$, the free energy is lowered by the quantity $\Delta\mu$ of Eq.(3).

As soon as $N > N/f$, the tube friction may be written as

$$\zeta_{tube} \sim n\zeta_1 , \tag{19}$$

and the rate of suction dn/dt is given by

$$\zeta_{tube} \frac{dn}{dt} = \Delta\mu . \tag{20}$$

The result is

$$\frac{n}{N} \sim \left(\frac{t}{\tau_s} \right)^{\frac{1}{2}} , \tag{21}$$

where the spreading time τ_s scales like

$$\tau_s = \tau_o f N^2 x^{-5} . \tag{22}$$

Eq.(22) holds when the hydrodynamic thickness ℓ is smaller than its value at saturation ($x < N^{-1/5}$). If we return to saturation, we have $\tau_s \sim \tau_o f N^3$, i.e., a normal reptation law with wall friction.

IV. DISCUSSION

The relative importance of the entry and spreading processes depends on three control parameters: the chain length N, the glassy friction factor f, and the level of saturation x. To locate the bottleneck we must compare the two times, as given by Eqs.(18) and (22). For instance, in conditions of starvation (with $1 >> x > N^{-1/5}$), we expect entry to be the bottleneck if

$$f < N x^{3.5} . \tag{23}$$

How can we hope to prove (or disprove) these ideas?

1. The Wageningen experiments[6] probably give us the spreading time τ_s. They can hopefully be performed at variable N, and also at variable x. However, the range ($1 >> x > N^{-1/5}$) in which our simple formulas hold is not very large, even for high N. Thus the x-dependence may not be a good check. The main difficulty is related to our poor knowledge (and control) of the glassy sublayer described by f. Our hope is that with certain highly flexible polymers (elastomers, polydimethylsiloxane..) adsorbed on suitably chosen surfaces, one can remove altogether the glassy features.

2. The rate equations for the growth of a layer[5] or for exchanges between adsorbed layers and solutions[3] may provide indirect checks for the present ideas. For instance, when the bottleneck is at entry, it is feasible to relate simply the macroscopic rate constants of Refs. 3, 4 to the entry time: this is done in the appendix. But when the bottleneck is the spreading process, the analysis appears more difficult.

3. Similar conclusions hold for the process of bridging between two grains. If the distance D between the grains is smaller than 2ℓ, the two adsorbed layers can interchange some molecules. The effective cut-off length is $\ell_{eff} = D/2$. When f is of order unity, and the entry process is

the bottleneck, the bridging time τ_B might tentatively be estimated from an adaptation of Eqs.(17) and (18)

$$\tau_B \sim \tau_o N^3 \left(\frac{D}{a}\right)^{\frac{1}{2}}, \tag{24}$$

but we need a systematic program on the one-layer systems before seriously approaching the bridging problem.

APPENDIX

Scaling discussion of exchange constants when the bottleneck is at entry In this appendix we discuss the transport currents J_{in} (from solution to layer) and J_{out} (from layer to solution) following the general framework of Ref. 3.

1) The inward current J_{in} (from a solution of concentration c_b) is expected to be of the form:

$$J_{in} = K_{in} c_b \tag{A.1}$$

$$= \frac{c_b}{c^*} \frac{1}{R_F^2} \frac{1}{\tau_e} . \tag{A.2}$$

Eq. (A.2) may be understood from the self-similar grid of Fig. 2. To have entry we must have a chain in a region of volume $R_F{}^3$ just above the grid: the probability for this is c_b/c^*, where $c^* = N/R_F{}^3$ is the overlap concentration. To get the number of such events per unit area we multiply by $R_F{}^{-2}$. The rate of entry is further proportional to $1/\tau_e$.

2) The outward current J_{out} is written as

$$J_{out} = K_{out} \Gamma \tag{A.3}$$

$$= \frac{\Gamma}{N} \frac{1}{\tau_e} exp\,(\Delta\mu/kT) . \tag{A.4}$$

In Eq. (A.4) the factor Γ/N gives the number of chains per cm^2. The factor $1/\tau_e$ gives the rate of exit. The last factor ($\exp \Delta\mu/kT$) ensures detailed balance: $\Delta\mu$ is the difference between the chemical potential (per monomer) in the layer and the reference potential in solution.

Eqs. (A.2,4) define the rate constants (except for numerical factors). But it may be useful to restate the (many) assumptions which are required to reach these simple forms:

a) the bottleneck is at entry (spreading is very fast);

b) the dynamics is dominated by reptation. This is probably true for an equilibrated chain, as discussed in Section I. Our situation here could be different, since just before entry our chain has zero contact with the wall, and is thus less entangled. However, a longer discussion of loop statistics for an entering chain indicates that the inner regions still impose reptation.

c) the first contact is taken to be through a chain end, not a hairpin. In fact, one can estimate the transmission factors for both situations: in Ref. 3 we argued that

$$\vartheta_{end\ point} \sim \frac{<\psi(R_F)>}{<\psi(a)>} \sim (a/R_F)^{\frac{1}{3}}, \quad (A.5)$$

while one can guess that

$$\vartheta_{hair\ pin} = \frac{<\Psi^2(a)>}{<\Psi^2(R_F)>} = \left(\frac{a}{R_F}\right)^{4/3}, \quad (A.6)$$

Here Ψ is the Edwards order parameter, such that $<\Psi^2>$ is the local concentration.

Thus transmission via chain-end processes is faster. But $N^{-1}\ \vartheta_{end\text{-}point} < \vartheta_{hair\text{-}pin}$: Thus, if all conformations were equally probable, hairpins would dominate; our (tentative) assumption is that the dynamics will favor reptation processes, and that the end-point process still dominates for this dynamical reason.

NOMENCLATURE

a	monomer size
$c^* = N/R_F^3$	concentration for the onset of overlap in bulk solutions
D	size of gap in two-plate experiments
$D_{\parallel}$	diffusion coefficient along the adsorbing wall
e_H	hydrodynamic thickness of adsorbed layer
$f = \zeta_1/\zeta_s$	reduced measure of the stickiness of the wall
f_n	weight distribution of loops
g_i	number of monomers inside subunit (i)
J_{in}, J_{out}	transport current
K_{in}, K_{out}	rate constants for solution ⟷ layer exchange
k	Boltzmann constant
ℓ	thickness of starved layer
N	number of monomers per chain
R_F	Flory radius: size of one coil in good solvent condition
$\mathbf{R}$	end-to-end distance of coil
ϑ	transmission factor
T	absolute temperature
U_{ci}	curvilinear velocity of subunit (i) on the self-similar grid
x	"starvation factor" = relative deviation from the equilibrium coverage
z_i	size of subunit (i)
Γ	number of monomers per cm^2
ζ_1	friction coefficient of one monomer near the adsorbing wall
ζ_i	friction coefficient of a subunit (i)

ζ_s	friction coefficient of one monomer inside solvent
ζ_{tube}	friction coefficient for a long-chain portion trapped in a tube
η_s	solvent viscosity
μ	chain mobility
μ	chemical potential
τ_e	entry time
τ_o	microscopic jump time
τ_s	spreading time
$\phi(z)$	concentration profile
Ψ	Edwards order parameter (defined, for instance, in Chap. 9 of "Scaling concepts in polymer physics" (P.G. de Gennes, Cornell Univ. Press, 2nd printing 1985).

REFERENCES

1. D.H. Napper, Polymer Stabilization of Colloidal Dispersions, Academic Press, New York (1983).
2. P.G. de Gennes, C.R. Acad. Sci. (Paris), II 306, 183-185 and II 306, 739 (1988).
3. P.G. de Gennes, C.R. Acad. Sci. (Paris), II 301, 1399-1403 (1985).
4. E. Pfefferkorn, A. Carroy and R. Varoqui, J. Polymer Sci., (Physics) 23, 1997-2008 (1985).
5. P.G. de Gennes, Proceedings of the Toyota Conference on Polymer Physics (1986) to be published.
6. M. Cohen Stuart and H. Tamai, to be published.
7. P.G. de Gennes, Macromolecules, 14, 1637-1648 (1981).
8. P.G. de Gennes, Adv. Colloid Interface Sci., 27, 189-209 (1987).
9. P. Pincus and G. Rossi, to be published.
10. M.A. Cohen Stuart, J. Keurentjes and B. Bonekamp, Colloids-Surfaces, 17, 1997-2002 (1985).
11. K. Kramer, J. Physique, 47, 1269-1271 (1986).
12. P.G. de Gennes, J. Chem. Phys., 55, 572-584 (1971).
13. P.G. de Gennes, C.R. Acad. Sci. (Paris), II.294, 1317-1320 (1982).

* The attractive and repulsive parts in $\Delta\mu$ are assumed to be comparable. We estimate the repulsive part, and take the overall $\Delta\mu$ to have the same scaling law, but with a reversed sign.

EQUILIBRIUM STRUCTURE OF THE FLUID INTERFACE

F.P. Buff

Department of Chemistry
University of Rochester
Rochester, New York 14627

M. Robert

Department of Chemical Engineering
Rice University, P.O. Box 1892
Houston, Texas 77251

ABSTRACT

Some recent results in the theory of the interface between continuous fluid phases at equilibrium are described. Emphasis is given to the role played by the external field in determining the microscopic structure of the interface, its anomalous effect on the critical behavior of fluid interfaces in two dimensions, the success of capillary wave theory and the failure of traditional van der Waals theory to describe not only transverse but also longitudinal interfacial correlations, as well as to account for the optical reflectivity of the interface of simple fluids and binary mixtures near the critical point.

I. FLUID INTERFACES IN ZERO EXTERNAL FIELD

I.1 Traditional point of view

It was long believed that the interface which separates two fluid phases at equilibrium has a well-defined structure even in the absence of a macroscopic external field such as gravity.[1] Gravity, which is generally very weak, was thought to have no effect on the interface apart from the obvious one of making it planar. It was assumed, in particular, that the density profile of the interface, the function which describes the continuous distribution of matter as one crosses the interface from one bulk phase to the other, has a finite thickness independent of gravity. Such an interface assumed to exist independently of any external field is called an intrinsic interface. As we will see below, this intrinsic interface does not exist and such a meaning of the word "intrinsic" must be modified in view of recent theoretical advances.

The thickness of the interface is typically of the order of the range of the intermolecular forces, except in the critical region where it increases considerably, ultimately diverging at the critical point itself. Strictly speaking, of course, this critical point divergence is prevented from occurring by the gravity-induced density gradients which are present in any fluid.[2]

The picture of a self-maintained interface forms the basis of the work of van der Waals[3] and culminates in its more recent extensions.[4a,b] In theories of the van der Waals type, a squared-density gradient term, which arises from the force imbalance induced by the density gradients, is incorporated in the free-energy functional describing the two-phase system. The density profile then follows by minimizing this functional. In its modern version, this approach yields a density profile of finite thickness which is proportional to the correlation length of the bulk spontaneous density fluctuations.[4b] While no external field such as gravity is necessary in van der Waals-type theories, such a field is readily incorporated into the field-free theory[5] and, except very close to the critical point, its effect is found to be small because normal gravity is itself weak.[6]

I.2 Modern point of view

The traditional picture of the fluid interface which was described above and which prevailed for over 150 years, has however recently been shown, on the basis of fundamental principles of statistical mechanics, to be incorrect.

It can indeed be shown[7] that for continuous fluids in three or fewer[8] dimensions of space in which the molecules interact via short-ranged forces, such as Lennard-Jones interactions, no density profile exists on the scale of the bulk correlation length in the absence of a macroscopic external field.

This essential role played by the external field had in fact first been suggested[9] in an alternative formulation of the interface problem based on a fluctuation theory employing capillary waves as relevant collective coordinates, which is described below in Section II.

The proof[7] of this result is based on exact bounds on the asymptotic behavior of the interfacial correlation functions, obtainable from the Bogoliubov inequality, and uses a *reductio ad absurdum*: a self-maintained interface is assumed to exist and it is then shown that this assumption leads to a contradiction. The key step in the demonstration consists in deriving and making use of the asymptotic behavior at large separations of the direct correlation function of Ornstein-Zernike, $c(\mathbf{r}, \mathbf{r}')$, defined in terms of the more familiar pair correlation function $h(\mathbf{r}, \mathbf{r}')$ which measures the probability of having a molecule at point $\mathbf{r}'$ given that there is one at point $\mathbf{r}$:

$$h(\mathbf{r},\mathbf{r}') = c(\mathbf{r},\mathbf{r}') + \int d^d\mathbf{r}''\, c(\mathbf{r},\mathbf{r}'')\rho(\mathbf{r}'')h(\mathbf{r}'',\mathbf{r}') , \qquad (1)$$

where $\rho(\mathbf{r}'')$ is the local density of the spatially nonuniform d-dimensional fluid at $\mathbf{r}''$.

For a planar fluid interface in three or fewer dimensions of space in which the vertical coordinate z is taken to be that along which the density $\rho(\mathbf{r}) = \rho(z)$ varies, the functions h and c are translational invariant in the transverse directions x and y, and we may write $h(\mathbf{r},\mathbf{r}') = h(z, z', |\mathbf{x}^{\perp}|)$ and $c(\mathbf{r},\mathbf{r}') = c(z,z', |\mathbf{x}^{\perp}|)$, where $|\mathbf{x}^{\perp}| = [(x-x')^2 + (y-y')^2]^{1/2}$ is the transverse separation between the points $\mathbf{r} = (x,y,z)$ and $\mathbf{r}' = (x', y', z')$. The Bogoliubov inequality, a unique tool to rule out spontaneous breaking of a continuous symmetry,[10] is first used[8] to obtain a bound on the asymptotic decay of $h(z, z', |\mathbf{x}^{\perp}|)$ for large values of $|\mathbf{x}^{\perp}|$, from which the following bound[7] on $c(z, z', |\mathbf{x}^{\perp}|)$ can be obtained:

$$c(z,z',|\mathbf{x}^{\perp}|)_{|\mathbf{x}^{\perp}|\to\infty} \sim \frac{m(z,z')}{|\mathbf{x}^{\perp}|^{d+\lambda}}, \quad 0 \le \lambda < 1 \quad for\, d \le 3 \,. \tag{2}$$

with m bounded.

It is instructive to rewrite Eq. (1) by Fourier transforming it in the d-1 transverse dimensions:

$$\hat{H}\cdot\hat{C} = \mathbb{1}, \tag{3}$$

where $C \equiv \rho^{-1}(z)\delta(\mathbf{r}-\mathbf{r}') - c(\mathbf{r},\mathbf{r}')$ and $H \equiv h(\mathbf{r},\mathbf{r}') + \rho(z)\delta(\mathbf{r}-\mathbf{r}')$.

The form of Eq. (3) illustrates an intimate connection between the problem of self-maintained interfaces and that of critical point fluctuations, for which

$$\hat{H}\cdot\hat{C} = 1 \,, \tag{4}$$

where the Fourier transform is now taken with respect to the full d-dimensional space. It follows from Eq. (4) and model calculations that at the critical point, both H and C, and therefore h and c, are slowly decaying at infinity;[11] likewise, it can be shown[7] that Eq. (3) implies that both h and c decay slowly in a self-maintained interface.

The final step in the proof is taken by making contact with the modern theory of surface tension[12,13] according to which the local work of deformation required to slightly distort an initially planar interface involves, in three dimensions:

$$\int d^2\mathbf{x}^{\perp}\,|\mathbf{x}^{\perp}|^2 c(z,z',|\mathbf{x}^{\perp}|)\,. \tag{5}$$

Eq. (2) implies that the integrand of Eq. (5) behaves, at large transverse separations, like:

$$|\mathbf{x}^{\perp}|^3 \,.\, c(z,z',|\mathbf{x}^{\perp}|) \sim m(z,z')\,.\,|\mathbf{x}^{\perp}|^{-\lambda}, \qquad for\; d = 3$$

and because $0 \le \lambda < 1$, this yields a nonintegrable decay at infinity. In other words, an infinite amount of work is required to distort a self-maintained planar interface in three or fewer dimensions, a clearly unacceptable result which concludes our *reductio ad absurdum*. It is important to recognize that Eq. (5) is an intermediate step in the derivation of the expression of surface tension in terms of the direct correlation function,[13] and that the present demonstration, unlike the original one given in Ref. 7, does not make use of the final expression for surface tension.

For intermolecular forces decaying slowly enough at infinity, Bogoliubov's inequality is inconclusive and the above proof is inapplicable. It is an open question whether such long-ranged forces can stabilize interfaces in three-dimensional continuous fluids, as they do in the mean-field limit of infinite interaction range. It is of interest to observe that the same situation prevails in the problem of crystalline or isotropic ferromagnetic order in two dimensions. The Bogoliubov inequality can again be used to show that such order is absent for short-ranged forces,[14] and while ferromagnetic order is restored in the presence of long-ranged forces which decay sufficiently slowly at infinity,[15] it is not known whether the same will occur for crystalline order.

II. CRITICAL BEHAVIOR OF FLUID INTERFACES IN TWO DIMENSIONS

II.1 Capillary wave and van der Waals theories

According to the traditional point of view described in Section I.1, the properties of fluid interfaces close to the critical point are directly related to those of the coexisting bulk phases and correspondingly share the universal critical properties the latter are well known to exhibit.[4a,b]

A different approach to the problem of fluid interfacial structure advocates capillary waves, which have no analogue in the bulk, as the proper modes of the interfacial fluctuations. While originally designed for low temperatures, capillary wave theory has been successfully extrapolated into the critical region of three-dimensional fluids.[9]

Quite surprisingly, these two completely different phenomenological theories make identical predictions[16] for the value of the critical exponent ω of the interfacial thickness:

$$L \sim (T_c - T)^{-\omega} ,$$

with T the temperature and T_c its critical value,[16] in all dimensions of space $d \geq 3$ where, moreover, both theories predict[16] the hyperscaling-like relation $\mu = (d-1)\omega$, with μ the critical exponent describing the vanishing of the surface tension $\gamma \sim (T_c - T)^{\mu}$ at the critical point.

In two dimensions,[10] however, the situation will prove to be very different from what it is in three or more dimensions: whereas van der Waals theory predicts $\omega = 1$, which follows at once from the postulated identification[4b] $\omega \equiv \nu$, with ν the critical exponent describing the divergence of the bulk correlation length $\xi \sim (T_c - T)^{-\nu}$, and the exact result $\nu = 1$ of the two-dimensional lattice gas (Ising) model,[17] the prediction of capillary wave theory is very different, as will now be seen.

In the two-dimensional version of capillary wave theory, we have an infinitely sharp fluctuating line separating two two-dimensional fluid phases. The probability of a single-valued distortion $z(x)$ is proportional to $\exp(-W(z)/k_BT)$, with:

$$W(z) = \tfrac{1}{2} m \Delta\rho \cdot g \int z^2(x)\,dx + \gamma_0 \int \{[1 + (\partial z/\partial x)^2]^{\frac{1}{2}} - 1\}\,dx \ , \qquad (6)$$

where k_B is Boltzmann's constant, m is molecular mass, $\Delta\rho$ is the difference between the densities of th vo coexisting phases, g is acceleration due to gravity, and γ_0 is the re surface tension, i.e., the tension of the fluctuating line.

Decomposing $z(x)$ into its Fourier components $z(x) = \sum_k a(k)\, e^{ikx}$ and assuming $\partial z/\partial x << 1$ turns Eq. (6) into:

$$W(z) = \tfrac{1}{2} m \Delta\rho \, . \, g L \sum_k a^2(k)(1 + \tfrac{1}{2}\ell^2 k^2) \,, \tag{7}$$

with periodic transverse edge length L and capillary length $\ell = (2\gamma_o / m\Delta\rho . g)^{\frac{1}{2}}$.

From Eqs. (6) and (7), the equilibrium average of $a^2(k)$ is found to be:

$$< a^2(k) > = \frac{1}{\beta m \Delta\rho . g . L} \, \frac{1}{1 + \frac{1}{2}\ell^2 k^2} \,, \tag{8}$$

with $\beta = 1/k_B T$, and the mean square of the displacement Z of the interface:

$$< Z^2 > = \sum_{o}^{k_{max}} < a^2(k) > \,,$$

with $k_{max} \sim$ (interfacial thickness)$^{-1} \equiv L^{-1}$,[18] is then calculated to be, at the thermodynamic limit:

$$< Z^2 > \sim \frac{1}{\beta m \Delta\rho g \ell} \cdot \arctan\left[\frac{\ell k_{max}}{\sqrt{2}}\right] . \tag{9}$$

As the critical temperature is approached from below, $\Delta\rho$ vanishes like $(T_c - T)^\beta$ and the capillary length ℓ vanishes like $(T_c - T)^{(\mu-\beta)/2}$. The critical behavior of $L = <Z^2>^{1/2}$ depends on that of ℓk_{max}. In the range of temperatures currently accessible to experiment $\ell k_{max} >> 1$, and we find from Eq. (9), setting $\arctan(y) \sim \pi/2 - 1/y$:

$$< Z^2 > \sim \frac{1}{m \Delta\rho . \ell} \sim (T_c - T)^{-(\mu + \beta)/2}$$

$$= (T_c - T)^{-9/16}$$

where the exact values $\mu = 1$ and $\beta = 1/8$ have been used. We thus have:

$$L \sim (T_c - T)^{-9/32} \,, \tag{10}$$

so $\omega = 9/32$, in strong contrast to the prediction $\omega = 1$ of van der Waals theory.

We finally observe that for fluids in four dimensions, capillary wave theory predicts:[16]

$$< Z^2 > \sim \frac{1}{m \Delta \rho \ell^2} \left[2 k_{max} - \frac{2\sqrt{2}}{\ell} \, arctan\left(\frac{\ell k_{max}}{\sqrt{2}} \right) \right] , \tag{11}$$

and therefore, in the regime $\ell k_{\max} >> 1$,

$$< Z^2 > \sim (T_c - T)^{\omega - \mu} ,$$

that is, since $<Z^2> \sim (T_c - T)^{-2\omega}$ by definition of the critical exponent ω:

$$\mu = 3\omega . \tag{12}$$

Comparison of Eq. (12) with the hyperscaling-like relation $\mu = (d-1)\nu$ reveals that $\omega = \nu$ for $d = 4$. It must be emphasized that the result of Eq. (12) only holds when $k_{\max}$ is taken to be proportional to L^{-1}, in accord with Ref. 9. It is then readily found[16] that for all $d \geq 3$, capillary wave theory predicts $\mu = (d-1)\omega$.

II.2 Singular behavior of effective mass density of capillary waves

In view of the striking disagreement between the predictions of capillary wave and van der Waals theories in two dimensions, it is useful to reexamine critically[18] the basic assumptions of capillary wave theory.

The Hamiltonian which describes capillary waves as collective modes has the form:[19]

$$H = 1/2 \sum_k \left[\frac{p_k^2}{m(k)} + m(k) \omega^2(k) a^2(k) \right] , \tag{13}$$

where the effective mass is given by:

$$m(k) \sim \frac{\rho_\alpha + \rho_\beta}{k} L^{d-1} , \tag{14}$$

with ρ_i the bulk phase densities, and where, according to the classical dispersion relation:[20]

$$\omega^2(k) = \frac{\rho_\alpha - \rho_\beta}{\rho_\alpha + \rho_\beta} g k + \frac{\gamma k^3}{\rho_\alpha + \rho_\beta} . \tag{15}$$

The total effective mass density per unit interfacial area, $\overline{m}$, is given by:

$$\frac{\int_o^{k_{max}} dk \, m(k) g(k)}{L^{d-1} \int_o^{k_{max}} dk \quad g(k)} ,$$

where $g(k)$, the density of modes, is found by a Rayleigh-Weyl enumeration to be proportional to $L^{d-1}\, k^{d-2}$. Consequently:

$$\overline{m} \sim (\rho_\alpha + \rho_\beta) \begin{cases} \dfrac{1}{k_{max}} & d > 2 \\ \\ \dfrac{1}{k_{max}}\, \ell n\, k \Big|_{o}^{k_{max}} & d = 2 \end{cases} \tag{16}$$

The case d = 2 is anomalous in that $\overline{m}$, which is regular for all d > 2, is logarithmically divergent for d = 2. The divergence of $\overline{m}$ in d = 2 occurs at all subcritical values of the temperature and is induced by the modes of long wavelengths.

Finally, we observe that in d = 2, the divergence of $\overline{m}$ corresponds, in a hydrodynamic description, to a diverging mean generalized inertia: $\overline{m}$ is proportional to the mean generalized inertia of a fluid in which a surface wave of wavelength k^{-1} carries with it a layer of fluid of depth approximately equal to $m(k)$.[21] Consequently, this diverging inertia inhibits the surface's excitations, making the interface less diffuse and yielding a weaker divergence of the interfacial thickness L in d = 2. This conclusion is of course consistent with the result of capillary wave theory.

II.3 Nonuniversal critical behavior in external fields

In order to elucidate the nature of the critical behavior of fluid interfaces in two dimensions and to better determine the special role played by the external field, it will prove instructive to analyze the effect of varying the external field[24] and test some of the current ideas on the nature of interfacial fluctuations, in particular that according to which capillary waves and intrinsic fluctuations should be combined and forced to coexist rather than exclude each other.[22]

In the original capillary wave theory of the planar interface,[9] one considers an infinitely sharp interface separating two incompressible phases. This infinitely sharp interface can, and indeed, according to current ideas,[22] should, be replaced by a diffuse interface of non-zero thickness, the intrinsic interface. The word "intrinsic" now acquires a different meaning from the one it had in the traditional point of view described in Section I. In such a picture, one attempts to describe the interfacial structure without loosing any aspects of the originally conflicting capillary wave and intrinsic structure (van der Waals) theories. It is hoped, in such an amalgamation, to recapture from a single picture the capillary wave aspects at large distances on the one hand and the traditional critical behavior of van der Waals theory on the other hand. As will now be seen, such a hope is fulfilled in three or more dimensions, but not in two dimensions.

In order to determine the thermal broadening of this new intrinsic profile, we will consider, for simplicity, a single component liquid-vapor phase equilibrium, in which the density profile of the intrinsic interface is denoted by $\rho_I(z)$. The energy change due to capillary wave fluctuations is assumed to be given by:[23]

$$W = \gamma_o \int \{ |1 + |\nabla\zeta(\mathbf{x})|^2]^{1/2} - 1 \} d\mathbf{x}$$

$$+ \int \rho_I(z - \zeta(\mathbf{x})) u_{ext}(z) d\mathbf{x} - \int \rho_I(z) u_{ext}(z) d\mathbf{x} , \quad (17)$$

where γ_0, the bare surface tension, is the surface tension of the intrinsic interface; u_{ext} is the external field taken to depend on the coordinate z only, and ζ is the instantaneous deformation of the intrinsic interface.

Expanding the distorted profile $\rho_I[z-\zeta(x)]$ to second order in $\zeta(\mathbf{x})$, assuming $\int\zeta(\mathbf{x})d\mathbf{x} = 0$ to preserve the average position of the profile and expanding $[1+ |\nabla\zeta(\mathbf{x})|^2]^{\frac{1}{2}}$ to second order in $|\nabla\zeta(\mathbf{x})|^2$ turns Eq. (17) into:

$$W = \frac{\gamma_o}{2} \int |\nabla\zeta(\mathbf{x})|^2 d\mathbf{x} + \tfrac{1}{2} \int \int \rho_I''(z) u_{ext}(z) \zeta^2(\mathbf{x}) dz d\mathbf{x}$$

$$= \frac{\gamma_o}{2} \int |\nabla\zeta(\mathbf{x})|^2 d\mathbf{x} - \tfrac{1}{2} \int \rho_I'(z) u'_{ext}(z) dz \int \zeta^2(\mathbf{x}) \, d\mathbf{x} , \quad (18)$$

where the symbol $'$ denotes differentiation with respect to z.

Assuming the fluctuations of the intrinsic interface as a whole to be capillary wave-like enables one to decompose them into decoupled surface waves: $\zeta(\mathbf{x}) = \sum_{\mathbf{k}} a(\mathbf{k}) \, e^{i\mathbf{k}.\mathbf{x}}$. Eq. (18) then assumes the form:[23]

$$W(a(\mathbf{k})) = \frac{\gamma_o L^{d-1}}{2} \sum_{\mathbf{k}} a^2(\mathbf{k}) |\mathbf{k}|^2 + \frac{K L^{d-1}}{2} \sum_{\mathbf{k}} a^2(\mathbf{k}) , \quad (19)$$

with L the periodic transverse edge length of the system and:

$$K \equiv - \int_{-\infty}^{+\infty} \rho_I'(z) u'_{ext}(z) dz . \quad (20)$$

K is positive, the derivatives in the integrand being of opposite signs.

A generalized capillary length ℓ may be defined by:

$$\ell \equiv \left(\frac{2\gamma_o}{K} \right)^{1/2} , \quad (21)$$

in terms of which Eq. (19) becomes:

$$W(a(\mathbf{k})) = \frac{K L^{d-1}}{2} \sum_{\mathbf{k}} a^2(\mathbf{k}) (1 + \tfrac{1}{2} \ell^2 . |\mathbf{k}|^2) . \quad (22)$$

We observe that if, following Ref. 9, we choose the intrinsic interface to be a step function, i.e., $\rho_I(z) = \rho_L - \Delta\rho . \theta(z)$, with $\theta(z)$ the unit step function, and the external field to be that of gravity, i.e., $u_{ext}(z) =$ mgz, then the constant K of Eq. (20) is equal to $mg \cdot \Delta\rho$, so that ℓ reduces to the ordinary capillary length $\ell = (2\gamma_0/mg\Delta\rho)^{\frac{1}{2}}$, in accord with Ref 9.

As in Section II.1, the equilibrium average of $a^2(\mathbf{k})$ is readily found to be:

$$< a^2(\mathbf{k}) > = \frac{1}{\beta K L^{d-1}(1 + \frac{1}{2}\ell^2 \cdot |\mathbf{k}|^2)} ,$$

leading to the mean-squared thickness of the interface:

$$< Z^2 > = \frac{1}{\beta K L^{d-1}} \sum_{\mathbf{k}} \frac{1}{1 + \frac{1}{2}\ell^2 \cdot |\mathbf{k}|^2} . \tag{23}$$

Eq. (23) enables us to determine the behavior of the interfacial thickness for several model intrinsic interfaces and several choices of the external field in any dimension of space. The case of an intrinsic interface of vanishing thickness and of a gravitational external field has already been treated for $d = 3$[9] as well as for general d.[16]

Here, we will consider a step-like external field of positive amplitude c:

$$u_{ext} = c \begin{cases} -1 & z < 0 \\ \\ +1 & z > 0 \end{cases} . \tag{24}$$

For such an external field, K of Eq. (20) is given by:

$$K = - \int_{-\infty}^{+\infty} \rho_I'(z) \cdot c\delta(z) dz$$

$$= - c\rho_I'(o) . \tag{25}$$

In two dimensions, we find for Eq. (23), at the thermodynamic limit $L \to \infty$ where the sum over $\mathbf{k}$ becomes an integral over d$\mathbf{k}$:

$$< Z^2 > \cong \frac{1}{\beta K \ell}$$

$$= \frac{1}{\beta(2\gamma_o K)^{1/2}} , \tag{26}$$

where Eq. (21) for the generalized capillary length has been used and where, as above in Eq. (9), the limit $\ell k_{max} >> 1$ has been taken.

As our first intrinsic density profile we choose the step-like profile $\rho_I(z) = \rho_L - \Delta\rho \cdot \theta(z)$. Such an intrinsic profile is found to support no thermal broadening: with u_{ext} given by Eq. (24), a vanishing total interfacial thickness obtains at all temperatures, in any number of dimensions. This result is restricted to continuous fluids; it does not hold for lattice models. (See Section III).

When one chooses next a more diffuse intrinsic profile such as an exponential profile with a decay length equal to the bulk correlation length ξ, K becomes simply:

$$K = \frac{c\Delta\rho}{2\xi} , \tag{27}$$

yielding from Eq. (26):

$$<Z^2> \sim \left(\frac{\xi}{\gamma_o \Delta\rho} \right)^{\frac{1}{2}} . \tag{28}$$

Consequently, the interfacial thickness $L \equiv <Z^2>^{\frac{1}{2}}$ diverges, as the critical point is approached, according to:

$$L \sim (T_c - T)^{-1/4(\mu + \beta + \nu)}$$

$$= (T_c - T)^{-17/32} , \tag{29}$$

where exact values of the critical exponents are used and it is assumed that the latter are not affected by the external field. While unproved, such an assumption is most likely to be correct, at least for $|u_{ext}|$ sufficiently small.

When a linear field such as gravity is chosen instead of the step-field, keeping the thickness of the intrinsic profile proportional to ξ, the total interfacial thickness L is found to diverge like:

$$L \sim (T_c - T)^{-9/32} , \tag{30}$$

which differs greatly from Eq. (29). As we saw in Eq. (10), a result identical to Eq. (30) obtains in ordinary capillary wave theory where the intrinsic profile is of vanishing thickness. Both results of Eqs. (29) and (30) disagree severely with the prediction of the generalized van der Waals theory according to which, as was seen in Section II.1, $L \sim (T_c - T)^{-1}$.

This strong non-universal critical behavior of the interfacial thickness in two dimensions prompts one to consider the same step-like external field in both three and four dimensions. We find from Eq. (23):

$$<Z^2> \sim \frac{1}{K\ell^2} \ell n(1 + \tfrac{1}{2}\ell^2 |\mathbf{k}^2_{max}|)$$

$$\sim \frac{1}{\gamma_o} , \tag{31}$$

in $d = 3$, while for $d = 4$:

$$<Z^2> \sim \frac{1}{K\ell^2}(2|\mathbf{k}_{max}| - \frac{2\sqrt{2}}{\ell} arctan(\ell . |\mathbf{k}_{max}|))$$

$$\sim \frac{1}{L\gamma_o} . \tag{32}$$

From Eqs. (31) and (32) it is seen at once that:

$$\omega = \begin{cases} \frac{\mu}{2} & for \ d = 3 \\ \\ \frac{\mu}{3} & for \ d = 4 \end{cases} . \tag{33}$$

As seen from Section II.1, these results are in full agreement with the prediction $\mu = (d - 1)\omega$ of ordinary capillary wave theory for the case of a gravitational field and an intrinsic profile of zero thickness,[16] as well as with the identical prediction[4b] of the generalized van der Waals theory. The critical behavior in $d \geq 3$ is thus insensitive to the choices of the intrinsic profile and external field.

In summary, assuming the equilibrium structure of the fluid interface to result from averaging capillary wave excitations on an intrinsic interface, it is found that while the external field does not affect the divergence of the interfacial thickness in the critical region of fluids in three or more dimensions (except, of course, extremely close to the critical point[2]), its effect is dramatic in two dimensions, where the critical behavior is found to be non-universal, i.e., depending on the external field. Consequently, the relation $\mu = (d\text{-}1)\omega$, which links the critical exponents of surface tension and interfacial thickness to the dimension of space and which is most probably correct in $d \geq 3$, appears to be incorrect in $d = 2$, since there ω, unlike μ, is strongly field-dependent.

III. INTERFACIAL CORRELATION FUNCTIONS: AN EXACTLY SOLUBLE MODEL

The above discussions concerned the microscopic quantity of prime interest, the thickness of the interface. Much more detailed information on the microscopic structure of the interface is contained in the pair correlation function. Unfortunately, unlike the density profile, the pair correlation function in the interfacial region is not accessible to experiment as readily as it is in the bulk phases, where a very large body of information is available from X-ray or neutron scattering.

Just as for the interfacial density profile itself, two very different points of view on the nature of interfacial correlations have been advocated, the first relying on local thermodynamic concepts in the spirit of the van der Waals theory, the second relying on capillary wave concepts.

While it has recently been recognized both from theory[25] and computer simulations,[26] that local thermodynamic concepts cannot account for the transverse correlations induced in directions parallel to the interfacial plane by an external field such as gravity, which are very long-ranged because gravity is so weak, very little is known about the nature of the longitudinal correlations, i.e., those in the direction of the external field. We describe here a model of an interface for which the longitudinal correlation functions can be determined exactly in closed form,[27] thus enabling one to test approximations based on either van der Waals or capillary wave concepts.

The model is the one-dimensional lattice gas (Ising) model with nearest-neighbor interactions, in which the separation of phases, which cannot occur spontaneously because of the linear character of the model and the short-rangedness of the interactions, is achieved by an external field that changes sign at the middle of the chain (see Fig. 1). The correlation functions of such a model can be calculated exactly by using standard transfer matrix techniques.[27]

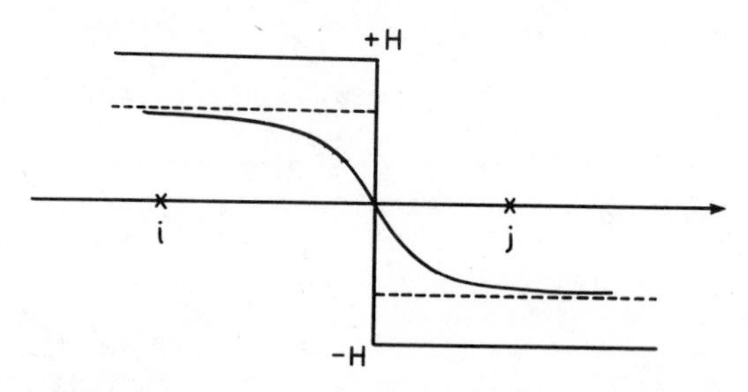

Fig. 1. External field and magnetization (density) profile.

In the magnetic transcription of the lattice gas, we have spins $s_i = \pm 1$ located at the lattice sites, and the spin-spin correlation function is found to be exactly given by:

$$\langle s_i s_j \rangle = \langle s_i \rangle \langle s_j \rangle + e^{-|i-j|/\xi} \tag{34a}$$

when sites i and j belong to different halves of the chain, and by:

$$\langle s_i s_j \rangle = \langle s_i \rangle \langle s_j \rangle - \langle s_j \rangle^2 e^{-|i-j|/\xi} + e^{-|i-j|/\xi} , \tag{34b}$$

when both sites i and j belong to the same half of the chain, j denoting in the latter case that site closest to the origin. In Eqs. (34a) and (34b), $\langle s_i \rangle$ is the magnetization profile and ξ is the correlation length of the spin fluctuations in either bulk phases, respectively given by:[28]

$$< s_i > = < s_\infty > sgn\ i (1 - e^{-|i|/\xi}),$$

and

$$1/\xi = \ell n \left(\frac{\cosh(H/k_BT) + [\sinh^2 (H/k_BT) + e^{-4J/k_BT}]^{\frac{1}{2}}}{\cosh(H/k_BT) - [\sinh^2 (H/k_BT) + e^{-4J/k_BT}]^{\frac{1}{2}}} \right),$$

with H the amplitude of the field and J that of the spin-spin interaction; $<s_\infty>$ is the well-known magnetization of the positively magnetized bulk phase found by Ising:[29]

$$< s_\infty > = \frac{1}{(1 + [e^{2J/k_BT} \sinh(H/k_BT)]^{-2\ 1/2})}.$$

In local thermodynamic theories of the fluid interface, the generally unknown pair distribution function $\rho^{(2)}(\mathbf{r}, \mathbf{r}') = \rho(\mathbf{r})\rho(\mathbf{r}')g(\mathbf{r},\mathbf{r}')$, with g related to the pair correlation function h by $g \equiv h + 1$, is approximated by either:

$$g(\mathbf{r},\mathbf{r}') = g_o \left[|\mathbf{r} - \mathbf{r}'|; \rho\left(\frac{\mathbf{r} + \mathbf{r}'}{2}\right)\right], \tag{35a}$$

$$g(\mathbf{r},\mathbf{r}') = g_o \left[|\mathbf{r} - \mathbf{r}'|; \frac{\rho(\mathbf{r}) + \rho(\mathbf{r}')}{2}\right], \tag{35b}$$

or:

$$g(\mathbf{r},\mathbf{r}') = \tfrac{1}{2} \left\{ g_o \left[|\mathbf{r} - \mathbf{r}'|; \rho(\mathbf{r})\right] + g_o \left[|\mathbf{r} - \mathbf{r}'|; \rho(\mathbf{r}')\right]\right\}, \tag{35c}$$

where g_o is the pair distribution function of a reference fluid of constant density, in which that constant density is simply related to the densities at $\mathbf{r}$ and $\mathbf{r}'$.

A completely different approximation, one that is based on dynamic capillary wave concepts rather than on static local thermodynamic ideas, was recently proposed[26] in which, like in the original capillary wave model, the two phases are assumed incompressible and uncorrelated, and the equilibrium interfacial structure arises from averaging over the displacements of an infinitely sharp dividing surface. In fluid language, this approximation reads:[26]

$$\rho^{(2)}(z_i,z_j) = \frac{1}{\rho_L - \rho_G} \Big[\rho_L^{(2)}(z_i,z_j)(\rho_{Min} - \rho_G)$$

$$+ \rho_L \rho_G (\rho_{Max} - \rho_{Min})$$

$$+ \rho_G^{(2)}(z_i,z_j)(\rho_L - \rho_{Max}) \Big], \tag{36}$$

where z_i denotes the distance of site i from the middle of the chain, $\rho_L(\rho_G)$ is the bulk liquid (gas) density, $\rho_L^{(2)}$ $(\rho_G^{(2)})$ is the pair distribution function of the bulk liquid (gas) phase and

$$\rho_{Min} \equiv Min[\rho(z_i), \rho(z_j)], \rho_{Max} \equiv Max[\rho(z_i), \rho(z_j)] \ .$$

The exact expressions for $\rho^{(2)}$ (which is related to $< s_i s_j >$ by $\rho^{(2)}(z_i, z_j) = 1/4\ (<s_i s_j>+<s_i>+<s_i> + 1\ ))$ obtained for the above model interface can be used to test these two different types of approximations. The results are shown in Fig. 2 (from Ref. 27), using the transcription $\rho(z_i) = (1 + <s_i>)/2$.

In view of the dramatic failure of the local thermodynamic approximations (35), it is natural to attempt to improve them by considering density functional expansions which take into account the nonlocal gradient correction terms. Standard density functional methods yield:[30]

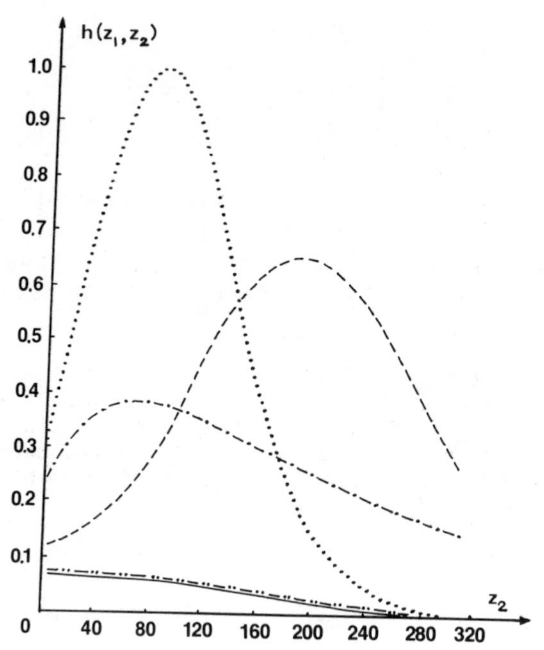

Fig. 2. Exact (______) and approximate (---,-·-, ···) (from Eqs. (35)) values of $h\ (z_1,z_2) \equiv g\ (z_1,z_2)-1$ as a function of z_2 and z_1 = 100 for H/k_BT = 0.01 and J/k_BT = 3.0. The curve (-··-) is the prediction of approximation (36).

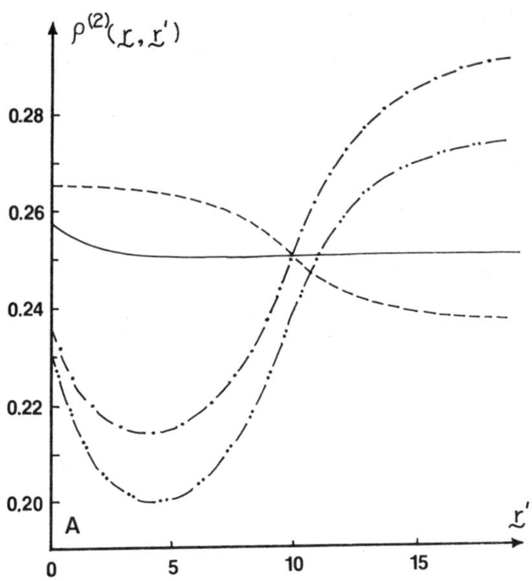

Fig. 3(a) Exact values (______), zeroth- (--), first (-·-·-), and second-order (-··-··) approximations of the pair distribution function $\rho^{(2)}$ $(\mathbf{r},\mathbf{r}')$ as a function of $\mathbf{r}'$ for $\mathbf{r}$ = -10, H/k_BT = 0.01, and J/k_BT = 0.5661. $\mathbf{r}$ and $\mathbf{r}'$ are on opposite sides of the origin.

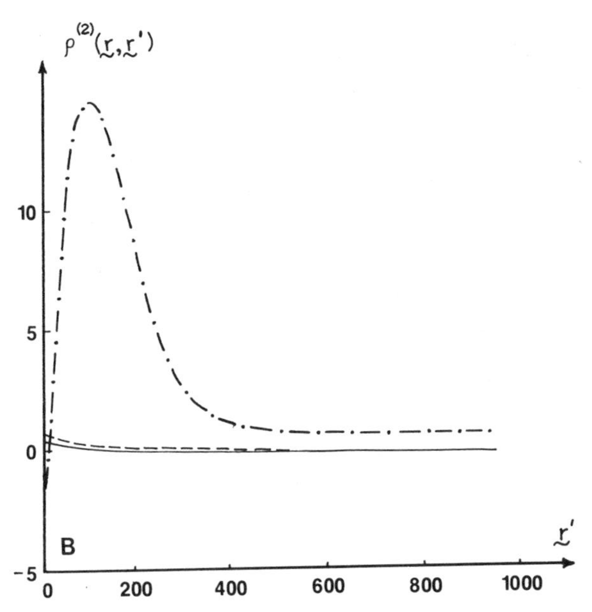

3(b) Exact values (______), zeroth- (--), and first-order (-·-·-) approximations of the pair distribution function $\rho^{(2)}$ $(\mathbf{r},\mathbf{r}')$ as a function of $\mathbf{r}'$, for $\mathbf{r}$ = -10, H/k_BT = 0.01, and J/k_BT = 3.0. $\mathbf{r}$ and $\mathbf{r}'$ are on opposite sides of the origin.

$$\rho^{(2)}(\mathbf{r},\mathbf{r}') = \rho_o^{(2)}(\mathbf{r},\mathbf{r}') + \int \frac{\delta\rho^{(2)}(\mathbf{r},\mathbf{r}')}{\delta\rho(\mathbf{r}_1)}\Bigg|_o [\rho(\mathbf{r}_1) - \rho_o(\mathbf{r}_1)d\mathbf{r}_1$$

$$+ \tfrac{1}{2}\int\int \frac{\delta^2\rho^{(2)}(\mathbf{r},\mathbf{r}')}{\delta\rho(\mathbf{r}_1)\delta\rho(\mathbf{r}_2)}\Bigg|_o [\rho(\mathbf{r}_1) - \rho_o(\mathbf{r}_1)][\rho(\mathbf{r}_2) - \rho_o(\mathbf{r}_2)]d\mathbf{r}_1 d\mathbf{r}_2 \tag{37}$$

$$+ \ldots ,$$

the subscript o indicating functions evaluated in a uniform reference fluid of constant density ρ_o.

For the present model interface, all terms in expansion (37) can be determined exactly in closed form,[31] offering a unique opportunity to test the accuracy of this method. However, as Figs. 3(a) and 3(b) clearly show, the functional expansion method is found to be not only incapable of reducing the disagreement with exact results of its zero-order term, which corresponds to Eqs. (35), but it even aggravates it considerably.

IV. NEW ANALYSIS OF OPTICAL REFLECTIVITY OF CRITICAL FLUIDS

The microscopic structure of fluid interfaces is most accurately determined experimentally by light reflectivity measurements.

The light reflectivity R at normal incidence for an infinitely sharp interface of vanishing thickness can be evaluated in terms of the refractive indices n_1 and n_2 of the bulk phases:

$$R_F = \left(\frac{n_2 - n_1}{n_2 + n_1}\right)^2 , \tag{38}$$

where the subscript F stands for Fresnel. Any departure of the reflectivity from Fresnel's equation (38) is a measure of the nonzero thickness of the interface.

The few available experimental data[32] had not enabled one to discriminate between the vastly diverging points of view advocated in the van der Waals and capillary wave theories. Although no reflectivity measurements have been performed very close to the critical point since the pioneering studies of Webb and his co-workers,[32] new measurements and analyses of the critical parameters (bulk correlation length[33] and surface tension)[34] of the systems originally studied by Webb et al., have been performed recently and very accurate values for both quantities, which differ appreciably from those used in earlier work, have recently become available.

Here we ask whether a reanalysis of these earlier reflectivity data, which is based on new more accurate values of the critical parameters, would not enable one to discriminate between the leading phenomenological theories.[35]

IV.1 Theory

We consider again, for convenience, a one-component fluid near its critical point, where the density profile $\rho(z)$ and the refractive index profile $n(z)$ can be assumed to be proportional to each other.

The specular reflectivity R is ordinarily expressed in units of the Fresnel reflectivity R_F given by Eq. (38), and for nearly normal incidence it can be written, in the Born approximation, as:[36]

$$\frac{R}{R_F} = \frac{1}{(\rho_1 - \rho_2)^2} \left| \int_{-\infty}^{+\infty} \frac{d\rho(z)}{dz} e^{2ikz} dz \right|^2 , \tag{39}$$

where $k = [2\pi n(z)]/\lambda$ is the wave vector of light, with λ the wavelength of incident light in vacuum.

In the temperature range considered in experiment, typical values of the interfacial thickness are much smaller than the wavelength of light, so that kz is small and Eq. (39) can be approximated, in the notation of Ref. 32(a), as:

$$\frac{R}{R_F} = e^{-(2k^2<\zeta^2>)/\pi} , \tag{40}$$

where $<\zeta^2>^{\frac{1}{2}} = (2\pi<Z^2>)^{\frac{1}{2}}$, with $<Z^2>^{\frac{1}{2}}$ a measure of the interfacial thickness, equal to the root-mean-square displacement of the bare interface in capillary wave theory, and proportional to the bulk correlation length in van der Waals theory. Reflectivity measurements therefore provide a direct determination[9] of $<Z^2>$.

It is convenient to estimate the root-mean-square displacement of the interface $(<Z^2>)^{\frac{1}{2}}$ in units of the bulk correlation length ξ. In the one-phase region, the correlation length behaves as $\xi_+ = \xi_+^0 t^{-\nu}$, with $t \equiv (T - T_c)/T_c$, $\nu = 0.63$, a universal exponent,[37] and ξ_+^0 a non-universal amplitude.

When in the original van der Waals theory the classical equation of state is replaced[4c] by a more accurate non-classical equation of state that agrees much better with experimental data, one finds:

$$L = 2.565\xi_+ . \tag{41}$$

Capillary wave theory, on the other hand, yields, in three dimensions:

$$<Z^2> = \frac{k_B T}{2\gamma_o} \ell n\left(\frac{1+\ell^2 k^2_{max}}{1+\ell^2 k^2_{min}}\right) \tag{42}$$

where:

$$k_{max} = \frac{\pi}{L} = \frac{\pi}{r\xi_+} , \tag{43}$$

the factor r being determined from experiment to be of order unity and where the bare surface tension γ_o is related to the observable surface tension γ by[9]

$$\gamma_o = \gamma + \frac{3}{16\pi} k_B T k^2_{max} . \tag{44}$$

Recent analyses[38] indicate that the scaling prediction $\gamma\xi^2 \sim k_BT$ is valid in the probed experimental range, where one finds:

$$\frac{k_BT_c}{\gamma} = R\,\xi_+^2\,, \tag{45}$$

R being predicted to be a universal number.

From Eqs. (43) and (44), the bare surface tension γ_0 can be expressed in units of the bulk correlation length

$$\gamma_o = \frac{k_BT}{\xi_+^2 R_{eff}} \tag{46}$$

where

$$R_{eff} \equiv \left(\frac{1}{R} + \frac{3\pi}{16r^2}\right)^{-1},$$

yielding

$$L_{cw}^2 = \frac{R_{eff}\,\xi_+^2}{2\pi}\,\ell n\left(\frac{1+\ell^2k_{max}^2}{1+\ell^2k_{min}^2}\right)$$

$$\simeq \frac{R_{eff}\,\xi_+^2}{2\pi}\,\ell n\left(k_{max}^2/k_{min}^2\right), \tag{47}$$

since both $\ell^2k^2_{max}$ and $\ell^2k^2_{min}$ are much larger than unity in the experimental temperature range.

If one assumes, following several authors, that interfacial fluctuations can be decoupled into short-wavelength spontaneous density fluctuations identical to those present in the bulk coexisting phases and into longer-wavelength capillary waves, neglecting any coupling between these different types of fluctuations as well as excluding any other surface waves, then one finds:[39]

$$L_{total}^2 = L_{int}^2 + L_{cw}^2\,. \tag{48}$$

IV.2 <u>Comparison with experiment</u>

Two systems have been studied by Webb <u>et al</u>., the binary mixture of cyclohexane and methanol,[32a] and the pure fluid sulfur hexafluoride.[32b]

For the cyclohexane-methanol system, the relevant data are: the bulk correlation length amplitude ξ_+^o = 3.24 Å, the critical temperature T_c = 318 K, and the interfacial thickness deduced from the reflectivity data, L_{exp} ≡ 12.4 $(-t)^{-0.66}$ [Å] ≃ $4\xi_+$. The corresponding effective ratio is R_{eff} = 2.29.

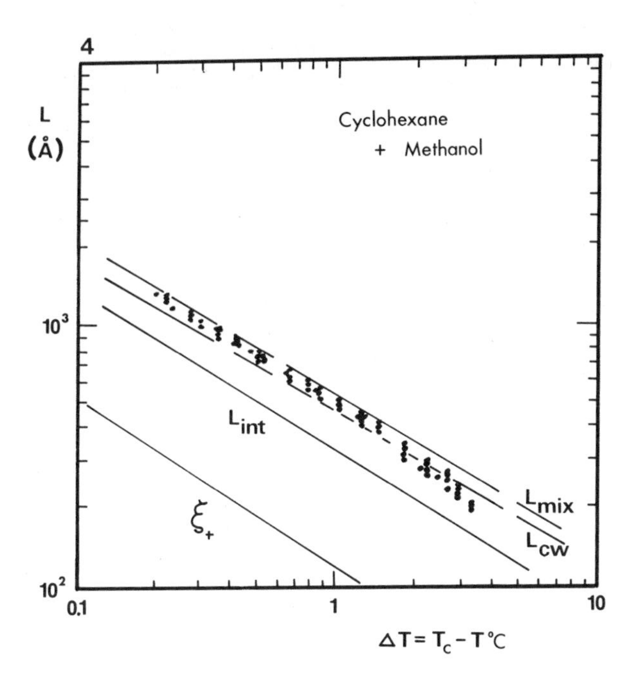

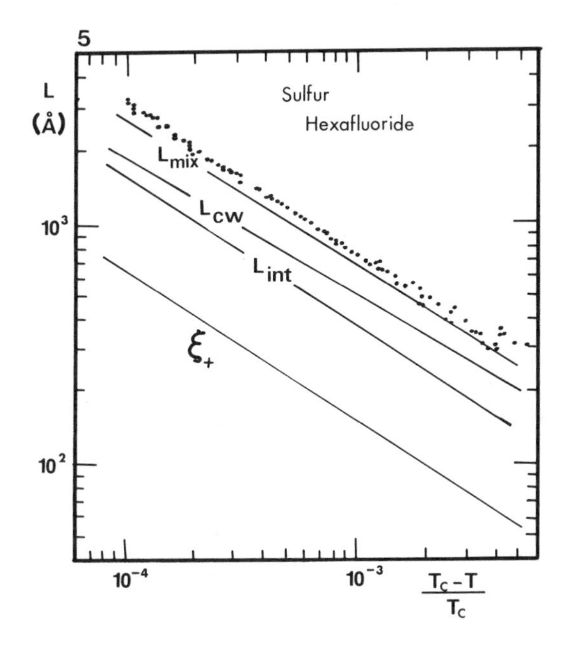

Figs. 4 and 5
Interfacial thickness near the critical point. ξ_+ = bulk correlation length in the one-phase region; L_{int} = interfacial thickness according to intrinsic structure theory; L_{mix} = interfacial thickness according to theory including both intrinsic structure and capillary waves. Fig. 4: The binary mixture of cyclohexane and methanol, from Ref. 32 (a). Fig. 5: The one-component fluid sulfur hexafluoride, from Ref. 32 (b).

For sulfur hexafluoride, the relevant data are:[38] ξ_+^o = 1.88[Å], T_c = 318.7 K and R = 2.6; the reflectivity measurements gave L_{exp} = $10.4(-t)^{-0.62}$ [Å] ≃ $5\xi_+^o$ yielding R_{eff} = 2.0.

In Figs. 4 and 5, we show the experimental interfacial thickness for each system, together with the predictions of theory. It is manifested in both figures that, as had been previously found[41] with mixtures of aniline and cyclohexane, intrinsic structure theory cannot account for the data; on the other hand, capillary wave theory is in very good agreement with the data for cyclohexane-methanol. Adding both intrinsic and capillary wave contributions is seen to significantly reduce the disagreement between theory and experiment for the case of sulfur hexafluoride, but does not improve the prediction of capillary wave theory for the mixture of cyclohexane and methanol.

ACKNOWLEDGEMENTS

The authors' work described in this article was supported in part by the National Science Foundation, the Shell Oil Company, the Welch Foundation and Donors of the Petroleum Research Fund administered by the American Chemical Society.

NOMENCLATURE

$c(\mathbf{r},\mathbf{r}')$	direct correlation function
d	dimension of space
$g(\mathbf{r},\mathbf{r}')$	$h(\mathbf{r},\mathbf{r}') + 1$
$g(k)$	density of modes
$h(\mathbf{r},\mathbf{r}')$	pair correlation function
H	field amplitude
J	coupling constant in lattice gas model
$\mathbf{k}$	wave vector with $k = \mid \mathbf{k} \mid$
k_B	Boltzmann's constant
ℓ	generalized capillary length
L	interfacial thickness or periodic edge length
T_c	critical temperature
β	critical index for bulk density difference of coexisting phases
γ	surface tension
$\zeta(\mathbf{x})$	instantaneous deformation of interface
λ	index for transverse asymptotic bound of $c(\mathbf{r},\mathbf{r}')$
μ	critical index for γ
ν	critical index for ξ
ξ	correlation length
$\rho(\mathbf{r})$	density
$\rho^{(2)}(\mathbf{r},\mathbf{r}')$	pair distribution function
ω	critical index for interfacial thickness L

REFERENCES

1. J.C. Maxwell, Scientific Papers, Vol. 2 (Dover, New York 1965), p. 561. For a review, see H. Minkowski, Kapillarität, in Enc. der Math. Wiss. (A. Sommerfeld, ed.) Vol. 5, Part 1 (Teubner Verlag, Leipzig 1906), p. 589.
2. J. Yvon, Fluctuations en Densité, Actualités Scientifiques et Industrielles, n°542, Herrmann et Cie, Paris 1937. This point is also emphasized in J.M.J. van Leeuwen and J.V. Sengers, Physica B8A, 1 (1986).
3. J.D. van der Waals, Verh. K. Ned. Akad. Wet. Afd. Natuurk, Reeks, 1, 8 (1893).

4. (a) F.P. Buff, in Proceedings of 38th National Colloid Symposium, Austin, (Texas) June 11-13, 1964; (b) Widom, J. Chem. Phys., 43, 3892 (1965); (c) S. Fisk and B. Widom, J. Chem. Phys. 50, 3219 (1969).
5. F.H. Stillinger and F.P. Buff, J. Chem. Phys. 37, 1 (1962); J.L. Lebowitz and J.K. Percus, J. Math. Phys., 4, 116 (1963).
6. J.V. Sengers and J.M.J. van Leeuwen, Intern. J. Thermophysics, 6, 545 (1985), Fig. 2.
7. M. Robert, in Proceedings of 13th Annual Conference on Statistical Physics, Oaxtepec (Mexico) January 3-6, 1984, Kinam 6A, 19 (1984) Phys. Rev. Lett., 54, 444 (1985).
8. The case of two dimensions of space was dealt with by M. Requardt, J. Stat. Phys., 31, 697 (1983).
9. F.P. Buff, R.A. Lovett and F.H. Stillinger, Phys. Rev. Lett., 15, 261 (1965).
10. For a lucid discussion, see: N.D. Mermin, J. Phys. Soc. (Japan) 26, 203 (1969) (Supplement.)
11. M.E. Fisher, J. Math. Phys., 5, 944 (1964).
12. R. Lovett, P.W. DeHaven, J.J. Vieceli, Jr., and F.P. Buff, J. Chem. Phys., 58, 1880 (1973).
13. F.P. Buff and R.A. Lovett, in Simple Dense Fluids, ed. by H.L. Frisch and Z.W. Salsburg (Academic Press, New York 1968).
14. N.D. Mermin, Phys. Rev., 176, 250 (1968). N.D. Mermin and H. Wagner, Phys. Rev. Lett., 17, 1133 (1966).
15. H. Kunz and C.E. Pfister, Commun. Math. Phys., 46, 245 (1976).
16. M. Robert, Phys. Rev., A30, 2785 (1984).
17. L. Onsager, Phys. Rev., 65, 117 (1944).
18. F.P. Buff and M. Robert, J. Stat. Phys., 41, 1037 (1985).
19. See, e.g., Ref. 18 and references therein.
20. See, e.g., H. Minkowski, in Ref. 1.
21. See, e.g., M. Lighthill, Waves in Fluids, Cambridge University Press, Cambridge, 1978, p. 213.
22. See e.g., B. Widom, in Phase Transitions and Critical Phenomena, ed. by C. Domb and M.S. Green, (Academic, New York 1972), Vol. 2; J.D. Weeks, J. Chem. Phys., 67, 3106 (1977); D.B. Abraham, Phys. Rev. Lett., 47, 545 (1981); D.J. Wallace and R.K.P. Zia, Phys. Rev. Lett. 43, 808 (1979).
23. J.K. Percus, in Liquid State of Matter: Fluids, Simple and Complex, E.W. Montroll and J.L. Lebowitz, eds. (North Holland 1982).
24. M. Knackstedt and M. Robert, J. Chem. Phys.,89, 3747, 1988).
25. M.S. Wertheim, J. Chem. Phys., 65, 2377 (1976).
26. M.H. Kalos, J.K. Percus and M. Rao, J. Stat. Phys., 17, 111 (1977).
27. M. Robert and R. Viswanathan, J. Chem. Phys., 86, 4657 (1987).
28. M. Robert and B. Widom, J. Stat. Phys., 37, 419 (1984).
29. E. Ising, Z. Phys., 31, 253 (1925).
30. See, e.g., J.K. Percus, in The Equilibrium Theory of Classical Fluids, edited by H.L. Frisch and J.L. Lebowitz, Benjamin, New York, 1984 p. II-33.
31. M. Robert, J-F. Jeng and R. Viswanathan, J. Chem. Phys., 88, 1983 (1988).
32. (a) J.S. Huang and W.W. Webb, J. Chem. Phys., 50, 3677 (1969); (b) E.S. Wu and W.W. Webb, Phys. Rev., A8, 2065 (1973).
33. C. Houessou, P. Guenoun, R. Gastaud, F. Perrot and D. Beysens, Phys. Rev., A32, 1818 (1985).
34. M.R. Moldover, Phys. Rev., A31, 1022 (1985).
35. D. Beysens and M. Robert, J. Chem. Phys., 87, 3056 (1987).
36. Ref. 9 and Ref. 32(a) (Eqs. (2) and (3)).
37. See, e.g., D. Beysens, in Phase Transitions, Cargese, 1980, edited by M. Levy, J.-C. Le Guillou and J. Zinn-Justin, Plenum, New York 1982.

38. H. Chaar, M.R. Moldover and J. Schmidt, J. Chem. Phys., 85, 418 (1986).
39. J. Meunier, C.R. Acad. Sci., 292, 1469 (1981); J. Meunier and D. Langevin, J. Phys. Lett., (Paris) 43, L185 (1982).
40. D.T. Jacobs, Phys. Rev., A33, 2605 (1986).
41. D. Beaglehole, Physica, B112, 320 (1982).

DISPLACEMENT OF HOMOPOLYMER FROM A SURFACE BY A BLOCK COPOLYMER

Hiroshi Watanabe,[†] Sanjay Patel* and Matthew Tirrell**

Department of Chemical Engineering and Materials Science
University of Minnesota
Minneapolis, Minnesota 55455

ABSTRACT

To interpret the force-distance profile of adsorbed vinylpyridine (PVP)-styrene (PS) diblock copolymer layers in a selective (toluene), we have been assuming that the soluble PS blocks are effectively end-grafted by the insoluble PVP blocks preferentially adsorbed to the substrate (mica). To examine this hypothesis, a surface-force device was used to measure the force-distance profiles for PS/toluene solutions before and after addition of PVP-PS copolymer. For PS/toluene solutions (before addition of PVP-PS), short-range repulsive interactions were observed. This repulsion indicates some PS was adsorbed on mica even from the good solvent toluene contrary to previous reports. However, after addition of PVP-PS, long-range repulsion prevailed and the force-distance profile was shifted toward that of pure PVP-PS solution. Thus, the adsorbed homo-PS chains were rapidly displaced by PVP-PS chains, suggesting the strong preferential adsorption of PVP blocks.

I. INTRODUCTION

Because of their inherent ability to straddle phase boundaries, amphiphilic molecules, such as block copolymers, find application as stabilizing agents for colloidal dispersions, compatibilizers in polymer blends and adhesion promoters or deterrents. Block copolymers afford tremendous potential for controlled manipulation of surface properties once a detailed understanding of how they arrange themselves at interfaces

---

† Permanent Address: Department of Macromolecular Science, Facult
Science, Osaka University, Toyonaka, Osaka 560, Japan.

* Current Address: AT&T Bell Laboratories, 600 Mountain Avenu
Hill, New Jersey 07974

** To whom all correspondence should be addressed.

is available. The conformations of the polymers govern the interactions between the surface and the surrounding environment, and in many practical applications one is interested in how the polymer affects this interaction. Often the interactions of interest are those between the polymer modified surfaces. For example, the nature (attractive or repulsive) and range of the interaction between polymer chains adsorbed on dispersed particles determines the stability of the colloidal suspensions.[1] The interactions between the adsorbed layers can be directly measured using a surface forces apparatus of the type developed by Israelachvili.[2,3] From these interactions, fundamental information such as the conformations of the polymers at the surface can be deduced.

In an effort to create model surfaces to correlate the conformations of the polymers with the measured interactions, we have been examinming the interactions between layers of 2-vinylpyridine(PVP)-styrene(PS) block copolymers adsorbed onto mica substrates (in the surface force apparatus).[4-7] In these systems, we observed repulsive forces at distances many (~10) times larger than the dimension of those polymers in solutions. In particular, we found that the distance at which the repulsive force becomes significant is a linear function of PS-block length.

This result suggests that these adsorbed block copolymers produce physical realizations of the dense, terminally adosrbed, or grafted layers analyzed by Alexander[8] and de Gennes.[9] These scaling theories showed that, when chains are end-grafted to a surface at a density high enough that the mean distance between the graft points is smaller than the mean radius of gyration of the free chain, solvent will try to penetrate this dense layer in order to dilute it, that is, to reduce the osmotic pressure within the layer. This favorable polymer-solvent mixing swells the layer, thereby stretching the polymer molecules. However, this stretching is countered by an elastic restoring force of entropic origin. The balance between osmotic swelling and entropic elasticity was shown by Alexander[8] and de Gennes[9] to yield a linear dependence of layer thickness on molecular weight. Halperin,[10] extending the Alexander-de Gennes analysis, has shown that this linear dependence of the layer thickness on the molecular weight of the terminally attached chains occurs independent of the quality of the solvent.

The solvent used, toluene, is a nonpolar and selective solvent, that is a good solvent for PS but a non-solvent for PVP. Thus, in toluene, the PVP-blocks (being much more polar and much less soluble than PS-blocks) are expected to exhibit compact configurations and have much stronger affinity (as compared to PS-blocks) to the polar surface of mica. In view of these facts, we formulated the following hypothesis for the adsorbed configurations.

Given the large difference in the relative affinity of the two blocks for the mica surface, the PVP-blocks would be strongly adsorbed, while the PS-blocks would not be adsorbed but end-grafted on mica by the PVP-blocks. Based on this **hypothesis** of preferential adsorption of PVP segments, we expected the block copolymer adsorbed on mica to form a layered structure with the densely packed PVP segments closer to the mica substrate and the toluene swollen PS segments extending away from the surface. Thus, the repulsive forces observed may be attributed to interaction between the PS-blocks end-grafted on two mica surfaces (in the surface force apparatus). urthermore, since the PVP-blocks, acting as the anchoring groups for PS-ocks, would exhibit compact configurations, we expected significant lap between the PS-blocks end-grafted on a mica surface. Then, the cks would be stretched into the solution phase because of the osmotic interaction, leading to very long-range repulsive forces, as observed.

In fact, based on the above molecular picture and applying the Alexander[8]-de Gennes[9] model, we could introduce reduced force and distance variables, both of which are essentially the quantities divided by the molecular weight of PS-block, and obtained universal force-distance profiles.[7] This result suggests that our hypothesis is valid, at least for PVP-PS diblock copolymers having sufficiently long PVP blocks.

However, some experimental check of our hypothesis is still desirable. Thus, we made the sequential measurements of the force in the surface force apparatus as follows: first, measurements were done on homo-PS/toluene solutions to check whether PS chains can be adsorbed on mica in our experimental condition. Then, a small amount of PVP-PS diblock copolymer was added to the system and the change in the force-distance profile was followed with time. The basic idea for this experiment is as follows:

If our hypothesis described above is **not** correct, that is, if the driving force for the adsorption of the PS segments is not much weaker than that for the PVP segments, we would not recover the force-distance profile for pure PVP-PS/toluene system even after the addition of PVP-PS copolymer. In other words, even if homo-PS chains are adsorbed on mica from solution in toluene, these PS chains would be displaced by PVP-blocks, and profile for pure PVP-PS system would be recovered upon addition of PVP-PS diblocks if the driving force for adsorption of PVP segments is much stronger than that for the PS segments. Therefore, the important factor is the competition of adsorption of PVP- and PS-segments determined by the relative magnitudes of the driving force for the adsorption of these blocks, and not the absolute magnitudes of their respective driving forces.

II. EXPERIMENTAL

The mica used in this study was Grade No. 4 ASTM V-2, clear and slightly stained Muscovite, ruby red mica obtained from Asheville-Schoonmaker Mica Co., Newport News, VA. Glass-distilled SPEC grade toluene supplied by EM Science (manufacture's specification for water content: 0.006%) was used as received (immediately after opening a new bottle) without further purification. The polystyrene homopolymer was obtained from Pressure Chemical Co. (weight average molecular weight M_w = 233,000; M_w/M_n = 1.06; lot# 50124) and was used as supplied. The PVP-PS diblock copolymers were synthesized by living anionic polymerization to yield blocks with narrowly distributed molecular weights. The synthesis and characterization techniques employed for these block copolymers are described elsewhere.[4]

III. RESULTS AND DISCUSSION

Figures 1 and 2 show the typical results of our experiments. Based on the Derjaguin approximations,[11] the data obtained from independent experiments (and different contact positions) are summarized as a plot of F/R (with F and R being the force and mean radius of curvature of mica surfaces, rspectively) against the separation between the mica substrates D. The measurements were carried out in toluene at 32°C. The incubation concentrations for both homo-PS (PS233) and PVP-PS (60-60 and 60-90) samples in the apparatus were approximately 3 µg/ml.

As can be seen from Figs. 1 and 2, short-range repulsive forces are observed for PS233/toluene system (filled circles), suggesting that homo-

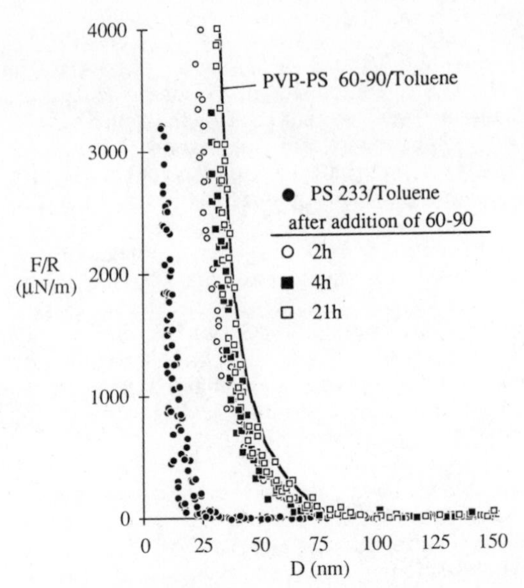

Fig. 1. Change in the force-distance profiles of PS233/toluene system at 32°C after addition of PVP-PS 60-90 diblock copolymer. Adsorption of PS233 and successive replacement by PVP-PS 60-90 were allowed to take place at widely (>2 mm) separated mica surfaces (in the surface force apparatus). The profile for pure PS233/toluene system (filled circles) was obtained after 24 hours (or more) adsorption of PS233k.

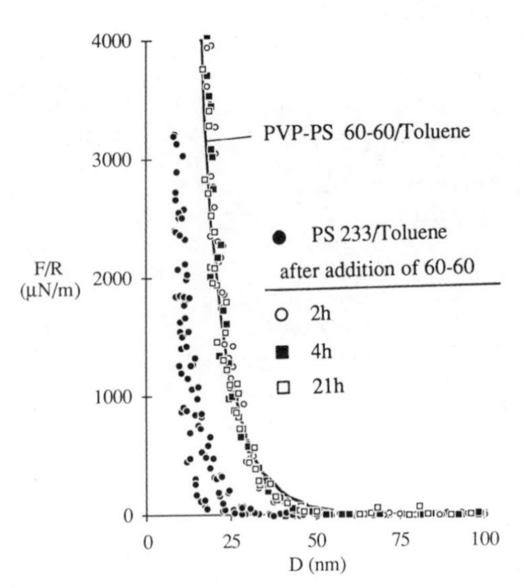

Fig. 2. Change in the force distance profiles of PS233k/toluene system at 32°C after addition of PVP-PS 60-60 diblock copolymer. The experimental conditions are the same as those in Fig. 1.

PS chains can be adsorbed on mica at our present (and also previous)[4-7] experimental conditions. This result is not the same as that reported by Luckham and Klein,[12] who observed no repulsive force for homo-PS/toluene systems. This difference may be presumably due to uncontrolled levels of some impurity in the solvent in one of the experiments. However, more importantly, we observe the shift of the force-distance profile to longer distances after addition of PVP-PS copolymer. After a sufficiently long time (~2h) the force-distance profile for PVP-PS/toluene systems (indicated by solid curves) appears to be attained, suggesting that the homo-PS first adsorbed on mica were driven out and replaced by PVP-blocks.

We also checked if some change in the profile was observed for the opposite case, that is, we first examined the pure PVP-PS 60-90/toluene system and then followed the change (if any) after addition of PS233. Although we do not show the figure here, the profile remained the same even 18 hours after the addition of PS233. The result, together with those shown in Figs. 1 and 2, appear to support our hypothesis of "preferential adsorption of PVP."* (Note that the PS-blocks in the 60-90 copolymer sample is significantly shorter than PS233 and would have less tendency to be adsorbed on mica as compared to PS233.)

Another finding of the present experiments concerns the rate of the change in the force-distance profile. As can be clearly seen from Figs. 1 and 2, the profile for pure PVP-PS system can be recovered (and presumably the homo-PS chains first adsorbed on mica are displaced by PVP-blocks) more rapidly for PVP-PS chains having shorter PS-blocks. Although we do not show the figure here, we also found that the recovery of the profile for the pure PVP-PS system was faster for shorter homo-PS chains (if the PVP-PS copolymer used is the same). These results may be explained: For the homo-PS chains adsorbed on the mica surface to be replaced by PVP-PS copolymer chains, copolymer chains in the solution phase should approach the surface. Then, the osmotic repulsion between the homo-PS chains and PS-blocks of copolymers hinders the displacement. If some of the homo-PS chains on mica surface have been already displaced by copolymer chains, the repulsion between the PS blocks end-grafter (by adsorbed PVP-blocks) on the surface and those in the solution phase also hinders the displacement. In both cases, the repulsion would be stronger for longer homo-PS and PS-blocks, leading to slower displacement.

IV. CONCLUSIONS

The exchange experiments show that although the PS segments can adsorb onto mica from solution in toluene, they do not seem to do so in the presence of PVP blocks. In other words, there is preferential adsorption of PVP segments (compared to PS segments) on the mica surface. A similar conclusion can be made based on recent X-ray photoelectron spectroscopy (XPS) experiments,[13] in which a layer-type structure of adsorbed PVP-PS block copolymers was observed (after evaporation of toluene). Both these measurements support our previous hypothesis that the adsorbed PVP-PS diblocks are akin to terminally attached PS chains.

---

* The system can be considered to be at equilibrium with respect to the adsorption of PVP-PS diblocks and PS "homopolymers" since the same force-distance profile was observed irrespective of the order of addition of the two polymers. From this point of view, our results suggest that the equilibrium constant for the adsorption of PS segments is highly decreased by the presence of PVP segments, that is, the adsorption is dominated by PVP.

REFERENCES

1. D.H. Napper, Polymeric Stabilization of Colloidal Dispersions Academic Press, New York (1983).
2. J.N. Israelachvili and G.E. Adams, Nature (London), 262, 774-776 (1976); J. Chem. Soc. Faraday Trans 1, 74, 975-1001 (1978).
3. J.N. Israelachvili, J. Colloid. Interface Sci., 44, 259-272 (1973).
4. G. Hadziioannou, S. Patel, S. Granick and M. Tirrell, J. Am. Chem. Soc., 108, 2869-2876 (1986).
5. S. Patel, G. Hadziioannou and M. Tirrell, in: Composite Interfaces (H. Ishida and J.L. Koenig, eds.), pp. 65-70, Elsevier, Amsterdam (1986).
6. M. Tirrell, S. Patel and G. Hadziioannou, Proc. Nat. Acad. Sci., 84, 4725-4728 (1987).
7. S. Patel, M. Tirrell and G. Hadziioannou, Colloids and Surfaces, 31, 157-179 (1988).
8. S. Alexander, J. Phys. (Paris), 38, 983-987 (1977).
9. P.-G. de Gennes, Macromolecules, 13, 1069-1075 (1980).
10. A. Halperin, Europhys. Letts., 4, 439-445 (1987).
11. B.V. Derjaguin, Koll Z., 69, 155-164 (1934).
12. P.F. Luckham and J. Klein, Macromolecules, 18, 721-728 (1985).
13. E. Parsonage, R. Nuzzo and M. Tirrell, in preparation.

THE MODIFICATION OF SURFACE FORCES BY GRAFTED POLYMER CHAINS

Hillary J. Taunton,
Chris Toprakcioglu

Cavendish Laboratory
Cambridge CB3 OHE England

Lewis J. Fetters

Exxon Research
Annandale, New Jersey 08801

Jacob Klein*

Weizmann Institute
Rehovot, 76100
Israel

ABSTRACT

We report the investigation of the forces between smooth solid surfaces bearing end-grafted chains in good solvent conditions. The forces are monotonically repulsive with a range about twice that for corresponding adsorbed chains. We observe no evidence of bridging attraction at low adsorbance, nor any hysteresis in the forces following strong compression of the grafted layers, both of which are effects characteristic of adsorbed chains. Our data are quantitatively in good accord with the scaling model of Alexander and de Gennes.

I. INTRODUCTION

A distinguishing feature of Pierre Gilles de Gennes' contributions to polymer physics[1] has been his application of new fundamental ideas to problems with practical implications. This paper deals with steric stabilization forces, an area where de Gennes has played an illuminating and stimulating role over the past several years.[2-5]

The role of polymers in stabilizing colloidal dispersions has been empirically known since long before the microscopic nature of either colloids or polymers was thought of. By adding the resin of the accacia

---

*Author to whom correspondence should be addressed.

tree to an aqueous dispersion of carbon black, the Egyptians, already at the time of the Pharaos, found that the resulting liquid would stay black for months, and could be used for writing and drawing. In the absence of such treatment, the carbon black particles would aggregate and sediment within a short time, the resulting liquid being quite useless as an ink.[6] Nowadays we realize that the accacia resin, consisting of a watersoluble, flexible polysaccharide of mean radius of gyration ca. 2000 Å, adsorbs on the solid carbon particles to provide a protective repulsive barrier against their adhesion in the primary van der Waals minimum. More generally, the field type forces, such as electrostatic double layer repulsion and dispersive van der Waals attraction, acting between solid surfaces, lead to interaction energies - either repulsive or attractive - which fall to ca. kT at surface separation of some 50-3000 Å.[7] This is a range very comparable with the size of flexible polymers: for this reason flexible polymeric chains have been used to adsorb onto surfaces to modify their mutual interactions in the range where Brownian collisions (energies $\sim kT$) are important. Such adsorbed polymer layers are effective in providing steric stabilization of colloidal particles in both aqueous, or ionic, and in organic, or non-ionic liquid media, and have been extensively studied for decades.[8] A recent[4] review on the structure of such adsorbed layers and their mutual interaction beautifully illustrates the main features.

Over the past few years the form of the forces between smooth solid surfaces in liquid media and their modification by adsorbed polymers has been studied directly using the mica force-balance method.[7,9] The results of force-profile studies for different solvency conditions of the adsorbed polymers are summarized in Figs. 1 and 2 for partial adsorbance and for equilibrium adsorbance of the polymers on the surfaces, respectively.

The main features are the range of the interactions (from $\sim 2\ R_g$ for θ solvents to $\sim 7\ R_g$ for good solvents), the equilibrium attraction in poor solvents compared with the equilibrium repulsion in good solvents (Fig. 2), and the kinetic features for the case of good solvents where finite times of order many minutes were required for relaxation of compressed adsorbed layers. The most remarkable (and perhaps surprising) feature to

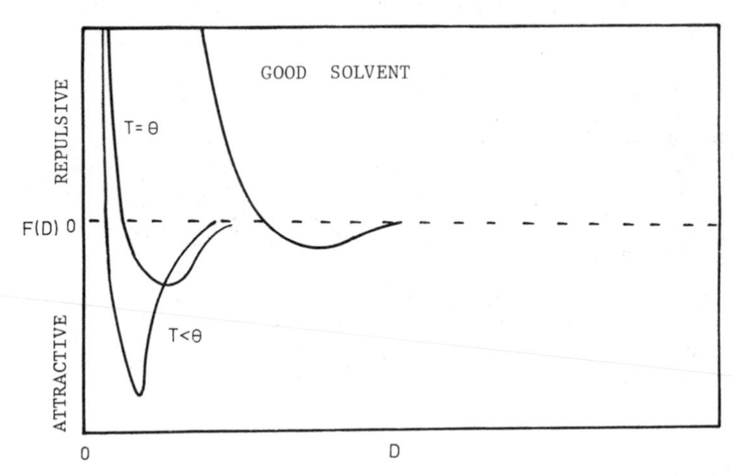

Fig. 1. Summary of force profiles between surfaces with low or undersaturated coverage of adsorbed chains in various solvency conditions. Based on Ref. 7.

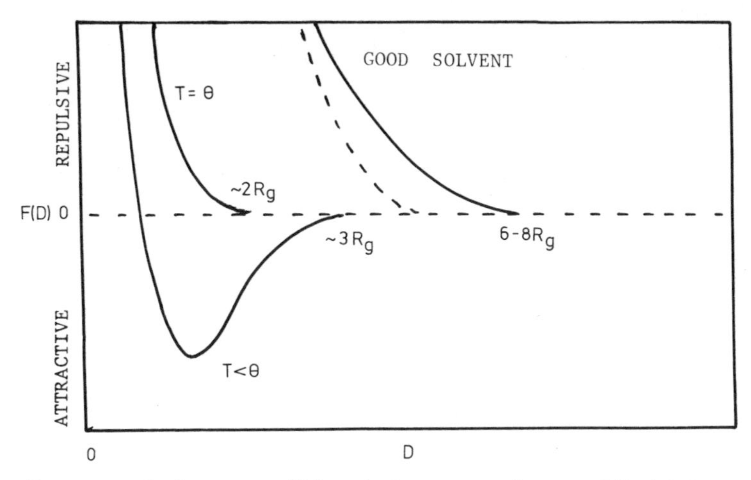

Fig. 2. Summary of force profiles between surfaces with high or saturated coverage by adsorbed polymers. The broken curve is the unrelaxed profile in the good solvent case. Based on Ref. 7.

emerge from these studies was the crucial importance of <u>bridging attraction</u> for the case when the surfaces were undersaturated (Fig. 1). Such attraction results from the simultaneous adsorption of given chains on both interacting surfaces, and is always attractive. The attraction is dominant whenever the surface coverage is low, and this applies even in the case of good solvents. This is empirically realized in the paint industry: when colloidal paint particles are immersed in a polymer solution under conditions where high adsorbance is rapidly obtained, steric stabilization results. Where, on the other hand, the adsorption process is slower, bridging and flocculation are frequently the case: this is because in the latter case the colloidal particles have time to aggregate in a bridging attractive well while they are still only sparsely covered with polymer, and once adhered no more polymer can access the intersurface gap. Thus the particles remain flocculated in the bridging minimum.

The problem of bridging may be overcome by the <u>grafting</u> of chains, as illustrated in Fig. 3. Here chains which have <u>no</u> propensity to adsorb on the surface are tied down, or anchored, at one end only. This can be achieved by either attaching a strongly 'adhesive' chemical group at one

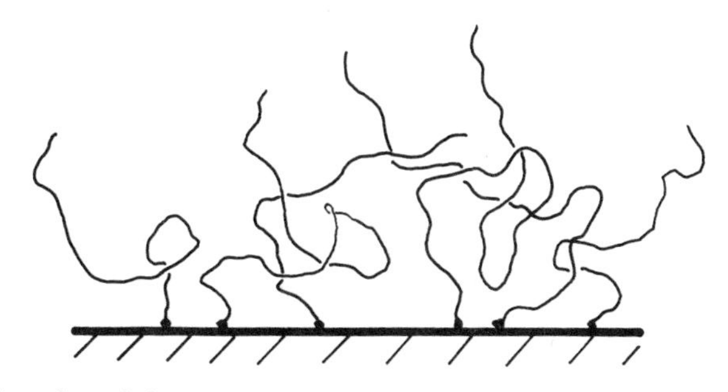

Fig. 3. Schematic illustration of grafted chains.

end of the non-adhering chain, or by using a diblock copolymer, one of whose components adsorbs while the other does not. We have carried out an extensive study of the forces between mica sheets bearing such layers in good solvents, both for end-functionalized polystyrene chains PS-X, where -X is a highly polar functional group, and for polystyrene poly(ethylene oxide) diblock copolymers, PS-PEO. PEO adsorbs strongly onto mica from toluene, while PS does not, so that a PS-PEO chain would be a good candidate to exhibit the end-anchoring features of grafted chains. In this paper we shall concentrate rather on the end-capped PS-X[10], though we note at this point that the broad features of the interactions with both types of end-anchored chains (PS-X and PS-PEO) are similar.

II. SURFACE FORCES WITH END-GRAFTED CHAINS[10]

Polystyrene is known from earlier work not to adsorb from toluene onto mica.[11] Polystyrene chains end-capped with the zwitterionic group $-N^{+}(CH_3)_2(CH_2)_3SO_3$ were prepared by the following route:

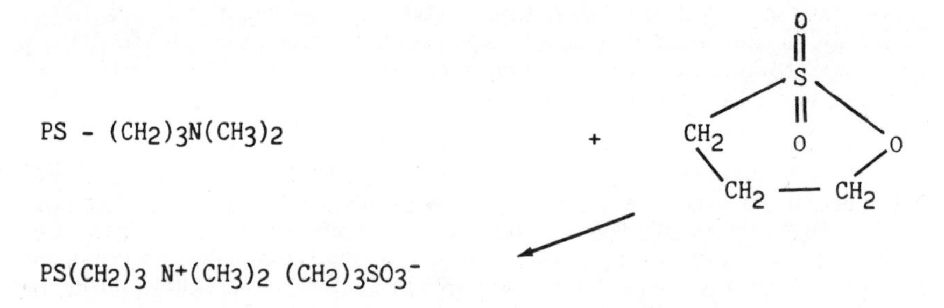

The molecular characteristics of both the regular PS used as control and the PS-X are given in Table 1.

In the mica force-balance studies, especially those involving polymers, it is essential to carry out control measurements at each stage of the investigation, to ensure that one is not measuring artefactual effects. This is especially important in view of the fact that the range of interest of interactions, some tens to perhaps thousands of angstroms, is comparable with the size of dust or microgel particles, organic contamination or - in the case of amphiphilic molecules - micelles. In the present study we measured first the forces between smooth mica sheets in polymer-free xylene (the results in toluene are identical) (Fig. 4, open circles): a short-range (<100 Å) jump into contact was observed, in agreement with earlier work, and probably due to water bridging in the undried solvent.[11] Following addition of regular uncapped PS (to a concentration 10^{-4} w/w) and a 12 hour incubation, the forces were again measured (Fig. 4, solid circles), and within the scatter show no difference to the polymer free case, indicating that no adsorption of the PS has taken place. This is again in accord with the earlier measurements.[11] Finally, PS-X was added to the cell to a concentration 10^{-4} w/w: within about an hour the short range jump into contact was replaced by a monotonic repulsion, which moved out within 2-12 hours (depending on rate of stirring or thermal agitation) to a limiting value (Fig. 4 squares). As the only microstructural difference between the PS and PS-X chains is the zwitterionic end-group, this shows without any ambiguity that the PS-X chains must be attached to the mica surface by the zwitterionic group - X. The attachment probably arises from the highly

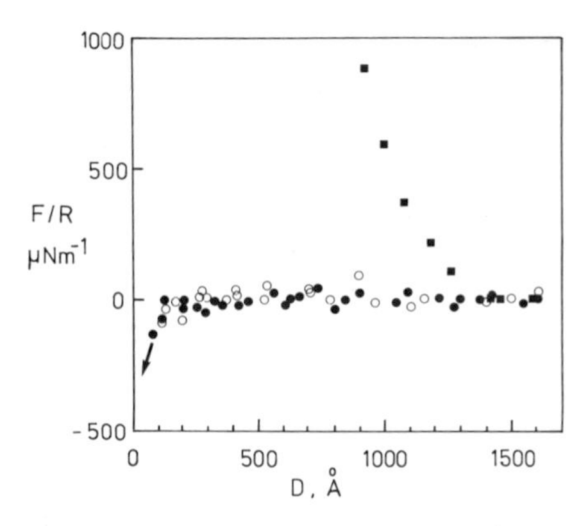

Fig. 4. Interaction between mica sheets in polymer-free xylene (O); in the same solvent with added PS following 12 hours incubation (●); and following addition of PS-X and 2 hr. incubation (■). The forces are plotted as *F(D)/R* where *F(D)* is the force at separation *D* and *R* is the mean radius of curvature of the mica sheets. This is proportional to the interaction energy per unit area *U(D)* between two flat parallel plates a distance *D* apart.[20]

polar nature of the zwitterion. Its magnitude is likely to be several *kTs* - we shall return to this point later. The essential points to note in Fig. 4 is the fact that the PS-X adsorbs and results in a long-ranged repulsion as the surfaces approach. We have observed no evidence of attraction even during the earliest stages of adsorption, where bridging attraction with adsorbed polymers was noted (Fig. 1).

We have also investigated the effect of slow and rapid compression-decompression cycles on the force-distance profiles, which shows a characteristic hysteretic behaviour for adsorbed chains in good solvents (dashed curve, Fig. 2). Figure 5 shows the results of a series of slow and rapid compression-decompression cycles with the end-grafted PS-X. Within the scatter, the profiles are the same, the effect of slow relaxation of the compressed layers being markedly absent.

The mica experiments allow one also to measure the refractive index of the gap separating the mica sheets, and hence to estimate directly the amount adsorbed during the force experiments themselves.[12] In these experiments the adsorbance of the PS-X was evaluated (from the refractive index data) as 3 ± 0.5 mg/m^2. Assuming each PS-X chain is anchored at a point to the mica surface, we may evaluate the mean inter-anchor spacing as ca. 85Å.

TABLE 1

Molecular Characteristics of the Polystyrene		
Sample	M_w	M_w/M_n
PS	1.31 x 10^5	1.03
PS-X	1.41 x 10^5	1.02

M values were determined using size exclusion chromatography.

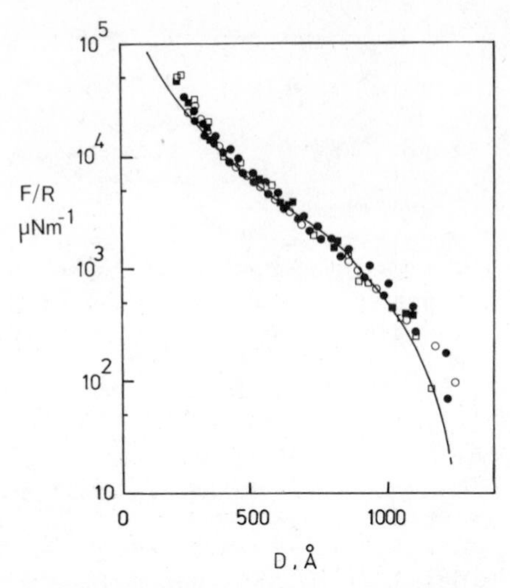

Fig. 5. Interaction between mica sheets bearing end-adsorbed PS-X for different rates of compression and decompression. The rates vary from compression following overnight incubation to a rapid (~5' in each direction) compression-decompression cycle. Different symbols correspond to different compression or decompression runs.

III. DISCUSSION

We note first the qualitative differences between our results with grafted polymers and the earlier studies with adsorbed chains (summarized in Figs. 1 and 2). The absence of bridging, suggested by the fact that no attraction was detected even at the earliest accessible stages of incubation (where bridging attraction with adsorbed chains was marked, even for good solvents - see Fig. 1) is consistent with our concept of dangling PS tails anchored by the zwitterionic moiety: such non-adsorbing tails will not form bridges. Secondly, the extended range for detection of onset of interactions, corresponding to some 12 Rg (unperturbed radius of gyration of the PS-X), compares with some 7 ± 1 Rg for adsorbed chains (covering the same M range - see also Fig. 2) in good solvents. We shall relate to this extended grafted-layer thickness later. The repulsion itself is due to the osmotic segment-segment interactions in the good solvent condition of the experiment. Finally, we note the absence of any hysteresis or detectable surface relaxation effect in the polymer layer following strong compressions, which was a characteristic feature with adsorbed chains; this had been attributed to the forcing onto the surface of adsorbing segments, which then require a finite time to desorb and permit the chains to attain their equilibrium configuration.[7,11] The absence of such a feature in the case of the non-adsorbing PS-X confirms our intuitive expectation that such chains do not undergo this 'sticking' process, but are purely elastic in their response to high compression (within the time scales of our experiments).

The reversible nature of the force profile following high compression deserves comments in view of the fact that the compressive energies involved (per chain) are very much larger than the few kT that are likely to be holding each PS-X down on the mica surface. We attribute this to kinetic factors: the molecules within the highly compressed (and thus entangled) grafted layers would diffuse only very slowly even if the anchoring points were forced off the surface, on the one hand; at the same time the very slow disentanglement process within the compressed layer results in an effective network, permanent on the time scales of our measurements, which is trapped between the mica surfaces.

Our results differ, both qualitatively and quantitatively, from an earlier study of forces between mica sheets in polystyrene-polyvinylpyridine (PS-PVP) diblock copolymer solutions in toluene,[13] and which had also been interpreted in terms of an adsorbing PVP moiety and a dangling PS tail. The reasons for this are considered in detail elsewhere.[14]

The fact that the force profile (Fig. 5) is independent of the compression rate suggests we may apply equilibrium theoretical models to understand the structure of the adsorbed layers and the form of the interaction. (Strictly speaking, we refer to a restricted equilibrium - one that is subject to a constant number of end-grafted chains on the surface). There have been several theoretical discussions of the problem of grafted chains.[15-18] Here we shall analyze our data in terms of some early and very simple ideas on non-adsorbing chains with adsorbing polar headgroups, due to Alexander[16] and extended by de Gennes[5] to the case of surface forces. This is illustrated in Fig. 6. For end-anchored chains with a mean interanchor spacing s, the total excess free energy of the grafted layer consists of two main parts. On the one hand, the osmotic repulsion between the segments (in the good solvent) tends to extend the polymers to minimize the overall repulsive energy. On the other hand, for chains that are too extended there is a high price to pay in terms of elastic energy increase. Minimization of the total energy with respect to the chain extension yields the equilibrium grafted-layer thickness, L. Alexander's treatment suggested that for the interanchor spacings $s << R_F$, (where R_F is the swollen end-to-end dimension of the chain in the good solvent), the grafted layer could be considered as a close packed array of 'blobs', each of size s (Fig. 6a), within which excluded volume interactions are effective, but where segments in different 'blobs' are 'screened' from each other. The predicted value was

$$L \simeq s[R_F/s]^{5/3}, \quad (1)$$

---

*Eq. (1) predicts that for a given surface coverage

$$L \simeq sR_F^{5/3} \propto N. \quad (A1)$$

However, when the anchoring moiety (-X in our experiments) has only a finite sticking energy, one might expect that <u>in equilibrium</u> longer chains would be more sparsely distributed, i.e., have a larger s. This is in order to increase their 'blob' size and reduce the number of blobs per chain in order for the excess adsorption energy (~ kT per blob) not to exceed the sticking energy. In the case of such equilibrium, it is very straightforward to show that the expected variation of L with N goes as

$$L \propto N^{3/5},$$

rather than (A1) above. For a <u>given</u> s value, however, Equation (1) in the text holds.

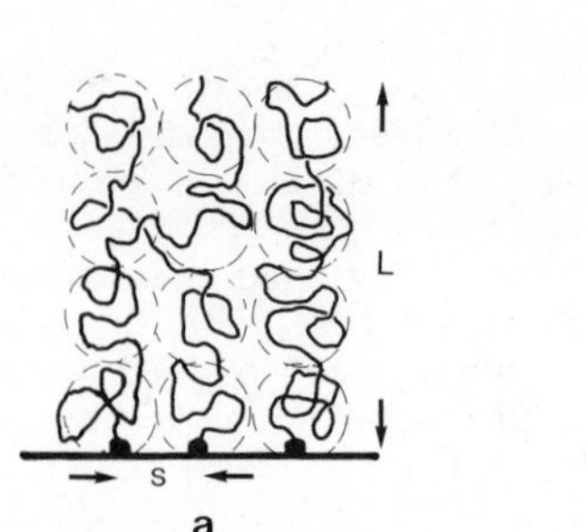

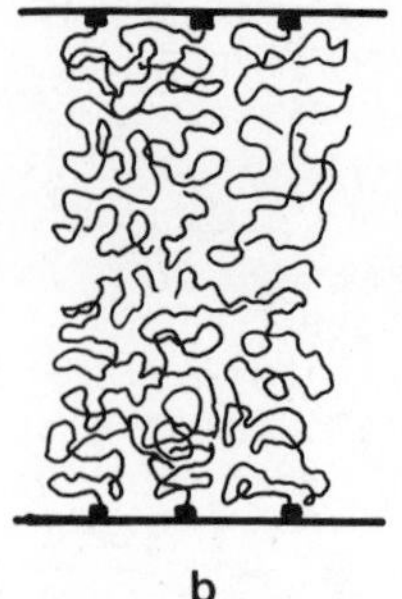

Fig. 6(a) Illustrating the Alexander picture of end-grafted chains as an array of close packed blobs.
6(b) Illustrating de Gennes' hypothesis of no interpenetration on compression of the grafted layers.

in this model.* Since this is a scaling result, a numerical prefactor is missing. It is interesting to compare our experimental determination of L, which we take as half the distance for detection of onset of interactions, with this prediction. We can do this since both s and R_F are independently measured, through the refractive index data as described above, and via intrinsic viscosity measurements,[20] respectively. Using these values (s = 85 Å, R_F = 320 Å) we find equation (1) to yield $L \simeq$ 740 Å, while from Fig. 5 we estimate $L \simeq$ 650 Å. The agreement is certainly reasonable.

We remark on two features here: the mean concentration of polymer in the grafted layers is around 4% w/w, compared with an overlap value c^* for this polymer in a good solvent of around 1-2%; secondly, R_F is considerably larger than s. Both inequalities suggest that we are in the correct regime for applying the scaling ideas of Alexander and de Gennes.

de Gennes extended Alexander's analysis to the case of interaction between two surfaces with grafted polymers.[5] The essential assumption here is that no interpenetration occurs; as the surfaces approach, the grafted chains are compressed (Fig. 6(b)). This changes the interaction energy in two ways: the mean concentration of polymer in the gap increases, thereby increasing the osmotic repulsion, and at the same time the overextended, stretched chains are compressed, resulting in a decrease of elastic energy. Overall, the force per unit area between two such layers is predicted as:

$$F(D) \simeq kT/s^3\,[(2L/D)^{9/4} - (D/2L)^{3/4}], \tag{2}$$

where $D < 2L$ is the distance between the mica sheets. Eq. (2) gives the force between two flat parallel plates, and to compare with our data (Figs. 4 and 5) it is necessary to convert to the corresponding interaction energy

$$U(D) = \int_{2L}^{D} F(D')\,dD'. \tag{3}$$

This variation is plotted as the solid line in Fig. 5 where all parameters in Eq. (2) and (3) are independently known. The fit is good, both as regards the spatial scale and the variation in energy. Since the scaling prediction (2) lacks a numerical prefactor, however, the curve has had to be shifted in the absolute energy scale. Knowing the other parameters explicitly, we find that this shift corresponds to a factor 1.8. In other words, the scaling prediction is quantitatively correct to within a factor 2.

IV. CONCLUSIONS

In summary, we find that surface forces with non-adsorbing end-anchored chains are monotonically repulsive with a range substantially greater than for corresponding adsorbed chains; we find no evidence of bridging attraction even at low adsorbance, and no indication of hysteretic effects following high compressions. Our data (for the polymer sample used) is well described by the scaling model of Alexander and de Gennes.

ACKNOWLEDGEMENTS

An SERC CASE studentship with Unilever (H.J.T.) and the partial support of the Israeli Academy, Basic Research Division and the Minerva Foundation (J.K.) are gratefully acknowledged. J.K. is the incumbent of the Herman Mark chair in Polymer Physics at the Weizmann Institute.

NOMENCLATURE

c	concentration
c^*	overlap concentration
D	distance between two mica sheets
$F(D)$	force per unit area
k	Planck constant
L	equilibrium grafted-layer thickness
N	degree of polymerization
R_T	swollen end-to-end dimension of the chain
R_g	radius of gyration
s	size of blob
T	temperature
$U(D)$	interaction energy

REFERENCES

1. P.G. de Gennes, Scaling Concepts in Polymer Physics, Cornell University Press, Ithaca, New York, 1979
2. P.G. de Gennes, Macromolecules, 14, 1637 (1981); 15, 492 (1982).
3. P.G. de Gennes and P. Pincus, J. Phys. (Paris) Lett., 45, L953 (1984).
4. P.G. de Gennes, Pour la Science (Scientific American-French Edition), January 1987, p.40.
5. P.G. de Gennes, Comptes Rendus Acad. Paris, 300, 839 (1985).
6. J. Klein, unpublished experiments.
7. J. Klein, in Proceedings of the First Toyota Conference, Elsevier, Holland 1988, p.333-352; M. Nagasawa, Editor.
8. D.H. Napper, Polymeric Stabilization of Colloidal Dispersions, Academic, London 1983.
9. (a) D. Tabor and R.H.S. Winterton, Proc. Roy. Soc., A312, 435 (1969); (b) J.N. Israelachvili and G. Adams, J. Chem. Soc. Faraday I, 74, 975 (1978).
10. A brief account of the main PS-X results has appeared, H.J. Taunton, C. Toprakcioglu, L.J. Fetters, J. Klein, Nature, 332, 712 (1988).
11. P.F. Luckham and J. Klein, Macromolecules, 18, 721 (1985).

12. J. Klein, J. Chem. Soc. Faraday I, 79, 99 (1983).
13. G. Hadziioannou, S. Granick, S. Patel and M. Tirrell, J. Amer. Chem. Soc., 108, 2869 (1986).
14. H.J. Taunton, C. Toprakcioglu, L.J. Fetters and J. Klein, Macromolecules, submitted.
15. A.K. Dolan and S.F. Edwards, Proc. Roy. Soc., A, 343, 427 (1975).
16. S. Alexander, J. Phys. (Paris), 38, 983 (1977).
17. S. Milner, M.E. Cates and T.A. Witten, Europhysics Lett., 5, 413 (1988); also S. Milner, private communication.
18. C.M. Marques, J.F. Joanny, L. Leibler, Macromolecules, 21, 1051 (1988).
19. (a) J.E. Roovers, and S. Bywater, Macromolecules, 5, 385-390 (1972).
    (b) Y. Einaga, Y. Miyaki and M. Fujita, J. Polym. Sci., 17, 2103 (1979).
20. B.V. Derjaguin, Kolloid Zh., 69, 155 (1934).

PERSISTENCE LENGTH IN MICROEMULSION SYSTEMS

O. Abillon,* B.P. Binks,* D. Langevin,* J. Meunier* and R. Ober**

*Laboratoire de Physique de l'Ecole Normale Supérieure
24, rue Lhomond, 75231 Paris Cedex 05
France

**Laboratoire de Physique de la Matière Condensée
Collège de France, place M. Berthelot
75231 Paris, Cedex 05, France

ABSTRACT

Microemulsions are dispersions of oil and water made with surfactant molecules. Their microstructure is closely dependent on the spontaneous curvature of the surfactant film. In some cases, the curvature can be very small, and the microstructure is locally lamellar. De Gennes pointed out that the macrostructure is disordered because of thermal fluctuations of the surfactant film. He introduced the persistence length ξ_k of the film, length over which the film remains locally flat, and related this length to the bending elastic modulus K of the film. He postulated that there should be a characteristic size of the oil and water microdomains, and that this size will be close to ξ_k. We will present measurements of the film elastic modulus K (obtained with ellipsometry) and of the characteristic size ξ_k (obtained with low-angle X-ray diffraction) done on several microemulsion systems confirming these ideas.

I. INTRODUCTION

Long linear polymer chains are locally rigid. They only behave as flexible coils on a scale larger than the persistence length ℓ_p of the chain. These coils are random-walk associations of elementary steps of size ℓ_p. ℓ_p can be much larger than the monomer size ℓ_0 if, for instance, the chains carry ionizable groups (polyelectrolytes). ℓ_p is always finite and has an Arrhenius temperature dependence: $\ell_p = \ell_0 \exp(\Delta\varepsilon/kT)$, where $\Delta\varepsilon$ is related to the bending energy of the chain.[1]

Two-dimensional flexible sheets like monolayers or bilayers of amphiphilic molecules can also be characterized by a persistence length ξ_k, which is basically the correlation length associated with order in the local normals to the surface:

$$\xi_k = a\, exp\,(2\,\pi K/kT)\ , \qquad (1)$$

where a is a molecular size and K the bending elastic modulus of the amphiphilic layer.[2] The surface is locally flat and crumpled on scales larger than ξ_k. The short wavelength undulations of the layer affect the value of the bending modulus which is scale-dependent.[3] Renormalization group calculations predict two different behaviors, according to the type of order in the layer. For liquid-like in-plane order, as for instance in L_α lamellar phases of lipid molecules, the undulations soften the macroscopic bending modulus.[4] The persistence length is always finite: the undulations destroy the long-range order in surface normals. For crystalline or hexadic in-plane order, like in L_β phases of lipids or in polymerized membranes, the situation is different. Because of the coupling between order in surface normals and positional order in the layer, the sign of the scale variation of the bending modulus can be sometimes reversed. A "crumpling" transition can thus occur between a state where the layer is flexible and crumpled (finite ξ) and a state where it is rigid and flat (infinite ξ).[5] This problem has close connections with elementary particle physics (string model).[6]

In multilayered systems, like lamellar phases, the interactions between the layers have also to be taken into account. This leads to another interesting type of transition for liquid-like layers. Thermal undulations are responsible for a steric repulsion between the layers inversely proportional to the bending modulus K.[7] Renormalization group calculations of the effective Hamiltonian for interacting layers shows the existence of a critical "unbinding" transition. According to the values of K or of an external pressure or constraint like the amount of solvent separating the layers, the effective interaction can be either attractive or repulsive at large distances. In the first case the layers are bound together, in the second they are completely separated. This might explain why certain lamellar phases can be swollen by an unlimited amount of solvent.[8] The unbinding transition is similar to the wetting transition and to the commensurate-incommensurate transition.

Microemulsions are mixtures of oil and water stabilized by surfactant molecules.[9] Frequently the medium is made of droplets either of oil or water, surrounded by a surfactant monolayer and dispersed in a continuous media water or oil respectively. The maximum droplet radius is close to the spontaneous radius of curvature of the surfactant monolayer R_o, which appears in the complete expression of the bending energy of the monolayer:[10]

$$F_c = \frac{K}{2}\,(c_1 + c_2 - 2\,c_o)^2 + \overline{K}\,c_1 c_2 \ \ per\,unit\,area\ , \qquad (2)$$

where K is the bending elastic modulus defined earlier, $\overline{K}$ the saddle-splay bending modulus, c_1 and c_2 the local principal curvatures of the layer and c_o the spontaneous curvature $c_o = 1/R_o$. The sign for c_o is by convention positive for oil droplets, negative for water droplets.

Although less frequent, a more original situation occurs when c_o is small and the structure locally lamellar. The bending elastic modulus in these systems is very small, comparable to kT. This means that the persistence length of the layers is microscopic: $a \sim 1$ Å and $K \sim kT$ gives $\xi_k \sim$ 500 Å. Long range order is destroyed and the medium is macroscopically isotropic: the structure is sponge-like, and bicontinuous in oil and water. Geometrical space filling models allow to calculate the mean size of the oil and water microdomains. De Gennes and Taupin proposed to use consecutive cubes of linear size ξ:

$$\xi = \frac{6\,\phi_o\,\phi_w}{c_s\,\Sigma}\,, \tag{3}$$

where ϕ_o and ϕ_w are respectively the oil and water volume fractions (the surfactant volume fraction has been neglected), c_s the surfactant concentration and Σ the area per surfactant molecule.[2] De Gennes and Taupin proposed also to take $\xi = \xi_k$. Recent theories taking into account the scale variation of K give more complete predictions for the ratio ξ/ξ_k.[11]

In the following, we will present measurements of the bending constant K. In order to suppress interactions between surfactant layers, we have studied a surfactant monolayer at a macroscopic oil-water interface. Due to the fact that such an interface possesses a finite surface tension γ, we had to consistently renormalize K and γ to interpret the measurements.[12] We have found:

$$\xi_k = a\,exp\,(\alpha\,K(a))/kT\,;\quad \alpha = \frac{4\,\pi}{3}\,(1 - e^{-1.5}) = 3.254\,, \tag{4}$$

where $K(a)$ is the bending constant at scale a.

II. EXPERIMENTAL PROCEDURE

We have studied three model systems of five-component mixtures: oil - water - surfactant - alcohol - salt. Three different ionic surfactants have been used: dodecyltrimethyl ammonium bromide (DTAB), sodium dodecyl sulfate (SDS) and sodium hexadecyl benzene sulfonate (SHBS). The alcohol used was butanol: its addition was necessary because the above surfactants alone could form microemulsions. The salt screened the electrostatic interactions between surfactant polar heads: it reduced c_0. In this way, a continuous structural evolution o/w - bicontinuous - w/o could be obtained by addition of salt. At each salinity, the size of the structural elements was the largest (maximum swelling: R_0 for droplets, ξ_k for bicontinuous structures) because the microemulsions coexisted with excess oil and/or water.

The compositions of the systems are (in wt.%):

System SHBS		System SDS or DTAB	
dodecane	38.19	toluene	47
brine*	56.93	brine*	47
SHBS	1.66	SDS or DTAB	2
butanol	3.32	butanol	4

* The brine is an aqueous solution of sodium chloride for SHBS and SDS and of sodium bromide for DTAB.

Table 1. Measured K and renormalized $K(a)$ values

system	K/kT	$K(a)/kT$	ξ_k(Å)	ξ(Å)
DTAB	0.41	0.55	62	97
SDS	0.75	1.10	375	250
SHBS	0.34	0.86	175	400

The bending elasticities have been measured using ellipsometry.[14] They are roughly independent of salinity. The measured and renormalized values K and $K(a)$ are given in Table 1. Sizes in bicontinuous microemulsions were measured with small-angle X-ray and neutron-scattering techniques in the SDS system by Auvray et al.[15] and de Geyer and Tabony,[16] in SHBS system in collaboration with Auvray[17] and in DTAB system by us.[18] A typical X-ray spectrum is shown in Fig. 1.

III. DISCUSSION AND CONCLUSIONS

In bicontinuous microemulsions, it can be shown that the peak of spectrum is related simply to ξ: $q_{max} = \pi/\xi$.[19] From the experimental data, we have made a comparison between ξ and ξ_k. The results are given in Table 1.

The data are in qualitative agreement with the theoretical predictions: the ratio ξ/ξ_k is of order unity although it varies from one system to another. Let us note that, although de Gennes took the condition $\xi = \xi_k$ as a rough assumption, it has been shown later that the ratio ξ/ξ_k should be about universal for microemulsions in the limit of coexistence with both water- and oil-excess phase: ξ/ξ_k variations may arise from differences between surfactant chemical potential in oil and water.[12] The origin of the observed variations of ξ/ξ_k may also arise from oversimplifications in the theories. Further improvement of these theories and possibly new experimental developments (measurement of K) are needed to achieve a complete understanding of the problem.

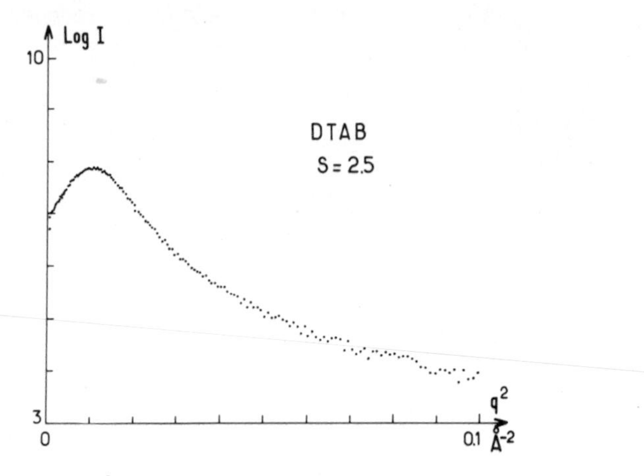

Fig. 1. X-ray spectrum for a bicontinuous microemulsion. DTAB system, brine salinity = 2.5 wt.%.

This problem is one aspect of the more general one of statistical mechanics of random surfaces. It is probably one of the most complicated because it falls in the range where the existing theories do not yet apply ($K \sim kT$, $\xi \sim \xi_k$). Many other systems have been intensively investigated in this field: lipid L_α and L_β lamellar phases, swollen lamellar phases, giant vesicles, polymerized lipid membranes, etc.[20] The two dimensional analog of the one-dimensional random-walk problem of linear polymer chains is therefore extremely rich both from a theoretical point of view where a variety of interesting behaviors has been predicted and from an experimental point of view, in particular in amphiphilic molecules systems. Interesting advances are expected in the coming years.

NOMENCLATURE

a	molecular size
c_0	spontaneous curvature of the surfactant layers
c_1,c_2	local principal curvatures
c_s	surfactant concentration in the bicontinuous microemulsion
ε	bending energy of the chain
I	scattered X-ray intensity
$\underline{K}$	splay elastic bending constant
$\overline{K}$	saddle-splay elastic bending constant
ℓ_p	persistence length of the chain
ξ	characteristic size in the bicontinuous microemulsion
ξ_k	persistence length of the surfactant layers
$\mathbf{q}$	scattering wavevector
Σ	area per surfactant molecule
ϕ_o	oil volume fraction in the bicontinuous microemulsion
ϕ_w	water volume fraction in the bicontinuous microemulsion

REFERENCES

1. P.G. de Gennes, Scaling Concepts in Polymer Physics, Cornell University Press, 1979.
2. P.G. de Gennes and C. Taupin, J. Phys. Chem., 86, 2294 (1982).
3. W. Helfrich, J. Phys., 46, 1263 (1985).
4. L. Peliti and L. Leibler, Phys. Rev. Lett., 54, 1690 (1985).
5. Y. Kantor and D. Nelson, Phys. Rev. Lett., 58, 2774 (1987).
6. J. Frölich, in Applications of Field Theory to Statistical Mechanics, L. Garido, ed., Lecture Notes in Physics, vol. 216, Springer, New York (1985).
7. W. Helfrich, Z. Naturforsch., 339, 305 (1978).
8. R. Lipowsky and S. Leibler, Phys. Rev. Lett., 56, 2541 (1986).
9. L.M. Prince, Microemulsions, Academic Press, New York (1977).
10. W. Helfrich, Z. Naturforsch., 28a, 693 (1973).; S.A. Safran, L.A. Turkevich and P. Pincus, J. Phys. Lett., 45, L-69 (1984).
11. S.A. Safran, D. Roux, M.E. Cates and D. Andelman, Phys. Rev. Lett., 57, 491 (1986).
12. J. Meunier, J. Phys., (Paris) 48, 1819 (1987).
13. D.A. Huse and S. Leibler, J. Phys., 49, 605 (1988).
14. J. Meunier, J. Phys. Lett., 46, L-1005 (1985).; B.P. Binks, J. Meunier, O. Abillon and D. Langevin, in preparation.
15. L. Auvray, J.P. Cotton, R. Ober and C. Taupin, J. Phys., 45, 913 (1984).
16. A. de Geyer and T. Tabony, Chem. Phys. Lett., 113, 83 (1985).
17. De Guest, D. Langevin and L. Auvray, J. Phys. Lett., 46, L-1005 (1985).

18. O. Abillon, B.P. Binks, C. Otero, D. Langevin and R. Ober, J. Phys. Chem., to appear.
19. S. Milner, S.A. Safran, D. Andelman, M.E. Cates and D. Roux, J. Phys., 49, 1065 (1988).
20. J. Meunier, D. Langevin and N. Boccara, editors, Physics of Amphiphilic Layers, Springer, New York (1987).

POLYMER ADSORPTION AT THE SOLID-LIQUID INTERFACE: THE INTERFACIAL CONCENTRATION PROFILE

I. Caucheteux, H. Hervet and F. Rondelez

Physique de la Matière Condensée, UA 792
Collège de France
11 place Marcellin-Berthelot
75231 Paris Cédex 05, France

L. Auvray and J.P. Cotton

Laboratoire Léon Brillouin (CEA-CNRS)
CEN-Saclay
91191 Gif-sur-Yvette
Cédex, France

ABSTRACT

Evanescent wave induced fluorescence (EWIF) and small angle neutron scattering (SANS) experiments have been performed at the polymer solution-solid interface, in the case of adsorption. The interfacial layer is characterized by its total surface excess Γ and by the monomer concentration profile $\Phi(z)$ in the direction normal to the interface.

EWIF has been used to study the adsorption of poly(methylmethacrylate) (PMMA) onto sapphire. Polymer molecular weights are 120,000 and 600,000 and some chains have been labelled with anthracene chromophores to allow fluorescence excitation by the optical evanescent wave. The surface excess is measured to be 1.47 mg/m^2. Moreover, the first moment of the monomer distribution, $\langle z \rangle$, can be evaluated. The observed value $\langle z \rangle$ = 73Å, is comparable to the chain radius of gyration, R_G, in good agreement with the theoretical models. The determination of higher moments of the distribution is not accessible with the present system. It would require the use of chains with molecular weights in the range 10^6-10^7.

SANS is not prone to this limitation. The full concentration profile has been measured for two different systems, namely poly(methylmethacrylate) (M_w = 265,000) adsorbed onto γ-alumina and poly(dimethylsiloxane) adsorbed onto porous silica. In both cases the results are in support of the scaling law behavior $\Phi(z) \propto (a/z)^{4/3}$. The profile is self-similar, as predicted by de Gennes, in the central region $Z_{min} < z < R_G$. Distances smaller than Z_{min} correspond to the proximal region. The measured width Z_{min} depends on the particular system. It is close to a monomer length $\simeq$ 2.5 Å in the case of the highly flexible poly(dimethylsiloxane) chains. It is significantly higher, $\simeq$ 11.5 Å, for poly(methylmethacrylate). This latter value is compatible with the PMMA chain persistence length. Finally, the surface excess measured by SANS, Γ = 1.7

mg/m^2 is in fair agreement with the value measured by EWIF on the PMMA-sapphire system.

In a separate experiment, the possibility of monitoring fast adsorption kinetics by EWIF is demonstrated.

I. INTRODUCTION

The presence of a sharply defined interface between a polymer solution and a solid wall leads to important modifications in the local polymer concentration with respect to the bulk concentration. These variations can be positive or negative depending on the sign of the interaction between the solid wall and the macromolecular chains immersed in the solvent. Attractive forces lead to adsorbed layers while repulsive forces lead to depletion layers. Due to their connection with important technological applications such as adhesives, protective coatings, microlithography, emulsion stabilizers, adsorbed polymer layers have been the subject of extensive theoretical studies.[1]

The ultimate goal was to predict the monomer concentration profile $\Phi(z)$ in the interfacial region as a function of the distance z away from the interface. For many years attempts have been made to derive the structure of polymer layers by some form of self-consistent field theory. Each chain is submitted to an average potential which is the combination of a short range attraction due to the wall and a long range repulsion proportional to the concentration profile $\Phi(z)$. A recent approach along these lines is due to Scheutjens and Fleer[2] who gave detailed numerical results on the repartition of the chain monomers between trains, loops and tails respectively. The trains correspond to chain segments in direct contact with the wall while the loops and tails extend into the solution over distances comparable to the chain radius of gyration in bulk solution. It has been progressively recognized, however, that the mean-field theory neglects important correlations, which, when properly taken into account, considerably modify the interfacial profile. The scaling approach, based on renormalization group theory, incorporates these correlations rigorously and predicts power law exponents: $\Phi(z) \propto (a/z)^{4/3}$ which are very different from the mean-field result $\propto (a/z)^2$. The above scaling prediction has been first derived by de Gennes in a seminal paper on polymer adsorption published in 1981.[3] Particularly striking is the simplicity with which the above result was obtained. The derivation is based on two essential ingredients:

1) Since the concentration is much larger than in the bulk, the chains overlap and can no longer be considered as non-interacting isolated objects. The adsorbed layer behaves locally as a semidilute solution. Under such conditions one defines a characteristic mesh size or correlation length ξ, which is the average distance between entanglement points and scales in good solvent as $\xi \sim \Phi^{-3/4}$.[4] It is typically in the range of 10-1000 Å.

2) There must be locally a single characteristic length in the problem. Consequently, the size ξ at any position within the interfacial layer has to be proportional to the distance z from the wall: the profile is thus self-similar.

The combination of these two ingredients:

$$\xi(\Phi(z)) \propto z ,$$

and

$$\xi(z) \propto \Phi(z)^{-3/4}$$

leads to the strikingly simple power law behavior

$$\Phi(z) \propto z^{-4/3}$$

This scaling approach does not give the precise numerical coefficients. Recently, however, des Cloizeaux[5] has shown that the prefactor A in the expression $\Phi(z) = A(z)^{-4/3}$ is determined as soon as the polymer mass and radius of gyration are known. The profile is then uniquely defined. Another approach to calculations of the prefactors is to use numerical methods such as Monte Carlo computations or direct enumerations on chains with finite lengths confined on lattice sites.

The above scaling law is only valid within a strict spatial range. For the usual case of strong adsorption (energy gain per monomer upon attachment to the solid surface comparable to kT) the lower limit Z_{min} for fully flexible chains is the monomer size, a; for more realistic chains it is the persistence length, p. The upper limit, Z_{max}, is the size, R_F of one polymer coil in bulk solution. Calling N the number of monomers per chain one has $Z_{max} \propto N^{3/5}$. As N is large in practice, the self-similar picture is therefore valid over a reasonably large domain of space. It was called the "central region" by de Gennes. The "proximal" region next to the wall extends to ($z < Z_{min}$) and the "distal" region is in contact with the bulk solution ($z > R_F$). The fact that the layer extends to R_F is by no means a trivial result. In the case of strong adsorption one would expect the chains to be adsorbed in a flat conformation with $Z_{max} \simeq a$. In reality Z_{max} is much larger because many chains compete for the same adsorption sites. Therefore, each chain attaches a limited number of its monomers to the wall.

The confrontation of this elegant prediction by experiment has taken many years to complete because all experimental techniques available at that time did not have the proper spatial resolution. Spectroscopic methods such as EPR,[6] NMR,[7] IR[8] are only sensitive to the fraction of monomers attached to the wall. Ellipsometry measures the first moment of the monomer distribution and yields an average length called the ellipsometric thickness $e_\ell \sim a^{1/3} R_F^{2/3}$.[9] Hydrodynamic methods, based on flow restric-tion in capillaries covered by the adsorbed layer, yield another average length related to the maximum extension of the adsorbed layer $e_H \sim R_F$.[10,11] To study concentration profiles at the solid-liquid interface, one has thus to devise special tools.

The first of these techniques is the EWIF method, an acronym for Evanescent Wave Induced Fluorescence. The power of EWIF has been first demonstrated in the case of depletion layers and the main results have been presented in several publications.[12,13] However, it is only very recently that the method has been applied to adsorbed layers.

The second technique of choice is SANS. Small angle neutron scattering is a classical tool for studying separately polymeric[14] and colloidal[15] systems. As first shown by Cosgrove et al,[16,17] it can also be used to observe polymers at interfaces. These authors were able to measure for the first time the concentration profile of poly(oxyethylene) chains adsorbed onto polystyrene colloidal grains. Their data were in qualitative agreement with the train-loop-tail description of Scheutjens and Fleer. However, these experiments were performed on too short chains to allow a valid comparison with scaling predictions to be made.

Still another approach is to use neutron reflectometry.[18] Results using this novel technique are presented at this conference by G. Jannink.[19]

In the following, we will present results obtained with the SANS and EWIF methods on several polymer-solvent systems. Experiments have been performed on chains in good solvents. Their molecular weights were chosen large enough ($M_w > 10^5$) to allow meaningful comparison with the scaling prediction. Poly(methylmethacrylate) and poly(dimethylsiloxane) were selected because they possess widely different glass transition temperatures and therefore different chain flexibility. The solid surfaces were silica glass and alumina. In one case the same system was studied by both methods, which enables one to see their respective advantages and disadvantages. We will start by the presentation of the EWIF results and then continue with the description of the SANS data.

II. RESULTS OBTAINED WITH EWIF:

The EWIF method allows one to measure the concentration profile in polymer solutions at the solid-liquid interface using optical illumination. It is based on the properties of evanescent waves. When a light beam propagating in a medium of high refractive index n_1 impinges on an interface with another medium of low refractive index n_2, the beam will experience total reflection if the angle of incidence exceeds the critical angle $\theta_c = \sin^{-1}(n_2/n_1)$. Under these conditions, Maxwell's equations show that an evanescent wave still penetrates in the optically rare medium. Its amplitude decays exponentially with the distance from the interface. The decrement of the exponential is the so-called penetration depth Λ. It depends on the incident wavelength λ_0, the refractive indices n_1 and n_2, and the angle of incidence according to the relation:

$$\Lambda = \frac{\lambda_o}{4\pi} \frac{1}{\sqrt{n_1^2 \sin^2\theta - n_2^2}} . \quad (1)$$

If the polymer chains bear some fluorescent groups, this evanescent wave can excite the chromophores contained in the illuminated volume. The total fluorescent intensity is simply given by

$$I(\Lambda^{-1}) = k \int_0^\infty \Phi(z) \exp - \frac{z}{\Lambda} dz , \quad (2)$$

where k is a prefactor in which have been lumped both the optical properties of the chromophores (adsorption characteristics, quantum yield, etc.) and those of the detection system (collection efficiency, photomultiplier response,...). This prefactor can be eliminated by performing ancillary measurements on a reference solution for which $\Phi(z)$ is known to be uniform. By making $\Phi(z)$ = constant = Φ_b in the expression for the total fluorescent intensity, one readily obtains:

$$I_F^{ref}(\Lambda^{-1}) = k\Lambda . \Phi_b , \quad (3)$$

Taking the ratio I_F/I_F^{ref}, the k factor cancels out and one gets $R(\Lambda^{-1})$ as a function of Λ, Φ_b and $\Phi(z)$ only:

$$R(\Lambda^{-1}) = \frac{1}{\Lambda \Phi_b} \int_o^{\infty} \Phi(z) e^{-z/\Lambda} \, dz \, . \tag{4}$$

Λ $R(\Lambda^{-1})$ is nothing but the Laplace transform of $\Phi(z)$. In principle, it is possible to extract $\Phi(z)$ by performing measurements at various Λ (or equivalently at various incidence angles) and then by taking the inverse laplace transform of Λ $R(\Lambda^{-1})$. This procedure however, does not converge easily and requires extreme data accuracy over the complete Λ range (0 to ∞). Moreover, the Λ range experimentally accessible has a natural cut-off at

$$\Lambda_{min} = \frac{\lambda_o}{4\pi(n_1^2 - n_2^2)^{1/2}} \, ,$$

corresponding to the glancing incidence angle. In most instances, it is therefore simpler to use a development in moments of the distribution $\Phi(z)$. Expansion of $R(\Lambda^{-1})$ in a power series of Λ^{-1} yields:

$$R(\Lambda^{-1}) = 1 + \frac{\Gamma}{\Phi_b} \frac{1}{\Lambda} - \frac{\Gamma < z >}{\Phi_b \Lambda^2} + \frac{\Gamma}{2\Phi_b} \frac{< z^2 >}{\Lambda^3} + o\left(\frac{1}{\Lambda^4}\right) , \tag{5}$$

where

$$< z^n > = \frac{\int_0^{\infty} z^n (\Phi(z) - \Phi_b) dz}{\int_0^{\infty} (\Phi(z) - \Phi_b) dz} \, . \tag{6}$$

This procedure has the advantage of being very general since it does not require any specific assumption about the concentration profile.

We can define the quantity Γ:

$$\Gamma = \rho \int_0^{\infty} (\Phi(z) - \Phi_b) dz \, , \tag{7}$$

where $\Phi(z)$ is the local volume fraction and ρ, the polymer density (in g/cm^3). The surface excess Γ is the total amount of polymeric material (in g/cm^2) contained in the interfacial layer, in excess of the bulk value. This surface amount is always positive in the case of adsorption.

The first moment $< z >$ gives the overall thickness of the adsorbed layer. For a square profile of width L, one gets $< z > = L/2$. For an exponential profile of characteristic length L, one gets $< z > = L$.

In order to extract the various moments of the distribution, it is sufficient to plot R as a function of Λ^{-1}. Close to the critical angle,

$\Lambda \rightarrow \infty$ and is much larger than $< z >$. The expression for $R(\Lambda^{-1})$ then simplifies to:

$$R(\Lambda^{-1}) = 1 + \frac{\Gamma}{\Phi_b} \Lambda^{-1} . \tag{8}$$

In this range, the variation of R with Λ^{-1} is linear and the slope yields directly the value of the surface excess Γ. When the experiments are performed at higher incidence angles, Λ decreases and becomes comparable to $< z >$. It is then no longer possible to neglect the higher-order terms in the distribution. The variation of $R(\Lambda)$ with Λ^{-1} starts to exhibit some curvature which contains the information on the layer structure.

This last condition is however not easily achieved in practice. Taking λ_o = 3500 Å, n_1 = 1.8, n_2 = 1.5 one finds Λ_{min} = 280 Å. By comparison the radius of gyration for a polystyrene chain of molecular weight 10^5 is only 10^2 Å. This shows one limitation of the method, which is best suited for very long chains. Unfortunately such chains are extremely difficult to synthesize with covalently bonded fluorescent groups.

The experimental system used consists of a dilute solution of poly(methylmethacrylate) (PMMA) in toluene (n_2 = 1.494) in contact with a solid surface of sapphire: γ-alumina, (n_1 = 1.79). Molecular weight of the polymer chain was 120,000 with a polydispersity index of 1.15. A small percentage of the chains were labelled with anthracene chromophores at a 1% mole concentration. The chromophores served as the fluorescent markers. Since they are located randomly along the chain, it was assumed that their distribution faithfully reflects the monomer concentration profile. PMMA of higher molecular weight was also used in some instances (M_w = 600,000). The reference solutions were anthracene molecules in pure toluene.

The sapphire surface was highly polished to make the surface roughness as small as possible (<30 Å rms), in order to minimize the possibility of stray light excitation of the bulk solution. Toluene was chosen not only because it is a good solvent for PMMA but also, and more importantly, because its refractive index is very close to that of PMMA (1.494 compared to 1.492). Under such conditions of optical matching, it is reasonable to assume that the refractive index of the solution is independent of the polymer concentration. This simplifies considerably the data analysis.

It was carefully checked that free anthracene molecules show no tendancy to absorb on the alumina surface. The most sensitive method is to compare the fluorescence spectra obtained from excitation of the bulk solution with the one obtained from evanescent wave excitation. Both spectra were found to be rigorously equivalent, using an excitation wavelength of 3630 Å. Obviously, the evanescent wave spectra was noisier because of the much smaller sample volume excited. However, the characteristic emission peaks at 3496 Å and 4162 Å were easily identified.

Figure 1 shows typical curves for the detected fluorescent intensity I_F as a function of the incident angle θ. Below θ_c, the light beam excites the whole sample, leading to a large signal. On the contrary, above θ_c, the evanescent wave sets in and there is a precipitous drop in the detected intensity. The solution is now excited over a thickness Λ much smaller than the cuvette thickness h (h/Λ ~ 500). The two curves correspond to a PMMA solution containing 5% of labelled chains and to a free chromophore solution respectively. The concentrations have been adjusted so that both intensities are approximately equal for angles smaller than θ_c (case of bulk excitation). This condition obviously is no longer satisfied in the evanescent mode regime. There is a difference by roughly two orders of magnitude in favor of the polymer solution. That means that the number of chromophores present in the volume illuminated by the evanescent wave is much larger than in the reference solution. This is a clear indication of an adsorbed polymer layer.

The same data have been replotted in Fig. 2 as $R(\Lambda^{-1}) = I_F/I_F^{ref}$ versus Λ^{-1} to enable the comparison with the moment expansion method. $R(\Lambda^{-1})$ starts from unity for $\Lambda = \infty$ and increases monotonically with decreasing Λ.

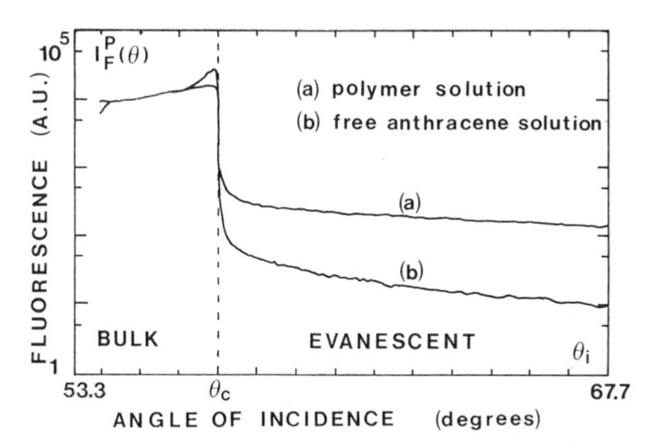

Fig. 1. Fluorescence intensity (in arbitrary units) versus incidence angle of the excitation optical beam. The critical angle for total reflection is $\theta_c = 56.6°$ in the case of γ-alumina-toluene solution interface. The two curves correspond to a solution of free anthracene molecules and to a solution of poly(methylmethacrylate) (PMMA) chains of molecular weight 120,000 bearing 1% of covalently-bonded anthracene moieties. The respective concentrations have been adjusted so to give identical signals for bulk excitation.

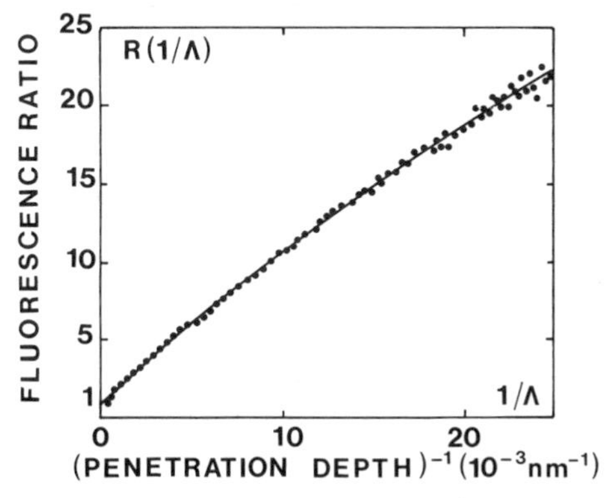

Fig. 2. Fluorescence intensity ratio R (see text for details) as a function of the inverse of the penetration length Λ for the optical evanescent wave. The solid curve represents the best fit to a first-order polynomial expansion in powers of Λ^{-1}. Polymer molecular weight is 120,000.

The surface excess of the adsorbed layer, Γ, given by the initial slope as $\Lambda \rightarrow \infty$, is equal to 1.47 ± 0.05 mg/m^2. The first moment $< z > = 73 \pm 8$ Å. Both of these results compare well with independent measurements. For instance, we have incubated γ-alumina spheres of diameter 3000 Å with the same PMMA solution and measured the adsorbed amount of polymer chain by the method of the supernatant. The remaining concentration in solution following the adsorption process was determined from the intensity changes of the IR adsorption band at 1735 cm^{-1}. We found $\Gamma = 3$ mg m^{-2}. In this last experiment, the surface-to-volume ratio had to be estimated from small-angle neutron scattering data, according to a procedure which is described in the second part of this paper.

The width of the adsorbed polymer layer is also consistent with the chain radius of gyration which can be calculated using the well-known formula $R_F = N^{3/5} a(\sqrt{7})^{-1/2}$. Taking $N = 1200$ and a = 2.52 Å for the PMMA chain, we obtain $R_G = 173$ Å. If the profile is assumed to be a step function of width R_G one than expects $< z > = R_G/2 = 86$ Å which is very close to our experimental finding of 73 ± 8 Å.

It may seem surprising that one uses a square profile and not the true profile $\Phi(z) \propto (a/z)^{4/3}$ predicted by the de Gennes scaling law. The reason is that our experiment does not allow us to test the concentration profile accurately enough, because the penetration distance of the evanescent wave $\Lambda_{min} = 2800$ Å is too large compared to the expected width of the adsorbed layer. Experiments on higher molecular weight samples are currently under way to get more insight on the profile.

Our absorbance data are also in excellent agreement with those already published in the literature. For instance, Takahashi et al.[9] have measured Γ to be between 1.0 and 1.2 mg m^{-2} for polystyrene in carbon tetrachloride adsorbed onto a chromium plate. Using ellipsometry, they were also able to measure an ellipsometric layer thickness of 200 Å for $M = 1.10 \times 10^5$. It is easy to show that $e_{ellips} = 2 < z >$. Therefore, this corresponds to a $< z >$ value of 100 Å which compares favorably to our own value of 68 Å.

As will be demonstrated in the second part, the SANS technique is more suited than the EWIF technique to investigate adsorbed polymer layers at short length scales ($0 < z < 300$ Å). Before turning to the description of the SANS experiments, we would like, however, to present kinetic experiments in which we have followed in real time the changes occurring in an adsorbed layer submitted to an external perturbation. After the adsorption equilibrium was established, a small quantity of water was allowed to sneak into the measuring cell. Water is a nonsolvent for PMMA and one therefore expects the polymer to precipitate out of the toluene solution. The adsorbed layer should then increase dramatically. Figure 3 shows the result of such an experiment.

We have plotted the ratio $R(\Lambda^{-1})$ versus Λ^{-1} at different times following the water uptake. For a given incidence angle, R increases with time. The initial slope which is proportional to the surface excess Γ gets steeper, indicating increased adsorption. Figure 4 shows the adsorption kinetics for two molecular weights of 120,000 and 600,000. It appears that smaller molecular weights are more rapidly adsorbed. After about 2 hours a new equilibrium is reached for Γ, which is about one order of magnitude larger than the initial equilibrium value in the pure toluene. With the 600,000 sample, the new equilibrium is not yet reached after 3 hours but it seems to converge towards the same final value as with the 120,000 sample.

This demonstrates the possibility of EWIF to perform adsorption kinetics measurements. The present time resolution is about 2 minutes if a complete $R(\Lambda^{-1})$ versus Λ^{-1} curve is needed and only a few seconds for data collection at fixed Λ. This is several orders of magnitude faster than SANS in which the low flux of neutrons requires a counting time of a few hours.

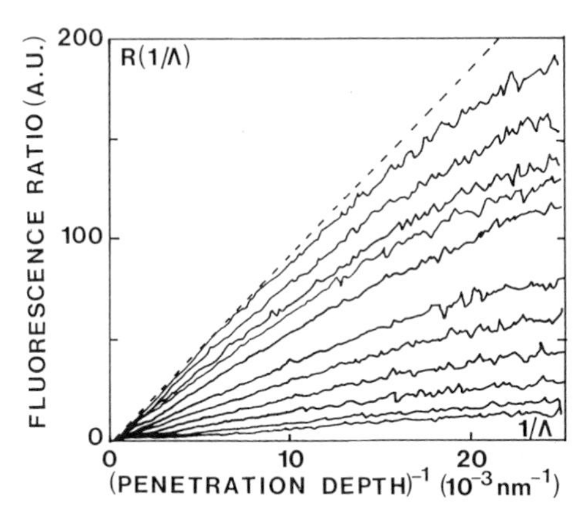

Fig. 3. Fluorescence intensity ratio R as a function of the inverse of the penetration length Λ for the optical evanescent wave. The various curves are taken at 12 successive time intervals after introducing a small quantity of a nonsolvent into the polymer solution. At fixed Λ, R markedly increases with time, indicating polymer accumulation in the adsorbed layer.

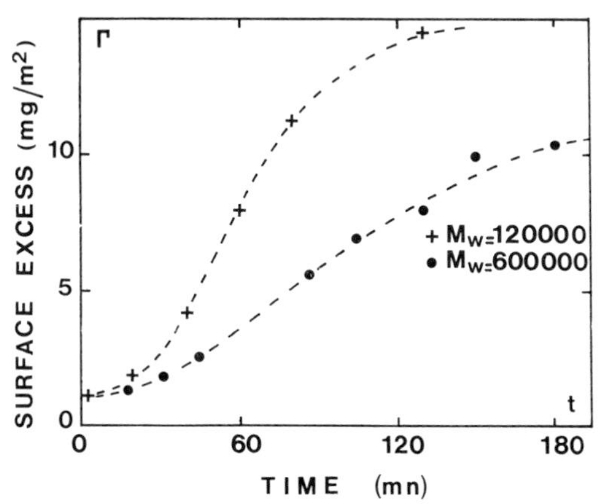

Fig. 4. Time evolution of the total surface excess Γ for poly(methylmethacrylate) chains adsorbed onto a γ-Alumina surface, following the introduction of a nonsolvent in the sample cell. Two different polymer molecular weights of 120,000 and 600,000 have been used.

III. NEUTRON-SCATTERING RESULTS

The second series of results that we present have been obtained by small-angle neutron scattering (SANS). Contrary to the EWIF method and to the neutron-reflectivity technique, SANS does not deal with a single, flat, well oriented, small surface but need a large amount of interface coated by polymer. The polymer substrate is therefore a porous medium or a colloidal suspension. For such samples, the scattered intensity $i(\mathbf{q})$ per unit volume (in cm^{-1}) is the sum of three terms resulting from the interferences between the amplitudes scattered by the solid and by the polymer.

$$i(\mathbf{q}) = (n_g - n_s)^2 S_{gg}(\mathbf{q}) - 2(n_g - n_s)(n_p - n_s) S_{pg}(\mathbf{q}) + (n_p - n_s)^2 S_{pp}(\mathbf{q}) ,$$

where n_g, n_s and n_p are the scattering length densities (in cm^{-2}) of the solid (g), solvent (s) and polymer (p) respectively. The scattering length density is the neutronic analog of the index of refraction. The scattering vector $\mathbf{q}$ is related to the scattering angle θ and the wavelength λ by $\mathbf{q} = (4\pi/\lambda) \sin \theta/2$. The partial structure factors S_{ij} (in cm^3) are proportional to the Fourier transform of the i-j density-density correlation function.

Because these structure factors depend on the particular geometry of the substrate, there is no simple relationship between them and the polymer density profile $\Phi(z)$. There is, however, one limiting case of interest, when the surface of the substrate is sharp and well-defined and when its curvature is smaller than the scattering vector $\mathbf{q}$, then the surface can be considered as flat at the scale $\mathbf{q}^{-1}$. In this regime, the partial structure factors reach an asymptotic limit depending only on the structure of the polymer layer and on the area per unit volume of the solid phase, S/V.[17,20]

The neutron-scattering results then becomes particularly informative. One can vary systematically the scattering length density of the solvent, n by isotopic substitution, and from the corresponding intensity measurements separate the different partial structure factors. This gives access to the detailed polymer profile.

- The first term of interest is the structure factor of the bare solid, $S_{gg}(\mathbf{q})$. In the asymptotic regime

$$S_{gg}(\mathbf{q}) = 2\pi \left(\frac{S}{V} \right) \mathbf{q}^{-4} . \tag{9}$$

This expression is known as the Porod's law and enables one to measure the specific area of the solid.

- The surface structure factor of the polymer layer, $S_{pp}(\mathbf{q})$ can be measured directly in a contrast matching experiment by choosing $n_s = n_g$. In principle, $S_{pp}(\mathbf{q})$ depends both on the average concentration profile $\Phi(z)$ and on the concentration fluctuations within the layer.[20] In the simple case, where the adsorbed layer can be assumed homogeneous, $S_{pp}(\mathbf{q})$ is related to the square of the Fourier transform of the profile $\Phi(z)$.

$$S_{pp}(\mathbf{q}) = \overline{S}_{pp}(\mathbf{q}) = 2\pi\left(\frac{S}{V}\right)\mathbf{q}^{-2}\left|\int_0^\infty dz\Phi(z)e^{i\mathbf{q}z}\right|^2 . \quad (10)$$

Until now, this expression, which can be inverted to yield $\Phi(z)$, and the underlying hypothesis of homogeneity of the adsorbed layer, have been the basic tools of investigation of polymer concentration profiles.[17,21,22]

- Finally, the contrast variation experiments yield the polymer-solid cross structure factor:

$$S_{pg}(\mathbf{q}) = 2\pi\left(\frac{S}{V}\right)\mathbf{q}^{-3}\int_0^\infty dz\Phi(z)\sin\,\mathbf{q}z . \quad (11)$$

It is remarkable that this expression is independent of the fluctuations in the layer.

The evaluation of the partial structure factors S_{pp} and S_{pg} in the framework of the scaling theory which predicts $\Phi(z) \propto (a/z)^{4/3}$ has been carried out in Ref. (23). In the case of strong adsorption and in the q-range $z_{min} < \mathbf{q}^{-1} < R$, one expects for a polymer layer in contact with pure solvent:

$$S_{pp}(\mathbf{q}) = 2\pi\frac{S}{V}\gamma^2\mathbf{q}^{-2}(1 - 4\sqrt{3}\frac{a}{\gamma}(\mathbf{q}a)^{1/3} + \lambda(\mathbf{q}a)^{2/3}) , \quad (12)$$

and

$$S_{pg}(\mathbf{q}) = 2\pi\frac{S}{V}(2a^4)(\mathbf{q}a)^{-8/3} , \quad (13)$$

where γ (in cm) is the adsorbed amount per unit surface, in unit of volume fraction.

$$\gamma = \int_0^\infty dz\Phi(z) ,$$

is related to Γ (in g/cm^2), the adsorbed amount in unit of weight, by $\Gamma = \gamma\rho$, where ρ is the polymer density (in g/cm^3). λ is an unknown constant.

We have studied two different systems:

- The samples of the first system are made of poly(dimethylsiloxane) (PDMS) (M_w = 270,000, M_w/M_n = 1.2) adsorbed on porous silica (pore diameter 3,000 Å, specific area = 2.5 m^2/cm^3) in presence of cyclohexane.

- Those of the second system are made of poly(methylmethacrylate) (PMMA) (M_w = 265,000, M_w/M_n = 1.12) adsorbed on porous γ-Alumina (pore diameter 800 Å, specific area 17 m^2/cm^3) in presence of pure benzene. This system is therefore analogous to the one studied by EWIF since γ-alumina and sapphire are two different names for the same material.

In both cases, the polymer is monodisperse and the solvent is a good solvent. The polymer chains are adsorbed from solution at a concentration close to the onset of the semidilute regime ($c = 3 \times 10^{-2}$ g/cm^3) and the supernatant is rinsed with pure solvent. Notice that the radius of the pores is always larger than the radius of gyration R_g of the polymers.

The neutron experiments have been made at Laboratoire Leon Brillouin on the spectrometer PACE. The range of scattering vector, 10^{-2} Å^{-1} < $\mathbf{q}$ < 10^{-1} Å^{-1}, is chosen to probe the inner structure of the layers, $\mathbf{q}R_g >> 1$. For both systems, the partial structure factors of the polymer are extracted from the difference between the signal of the polymer-coated solid and that of the bare solid, observed at three different contrasts.

The results concerning the first system with a complete description of sample preparation and data analysis are already published in Ref. (24). Those concerning the second system are new.

We start by describing the structure of the PDMS chains adsorbed on silica.

- The structure factor of the layer, $S_{pp}(\mathbf{q})$

The theoretical expression (12) of the structure factor of the polymer layer, $S_{pp}(\mathbf{q})$, suggests a plot of the product $\mathbf{q}^2 S_{pp}(\mathbf{q})$ as a function of $\mathbf{q}^{1/3}$. This is done in Fig. (5). We observe a linear variation:

$$\mathbf{q}^2 S_{pp}(\mathbf{q}) = Y(1 - R\,\mathbf{q}^{1/3}) \ ,$$

with $Y = 0.24 \pm 0.02$ Å and $R = 1.9 \pm 0.2$ Å$^{1/3}$.

This dependence is in agreement with the prediction of the scaling theory. Note, however, that we do not observe the third term of Eq. (12), which varies as $\mathbf{q}^{-4/3}$ and is the sum of a contribution of the average profile and the contribution of the transverse fluctuations. In the $\mathbf{q}$-range observed, these fluctuations are not detected.

By comparison with Eq. (12), the values of the coefficients Y and R yield the adsorbed amount γ (in Å) or Γ (in g/cm^3) and the monomeric length a of the self-similar profile. We obtain

$$\gamma_{PDMS/SiO_2} = 12 \pm 1 \text{ Å} \ ; \ \Gamma = 1.2 \pm 0.1 \ mg/m^2$$

$$a_{PDMS} = 2.5 \pm 1 \text{ Å} \ .$$

- The polymer-solid cross structure factor, $S_{pg}(\mathbf{q})$

The experimental results obtained for the cross-term $S_{pg}(\mathbf{q})$ are displayed in Fig. (6) in a log-log representation. Using the first 20 experimental points for which the signal-to-noise ratio is sufficiently high, we observe that $S_{pg}(\mathbf{q})$ decreases as a power law of $\mathbf{q}$:

$$S_{pg}^{exp}(\mathbf{q}) = C\,\mathbf{q}^{-b}$$

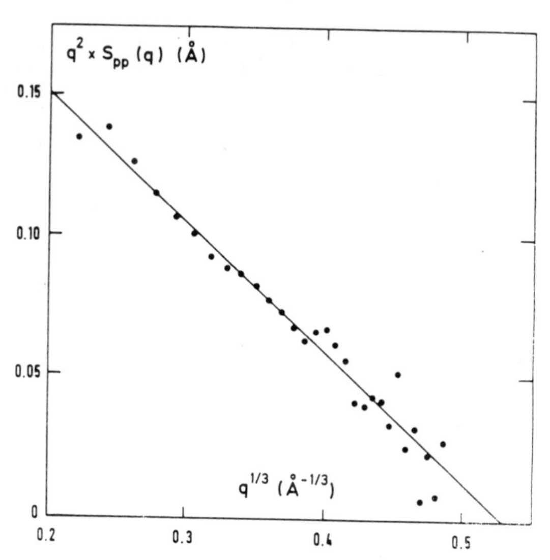

Fig. 5. Structure factor $S_{pp}(\mathbf{q})$ of the poly(dimethylsiloxane) (PDMS) layer adsorbed on silica in presence of cyclohexane, in the representation $\mathbf{q}^2 S_{pp}(\mathbf{q})$ as a function of $\mathbf{q}^{1/3}$.

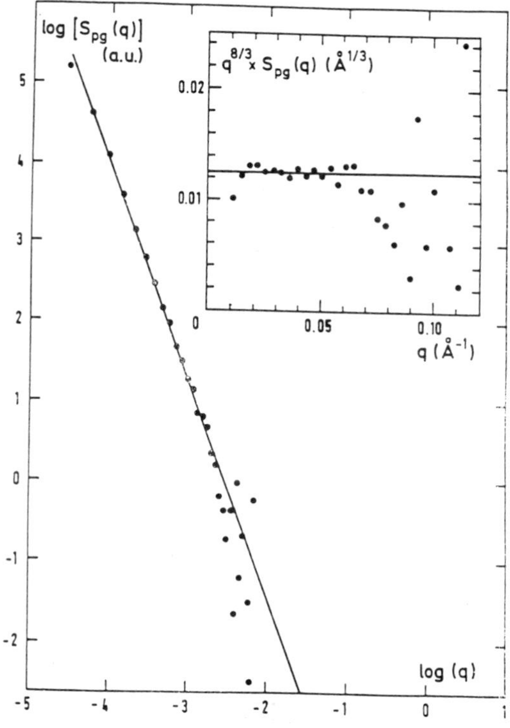

Fig. 6. The polymer-solid cross-structure factor of the system PDMS/silica/cyclohexane. The slope of the continuous straight line in the log-log plot is -2.65.

The observed value of the exponent, $b = 2.65 \pm 0.1$, is equal, within the experimental uncertainties, to the value predicted by the scaling theory. The value of the prefactor $C = (1.25 \pm 0.1)\ 10^{-2}\ \text{Å}^{1/3}$ enables us to evaluate the length a. One finds $a_{PDMS} = 3 \pm 1$ Å. This value is entirely consistent with the previous, independent evaluation obtained from S_{pp}.

The study of the system PDMS-silica-cyclohexane provides the first direct evidence for the de Gennes description of polymer adsorption from good solvent. A difficulty of this system is, however, that the contrast between PDMS and silica is relatively small.

We will now turn to the description of the structure of the PMMA chains adsorbed on alumina. This system offers a much better contrast for neutrons, which makes it experimentally very convenient. Moreover, it enables one to test the universality of the observed laws and to compare with the results obtained by EWIF.

- The structure factor of the polymer layer

As for the first system, we plot the structure factor of the polymer layer $S_{pp}(\mathbf{q})$ in the representation $\mathbf{q}^2\ S_{pp}(\mathbf{q})$ as a function of $\mathbf{q}^{1/3}$ (Fig. 7). Here also we obtain a straight line:

$$\mathbf{q}^2 S_{pp}(\mathbf{q}) = Y(1 - R\,\mathbf{q}^{1/3}) \ .$$

A self-similar profile decaying as $z^{-4/3}$ is thus coherent with the experimental data. As in the first case, we do not observe the transverse fluctuations of the layer. With the experimental values $Y = 2.13$ Å and $R = 1.38\ \text{Å}^{1/3}$, we obtain the three quantities:

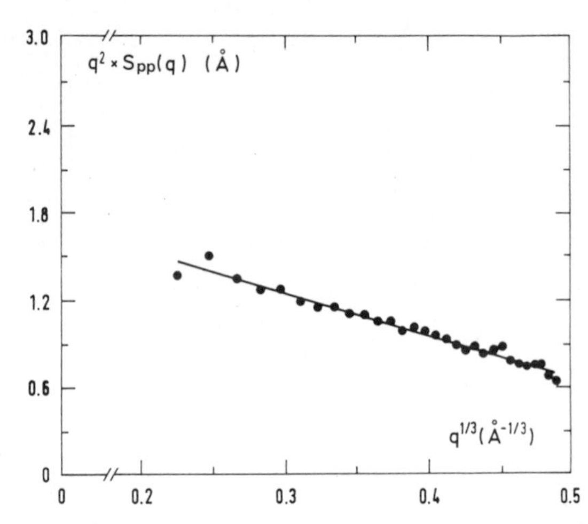

Fig. 7. Structure factor $S_{pp}(\mathbf{q})$ of the PMMA layer adsorbed on γ-alumina in presence of benzene, in the representation $\mathbf{q}^2 S_{pp}(\mathbf{q})$ as a function of $\mathbf{q}^{1/3}$.

$$\gamma_{PMMA/Al_2O_3} = 14 \pm 1 \text{Å} \quad ; \quad \Gamma_{PMMA/Al_2O_3} = 1.7\, mg/m^2$$

$$a_{PMMA} = 2.2 \text{Å}$$

The adsorbed amount $\Gamma_{PMMA/A\ell_2O_3}$ is relatively high and is of the same order of magnitude as the one measured on the monocrystalline sapphire. We note also that the cut-off lengths a for the two polymers, PDMS and PMMA, are comparable.

- The solid-polymer cross-structure factor

Since the structure factor of the PMMA polymer layer is well described by the scaling theory, we expect the cross-structure factor to decrease as $\mathbf{q}^{-8/3}$, as was indeed observed for PDMS on silica. However, it turns out, as shown in Fig. (8), that the product $\mathbf{q}^{8/3}\, S_{pg}(\mathbf{q})$ is not constant but increases with the scattering vector $\mathbf{q}$.

An explanation of this behaviour is possible if one takes into account the existence of a dense, thin proximal region in the adsorbed layer. In this region of width Z_{min}, the polymer concentration profile strongly depends on the particular interactions with the solid and is expected to be less universal than in the central self-similar region of the layer.

Because the $z^{-4/3}$ profile is singular at the origin, this proximal zone plays no role in the structure factor of the layer, $S_{pp}(\mathbf{q})$. As soon as $\mathbf{q}^{-1}$ is larger than Z_{min} its contribution is automatically included in the $\mathbf{q}^{-2}$ term related to the total adsorbed amount Γ. On the contrary, its existence modifies the cross-structure factor $S_{pg}(\mathbf{q})$. Going back to the calculation of the Sine-Fourier transform of the profile $\Phi(z)$, we can separate the contribution of the proximal region from that of the central zone:

$$\int_0^\infty dz\Phi(z)\sin\,\mathbf{q}z = \int_0^{Z_{min}} dz\Phi(z)\sin\,\mathbf{q}z + \int_{Z_{min}}^\infty dz\,\Phi(z)\sin\,\mathbf{q}z$$

Assuming that $\mathbf{q}\, Z_{min}$ is smaller than 1 and that the dominant contribution of the central region varies again as $\mathbf{q}^{1/3}$, we obtain $S_{pg}(\mathbf{q})$ as the sum of two terms:

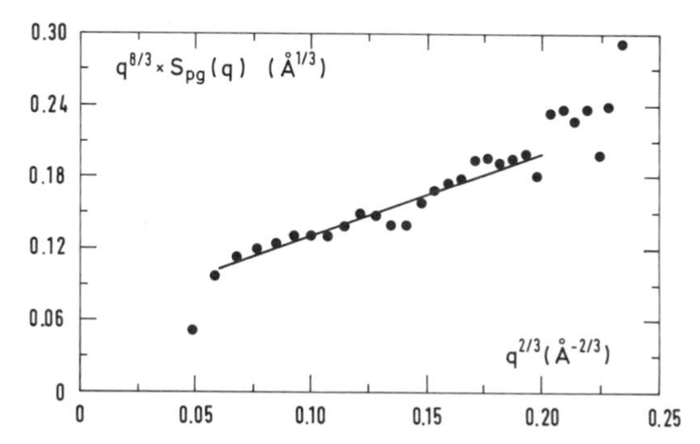

Fig. 8. The polymer-solid cross-structure factor of the system PMMA/alumina/benzene, in the representation $\mathbf{q}^{8/3}\, Spg(\mathbf{q})$ as a function of $\mathbf{q}^{2/3}$.

$$S_{pg}(\mathbf{q}) = \alpha \mathbf{q}^{-2} + \beta \mathbf{q}^{-8/3}.$$

The prefactors β and α are given by $\beta = 2\pi S/V(2a^{4/3})$ as in Eq. (13) and by

$$\alpha = 2\pi \frac{S}{V} \frac{Z_{min}^2}{2}$$

if the adsorbed layer is dense in its proximal part ($\Phi(z) = 1$ for $z < Z_{min}$).

A natural plot is then $q^{8/3} S_{pg}(q)$ as a function of $q^{2/3}$. We expect a straight line. As shown in Fig. (8), the agreement between the model and the experimental results is good. We get $\alpha = 0.7$ Å and $\beta = 6 \cdot 10^{-2}$ $Å^{1/3}$.

From these values, we deduce an estimation of the thickness Z_{min} of the proximal region, $Z_{min} = 11.5$ Å, and another determination of the microscopic length $a_{PMMA} = 2.2$ Å. This latter value is the same as the one previously determined. On the other hand, the proximal region is rather thick, of the order of the PMMA persistence length. Assuming a dense packing, one calculates that 80% of the polymer mass is contained in the proximal region.

IV. CONCLUSIONS

The SANS data presented here give strong support to the de Gennes main predictions for adsorbed polymer layers: the monomer concentration profile is self-similar and scales as $\Phi(z) \propto (a/z)^{4/3}$ for polymer chains in good solvent. This particular profile is observed over a spatial scale which extends between the chain persistence length, p, and the chain radius of gyration, R_F, in bulk solution. The width of the proximal layer in the immediate vicinity of the solid wall depends on the polymer system. In the case of poly(dimethylsiloxane) on silica, it is of the order of a few Å and comparable to the monomer length. For poly(methylmethacrylate) chains on alumina, it is of the order of 10 Å, and comparable to the persistence length of the chain backbone.

The SANS observations also show the interest of performing full contrast variation experiments. Indeed, the information given by the two partial structure factors of the polymer are not equivalent because of the very singular nature of the profile at close distances from the wall. The experimental data yield values of the surface excess Γ. They correspond roughly to one full monolayer of dense polymer. It must be stressed, however, that the chains are not adsorbed in a flat configuration on the solid surface, as demonstrated as well by the EWIF method.

In the case of PMMA adsorbed onto γ-alumina, a direct comparison can be made between EWIF and SANS since the same system was investigated by both techniques. The measured surface excesses are within 50% of each other. The origin of this difference may reside in slight variation in the composition or roughness of the two surfaces.

EWIF yields readily the total surface excess and the average adsorbed layer thickness but fails to obtain the power law exponent describing the concentration profile. This is due to the relatively long penetration distance of the evanescent electromagnetic field into the polymer solution. In order to extract relevant information, it would be necessary to work with penetration lengths Λ_{min} much smaller than the chain radius of gyration. Since Λ_{min} is more or less fixed to a few hundreds of Angströms by the optical properties of the materials used, it will be necessary in the future to use much longer polymer chains. Semi-flexible chains with long persistence lengths, such as some of the water-soluble polysaccharides, are good potential candidates.

There is one domain, however, in which EWIF is probably unsurpassable. Due to the high sensitivity of fluorescence measurements, significant data can be obtained in much less than one minute. Therefore, the kinetics of formation of the adsorbed layer can be monitored almost in real time. Schemes in which the incident beam is kept fixed but where the detection is performed at various angles should further speed up the measurements especially if the single channel photomultiplier is replaced by a linear position sensitive detector.

Dynamical effects in adsorbed layers have not yet been investigated experimentally. It is being progressively recognized, however, that adsorbed chains are by no means immobile. There are breathing collective modes which modulate the thickness of the adsorption layer.[25] There are also individual crawling motions of the chains along the solid surface.[25] Measurements of these dynamical properties will no doubt represent a real challenge for experimentalists, because they are on the borderline of the present capabilities, even with the most sensitive optical techniques such as Forced Rayleigh Scattering[26] and Fluorescence Photobleaching Recovery.[27]

NOMENCLATURE

a	monomeric length
I_F	intensity of fluorescence of the polymer solution
I_F^{ref}	intensity of fluorescence of the reference solution
$i(\mathbf{q})$	neutron scattering intensity at scattering vector $\mathbf{q}$
N	number of monomers per polymer chain
n_1	optical index of refraction of medium 1
n_2	optical index of refraction of medium 2
n_i	scattering length density (cm^{-2}) of medium i for neutrons
$\mathbf{q}$	scattering vector
$R(\Lambda^{-1})$	ratio I_F/I_F^{ref}
R_F	radius of gyration of the polymer chain
S	surface of the solid particles
$S_{gg}(\mathbf{q})$	structure factor of the bare solid
$S_{pp}(\mathbf{q})$	structure factor of the polymer layer
$S_{pg}(\mathbf{q})$	polymer-solid cross-structure factor
V	volume of the solid particles
γ	adsorbed amount in unit of volume fraction
Γ	adsorbed amount in unit of mass (g/cm^2)
θ	angle of incidence
θ_c	critical angle of incidence
λ	a constant in SANS analysis
λ_o	wavelength of light in vacuum
Λ	penetration depth of the evanescent wave
Λ_{min}	minimum penetration depth
ξ	average distance between entanglements in a semi-dilute polymer solution
Φ_b	monomer volume fraction in the bulk solution
$\Phi(z)$	monomer volume fraction at a distance z from the solid surface

REFERENCES

1. See the recent review by M. Cohen Stuart, T. Cosgrove and B. Vincent, Adv. Colloid Interface Sci., 24, 143 (1986).
2. J.M.H.M. Scheutjens and G.J. Fleer, J. Phys. Chem. 83, 1619 (1979), J. Phys. Chem. 84, 178 (1980), Adv. Colloid Interface Sci., 16, 341 (1982).
3. P.G. de Gennes, Macromolecules, 14, 1637 (1981).
4. P.G. de Gennes, Scaling Concepts in Polymer Physics, Cornell University Press, 2nd Ed. (1985).
5. J. des Cloizeaux, J. Phys., (Paris) 49, 699 (1988).
6. I.D. Robb and R. Smith, Europ. Polym. J., 10, 1005 (1974).
7. T. Cosgrove and B. Vincent, Macromolecules, 14, 1018 (1981).
8. M. Cohen-Stuart, J. Colloid Interface Sci., 90, 321 (1982).
9. M. Kawaguchi, K. Hayakawa and A. Takahashi, Macromolecules, 16, 631, 1465 (1983).
10. Z. Priel and A. Silberberg, J. Polym. Sci., 16, 1917 (1978).
11. R. Varoqui and P. Dejardin, J. Chem. Phys., 66, 4395 (1977).
12. C. Allain, D. Aussèrré and F. Rondelez, Phys. Rev. Lett., 49, 1694 (1982).
13. D. Aussèrré, H. Hervet and F. Rondelez Phys. Rev. Lett., 54, 1948 (1985), Macromolecules, 19, 85 (1986).
14. J. des Cloizeaux and G. Jannink, Les Polymères en solution, Les Editions de Physique, Les Ulis (1987).
15. B. Cabane in Colloïdes et Interfaces, p. 101,, A.M. Cazabat and M. Veyssié eds. Les Editions de Physique, Les Ulis (1984).
16. K.G. Barnett, T. Cosgrove, B. Vincent, A.W. Burgess, T.L. Crowley, T. King, J.D. Turner and Th.F. Tadros, Polym. Commun., 22, 283 (1981).
17. T. Cosgrove, T.L. Crowley, B. Vincent, K.G. Barnett and Th.F. Tadros, Faraday Symp. Chem. Soc., 16, 101 (1982).
18. X.D. Sun, E. Bouchaud, A. Lapp, B. Farnoux, M. Daoud and G. Jannink, Europhys. Lett., 6, 207 (1988).
19. X.D. Sun, B. Farnoux, J. des Cloizeaux and G. Jannink, This volume of Proceedings.
20. L. Auvray, C.R. Acad. Sci. (Paris) Ser. 2, 302, 859 (1986).
21. T. Cosgrove, T.G. Heath, K. Ryan and B. Van Lent, Polym. Commun., 28, 64 (1987).
22. T. Cosgrove, T.G. Heath, K. Ryan and T.L. Crowley, Macromolecules, 20, 2879 (1987).
23. L. Auvray and P.G. de Gennes, Europhys. Lett., 2, 647 (1986).
24. L. Auvray and J.P. Cotton, Macromolecules, 20, 202 (1987).
25. P.G. de Gennes, C.R. Acad. Sci. (Paris), 302, II, 55 (1986), 302, II, 765 (1986), 301, II, 1399 (1985).
26. H. Hervet, L. Leger and F. Rondelez, Phys. Rev. Lett., 42, 1681 (1979).
27. J. Davoust, P.F. Devaux and L. Leger, Embo. J., 1, 1233 (1982).

CHIRAL DISCRIMINATION IN MONOLAYER PACKING OF HEXADECANOL-THIOPHOSPHORYL-2-PHENYLGLYCINOL WITH TWO CHIRAL CENTERS IN THE POLAR HEAD-GROUP

M. Dvolaitzky and M.A. Guedeau-Boudeville

Laboratoire de Physique de la Matière Condensée
Collège de France
75231 Paris Cedex 05
France

ABSTRACT

The four possible stereomers of a chiral surfactant with two asymmetric centers within the polar head group have been synthesized and their absolute configuration determined by X-ray diffraction. One of the diastereomers exhibits a chiral discrimination when spread on water interface : the monolayer racemic film undergoes a phase transition from a liquid-expanded towards a liquid-condensed phase upon compression, while the pure enantiomers only have a liquid-expanded phase, as revealed by the measured pressure-area isotherms. The transition pressure-composition diagram indicates that heterochiral interactions are favored. Our results are compared to predictions of Andelman and de Gennes[1] based upon a statistical model.

I. INTRODUCTION

The effect of chirality on molecular interactions in membranes has recently received increasing attention. Although chiral monolayers at the air-water interface might be used as models of biological membranes where most of the components are chiral, there are only few studies which show that surface properties of monolayers may be sensitive to stereochemistry. Since the first observation of a chiral discrimination in a monolayer in 1956 by Zeelen,[2] few examples have been reported in the literature; they have been reviewed by Stewart and Arnett in 1982.[3] More recently, Bouloussa and Dupeyrat have described a very large chiral discrimination in monolayers of tetradecanoyl-alanine.[4] Chiral discrimination in monolayers can be investigated by comparing the surface-active properties of the enantiomerically pure surfactants with those of the corresponding racemic mixtures, for instance, using a Langmuir balance to obtain the pressure-area isotherms. In most of the described cases which exhibit a chiral discrimination, the racemic film has less tendency to condense than the enantiomeric one. No significant difference was detected in dipalmitoyl phosphatidylcholine between the racemic mixture and the natural enantiomer.[5] In this latter case, the asymmetric carbon atom, in the glycerol backbone, is probably hidden from its neighbours in the film.

Our purpose was to improve the understanding of molecular interactions in chiral monolayers by introducing a second chiral center in the polar head, with the hope to obtain a better anchorage at the interface and an increase of the chiral discrimination. Moreover, four possible stereomers may be obtained which will allow a more complete study of the relationships between the molecular packing and configurations.

II. SYNTHESIS OF THE SURFACTANTS AND THEIR ABSOLUTE CONFIGURATIONS

By analogy with the works of Tsai et al.[6] on thiophospholipids chiral at phosphorous, we chose to prepare the hexadecanol-thiophosphoryl-phenyl-2-glycinols 3. The synthesis of the two diastereomers 3a and 3b, from D-(-)-(R)-phenylglycinol 1 is outlined in the Scheme (Fig. 1). The enantiomers 3c and 3d were similarly prepared from the L(+)-(S)-phenylglycinol.

As the NMR spectra of the phospholidine-thiones 2a and 2b do not permit to assign the configuration of the phosphorous chiral center,[7] we have tried to grow crystals suitable for X-ray diffraction. We were successful only for a homologue of 2a with a shorter chain; the crystal structure of the major (2-butanoxy-4-phenyl)1,3,2-oxazaphospholidine-2-thione (2a, R = C_4H_9), shows the relative S, R configurations on positions 2 and 4.[8] As the acid cleavage is expected to proceed with inversion of configuration,[9] it is reasonable to assign the P(R), C(R) configuration to the major surfactant 3a and the P(S), C(R) one to the minor isomer 3b (see Fig. 1).

Since the infrared spectra in the solid state of the racemates are different from those of the enantiomers for both diastereomers, the racemates in tri-dimensional arrangement are probably racemic compounds.[10] These observations were confirmed by differential scanning calorimetry (DSC), which shows that the racemates melt at higher temperature than the enantiomers, and by the phase-composition diagram of the major diastereomer (Fig. 2).

III. CHIRAL MONOLAYERS AT THE AIR-WATER INTERFACE OF THE ENANTIOMERIC AND RACEMIC THIOPHOSPHORYLPHENYLGLYCINOLS

The spreading experiments have been done first on pure water at room temperature

The racemate and enantiomers of the minor diastereomer give identical π - A isotherms of a liquid-expanded phase, and therefore show no chiral discrimination (Fig. 3). On the other hand, isotherms for the major diastereomer show an interaction between the two enantiomers in the racemic film, since a phase-transition is observed, which is not present in the pure enantiomeric film. However, because the reproducibility was not good on pure water subphase, an attempt was made to measure the equilibrium spreading pressure (ESP) by Wilhelmy plate method, disposing crystals of the surfactants on the water surface. As shown in Fig. 4, no spreading pressure is developed on pure water; a surface active behaviour appears only on acid subphase (2N H_2SO_4) and the ESP increases with the acid concentration. Moreover, as has been shown by Arnett et al.[11] in the case of N-methyl-benzylstearamide, the racemate exhibits a higher surface pressure than the enantiomer on 6N sulfuric acid for the same area/molecule. The pressure-area curves were then again measured on 0,3 N H_2SO_4. The results are now reproducible (Fig. 5) : until about 40 $Å^2$/molecule, the enantiomers and the racemate give the same expanded

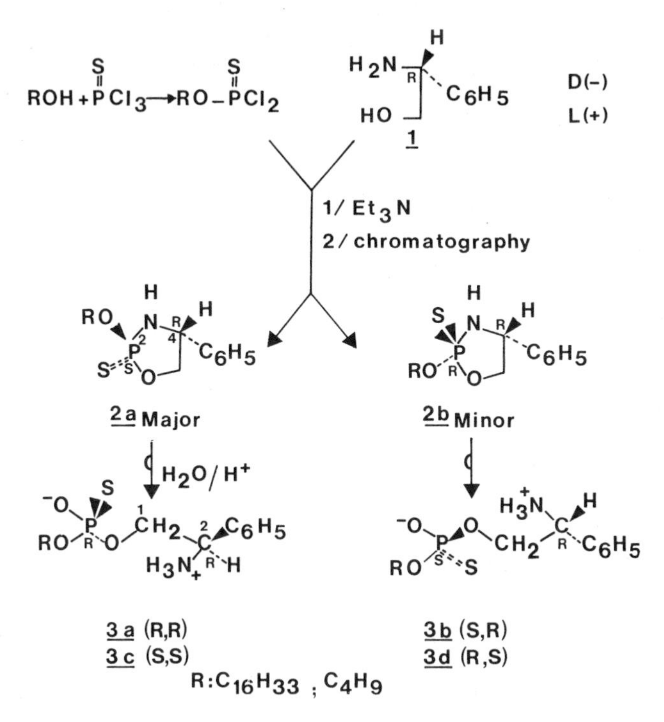

Fig. 1. Scheme of Stereomer Formation.

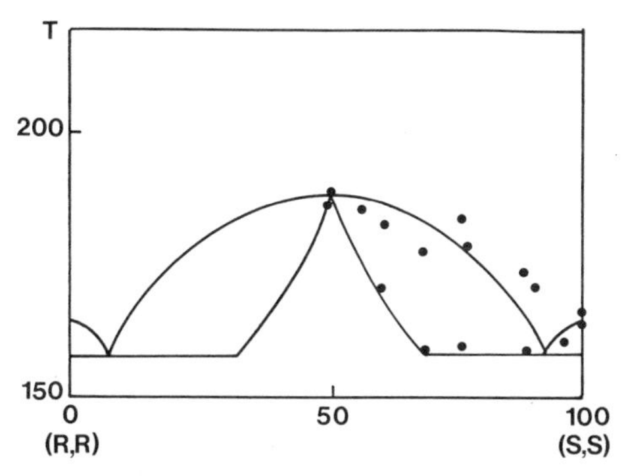

Fig. 2. Phase diagram of the major diastereomer 3a and 3b, (R,R and S,S) in crystalline state.

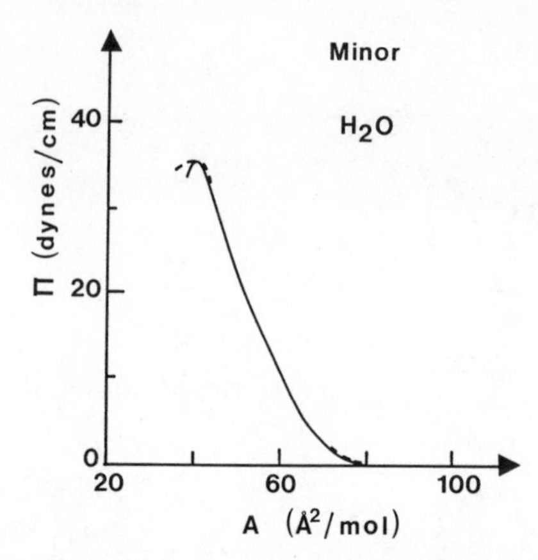

Fig. 3. π - A isotherms of the minor diastereomer, obtained by compression at the air-water interface at 20° C; (---)pure enantiomers; (——)racemic mixture.

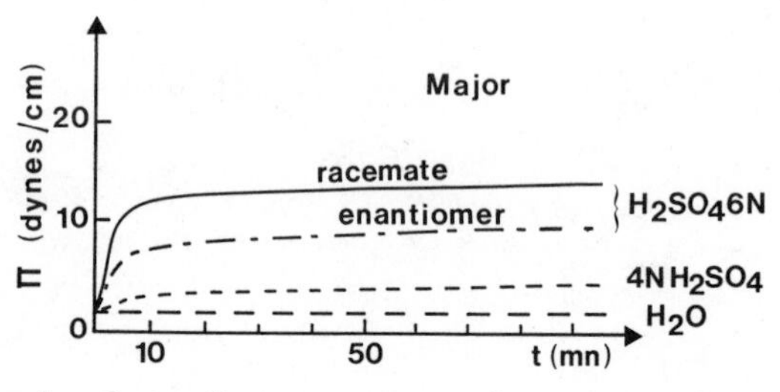

Fig. 4. Development of surface pressure for the major diastereomer with the acid concentration of the sub-phase: (---)crystals of pure enantiomer (R,R) and (——) crystals of the racemic.

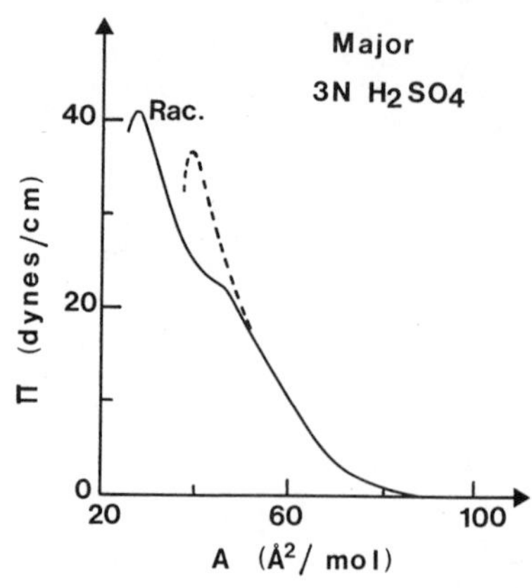

Fig. 5. π - A isotherms of the major diastereomer, obtained by compression at the air-0.3N H_2SO_4 interface;(---)pure enantiomers, (——)racemic mixture.

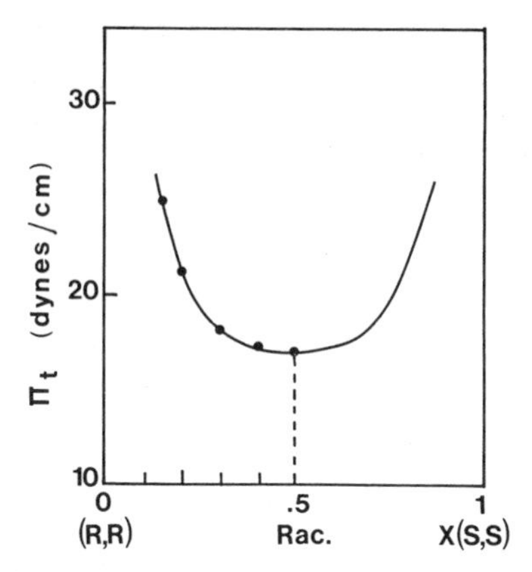

Fig. 6. Plot of transition pressure versus composition, for the major diastereomer.

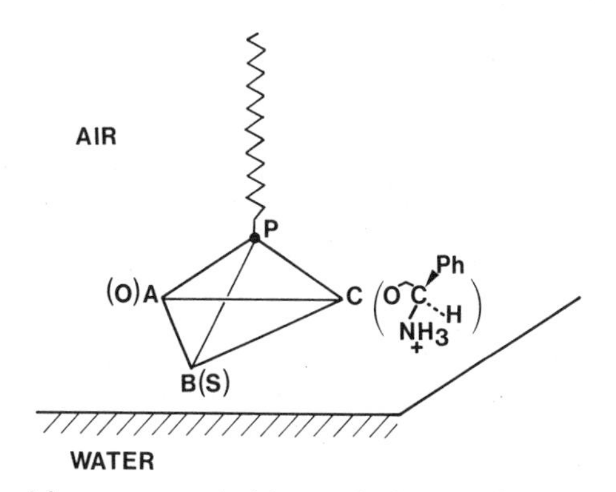

Fig. 7. Schematic representation of the surfactant 3 as a tripod.

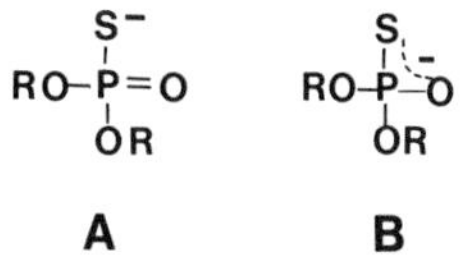

S^- | $RO-P=O$ | OR

A

S | $RO-P-O$ | OR

B

Fig. 8. Representation of a phosphothiate anion.[13]

phase; then, for smaller areas, the racemic film exhibits a condensing effect with a phase-transition towards a more condensed phase, and its collapse occurs at a higher pressure, thus suggesting that the racemic film is more stable. By measuring the π - A isotherms of mixed monolayers of various compositions, it is possible to plot the transition pressure versus composition as represented in Fig. 6 : this diagram is characterized by a minimum point for the racemic composition, indicating that the mutual interactions between the two enantiomers in the monolayer are stronger than the interactions between the pure enantiomeric molecules themselves.[12] In other words, the heterochiral interactions are favoured in the film. This result is, to our knowledge, the first observation of preferred heterochiral interactions in a monolayer.

IV. CONCLUDING REMARKS

Was it possible to predict this result by considering the molecular structure? Andelman and de Gennes have proposed a model[1] which permits one to predict the homo- or heterochiral case for surfactants with a tripod shape, containing three functional groups lying on the water surface and bounded to an asymmetric atom. Two neighboring molecules are assumed to associate via two intermolecular bonds (ij and $i'j'$). Each of these bonds is associated with an energy which is negative (attractive) if i and j tend to associate; if $Z{++}$ and $Z{+-}$ are the partition functions for a pair of molecules with the same or opposite chirality, respectively, a positive difference $\Delta = Z{++} - Z{+-}$ will yield a preferred homochiral interaction (HOC) and a negative Δ a preferred heterochiral one (HEC). The studied surfactants 3 (R,R and S,S) are not really tripods, but may be considered as such if the three groups lying at the interface are O, S and NH_3 (Fig. 7). The possibility of preferred heterochiral interactions depends on the relative strengths of the intermolecular bonds between the various groups (NH_3, S, O ...) which are difficult to estimate. If we assume, as recently reported,[13] that the charge distribution in the phosphothiate group is intermediate between a structure similar to A that exists in water, and a structure B which seems to be the best representation for two-dimensional crystalline arrays (Fig. 8), then, we obtain $\Delta > 0$ (HOC) with the model[1] and our assumed tripod molecule (Fig. 7). This disagrees with our phase diagram results that indicate a HEC case. Note that the interpretation of our experimental results lies heavily on the way we assigned the charge distribution to the various groups, and on our interpretation of the surfactant molecule as a tripod. Also, theoretically, the case of two negative and one positive charges cannot be simply interpreted because simple packing cannot avoid having identical charges that oppose each other in nearest neighbors. In the tripod representation of Fig. 7, we have neglected the second chiral center; if we had considered the three groups bounded to the asymmetric carbon atom (A = H = aliphatic, B = C_6H_5 = aromatic and C = NH_3 = charged group), then the model would have predicted heterochiral interactions. Thus, we hope that synthesis of true tripodal surfactants, currently in progress, will provide a better way to check the validity of the theoretical model.

REFERENCES

1. D. Andelman and P.G. de Gennes, C.R. Acad. Sci., (Paris) Ser. 2, 307 (3) 233-237 (1988).
2. F.J. Zeelen, Doctoral Thesis Stearoyl-Aminozuren : Synthese, Spreiding, en Photochemie, Leiden, Netherlands, 1956.
3. M.V. Stewart and E.M. Arnett, Topics in Stereochemistry, N.L. Allinger, E.L. Eliel and S.H. Wilen Eds. Vol. 13, pp.195-262, Wiley, New York (1982).

4. O. Bouloussa and M. Dupeyrat, Biochim. Biophys. Acta, 938, 395-402 (1988).
5. E.M. Arnett and J.M. Gold, J. Amer. Chem. Soc., 104, 636-639 (1982).
6. K. Brusik and M-D. Tsai, J. Amer. Chem. Soc., 104, 863-865 (1982).
7. D.B. Cooper, C.R. Hall, J.M. Harrison and T.D. Inch, J. Chem. Soc., Perkin I, 1969-1980 (1977).
8. X-ray Structure determination have been performed by M. Cesario and C. Pascard, Institut de Chimie des Substances Naturelles, Gif-sur-Yvette, France.
9. S.J. Abbot, S.R. Jones, S.A. Weinman and J.R. Knowles, J. Amer. Chem. Soc., 100, 2558-2560 (1978).
10. J. Jacques, A. Collet and S.H. Wilen, Enantiomers, Racemates and Resolutions; J. Wiley and Sons, New York, Brisbane, Toronto p.18 (1981).
11. E.M. Arnett, J. Chas, B.J. Kinzig, M.V. Stewart, O. Thomson and R.J. Verbiar, J. Amer. Chem. Soc., 104, 389-400 (1982).
12. H. Matuo, K. Motomura and R. Matuura, Chem. Phys. Lipids, 30, 353-365 (1982).
13. J. Baraniak and P.A. Frey, J. Amer. Chem. Soc., 110, 4059-4060 (1988).

Discussion

On the Paper by P.G. de Gennes

Stuart L. Cooper: (University of Wisconsin) Have you considered desorption of polymer from surfaces into pure solvent, as opposed to a solution of polymer and solvent? Studies of protein adsorption show the process to be somewhat irreversible, especially when the protein solution is replaced by pure solvent (water). For example, it has also been shown using radio-labelled protein that adsorbed protein will exchange with protein in solution, but that very little will desorb into pure solvent.

P.G. de Gennes: The experiments of Varoqui et al. with tritiated PS show that, for this polymer, the rate of exchange is proportional to the bulk concentration C_B, and vanishes when C_B=0. Thus, proteins and flexible polymers seem to behave similarly in that respect.

T.G.M. van de Ven (Paprican/McGill University): Regarding the kinetics of polymer entry in an adsorbed polymer layer, you mentioned that it is more likely for the ends of the polymer to attach to the substrate than for loops or "hairpins." However, there are far more "hairpins" than ends. Why do you dismiss the possibility of the attachment of "hairpins"?

P.G. de Gennes: This question is still open. For dense systems (i.e., = polymer melts) you can argue that hairpins of length n have a very small weight ($\sim e^{-an}$). But for adsorbed layers, this weight decreases only as a power law. More precisely, in terms of an order parameter $\Psi(z)$ (such that Ψ^2 = conservation) the barrier factor for end entry is

$$\tau_1 = \frac{<\Psi(R_G)>}{<\Psi(a)>} \sim \left(\frac{a}{R_F}\right)^{1/2} \sim \frac{1}{N^{0.3}}$$

The barrier factor for hairpin entry should be

$$\tau_2 = \frac{<\Psi^2(R_F)>}{<\Psi^2(a)>} \sim \left(\frac{a}{R_F}\right)^{4/3} \sim \frac{1}{N^{0.75}}$$

Since $N\,\tau_2 > \tau_1$ it may well be that hairpins dominate. I have not yet worked out the impact of this on rate equations.

Christopher Lantman (University of Massachusetts): Do you expect the argument of a self-similar structure to be found in a poor solvent as well as in a good solvent? Will a different fractal exponent be seen or must a different structure be envisioned?

P.G. de Gennes: This case has been analyzed by Pincus and Klein. One will expect self-similarity in complete thermodynamic equilibrium. The fractal dimension should be 2 and the concentration profile should scale like $\phi(z) \sim z^{-1}$. However, a recent mean-field calculation by SCHEUTENS give a more complicated profile, which is not yet understood.

K.J. Mysels (8327 LaJolla Scenic Drive, La Jolla, CA 92037): Energetics require that only exchange occurs stepwise and within the adsorbed layer. This must affect the kinetic processes.

P.G. de Gennes: Exchanges near the surface are slowed down by two processes: a) direct binding of monomers and b) possible existence of a glassy polymer layer near the wall. In my crude discussion, effects (a) and (b) are lumped into one same factor, f.

G. Slater (Xerox Research Center of Canada): What is the effect of surface roughness on the self-similarity idea?

P.G. de Gennes: This is discussed in the paper by F. Brochard. In conditions of complete thermal equilibrium, the crucial point amounts to compare the fractal dimension D_p of the polymer (here $D_p = 5/3$) and the fractal dimension of the absorbing surface D_S ($2 < D_S < 3$). Whenever $D_p < D_S$ the polymer should cover all pores (even those which are smaller than the coil size R_F). But, in many practical cases, there is a kinetic barrier = adsorbed polymers do not enter easily in small pores, and the large pores (of size $> R_F$) are the only accessible pores.

On the Paper by H. Watanabe, S. Patel and M. Tirrell

J. Emert (Exxon Chemical): Were the force-distance curves the same when the polymer-coated surfaces were brought closer together as when they were moved apart?

S. Patel

Yes.

Stuart L. Cooper (University of Wisconsin): The authors should be aware of the possibility of surface mobility confounding angular-dependent XPS measurements. In the vacuum of the XPS apparatus hydrophobic chain segments may rapidly diffuse to the surface, in contrast to where they may reside in a media.

On the Paper by M. Dvolaitzky and M.A. Guedeau-Boudeville

Jane E.G. Lipson (Dept. of Chemistry, Dartmouth College, Hanover, N.H.): Would it be possible to measure diffusion across (perpendicular to) these kinds of monolayers? Could one discriminate between the different kinds of enantiomers. For example, in biological systems perhaps the different enantiomers of a small diffusing molecule could be discriminated against.

M. Dvolaitzky (College de France): Indeed, the permeability of, say, an *L* monolayer, should be different for two enantiomers (ℓ) and (d) of any chiral molecule.

F. Brochard-Wyart (P & M Curie University, Paris): You study (*RS* and *SR*), and (*SS* and *RR*) racemates - Is it not easier to understand first (*RS* and *SS*) or (*RS* and *RR*) couples?

M. Dvolaitzky: Yes, but this is not a chiral discrimination but a diastereomeric one. It will be interesting to study it too.

PART TWO:

ADHESION, FRACTAL AND WETTING OF POLYMERS

NEW PERSPECTIVES ON POLYMER ADHESION MECHANISMS

Lieng-Huang Lee

Webster Research Center
Xerox Corporation
800 Phillips Road
Webster, New York 14580

ABSTRACT

This paper reviews new perspectives on polymer adhesion mechanisms. Professor P.G. de Gennes' theories about polymer melts and networks have enlightened us about polymer adhesion mechanisms. Both reptation theory and related experiments have demonstrated the existence of either self-diffusion or interdiffusion of polymer at the interfaces. Thus, adhesive strength, green strength and tack of elastomeric adhesives can be at least partially explained on the basis of the diffusion mechanism. Besides diffusion, Dr. de Gennes and his colleagues' studies on the kinetics of spreading and wetting of polymers have provided us with new understandings about the adsorption mechanism. Without wetting, polymer molecules can not achieve adequate adsorption at the interfaces to result in van der Waals attraction and/or acid-base (or donor-acceptor) interaction. The latter has not been explored by the French school. In addition to the above mechanisms, we shall mention briefly the mechanisms of mechanical interlocking, chemical bonding and electronic adhesion.

I. INTRODUCTION

In 1967, Lee published two papers on "Adhesion of High Polymers"[1,2] on the basis of Buche-Cashin-Debye's equation:[3]

$$D\eta = (A \rho k T/36)(R^2/M), \tag{1}$$

where D is the molecular diffusion constant, η the bulk viscosity, A Avogadro's number, ρ the density, k the Boltzmann's constant, T the absolute temperature, M the molecular weight, and R^2 the mean square end-to-end distance of a single polymer chain. It was concluded that the physical state of polymer determined the major adhesion mechanism involved. Polymer adhesion can be subdivided into rubbery polymer-rubbery polymer adhesion (*R-R* adhesion), rubbery polymer-glassy polymer adhesion (*R-G* adhesion), and rubbery polymer-nonpolymer adhesion (*R-S* adhesion). Diffusion, which depends to a great extent on the physical state of a polymer, is actually a limited selective process. Thus, diffusion of rubbery polymers can take place at the interface, but diffusion of a glassy polymer at a viscosity of 10^{13} poise or a diffusion constant of

10^{-21} cm^2/sec appears to be nearly impossible. On the other hand, physical adsorption is common to all three types of the above adhesion systems.

Since 1971, polymer physics has taken a new turn with a remarkable observation by Professor Pierre-Gilles de Gennes[4] that there is a correspondence between self-avoiding random walks and the phase-transition properties of an **n**-vector model of ferromagnetism for **n** = 0. (In the **n**-vector model, **n** denotes the number of degrees of freedom of electron spins; **n** = 1 corresponds to the usual Ising model.) Subsequently, this discovery has opened a new path for polymer physicists to apply powerful scaling[5] and renormalization group methods to solve polymer problems, e.g., entanglement in polymer solutions[6] and melt,[7] diffusion,[8,9] welding (or healing),[10] and adsorption at interface.[11,12]

In recent years, in addition to theoretical developments, e.g., reptation[4] and "tube" theories[13,14] there have been many new analytical techniques,[15] e.g., electron-induced X-ray fluorescence, NMR field gradient method, forced Rayleigh scattering(FRS), forward recoil spectrometry(FRES), photon correlation spectroscopy, Rutherford back scattering, small-angle neutron scattering(SANS), etc. All of these techniques have been used to verify theoretical predictions about polymer dynamics. In comparison to the 60's, we now know more about diffusion, welding, polymer interfaces, adsorption, chemical bonding, and other mechanisms related to adhesion.

For an historical review of adhesion mechanisms published in the literature, the readers should consult two articles[16,17] and a book[18] by Kinloch. In the following, we shall summarize briefly some of the major developments on adhesion mechanisms in recent years and some of the studies that are still in progress.

II. DIFFUSION MECHANISM

Voyutskii[19] first proposed that diffusion was the major driving force for polymer autohesion and heterohesion. Since the introduction of the reptation theory, there have been many publications devoted to self-diffusion and inter-diffusion in polymer solutions and melts. Here, we only want to demonstrate with several examples in the following subsections.

II.1. Viscoelastic Properties of Polymers

For a certain linear, flexible polymer, the viscoelastic properties[7] display a certain characteristic (relaxation) time τ, which increases rapidly with the molecular weight M, (or equivalently with the number N monomers per chain). Thus,

$$\tau = \tau_o N^a , \tag{2}$$

where τ_o is a microscopic time (of the order of 10^{-10} sec in melts) and a is an exponent of the order of 3.2-3.4. Since N is generally in the range of 10^4-10^5, the time τ may be extremely long (minutes). Thus when $t > \tau$, the chains appear to be tied up in knots. Conversely, $t < \tau$, the chains flow.

As the molecular weight further increases passing a critical point, M_c (or N_c), the entanglement of flexible chains takes place. The steady-flow viscosity η at low shear rate abruptly increases at N_c as shown in Fig. 1:[20]

$$\eta \propto N, (N < N_c) \; ; \tag{3}$$

$$\eta \propto N^{3/4}, (N > N_c) \; . \tag{4}$$

Generally, N_c is between 300 and 600 in an undiluted polymer (or melt).

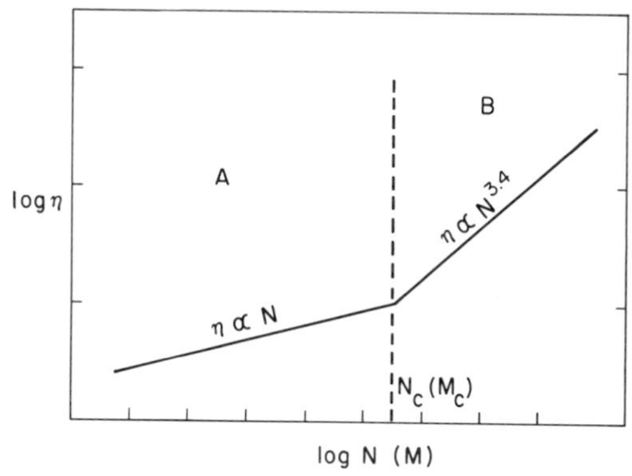

Fig. 1. Variation of the steady-flow viscosity with degree of polymerization N for a linear polymer in solution at fixed polymer concentration $c(>>c^*$, the overlap concentration). (From J. Klein, Macromolecules, 11, 853 (1978), reprinted with permission of the author).

II.2. Theories of Self-diffusion

The self-diffusion of polymers[8] in concentrated and undiluted solutions has been described by at least three theories: entanglement coupling, reptation, and cooperative.

In the entanglement coupling theory,[8] the self-diffusion constant D_s is related to N as

$$D_s \propto N,^{-1} \tag{5}$$

on the basis of the free draining concept. It was postulated that above a critical value of N ($N > N_c$) coupling of chains takes place, resulting in the increase of frictional drag. As a result, the viscosity increases as shown in Fig. 1, and the self-diffusion decreases. However, the original theory did not consider the spatial or the topological constraints along the contour of a polymer chain when it is heavily penetrated and surrounded by other chains.

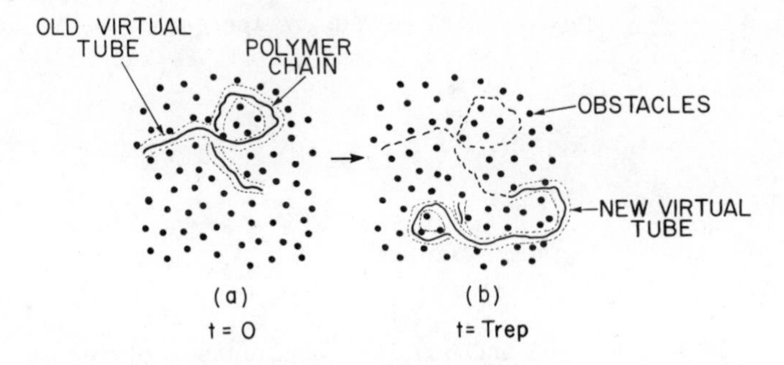

Fig. 2. Schematics of reptation of polymer chains.

On the other hand, the reptation theory proposed by de Gennes[4] assumes that a flexible chain diffusing in a fixed three-dimensional mesh of obstacles which the chain can not cross (Fig. 2). Thus, the chain would be topologically constrained to move by a curvilinear, or snake-like motion alone. This motion has been named as reptation (from "reptile"). One can visualize that the flexible chain is reptating by a Brownian diffusion within a "tube" surrounded by obstacles, but motions proceed perpendicular to the axis if the tube is blocked. For a chain made from N monomers of size a, the coefficient of the curvilinear diffusion, D_t, along the tube is

$$D_t = \frac{kT}{\eta a N}, \tag{6}$$

where η is the viscosity or the monomer-monomer friction coefficient.

At each instant, the ends of the chain find their ways randomly among the obstacles formed by other chains, and progressively define a new tube. The time required to completely recover the conformation of the initial tube is the reptation time, T_{rep} such that

$$D_t T_{rep} = L_t^2, \tag{7}$$

where $L_t = [(N/N_e)^2 N_e a^2]^{1/2}$ is the curvilinear length of the tube, and N_e is the average number of monomers between two entanglements. For a common polymer, N_c is a phenomenological parameter equal to $\simeq$ 100. The average diameter of the tube is therefore $(N_e a^2)^{1/2}$. From the above two equations, one can deduce that

$$T_{rep} \simeq N^3. \tag{8}$$

Since the tube has a random conformation, the experimentally determined diffusion coefficient in a given direction is not D_t but rather D_s (self-diffusion coefficient) such that

$$D_s T_{rep} \simeq R_G^2 \simeq N a^2 , \tag{9}$$

which is the random walk of step time T_{rep} and of step length equal to the radius of gyration of the chain, R_G ($\sim N^{1/2}$), thus

$$D_s = a^2 N^{-2} , \tag{10}$$

or

$$D_s \sim N^{-2} . \tag{11}$$

This diffusion process is very slow and experimental measurements reveal that for $a \sim 3 \times 10^{-8}$, $N \sim 10^3$, $D_s \sim 10^{11} cm^2/sec$ (for most rubbery polymers). In the melt, Klein and Briscoe[21] were able to obtain a $D_s \sim M_w^{-2 \pm 0.1}$ for a polymer of large polydispersity ($M_w/M_N \sim 20$). This result is very close to the value predicted by the reptation model.

The power law in N^{-2} for D_s is a characteristic for the reptation model and is rather well verified experimentally, although refinements of the model are required to take into account the fact that the topological constraints imposed by other chains are not fixed.[20]

The third theory is cooperative diffusion, which was proposed by Edwards and Grant.[22] In this model, the virtual "tube" is composed of more than one chain. As those chains move, so does the tube. With difficult mathematics, they could describe that one chain moves due to its cooperative motion with its neighbors or even in the presence of intermingling or knot-forming with other chains. Though the description of this model is more exact, the curvilinear diffusion (reptation) is not taken into account. Moreover, the diffusion coefficient $D(\omega)$ and the characteristic relaxation $T(\omega)$ depend on the mode of motion characterized by a frequency ω. The calculations also exclude the diffusion coefficient corresponding to the center-of-mass and the overall molecular relaxation time corresponding to the long-range, long-time behavior at $\omega = 0$. Thus, this model does not represent the actual situation of a real polymer melt.

II.3. Interdiffusion, Healing and Welding

Through diffusion, the cracks in a polymer can be healed, and two different parts of a polymer can be rejoined or welded together. In the following, we shall discuss briefly the application of the reptation model for the study of the healing and welding processes.

II.3.1. Number of bridges crossing the interface, $p(t)$

A typical experiment is shown schematically in Fig. 3. Two samples of the same polymer are put in contact at time $t = 0$, when no chain is crossing the interface. Eventually, the chains start to wiggle, resulting in the change of conformation by Brownian diffusion. Hence at time t, $p(t)$ chains have crossed the interface to form bridges, which can be expressed as[23,24,25]

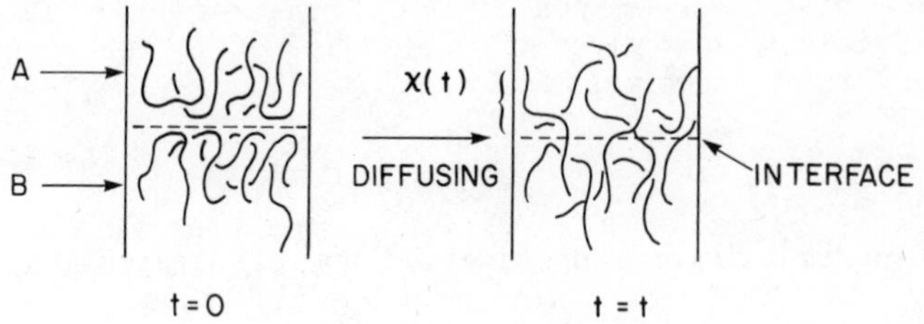

Fig. 3. Interdiffusion across the interface.

$$p(t) \sim t^{1/2} N^{-3/2}, \tag{12}$$

$$p_{\infty} \sim N^{\circ}. \tag{13}$$

where p_{∞} is the number of bridges at the virgin state.

The relation for $p(t)$ was first derived by de Gennes.[23] It is equivalent to the "crossing density" calculated by Roger and Tirrell.[10] Wool[26] further developed the model for the study of the healing process of polymers.

II.3.2. Average monomer interpenetration depth, $x(t)$

The average monomer interpenetration depth of segments on the interdiffusing chains, $x(t)$, is obtained from an integration over the Gaussian segment density profiles of the interdiffused minor chain population[21], such that

$$x(t) \sim t^{1/4} N^{-1/4}, \tag{14}$$

$$x_{\infty} \sim N^{1/2}. \tag{15}$$

where x_{∞} is the average monomer penetration depth at the virgin state.

The term of $x(t)$ has also been defined as the thickness of the interdigitated region,[23] which can be calculated to be:

$$x(t) \simeq R_G \left(\frac{t}{T_{rep}} \right)^{1/4}, \tag{16}$$

where $R_G = (La)^{1/2}$ is the coil size. When $t < T_{rep}$, $x(t)$ is smaller than R_G. In this short distance regime, where the reptation model applies, the macroscopic diffusion equations developed by Vasenin[28] for adhesion appear to be inapplicable. This is one of the major advances in the diffusion theory related to adhesion, since the reptation model enters into the picture.

II.3.3. Number of monomers crossing the interface, $L_o(t)$

The number of monomers that have crossed from one side of the interface to the other is equivalent to the total interpenetration contour length, $L_o(t) = n(t)\ \ell(t)$. Summing over the segment density profiles of the interdiffused minor chains, one obtains

$$L_o(t) \sim t^{3/4} N^{-7/4}, \tag{17}$$

$$\left(L_o\right)_\infty \sim N^{1/2}. \tag{18}$$

where $(L_o)_\infty$ is the number of monomers crossing the interface at the virgin state.

II.3.4. Scaling law for a polymer-polymer interface

Wool[25] described the above properties and others listed in Table 1 in a convenient scaling law that relates the dynamic properties, $H(t)$, to the static properties, H_∞, with the reduced time, t/T_r, by

$$H(t) = H_\infty (t/T_r)^{r/4}, \tag{19}$$

Table 1

Molecular Aspects of Interdiffusion at a Polymer-Polymer Interface

Molecular Aspect	Symbol	Dynamic Relation $H(t)$	Static Relation H_∞	r	s
General Property	$H(t)$	$t^{r/4} N^{*-s/4}$	$N^{(3r-s)/4}$	r	s
No. of Chains	$n(t)$	$t^{1/4} N^{-5/4}$	$N^{1/2}$	1	5
No. of Bridges	$p(t)$	$t^{1/2} N^{-3/2}$	N^{0}	2	6
Ave. Monomer Depth	$x(t)$	$t^{1/4} N^{-1/4}$	$N^{1/2}$	1	1
Total Monomer Depth	$x_o(t)$	$t^{1/2} N^{-3/2}$	N°	2	6
Ave. Contour Length	$\ell(t)$	$t^{1/2} N^{-1/2}$	N	2	2
Total Contour Length, No. of Monomers	$L_o(t)$	$t^{3/4} N^{-7/4}$	$N^{1/2}$	3	7
Ave. Bridge Length	$\ell_p(t)$	$t^{1/4} N^{-1/4}$	$N^{1/2}$	1	1
Center of Mass Depth+	x_{cm}	$t^{1/2} N^{-1}$	$N^{1/2}$	2	5

+ This equation does not apply to chain whose center of mass is within a radius of gyration of the surface.

* The original table uses M instead of N

(Ref. R.P. Wool, Rubber Chem. and Technol. 57, 307 (1984), and in Fundamentals of Adhesion, Ed. L. H. Lee, Chapt.7, Plenum, New York (1990).

$$H_{\infty} \sim M,^{(3r-s)\,4} \tag{20}$$

$$where\ r, s = 1, 2, 3 \ldots . \tag{21}$$

Several of those parameters listed in Table 1 have also been derived by de Gennes[23] and by Prager and Tirrell.[10]

II.3.5. Modulus at the interface, $E(t)$

At time t, $p(t)$ chains have crossed the interface and therefore contribute to the elastic modulus $E(t)$.[29] From the thickness $x(t)$, it is possible to evaluate the number of monomers within this slice $x(t)$ originating from chains initially located below the interface. Then one finds

$$p(t) \simeq E(t) . \tag{22}$$

However, the strong molecular weight dependence of $E(t)$ ($\simeq N^{-3/2}$) has not been verified experimentally.

In the case where the two polymers in contact are chemically different but compatible,[29] the kinetic of interdiffusion could be more rapid than that between two identical polymers. Owing to the attraction between the monomers of each species, one may expect rather

$$p(t) \simeq E(t) \simeq t . \tag{23}$$

II.3.6. Fracture energy of the healed polymer

The healing and welding of a polymer results not only in the increase in the thermodynamic work of adhesion W_A but also in the fracture energy G_c or fracture toughness K_c. It is well known that both G_c and K_c ($\simeq (EG_c)^{1/2}$) are much greater than W_A because the fracture energetic terms are also functions of deformation.[30,31] In other words, after healing, the polymer becomes whole and is completely recovered to its full strength as measured by the fracture energy.

The above contention is borne out by the work of Jud et al.[32] They studied the rehealing[33] of PMMA-PMMA, SAN-SAN, and PMMA-SAN (PMMA-polymethyl methacrylate; SAN-styrene-acrylonitrile copolymer). At temperatures above T_g, they have observed that the fracture toughness at the interface $(K_I)_i$ increases with the contact time, t, as $(K_I)_i \propto t^{1/4}$. In terms of the fracture energy $(G_I)_i$, the relation becomes $(G_I)_i \propto t^{1/2}$ as predicted by de Gennes with the reptation model.[23] In addition, the depth of the interdigitation $x(t)$ is calculated to be 2 to 3 nm. Since R_G is 5 nm, this value of $x(t)$ is rather reasonable.

II.3.7. Self-diffusion coefficient

Though the diffusion model predicts very well the time-dependence of the rehealing experiments, there are still many unanswered questions, such as the absolute value of D, the roles of chain-ends and of molecular weight, the influence of the relaxing fibrils derived from the fracture event, and the nature of the physical links. The relationships of D to viscoelastic and structural parameters still rely on the macroscopic formulation such as the Buche-Cashin-Debye's equation[3] as in Eq.(1), which was derived from the free draining model for both entangled and nonentangled systems. For example, the D_S for polyethylene (M_w = 11,000) at 176°C was determined by Klein et al.[21] to be 3.7 x 10^{-9} cm^2s^{-1}. In contrast, the D_s calculated from Eq.(1) is 1.4 x 10^{-10} cm^2s^{-1}. The agreement between the experiment and the calculated value of D_S is good.

In terms of critical molecular weight, M_c, Graessley[34] has introduced a more sophisticated equation for D_S on the basis of the work by Doi and Edwards:[13]

$$D_s = \frac{G_N^o}{135}\left(\frac{\rho R_g T}{G_N^o}\right)^2\left(\frac{<R^2>}{M}\right)\left(\frac{M_c}{M^2\eta_o(M_c)}\right), \tag{24}$$

where G_N° is the plateau modulus and η_o is the zero shear viscosity. Here R_g is the universal gas constant, and c is the polymer concentration (wt/vol). This equation has been shown to give better results than that derived from the uniformly effective friction model.

For polystyrene, the value D obtained from Eq.(1), however, agrees very well with that based on the tube model, although Eq.(1) was presumably derived from the free draining model.

II.3.8. Interdiffusion coefficient

Interdiffusion between two compatible polymers, e.g., PMMA and PVF_2 (polyvinylidene fluoride), has been studied by Wu et al.[35] In this case, the two diffusion coefficients and rates are unequal, and D_i is proportional to N^{-2}. In addition, the interfacial thickness $x(t)$ grows by $t^{1/2}$, hence

$$x(t) \sim N^{-1} t^{1/2} . \tag{25}$$

This is in accordance with the reptation model. Furthermore, the adhesive bond strength $\sigma(t)$ at the interface can be calculated to be

$$\sigma(t) \propto t^{1/4} N^{-1/4} . \tag{26}$$

II.4. Tack and Green Strength

The application of the reptation model to adhesion is not limited to the adhesive bond strength at the interface. The reptation model can also be used to study tack and green strength of polymers.[24]

Tack[35] is the ability of two materials to resist separation after bringing their surfaces into contact for a short time under a light pressure. There are two types of tack; namely, autohesive tack from a pair of similar materials and adhesive tack from a pair of dissimilar materials. On the other hand, the green strength of an elastomer is its resistance to deformation and fracture before vulcanization. We shall discuss tack and green strength on the basis of diffusion mechanism. It is well known that not all types of diffusion can result in the enhancement of adhesion. Only the diffusion involving polymer entanglement can improve tack and green strength.

From microfracture criteria, tack in the unaxial tension in the absence of a large stress is identical to $\sigma(t)$ in Eq.(26). Thus, tack is

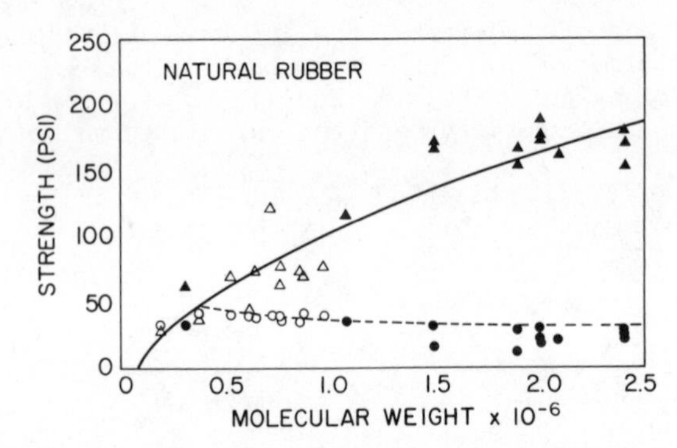

Fig. 4. Tack (circles) and green strength (triangles) as a function of M_v for fractionated samples of natural rubber (data of W.G. Forbes and L.A. McLeod, Trans Inst. Rubber Ind. 30(5), 154 (1958)). The green strength was evaluated at an uniaxial test speed of 26cm/min at 25°C, and the tack was evaluated at a contact time of 30 s for each sample. The solid lines are the calculated values based on the cited equations in the text. (From R.P. Wool, Rubber Chem. and Technol. 57, 307 (1984)).

$$\sigma(t) \sim t^{1/4} N^{-1/4} . \tag{27}$$

and the green strength is

$$\sigma_{\infty} \sim N^{1/2} . \tag{28}$$

The above two relations have been obtained independently by Forbes and McLeod[36] as shown in Fig. 4.

II.5. Fractals at Interface

The structure of interface portrayed in the literature and that in this section appears to be a sharp boundary. In reality, this is not so. An interesting work by Wool[37] indicates that the actual, interdiffused interface is nothing but fractals.[38] Indeed, it has been suggested that all polymers can be described as r-fractals.[39] Here r is defined by the scaling of the maximum radius of gyration with the total mass in fractal N, i.e., $r \sim N^{-1}$ where r is assumed to be a superuniversal exponent, i.e., independent of the Euclidean space dimension d. The inverse of r has also been suggested to be the spectral dimension itself.[40] We believe that the entire subject of fractal should have some impact in the future studies of polymer adsorption,[41] wetting,[42] and adhesion.

III. LIFSHITZ-VAN DER WAALS INTERACTIONS

As stated in the Introduction, physical adsorption is common to all three types of adhesion systems. For good adsorption, wetting is necessary and essential. Wetting and wettability of polymers have been reviewed by Zisman,[43] Mittal,[44] etc. In the past, the efforts have been focused on the thermodynamic aspects of wetting because wetting is the only way to achieve maximum thermodynamic (Lifshitz-van der Waals) work of adhesion W_A^{LW} according to Dupree's Equation,[45]

$$W_A^{LW} = F_{LV} + F_{SV} - F_{SL}, \tag{29}$$

where F_{LV}, and F_{SV} are the surface free energy of the liquid and solid, respectively, and F_{SL} is the interfacial energy. Instead, surface tension γ is commonly used to express surface free energy, thus $W_A^{LW} = \gamma_{LV} + \gamma_{SV} - \gamma_{SL}$. Throughout this discussion, we shall put emphasis on W_A instead of on adhesive strength, which also involves the work of deformation. We do not intend to discuss the Lifshitz-van der Waals interaction in this paper, but recommend readers to refer to recent articles by Good, et al.[46]

In the past decade, one of the important developments in the wetting studies is the kinetic aspect. Several reviews by Marmur,[41] de Gennes,[48,49] and Cazabat[50] should serve to highlight many careful studies carried out in recent years. In the following we shall briefly summarize the important work by de Gennes and his colleagues on the kinetics of wetting and spreading of a nonvolatile liquid or polymer. In addition, Leger et al.,[51] have reported the recent experimental study of polymer and Cazabat et al.[52] have highlighted recent work on the spreading of a liquid drop.

III.1. Partial Wetting of Liquid ($\theta_e > o$)

III.1.1. Young's equation

According to Young's equation,[53] the surface tension of the liquid, γ_{LV}, the surface tension of the solid, γ_{SV}, and the interfacial tension, γ_{SL}, also represent the three forces by unit length applied on the line of contact. In the case of partial wetting, they are related to the equilibrium contact angle, θ_e (Fig.5) as follows:

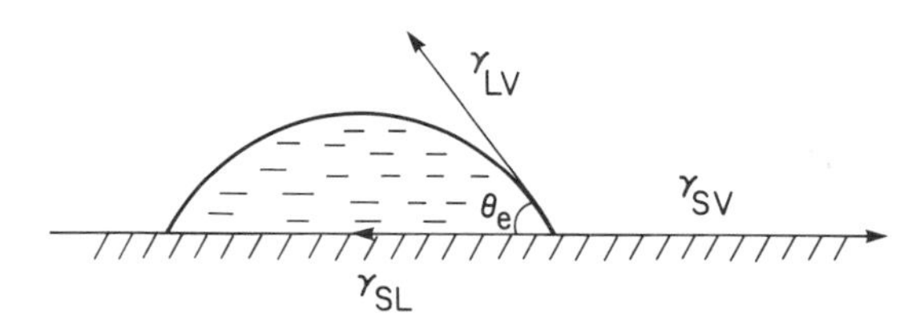

Fig. 5. Contact angle of a sessile drop on a solid surface.

$$\gamma_{LV} \cos\theta_e = \gamma_{SV} - \gamma_{SL} \, . \tag{30}$$

This equilibrium also requires a balance of vertical forces: capillary forces along the contact line ($\gamma \sin\theta_e$), Laplace pressure, and drop weight.[54]

III.1.2. Effect of Drop Size

If the drop is sufficiently small that the difference in hydrostatic pressure between the top and the bottom of the drop is negligible compared to the Laplace pressure associated with the curvature of the liquid air interface (i.e., small size comparable to the capillary length (height) $k^{-1} = (\gamma/\rho g)^{1/2} \simeq 1$ mm, where ρ is the density of the liquid and g is the gravitational acceleration), the drop takes the shape of a spherical cap and is entirely characterized by θ_e.

The size of the drop is determined by the radius of the wetted spot R, its relation with k^{-1} results in two following power laws:

$$R << k^{-1} : \; R \sim t^{0.1} \, ; \tag{31}$$

$$R >> k^{-1} : \; R \sim t^{1/8} \, . \tag{32}$$

For a very small drop ($R << k^{-1}$)[55] gravity can be neglected, and the drop maintains a spherical cap. For a large drop ($R >> k^{-1}$),[56] gravity tends to affect the shape of the drop. Then the kinetics becomes

$$R(t) \sim \Omega^{3/8} \left(\frac{\rho g}{\eta} t \right)^{1/8} . \tag{33}$$

Experiments carried out by Cazabat and Cohen Stuart[57] have confirmed the relation in Eq.(33), in which Ω is the volume of the drop.

III.1.3. Spreading coefficient

For spreading, another parameter, spreading coefficient,[58] $S = \gamma_{SV} - \gamma_{SL} - \gamma_{LV}$, appears to be important in classifying liquids that have a tendency to form good films on a given substrate. In general, the larger and the more positive the S the more energy is gained by intercalating a liquid film between the solid and air. Thus,

$S > o$;	spontaneous spreading
$S < o$.	not spontaneous spreading

Though the condition $S > o$ is necessary for a liquid to spread spontaneously on a solid, it is insufficient to describe the final state of the film. According to Joanny and de Gennes,[59] the final state of the film is controlled by long-range forces existing within the liquid. If one deposits on the solid a liquid film whose thickness is much larger than the range of interaction in the liquid, the energy associated with the creation of the film is that required to form two solid-liquid and liquid-air interfaces; that is, $\gamma_{SL} + \gamma_{LV}$.

III.1.4. Disjoining pressure

On the other hand, if the film thickness is smaller than the range of long-range forces, the energy contains an interaction energy term $P(z)$:

$$E = \gamma_{SL} + \gamma_{LV} + P(z) \, , \tag{34}$$

where $P(z)$ is directly related to the disjoining pressure (or force) $\Pi(z) = -dP(z)/dz$ introduced by Deryagin.[60-62] Thus, the film tends to thicken

$$(\lim_{z \to \infty} P(z) = 0)$$

resulting from the long-range interactions. The quantity $\Pi(z)$ is the pressure applied to the film for maintaining it at thickness z. For the non-retarded *VDW* interactions, $\Pi(z)$ can be expressed in terms of the thickness and the Hamaker constant, A_H:

$$\Pi(z) = \frac{A_H}{6 \pi z^3}, \tag{35}$$

and $P(z) = A_H / (12 \pi z^2)$. For complete wetting, A_H is positive.

III.1.5. Equilibrium thickness of "pancake"

For "dry" liquid (nonvolatile), the film spreads at equilibrium like a pancake with an equilibrium thickness, $\dot{e}$ (Fig.6). This is one of the major findings by Joanny and de Gennes.[59] They have calculated this thickness by taking into account the counter effects of the spreading term in thinning of the film and the disjoining pressure in thickening of the film. Thus, for non-retarded *VDW* interactions,

$$\dot{e} = a \left[3\gamma / 2S \right]^{1/2} . \tag{36}$$

where the length $a = (A_H/6 \pi \gamma)^{1/2}$ is of the order of the molecular size.

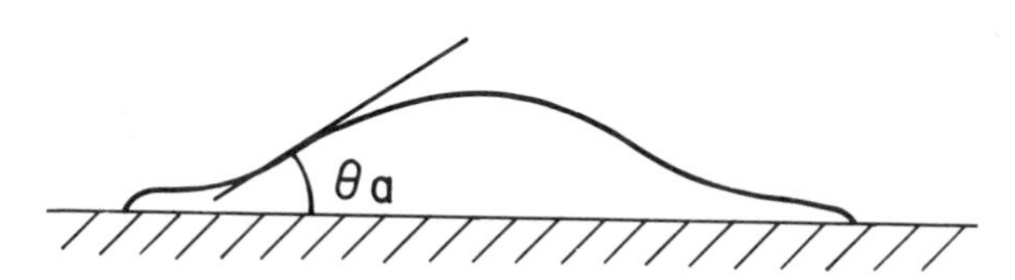

Fig. 6. Equilibrium film on a solid surface resembling a "pancake".

From the above discussion, one can see that it is not necessary to look for a large value of S to realize a "good film." If S is large, $\dot{e}$ will be of the order of the molecular thickness, a, i.e., very thin. On the other hand, if γ_{LV} is close to γ_c of the solid and $S \approx$ o, $\dot{e}$ can become much larger than a and reach 100 to 500 Å.

III.2. Complete Wetting of Liquid ($\theta_e \rightarrow$ o):

III.2.1. Macroscopic region-spherical cap

For complete wetting, there is no balance of horizontal forces, and the spreading parameter $S \geq 0$. The drop of a "dry" liquid spreads continuously (under a cap) and one follows the kinetics of wetting by measuring the radius $R(t)$ and the apparent contact angle $\theta_a(t)$ as a function of time t. After a short interval, the measured values of R and $Cos\theta_a$ are following simple Tanner's scaling law[55] for small drops:

$$R(t) \propto t^{0.1} \Omega^{0.3}; \tag{37}$$

$$\theta_a(t) \propto t^{-0.3}, \tag{38}$$

where Ω is the volume of the drop, and $\Omega = \pi R^3\theta/4$ = const. It is important to point out that this universal behavior does not appear to depend on S, as long as S is positive. In other words, the exact nature of the solid surface appears to play no role in the kinetics of spreading.

III.2.2. Microscopic region-precursor film

A precursor film is progressively spreading in front of the macroscopic drop (Fig.7). The development of the precursor film results from two factors: the drift velocity $dR/dt = U$ of the macroscopic edge, and the expansion of the film itself, which takes place even if the macroscopic edge has stopped. In the latter case, the cause of expansion is clear: the unbalanced capillary force S pulls out of the drop a film whose thickness is controlled by the disjoining pressure. In general, both drift and expansion take place simultaneously.

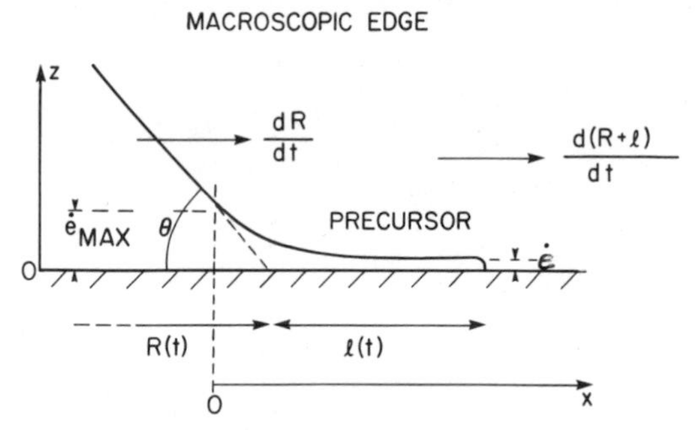

Fig. 7. The precursor film precedes the macroscopic edge of the drop, which advances with the velocity $U = dR/dt$. From this edge, the length of the precursor is ℓ. At the tip, the velocity is $d(R + \ell)/dt$. (From A. Cazabat, Contemp. Phys., 28(4), 347 (1987), reprinted with permission of the author).

In the film, there are apparently two zones:[59] the adiabatic zone and the diffusive zone (Fig.8). In the adiabatic zone, the precursor thickness is a function of the horizontal distance X to the macroscopic drop: $Z(X)$ decreases as X^{-1} between its maximum value $e_{max} = a/\theta$ at the drop and the limit thickness $\dot{e}$ (Fig.8).

When the drift is negligible, the "diffusive" phase starts because after a period of time when the thickness of the film is almost e at different points, its length becomes

$$\ell = \sqrt{Dt}\,, \tag{39}$$

with

$$D = \frac{2}{3}\,\frac{\dot{e}S}{\eta} = \frac{1}{3\eta}\sqrt{\frac{AS}{\pi}}\,. \tag{40}$$

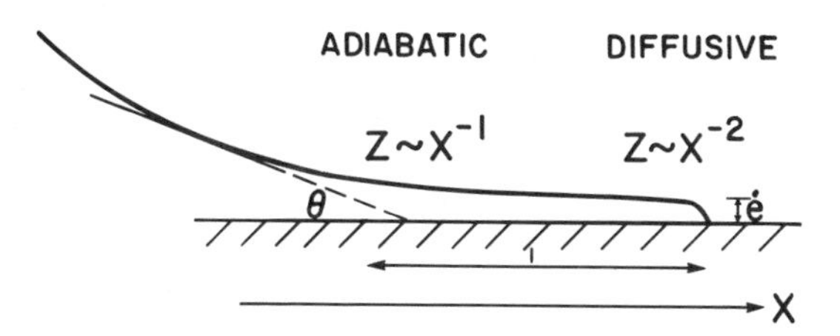

Fig. 8. Adiabatic and diffusive regimes of the precursor film on a solid surface (From A. Cazabat, Contemp. Phys., 28(4), 347 (1987), reprinted with permission of the author).

For sessile drops, the diffusive zone occurs over a long period of time. In the crossover regime, an X^{-2}-profile develops at the tip of the film. (Fig.8)

The above descriptions of the kinetics of spreading are summarized in Table 2. Most of above discussions about the spreading of the "dry" liquid also apply to the case of the "moist" liquid (or volatile liquid). However, the length of the "moist" precursor film is much shorter. In general, the kinetics of spreading of the macroscopic part of the drop is independent of the spreading coefficient S, while the shape and the spatial extent of the microscopic precursor film are determined by S. Consequently, all the gains in the interfacial energy during the spreading are dissipated by friction in the precursor film.

Table 2 The Kinetics of Dry Spreading of Liquid and Polymer

Region of Drop	**Spherical Cap**		**Macroscopic Foot**		**Microscopic Precursor Film**	
Type of liquid	Liquid	Polymer	Liquid	Polymer	Liquid	Polymer
Border of Region	$Z(r) > Z_o$	$Z(r) > b$	----	$b > Z(r) > Z_o$	$Z_o > Z(r)$	$Z_o > Z(r)$
Driving Forces	VDW+Vis	Vis.flow	----	Cap. pres plug flow.	VDW+Vis	VDW plug flow
Spreading Coefficient, S	Independent	Independent	----	Independent	Dependent	Dependent
Contact Angle, $\theta_a(t)$	$\theta_a(t) \sim t^{-0.30}$	$\theta_a(t) \sim t^{-0.30}$	----			
Radius of Cap, $R(t)$	$R(t) \sim t^{0.10}$	$R(t) \sim t^{0.10}$	----			
Velocity, $U(=dR/dt)$	$U \sim \theta_a^3(t)\gamma/\eta$	$U \sim \theta_a^3(t)\gamma/\eta$	----			
Thickness (height)				Entangled polymer: $(N > N_e)$ $b \sim N^3 N_e^{-2} a$ Nonentangled polymer $(N < N_e)$ $b \sim Na$	Adiabatic: $Z(X) \sim X^{-1}$ Diffusive (at long time): $Z(X) \sim X^{-2}$	$Z^2(X) \sim \frac{A}{4\pi k u}\left(\frac{1}{X+X_1}\right)$ where: $k = \eta/b = \eta_0/a$
Length f = width of foot, $\ell(t)$=length of the precursor film	---------- Details see Ref.(50).	---------- Details see Ref.(63).	-------	$f \sim \left(\frac{3b^2\lambda}{8}\right)^{1/3} \sim b\theta_a^{-1}(t)$	$\ell(t) \sim \left[\frac{1}{3\eta(dR/dt)}\right] \times \left(\frac{AS}{\pi}\right)^{1/2}$ Diffusive $\ell(t) \sim \frac{1}{3\eta}\left(\frac{AS}{\pi}\right)^{1/2}$	$X_1 \sim X_o \sim (ab)^{1/2}\theta_a^{-3/2}$

III.3. Complete Wetting of Polymer Melt

The spreading of polymer droplets has been studied by Brochard and de Gennes.[63] If the polymer droplet height h is larger than b, we are considering a case of $h>>b>>Z$. Then there will be three regions as shown in Fig. 9. The one that is not present in the case of liquid is the "foot," which has been previously reported by Schonhorn, et al.[64] The three regions of a polymer droplet are:

a) a spherical cap ($Z(r)>b$): Here normal viscous flow takes place, the slip is negligible, and the kinetics follows the Tanner laws;[55]

b) a foot ($b>Z(r)>Z_0$): In the foot there are plug flows driven by the capillary pressure, and

c) a precursor film ($Z>(r)>Z_0$). In this case, it has a plug flow driven by van der Waals forces.

The comparison between the dry spreading of a liquid and that of a polymer is given in Table 2. Some parameters in Table 2 are still being examined currently by others. Recent experiments[65a] on a model system (silicone oil on a potential surface of silicon oxide) and the latest work by Dallant et al.[65b] with X-ray reflectivity have confirmed the theory proposed by Joanny and de Gennes[59] and the existence of a thin precursor film.

III.4. Dry Spreading of Polymer Solution

The dry spreading of polymer solutions[66] is more complex than that of either the pure liquid or a polymer melt. Assuming the polymer does not adsorb on the surface, the dry spreading of a semidilute solution (from a good solvent) of neutral, flexible chains on a solid surface can be summarized as follows:

The polymer solute has two effects: 1) it modifies the spreading coefficient of the liquid, and 2) it introduces a new contribution to the disjoining pressure, which, as pointed out by de Gennes,[48] is present only when the polymer cannot exchange with a reservoir. After the free energy of a completely spread droplet is analyzed, a phase diagram can be constructed involving three different states: (1) the bulk droplet of the

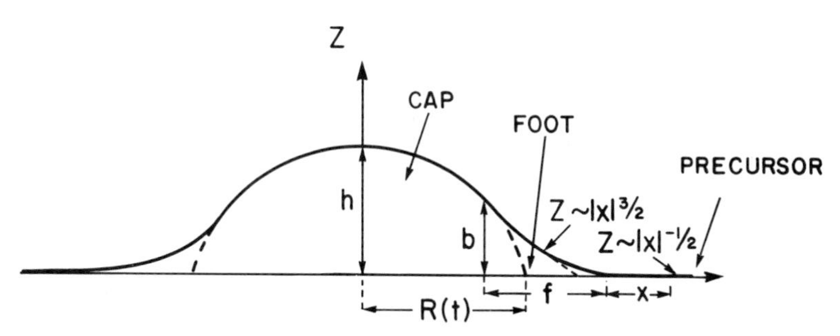

Fig. 9. The foot of a polymer droplet on a solid surface (From F. Brochard and P.G. de Gennes, J. Phys. Lett., (Paris) 45, L-597 (1984), reprinted with permission of the authors).

solution, (2) the solution film containing the polymer, and (3) the film without the polymer. In particular, some polymer solutions will not spread on a solid surface, although the pure solvent does spread. Then, a film of pure solvent will leak out of the solution droplet. Consequently, the concentration, the spreading coefficient, and the Hamaker constant related to the VDW interactions will control various cases of wetting and spreading. In some cases, two precursor films were found: one from the pure solvent and the other from the polymer.

The above brief summary of recent studies on wetting and spreading of liquids and polymers clearly indicate the contributions of Professor de Gennes and his colleagues in France. To us the importance of these studies is in the better understanding of the adsorption mechanism based on the Lifshitz-van der Waals interactions.

IV. Molecular Interactions

Besides the Lifshitz-van der Waals interactions, there are short-range (<0.2 nm) forces due to the donor-acceptor interaction[67-69] or the acid-base interaction.[70] The role of the acid-base interaction in polymer adhesion has been studied by Fowkes[71,72] and Bolger et al.[73] Recently Good et al.[74] have examined the role of the interfacial hydrogen bond in enhancing the adhesion through the acid-base interaction. If there is a major aspect of the adhesion mechanism that the French school, led by Prof. de Gennes, has not been engaged in, it is the acid-base interaction. Since the French school is dominated by physicists, it is understandable that the conventional subjects for chemists, such as the acid-base interaction, have not been covered in their recent studies. However, there may be a few exceptions. Boiziau and Lecayon[75] at Saclay have applied the concept of the acid-base interaction to the study of polymer-metal adhesion involving the grafting of acrylonitrile.

IV.1. Enthalpy of the Acid-Base Interaction

Now let us discuss briefly the acid-base interaction. The generalized acid-base interaction can be represented[70] as

$$\underset{\text{(Acid)}}{A} + \underset{\text{(Base)}}{:B} \rightarrow \underset{\text{(Acid-Base Complex)}}{A\!:\!B} . \tag{41}$$

This interaction actually involves both covalent (homopolar) and ionic (heteropolar) factors. Thus, Drago et al.[76] introduced four parameters for the prediction of the enthalpies of the acid-base interactions. For an A-B pair, the enthalpy or molar energy of the adduct formation can be expressed by the following empirical relation:

$$-\Delta H^{AB} = E_A E_B + C_A C_B , \tag{42}$$

where E_A and E_B are the susceptibilities of the acid (A) and base (B), respectively, to undergo the electrostatic interaction, and C_A and C_B are those to undergo covalent interaction. Several examples of Drago's parameters for acids and bases are listed in Table 3.

Table 3 Drago's Parameters for Acids and Bases

Acids	C_A	E_A	Bases	C_B	E_B
Iodine	1.00	1.00	Pyridine	6.40	1.17
Phenol	0.44	4.33	Ammonia	3.46	1.36
Boron trifluoride	3.08	7.96	Ethyl acetate	1.74	0.98
Sulfur dioxide	0.81	0.92	Acetone	2.33	0.99
Chloroform	0.15	3.31	Benzene	0.71	0.49
Water	2.45	0.33	Triethylamine	11.09	0.99

IV.2. Work of Adhesion (the Acid-Base component)

From the above discussion of Drago's equation, the work of acid-base interaction for a pair of molecules is simply the enthalpy of the interaction

$$W_{int}^{a-b} = -\Delta H^{a-b} . \tag{43}$$

In terms of the work of adhesion W_A due to the acid-base interaction, we can equate it to the work of acid-base interaction:

$$W_A^{a-b} = W_{int}^{a-b} . \tag{44}$$

However, in the case of a solid-solid interaction, the interaction area can be very much localized. Therefore, we need to determine the surface fraction or the population of the interaction n^{a-b} in terms of acid-base pair per unit area.[72] As a result, the work of adhesion, $W_A{}^{a-b}$ can be expressed as

$$W_A^{a-b} = -fn^{a-b}\Delta H^{a-b} , \tag{45}$$

where f is a correction factor, which is close to one.

IV.3. Total Work of Adhesion

When a solid surface involves both the Lifshitz-van der Waals and the acid-base interactions, the total work of adhesion should be the sum of the following two components:

$$W_A = W_A^{LW} + W_A^{a-b} . \tag{46}$$

According to Fowkes,[72] the first term can be obtained from the dispersion component of the work of adhesion between two materials:

$$W_A^{LW} \simeq W_{12}^d , \qquad (47)$$

where

$$W_{12}^d = \left(W_{11}^d W_{22}^d \right)^{1/2}$$

$$= 2\left(\gamma_1^d \gamma_2^d \right)^{1/2} . \qquad (48)$$

It is important to note that though the first term can be obtained from the geometric mean, the second term $W_A{}^{a\text{-}b}$ can not be obtained in the same manner. Thus, $W_A{}^{a\text{-}b}$ can only be calculated from Eq.(45) despite the assumption by others to use the geometric mean approach, instead.

IV.4. Extension of the Hard-Soft Acid-Base (HSAB) Principle to Solid Adhesion

The hard-soft acid-base (HSAB) principle originally proposed by Pearson[77] has been well tested in many organic and inorganic reactions in solutions.[78] The HSAB principle states:

1. About equilibrium: Hard acids prefer to coordinate with hard bases, and soft acids with soft bases.

2. About kinetics: Hard acids react readily with hard bases, and soft acids with soft bases.

Recently, Lee[79] extended the original HSAB principle to the solid adhesion and surface interactions. With this new attempt, we start to visualize that many solid interactions can now be explained by the HSAB principle. Solid adhesion between polymers and metals can also be achieved presumably by a proper matching of acid and base at the interface.

V. ADSORPTION MECHANISM - POLYMER AT INTERFACE

V.1. Adsorption profile

The above discussion on the two components of W_A should lead to a better understanding of physical adsorption. Theoretically, polymer adsorption[80] can be treated by the Scheutjens-Fleer (SF)[81] mean field theory, the Monte Carlo (MC) method,[82] or the scaling approach.[83] In Fig. 10, two profiles are given to the cases of adsorption (χ=1) and depletion (χ=0) using the SF theory, where χ is the Flory-Huggins interaction parameter[84] between a polymer and a solvent with respect to pure components. The polymer coil expands if $\chi < 0.5$ and contracts if $\chi >$ 0.5. These two cases are referred to as "good" and "poor" solvents, respectively. From the volume fraction profile $\phi(z)$, we can calculate other adsorption parameters, e.g., Γ, the adsorbed amount:

$$\Gamma = \sum_{z=1}^{M} \phi(z) \qquad (49)$$

where z is the normal distance from the interface, and M is the number of layers parallel to the interface. The adsorbed amount defined in Eq.(49)

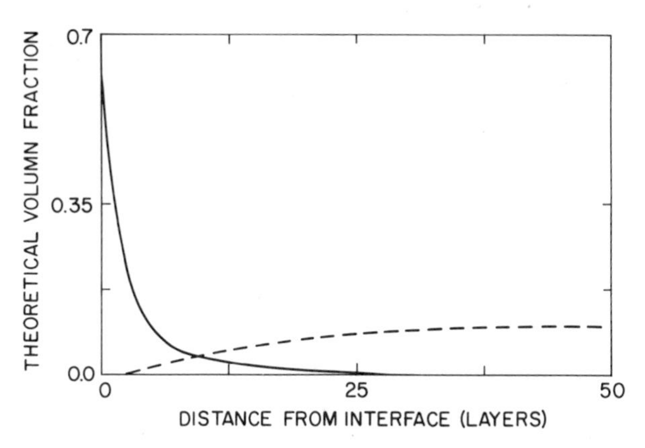

Fig. 10. Volume fraction profiles for adsorption (χ_s = 1; solid line) and depletion (χ_s = 0; dashed line and scale multiplied by 1000). Chain length r = 1000 segments. χ = 0.5, C_{eq} = 1000 ppm. (From T. Cosgrove, Chem. and Ind. 45, January, 1988), reprinted with permission of the author).

is the number of adsorbed segments per surface site: p, the average bound fraction of segments per polymer chain at the interface (layer 1),

$$p = \phi(1)/\Gamma \; . \tag{50}$$

The thicknesses of the adsorbed layer can be expressed in terms of the root-mean-square layer thickness δ_{RMS},

$$\delta_{RMS} = \left\{\left(\sum_{z=1}^{M} \phi(z)z^2\right)/\Gamma\right\}^{1/2} . \tag{51}$$

or the hydrodynamic thickness δ_H, which can be defined in terms of the solvent flux with or without a polymer layer, but is not easy to calculate.

V.2. Loops, Trains and Tails

At the microscopic level, it is possible to visualize the adsorbed chain in the fraction of segments; in loops, trains and tails. Figure 11 shows the distribution of three fractions as generated by the MC method. For the adsorbed layers, the surface appears to be dominated with tails.[85] This may be important in terms of adhesion. By tailor-making the kind of tails, we can perhaps achieve a certain type of adhesion to a selected surface.

For some type of adsorption without being bound tightly to the surface, the scaling technique[82] can also be used, but it is experimentally more difficult to verify. For example, the scaling theory[34] predicts a profile of the form $\phi(z) = (a/z)^{4/3}$, which is a very slow decay function not readily observed experimentally. Recent works by Bouchaud et al.[86] and Caucheteux et al.[87] support this equation derived by de Gennes. However, the scaling theory involves the concept of "self-similarity,"[88] which may imply that the interface could also be fractal.[38,89] Fractal geometry reflects "irregularity," and this may well be the case for the polymer-adsorbed interface as on the electrodeposited surface.

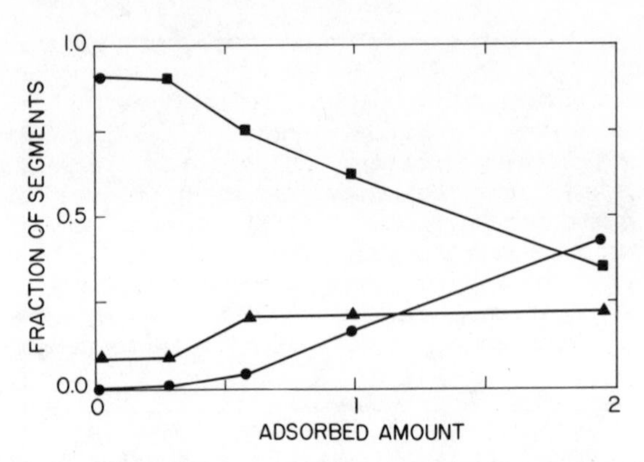

Fig. 11. Fraction of segments in loops, ▲, trains, ■ and tails, ● as a function of the adsorbed amount. $\chi_s = 0.5$, $\chi_s = 1$, chain length 101 segments, generated by the Monte Carlo method. (From T. Cosgrove, Chem. and Ind. 45, January, 1988), reprinted with permission of the author).

V.3. Surface Energetics of Adsorption

One of the important questions about polymer adsorption is how it takes place. We now have some clues about the answer from surface energetics and surface entropic considerations discussed by various researchers and in the above section:

1) Adsorption can take place when $\gamma_p < \gamma_\ell$[82] (where γ_ℓ is the surface energy of the pure solvent and γ_p that of the polymer).

2) Adsorption is in progress when the surface tension of the solution first drops sharply and then gradually with the volume fraction ϕ of the solution.[82]

3) Depletion[11,90] is imminent when the surface tension of the solution increases very slowly with ϕ, and at low ϕ, the polymer never touches the surface.[82]

4) Polymer adsorption from solution involves a decrease in entropy (or $\Delta S^{ads} < 0$) because the adsorbed polymer has fewer possible configurations in the adsorbed state than in solution.[72]

5) Adsorption will occur only when ΔH^{ads} is sufficiently negative (exothermic) to overcome the positive $T\Delta S^{ads}$ term.[72] In other words, adsorption is facilitated by the acid-base interaction at the interface.[91]

V.4. Mean field versus Scaling Theories

The study of adsorption is a very fast moving field, and it is impossible for this brief discussion to cover the entire spectrum of various theories. Specifically, we still do not have a total picture of the adsorption in solution and its effect on adhesion.

A recent review by Fleer et al.[92] has summed up this subject of adsorption according to their findings. They believe that scaling methods are only adequate for weakly overlapping, long flexible chains in good solvents where the mean field model produces incorrect results. However, for the overall adsorption problem, scaling is, in most cases, not a useful alternative because the densities encountered in adsorbed layers are often an order of magnitude too high for scaling laws to apply. Indeed, this has been disputed by the French school led by de Gennes. A recent work by Caucheteux et al.[87] is a good example supporting the scaling theory. However, in most situations, the mean field theory appears to give a better description, which compares well with experimental data.

Most theoretical treatments have neglected end effects, e.g., by assuming one dominating eigenfunction. Although this helps one find analytical solutions, this approximation is not suitable for polymers adsorbing from solutions of finite solutions. One must then apply numerical models, of which the Scheutjens-Fleer theory appears to be rather detailed and adequate. The beauty of this model is that it is valid for any chain length and any solution concentration and considers all possible configurations, including those with tails. Tails have been shown to be very important, e.g., for the hydrodynamic thickness of the adsorbed layers, for steric stabilization, and for bridging flocculation. The proponents also believe that another advantage of this theory is that it can be readily extended to a variety of systems, such as polydisperse samples, polymer mixtures, grafted chains, non-adsorbing polymer, and even polyelectrolytes, etc., not only at one surface but also between two surfaces and for other geometries (using spherical or cylindrical lattices). Thus, so far the mean field theory proposed by Scheutjens and Fleer is more universal than the other two alternatives.

V.5. Role of Adsorption in Adhesion

Since adsorption is believed to be one of the important mechanisms[1] in achieving adhesion, diffusion and wetting are merely kinetic means to attain good adsorption of polymer at the interface. By diffusion and wetting, the polymer molecules can reach an intimate contact that either Lifshitz-van der Waals interaction (long-ranged) or the acid-base interaction (short-ranged) or both may take place. The key for achieving the short-range interaction is to keep the molecules within the approximity of 2 Å.

Recent contributions of the French school on diffusion, wetting and adsorption have certainly made some impact in the field of adhesion. However, the entire field is still wide open for new ideas to germinate.

VI. MECHANICAL INTERLOCKING MECHANISM

The above discussions on diffusion, wetting and adsorption are all on the microscopic scale. For the macroscopic scale, in order to build a strong adhesive joint, mechanical interlocking can be applied, such as the surface treatments of metals to provide various topologies. However, mechanical interlocking is not one of the adhesion mechanisms, at least from the molecular level. It is only a technological means in achieving adhesive bonding as in the case of structural adhesives.

For those interested in mechanical interlocking, one may refer to the papers by Venables,[93] Clearfield et al.[94] Minford,[95] Brockmann,[96] and Kinloch.[18] It is important to point out that the ultimate goal of mechanical interlocking is also to achieve optimum adsorption.

VII. CHEMICAL BONDING MECHANISM

Physical adsorption mechanisms depend on van der Waals forces or the acid-base interaction, while the chemical bonding mechanism is based on the primary covalent bond at the interface. One of the most significant findings in the last several years is the confirmation of the chemical bonding mechanism. Chemical bonding is one of the strongest kinds one can hope for for the polymer adhesion. For example, the chemical reactions created by coupling agents at the interface have been established.[39] Two recent reviews by Plueddemann[97] and Miller and Ishida[98] should be helpful for those interested in designing strong interfaces through the use of coupling agents.

Chemical interactions between metals and common polymers have been reviewed by Buckley[49] and Lee.[100] Recent progress in the work of metal and polyimide bonding has been reported by Ho _et al_.[101] and R. Haight _et al_.[102] and reviewed by Lee.[100]

A well known method, grafting, has been used to improve adhesion of polypropylene,[105] and polyethylene[106] to metals. Schultz _et al_.[106] show that the addition of acrylic acid can enhance the adhesion of polyethylene to aluminum.

In the case of polyimides, Cr has been found to form a complex with the π electron on the PMDA unit of PI.[103,104] It appears to be also an acid-base interaction, where Cr is a soft acid and PMDA is a soft base.

VIII. ELECTROSTATIC MECHANISM

The last, but not the least, is the electrostatic mechanism. The electrostatic (or electronic) mechanism proposed by Deryagin[107] is especially important in particle adhesion.[108] Though the electronic adhesion has been claimed to be first investigated by Deryagin, some early work by Skinner, Savage and Rutzler[109] reported in 1953 was first known to the west and was carried out without the knowledge of the Russian work. Skinner[110] actually first laid the theoretical foundation for electronic adhesion based on the electrical double layer (*EDL*) theory, despite some errors in the original derivation noted by the Russian school published in English[111] twelve years after Skinner's publication.

There have been at least three controversies surrounding the electronic adhesion theory: 1) the *EDL* could not be identified without the separation of the adhesive bond, 2) the effect of *EDL* on the adhesive bond strength was exaggerated, and 3) the Russian school later attempted to encompass the donor-acceptor interaction within the *EDL* electronic theory.[69] Several recent works outside the U.S.S.R. have shed new light on the original electronic adhesion theory. We shall attempt to summarize these new developments.

VIII.1. Direct Evidence of the Electrical Double Layer

In the past, it was very difficult to prove the existence of an electrical double layer (*EDL*) without the separation of an adhesive bond. In 1984, an interesting experiment carried out by Possart and Röder[112] used SEM for determining the potential distribution at the polymer-metal interface without separation. The existence of a double layer was thus confirmed. One of the remaining problems is to establish the extent of the *EDL* and its effect on adhesive bonding.

VIII.2. Work of Electrical Double Layer Versus Peel Work

Recently Possart[113,114] has attempted to determine the actual charge density and the work of the electrical double layer without breaking the bond. According to him, the maximum charge moving in the whole measuring interval reaches 2.7 x 10^{-2} C/m^2. His calculations show that the electrostatic component of W_{el} due to *EDL* is only 1.71 x 10^{-3} J/m^2. Here W_{el} is the interaction energy between the space charges in the undisturbed *EDL*.

On the other hand, the actual peeling of the LDPE film from the aluminum foil yields a specific peel work of 1 J/m^2, which is about 600 times that of the W_{el}. Thus, the specific peel energy is much larger than the stored electrostatic energy due to the presence of the *EDL*. This is an indication that the electrostatic component is only a fraction of the total adhesive strength. Thus, this result still can not serve as definitive proof for the Deryagin's hypothesis on the dominant role of *EDL* in adhesion.

In the literature, there has been other work that disputes Deryagin's original claim. Roberts, in his studies of rubber adhesion,[115] has indicated that the contribution of the electrostatic component is less than 10%, usually only 0.1-1%.

VIII.3. Elimination of Double Layer by Surface Modification

Since the *EDL* adhesion is not equal to the acid-base (or donor-acceptor) interaction, it is possible that in some systems both interactions can coexist. However, the increase of the acid-base interaction on the same surface can reduce the *EDL* interaction due to the decrease of the surface charge density[69]

$$\sigma_s = e(n^a - n^b) \ , \qquad (52)$$

where n^a is the surface concentration of the acidic component (or acceptor) and n^b that of the basic component. When $n^a = n^b$, $\sigma_s \simeq 0$.

Hence, to eliminate the *EDL*, either acidic or basic properties must be imparted to the surface by surface modification. Thus the application of any of the following donor-acceptor series containing various functional groups can change the surface charge density.

Donor (base) $-NH_4$ > - OH > - OR > - COOR

> - CH_3 > - C_6H_5 > X > -C=O > - CN. Acceptor (acid)

VIII.4. Electrostatic Adhesion Between Anion-Cation Pairs

Though the Russian school has been credited for the introduction of the *EDL* electronic adhesion theory,[107] the idea of donor-acceptor (or acid-base) interaction did not originate from their school.

Recently, Hays[116] discussed the electrostatic adhesion of two planar materials (donors and acceptors) between anions and cations. He considers two cases for two insulators: Case 1: $r << a'$, and case 2: $r >> a'$, where r is the average distance between anionic and cationic groups and a' is the distance between the two identical groups.

Case 1: $r << a'$

$$Electrostatic\ force:\ F_{el} = e^2/(4\pi k \varepsilon_o r^2); \tag{53}$$

$$Electrostatic\ force\ per\ unit\ area:\ P_{el} = nF_{el} = ne^2/(4\pi k \varepsilon_o r^2), \tag{54}$$

where k is the dielectric constant, ε_o is the permittivity in the free space and n is the density of anion-cation pairs per unit area.

Case 2: $r >> a'$

$$Electrostatic\ force:\ F_{el} = ne^2/(2k\varepsilon_o); \tag{55}$$

$$Electrostatic\ force\ per\ unit\ area:\ P_{el} = n^2e^2/(2k\varepsilon_o), \tag{56}$$

The calculated results for P_{el}'s for both cases are shown in Table 4 for various densities of the anion-cation pairs. It can be seen that the electrostatic component due to the anion-cation interaction is comparable to or greater than the van der Waals component ($P_{LW} = A/(6\pi Z^3)$; when $A = 1 \times 10^{-19}$ J and separation, Z = 0.4 nm, P_{LW} = 83 MPa) when the density of the anion-cation pairs is approximately $10^{17}/m^2$ or greater. Incidentally, a surface charge density of $10^{17}/m^2$ actually corresponds to a charge density of $1.6 \times 10^{-6} C/cm^2$ or 16,000 uC/m^2. Since a monolayer of atoms is approximately $10^{19}/m^2$, the anion-cation interaction is significant when the density of the pairs is greater than 1% of a monolayer.

VIII.5. Particle Adhesion

For particle adhesion, the total forces[108] consist of Lifshitz-van der Waals forces, F_{LW}, the electrostatic induced image forces, F_{im}, the capillary force, F_{ca}, the chemical forces, F_{ch} (such as the acid-base interaction), and the double layer force, F_{ed}:0

Table 4 Electrical component of adhesive pressure in units of MPa

$P_{el}\ (r<<a') = 7.6\times10^{-17}(n/r^2)$; $P_{el}\ (r>>a') = 4.6\times10^{-34}(n^2)$

n (m^{-2})	a' (nm)	r(nm)				
		0.3	0.5	1	10	100
10^{14}	100	8.4×10^{-2}	3.0×10^{-2}	7.6×10^{-3}	7.7×10^{-5}	4.8×10^{-6}
10^{15}	31	8.4×10^{-1}	3.0×10^{-1}	7.6×10^{-2}	9.5×10^{-4}	4.7×10^{-4}
10^{16}	10	8.4	3.0	7.7×10^{-1}	4.8×10^{-2}	4.6×10^{-2}
10^{17}	3.1	85.0	31.0	9.5	4.7	4.6
10^{18}	1.0	1.0×10^{3}	5.8×10^{2}	4.8×10^{2}	4.6×10^{2}	4.6×10^{2}
10^{19}	0.31	4.8×10^{4}	4.7×10^{4}	4.7×10^{4}	4.6×10^{4}	4.6×10^{4}

(Ref.: D. Hays, in Fundamentals of Adhesion, Ed. L.H. Lee, Chapt. 8, Plenum, New York, (1990).

$$F = F_{LW} + F_{im} + F_{ca} + F_{ch} + F_{ed} \; . \tag{57}$$

According to Krupp[117] in general F_{LW} dominates, and the first term is:

$$F_{LW} = \frac{\hbar \omega R}{8 \pi Z^2} \; , \tag{58}$$

where R is the radius of the sphere, Z the distance between the sphere and the plane, $\hbar\omega$ the Lifshitz-van der Waals constant. The second term, the image force is

$$F_{im} = K_{im} \left(q^2 / R^2 \right) \tag{59}$$

where K_{im} is the constant, which depends on the roughness of the spherical particle and the density of the particle layer.[118]

The third term F_{ca} in ambient condition is generally negligible. The fourth term F_{ch} can be significant if the donor-acceptor (or anion-cation) interaction is involved. This was demonstrated in the preceding section. Finally, the fifth term, F_{ed}, is the electrical double layer force proposed by Deryagin,[119] and it is expressed as

$$F_{ed} = 2 \pi \sigma^2 S_d \, , or \;\; \sigma^2 S_d / 2 \varepsilon_0 \;\; (in\ S.I.\ unit\ as\ in\ Eq. 55) \tag{60}$$

where σ is the surface charge density and S_d is the area. Thus, even in the case of particle adhesion, the *EDL* is only a part of the total forces of attraction. Since Uber et al.[120] have reported that the particle adhesion is not affected by the charge transfer, the *EDL* force in the fifth term is thus insignificant or equal to zero.

The dominating role of the Lifshitz-van der Waals component on the adhesion of particles (<50 μm) has also been shown on a semiconductor surface.[121] These *LW* forces can increase with time due to particle and/or surface deformation, which increases the contact area. Micron size particles can be held to surfaces by forces exceeding 100 dynes, which correspond to a pressure of 10^9 dyne/cm^2 or more. Total forces of adhesion of micron size particles exceed the gravitational force on the particle by factors greater than 10^6. Electrostatic forces only become important and predominant for particles larger than 50 μm.

In the case of *pre-charged* particles, e.g., xerographic toners, the electrostatic (Coulombic) component is more dominant[122-125] than the van der Waals component. The contact-charging and adhesion of *pre-charged* particles are easier to understand than those of neutral particles.

Since new questions have been raised about the mechanisms of contact charging,[126] we are unable to infer one way or the other whether the controversies of the electronic adhesion theory have been put to rest. Once we know exactly how charges are being transferred at the interface, we should be able to draw a better picture about the electrical double layer and its effect on adhesion.

IX. SUMMARY

In the past two decades, the progress in polymer dynamics has been phenomenal. Since the introduction of the scaling concept by Professor P.G. de Gennes of the College de France, several problems related to polymer interface and adhesion have been widely investigated.

This paper was intended to review recent developments in the studies on mechanisms of polymer adhesion. For diffusion mechanism, the reptation model has been applied successfully to study tack, green strength, healing, and welding of polymers. For adsorption mechanism, new findings by the French school related to the kinetics of wetting and spreading of polymer melt provide insights in achieving good adhesive bonding. The theoretical development of the adsorption mechanism reveals the roles of surface energetics and acid-base interaction on optimum adsorption. However, it appears that the mean field theory is more applicable to adsorption than the scaling theory.

In addition to the above mechanisms, we also briefly examined the mechanical interlocking, the chemical bonding, and the electronic adhesion mechanisms. We noted that the influence of the French school on these mechanisms was minimal.

Nomenclature

A	Avogadro's number
A_H	Hamaker constant
a	molecular size
a'	distance between two identical ionic groups
b	height of polymer-drop foot
c	concentration of polymer
c^*	polymer overlap concentration
C	susceptibility of covalent interaction
D	diffusion coefficient
D_i	interdiffusion coefficient
D_s	self-diffusion coefficient
D_t	curvilinear diffusion coefficient along the tube
$\dot{e}$	thickness of spreading film
e	electronic charge
$E(t)$	modulus of elasticity
E	susceptibility of electrostatic interaction
F	surface free energy
F_{ca}	capillary force
F_{ch}	chemical force
F_{ed}	electrical double layer force
F_{el}	electrostatic force
F_{im}	image force
f	correction factor
G_c	fracture energy
$(G_I)_i$	interfacial fracture energy (Mode I)
$G_N{}^\circ$	plateau modulus
g	gravitational acceleration
$H(t)$	dynamic properties
$H(\infty)$	static properties
$\Delta H^{a\text{-}b}$	enthalpy of the acid-base interaction
ΔH^{ads}	enthalpy of adsorption
$\hbar$	Planck constant $h/2\pi$
h	drop height
K_c	fracture toughness
$(K_I)_i$	interfacial fracture toughness (Mode I)
k	capillary length (height)

$L_o(t)$	number of monomers crossing the interface
$L(t)$	length of the tube
ℓ	diffusive length
$\ell(p)$	average bridge length
$\ell(t)$	average contour length
M	molecular weight
M_c	molecular weight at critical point
M_N	number-average molecular weight
M_W	weight-average molecular weight
N	degree of polymerization or number of monomer units per chain
N_c	N at critical point
N_e	N between two entanglement points
$\mathbf{n}$	**n**-vector
n	surface population fraction
$n(t)$	number of chains
P_{LW}	Lifshitz-van der Waals pressure
$P(e)$	related to disjoining pressure
P_{el}	electrostatic force per unit area
p	average bound fraction of segments of polymer chain
$p(t)$	number of bridge crossing the interface
q	point charge
R	radius of particle, and radius of wetted spot
R^2	root-mean-square end-to-end distance
R_g	universal gas constant
R_G	radius of gyration
r	scaling of the maximum radius of gyration
S	spreading coefficient
S_d	charged area
ΔS^{ads}	entropy of adsorption
T	temperature (Kelvin)
T_g	glass temperature
T_r, T_{rep}	reptation time
$T(\omega)$	characteristic relaxation at frequency ω
t	relaxation time, contact time
U	drift velocity
W_A	work of adhesion
$W_A{}^{a\text{-}b}$	the acid-base component of W_A
$W_A{}^d$	the dispersion component of W_A
$W_A{}^{LW}$	the Lifshitz-van der Waals component of W_A
W_{el}	electrostatic double layer work
$W_{int}{}^{a\text{-}b}$	work of the acid-base interaction
W_P	peel work
X	horizontal distance of spreading
x_{cm}	center-of-mass depth
$x(t)$	average monomer interpenetration depth, or thickness of the interdigitated region
Z	separation between the spheres
$Z(r)$	height of drop

Greek letters

γ	surface tension
γ^d	dispersion component of γ
γ_ℓ	γ of pure solvent
γ_p	γ of polymer
δ_H	hydrodynamic thickness
δ_{RMS}	root-mean-square thickness
ε_o	permittivity at free space
η	bulk viscosity, also monomer-monomer friction coefficient
η_o	zero shear viscosity
θ	contact angle

θ_a	apparent θ
θ_e	equilibrium θ
K	Boltzmann's constant
$\pi(e)$	disjoining pressure
ρ	density
σ_s	surface charge density
$\sigma(t)$	adhesive bond strength
Γ	the adsorbed amount of polymer
τ_o	microscopic relaxation time
τ	characteristic relaxation time
$\phi(Z)$	volume fraction profile of polymer
χ	interaction parameter
ω	Lifshitz-van der Waals constant frequency
Ω	volume of drop

References

1. L.H. Lee, J. Polym. Sci. A-2, 5, 751 (1967).
2. L.H. Lee, J. Polym. Sci. A-2, 5, 1103 (1967).
3. F. Bueche, W.M. Cashin and P. Debye, J. Chem. Phys. 20, 1956 (1952).
4. P.G. de Gennes, J. Chem. Phys. 55 (2), 572 (1971).
5. P.G. de Gennes, Scaling Concepts in Polymer Physics, Cornell University Press (1975).
6. P.G. de Gennes, Phys. Today, 33, June, 1983.
7. P.G. de Gennes and L. Leger, Ann. Rev. Phys. Chem. 33, 49-61 (1982).
8. J. Klein, Contemp. Phys. 20, (6), 611 (1979).
9. M. Tirrell, Rubber Chem. and Technol., 57, 523 (1984).
10. S. Prager and M. Tirrell, J. Chem. Phys., 75, (10), 5195 (1981).
11. P.G. de Gennes, Macromolecules, 14, 1637 (1981).
12. P.G. de Gennes, Macromolecules, 15, 492 (1982).
13. M. Doi and S.F. Edwards, J. Chem. Soc. Faraday Trans. II, 74, 1789 (1978), and 74, 1802 (1978).
14. M. Doi and S.F. Edwards, The Theory of Polymer Dynamics, Clarendon Press, Oxford, London (1986).
15. F. Brochard, in Fundamentals of Adhesion, Ed. L.H. Lee, Chapt. 6, Plenum, New York, to be published in 1990.
16. A.J. Kinloch, J. Mater. Sci., 15, 2141 (1980).
17. A.J. Kinloch, J. Mater. Sci., 17, 617 (1982).
18. A.J.Kinloch, Adhesion and Adhesives, Science and Technology, Chapt. 3, Chapman and Hall, London, New York (1987).
19. S.S.Voyutskii, Autohesion and Adhesion of High Polymers, Interscience, New York (1963).
20. J. Klein, Macromolecules, 11(5), 852 (1978).
21. J. Klein and B.J. Briscoe, Proc. R. Soc. London Ser. A, 365, 53 (1979).
22. S.F. Edwards and Grant, J. Phys. A, 6, 1186 (1973).
23. P.G. de Gennes, in Microscopic Aspects of Adhesion and Lubrication, J.M. Georges, Ed., Elsevier, Amsterdam, Oxford, New York, 1982, p.355.
24. R.P. Wool, Rubber Chem. and Technol. 57, 307 (1984).
25. R.P. Wool, in Fundamentals of Adhesion, Ed. L.H. Lee, Chapt. 7, Plenum New York, to be published in 1990.
26. R.P. Wool and K.M. O'Connor, J. Polym. Sci., Polym. Lett., 20, 7 (1982).
27. Y.-H. Kim and R.P. Wool, Macromolecules, 16, 1115 (1983).
28. R.M. Vasenin, in Adhesion, Fundamentals and Practice, Ministry of Technology, U.K., Gordon and Breach, 1969, p.29.
29. L. Léger, Ann. Chim. Fr., 12, 175 (1987).

30. L.H. Lee, in Physicochemical Aspects of Polymer Surfaces, K.L. Mittal, Ed., Vol. 1, Plenum, New York, 1983, p.523.
31. D. Maugis, in Adhesive Chemistry-Developments and Trends, L.H. Lee, Ed. Plenum, New York, 1984, p.63, also in "Fundamentals of Adhesive Bonding, L.H. Lee, Ed., Chapt. 10, Plenum, New York, to be published in 1990.
32. K. Jud and H.H. Kausch, Polym. Bull., 1, 697 (1979).
33. K. Jud, H.H. Kausch and J.G. Williams, J. Mater. Sci., 16, 204 (1981).
34. W.W. Graessley, J. Polym. Sci. Polym. Phys., 18, 27 (1980).
35. G.R. Hamed, Rubber Chem. and Technol., 54, 576 (1981).
36. W.G. Forbes and L.A. McLeod, Trans. Inst. Rubber Ind. 30 (5), 154 (1958).
37. R.P. Wool, in New Trends in Physics and Physical Chemistry of Polymers, L.H. Lee, Ed. Plenum, New York, 1989.
38. B.M. Mandelbrot, The Fractal Geometry of Nature, Freeman, New York, 1982.
39. D. Lhuillier, J. Phys. (France), 49, 705 (1988).
40. T.A. Vilgis, J. Phys. (France), 49, 1481 (1988).
41. C.M. Marques and J.F. Joanny, J. Phys. (France), 49, 1103 (1988).
42. P.G. de Gennes, in Physics of Disordered Materials, D. Adler, H. Fritzche, S.R. Ovshinsky and N.F. Mott, Eds., Plenum, New York, 1985, p.227.
43. W.A. Zisman, in Adhesion Science and Technology, Ed. L.H. Lee, Vol. A, 55, Plenum, New York (1975).
44. K.L. Mittal, in Adhesion Science and Technology, Ed. L.H. Lee, Vol. A, 129, Plenum, New York (1975).
45. A. Dupree, Theorie Mechanique de la Chaleur, Gauthier-Villars, Paris (1869), p.393.
46. R.J. Good and M.K. Chaudhury, in Fundamentals of Adhesion, L.H. Lee, Ed., Chapt. 3, Plenum, New York, to be published in 1990.
47. A. Marmur, Adv. Coll. and Interface Sci., 19, 75 (1983).
48. P.G. de Gennes, Rev. Mod. Phys., 57(3), Part 1, 827 (1985).
49. P.G. de Gennes, in Fundamentals of Adhesion, L.H. Lee, Ed., Chapt. 5, Plenum, New York, to be published in 1990.
50. A.-M. Cazabat, Contemp. Phys., 28,(4), 374 (1987).
51. L. Léger, A.M. Guinet-Picard, H. Hervet, D. Ausserré and M. Erman, in New Trends in Physics and Physical Chemistry of Polymers, L.H. Lee, Ed. Plenum, New York, 1989.
52. A.M. Cazabat, F. Haslot and P. Levinson, in New Trends in Physics and Physical Chemistry of Polymers, L.H. Lee, Ed. Plenum, New York, 1989.
53. T. Young, Philos. Trans., Roy. Soc. London, 95, 65 (1905).
54. A.W. Adamson, Physical Chemistry of Surfaces, Ed., Wiley, New York, 1982.
55. L.H. Tanner, J. Phys. D, 12, 1473 (1979).
56. J. Lopez, C.A. Miller and E. Ruckenstein, J. Colloid Interface Sci., 56, 460 (1976).
57. A.M. Cazabat and M.A. Cohen Stuart, J. Phys. Chem., 90, 5845 (1986).
58. W. Cooper and W. Nuttal, J. Agr. Sci., 7, 219 (1915).
59. J.F. Joanny and P.G. de Gennes, J. Phys., (Paris) 47, 121 (1986).
60. B.V. Deryagin, Zh. Fiz. Khim., 14, 137 (1940).
61. B.V. Deryagin, Kolloidn. Zh., 17, 191 (1955).
62. B.V. Deryagin and N.V. Churaev, Kolloidn. Zh., 38, 438 (1976).
63. F. Brochard and P.G. de Gennes, J. Phys. Lett., (Paris) 45, L-597 (1984).
64. H. Schonhorn, H.L. Frisch, and T.K. Kwei J. Appl. Phys., 37, 4967 (1966).
65.(a) D. Ausserré, A.M. Picard and L. Léger, Phys. Rev. Lett., 57, (21), 2671 (1986).

65.(b) J. Daillant, J.J. Benattar, L. Bosio and L. Léger, Europhys. Lett., 6,(5), 431 (1988).

66. M. Boudoussier, J. Phys., (Paris) 48, 445 (1987).

67. F.M. Fowkes, J. Adhes., 4, 155 (1972), also in Recent Advances in Adhesion, L.H. Lee, Ed. Gordon and Breach, New York and London (1973), p.39.

68. V. Gutman, The Donor-Acceptor Approach to Molecular Interactions, Plenum Press, New York, 1978.

69. B.V. Deryagin, N.A. Krotova and V.P. Smilga, Adhesion of Solids, Eng.Ed. Translated by R.K. Johnson, Plenum, New York, 1978, p 143.

70. W.B. Jensen, Chem. Rev., 78,(1), 1 (1978).

71. F.M. Fowkes, in Physicochemical Aspects of Polymer Surfaces, K.L. Mittal, Ed. Vol. 2, Plenum, New York, 1981, p.583.

72. F.M. Fowkes, J. Adhes. Sci. Technol., 1,(1) 7 (1987).

73. J.C. Bolger and A.S. Michaels, in Interface Conversion, P. Weiss and D. Cheevers, Eds. Chapt. 1, Elsevier, New York (1969).

74. R.J. Good, M.K. Chaudhury and C.J. van Oss, in Fundamentals of Adhesion, L.H. Lee, Ed., Chapt. 4, Plenum, New York, to be published in 1990.

75. C. Boiziau and G. Lecayon, Surf. and Interface Ana., 12, 475 (1988).

76. R.S. Drago, G.C. Vogel and T.E. Needham, J. Am. Chem. Soc., 93, 6014 (1971).

77. R.G. Pearson, J. Am. Chem. Soc., 85, 3533 (1963).

78. R.G. Pearson, Hard and Soft Acids and Bases, Dowden, Hutchinson and Ross, Stroudsburg, PA, (1973).

79. L.H. Lee, in New Trends in Physics and Physical Chemistry of Polymers, L.H. Lee, Ed. Plenum, 1989, also presented to the Fifth International Congress in Tribology, June, 1989.

80. T. Cosgrove, Chem. and Ind., 18, 45, Jan. 1988.

81. J.M.H.M. Scheutjens and G.J. Fleer, J. Phys. Chem., 83, 1619 (1979), 84, 178 (1980).

82. P.G. de Gennes, Adv. Colloid and Interface Sci., 27, 189 (1987).

83. T. Cosgrove, T.G. Heath, B. van Lent and J.M.H.M. Scheutjens, Macromolecules, 20, 1692 (1987).

84. P.J. Flory, Principles of Polymer Chemistry, Cornell University Press, Ithaca, New York (1953).

85. J.M.H.M. Scheutjens, G.J. Fleer and M.A. Cohen Stuart, Colloids and Surfaces, 21, 285 (1986).

86. E. Bouchaud and M. Daoud, J. Phys., (Paris) 48, 1991 (1987).

87. I. Caucheteux, H. Hervet, F. Rondelez, L. Auvray and J.P. Cotton, in New Trends in Physics and Physical Chemistry of Polymers, Plenum, (1989).

88. L. Auvray and J.P. Cotton, Macromolecules, 20, 202 (1987).

89. R. Orbach, Science, 231, 814 (1987).

90. J.F. Joanny, L. Leibler and P.G. de Gennes, J. Polym. Sci. Polym. Phys. 17, 1073 (34) (1979).

91. F.M. Fowkes and M.A. Mostafa, Ind. Eng. Chem. Prod. Res. Dev., 17(1), 3, (1978).

92. G.J. Fleer, J.M.H.M. Scheutjens and M.A. Cohen Stuart, Colloid and Surfaces, 31, 1 (1988).

93. J.D. Venables, J. Mater. Sci, 19, 2431 (1984).

94. H.M. Clearfield, D.K. McNamara and G.D. Davis, in Fundamentals of Adhesive Bonding, L.H. Lee, Ed., Plenum, New York to be published in 1990.

95. J. Minford, in Fundamentals of Adhesive Bonding, L.H. Lee, Ed. Chapt. 9, Plenum, New York, to be published in 1990.

96. W. Brockmann, O.D. Hannemann, H. Kollek and C. Matz, Int. J. Adhes. and Adhes., 6(3), 115 (1986).

97. E.P. Plueddemann, in Fundamentals of Adhesion, L.H. Lee, Ed., Chapt. 9, Plenum, New York to be published. in 1990.

98. J.D. Miller and H. Ishida, in Fundamentals of Adhesion, L.H. Lee, Ed., Chapt. 10, Plenum, New York, to be published in 1990.
99. D.H. Buckley, Surface Effects in Adhesion, Friction, Wear and Lubrication, Elsevier, Amsterdam (1981).
100. L.H. Lee, in Fundamentals of Adhesion, L.H. Lee, Ed. Chapt. 1, Plenum, New York, to be published in 1990.
101. P.S. Ho, P.O. Hahn, J.W. Bartha, G.W. Rubloff, F.K. LeGouses and B.D. Silverman, J. Vac. Sci. Technol., A 3(3), 739 (1985).
102. R. Haight, R.C. White, B.D. Silverman and P.S. Ho, J. Vac. Sci. Technol., A6(4) 2188 (1988).
103. P.S. Ho, B.D. Silverman, R.A. Haight, R.C. White, P.N. Sanda and A.R. Rossi, IBM J. Res. Develop., 32(5), 658 (1988).
104. P.S. Ho, R.A. Haight, R.C. White, B.D. Silverman and F. Faupel, in Fundamentals of Adhesion, L.H. Lee, Ed. Chapt. 13, Plenum, New York, to be published in 1990.
105. A.F. Lewis and L.J. Forrestal, in Symposium on Recent Developments in Adhesion Science, American Society for Testing and Materials, 1963, p.59.
106. J. Schultz, A. Carre and C. Mazeau, Int. J. Adhes. and Adhes, 4(4), 163 (1984).
107. B.V. Deryagin and N.A. Krotova, Adgeziya (Adhesion), Acad. Sci. Publ., Moscow, 1949.
108. A.D. Zimon, Adhesion of Dust and Powder, Eng. Ed. Plenum, New York (1969).
109. S.M. Skinner, R.L. Savage and J.E. Rutzler, Jr., J. Appl. Phys., 24(4), 438 (1953).
110. S.M. Skinner, J. Appl. Phys., 26(5), 509 (1955).
111. B.V. Deryagin and V.P. Smilga, J. Appl. Phys., 38(12), 4609 (1967), also in Adhesion-Fundamentals and Practice, Ministry of Technology, published by Gordon and Breach, New York, 1969, p.152.
112. W. Possart and A. Röder, Phys. Stat. Sol., (a) 84, 319 (1984).
113. W. Possart and I. Müller, Phys. Stat. Sol. (a) 106, 525 (1988).
114. W. Possart, Int. J. Adhes. and Adhes., 77, April (1988).
115. A.D. Roberts, J. Phys. D.10, 1801 (1977).
116. D. Hays, in Fundamentals of Adhesion, L.H. Lee, Ed. Chapt. 8, Plenum, New York to be published in 1990.
117. H. Krupp, Adv. Colloid and Interface Sci., 1(2), 111 (1967).
118. N.S. Goel and P.R. Spencer, in Adhesive Science and Technology, Ed. L.H. Lee, Vol. B, Plenum, New York (1975), p. 763.
119. B.V. Deryagin, I.N. Aleinikova and Yu. P. Toporov, Powder Technology, 2, 154 (1968/69).
120. A.E. Uber, J.F. Hoburg and G. Penney, IEEE Trans. Ind. Appl. IA-20(1), 148 (1984).
121. R. Allen Bowling, J. Electrochem. Soc.:Solid-State Sci. and Technol., 132(9), 2208 (1985).
122. D.K. Donald and P.K. Watson, Phot. Sci and Eng., 14,(1), 36 (1970).
123. D.K. Donald in Recent Advances in Adhesion, L.H. Lee, Ed., Gordon and Breach, New York (1973), p.129.
124. M.H. Lee and J. Ayola, J. Imaging Technol., 11(6), 279 (1985).
125. L.H. Lee, Phot. Sci. and Eng., 22,(4), 228, (1978).
126. A.R. Akande and J. Lowell, J. Phys. D, 20, 565 (1987).

DYNAMICS AND FRACTAL STRUCTURE OF POLYMER INTERFACES

R.P. Wool

Department of Materials Science and Engineering
University of Illinois
Urbana, Illinois 61801

ABSTRACT

Interdiffusion at polymer-polymer interfaces is analyzed using the reptation model proposed by P.-G. de Gennes for the dynamics of linear polymer melts. Correlated motion effects unique to the reptation model are explored by computer simulation and analysis of concentration profiles of symmetric amorphous interfaces at times less than the reptation time, T_r, and diffusion depths less than random coil diameter. Results are compared with the scaling laws for interdiffusion given by $H(t)/H_\infty \approx (t/T_r)^{r/4}$, where $r = 1, 2, 3$. The diffusion front of the highly ramified interface was analyzed on a square latice. The front was found to have fractal characteristics for $t > T_r$ but was not fractal for $t < T_r$. A fractal diffusion front for a polymer/melt interface (polyimide/Ag) is demonstrated.

I. POLYMER-POLYMER INTERDIFFUSION

The reptation model as proposed by de Gennes[1] to describe the dynamics of linear polymer melts was used to determine the molecular properties and structure of an interdiffused symmetric polymer-polymer interface. The interface is created when two amorphous polymers achieve good contact in the melt. Wetting first occurs followed by interdiffusion. The extent of interdiffusion of polymer segments is an important problem and controls the strength development of the interface. Solutions to this problem have application to polymer processing, adhesion, tack, composite lamination, welding, coatings, etc.

The minor chain (MC) model of reptating chains as shown in Figure 1 was proposed by Kim and Wool to analyze interdiffusion in polymer melts.[2] Only those parts of the chains which have escaped by reptation from their initial tubes (the minor chains) at the time of contact can contribute to interdiffusion. Using this model, the average molecular properties of the interface were derived[2,3] and are summarized in Table 1. The molecular properties have a common scaling law which relates the dynamic properties, $H(t)$, to the static equilibrium properties, H_∞, via the reduced time, t/T_r, by[3,4]

$$H(t) = H_\infty (t/T_r)^{r/4} ,$$

$$H_\infty \sim M^{(3r-s)/4} , \tag{1}$$

where r,s = 1, 2, 3, ...

In these relations the reptation time, T_r, is related to the molecular weight, M, via $T_r \propto M^3$. Values for r and s are given for the specific molecular properties in Table 1. The properties can be used to explore various molecular models proposed for welding and fracture of polymers. In this paper we investigate the scaling laws for interdiffusion based on the reptation model and examine the fractal nature of polymer interfaces.

II. CONCENTRATION PROFILE

The concentration profile, $\Phi(X,t)$, for polymer melts with reptation has non-Fickian characteristics[5] for diffusion times $t < T_r$ and diffusion depts $X < R$, where R is the end-to-end vector of the random coil chains. Deviations from normal Fickian diffusion arise by correlated motion effects of reptating chains such that the average monomer motion has a different time dependence than the center-of-mass motion of the entire chain. At $t > T_r$, normal Fickian diffusion behavior is obtained when the correlations have decayed. The concentration profiles have been determined by Tirrell, Adolf and Prager,[6] and by Zhang and Wool,[7] using the MC model in the latter case. The concentration profile is obtained, assuming a uniform spatial distribution of chain ends, from the integral of the MC Gaussian conformations as a function of depth, X, at $t < T_r$ as[7]

$$\Phi(X,t) = \left(\frac{\ell(t)}{L} + \frac{X^2}{Lb} \right) erfc \left[\frac{X}{\sqrt{(2n)}b} \right] - \frac{\sqrt{(2n)}}{\sqrt{\pi}} \frac{X}{L} e^{-\left(\frac{X^2}{2nb^2}\right)} \tag{2}$$

where b is the bond length, n is the number of segments in the minor chain of length $\ell(t)$ and L is the contour length of the chain. The time dependence of $\ell(t)$ (and n) is given by Kim and Wool as,[2]

$$\ell(t) = \{16 D_1 t/\pi\}^{1/2} , \tag{3}$$

where $D_1 \propto 1/M$ is the one-dimensional curvilinear diffusion coefficient.

The concentration profile as derived by Zhang and Wool[7] is shown in Figure 2 at times less than the reptation time and diffusion distances of the order of the radius of gyration. The discontinuity in $\Phi(X,t)$ at the original interface plane at X = 0, represents one of the peculiar non-Fickian features predicted by the reptation model, and can be evaluated from (Eq. 2) as,

$$\Phi(0,t) = \ell(t)/L . \tag{4}$$

Substituting from Table 1 for $\ell(t)$, $\Phi(0,t)$ scales as 1/2 $t^{1/2}M^{-3/2}$ which has the same scaling law as the number of bridges and crossing density determined by de Gennes[5] and Prager and Tirrell,[8] respectively. At $t = T_r$, $\ell(t) = L/2$ and $\Phi(0,t)$ = 1/2 as predicted for normal Fickian diffusion. It is also worth noting that the chains contributing to the

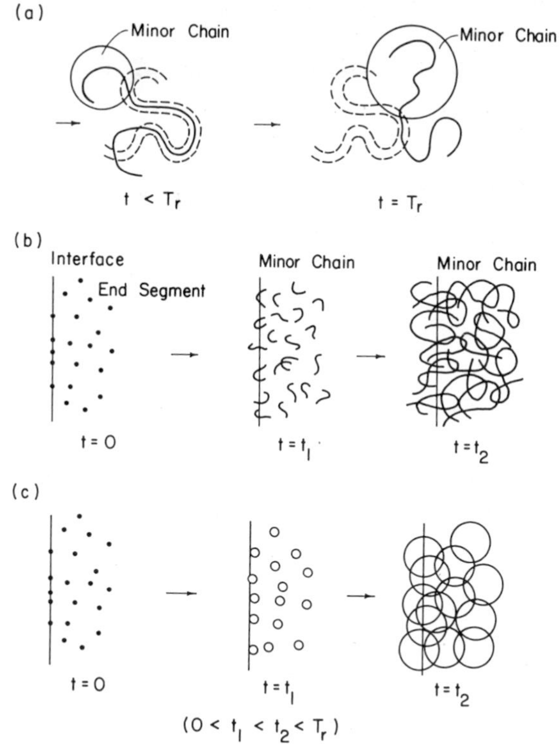

Fig. 1. The Minor-Chain Reptation model for diffusion at a polymer-polymer interface is shown in terms of (a) the evolution of the minor chain ends by reptation, (b) the behavior of the minor chains at the interface and (c) the minor chain most probable spherical envelopes. One side of the interface is shown for clarity. (Kim and Wool)

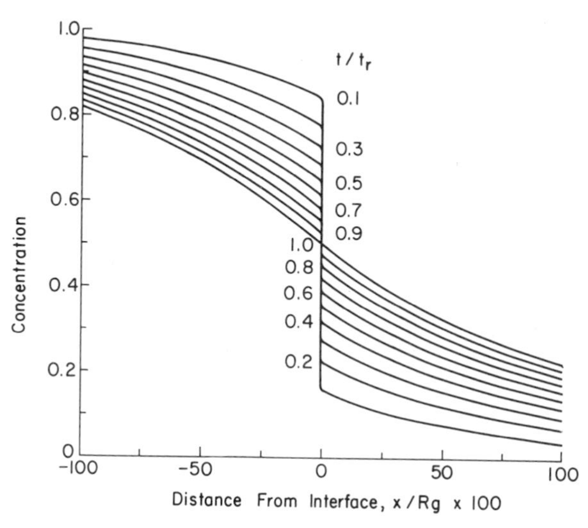

Fig. 2. The normalized monomer concentration depth profile at a polymer-polymer interface is shown at times $t \leq T_r$ and percent depths, X/R_g. (Zhang and Wool)

Table 1. Molecular Aspects of Interdiffusion at a Polymer-Polymer Interface

Molecular aspect	Symbol	Dynamic relation $H(t)$	Static relation H_∞	r	s
General property	$H(t)$	$t^{r/4}\ M^{-s/4}$	$M^{(3r-s)/4}$	r	s
No. of chains	$n(t)$	$t^{1/4}\ M^{-5/4}$	$M^{-1/2}$	1	5
No. of bridges	$p(t)$	$t^{1/2}\ M^{-3/2}$	M^{o}	2	6
Ave. monomer depth	$X(t)$	$t^{1/4}\ M^{-1/4}$	$M^{1/2}$	1	1
Total monomer depth	$X_o(t)$	$t^{1/2}\ M^{-3/2}$	M^{o}	2	6
Ave. contour length	$\ell(t)$	$t^{1/2}\ M^{-1/2}$	M	2	2
Total contour length, no. of monomers, N	$L_o(t)$	$t^{3/4}\ M^{-7/4}$	$M^{1/2}$	3	7
Ave. bridge length	$\ell_p(t)$	$t^{1/4}\ M^{-1/4}$	$M^{1/2}$	1	1
+Center of mass depth	X_{cm}	$t^{1/2}\ M^{-1}$	$M^{1/2}$	2	5
Diffusion front length	N_f	$t^{1/2}\ M^{-3/2}$	M^{o}	2	6

+ This equation applies to chains in the bulk and does not apply to chains whose center of mass is within a radius of gyration of the surface.

interface structure have a highly non-equilibrium compressed conformation at the surface at $t = 0$. The reduced entropy of the surface chains contributes to finite interdiffusion for incompatible asymmetric interfaces but permits little center-of-mass motion during the first relaxation time for symmetric interfaces.[2]

Another interesting characteristic of the reptation model is that the number of monomers crossing the interface as determined by the integral of the concentration profile behaves as[3,7]

$$N \propto t^{3/4} M^{-7/4}. \qquad (t < T_r) \tag{5}$$

For $t > T_r$, N should behave as $N \propto t^{1/2}$ and the crossover from $t^{3/4}$ to $t^{1/2}$ should be observable.

Experiments to explore the concentration profiles at $t < T_r$ and $X < R$ require a depth resolution of 100 Å, or better, and are currently being attempted by Whitlow and Wool using Sputtered Neutral Atom Mass Spectrometry (SNMS) and Secondary Ion Mass Spectrometry (SIMS).[9] These techniques are being used to evaluate the depth profile of a thin deuterated polystyrene (DPS) film diffusing into a thick protonated polystyrene (HPS) substrate. The SIMS technique was recently used to measure the concentration profile over a 2000 Å depth for a DPS/HPS interface; the self-diffusion coefficient was determined as $D_S = 5.6 \times 10^{-16}$ cm^2/sec for M(DPS) = 110,000 and M(HPS) = 100,000 (monodisperse) at 125°C.[6] The SIMS data were obtained with a depth resolution of 135 Å which for this molecular weight (random coil diameter of about 300 Å), was not sufficient resolution to evaluate the discontinuity in the concentration profile. This study is being continued with higher molecular weights of 500,000 and 1,000,000.

Experiments involving relaxation of centrally deuterated polystyrene chains in a higher molecular weight matrix recently provided support for the minor chain reptation model.[10] Using infrared dichroism studies of step-strained PS films, the protonated chain-ends were found to relax faster than the centrally deuterated fraction. This is consistent with the MC model shown in Fig. 1(a) where the chain-ends first lose memory of their initial orientation.

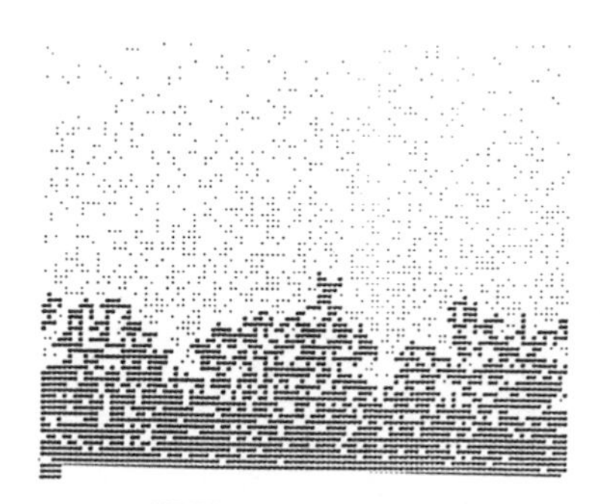

MONOMER INTERFACE

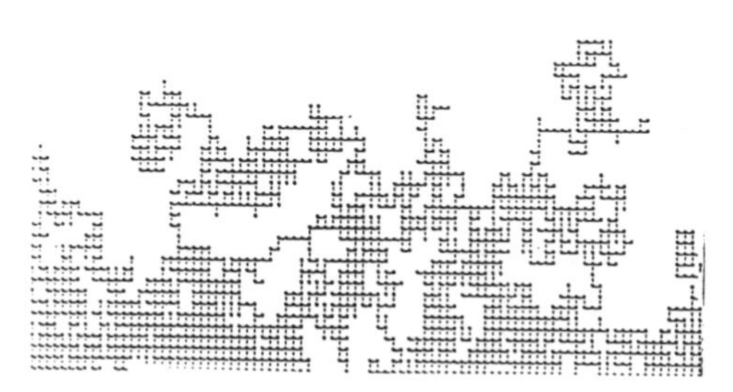

POLYMER INTERFACE

POLYMER-METAL INTERFACE

Fig. 3. The ramified fractal nature of diffuse interfaces is shown for (top) a computer simulated 2-d monomer-monomer interface where the heavy region represents the connected monomers on one side, (middle) a simulated 2-d random coil polymer interface at the reptation time, and (bottom) electrochemically deposited Silver diffusing in polyimide with the unconnected metal atoms removed to show the fractal diffusion front of the connected metal atoms. (Wool and Long)

III. FRACTAL INTERFACES

While the concentration profile due to diffusion varies smoothly as shown in Fig. 2, the diffusion field when viewed in two or three dimensions has a very "rough" nature. The highly ramified character of an interface produced by diffusion was analyzed using the techniques suggested by Sapoval, et al.[11] Figure 3 shows monomer, polymer, and polymer-metal interfaces simulated and analyzed by Wool and Long.[12] Monomer diffusion was simulated by placing particles along points on a lattice line at depth X with a probability $p(X)$, compared with the theoretical error function concentration profile for Einstein diffusion, $\Phi(X)$. The computer picks a random number, $0 < p(X) < 1$, and if $p(X) < \Phi(X)$, a particle is placed at that point in the two dimensional lattice. The boundary conditions were maintained such that $\Phi = 1$ at $X = 0$ (bottom of Figure 3).

The diffusion front for an A/B interface is defined in two dimensions for the A-monomers on the B-side by the percolation condition that each A-monomer on the front be connected via other A-monomers to the diffusion source at $X = 0$, and also be in contact with B-molecules connected to their diffusion source. The front can be considered as the ramified leading edge of the connected portion of the diffusion field. Figure 3 shows only the A-monomer diffusion for clarity. The diffusion front (or "seashore" of each interface (in two dimensions) was found to be fractal such that the mass of the front obeys the relation, $N_f \propto (\text{Radius})^{D_f}$, where the fractal dimension $D_f = 7/4$. The mass of the front, which is a relative measure of the roughness of the interface, increases with diffusion depth X, according to $N_f \propto X^{0.43}$.[11] Thus, the roughness of the interface increases with diffusion depth. The mean position of the diffusion front occurs at a normalized concentration in the depth profile equivalent to the scalar percolation threshold, P_c, for the system.

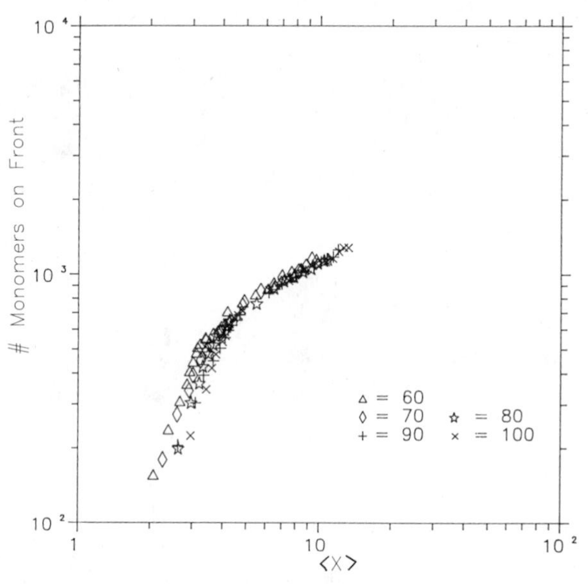

Fig. 4. The number of monomers on the diffusion front of a simulated 2-d diffusion of reptating random coil chains versus the average monomer diffusion depth $<X>$ is shown for molecular weights 60-100 times the entanglement molecular weight.

The fractal characteristics of a polymer diffusion front and the scaling laws, $H(t)$, for interdiffusion were analyzed by a computer simulation of reptating chains. A reptation algorithm was used in which monomers were randomly added and subtracted from the ends of a Gaussian chain of N-steps in two dimensions. The simulations were done on an IBM-AT microcomputer attached to a Cray supercomputer. For $t < T_r$, the concentration profile is discontinuous near the origin for reasons discussed above and illustrated in Fig. 2. Consequently, the condition of self-similarity is not obeyed and the diffusion front, though highly ramified, is not fractal during the first reptation time.

A measure of the ramified nature of the diffusion front as shown in Fig. 3 is given by the number of monomers per unit length on the front, N_f. When correlated motion effects are important at $t < T_r$, N_f can be approximated by

$$N_f \propto n(t)\sqrt{\ell(t)}\ , \tag{6}$$

such that the front is largely constituted by individual minor chain contributions, $\sqrt{\ell(t)}$, from $n(t)$ chains. Using Table 1, we can relate N_f to the monomer interdiffusion distance, X, via $\sqrt{\ell(t)} \sim X$ and $n(t) \sim X/M$, as

$$N_f \propto X^2/M\ . \qquad (t<T_r) \tag{7}$$

When $t > T_r$, the polymer diffusion front becomes fractal and the monomer diffusion result is obtained such that $N_f \propto X^{0.43}$. Figure 4 shows a plot of N_f vs. average monomer diffusion depth, X, for dense reptating chains of varying molecular weights ranging from 60 to 100 times the entanglement molecular weight. At $t < T_r$, a slope of 2 is obtained with a molecular weight dependence of M^{-1} in agreement with the above analysis. At $t >> T_r$, the slope becomes independent of molecular weight and decreases towards 0.43. Thus, at long times and diffusion distances $X >> R$, the fractal properties of a polymer-polymer interface are similar to a monomer/monomer interface.

The scaling laws for diffusion outlined in Table 1 were investigated by this computer simulation and found to be correct within the framework of the reptation model. They were also derived from the concentration profile given by Eq. (2) by Zhang and Wool.[7] The scaling laws are useful in developing relations for welding, adhesion, fatigue and fracture of polymer materials.[13]

The polymer-metal interface shown in Fig. 3 was derived from an electron micrograph obtained by Mazur and Reich.[14] They electrodeposited silver from a silver ion solution diffusing through a polyimide film. Particles not connected to the diffusion source were removed by computer analysis. The deposited silver particles essentially "decorate" the concentration profile and permit the diffusion front to be observed. A 1000-Å thin slice was used to aproximate two dimensional diffusion. The fractal dimension of this interface was determined by computer analysis to be approximately 1.7.[12] Similar ramified interface fronts are created by vapor deposition of metal atoms on polymers and by certain ion bombardment treatments of polymer surfaces. The fractal front is fairly insensitive to the details of the concentration profile. However, strong chemical potential gradients in asymmetric interfaces may promote a more planar, less ramified structure. The fractal characteristics of polymer interfaces are important in determining their mechanical, thermal and electrical properties.

ACKNOWLEDGEMENTS

The author is grateful to the National Science Foundation, Grant DMR 86-11551; the Army Research Office, Grant DAAL03-86-K-0034; and IBM for financial support of this work. Support from the Materials Research Laboratory was received for the SIMS and SNMS facility and Fractal Interface study, NSF Grant DMR 86-12860. The author particularly appreciates the helpful comments from Professor de Gennes on this work.

NOMENCLATURE

b	bond length
D_1	curvilinear diffusion coefficient
D_s	self-diffusion coefficient
D_f	fractal dimension
$\Phi(X,t)$	concentration profile
$H(t)$	interface dynamic property
H_∞	interface equilibrium property
$\ell(t)$	minor chain length
L	chain length
M	molecular weight
n	number of segments of length b in minor chain
$n(t)$	number of chains intersecting interface
N	number of monomers diffused
N_f	number of monomers on diffusion front
$p(X)$	probability of placing atom at diffusion depth, X
p_c	percolation threshold
r	1, 2, 3, ...,
R	end-to-end vector of Gaussian chain
s	1, 2, 3, ... ,
t	time
T_r	reptation time
X	diffusion distance

REFERENCES

1. P.-G. de Gennes, J. Chem. Phys., 55, 572-579 (1971).
2. Y.-H. Kim and R.P. Wool, Macromolecules, 16, 1115-1120 (1983).
3. R.P. Wool, Rubber Chem. Technol., 57, (2), 307-318 (1984).
4. R.P. Wool, J. Elast. and Plastics, 17, 106 (1985).
5. P.-G. de Gennes, C.R. Acad. Sci., Paris, 292, (2), 1505 (1981).
6. M. Tirrell, D. Adolf and S. Prager, IMA Vols in Math and Its Applications, Springer-Verlag, 5, (1986).
7. Huanzhi Zhang and R.P. Wool, Macromolecules, in press (1989).
8. S. Prager and M. Tirrell, J. Chem. Phys., 75, 5194 (1981).
9. S.C. Whitlow and R.P. Wool, Macromolecules, in press; Bull. Amer. Phys. Soc., 33, (3) 640, New Orleans (1988).
10. Andre Lee and R.P. Wool, Macromolecules, 20, 1924-1927 (1987).
11. B. Sapoval, M. Rosso and J.F. Gouyet, J. Phys., Lett., 46, 149-156 (1985).
12. R.P. Wool and J.M. Long, Paper presented at the Materials Research Society, Boston, December 1987, to be published.
13. R.P. Wool, "Welding, Tack and Green Strenth of Polymers", in Fundamentals of Adhesion, L.-H. Lee (Ed.), Plenum Press, to be published 1989.
14. S. Mazur and S. Reich, J. Phys. Chem. 90, 1365-1372 (1986).

SIZE EXCLUSION CHROMATOGRAPHY AND SURFACE EFFECTS IN POROUS FRACTALS

F. Brochard-Wyart

Structure et Réactivité aux Interfaces
Universit Pierre & Marie Curie
4 place Jussieu, 75231 Paris
Cedex 05, France

M. Le Maire

Centre de Génétique Moléculaire
C.N.R.S., 91190 Gig-sur-Yvette
France

A. Ghazi

Laboratoire des Biomembranes
Université Paris XI, 91450
Orsay Cedex, France

M. Martin

E.S.P.C.I.
Laboratoire d'Hydrodynamique Physique
10 rue Vauquelin, 75231 Paris
Cedex 05, France

ABSTRACT

Certain porous media have been claimed to be fractals in a broad range of scales $a < \ell < L$. The fractal dimension D_f influences various surface effects characterized by a penetration length ℓ_s. If ℓ_s can be varied (by suitable external agents) through the interval (a,L) some information on D_f can be extracted. We construct a tentative list of physical processes leading to the variable ℓ_s:

a) Depletion layers in a macromolecular solution, where the molecules are repelled by the wall, with specific implications for size exclusion chromatography.

b) Preferential adsorption of one species in a binary mixture: whenever the fractal dimension of the adsorbed molecule is larger than D_f, the excess adsorption may detect fractality.

c) Capillary condensation; a fractal grain exposed to a vapor becomes surrounded by a "cocoon" of liquid provided that $D_f > 2$.

d) Electrophoresis and electro-osmosis, where electrical boundary conditions play an essential rôle: the mobility of a dilute fractal aggregate is sensitive to D_f only if the charge per grain is maintained during aggregation.

I. POROUS FRACTALS

Porous media with high specific areas play an important rôle in many fields: hydrology, oil sciences, heterogeneous catalysis, chromatography, absorbers electrochemistry, electrophoresis, etc.,. Sometimes the high-area systems have a relatively narrow distribution of pore sizes (say, over less than a decade). In other cases, the distribution of pore sizes is extremely broad (say, over three decades). Certain porous media of the second group can be described as fractals in a certain range of lengths $a < \ell < L$.

I.1 What is a Porous Fractal?

I.1.1 Selfsimilarity The word fractal was introduced by B. Mandelbrot[1] to describe wrinkled objects whose rugosities show up in a large range of scales. Fractal structures are statistically selfsimilar. In a more concrete term, this means that it is not possible to discriminate between two photographs of the same medium, taken with two different frame sizes (ℓ_1, ℓ_2) (with ℓ_1, ℓ_2 in the fractal interval a,L). We show in Fig. 1 a rigourous and a statistically selfsimilar structure. For our purposes we shall classify the porous fractals into two groups:

I.1.1a. Fractal Aggregates (Fig. 1): There are many examples of fractal aggregates in nature formed by aggregation of colloidal particles or aerosols (dust, smoke...)[2,3,4] or by chemical bonding of multifunctional groups (branched polymers).[5] A single polymer chain in solution,[6] the micro-droplets of one A species in an A/B binary mixture are also fractal objects.[7] The natural cut-offs are the typical size a of the individual particle and the overall size L of the cluster.

I.1.1b. Surface Fractals (Fig. 2): Grains or pores in a porous media may have a smooth surface or a corrugated surface. As the irregularity of the surface increases, the fractal dimension of the surface increases from 2 to 3. The value $D_f = 2$ corresponds to a smooth surface. On the other hand, values close to $D_f = 3$ correspond to a highly disordered volume (silica gels). In Fig. 2 several models for fractal pores are presented.

Various experiments suggest the existence of fractal structures in certain porous media, e.g., sandstones,[8] industrial aluminas or silicas,[9,10] coals.[11] Some of these experiments will be reviewed later in this section after a brief description of the fractal dimension D_f.

I.1.2 Definition of D_f

I.1.2a. Aggregates: If one counts the number of particles n within a sphere of radius r around a particle near the center of the aggregate, one finds that n varies as a power law

$$n = k r^{D_f} , \qquad (1)$$

where D_f is the fractal dimension of the aggregate. For a latex or a glass sphere, the density is constant and $D_f = 3$. For a flexible polymer in dilute solution,[6] $D_f = 5/3$ in a good solvent and $D_f = 2$ a θ-solvent. For a cluster of A molecules in an AB critical mixture, $D_f \simeq 2.5$.[7] The fractal dimension of certain branched polymers in dilute solution is $D_f = 2$.[5] D_f may be measured from small-angle neutron scattering or light scattering: the scattered intensity at wave vector $\mathbf{q}$ is proportional to

$n_{\mathbf{q}}$, the number of coherent scatters in a volume of size $\mathbf{q}^{-1}$. From Eq. (1), one finds that the scattering structure factor decreases as

$$S(\mathbf{q}) \sim \mathbf{q}^{-D_f} . \tag{2}$$

I.1.2b <u>Surface Fractals</u> Here we define D_f as the fractal dimension <u>of the interface</u> ($2 < D_f < 3$). D_f can be defined in the following way: we generate spheres of radius ε with their centers on the interface of one grain (size L) (Minkowski sausage). The volume spanned by the spheres per grain is

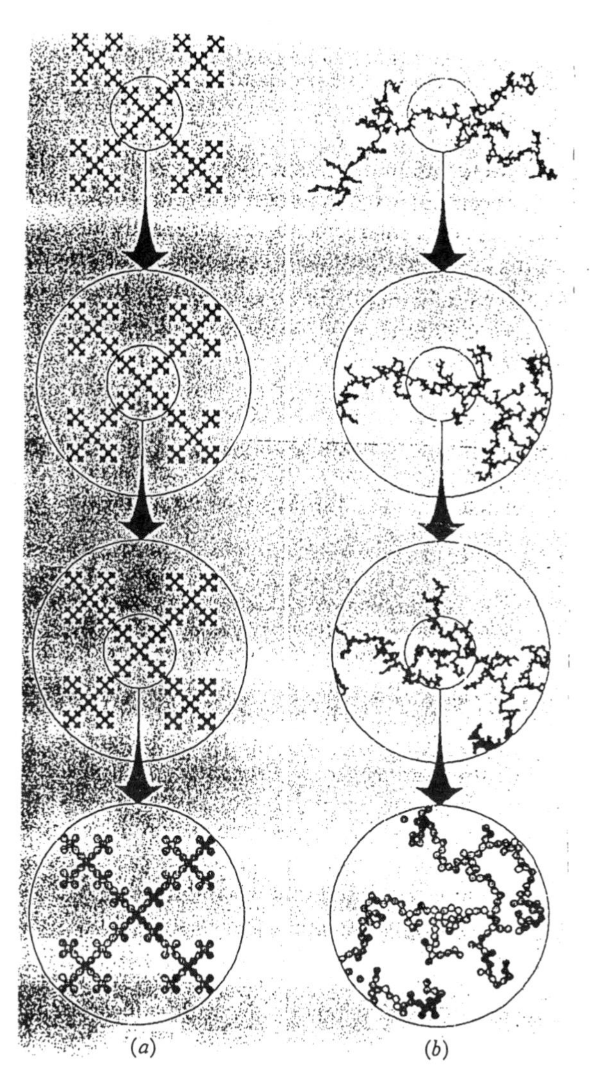

Fig. 1. Illustration of the selfsimilarity property in the case of (a) a deterministic and (b) a disordered cluster-cluster[3] <u>fractal aggregate</u>. Both aggregates have almost the same fractal dimension ($D_f \simeq 1.5$, $d = 2$). The top frame shows the original aggregate. The frames below show successive enlargements of the central region (from R. Jullien, <u>et al.</u>[4] (1987)).

$$\Omega(\varepsilon) = L^{D_f} \varepsilon^{d - D_f} , \tag{3}$$

where $d = 3$ is the space dimension. This definition can be immediately verified on points, lines and planes ($D_f = 0,1,2$).

For many qualitative purposes, we may present this differently: instead of using a continuum of spheres, we can restrict our attention to a discrete set of spheres in contact with each other: $(L/\varepsilon)^{D_f}$ is then the number of spheres. A related qualitative picture, used in Ref. (12), is the following: the interface is covered with "tiles" of size ε. The surface $\Sigma(\varepsilon)$ covered by the tiles depends on their size: large tiles cannot follow the details of the irregular surface. Then

$$\Sigma(\varepsilon) = L^{D_f} \varepsilon^{(d-1) - D_f} = \frac{\Omega(\varepsilon)}{\varepsilon} . \tag{4}$$

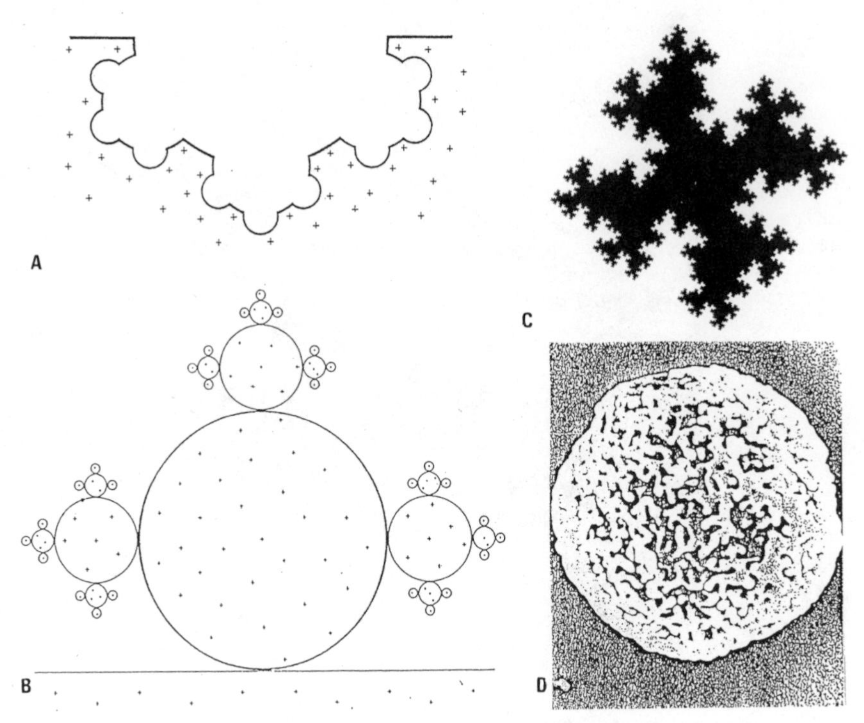

Fig. 2. Surface fractals: Illustration of three models for fractal pores and a real porous grain used in size exclusion chromatorgaphy.
A) Iterative pits (from P.G. de Gennes[27] (1985)).
B) Iterative flocs (from P.G. de Gennes[27] (1985)).
C) Koch island (from B. Mendelbrot[1] (1981)).
D) Scanning electron micrograph of a Li Chrospher Si 4000 particle (magnification X 8000) (from K.K. Unger, et al.[16] (1980)).

I.2 Evidence for Surface Fractal Porous Structure

I.2.1 Small Scales (1 Å < r < 100 Å) Adsorption data for small molecules on silica absorbers have been interpreted by Pfeifer and Avnir[9] in terms of the size r of the absorbed molecules. The number n of molecules absorbed per unit volume decreases as:

$$n \sim r^{-D}. \quad (D \sim 2.7) \tag{5}$$

They interpret D as the fractal dimension of the interface. Indeed according to Eq. (3), if the molecules pack densely on the solid, the number of molecules n_g adsorbed per grain is

$$n_g = \frac{\Omega(r)}{r^3} = \left(\frac{L}{r}\right)^{D_f} \sim r^{-D_f}. \tag{6}$$

Some critiques can be raised, however:

(i) r was varied only by a factor of ten; (ii) the molecules were not chemically identical.

To cover a larger length range, and to use chemical analogues, the same group later reinterpreted some data on polystyrene adsorbed from a solvent on porous alumina.[10] They find the same law Eq. (5) by taking for r the radius of gyration R_G of the free polymer chains. The equilibrium adsorption of flexible polymers on a fractal surface has been discussed theoretically in Ref. (19). The result is that n_g is independent of R_G as soon as $D_p < D_f$. The interpretation of the adsorption data for PS is thus not obvious. They may well be dominated by kinetic effects: chains of size R_G cannot crawl into pores smaller than R_G, although they would lower their free energy by increasing their number of contacts with the pore.

I.2.2. Intermediate scales (10 Å < r < 1000 Å) Small angle X-ray or neutron scattering data[11] from a lignite coal were interpreted by Bale and Schmidt in terms of the fractal dimension of the pore surface. The structure factor of the coal scatterers is $S_{\mathbf{q}} \sim \mathbf{q}^{D_f-6}$. Witten[14] gives a simple derivation of this result: one fills the space with spheres of size $\mathbf{q}^{-1}$. The spheres are either empty, partially empty or completely full. The scattering, due to density fluctuations, is entirely due to spheres which are partially full. The number of such spheres per grain is, according to Eq. (3),

$$n_g(\mathbf{q}^{-1}) = \Omega(\mathbf{q}^{-1})\mathbf{q}^3 = (L\mathbf{q})^{D_f}. \tag{7}$$

The number of scatterers per sphere is of the order of $(\mathbf{q}\,a)^{-3}$. The scattering structure factor is then

$$S(\mathbf{q}) \simeq n_g(\mathbf{q}^{-1})(\mathbf{q}\,a)^{-6} \simeq L^{D_f}\mathbf{q}^{D_f-6}. \tag{8}$$

For smooth surfaces ($D_f = 2$), Eq. (8) reduces to the classical Porod Law q^{-4}. The data give $D_f \simeq 2.56 \pm 0.03$ for lignite.

I.2.3. Large scales (10 Å < r < 100 μm) A number of sandstones have been studied by Katz and Thomson.[8] They analysed sections of the rocks by scanning electron (SEM) or optical microscopy. They claim that the rock is fractal over five decades of sizes (extending from 10 Å to 100 μm, the size of the sandgrains). D_f varies slightly from sample to sample ($2.57 < D_f < 2.87$).

I.3. Physical processes which are sensitive to D_f

Whenever a physical process (near the surface) depends on an adjustable characteristic length ℓ_s, this process may allow to probe the fractal interface with a variable yardstick ℓ_s.[12,13] However, in many instances, the measurements will be sensitive to various features of the surface, not only to D_f. We shall discuss here certain statistical and dynamical measurements, which should be mainly sensitive to D_f.

I.3.1. Adsorption of large objects We study the coating of a fractal with a) latex spheres, b) rigid or flexible macromolecules, c) a layer of preferential adsorption of one species in a binary mixture (near criticality). In all these cases, on a "surface" fractal (D_f) another fractal object of fractal dimension D_p and size R was adsorbed. For macromolecules, D_p ranges from 3 for globular proteins to unity for rods.

We shall see that adsorption can detect fractality only if $D_p > D_f$. A compact object ($D_p > D_f$) cannot penetrate in the smaller pores and one can picture the surface coverage as a tiling of the surface with objects of size R. The adjustable length ℓ_s is the size R. On the other hand, if $D_p < D_f$, the fractal object adjust easily to the surface heterogeneity and ℓ_s is not equal to R, but to the size "a" of a subunit.

I.3.2. Depletion and size exclusion chromatography Here we take the walls of the porous fractal to be repulsive: the macromolecules are repelled from the solid surface. A "depletion layer" of size $\ell_s = R$ surrounds the fractal: depletion can detect fractality.

Size exclusion chromatography measures directly this excluded-volume. Silica beads used in SEC (Fig. 2-c) have been claimed[16] to be "fractal." We analyse chromatographic data using this picture and shall see in section III that in many cases, it is possible to characterize a chromatographic column by a fractal dimension D_f.

I.3.3. Capillary condensation on a fractal object Let us assume that a colloidal aggregate is exposed to a slightly unsaturated vapor at pressure p. The grain becomes surrounded by a "cocoon" of liquid of volume Ω $(,p)$ dependent on the pressure. We construct the scaling form of Ω (p) in section III.

I.3.4 Electrophoresis An electrically charged aggregate is suspended in water (with a controlled ionic strength). An electric field E is applied, and the aggregate moves with a velocity $V = \mu E$. How does μ depend on the salt concentration c? We shall discuss this in section V.

II. DETERMINATION OF D_f BY ADSORPTION OF LARGE OBJECTS

The object is characterized by a fractal dimension D_p and we shall show that D_p has to be larger than D_f to test fractality.

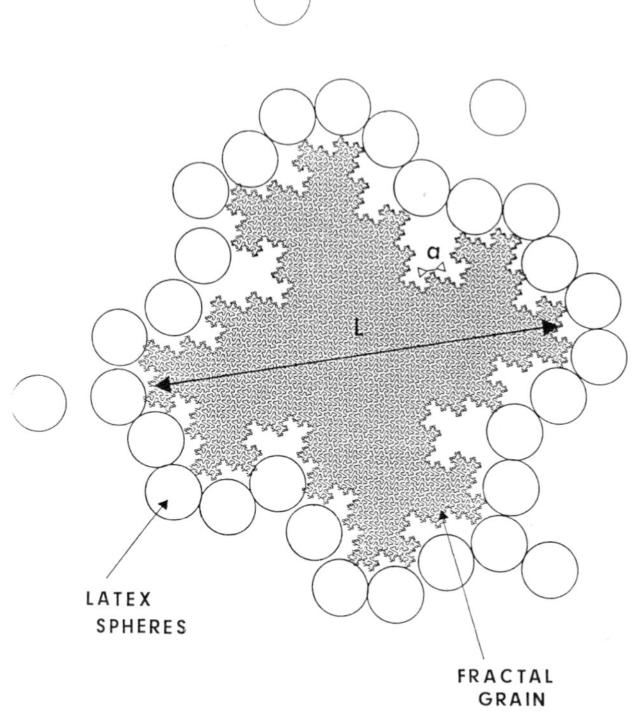

Fig. 3. Latex or glass spheres adsorbed on a fractal porous grain: the spheres cannot penetrate the small pores. The shape of the grain is copied from a "Koch island" in the book by B. Mandelbrot.[1]

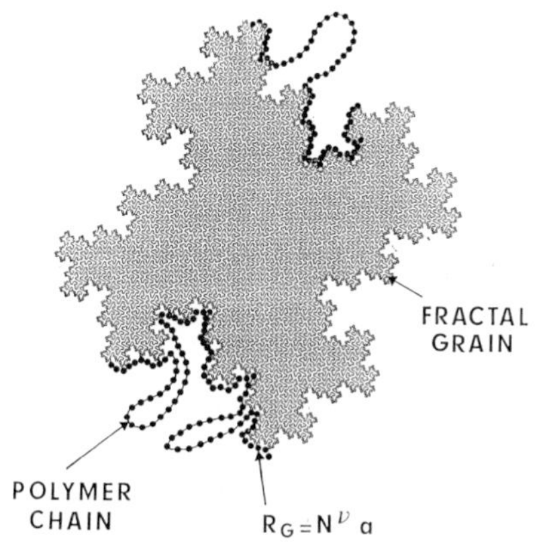

Fig. 4. Polymer chain adsorbed on a fractal grain: the *P*-chain can follow the irregularities of the surface.

II.1. Coating with latex spheres ($D_p = 3$) (Fig. 3) It is possible to prepare latex or glass particles of controlled size between 10 Å[17] and few microns.[18] The suspension is stabilized by two possible means: a) electric charges (in water); b) steric repulsion by covering the sphere with a polymer. The porous solid is incubated with the suspension. For case a), the sign of the sphere charges must be opposite to the sign of the surface charges, to provide attraction. We also want the superficial charge density of the spheres to be smaller (or at least comparable) in absolute value than that of the solid surface. In this way, the repulsion between adsorbed spheres is weak when compared to the attraction toward the solid and we expect a good coverage of the interface. For case b), we must choose a polymer which is strongly attracted by the interface.

The number of spheres of size b attached per grain is then

$$n_g = \left(\frac{L}{R}\right)^{D_f} .$$

The quantity adsorbed per unit volume will be :

$$C_B = \left(\frac{L}{R}\right)^{D_f} \cdot \left(\frac{\rho_B R^3}{L^3} \right) = \left(\frac{L}{R}\right)^{D_f - 3} \rho_B . \tag{9}$$

One must wash the porous solid with pure solvent to eliminate all free particles. Then one can either measure in terms of density or UV absorbance. It may also be possible to use various diffraction methods (neutrons, X-rays or light scattering).

A difficulty of the experiment is that it cannot probe pockets of pore space with small openings or fractals with nesting difficulties.[12]

II.2. Adsorbed polymer chains ($D_p = 5/3$) (Fig. 4) The porous fractal is exposed to a dilute polymer solution. This case is studied in detail in ref. (19). For polymer adsorbed by a planar surface, de Gennes pointed out that the profile is selfsimilar and given by the simple law

$$\xi(C(r)) = r , \tag{10}$$

where $\xi(C) \propto C^{-3/4}$ is the semidilute correlation length for a polymer solution at concentration C. Eq. (10) can be written as

$$a^3 C(r) = \left(\frac{a}{r}\right)^{3-D_p} ; \qquad r < R_G$$

$$C(r) \simeq 0 ; \qquad r < R_G \tag{11}$$

where D_p is the fractal dimension of the coils (D_p = 5/3 in good solvent) and R_G the chains radius of gyration. The concentration profile has been well-confirmed experimentally.[20]

The total number of adsorbed monomers per grain in full equilibrium is

$$n_{mono} = \int_a^{R_G} C(r)\Sigma(r)dr \,. \tag{12}$$

Inserting Eq. (11) into Eq. (12) this gives

$$n_{mono} = \left(\frac{L}{a}\right)^{D_f} \int_0^{R_G} \frac{dr}{r}\left(\frac{a}{r}\right)^{D_f - D_p} . \tag{13}$$

The integral is dominated by the small length limit and n_{mono} = $(L/a)^{D_f}$. The polymer forms a monolayer of size a, independent upon the degree of polymerization N. The probe size is just a monomer size (ℓ_s = a)! This is due to the fact that polymer chains are not compact and adjust easily to surface heterogeneity.

However, experimental data do show a dependence upon N. It is possible that equilibrium times are exceedingly long. The chains are first adsorbed as spheres of size R_G and it takes a long time for them to creep and fill the small structures. This will lead to

$$n = \left(\frac{L}{R_G}\right)^{D_f} N = \left(\frac{L}{a}\right)^{D_f} R_G^{\,5/3 - D_f} \tag{14}$$

which is in better agreement with experimental data.[9]

II.3. <u>Critical binary mixture (D_p= 2,5)</u> In contact with a solid wall, a binary mixture has a concentration C_s at the wall different from the bulk concentration C. If the mixture is near critical, this modification is not restricted to the interface, but extends over a length

$$\xi(T) = \xi_0 t^{-\nu} \,, \quad t = \frac{|\Delta T|}{T_c} \,. \tag{15}$$

ξ_0 ~ 1 Å for small molecules and ξ_0 ~ 100 Å for polymer/solvent mixtures.

At scales $r < \xi$, it is useful to introduce a correlation length which depends only upon the increase ΔC of a concentration C above the critical concentration C_c.[7]

$$\xi(C) = \xi_0 (C - C_c)^{-\nu/\beta} \,. \tag{16}$$

The increase of one species A in a volume $\xi(C)$ is then

$$n_A = (C - C_c)\xi^3 \sim \xi^{D_p} , \tag{17}$$

where $D_p = 3 - \beta/\nu \simeq 2.5$ is the fractal dimension of the fluctuation concentration or "cluster."

The profile as for the polymer solution is also given by Eq. (10)

$$\xi(C) = r ,$$

$$\text{i.e., } a^3(C - C_c) = \left(\frac{a}{r}\right)^{3 - D_p} . \tag{18}$$

From equation (18), one can derive the increase of A molecules near the interface

$$n_A = \left(\frac{L}{a}\right)^{D_f} \int_a^{\xi(T)} \frac{dr}{r} \left(\frac{a}{r}\right)^{D_f - D_p} . \tag{19}$$

<u>$D_f > D_p$</u>: the integral is dominated by the limit $r = a$:

$$n_A = \left(\frac{L}{a}\right)^{D_p} . \tag{20}$$

$\ell_s = a$ and we expect no critical adsorption.

<u>$D_f > D_p$</u>: n_A is dominated by the limit $r = \xi$:

$$n_A = \left(\frac{L}{a}\right)^{D_f} \left(\frac{\xi(T)}{a}\right)^{D_p - D_f} , \tag{21}$$

which can also be written as

$$n_A = \left(\frac{L}{\xi}\right)^{D_f} \Delta C(\xi)\, \xi^3 . \tag{22}$$

II.3.1. <u>Conclusion</u>

If $D_p > D_f$, the clusters cannot penetrate into the smaller pores and we may understand the surface coverage as a tiling of the surface with objects of size ξ and inner concentration $\Delta C(\xi) \sim \xi^{-\beta/\nu}$. This case is similar to the adsorption of "spheres" (corresponding to $D_p = 3 > D_f$).

If $D_p < D_f$, the clusters are more irregular and can penetrate into the smaller pores: the dominant clusters are then comparable in size to the lower cut-off (a). This case is similar to the adsorption of polymer ($D_p < D_f$).

Thus, experimental measurements might discriminate between fractals of dimensionality smaller or larger than $D_p \simeq 2.5$.

III. DETERMINATION OF D_f BY DEPLETION OF MACROMOLECULES - APPLICATION TO G.P.C. (SIZE EXCLUSION CHROMATOGRAPHY)

III.1. Polymer Depletion at the Solid/Solvent Interface (Fig. 5) With suitable choice of solvent and of surface treatment, the polymer is repelled by the interface and a depletion layer is expected to build up.[21] The thickness of the depletion layer is R_F - the chain radius of gyration - for dilute solution and $\xi(C)$ - the pair-correlation length - for semidilute solution. Thus, repulsive surfaces allow to probe the interface with a variable yardstick

$$\ell_s (\ell_s = R_{G'} C < N/R_G^3 = C^*, \ell_s = \xi(C) = a(Ca^3)^{-3/4} C > C^*).$$

If the porous solid is saturated with a polymer solution, the volume fraction occupied by the chains will be

$$\frac{V}{PL^3} = 1 - \frac{\Omega(\ell_s)}{L^3} , \tag{23}$$

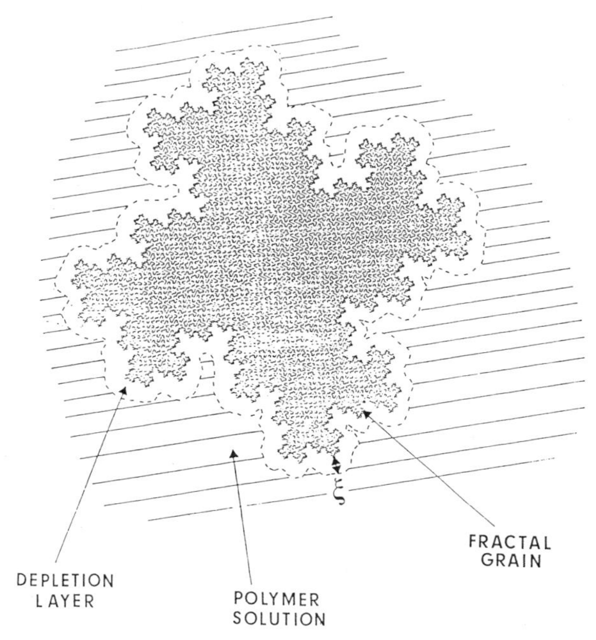

Fig. 5. Depletion layer of a polymer solution around fractal porous grain.

where P is the porosity.

$$\ell_s = R_G; \quad C < C^*$$

$$\ell_s = \xi . \quad C > C^*$$

The concentration C_i per unit volume of the porous solid is then

$$C_i = C_e P \left[1 - \frac{\Omega(\xi)}{L^3} \right] . \tag{24}$$

This repulsive case is theoretically clean, but the difference

$$\delta = \frac{C_i}{C_e P} - 1$$

is small and hard to detect $\delta \sim (\xi/L)^{3-D_f} \sim 1/50$. As pointed out to us by H. Benoit,[22] G.P.C. leads directly to V in Eq. (23).

III.2. Size Exclusion Chromatography (or G.P.C.) Gel Permeation Chromatography is a method widely used for the separation of molecules by size. The separation is carried out on columns packed with a porous material. Large molecules, which are excluded from the pores, are eluted first. If the wall is repulsive, a species is eluted at a volume exactly equal to the volume available to it in the column. For very large molecules, the elution volume V_e is equal to the interstitial volume V_0. For very small molecules, $V_e = V_0 + V_i$, where V_i is the internal pore volume. For intermediate size molecules, $V_e = V_0 + K_d V_i$, where K_d is the partition coefficient. K_d is equal to the ratio of accessible pore volume to V_i.

We estimate K_d for a column packed with a porous fractal material. In our notation, $V_i = PL^3$, and from Eq. (23)

$$K_d = 1 - \frac{\Omega(R)}{L^3} = 1 - \left(\frac{R}{L}\right)^{3-D_f} . \tag{25}$$

where R is the radius of the molecule.

In biochemistry, in particular, G.P.C. is used either to purify macromolecules, especially proteins, or to determine their size using properly calibrated columns. Two types of column materials are used: classical gels and High Performance Liquid Chromatography (HPLC) gels. Classical gels consist in polymers (for example, agarose in the case of Sepharose and allyl dextran crosslinked with N, N'-methylene bisacrylamide in the case of Sephacryl). HPLC gels, which allow a much better resolution, are made of small well-defined porous particles (for example, silica-based support for TSK SW, polyvinyl-based support for TSK PW).

Classical gels (Sepharose and Sephacryl) as well as HPLC gels (TSK SW and TSK PW) have been calibrated: the elution position, and hence the K_d, have been determined for a number of globular proteins of known sizes (refs. 23 and 24). The size of each protein (R) is determined independently by a combination of sedimentation equilibrium and sedimentation velocity experiments so that this parameter is in fact the Stokes radius of the protein (R_S). The range which was covered was 20 to 90 Å.

We have used these calibrations to test the above formula. The plot of $\ell n(R_S)$ against $\ell n(1 - K_d)$ should give a straight line of slope $1/(3 - D_f)$ and intercept at the origin $\ell n(L)$. This prediction is poorly fulfilled for the classical gels that we have tested (Sepharose and Sephacryl) as shown in Fig. 6. On the contrary, linearity is observed in a wide range of protein size for the HPLC gel TSK 3000 SW (Fig. 7). In the case of the HPLC gel TSK 4000 PW, the fit is even better and cover the whole range of calibration (Fig. 8). The values of D_f that are obtained for the TSK 3000 SW and the TSK 4000 PW gels are 2.22 and 2.45 respectively, whereas the values of L are 87 Å and 143 Å. For the HPLC gel TSK 3000 PW, D_f is 2.58 and L is 70 Å (data not shown).

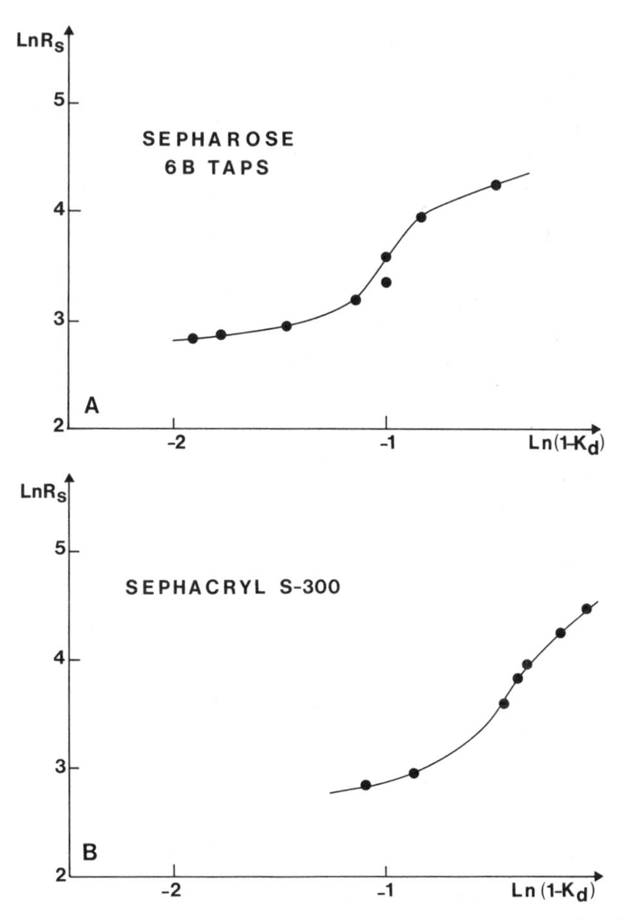

Fig. 6. Calibration curve for classical gels: Sepharose 6B (Fig. 6-a) and Sephacryl S-300 (Fig. 6-b). The data are taken from ref. (23).

Thus, it seems that HPLC gels can be described as fractals in the range of protein sizes. For classical gels, whose structure is different from HPLC gels and more heterogeneous, the model does not seem to apply.

This finding has some practical applications. First of all, the existence of a linear relationship between $\ell n(R_s)$ and $\ell n(1 - K_d)$ is of interest since it simplifies the calibration procedure in that only two standards are needed in principle. It is noteworthy that such linear relationships have been searched in vain in the past (refs. 23-25). Secondly, the characterization of a gel by two simple parameters (D_f and L), might be of some help in the design of new gels. A low D_f will result in a better resolution whereas a large D_f will permit a separation of proteins in a wide range of size.

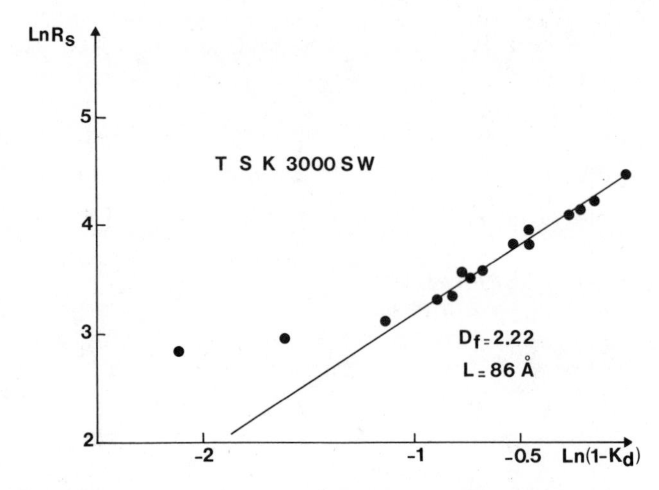

Fig. 7. Calibration curve for the HPLC gel TSK 3000 SW. The data are taken from ref. (24).

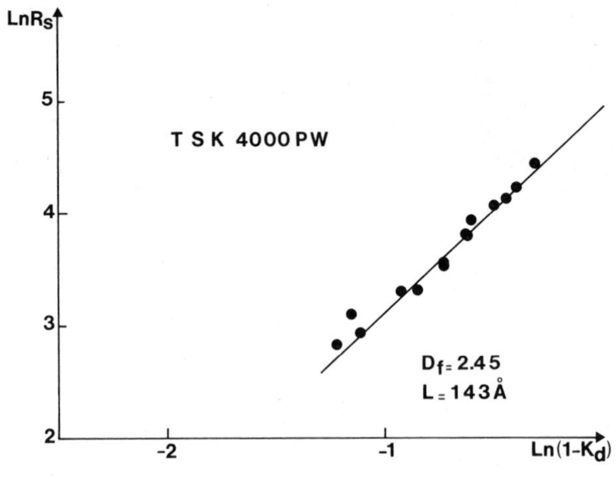

Fig. 8. Calibration curve for the HPLC gel TSK 4000 PW. The data are taken from ref. (24).

IV. CAPILLARY CONDENSATION ON A FRACTAL GRAIN

We have discussed up to now various effects on detecting fractality at scales less than one micron. Lenormand[26,29] has shown that one can characterize "giant" fractal structures ($\mu m < \ell_s <$ mm) by capillary condensation: the fractal object is exposed to two immiscible liquids A and B and prefers A. When the grain is immersed in B at a depth h below A, it becomes surrounded by a cocoon of A (Fig. 9). One can also expose the fractal grain to an unsaturated vapor. For instance, the vapor may be taken at height h above a liquid reservoir (A = liq, B = vapor).

The Laplace equation $2\gamma/e = \rho g h$ ($\rho = \rho_B - \rho_A$, $\gamma = \gamma_{AB}$ is the liq. A/liq. B interfacial tension) determines the local curvature $1/e$ of the cocoon surface. Then the "pits" of size smaller than e are invaded by the wetting liquid. One can vary e on several decades, by varying either h or ρ.

De Gennes[27] has studied the capillary condensation for the two strictly selfsimilar porous fractals "iterative pits" and "iterative flocs" shown in Fig. 2 A,B.

We have generalized[28] this analysis to statistically selfsimilar structures. We will assume that the spreading coefficient $S = \gamma_{SB} - (\gamma_{SA} + \gamma)$ is negative. For this case, the liquid A wets only partially a smooth surface made of the grain material. Then the cocoon cannot be formed on a smooth surface ($D_f = 2$). On the other hand, in the case of complete wetting ($S > 0$), an ultra thin film controlled by long-range van der Waals forces will be formed.[29]

Fairbridge[30] has interpreted mercury porosimetry data using a similar description and found a fractal dimension for coke particles in good agreement with the result obtained from adsorption data.

IV.1. Free Energy of the "Cocoon" Consider a fractal object immersed in a binary mixture at a depth h below the A/B interface, a sheath of liquid A and thickness e is maintained. The thickness e is again resulted from a balance between the gain of interfacial energy of the coated grain and the loss of gravitational energy. The free energy difference between the coated grain and the dry grain is given by

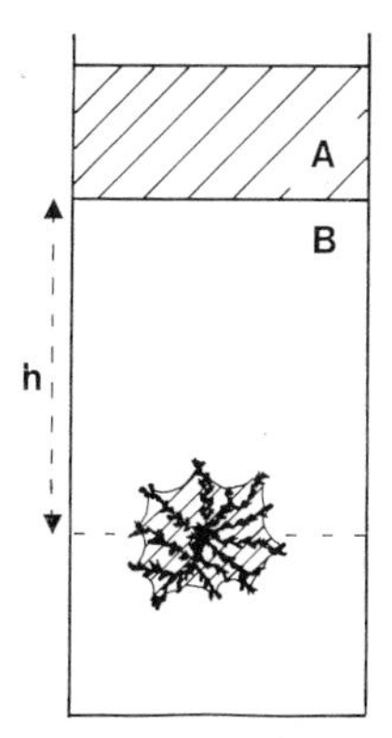

Fig. 9. Capillary condensation on a fractal structure of fractal dimension D_f. If $D_f > 2$, a "cocoon" of liquid A is expected to appear at equilibrium provided that $h < h_c$.

$$\Delta F = (\gamma_{SA} - \gamma_{SB})\Sigma(a) + \Sigma(e) + \Omega(e)\rho g\, h\ , \tag{26}$$

$\Sigma(e)$ is the surface of the fractal measured at scale e and $\Omega(e)$ the volume of the cocoon given by Eq. (5-4) respectively.

The first term corresponds to the decrease of the grain surface energy coated by A; the second term the A/B interfacial free energy, and the last term the gravitational contribution.

The thickness of the cocoon is obtained by the minimization of ΔF

$$\frac{\partial}{\partial e}\Delta F \sim \alpha_1(2 - D_f)e^{1-D_f} + \alpha_2 e^{2-D_f} h\,\kappa^2 = 0\ , \tag{27}$$

where α_1, α_2 are positive numerical constants and $\kappa^2 = \rho g/\gamma(\kappa^{-1}$ is the capillary length). This gives

$$e = cte\,(D_f - 2)\kappa^{-1} h\ . \tag{28}$$

For $\kappa^{-1} = 1$ mm ($\rho = 1$ g/cm^3), $e \simeq 1\ \mu$ for $h = 1$ m.

The volume of A liquid trapped in the fractal grain is

$$\Omega(e) \sim L^{D_f} e^{3-D_f}\ , \tag{29}$$

which is in agreement with the result found by de Gennes.[27]

IV.2. Dewetting Transition The cocoon is stable only if $\Delta F < 0$, $F = 0$ defines a threshold depth h_c

$$h_c a \kappa^2 = \left(\frac{\gamma_{SB} - \gamma_{SA}}{\gamma}\right)^{1/(D_f - 2)}\ . \tag{30}$$

If $\cos\theta_e \sim 1$, h_c is large ($h_c \sim 10$ kms), but if $(\gamma_{SB} - \gamma_{SA})/\gamma \simeq 10^{-2}$ ($\theta_e \sim 90°$), $h_c \sim 1$ m for $D_f = 2.5$. For $h = h_c$, the cocoon becomes unstable and one expects a first-order dewetting transition. For $h > h_c$, the fractal grain is "dry", i.e., in direct contact with liquid B.

IV.3 Conclusions One could test the above prediction by using large aggregates made of colloidal particles of size $a \sim 1000$ Å to 10 µm or surface porous fractals. However, the situation is complicated by two effects:

IV.3.1 Hysteresis Effects The energy barrier ($\sim \gamma e^2$) to fill or to empty a pore is much larger than kT. Capillary equilibrium may be facilitated by strong sonification. Gradual imbibition from a vapor will be a better way to lead to equilibrium.

IV.3.2 Collapse of Fragile Structures If the aggregate is flexible, it will shrink to decrease both gravitational and A/B surface energies. The fractal dimension of the aggregate may increase by capillary condensation. We have studied[33] the condensation of a vapor on a flexible polymer chain (assuming that the liquid is a good solvent of the polymer). At small pressure, the chain is collapsed (D_f = 3). At higher pressure, the chain is only partially collapsed (D_f = 2,5).

V. CHARGED FRACTAL GRAINS: ELECTROPHORESIS

A charged fractal grain is surrounded by the diffuse electrical double layer. The thickness of the double layer is controlled by the Debye length

$$\kappa_e^2 = \sum_i 8\pi C_i Z_i^2 / \varepsilon k T ,$$

where C_i is the concentration of ions of charge Z_i. Screening effects can be varied easily from thousand to few angströms by varying the ionic strength.

We discuss several experiments which are sensitive to the fractal character of the double layer:

V.1. The negative adsorption of co-ions[30,31] determines the excluded-volume per grain

$$\Omega(\kappa_e^{-1}) = \kappa_e^{-1} \sum(\kappa_e^{-1}) = (L\kappa_e)^{D_f} \kappa_e^{-3} . \tag{31}$$

V.2 Capacitance Measurements If the grains are conductors and stacked to form an electrode, one can determine the capacitance per grain. The surface potential ζ of the grain is proportional to the charge $Q\kappa_e^{-1}$) in a "blob" of size κ_e^{-1}, because κ_e^{-1} measures the screening of the Coulomb interactions

$$\zeta = \frac{Q(\kappa_e^{-1})}{\varepsilon \kappa_e^{-1}} , \tag{32}$$

where ε is the dielectric constant.

The number of "blobs" of size κ_e^{-1} per grain is $(\kappa_e L)^{D_f}$ and the total charge Q of one grain is

$$Q = (\kappa_e L)^{D_f} \varepsilon \kappa_e^{-1} \zeta = \varepsilon \zeta \kappa_e \Sigma(\kappa_e^{-1}) . \tag{33}$$

The capacitance

$$C = \varepsilon \zeta \kappa_e \Sigma(\kappa_e^{-1}) . \tag{34}$$

leads to a direct derivation of the surface of the grain measured at a scale κ_e^{-1}.

V.3. Electrophoresis We have discussed[32] the mobility of dilute charged fractal objects, e.g., colloidal aggregates, or polyelectrolytes and have shown that the electrical boundary conditions play an essential rôle.

V.3.1 Mobility of One Particle (size a) A charged particle of size a in suspension in water has a velocity V in response to an external applied electrical field E. The mobility $\mu = V/E$ is given by

$$\mu = \frac{\varepsilon \zeta}{\eta} cte \ , \tag{35}$$

where η is the water viscosity. Equation (35) is valid for $\kappa_e a > 1$ or < 1 (the numerical factor is modified) and whatever the shape of the particle (a sphere or a rod).

V.3.2. Mobility of an Aggregate of Particles If $\kappa_e a > 1$, aggregation does not modify the flows and Eq. (35) remains valid. We want to discuss what happens in the range $a < \kappa_e^{-1} < R$. If one builds an aggregate with particle of size a, one must know if one works at constant surface potential ζ on the grains, or if the charge Q_1 per grain is maintained during aggregation.

1) If the surface potential ζ remains constant, one can always use Eq. (35). One can illustrate this result with a rather non-physical example of fractal structure, formed of an object elongated in the direction z of the field but with a cross section (in the XY plane) represented by a fractal curve. The action of the electrical field E on the counter-ions give rise to a velocity field V_z (x, y) which satisfies the following condition

$$\eta \nabla^2 V_z = - \rho E = \frac{\varepsilon}{4\pi} E \nabla^2 \psi \ , \tag{36}$$

where ρ is the local charge of the counter-ions and ψ the electrostatic potential. The boundary conditions at the surface of the object are $V_z = V$ and $\psi = \zeta$ and $V_z = 0$ and $\psi = 0$ at infinity. Thus Ψ and V_z are proportional. Equation (35) then leads to the Smoluchovski relation

$$\mu = \frac{V}{E} = \frac{\varepsilon \zeta}{4\pi\eta} \ . \tag{37}$$

If ζ is independent on κ_e^{-1}, μ is constant and not sensitive to the fractal dimension.

Remark: In the case of electro-osmosis, the flow boundary conditions are $V = 0$ at the surface and $V(\infty) = V$. Equation (36) with $\psi = \zeta$ at the surface shows that the current "J" of solvent is strongly reduced in pores of size $b < \kappa_e^{-1}$. "J" will slightly depend on κ_e^{-1}, $J(\kappa_e^{-1}) = J(a)\ (1 - \Omega\ (\kappa_e^{-1}))$.

2) We assume now that the charge Q_1 per grain remains constant during aggregation: this may apply if the surface has strongly ionized groups (sulfonate...) or for branched polyelectrolytes.[32] Here also the flows

are induced in a region of size κ_e^{-1} coating the fractal grain, and the viscous dissipation can be written as

$$TS^o \cong \kappa_e^{-1} \Sigma(\kappa_e^{-1}) \eta (\kappa_e V)^2 . \tag{38}$$

The dissipation is equal to the electrical power QEV, where Q is the total charge of the fractal

$$Q = Q_1 \left(\frac{R}{a} \right)^{D_f} . \tag{39}$$

It leads to

$$\mu = Q_1 \eta^{-1} \kappa_e^{1-D_f} a^{-D_f} , \tag{40}$$

In this case, μ depends on the ionic strength and is sensitive to D_f. One can also notice that Eqs. (37) and (40) are compatible. The capacitance is $C = \varepsilon \Sigma (\kappa_e^{-1}) \kappa_e$ and the surface potential is

$$\zeta = C^{-1} Q = \varepsilon^{-1} Q_1 \kappa_e^{1-D_f} a^{1-D_f} . \tag{41}$$

VI. CONCLUSIONS

We have reviewed physico-chemical methods of investigating the fractal dimension D_f of a porous solid. In all cases, a penetration length ℓ_s can be varied by suitable external conditions (temperature, concentration, molecular weight...) through the interval (a,L), and some information on D_f can be extracted.

However, several difficulties may appear: 1) Hysteresis effects may be important in the case of adsorption of macromolecules and for wetting phenomena. 2) The fractal dimension can be modified by external conditions: a flexible aggregate can collapse during the capillary condensation of a vapor, the fractal dimension of a charged aggregate may be modified by varying the ionic strength. However, we have shown[32] that the fractal dimension of dilute, branched polyelectrolytes remains equal to 2 at high salt concentration (neutral) or in the absence of salt. 3) Fractals may have low accessibility: it is mathematically possible to generate fractal objects in which the wiggly surface is inside a nearly closed pore. Surface of this type cannot be probed by adsorption or depletion of large objects. 4) Fractal with nesting difficulties: when D_f is large, it may become difficult to nest a large sphere which is tangent to the surface at one point M without having the sphere intersecting some other portion of the surface.

In the case of adsorption measurements an object of fractal dimension D_p can adsorb on a surface fractal (D_f). If $D_p > D_f$, the porous solid is covered with tiles ℓ_s and one measures $\Sigma(\ell_s)$. On the other hand if $D_p < D_f$ one measures only $\Sigma(a)$.

In the case of depletion, we have applied our results to interpret certain gel permeation chromatography data. We showed the HPLC gels could be described as fractals in the range of protein sizes. We have shown a linear relationship between $\ell n R_s$ and $\ell n(1 - K_d)$ (where K_d is the partition coefficient), such linear relationships have been searched in vain in the past.

For the wetting of fractal structures, Lenormand has shown that hysteresis effects may be important in the case of the wetting by two immiscible liquids. But capillary condensation of a vapor may allow to avoid this difficulty.

For charged porous fractals, we have shown that depletion of counterions and capacitance measurements may lead to a simple derivation of D_f. For dilute aggregates, the electrophoretic mobility should be sensitive to the fractal dimension only if the charge Q_1 per grain (or per monomer) is maintained during aggretation.

NOMENCLATURE

γ, γ_{SB}, γ_{SA}	liquid/air, solid liquid A,B interfacial tensions
C	concentration
C_c	critical concentration of a binary mixture
ξ	correlation length
ν,β	critical exponents associated to ξ and C
D_f	fractal dimension
D_q	fractal dimension of the adsorbed object
ε	dielectric constant
g	gravity acceleration
h	altitude above the liquid revervoir
κ_c	capillarity length
ζ	surface potential
k	Planck constant
k_e	electrical Debye length
K_d	partition coefficient in S.E.C.
ℓ_s	penetration length
L,a	upper and lower limit for fractal behaviour
n	number of particles
n_g	number of particles per grain
$\mathbf{q}$	scattering wave vector
R_G	radius of gyration
R_s	hydrodynamic radius
ρ	mass per unit volume
$\Sigma(\varepsilon)$	surface spanned by tiles of size ε per fractal grain
V_e	elution volume
V_o	interstitial volume
V_i	internal pore volume
$\Omega(\varepsilon)$	volume spanned by spheres of radius ε surrounding the fractal grain

REFERENCES

1. B.M. Mandelbrot, Fractal Geometry of Nature, Freeman, San Francisco (1981).
2. T.A. Witten and L. Sander, Phys. Rev. Lett., 47, 1400 (1981).
3. M. Kolb, R. Botet and R. Jullien, Phys. Rev. Lett. 51, 1123 (1983).
4. R. Jullien, R. Botet and M. Kolb, La Recherche, 16, 1334 (1985).
5. E. Bouchaud, M. Delsanti, M. Adam, M. Daoud and D. Durand, J. Phys. (Paris), 47, 1273 (1986).
6. P.G. de Gennes, Scaling Concepts in Polymer Physics, Ithaca, N.Y., Cornell University Press, 1979 (second impression 1985).
7. M.E. Fisher and P.G. de Gennes, C.R. Acad. Sci.(Paris), 287, 207 (1978).
8. A. Katz and H. Thomson, Phys. Rev. Lett., 55, 10 (1985).
9. D. Avnir and P. Pfeifer, Nouv. J. Chim., 7, 71-71 (1983).

10. D. Avnir, P. Pfeifer and D. Farin, Nature, 308, 261-263 (1983).
11. H. Bale and P. Schmidt, Phys. Rev. Lett., 53, 596 (1984).
12. F. Brochard, J. Phys. (Paris), 46, 2117 (1985).
13. E. Guyon, Chance and Matter, Les Houches (1986), edited by J. Souletie, J. Vannimenus and R. Stora, North-Holland (1987).
14. T.A. Witten, private communication.
15. P.G. de Gennes, C.R.Acad. Sc. (Paris), 295, 1061 (1982).
16. K.K. Unger and M.G. Gimpel, J. of Chromat., 180, 93 (1980).
17. W.V. Smith and R.M. Ewart, J. Chem. Phys., 16, 592 (1948).
18. Y. Leong and F. Candau, J. Phys. Chem., 86, 2269 (1982).
19. P.G. de Gennes, C.R.Acad. Sc. (Paris), 299, 913 (1984).
20. L. Auvray and P.G. de Gennes, Europhys. Lett., 2, 647 (1986); L. Auvray and J.P. Cotton, Macromolecules, 20, 202 (1987).
21. J.F. Joanny, L. Leibler and P.G. de Gennes, J. Polym. Sci. Polym. Phys Ed., 17, 1073 (1979).
22. I am grateful to H. Benoit for suggesting this idea.
23. M. Le Maire, E. Rivas and J.V. Moller, Anal. Biochem., 106, 12-21 (1980).
24. M. le Maire, L.P. Aggerbeck, C. Monteilhet, J.P. Andersen and J.V. Moller, Anal. Biochem., 154, 525-535 (1986).
25. M. Le Maire, A. Ghazi, J.V. Moller and L.P. Aggerbeck, Biochem. J., 243, 399-404 (1987).
26. G. Daccord and R. Lenormand, Nature, 325, 41 (1987); R. Lenormand, A. Soucemarianadin, E. Touboul and G. Daccord, Phys. Rev. A, 36, 1855 (1987).
27. P.G. de Gennes, Physics of Disordered Materials, edited by D. Adler, H. Fritzche and S.R. Ovshinsky, Plenum Publishing Corporation (1985).
28. F. Brochard, C.R.Acad.Sci. (Paris), 304, 785 (1987).
29. R.K. Schafield and H.R. Samson, Discuss. Faraday Soc., 18, 135 (1954).
30. S.H. Ng. Fairbridge and R.B.H. Kaye, Langmuir, 340, (1987).
31. H.J. Van der Hul and J. Lyklema, J. Colloid Interface Sci., 23, 500 (1967); J. Amer. Chem. Soc., 90, 3010 (1968).
32. F. Brochard and P.G. de Gennes, C.R.Acad.Sc. (Paris), 307 Série II, p.1497 (1988).
33. F. Brochard et A. Halperin, C.R. Acad. Sci. (Paris) 302 ρ = °(17), 1043 (1986).

DRY SPREADING OF POLYMER LIQUIDS ON SOLID SURFACES : ROLE OF LONG RANGE FORCES, PRECURSOR FILM PROFILES AND SPECIFIC POLYMERIC EFFECTS

L. Léger, A.M. Guinet-Picard, H. Hervet and D. Ausserre

Laboratoire de Physique de la Matière Condensée*, U.A. CNRS 792
Collège de France
11 place Marcelin-Berthelot
75231 Paris Cedex 05
France

M. Erman

Laboratoire d'Electronique et de Physique appliqúee**
3, avenue Descartes, B.P. 15
94451 Limeil-Brévannes, Cedex
France

ABSTRACT

We present experimental investigations of the spreading of nonvolatile liquids (polydimethylsiloxane drops) on smooth horizontal silicon wafers. We distinguish the macroscopic part of the drop, which can be seen by bare eye (or through a microscope) from the microscopic part, or precursor film, which progressively develops, like a liquid tongue (thickness smaller than 1000 Å) all around the drop. The spreading kinetics of the macroscopic drop appears independent of the spreading parameter, in contrast to the precursor film which is deeply influenced by the surface energies. We have characterized this precursor film (both its profile and its evolution) by optical techniques. The results are compared with the recent de Gennes and Joanny' theoretical predictions, and specific polymeric effects are pointed out.

I. INTRODUCTION

The spreading of a liquid on a solid surface is of obvious importance in a number of practical situations (paints, textile dying, metal or glasses anticorrosive coating, lubrication, gluing, treatments of plants). The underlying mechanisms are, however, only poorly understood. Recent theoretical approaches have been developed by Scriven *et al.*,[1] de Gennes[2]

---

* Associated with CNRS, UA 792

** LEP is a member of the PHILLIPS RESEARCH ORGANIZATION

and Joanny,[3] which emphasize the role of long-range cohesive forces in the liquid on the spreading process. Following these approaches, the behaviour of volatile and nonvolatile liquids become clearly different : a volatile liquid is always in equilibrium with its vapour; molecules of the liquid can be efficiently transported through the vapor phase very far from the liquid front. Thus, when a drop of a volatile liquid is deposited on a solid, in a situation in which the liquid wets the solid, the surface is very rapidly covered by a thin layer of recondensed liquid molecules coming from the vapour. On the contrary, if the liquid has a very low vapour pressure, the only efficient way of covering the solid surface by a liquid film is by the flow of the liquid. The kinetics of the process, due to viscous dissipation in the thin liquid film formed, will be strongly slowed down compared to the volatile case. Moreover, not only the kinetics, but also the final equilibrium state of spreading will be quite different in the two situations. One can guess that the final state of spreading is a thin flat liquid layer. If the thickness of the liquid film becomes smaller than the range of the cohesive interactions in the liquid, the two interfaces solid-liquid and liquid-vapor become correlated: if one wants to make the liquid still thinner, one has to break cohesive interactions on both sides, and the energy of the thin film is larger than that of a film having a thickness larger than the range of these interactions. A way of taking this effect into account is to introduce a pressure, the disjoing pressure $\pi(z)$,[4] which increases when z goes to zero. For a volatile liquid, the final state of spreading corresponds to an equilibrium between the molecules in the film (at the pressure $\pi(z)$) and in the vapour phase. The equilibrium thickness of the film is thus fixed by an external parameter, the degree of saturation of the vapour. On the contrary, when the liquid is nonvolatile, the total volume of the liquid is fixed; make the film thinner means that a larger area of the solid surface will be covered, and the equilibrium film thickness is the result of a balance between two antagonist effects : disjoining pressure which tends to thicken the film and spreading forces which tend to make it thinner.

The spreading of a nonvolatile liquid thus appears more directly linked to the intrinsic properties of the solid/liquid system, and a detailed description of the process has been proposed by Joanny and de Gennes.[2,3] It is tempting to compare these predictions with the large number of available experimental work performed on the subject, owing to its practical importance.[5] However, many of these experiments have been performed on volatile liquids or, when silicone oils were used, on not well controlled surfaces. For this reason, we have undertaken systematic spreading experiments on a model system (polydimethylsiloxane (PDMS) deposited on smooth silicon wafers), using complementary techniques in order to characterize the drop behaviour at all length scales. In part II we shall present the experimental system and the different techniques used. In part III the present results will be gathered both for large drops where macroscopic and microscopic (precursor film) regimes coexist, even at very long spreading times, and for tiny drops which are rapidly thin enough to be entirely affected by long-range forces. In part IV, these results will be discussed and compared with theoretical predictions.

II. THE EXPERIMENTAL SYSTEM AND TECHNIQUES

II.1 The Liquid

We have used a polymer melt, the polydimethylsiloxane, which is well into its liquid phase at room temperature (glass transition temperation T_g = - 120°C). The choice of a polymer is convenient : it ensures the low volatility of the liquid and allows to vary the viscosity in an enormous range (0.5 - 25.10^3 P in the present experiments) without affecting the interactions. However, polymers are always polydisperse, especially silicon oils, which contain a large amount of low molecular weight oligomers. To avoid or limit as much as we could a possible contamination of our results by these volatile low molecular weight oligomers, we have used narrow molecular weight fractions kindly given to us by C. Strazielle (Institut Charles Sadron, Strasbourg, France). Their characteristics are reported in Table I, along with the viscosities, evaluated from the $\eta(M)$ data of the literature.[6] The surface tension of the polymer is $\gamma \sim 22$ mN/m at the working temperature.

II.2 The Solid

Silicon wafer surfaces have been chosen because of their high reproducible chemical quality and very small residual roughness. A silicon wafer was a slab of a silicon monocrystal (cut in our case along the [1,1,1] plane), covered by a 20 Å thick SiO_2 layer. The residual roughness, characterized by X-rays reflectivity measurements,[7] was of the order of 4 to 4.5 Å.

Table 1. Characteristics of the polymer samples

M_w	M_w / M_n	η(Poises)
6,500	2.0	0.5
79,000	1.15	350
160,000	1.11	4,260
280,000	1.19	24,500

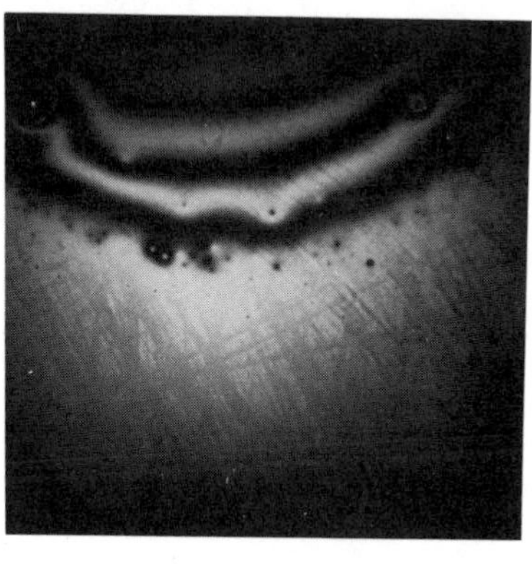

Fig. 1. Aspect of drop, as observed through a microscope, by ellipsocontrast techniques. The precursor film appears like a bright halo all around the macroscopic drop. This halo is uniform when the drop is deposited on a bare wafer, but when a silanated wafer is used like in the picture, irregularities of the spreading power of the surface are revealed by the presence of the film, and attributed to inhomogeneities in the grafted monolayer.

In order to test the influence of the spreading power of the solid, two surface states have been achieved:

–a high energy surface was obtained after a strong oxidation of all the adsorbed organic impurities, by oxygen flow under UV irradiation;[8] this is an efficient cleaning procedure which does not affect the surface roughness.

–a low energy surface was obtained by chemical grafting of a monolayer of octadecyltrichlorosilane, following the Sagiv's procedure.[9]

The spreading power of the surface can be quantified by the spreading parameter $S = \gamma_{SV} - \gamma_{SL} - \gamma_{LV}$ with γ_{SG}, γ_{SL} and γ_{LV} the solid-vapor solid-liquid, liquid vapor interfacial tensions respectively. S represents the energy gained by covering one unit area of the solid by a thick liquid film. In the first case (clean wafer) water spreads on the surface, which has thus a critical surface tension γ_C[10] larger than 72 mN/m, leading to large S values (an estimation of S is $S = \gamma_C = \gamma_{LV} > 50$ mN/m). On the contrary γ_C deduced from contact angle measurements of alkanes on the grafted surface is of the order of 22 to 23 mN/m and S is smaller or close to 1.

II.3 A typical experiment was performed in the following way : a small drop (volume Ω varying between 10^{-7} and 10^{-4} cm^3) was deposited on the wafer which was immediately enclosed in a sealed box equipped with a glass window in order to prevent further contamination by dust or atmosphere impurities during the spreading period which could last for several months. Up to now, the atmosphere around the drop was not further controlled (no special care was taken to avoid humidity, or work under inert nitrogen atmosphere). The size R of the macroscopic drop and its apparent contact angle θ_a were measured periodically by direct observation through a microscope (size and small contact angle measurements through equual thickness fringes spacing) or using a recently proposed method in which the drop was used as a convex mirror, the cone inside which the light was reflected having an angle $2\ \theta_a$.[11] At the same time, a careful examination of the system, through a microscope in polarized reflected light, revealed a bright halo all around the macroscopic drop (Fig. 1). As time went on, this halo extended further and further from the visible edge of the drop (last black interference fringe) and at the same time became fainter. We have attributed this special contrast to the presence of a thin liquid film on the top of the silicon wafer surface. At the wavelength used (green or white light), the silicon acted as a metallic mirror. It was covered by a 20 Å thick dielectric oxide layer plus eventually a liquid film. The optical properties of such a layered thin film are classical and can be calculated using the Fresnel formula.[12] The important point is that the reflection coefficient is different for a polarization in or out of the plane of incidence ($r_{\parallel}$ and $r_{\perp}$ respectively), a property classically used in ellipsometry to gain information on the dielectric layer. In the present observation through a microscope, the direction of polarization of the incident light is fixed with respect to the sample, but all beams contained inside the cone of aperture of the objective are accepted to form the image. To calculate how the collected intensity is affected by the liquid film, one has to perform a double integration on all angles of incidence contained in that cone and, at fixed angle of incidence, on all possible orientations of the plane of incidence with respect to the direction of polarization. The resulting contrast thus calculated, for the optical parameters of our system is reported in Fig. 2 as a function of the PDMS thickness. Observation in polarized reflected light, through a microscope, reveals itself a useful

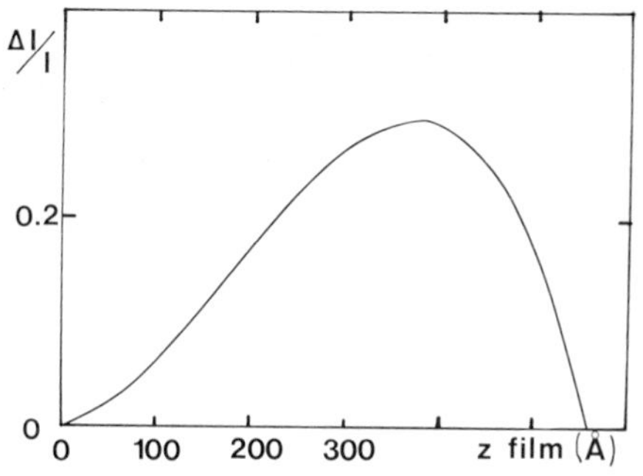

Fig. 2. Evaluation of the contrast introduced by the film when observed in polarized reflected light, through a microscope, as a function of the film thickness. The oxide layer is assumed to be 20 Å thick. The postulated optical constants of the system are: Si : n = 4 - i x 0.03; SiO_2 : n = 1.46; PDMS : n = 1.43, and the wavelength λ = 0.5 μm. The incident angle is 20°.

tool to study inhomogeneous thin films, a method that we have called ellipsocontrast.[13] It appears quite sensitive for thicknesses in the range 100 Å - 500 Å (for the optical constants of our system), and has the strong advantage of a high spatial resolution (1 μm). The sensitivity rapidly decreases for smaller thicknesses, in contrast to conventional ellipsometry, because the aperture of the microscope is not large enough to allow for incidence angles close to Brewster's angle.

In order to extend the range of thicknesses investigated, we have also used ellipsometry, with a high performance ellipsometer, built by one of us (M.E.) and J.P. Theeten,[14] in the Labortoire d'Electronique et de Physique Appliquée (LEP), and particularly designed to have a good spatial rsolution.[14] The size of the illuminated area on the sample was 10 x 25 μm. A light beam of fixed polarization and fixed incidence angle was shone on the sample. The reflected intensity was collected through a rotating analyzer, and the two ellipsometric parameters cos Δ and tan ψ, with $r_{\|}/r_{\perp}$ = tan ψ $e^{i\Delta}$, extracted from the amplitude and phase of the component of the reflected intensity modulated at twice the rotation frequency of the analyzer.[14] From tan ψ and cos Δ, the index of refraction and the thickness of a uniform dielectric layer can classically be extracted.[15] In the present experiments, we have assumed the thickness locally uniform in the illuminated window, treated the oxide layer (index of refraction n = 1.46) and the PDMS (n = 1.43) as the same optical medium, and extracted the film thickness z as a function of the position through the drop from the cos Δ, tan ψ data. Such an inversion of ellipsometric data, even if classical, is not totally obvious, and the procedure used - with its limitations - is discussed in detail in ref. (16).

III. RESULTS

III.1 Macroscopic Spreading Kinetics

The time evolution of the size and of the apparent contact angle on bare and grafted wafers has yet been reported.[13] We just recall here the main results which are illustrated in Fig. 3. All the drops studied are well inside the capillary regime ($R < K^{-1}$ with the capillary length K^{-1} = $(2\gamma/\rho_g)^{1/2}$ ~ 2.2 mm for our system). Both the size and the apparent contact angle follow universal scaling laws with time :

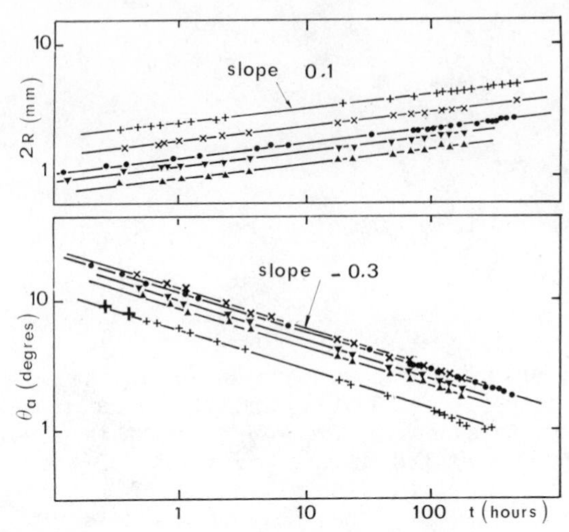

Fig. 3. Spreading kinetics of the macroscopic part of the drops, characterized by their size R and their apparent contact angle θ_a, for a wide range of drop characteristics, and two spreading parameter values.

o	Clean wafer,	$M_w = 1.6\ 10^5$,	$\Omega = 6.5\ 10^{-5}\ cm^3$
x	Silanated wafer,	$M_w = 1.6\ 10^5$,	$\Omega = 1.8\ 10^{-4}\ cm^3$
∇	Silanated wafer,	$M_w = 1.6\ 10^5$,	$\Omega = 3.5\ 10^{-5}\ cm^3$
Δ	Silanated wafer,	$M_w = 1.6\ 10^5$,	$\Omega = 1.5\ 10^{-5}\ cm^3$
+	Silanated wafer,	$M_w = 7.9\ 10^4$,	$\Omega = 2.1\ 10^{-4}\ cm^3$

$$R(t) \sim t^{0.1 \pm 0.01}\ ;\ \theta_a(t) \sim t^{-0.3 \pm 0.01}\ . \qquad (1)$$

The exponents are independent of the spreading parameter, and in very good agreement with Tanner's law.[17]

For very small drops, an acceleration of the spreading process is observed ($R(t) \sim t^{\alpha}$ with $\alpha > 0.1$), but $R^3\ \theta_a$ remains constant.

III.2 Precursor Film Characterization

Figure 1 is a visualization, by ellipsocontrast, of a precursor film developed on a grafted wafer. Strong irregularities of the film thickness are immediately visible, and revealed by the presence of the film (they were totally invisible before the film deposition, even using sensitive Nomarski's techniques).

In Fig. 4 an analogous illustration is obtained through a cartography performed by ellipsometry. On b, the lines are equal cos Δ lines, i.e., equal thickness lines. Again strong irregularities are clearly visible. These irregularities are not present when a bare wafer is used. For this reason, a detailed quantitative analysis of precursor film profiles has only been conducted, at the present time, on bare wafers.

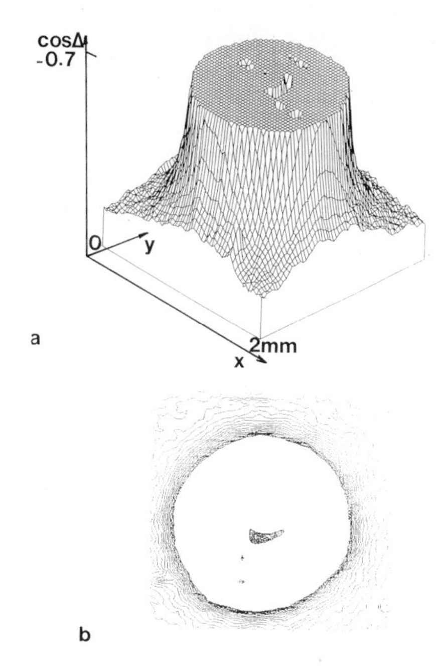

Fig. 4. Ellipsometric cartography of the thin precursor film zone, for a drop deposited on a grafted wafer. a) cos Δ as a function of the position (x,y) in the wafer plane; b) Equal cos Δ lines (i.e., equal thickness lines), displaying an overall axial symmetry, plus strong fluctuations, related to the inhomogeneities of the grafted silane monolayer. For a better visualization, the cartography is truncated at Δ = - 0.7, i.e., thicknesses of order 250 Å.

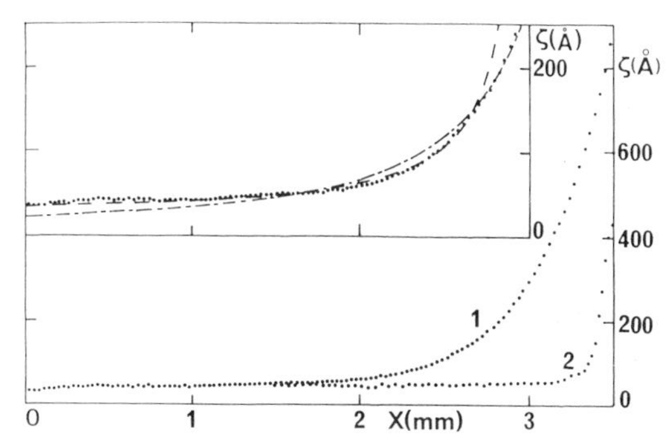

Fig. 5. Precursor film profiles obtained by spatially resolved ellipsometry on two PDMS drops of very different characteristics: (1) : M_w=6500, Ω=1.34 10^{-5} cm^3, viscosity η=0.53 P., spreading time 75 hours, θ_a=0.055°. (2) : M_w = 280,000, Ω=6.11 10^{-7} cm^3, η=24500 P., spreading time 2856 hours, θ_a=0.914°. The two profiles are cut at the transition towards the macroscopic drop. In the insert (vertical scale expanded x 2) the dotted line is the best adjustment of the tail of (1) on the asymptotic form $z \sim 1/(x-x_0)^\alpha$ with α=1 (predicted by ref. (2) and (3)), while the semi-dotted line is for α = 2.

In Fig. 5 two precursor film profiles obtained on very different drops (characteristics detailed in the figure caption) are reported. The profiles are cut at the transition towards the macroscopic drop, a point which can be precisely located, as the slope of the free surface there becomes large enough to prevent the ellipsometric inversion (assuming a locally uniform film) to be valid. We can notice in Fig. 5 that the precursor film appears much more developed (thickness and extension) for the drop having the smaller apparent contact angle. No extremity of the film can be located on these profiles.

In Fig. 6 a rather fresh precursor film for a high molecular weight PDMS is reported, and shows a non-monotonic decrease of thickness. This "bump", which appears for thicknesses comparable to the radius of gyration of the chains, slowly disappears when the film evolves and develops.

III.3 Very Small Drops

An ellipsometric profile of a very small drop is shown in Fig. 7. The maximum thickness is of the order of 200 Å, which means that long-range forces are important everywhere in this drop, and no macroscopic regime remains. Such drops are totally invisible by bare eye. As time goes they progressively flatten down, and evolve towards a thin liquid film of uniform thickness.[7]

These last stages of spreading have been characterized by X-rays reflectivity, and appear to strongly depend on the spreading parameter.[7]

IV. DISCUSSION

All the results presented above seem to agree qualitatively well, with de Gennes and Joanny's predictions : small spreading drops of a nonvolatile liquid indeed separate into two parts having very different characteristics. The macroscopic part evolves independently of the spreading power of the surface. On the contrary, the thin precursor film

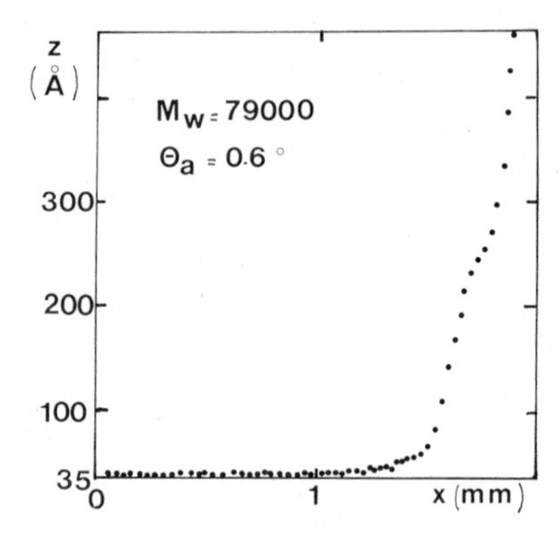

Fig. 6. Precursor film profile, obtained by ellipsometry, of a PDMS drop of molecular weight 79000. When the film is rather young, a "bump" appears for a thickness comparable to the radius of gyration of the polymer chains.

which slowly develops ahead of the macroscopic drop appears deeply influenced by the value of the spreading parameter, as illustrated by Figs. 1 and 4. Even if very preliminary, the X-rays reflectivity results on the final state of spreading,[17] seem to show that the picture of a "pancake" with a thickness strongly dependent on S is not very different from reality. However, these experiments demonstrate the difficulty of a complete quantitative characterization of the final state of spreading : if S is large, the final pancake is so thin that roughness becomes a leading effect. Small S values have to be achieved to increase the equilibrium thickness of the film, but then, it becomes highly sensitive to small S variations.

A more quantitative analysis has been conducted with the profiles of Fig. 5,[16] and we just recall here the results. The analysis of the precursor film profile performed by de Gennes and Joanny[2,3] assumes that the liquid is injected into the film at a velocity $U = dR/dt$. The precursor film profile is then obtained, assuming that the incoming liquid flows under the action of the Laplace and disjoining pressure gradients in the film arising from curvature and thickness variations. Assuming a Poiseuille-type flow, with a vanishing velocity at the solid wall, leads to a differential equation for the film profile (for van der Waals interactions).

$$\frac{3\eta U}{z^2} = \frac{d}{dx}\left[-\gamma \frac{d^2 z}{dx^2} - \frac{A}{6\pi z^3} \right] , \qquad (2)$$

where A is the effective Hamaker constant of the problem.

If the curvature is negligible (small thickness region), Eq. (2) has a simple solution

$$z \sim \frac{a}{(x - x_0) w} ,$$

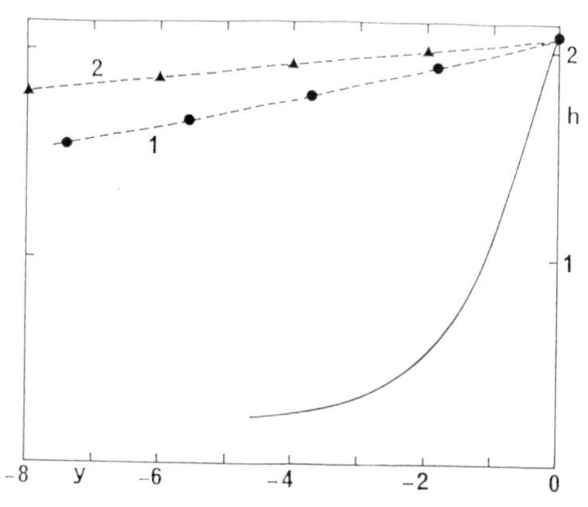

Fig. 7. Comparison of the experimental profiles of Fig. 5 with the numerical solution (full line), in the reduced adimensional variables $h(y)$ of ref. (18). The experimental profiles develop much more rapidly than predicted, and the discrepancy is more pronounced for the higher molecular weight.

where $a = (A/6\pi\gamma)^{1/2}$, $w = \eta U/\gamma$, and x_0 is an integration constant to be determined by a suitable connection to the full profile. The dotted line in the insert of Fig. 5 is the best least square fit of the tail of the profile 1 on such a function. The agreement appears satisfactory. However, adjustment of the full profile on the calculated form (taking curvature terms into account), discussed in detail in ref. 16, appears to be impossible : Eq. (2) has been solved numerically by H. Hervet and P.G. de Gennes.[19] It has only one solution which can be related to the macroscopic drop on one side (zero curvature of the free surface) and going to zero like $1(x-x_0)$ on the other side. This solution, called the maximum film, is compared to our experimental data (in normalized units) (Fig. 7). One can see immediately that the experimental precursor films develop much more rapidly than predicted. The departure is more pronounced for the larger molecular weight. One could think that this is due to a trivial polydispersity effect, but the polydispersity of the high molecular weight sample is much smaller than that of the low molecular weight sample (see Table 1).

We rather think that this discrepancy between predicted and observed behaviour reveals specific polymeric effects. An indication that specific polymeric effects exist is the transient "bump" observed for $z \sim R_G$ (Fig. 6). Specific polymeric effects on the spreading behaviour have been predicted by de Gennes and Brochard,[19] noticing that the flow boundary condition at the solid wall for an entangled polymer should rather be a finite velocity.[20] Such slipping at the wall should modify both the macroscopic drop profile (appearance of a foot) and the precursor film profile which, far from the macroscopic drop, should rather decay like $z \sim 1/\sqrt{x-x_0}$. Our experiments do not show any of these behaviours. Many complications could occur and prevent slipping at the wall, as for example an anomalous structuration of the liquid close to the wall due to the high disjoining pressure,[21] but this seems hardly compatible with the very thin final thickness we observe, and with the apparent acceleration of the spreading process. An alternative explanation could be that the long-range forces are not correctly modeled by a simple van der Waals forces, and/or that the viscosity in such a thin polymer film (thinner than the radius of gyration of the chains) is not the bulk liquid viscosity. We are presently trying to test these hypothesis.

An interesting result, and apparently simple one, is illustrated in Fig. 8 where the dotted line is the best least square fit of the experimental profile of the tiny drop on a Gaussian curve. The agreement is quite good and it was impossible to obtin similar results with the predicted Lorentzian shape.[22]

V. CONCLUSIONS

We have presented a series of experiments developed to understand the spreading behaviour of small drops of nonvolatile liquid deposited on smooth solid surfaces. Several complementary techniques have been used to completely characterize the drop profile at all length scales. We have paid special attention to the thin precursor film which develops all around the drop, first because precursor film studies on well-controlled systems are not numerous, and second, because this precursor film is of fundamental importance to understand the role of long-range forces on spreading.

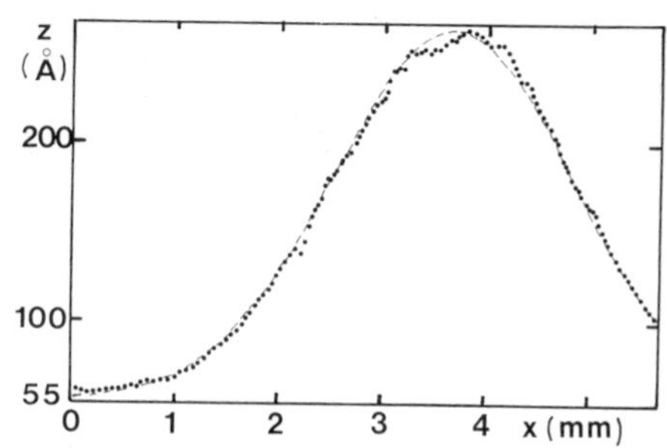

Fig. 8. Ellipsometric profile of a very small drop. After a few days of spreading, the thickness of a drop of volume of order 5.10-7 cm^3 and of molecular weight 6500, is small enough for producing a drop entirely sensitive to long-range forces. The macroscopic drop has disappeared. The profile of such a small drop is well described by a Gaussian (dotted line).

Our present results are in qualitative agreement with the theoretical predictions of Joanny and de Gennes : the spreading drop behaves in a very different way depending on the length scale. For thicknesses larger than the range of the interactions, it is in the macroscopic drop range whose spreading kinetic is independent of the surface. For thicknesses smaller than the interaction range, it enters into the precursor film which is deeply influenced both by the state of the macroscopic drop and by the spreading parameter. At fixed spreading parameter, the precursor film is more developed when the apparent contact angle of the macroscopic drop is smaller. For small spreading parameter values, the precursor film is highly sensitive to local inhomogeneities of the surface, and may be used, along with ellipsocontrast observation techniques, as a characterization tool of surface imperfections. When the precursor film thickness becomes small, roughness drives the spreading.

Quantitative discrepancies remain between the observed and predicted precursor film profiles which may reveal either an incorrect description of the long-range interactions by simple van der Waals terms or an anomalous viscosity for liquid thicknesses smaller than the chain radius of gyration. Complementary experiments and analysis are presently underway to test these problems.

NOMENCLATURE

A effective Hamaker constant
γ interfacial tension
γ_{SG} interfacial tension of the solid-gas interface
γ_{SL} interfacial tension of the solid-liquid interface
γ_C critical surface tension of the substrate
Δ ellipsometric parameter
g acceleration of gravity
η viscosity of liquid
θ_a apparent contact angle
K^{-1} capillary length
n refractive index (eventually complex)
$\pi(z)$ disjoining pressure
$r_{\parallel}$ reflection coefficient for an incident beam polarized in the plane of incidence
$r_{\perp}$ reflection coefficient out of the plane of incidence

R size of macroscopic drop
ρ density of liquid
S spreading parameter
t time
T_g glass transition temperature
U velocity of injection, $U = dR/dt$
ψ ellipsometric parameter
χ_o integration constant
Z film thickness
Ω volume of the drop

REFERENCES

1. G.F. Teletzke, H.T. Davis and L.E. Scriven, J. Colloid Interface Sci., to appear.
2. P.G. de Gennes, Review of Modern Physics, 57, 827 (1985).
3. J.F. Joanny, Ph.D Thesis, Université Paris VI, 1985; and J.F. Joanny, J. of Theor. and Appl. Mech. 0750-7240, 249 (1986).
4. B. Deryagin, Zh. Fiz. Khim., 14, 137 (1940).
5. A. Marmur, Adv. Colloid and Interface Sci., 19, 75 (1983).
6. R.R. Rahalker, J. Lamb, G. Harrison, A.J. Barlow, W. Hawthorn, J.A. Semlyen, A.M. North and R.A. Pethrick, Proc. R. Soc. London, A394, 207 (1984).
7. J. Daillant, J.J. Benattar, L. Bosio and L. Léger, Europhys. Lett., 6, 431 (1988).
8. J.R. Vig. J. Vac. Technol., A3, 1027 (1985).
9. J. Gun and J. Sagiv, J. of Colloid and Interface Sci., 112, 457 (1986).
10. E.G. Shafrin and W.A. Zisman, J. Colloid Sci., 7, 166 (1952).
11. C. Allain, D. Ausserré and F. Rondelez, J. Colloid Interface Sci., 107, 5 (1985).
12. M. Born and E. Wolf, Principle of Optics, Pergamon, Oxford 1980.
13. D. Ausserré, A.M. Picard and L. Léger, Phys. Rev. Lett., 57, 2671 (1986).
14. M. Erman and J.B. Theeten, J. Appl. Phys., 60, 859 (1986); and M. Erman, Ph.D thesis, Université Paris VI, 1986.
15. R.M.A. Azzam and N.M. Bashara, Ellipsometry and Polarized Light, North Holland Publishing Company, Amsterdam (1977).
16. L. Léger, M. Erman, A.M. Guinet-Picard, D. Ausserré and C. Strazielle, Phys. Rev. Lett., 60, 2390 (1988).
17. L.H. Tanner, J. Phys., D 12, 1473 (1979).
18. H. Hervet and P.G. de Gennes, C.R. Acad. Sci. Paris, 299II, 499 (1984).
19. F. Brochard and P.G. de Gennes, J. Phys. Lett., 45, L597 (1984).
20. P.G. de Gennes, C.R. Acad. Sci. Paris, B298, 219 (1979).
21. R.G. Horn and J.N. Israelachvili, Macromolecules, 21, 2836 (1988).
22. P.G. de Gennes, C.R. Acad. Sci. Paris, 298II, 475 (1984).

WETTING PHENOMENA

A.M. Cazabat, F. Heslot and P. Levinson

Collège de France, Physique de la Matière Condensée
11 Place Marcelin-Berthelot
75231 Paris Cedex 05
France

ABSTRACT

Wetting phenomena are very commonly observed in everyday life and have been the subject of many experimental investigations. However, as they usually involve a wide variety of processes, finding the clue for a general analysis is not straightforward.

Recently, de Gennes showed that the clue to this analysis was the role of the long-range forces. Many experimental situations appeared to be described by simple, universal laws which can also be used as guides for describing more complex phenomena. These laws result from interactions taking place at a microscopic scale, between 10 and 1000 Å.

This paper is an introductory review of wetting phenomena, presented mainly within the framework of de Gennes' theories.

I. INTRODUCTION

The spreading of liquids on solid surfaces or into porous media is of primary importance in fields as different as soil science, oil recovery, paper industry, adhesives.... A large amount of experimental and theoretical work[1-5] has been done for understanding how the water penetrates into the soil,[5] what are the mechanisms for oil trapping and migration in sedimentary rocks, and how to write properly with ink on paper.[4]

In the fields of adhesion and lubrication, the interesting liquids are often polymeric oils,[6,7] whose spreading on smooth or rough surfaces must be precisely controlled. However, wetting processes usually involve many parameters: the chemical[2] and geometrical[4] structure of the solid and the liquid, the possible presence of impurities[8] solubilized in the bulk liquid or adsorbed at interfaces, etc.

In the following, we shall discuss only the case of smooth or moderately rough[9] solid surfaces and focus our attention on the spreading of pure polymeric liquids. Spurious effects due to impurities or surface contamination will be only briefly mentioned.

First, general considerations on wetting in (II) will show the main problems to discuss and the theoretical answers will be given in (III). Comparison with experimental results will follow in (IV).

II. SOME GENERALITIES ON WETTING

II.1. Basic definitions

Let us consider a drop of liquid on a smooth horizontal solid surface. Its behaviour depends on the balance of capillary forces at the contact line, i.e., on the interfacial tensions between solid, liquid and the surrounding gas.

Let us assume that this gas phase is an inert atmosphere, and does not adsorb on the solid or liquid surfaces.[2,10-11] Let us assume further that the liquid is hardly volatile, i.e., that the transport of molecules through the gas phase can be neglected within the time of the experiments.

Then the interfacial tension between liquid and gas phase is very close to the interfacial tension γ between liquid and vacuum. Similarly, within the time of the experiments, the interfacial tension betwen solid and gas does not differ from the interfacial tension γ_S (or γ_{SV}) between solid and vacuum.

In this case, the quantity which controls spreading is the spreading coefficient S:

$$S = \gamma_S - \gamma - \gamma_{SL} . \tag{1}$$

Let us note that S is not a true thermodynamic equilibrium quantity, but corresponds to a metastable situation.

Let us consider a solid-gas interface: the surface free energy is γ_S per unit area. If the same solid is covered by a thick liquid film, the free energy becomes $\gamma + \gamma_{SL}$ per unit area.

*A positive value of S means that the latter situation is the more favourable, i.e., the liquid will spread on the surface. It is the complete wetting case.

*A negative value of S means that the former situation is obtained: it is the nonwetting case. A metastable equilibrium is reached, with a non-zero value for the contact angle θ_e. The value θ_e obeys Young's equation, which expresses the balance of the horizontal components of the capillary forces (Fig. 1).

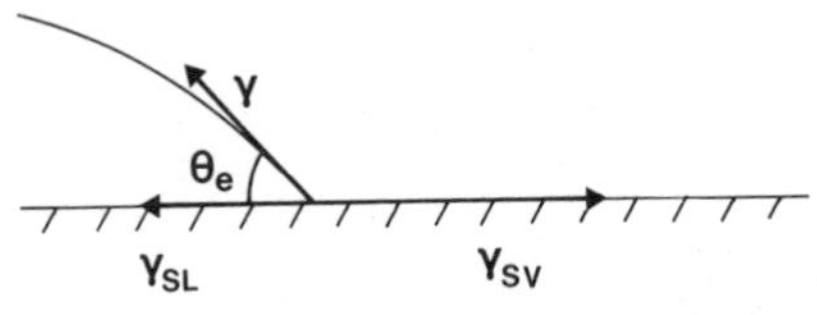

Fig. 1. Capillary forces at the edge of a macroscopic drop. θ_e is the (metastable) equilibrium contact angle.

$$\gamma_S - \gamma \cos\theta_e - \gamma_{SL} = 0 \,, \tag{2}$$

$$\text{or} \qquad \cos\theta_e = \frac{\gamma_S - \gamma_{SL}}{\gamma} = 1 + \frac{S}{\gamma} \,. \qquad (S < 0) \tag{3}$$

Polymer liquids, which are the topics of this volume are good examples of nonvolatile liquids for which the preceding assumptions hold. This case is usually referred to as "dry" wetting, because the part of the solid which is not covered by the liquid is dry. The opposite case which is not discussed here corresponds to "moist" wetting, where a pre-existing film is adsorbed on the solid surface.[11]

II.2 Some observations for $S > 0$

Let us now observe a drop spreading on a horizontal solid surface.

When the drop is just put on the surface, it takes rapidly a well defined shape, a spherical cap for small drops, or a more flattened shape for larger ones, with a fairly low contact angle (say, of the order of a few degrees). Then the drop goes on spreading more slowly with only slight changes in its shape.

During this latter stage, the only one we shall discuss here, the shape of the drop is practically the same as it is of a static drop with the same contact angle, except very close to the edges.[12,13] This static shape results from the balance of vertical capillary forces and gravity. It is usually close to a spherical cap and its exact dependence is controlled by the ratio of the radius R of the wetted spot to the capillary length κ^{-1}:

$$\kappa^{-1} = \sqrt{\frac{\gamma}{\rho g}} \,. \qquad (\kappa^{-1} \sim mm) \tag{4}$$

Here ρ is the liquid density and g the gravitational acceleration.

Tabulated static shapes can be found in the literature.[14]

The radius R of the wetted spot increases with time t according to simple power laws[12,15,16]

$$R \sim t^{1/10} \qquad \text{for} \qquad R << \kappa^{-1}, \quad (\text{spherical caps}) \tag{5}$$

$$R \sim t^{1/8} \qquad \text{for} \qquad R >> \kappa^{-1}, \quad (\text{flattened drops}) \tag{6}$$

For a given liquid, the dynamics of spreading is found to be independent of the solid, as soon as complete wetting is achieved ($S > 0$). This is very surprising as the (horizontal) unbalanced capillary force for spreading is now

$$F_c = \gamma_S - \gamma_{SL} - \gamma \cos\theta = S + \gamma(1 - \cos\theta) \,.$$

and does contain S. Here, θ is the dynamic (instantaneous) contact angle.

The observed power laws can be found theoretically by hydrodynamical calculations[12,16] in the $S = 0$ case. For strictly positive values of S and volatile liquids, one might argue that the drop spreads on a pre-existing liquid film and that the true value of S is zero.[16] But the paradox remains in the case of nonvolatile liquids. Why is the dynamics of spreading independent of S?

This paradox has its counterpart in the theoretical calculations: when the hydrodynamic equations for spreading are solved, one obtains a divergence of the viscous dissipation at the edge of the drop.[16,17] Ad hoc cut-off lengths or modified limit conditions must be introduced to remove this divergence.

II.3 The precursor film The clue to the riddle is an experimental study of Hardy[18] who noticed the presence of a very thin film preceding the spreading drops. This film exists even for nonvolatile liquids and has been called the precursor film. It means that the shape of the drop's edge at the macroscopic scale is not just a magnification of its shape at the microscopic scale. This is due to the presence of interactions whose range is below the macroscopic scale.

Taking into account these interactions in the theoretical analysis will solve the paradox of a S-independent spreading dynamics and the mathematical problems of diverging dissipation.[11,16,17,19,20]

III. THEORY FOR COMPLETE, DRY WETTING

III.1 Interactions at the microscopic scale

The interactions which control the behaviour of thin and ultrathin films belong to two main classes:

- the "long-range" interactions: the most important ones are van der Waals forces due to induced dipolar interactions in nonionic systems, and electric double layer repulsion in ionic systems.[21] The corresponding forces laws have quite general expressions with a range of a few thousand Angströms.

- the "short-range" interactions: they depend explicitly on the molecular structure[22] in the first layers of the interacting media. No general behaviour is expected for the force laws whose range is of the order of a few Angströms.

Let us consider a solid surface (medium 1) covered by a film of thickness z (medium 3) in contact with air (medium 2). For large z (thick, macroscopic film) the surface free energy of the system is just

$$E(\infty) = \gamma + \gamma_{SL} , \qquad (8)$$

per unit area.

For thinner films, with a thickness in the microscopic range (a few thousand Angströms or less), an additional term must be added. It accounts for the interaction between the liquid/vapour and the liquid/solid interfaces. Now, E depends on z:

$$E(z) = \gamma + \gamma_{SI} + P(z) \; ; \tag{9}$$

$$P(z) \to 0 \; . \; (z) \to \infty$$

The interaction energy $P(z)$ results from various short- and long-range contributions and has a large variety of possible shapes.[1,11,22-24]

Let us note that for $z \to 0_+$ (close contact) we get the dry situation, i.e., lim $P(z) = \gamma_S - \gamma - \gamma_{SL} = S$, when $z \to 0_+$.

In the following, we shall focus our attention on nonionic polymers, where van der Waals interactions dominate the long-range behaviour of P(z). Also it can be expected that the exact shape of the short-range interactions will have little influence on the results. Thus, even in the field of wetting, universal laws are expected for polymer systems.

III.2 Van der Waals interactions

In nonionic media, van der Waals interactions between molecules are usually the dominant long-range contribution.

Van der Waals forces between molecules in vacuum are attractive. As a result, two infinite media 1 and 2 interacting across vacuum attract each other. If they are separated by a gap of constant thickness z, the interaction energy per unit area can be written as[22]

$$P(z) = -\frac{A_{12}}{12\pi z^2} \; . \tag{10}$$

The strength of this interaction is characterized by the Hamaker constant A_{12}, which can be calculated in terms of the dielectric properties of the media 1 and 2. A_{12}, is a positive quantity, i.e., the interaction $P(z)$ tends to stick the media together.

The situation is somewhat different if the gap between the media 1 and 2 is filled by a third medium 3 instead of vacuum. Now, the interaction $P(z)$ between 1 and 2 across 3 can be written

$$P(z) = -\frac{A_{132}}{12\pi z^2} \; . \tag{11}$$

The effective Hamaker constant A_{132} is a combination of the various Hamaker constants A_{13}, A_{12}, A_{23}, A_{23} and can be positive or negative. A comprehensive review on the calculation of Hamaker constants and retardation effects is out of the scope of this introductory review. It can be found in Ref. 22.

If A_{132} is positive, $P(z)$ is negative, i.e., there is an attraction between 1 and 2: the interaction tends to thin the film which will ultimately disappear: it is a nonwetting situation.*

---

* If the short-range interactions are strongly positive, a monomolecular film can be left on the surface.[10,11]

If A_{132} is negative, the interaction is repulsive and tends to thicken the film, A_{132} leading to a wetting situation. It is the case which will be considered in the following. Let us just note

$$A = -A_{132} . \tag{12}$$

The van der Waals interaction can be written as

$$P(z) = \frac{A}{12\pi z^2} , \qquad A > 0$$

$$where \quad z \gtrsim 10\,Å . \tag{13}$$

The short-range part of the interaction can be positive or negative, and possible shapes of $P(z)$ are schematized in Fig. 2.

The shape (b) can be obtained for surfaces covered by a hydrophobic grafted layer of thickness δ.[24]

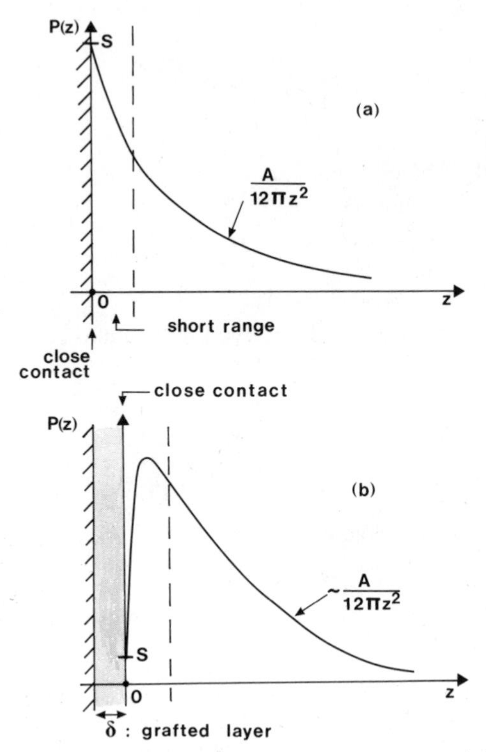

Fig. 2. Possible shapes of the interaction energy $P(z)$, z being the film thickness. The shaded area is the solid. The dotted line represents the range of the short ranged interactions.

(a) bare high energy surface,
(b) surface with a grafted layer of thickness δ.

In this case, the grafted layer is a fourth medium 4 whose properties will play a role for $z < \delta$. More precisely, the interaction $P(z)$ can be written as[24]

$$P(z) = -\frac{A_{132} - A_{432}}{12\pi(z+\delta)^2} - \frac{A_{432}}{12\pi z^2}$$

where 1 denotes the solid,

2 denotes the gas phase,

3 denotes the liquid (thickness z) , and

4 denotes the grafted layer (thickness δ).

For 10 Å $< z << \delta$, the medium 1 plays no role

$$P(z) \approx -\frac{A_{432}}{12\pi z^2} .$$

For $z >> \delta$, the grafted layer is not seen

$$P(z) \approx -\frac{A_{132}}{12\pi z^2} .$$

III.3. The equilibrium situation

The equilibrium thickness e_c of a film of a nonvolatile wetting liquid on a solid surface can be obtained by minimizing the surface free energy $E(z)$ for a given volume of liquid.* An implicit relation is obtained[11,19,20]

$$S = P(e_c) - e_c\left(\frac{dP}{dz}\right)_{z=e_c} . \tag{14}$$

Let us assume that $P(z)$ is dominated by its van der Waals contribution except for $z \to 0$ ($P(0) \to S$). One obtains:[19,20]

$$e_c = \frac{1}{2}\sqrt{\frac{A}{\pi S}} = a\sqrt{\frac{3\gamma}{2S}} , \tag{15}$$

with $a^2 = A/(6\pi\gamma)$. For polydimethylsiloxanes on silica, $A \sim 10^{-20}$ J and $\gamma \sim 20 \times 10^{-3}$ Jm^{-2}, so that a ~ 1.6 Å.

---

* For a volatile liquid, the condition of volume conservation is replaced by a condition of equilibrium with the vapour pressure.[11]

* In the case of Fig. 2(a), S is of the order of γ and e_c is of the order of a.

* In the case of Fig. 2(b), S can be made very small by a convenient grafting and larger values of e_c can be obtained.[24] Another interesting situation will be in the vicinity of a wetting transition[25] ($S \to 0$).

In contrast to our instuitive feeling, the final equilibrium situation is not necessarily a monomolecular layer of liquid. It is rather a "pancake"[19] whose thickness results from the balance between spreading parameter S and long range interactions (A).

This pancake thickness e_c is the minimum thickness for a liquid film covering the solid surface.

III.4. Nonequilibrium situations Let us now discuss the nonequilibrium situations, i.e., the spreading dynamics.

The hydrodynamic equations[12,16] describing the motion of a liquid edge on a surface hold only for macroscopic thicknesses of the liquid. Very close to the edge, the interaction $P(z)$ between the interfaces must be taken into account.[11,19,20] The complete equations have been discussed by de Gennes and Joanny in the dry wetting case.[19,20]

We shall only recall briefly the main results of the theory, in the case of spreading drops:

* For drops whose maximum thickness is in the macroscopic range, the macroscopic "cap" and the precursor can be treated separately.

In the driving force (Eq. 7):

$$F_C = S + \gamma\,(1 - \cos\theta)\ ,$$

the first term balances the viscous friction in the precursor, i.e., at the microscopic scale. At the macroscopic scale, the driving force is just γ (1 - cosθ) and does not contain S, what solves the paradox of the $II.2.: actually, the macroscopic dynamics is independent of the spreading parameter and a hydrodynamic calculation with $S = 0$[12,16] will lead to correct results. They are[19,20]

For $\kappa R << 1$ (cf. 5):

$$R\,(t) \sim \Omega^{\,3/10}\left(\frac{\gamma\, t}{\eta}\right)^{1/10}, \tag{16}$$

Ω being the (constant) drop volume and η the viscosity of the liquid. For $\kappa R >> 1$ (cf. 6):

$$R\,(t) \sim \Omega^{\,3/8}\left(\frac{\rho\, g\, t}{\eta}\right)^{1/8}. \tag{17}$$

At the microscopic scale, explicit results have been obtained for dominating van der Waals forces. The length of the precursor is given by:

$$\ell(t) = \frac{1}{3\eta \dfrac{dR}{dt}} \sqrt{\frac{AS}{\pi}} \quad . \tag{18}$$

Its thickness z is a function of the horizontal distance x to the drop edge and decreases as $1/x$ between its maximum value at the edge:

$$z_{max} = \frac{a}{\theta} \, , \qquad (a \sim 1.6\,\text{Å}) \tag{19}$$

and the minimum thickness:

$$z_{min} = e_c = \frac{1}{2} \sqrt{\frac{A}{\pi S}} \quad .$$

According to the values of drop volume, liquid viscosity, and minimum thickness, the development of a precursor of length $\ell \sim 1$ mm can take minutes or days:

$$\ell(t) \approx 10 \frac{a^2}{e_c} \, \Omega^{-3/10} \left(\frac{\gamma \, t}{\eta} \right)^{9/10} \quad . \tag{20}$$

* For very flat drops no clear distinction can be made between a macroscopic part and a microscopic precursor.[20] Moreover, when the edge velocity dR/dt becomes very low, the preceding formulae for the precursor length and profile are no longer valid: a diffusive behaviour is expected with, ultimately[11,26]

$$\ell(t) \sim (D\,t)^{1/2} \; ,$$

$$D = \frac{1}{3\,\eta} \sqrt{\frac{AS}{\pi}} \quad . \tag{21}$$

At long times, both phenomena will occur simultaneously for drops and the resulting behaviour might be rather intricate.

III.5. Discussion

These theoretical predictions contain some hypotheses:

- the film is assumed to have the bulk properties of the liquid and is treated as a continuous medium,

- molecular diffusion of liquid molecules on the solid surface is neglected.

These assumptions are certainly valid for simple liquids when the film thickness is much larger than the molecular size.[11]

In the case of polymers, the discussion is less obvious. The monomer size is one of the characteristic lengths, but the slipping length in the velocity profile[27] and the radius of gyration of the coil[28] are also good candidates.

Now, the experiments have to bring further information

IV. EXPERIMENTAL STUDIES AND COMPARISONS WITH THEORY

IV.1. The macroscopic scale

Many experimental studies on the spreading of the drops of nonvolatile liquids in an inert atmosphere (clean air) have been performed.[12,15] A good agreement with the theoretical power laws was always obtained, both in the $kR << 1$ (Eq. 16) and in the $kR >> 1$ (Eq. 17) ranges.[29] No specific polymeric effect was observed for PDMS oils[30] with molecular weight up to 2.8×10^5 in the $kR << 1$ as well as in the $kR >> 1$ range.[31] It suggests that slipping effects[27,28] do not play a major role in the spreading of polymer drops.

It is worth noting that very good precision and reproducibility of macroscopic spreading experiments with PDMS were obtained. They are practically insensitive to surface roughness (at least below 1 µm) and atmospheric contamination (except by good solvents of the silicone oil).[6,32]

IV.2. The microscopic scale

Various techniques have been used for the study of precursor films (interferometry,[33] scanning electron microscopy[34] and even tunneling microscopy).[35] However, the standard technique is ellipsometry[6] which allows to measure thicknesses in the Å range with a spatial resolution between 0.5 mm (usual) and 20 µm.[36] Spectroscopic ellipsometry is also available.[37]

In contrast with the macroscopic case, the spreading experiments are highly sensitive to roughness[34,38] and contamination.[6] However, ellipsometric studies with controlled atmosphere and high-purity liquids have undoubtedly evidenced a precursor film,[6] some tens of Å thick, preceding advancing liquid edges. Systematic studies of the length and the profile of such films are developing now.[39-41]

In the field of polymers, a similar study was presented at this conference.[41]

The importance of these experiments is obvious: if <u>the existence and the role of the precursor film</u>[19] <u>are now accepted facts</u>, then the predictions about the film properties are still questionable.

As a matter of fact, thin films can have peculiar <u>structures</u>. Force measurements in PDMS suggest that the molecules in the first molecular layer are <u>lying flat</u> on the surface,[42] and this is strongly supported by recent experimental results.[39] The first layer of molecules should behave as a solid on which the rest of the liquid advances with the bulk viscosity.[42] A modified precursor dynamics would result.

The second point is the <u>molecular diffusion</u> which obviously plays a role at the tip of the precursor:[11] thicknesses below that of PDMS molecules are commonly measured:[39] it means that the ellipsometry gives an averaged value over isolated molecules which are diffusing over the surface, and that the "pancake" is only a transient situation which eventually disappears.[39]

V. CONCLUSION

Taking into account long range forces, de Gennes was able to describe many experimental results in a unifying framework, and to solve the paradoxes occurring in the wetting dynamics.[19]

For practical applications like lubrication and adhesion, a precise knowledge of thin film properties is required now. A systematic study of the precursor film is then of the highest importance to lay the foundation for forthcoming applications, which are expected to be especially fruitful in the field of liquid polymers.

ACKNOWLEDGEMENTS

Fruitful discussions on wetting theories with P.G. de Gennes and J.F. Joanny are gratefully acknowledged.

NOMENCLATURE

a	$\sqrt{A/6\pi\gamma}$
A	Hamaker constant
A_{132}	1 and 2 interacting across 3
γ	surface tension of liquid/vacuum interface
γ_S	surface tension of solid/vacuum (or γ_{SV}) interface
γ_{SL}	surface tension of solid/liquid interface
δ	thickness of a grafted layer
e_f	pancake equilibrium thickness
E_z	surface free energy
F_c	capillary force
g	gravitational acceleration
η	viscosity of liquid
θ_e	metastable equilibrium contact angle
θ	dynamic (instantaneous) contact angle
k^{-1}	capillary length = $\sqrt{\gamma/\rho g}$
ℓ	precursor length
$P(z)$	interaction energy
ρ	density of liquid
R	radius of the wetted spot
S	spreading coefficient
t	time
Ω	volume of the drop
z	film thickness

REFERENCES

1. B.V. Derjaguin and N.V. Churaev, Wetting Films, Nauka, Moscow (1984) in Russian.; B.V. Derjaguin, N.V. Churaev, V.M. Muller, Surface Forces, Consultant Bureau New York and London, 1987; Adv. Colloid and Interface Sci., 28, p.197 (1988) and references.
2. W.A. Zisman, in: Adhesion Science and Technology, L.H. Lee, editor, Vol. 9, p.55, Plenum, New York (1975).
3. A.W. Neumann and R.J. Good, J. Colloid Interface Sci., 38, p.341 (1972).
4. J.F. Oliver and S.G. Mason, J. Mater. Sci., 15, p.431 (1980).
5. W.B. Haines, J. Agric. Sci., 20, p.97 (1930); P.G. de Gennes, in:

Physics of Disordered Materials, D. Adler, editor, p.227, Plenum (1985).
6. R.L. Cottington, C.M. Murphy and C.R. Singleterry, in: Contact Angle Wettability and Adhesion, R.F. Gould, editor, Adv. in Chemistry, Series 43, p.341, American Chemical Society, Washington, D.C. (1964); W.D. Bascom, R.L. Cottington, C.R. Singleterry, same reference, p.355.
7. M.K. Bernett and W.A. Zisman, ref. (6), p.332.
8. E. Bayramli, T.G.M. Van de Ven and S.G. Mason, Can. J. Chem., 59, p.1954 (1981).
9. R.E. Johnson and R.H. Dettre, ref. (6), p.112.; R.H. Dettre and R.E. Johnson, ref. (6), p.136; L.H. Schwartz and S. Garoff, Langmuir, 1, p.219 (1985), see also ref. (4).
10. A.W. Adamson, Physical Chemistry of Surfaces, John Wiley Ed., 4th edition, New York (1982); see also ref. (2) and ref. (6), p.1 (review papers by W.A. Zisman).
11. G.F. Teletzke, thesis, University of Minnesota (1983); G.F. Teletzke, H.T. Davis and L.E. Scriven, to appear in Revue de Physique Appliquée.
12. L.H. Tanner, J. Phys. D, 12, p.1478 (1979).
13. F. Brochard and P.G. de Gennes, J. Phys. Lett., 45, L-597 (1984).
14. F. Bashforth and J.C. Adams, An Attempt to Test the Theories of Capillary Action, University Press, Cambridge, England (1983). see also ref. (10).
15. A. Marmur, Adv. Colloid Interface Sci., 19, p.75 (1983).
16. E. Ruckenstein and P.S. Lee, Surf. Science, 50, p.597 (1975); J. Lopez, C.A. Miller and E. Ruckenstein, J. Colloid Interface Sci., 56, p.460 (1976).
17. E.B. Dussan V., Ann. Rev. Fluid, Mech., 11, p.371 (1979).
18. H.W. Hardy, Philos, Mag., 38, p.49 (1919).
19. P.G. de Gennes, Rev. Mod. Phys., 57, p.827 (1985).
20. J.F. Joanny, thesis, University of Paris (1985).
21. J. Lyklema, in: Colloides et Interfaces, A.M. Cazabat and M. Veyssié, editors, p.1, Les Editions de Physique (1984) in English.
22. J.N. Israelachvili, Intermolecular and Surface Forces, Academic Press, New York (1985).
23. G.F. Teletzke, H.T. Davis and L.E. Scriven, J. Colloid Interface Sci., 87, p.550 (1982); see also ref. (11).
24. G.J. Hirasaki, S.P.E., submitted for publication. For dominating van der Waals forces, the precursor thickness e_c is approximately given by

$$e_c + \delta \approx \frac{1}{2}\sqrt{\frac{A}{\pi S}} \quad \textit{(author's calculation)}.$$

25. J.W. Cahn, J. Chem. Phys., 66, p.3667 (1977); M. Moldover and J.W. Cahn, Science, 207, p.1073 (1980); M. Moldover and R. Gammon, J. Chem. Phys., 80, p.528 (1983).
26. J.F. Joanny and P.G. de Gennes, J. Phys. (Paris), 47, p.121 (1986). see also ref. (11).
27. P.G. de Gennes, C.R. Acad. Sci. B, 288, p.219 (1979).
28. P.G. de Gennes, Scaling Concepts in Polymer Physics, 2nd Edition Cornell University Press, Ithaca, N.Y., (1985).
29. A.M. Cazabat and M.A. Cohen Stuart, J. Phys. Chem., 90, p.5845 (1987).
30. D. Ausserré, A.M. Picard and L. Léger, Phys. Rev. Lett., 57, p.2671 (1986).
31. P. Levinson, A.M. Cazabat, M.A. Cohen Stuart, F. Heslot and S. Nicolet, to appear in Revue de Physique Appliquée.
32. P. Carles and A.M. Cazabat, submitted.

33. G. Sawicki, in: Wetting, Spreading and Adhesion, J. Padday, editor, p.361, Acad. Press,(1978).
34. J.F. Oliver and S.G. Mason, J. Colloid Interface Sci., 60, p.480 (1977).
35. D. Pohl, private communication.
36. M. Erman, thesis, University of Orsay, France (1986).
37. E. Drévillon, private communication.
38. F. Meyer and C.J. Loyen, Acta Electronica, 18, p.33 (1975).
39. F. Heslot, A.M. Cazabat and P. Levinson, submitted.
40. D. Beaglehole, private communication; S. Garoff, private communication.
41. L. Léger, A.M. Picard, H. Hervet, D. Ausserré, M. Erman, F. Brochard and J.F. Joanny, this volume.
42. H.K. Christenson, D.W.R. Gruen, R.G. Horn and J.N. Israelachvili, J. Chem. Phys., 87, p.1834 (1987) and R.G. Horn, J.N. Israelachvili, Macromolecules 21, 2836 (1988).

APPLICATIONS OF HARD-SOFT ACID-BASE (HSAB) PRINCIPLE TO SOLID ADHESION AND SURFACE INTERACTIONS BETWEEN METALS AND POLYMERS

Lieng-Huang Lee

Webster Research Center
Xerox Corporation
Webster, New York 14580

ABSTRACT

We attempt to extend the Hard-Soft Acid-Base (*HSAB*) principle for the reactions in solutions to interactions in solids. First we point out the important link between the absolute hardness of acid-and-base and the average energy gap. Then we discuss the electronic band structures of various solids, e.g., metals, semimetals, semiconductors and insulators. On the basis of energy gaps, we elaborate various consequences of the acid-base interactions in solids. The applications of *HSAB* principle and the frontier orbital concept to the solid adhesion and surface interactions between metals and polymers will be verified by experimental results reported in the literature. The new findings reported in this paper should be beneficial to those who are carrying out research in or processing thin-film microelectronic devices or thick-film multilayer structures.

I. INTRODUCTION

The donor-acceptor interaction[1,2] and the acid-base interaction[3] have been reviewed. In many occasions, the two terms, though different, have been used interchangeably to describe the interactions involving the exchange of electrons between a donor and an acceptor. For polymer adhesion, Fowkes[4,5] and Bolger et al.[6] have pointed out the important role of the acid-base interaction in the formation of an adhesive bond.

For the acid-base interaction in solutions, in 1963, Pearson[7] proposed the hard-soft acid-base (*HSAB*) principle to describe some basic rules about the kinetics and equilibrium of the reaction. In this paper, we attempt to apply the *HSAB* principle to solid interactions with the aid of the frontier orbital method. We shall first describe the *HSAB* principle as it has been evolved in recent years[8-11] and then the band structures of solids. After we demonstrate the compatibility between the *HSAB* principle and the band structures in the solid state, we then illustrate with several examples of adhesion and tribointeractions between metals and polymers. We are hoping that, in so doing, we can further understand the dynamics and energetics ofollowing tribointeractions at the molecular level.

II. HSAB PRINCIPLE

II.1 HSAB PRINCIPLE FOR INORGANIC REACTIONS

In an inorganic reaction, the generalized Lewis acid-base interaction can be represented:

$$\underset{\text{(Acid)}}{A} + \underset{\text{(Base)}}{B} \rightleftharpoons \underset{\text{(AB Complex)}}{A{:}B} \quad (1)$$

Pearson[7] has suggested for the above interaction to classify acids and bases according to the absolute hardness, which will be defined later:

- A hard acid contains an acceptor atom of high positive charge and small size. It does not have easily excitable outer electrons and is not polarizable.
- A soft acid contains an acceptor atom of low positive charge and large size. It has several easily excitable outer electrons and is polarizable.
- A hard base contains a donor atom of low polarizability that is hard to reduce. It is associated with empty orbitals of high energy and hence inaccessible.
- A soft base contains a donor atom of high polarizability and low electronegativity. It is easily oxidized and associated with unoccupied low-lying orbitals.

There are two rules governing the acid-base interactions in solutions:

1. About equilibrium: Hard acids prefer to coordinate with hard bases, and soft acids to soft bases.
2. About kinetics: Hard acids react readily with hard bases, and soft acids with soft bases.

The above principle has been well tested in inorganic reactions.[11]

II.2 HSAB PRINCIPLE FOR ORGANIC REACTIONS AND THE FRONTIER ORBITAL APPROACH

For an organic reaction, the *HSAB* principle can also be applied for the study of kinetics and equilibrium, and the frontier orbital method can be used to illustrate the electrophilic and nucleophilic interactions[12] (Fig. 1):

- A hard electrophile (or acid) has a high-energy *LUMO* (lowest unoccupied molecular orbital) and usually has a positive charge.
- A soft electrophile (or acid) has a low-energy *LUMO* but does not necessarily have a positive charge.
- A hard nucleophile (or base) has a low-energy *HOMO* (highest occupied molecular orbital) and usually has a negative charge.

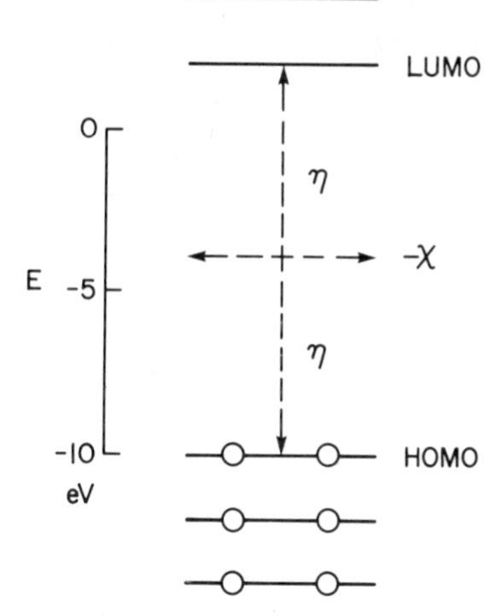

Fig. 1. Orbital energy diagram for a molecule. *HOMO*, highest occupied *MO*; *LUMO*, lowest unoccupied *MO*. (After R.G. Pearson, J. Chem. Educ. 64 (7), 563 (1987), reprinted with permission of the author).

- A soft nucleophile (or base) has a high-energy *HOMO* but does not necessarily have a negative charge.

Furthermore, according to Koopmans' theorem,[13] the frontier orbital energies are given by

$$-E_{HOMO} = I; -E_{LUMO} = A \quad . \tag{2}$$

where I is the ionization potential, and A is the electron affinity.

II.3 PERTURBATION EQUATION

Hudson and Klopman[14] proposed an equation to describe the effect of orbital perturbation of two molecules on chemical reactivity. Their equation for the interaction energy ΔE can be simplified by using the *HOMO* of a nucleophile and the *LUMO* of an electrophile:

$$\Delta E = -\frac{Q_N Q_E}{\varepsilon R^2} + \frac{2(C_N C_E \beta)^2}{E_{HOMO} - E_{LUMO}} , \tag{3}$$

The Coulombic term   The frontier orbital term

where Q_N, Q_E are the total charges for the nucleophile and electrophile, respectively; C_N and C_E are the coefficients of the atomic orbitals N and E, respectively, in the molecular orbital; β is the resonance integral; ε is the permittivity and R is the distance between N and E. E_{HOMO} is E_V, the energy of the valence band edge and E_{LUMO} is E_C, the energy of the conduction band edge.

On the basis of the above equation, the *HSAB* rules can be restated as follows;

- A hard-hard interaction (or reaction) is fast because of a large Coulombic attraction as described by the first term of Eq. (2).

- A soft-soft interaction (or reaction) is fast because of a large orbital interaction between the *HOMO* of the nucleophile and the *LUMO* of the electrophile as described by the second term of Eq. (2).

II. ELECTRONEGATIVITY AND ABSOLUTE HARDNESS

In the beginning when the *HSAB* principle was just introduced, the meaning of hardness was not easy to understand. Certainly, it is not the hardness that measures the resistance against deformation. Then, what is it? It was not until the last several years that the real definition of absolute hardness[8,9] received a quantitative and theoretical backing.[10]

First, let us go back one step to define the absolute electronegativity, χ. According to Mulliken,[15] for atomic species, the electronegativity

$$\chi = 1/2(I + A) = -\mu \ , \tag{4}$$

where μ is the chemical potential. However, this relation is only applicable to a single atom, but not to a molecule. If the total electronic energy of an atom (or molecule) is plotted against the total number of electrons, a curve similar to that in Fig. 2 is obtained. The first ionization potential for the species will be much larger than the

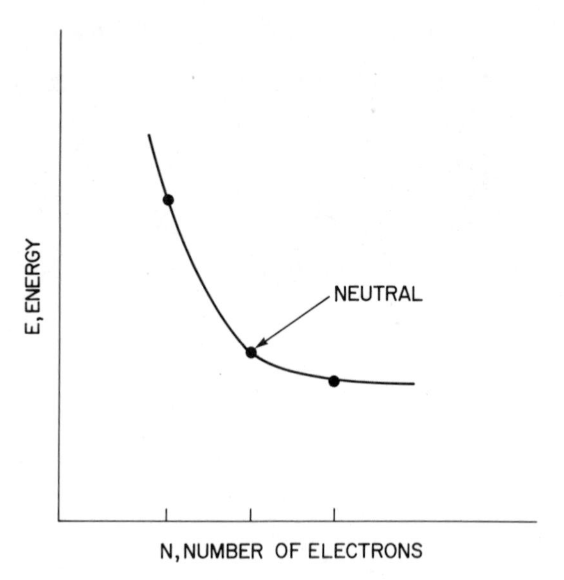

Fig. 2. Plot of electronic energy vs. number of electrons for a fixed collection of nuclei (After R.G. Pearson, J. Chem. Educ. 64 (7), 563 (1987), reprinted with permission of the author).

electron affinity and much smaller than the second ionization potential. Assuming a differential curve, the electronegativity may be defined as

$$\chi = -(\partial E/\partial N)_Z , \tag{5}$$

where Z is the total number of nuclear charges in the atom (or molecule).

The second derivative of the same is defined to be absolute (or chemical) hardness, η, the quantity:

$$\eta = 1/2(\partial^2 E/\partial N^2)_Z . \tag{6}$$

The corresponding operation definition is the corresponding finite difference formula

$$\eta = 1/2(I - A) . \tag{7}$$

Hence, the absolute hardness η is the resistance of the chemical potential to change in the number of electrons. The opposite of η is the absolute softness, σ.

According to the frontier orbital method,[16] the relationship of η with the energies of *LUMO* and *HOMO* is shown in Fig. 1. Thus, from Fig. 1,

$$\eta = -1/2(E_{HOMO} - E_{LUMO}) . \tag{8}$$

Here χ is the horizontal broken line in the middle of the energy gap. Hence, the energy gap E_g is twice the absolute hardness η. This is an important link between chemistry and physics in the solid state.

III EXTENSION OF THE HSAB PRINCIPLE TO SOLID INTERACTIONS

In the literature, we have not found any formal application or extension of the *HSAB* principle to solid interactions. In this paper, we shall demonstrate that the extension of the *HSAB* principle to solid interactions is feasible in view of the electronic band structures of solids. We shall discuss the physical meaning of the absolute hardness in terms of the average energy gap.

III.1 ELECTRONIC BAND STRUCTURES OF SOLIDS

Solids can be classified as metals, semimetals, intrinsic semiconductors and insulators. The band structures of solids can be illustrated in Fig. 3. Monovalent metals,[17] e.g., Na°, have a partially filled valence band, the lower half of which is occupied. The Fermi level is in the valence band but at the top of the occupied orbitals. Furthermore, there is still an energy gap between the valence band and the conduction band (unoccupied *MO*). In some metals, such as the bivalent metals, the valence band is full but overlaps a higher unoccupied conduction band. In this case, the Fermi level is in the conduction band and the overlapped valence band. Thus, the electrons close to the Fermi level are still free to move as the extra bands supply the unoccupied states. In the latter case, there appears to be no minimum energy gap, $E_g°$, which is generally reported in the literature. However, it is not

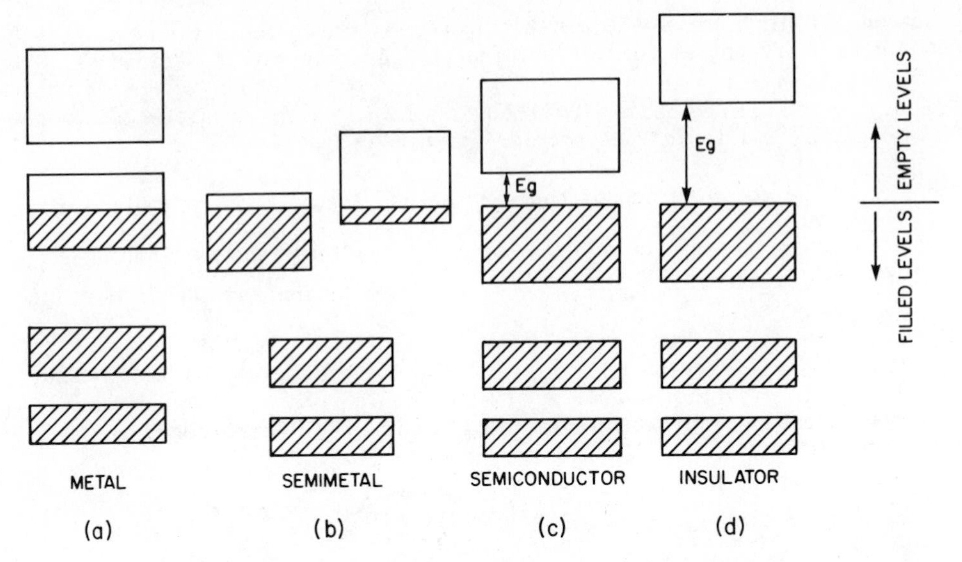

Fig. 3. Band structures of solids: (a) metal, (b) semimetal, (c) semiconductor and (d) insulator (Ref. A.J. Epstein and J.S.Miller, Sci. Amer. 241 (4) 52 (1979)).

obvious that for each metal atom there will always be an average energy gap E_g^{Av}, which will be discussed later.

In addition to metals, there are semimetals,[18] such as graphite, whose valence band and conduction band can overlap. In general, their minimum energy gaps are very narrow. The third class of solids is the intrinsic semiconductor; its minimum energy gap $E_g°$ is generally below 3 eV. Thus, thermal excitation alone can create an electron-hole pair to enhance conduction. The Fermi level of the intrinsic semiconductor[17] lies between the valence band (*HOMO*) and the conduction band (*LUMO*). Hence,

$$E_F = 1/2(E_{HOMO} + E_{LUMO}) = 1/2(E_c + E_v) , \tag{9}$$

where E_F is the energy of the Fermi level. The fourth class of solids is the insulator, which includes most polymers. The minimum energy gap of an insulator is generally above 3 eV. Therefore, thermal excitation alone can not enhance the conduction of electricity.

III.2 AVERAGE ENERGY GAP AND ABSOLUTE HARDNESS

The energy gaps generally reported in the literature are minimum energy gaps, $E_g°$. On the other hand, the average energy gap E_g^{Av} [19,20] is defined as the difference in the energy between the conduction and valence bands.

$$E_g^{Av} = -(E_{HOMO} - E_{LUMO}) = -(E_v - E_c)\,. \tag{10}$$

According to Phillips,[19] E_g^{Av} should contain both covalent (homopolar) component E_h and ionic (heteropolar) component C:

$$(E_g^{Av})^2 = (E_h)^2 + (C)^2\,. \tag{11}$$

For many $A^N B^{8-N}$ compounds, E_g^{Av} can be determined from dielectric measurements.

For comparison, both $E_g°$ and E_g^{Av} for several elements are listed in Table 1. In general, E_g^{Av} is always larger than $E_g°$. From Table 1, the η values calculated from E_g^{Av} are rather close to those calculated from the A-I data; however, the agreement is still not good enough.

Table 1

Minimum and Average Energy Gaps and Absolute Hardness Values for Atoms (eV)

	$E_g°$ (1)	E_g^{Av} (1)	η (2) (cal'ed)	η (3)
C	5.4	13.6	6.8	5.0
Si	1.1	4.8	2.4	3.4
Ge	0.7	4.3	2.2	3.4
Sn	0	3.1	1.6	3.1

Note: 1. $E_g°$ and E_g^{Av} from Burstein, Pinczuk and Wallis (Ref. 20).
2. η^{Av} is calculated from E_g^{Av}.
3. η is the value obtained by Pearson (Ref. 9).

We also attempt to calculate E_g^{Av} from η for several transition metal atoms (Table 2). The η values that fall between 3 to 4 indicate that most of these metal atoms are rather "soft." Since there are not many E_g^{Av} values in the literature for comparison, we assume that they are reasonably good. We shall discuss the applications of the *HSAB* principle based on these η values. In the meantime, it is important to note that bulk metals are different from metal atoms. The former are generally softer than the latter.

Table 2

Average Energy Gap and Hardness Values for Transition Metal Atoms (eV)

Metal Atom	$I^{(1)}$	$A^{(2)}$	χ	$\eta^{(3)}$	$E_g^{Av(4)}$ (cal'ed)
Ti	6.82	0.08	3.45	3.4	6.8
V	6.7	0.5	3.6	3.1	6.2
Cr	6.77	0.66	3.72	3.1	6.2
Mn	7.44	0.	3.72	3.7	7.4
Fe	7.87	0.25	4.06	3.8	7.6
Co	7.8	0.7	4.3	3.6	7.2
Ni	7.64	1.15	4.4	3.3	6.6
Cu	7.73	1.23	4.48	3.3	6.6
Mo	7.10	0.75	3.9	3.2	6.4
Ru	7.40	1.5	4.5	3.0	6.0
Pd	8.34	0.56	4.45	3.9	7.8
Ag	7.58	1.30	4.44	3.1	6.2
Pt	9.0	2.1	5.6	3.5	7.0
Au	9.23	2.31	5.77	3.5	7.0

Note: 1. C.E. Moore, Natl. Stand. Ref. Data. Ser. NSRDS-NBS 34, 1970.
2. H. Hotop and W.C. Lineberger, J. Phys. Chem. Ref. Data, 14, 731 (1985).
3. R.G. Pearson, Inorg. Chem. 27, 734 (1988).
4. E_g^{Av} is calculated from η obtained from A-I values.

III.3 ABSOLUTE HARDNESS OF SOLID

Now let us examine again the band structures of solids in Fig. 3. In Fig. 3a, the structure of metals with zero or low E_g signifies that all metals are "soft." However, they do not appear to be as "soft" as the η values calculated from the *A-I* data (Table 2). Indeed, metals have been classified as amphoteric materials. Most of the metals are "soft" acids, and some of them "soft" bases. When two metals are brought together into close contact, one of them assumes to be an acid while the other a base. In a recent paper by Cain et al.,[21] the interaction at the Cu/Cr interface has been treated as an acid-base interaction. In this case, a soft base (Cu) and a soft acid (Cr) reacts preferentially. In fact, Cr (η = 3.1 eV) is very slightly softer than Cu (η = 3.3 eV) (See Table 2).

Figures 3b and 3c show that both semimetals and semiconductors should be rather "soft" because of relatively low E_g^o and E_g^{Av} as well (most of η's < 3.0 eV). Semiconductors, generally, react readily with metals even at ambient temperatures.[22] Thus, the interaction takes place as a "soft" acid (metal) with a "soft" base (semiconductor). Both chemical reaction and inter-diffusion,[23] jointly create a diffused "interphase" instead of an abrupt "interface."

In Fig. 3d, it is noted that the band structure of an insulator contains a large gap. Thus, most insulators, regardless of organic or inorganic, are considered to be rather "hard" on the basis of the energy gap alone. Generally, they are harder than metals, semimetals and semiconductors.

It is important to remember that the hardness of one material with respect to the other depends on its η. The one with a lower η value is "softer" than the one with a higher η though they both can be "hard." Even in the same material, one functional group can be harder than the other. In this case, a new term, local hardness, $\tilde{\eta}$,[10b] has been introduced to signify the relative hardness of each functional group.

IV. APPLICATIONS OF THE HSAB PRINCIPLE TO ADHESION AND SURFACE INTERACTIONS BETWEEN METALS AND POLYMERS

The interactions between metals and polymers are good examples of the applications of the *HSAB* principle to solids. When Cr is deposited on pyromellitic dianhydride oxydianiline polyimide (PMDA-ODA *PI*), there appears to be some chemical bonding between Cr and *PI*. In this case, Cr is a "soft" acid, and *PI* is "hard" base. How can the bonding take place? According to Ho et al.,[24,25] indeed the reaction does not proceed in the manner as the Cr atom attacking one of the carbonyl groups (nucleophiles). What really happens is that the Cr atom delocalizes and forms a charge transfer (or acid-base) complex with the *PMDA* unit of the *PI* as shown in Fig. 4. In the frontier orbital terminology, the stabilization of the complex is achieved through the transfer of electronic charge from the metal *d*-states of Cr (*HOMO*) to the *LUMO* of the π system of the *PMDA* unit of *PI*. In this manner, *PI* acts as a "soft" acid by accepting the electron and Cr (η = 3.1 eV) becomes a "soft" base by donating the electron. Other transition metals, e.g., Ti (η = 3.4 eV),[26] are also very reactive with *PI*. The general reaction presumably follows the same path of the acid-base interaction through a charge-transfer complex, which should involve the polarization and the partial transfer of electrons.

One of the exceptions is Cu,[27] which does not interact with *PI* after the deposition, and gives rise to a much weaker complex. Recently, Cain and Matienzo[28] have used the tight binding calculations of the extended Hückel type to find the relative acid strengths of Cu compounds in the following decreasing order: $CuF_2 > CuO > CuF \approx Cu_2O > Cu$. Indeed, the Cu ions react faster than Cu with functional polymers. The Cu^{2+} ion has a Fermi level lying below the top of the Cu 3*d* band, and some of the Cu 3*d* orbitals are unoccupied and able to accept electron pairs by acting as an acid. In the *HSAB* terminology, Cu becomes a "hard" acid in the form of Cu^{2+} ion because the hardness for Cu is 3.3 eV, and that for Cu^{2+}, 8.3 eV.

Generally, many transition metals react well with polymers, ceramics,[29] etc., partly because of the ease of oxidation and partly because of the availability of *d*-orbital electrons. Buckley[30] have found that metals reacted with *PTFE* and *PI* during his pin-and-disc experiments. With *PTFE*, the adhesive forces are three times the applied load. Now, we can also attribute these reactions to the *HSAB* interactions. In the literature, there should be numerous examples of the surface interactions that can be classified as the acid-base interaction. However, we are unable to discuss them all in this paper.

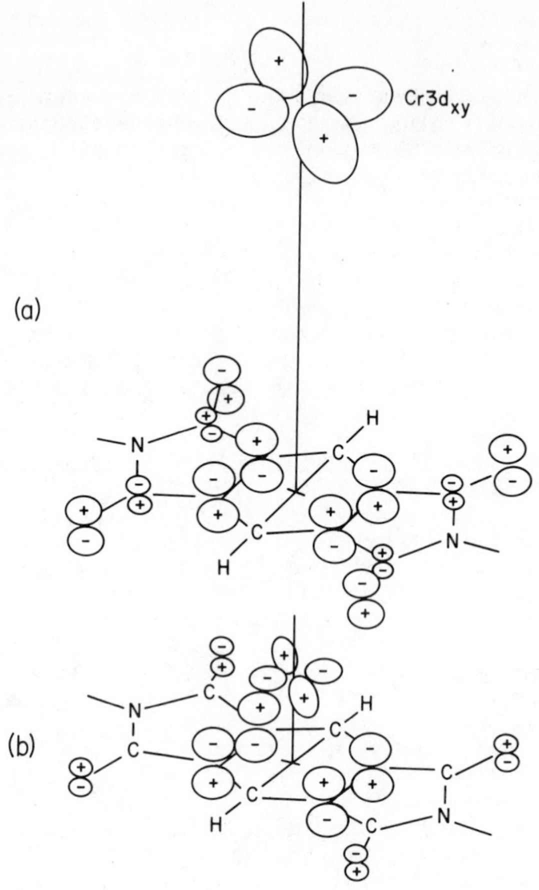

Fig. 4. Scheme of Cr-polyimide complex formation (a) *LUMO* of *PMDA* fragment and Cr 3 d_{xy} orbital (occupied), when Cr is at a distance; and (b) Cr over the 6-member central ring of the *PMDA* (After P.S. Ho, et al., IBM Res. Develop. 32 (5), 658 (1988), reprinted with permission of the authors).

V. CONCLUSIONS

The extension of the *HSAB* principle and the frontier orbital concept to solid interactions have been demonstrated. For the *HSAB* principle, the absolute hardness was discussed alongside with the average energy gaps in solids. The mechanisms of interactions between different types of solids were elaborated in light of the *HSAB* principle. Both the *HSAB* principle and the frontier orbital concept were found to be beneficial in explaining the mechanisms for several metal-polymer adhesion and surface interactions encountered in tribological environments. In practice, this new finding should help those in carrying out research in or processing thin-film microelectronic devices and thick-film multilayer structures.

ACKNOWLEDGMENTS

The author would like to thank Prof. R.G. Pearson of the University of California at Santa Barbara for his comments and for pointing out one paper by W. Yang and R.G. Parr on metals and catalysts (Proc. Natl. Acad. Sci. U.S.A. 82, 6723, 1985), Prof. R.J. Good of the State University of New York at Buffalo and Dr. L.J. Brillson of our Center for comments on the original manuscript.

NOMENCLATURE

A	electron affinity
C	ionic (heteropolar) component of the average energy gap
C_E	coefficient of the atomic orbital of the electrophile
C_N	coefficient of the atomic orbital of nucleophile
ΔE	interaction energy
E	electronic energy
E_c	energy of the conduction band edge
F_F	energy of the Fermi level
$E_g{}^{\circ}$	minimum energy gap
$E_g{}^{Av}$	average energy gap
E_h	covalent (homopolar) component of the average energy gap
E_{HOMO}	energy of the highest occupied molecular orbital
E_{LUMO}	energy of the lowest unoccupied molecular orbital
E_V	energy of the valence band edge
I	ionization potential
N	number of electrons for a fixed collection of nuclei
Q	total electron charge
R	distance between the electrophile and nucleophile
Z	total number of nuclear charge in the atom
β	resonance integral
ε	permittivity
η	absolute hardness
$\tilde{\eta}$	local hardness
μ	chemical potential
σ	absolute softness
χ	electronegativity (Mulliken)

REFERENCES

1. V. Gutman, The Donor-Acceptor Approach to Molecular Interactions, Plenum Press, New York, 1978.
2. B.V. Deryagin, N.A. Krotova and V.P. Smilga, Adhesion of Solids, Engl. Ed. Translated by R.K. Johnson, Plenum, New York, 1978.
3. W.B. Jensen, Chem. Rev., 78 (1), 1 (1978).
4. F.M. Fowkes, J. Adhes., 4, 155 (1972), also in Recent Advances in Adhesion, L.H. Lee, Ed. Gordon and Breach, New York and London (1973), p.39.
5. F.M. Fowkes, J. Adhes. Sci. Technol., 1 (1) 7 (1987).
6. J.C. Bolger and A.S. Michaels, in Interface Conversion, P. Weiss and D. Cheevers, Eds. Chapt. 1, Elsevier, New York (1969).
7. R.G. Pearson, J. Am. Chem. Soc., 85, 3533 (1963).
8. R.G. Pearson, J. Chem. Educ., 64, (7), 563 (1987).
9. R.G. Pearson, Inorg. Chem., 27, 734 (1988), also J. Am. Chem. Soc., 110, 7684 (1988).

10(a). R.G. Parr and R.G. Pearson, J. Am. Chem. Soc., 105, 7512 (1983).
10(b). M. Berkowitz, S.K. Ghosh and R.G. Parr, J. Am. Chem. Soc., 107, 6811 (1985).
10(c). M. Berkowitz and R.G. Parr, J. Chem. Phys., 88 (4), 2554 (1988).

11. R.G. Pearson, Hard and Soft Acids and Bases, Dowden, Hutchinson and Ross, Stroudsburg, PA, 1973.
12. N.S. Isaacs, Physical Organic Chemistry, Longman Scientific and Technical, with Wiley, New York (1987).
13. T. Koopmans, Physica, 1, 104 (1933).
14. R.F. Hudson and G. Klopman, Tetrahedron Lett., 12, 1103 (1967); also Theo. Chim. Acta, 8, 165 (1967).
15. R.S. Mulliken, J. Chem. Phys, 3, 573, (1935), also 586 (1935).
16. K. Fukui, Science, 218 (4574), 747 (1982).
17. F. Gutmann and L.E. Lyons, Organic Semiconductor, Wiley, New York (1967), p.17.

18. D.L. Carter and R.T. Bate, Eds. The Physics of Semimetals and Narrow-Gap Semiconductors, Pergamon Press, Oxford (1971).
19. J.C. Phillips, Rev. Mod. Phys. 42 (3), 317 (1970).
20. E. Burstein, A. Pinczuk and R.F. Wallis, The Physics of Semimetals and Narrow Gap Semiconductors, D.L. Carter and R.T. Bate, Eds. Pergamon Press, Oxford (1971), p.251.
21. S.R. Cain, L.J. Matienzo and F. Emmi, J. Phys. Chem. Solids, 50, (1), 87 (1989).
22. L. Braicovich, The Chemical Physics of Solid Surfaces and Heterogen-eons Catalysts, Ed. D.A. King and D.P. Woodruff, Chapt. 6, Elsvier, Amsterdam (1988).
23. L.J. Brillson, Surf. Sci. Rep., 2, 123 (1982).
24. P.S. Ho, B.D. Silverman, R.A. Haight, R.C. White, P.N. Sanda and A.R. Rossi, IBM J. Res. Develop, 32, (5), 658 (1988).
25. P.S. Ho, R. Haight, R.C. White, B.D. Silverman and F. Faupel, in Fundamentals of Adhesion, L.H. Lee, Ed. Plenum, to be published in 1990.
26. F.S. Ohuchi and S.C. Freilich, J. Vac. Sci. Technol., A4 (3), 1039 (1986).
27. R. Haight, R.C. White, B.D. Silverman and P.S. Ho, J. Vac. Sci. Technol., A6 (4) 2188 (1988).
28. S.R. Cain and L.J. Matienzo, J. Adhes. Sci. Technol., 2, (5) 395 (1988).
29. D.H. Buckley, Surface Effects in Adhesion, Friction, Wear and Lubri-cation, Elsevier, Amsterdam (1981).
30. D.H. Buckley and W.A. Brainard in Advances in Polymer Friction and Wear, L.H. Lee, Ed. Vol A, Plenum, New York (1975), p.315.

APPENDIX

HARD AND SOFT ACIDS AND BASES (FOR SOLUTIONS)

REF. R.G. PEARSON J. Am. Chem. Soc. 85, 3533 (1963)

TABLE 1

Classification of Lewis Acids

HARD	SOFT
H^+, Li^+, Na^+, K^+	Cu^+, Ag^+, Au^+, Tl^+, Hg^+, CS^+
Be^{2+}, Mg^{2+}, Ca^{2+}, Sr^{2+}, Mn^{2+}	Pd^{2+}, Cd^{2+}, Pt^{2+}, Hg^{2+} CH_3Hg^+, $Co(CN)_5{}^{2-}$, Pt^{4+}, Te^{4+}
Al^{3+}, Sc^{3+}, Ga^{3+}, In^{3+}, La^{3+}, Gd^{3+}, Lu^{3+}	Tl^{3+}, $Tl(CH_3)_3$, BH_3, $Ga(CH_3)_3$
Cr^{3+}, Co^{3+}, Fe^{3+}, As^{3+},	$GaCl_3$, GaI_3, $InCl_3$,
Si^{4+}, Ti^{4+}, Zr^{4+}, Th^{4+}, U^{4+}, Pu^{4+}, Ce^{3+}, Hf^{4+}	RS^+, RSe^+, RTe^+
$UO_2{}^{2+}$, $(CH_3)_2Sn^{2+}$, VO^{2+} MoO^{3+}	I^+, Br^+, HO^+, RO^+
$BeMe_2$, BF_3, BCl_3, $B(OR)_3$	I_2, Br_2, ICN, etc.
$Al(CH_3)_3$, $AlCL_3$, AlH_3,	Trinitrobenzene, etc.
$RPO_2{}^+$, $ROPO_2{}^+$,	Chloranil, quinones, etc.
$RSO_2{}^+$, $ROSO_2{}^+$, SO_3	Tetracyanoethylene, etc.
I^{7+}, I^{5+}, Cl^{7+}, Cr^{6+}	O, Cl, Br, I, N
RCO^+, CO_2, NC^+	M° (metal atoms)
HX (hydrogen bonding molecules)	Bulk metals
	CH_2, carbenes

Borderline
Fe^{2+}, Co^{2+}, Ni^{2+}, Cu^{2+}, Zn^{2+}, Pb^{2+}, Sn^{2+}, Sb^{3+}, Bi^{3+}, Rh^{3+}, Ir^{3+}, $B(CH_3)_3$, SO_2, NO^+, Ru^{2+}, Os^{2+}, R_3C^+, $C_6H_5{}^+$, GaH_3

TABLE 2

Classification of Lewis Bases
The symbol R stands for an alkyl group such as CH_3 or C_2H_5

HARD	SOFT
HO_2, HO^-, F^-	R_2S, RSH, RS^-
$CH_3CO_2{}^-$, $PO_4{}^{3-}$, $SO_4{}^{2-}$	I^-, SCN^-, $S_2O_3{}^{2-}$
Cl^-, $CO_3{}^{2-}$, $ClO_4{}^-$, $NO_3{}^-$	R_3P, R_3As, $(RO)_3P$
ROH, RO^-, R_2O	C_2H_4, C_6H_6
NH_3, RNH_2, N_2H_4	H^{2-}, R^-

Borderline
$C_6H_5NH_2$, C_5H_5N, $N_3{}^-$, Br^-, $NO_2{}^-$, $SO_3{}^{2-}$

Discussion

On the First Paper by L.H. Lee

A. Silberberg: (Weizmann Institute, Israel): In terms of the strength of adhesion, what is more important, the equilibrium bond energy or the activation energy?

L.H. Lee (Xerox Corp.): For the bond-formation, the thermodynamics of adhesion is the most important. This is an equilibrium property which will be a factor determining the final strength for the bond-separation. For example, the peel strength has been shown by Gent and Schultz (A.N. Gent and J. Schultz in Recent Advances in Adhesion, L.H. Lee, Ed., Gordon and Breach, New York and London (1973), p.253). to be the product of the equilibrium work of detachment given by the thermodynamic considerations and a numerical factor representing the enhancement of strength when the adhesive is perfectly elastic.

For the bond-separation, the process is similar to an absolute rate process with an activation energy. Then the activation energy is a determining factor. It is assumed that the bonding and debonding are rate processes and do not occur instantaneously to result automatically in a time-dependent failure. The general implications of the absolute rate theory are that 1) loads below a critical force will never produce failure, 2) the thermodynamic work of adhesion will have a profound effect on the time to breakage for a given load, although it represents only a small fraction of the energy expanded, and 3) the free energy of activation for viscoelastic flow of the adhesive also has a profound effect on the time to breakage for a given load. (Ref. M.R. Hatfield and G.B. Rathmann, J. Phys. Chem. 60, 957, 1956).

On the Paper by R.P. Wool

P.G. de Gennes (Collège de France): Just a short remark on the nonfractal structure of the diffusion front at times t shorter than the reptation time, T_{REP}. I once had to consider the kinetics of termination reactions in entangled melts and found that each chain end performed a certain type of compact exploration at $t < T_{REP}$. Do you agree that compact exploration and nonfractal fronts are two facets of the same physical picture?

R.P. Wool (University of Illinois): Yes. The anisotropic motion of monomers in a melt permits the chain to relax its configuration first at its ends. At $t < T_{REP}$, minor chains of length $\ell(t) \sim t^{1/2}M^{-1/2}$ escape from each end as shown in Figure 1(a), leaving a legth $L-2\ell(t)$ in the tube. When viewed by computer graphics, the minor chains with Brownian motion appear as balanced expanding/contracting spheres centered at the ends of the unrelaxed portion of the tube and the chain ends explore a region of radius $X \sim \sqrt{\ell(t)}$. At an interface, the minor chains diffuse while the fraction of chains remaining in the tube causes the discontinuity in the concentration profile shown in Figure 2. Thus, at $t < T_{REP}$, the diffusion front appears as a region with expanding/contracting (minor chain) semispheres of varying radii randomly distributed on a plane (Figure 1) and is nonfractal due to the lack of self-similarity. At $t >> T_{REP}$, the chain end exploration is noncompact and the diffusion front is fractal.

Michael Rubenstein (Eastman Kodak Company): Can one use the reptation diffusion coefficient $D \sim M^{-2}$ to describe polymer motion at times much shorter than the relaxation time, T_{REP}? At very short times, different sections of the chains do not communicate with each other. At times $t < T_{REP}$, other modes may be important.

R.P. Wool: If we consider a chain in the melt as a Rouse chain in a tube, then the following comments can be made. For times $T_{REP} > t > T_{RO}$, where $T_{RO} \sim M^2$ is the Rouse relaxation time, the monomer diffusion behaves as $<X^2> \sim t/M$. However, since the chain ends are free to move randomly in an uncorrelated manner, the center of mass motion is uncoupled from the monomer motion and the self-diffusion coefficient behaves as $D \sim M^2$. This point was nicely made by Sam Edwards, et al. For a chain at an interface, the situation is more complicated. At $t = 0$, the chain has a nonequilibrium configuration due to the reflecting boundary condition imposed by the surface. As the minor chains diffuse, the non-Gaussian configuration relaxes but very little change occurs in the center of mass position of the whole chain and the use of $D \sim M^2$ is not accurate at $t < T_{REP}$ for chains whose center-of-mass was a distance less than the radius of gyration from the surface. For $t < T_{RO}$ we need to consider higher order Rouse modes and fluctuations. Referring to Sam Edwards' analysis for the effect of Rouse modes at $t < T_{RO}$, the effect of chain-end fluctuations at $t << T_{REP}$ is to produce an initial small amount of minor chain diffusion to contour lengths less than the entanglement spacing. Due to the expanding-contracting nature of the minor chains, the contributions from chain-end fluctuation would be averaged to zero at $t \sim T_{RO}$.

On the Paper by F. Brochard

A. Silberberg (Weizmann Institute): Would you not expect that the fractal dimensions of the particles to be separated is not of importance for the separation in a porous media?

F. Brochart-Wyart (P & M Curie University): The thickness of the depletion layer is controlled by the largest site. If the macromolecule, whatever its fractal dimension (for rods (D_p = 1) as wide as for globular proteins (D_p = 3))= the variable yardstick is then $\ell_s = R$ whatever D_p. However, the concentration profile depends upon D_p and thus do lead to small differences in numerical coefficients. On the other hand, for the case of adsorption, $\ell_s = R$ only if $D_p > D_f$, the fractal dimension of the interface.

On the Paper by L. Léger, A.M. Guinet-Picard, H. Hervet, D. Aussere and M. Erman

R.P. Wool (University of Illinois, Urbana): Optical properties of thin layers (ca 10 Å) could become anisotropic, perhaps revealing interesting details of conformational properties of adsorbed molecules on substrate and structure of film? Have you observed/considered this aspect?

Liliane Léger (College de France): We have been looking for such structure of the very thin films, which should lead to points following different ellipsometric trajectories in the $\tan\Psi$, $\cos\Delta$ plane, than the points for thicker films. Up to now, we have not been able to put anything clearly into evidence.

T.P. Russel (IBM Almaden Research Center): I have three questions. First, what was the density that you used to calculate the reflectivity profiles for the 8 Å films? Second, what was the roughness of the 8 Å films? Third, would you comment on the waviness, which is different from the roughness, that you used to calculate the reflectivity profiles?

L. Léger:

1) In the case of the 8 Å "pancake", the electronic density that we used to analyze the X-ray reflectivity data was very close to that of the bulk polymer, slightly smaller.

2) The residual roughness of the 8 Å pancake was only 3 Å while the substrate roughness was 6 to 7 Å.

3) The roughness determined in the present X-ray reflectivity measurements is an averaged, integrated quantity - Up to now we can not say anything on the wavelength range of this roughness. Scattering experiments at varying wavevectors would be necessary to answer this point.

PART THREE:

DYNAMICS AND CHARACTERIZATION OF POLYMER SOLUTIONS

DIFFUSION OF POLYMER IN BINARY AND TERNARY SEMIDILUTE SOLUTIONS

Zhulun Wang and Benjamin Chu*

Chemistry Department
State University of New York
at Stony Brook
Long Island, New York 11794-3400

Qinwei Wang

Chemistry Department
Peking University
Beijing,
People's Republic of China

Lewis Fetters
Exxon Research & Engineering Co.
Annandale, New Jersey 08801

ABSTRACT

Dynamic properties of PS/TOL, PS1/PS2/TOL and PS1/PMMA2/TOL semidilute solutions (in which PS, PMMA and TOL denote polystyrene, poly(methyl methacrylate), and toluene, respectively; "1" denotes the matrix polymer and "2", the probe polymer) have been studied systematically by using dynamic light scattering. Unimodal characteristic linewidth distribution was observed for two narrow PS/TOL binary solutions consisting of high molecular weight and narrow molecular weight distribution(MWD) polystyrenes with M_w = 8.6 x 10^6, $M_w/M_n \leq 1.17$, and M_w = 10 x 10^6, $M_w/M_n \leq 1.20$. But two modes in the characteristic linewidth distribution appeared in a polydisperse PS/TOL binary solution with M_w = 23 x 10^6, $M_w/M_n \sim 2.0$, which indicated that polydispersity could lead to the observed bimodal behavior in the characteristic linewidth distribution even at small scattering angles where $KR_g < 1$, with K and R_g being the scattering vector and the root-mean-square z-average radius of gyration. Two modes also existed in a ternary solution composed of a high molecular weight component PS (M_w = 8.6 x 10^6, $M_w/M_n \leq 1.17$) and a low MW component PS (M_w = 2.3 x 10^5, $M_w/M_n \sim 1.06$). This observation agreed with the theory on ternary mixtures proposed recently by Benoît and his coworkers, and confirmed the effect of polydispersity on dynamic behavior of semidilute polymer solutions. The fast mode was interpreted as the cooperative diffusion of entangled chains and the slow mode could be attributed to the coupling of the matrix polymer with the probe polymer. This coupling was further ascertained by the bimodal behavior in two PS1/PMMA2/TOL ternary solutions, in which the probe polymer PMMA was isorefractive with the solvent TOL. The effect of star configuration and the extent of entanglements in two PS1/PS2/TOL ternary solutions with 4-arm and 12-arm star PS as probe polymers have also been examined.

*Author to whom all correspondence should be addressed

I. INTRODUCTION

The dynamics of polymer chains in semidilute solutions has been one of the central topics in polymer physics for more than ten years. Inspired by the scaling concepts advanced so eloquently by de Gennes[1] and subsequent refinement of theory based on more rigorous calculations,[2] many experiments have been reported. Unfortunately, several inconsistent conclusions have been drawn from many such experiments. It seems that the dynamic behavior of polymer semidilute solutions requires further investigation, at least from the experimental viewpoint.

Recently, based on the random phase approximation, Benoît and his coworkers[3] (BBA) proposed that any mixture of two monodisperse polymers could exhibit two relaxation modes, which was confirmed by their tests on a polystyrene-poly(methyl methacrylate)-toluene ternary mixture.[4] Their formalism was based on the assumption that there existed only one translational mode in monodisperse polymer solutions. Nemoto et al.[5] has also reported that two modes were observed in one of several semidilute solutions of polymer-polymer-solvent ternary systems, but only one mode in polymer-solvent binary solutions, which agreed with the BBA theory.

However, two modes have sometimes been observed in polymer-solvent binary solutions.[6-16] Furthermore, the source of the slow mode is still a controversy. For example, Amis and Han[11] observed two modes in semidilute solutions of polystyrene in tetrahydrofuran, in which the slow mode was attributed to self-diffusion. Brown[13] reported a slow relaxation mode which was 1 ~ 2 orders of magnitude slower when compared with that based on the self-diffusion concept. He proposed that the slow mode reflected the translational diffusion of intermolecularly entangled coils in polymer semidilute solutions. Mathiez et al.[7] found that the slow mode was independent of molecular weights, and suggested that it was knots of higher concentrations in polymer solutions that had caused the slow mode. Nishio and Wade[8] attributed the slow mode to the diffusion of the "whole polymer" since it became slower as the concentration and solution viscosity increased. On the other hand, Chu[17] pointed out that two modes could be detected by dynamic light scattering in the crossover regime from the dilute to the semidilute concentration range. He considered that these two modes were related to the cooperative diffusion of the polymer chains and the motion of the entire polymer coils which had not yet been entangled because of their smaller sizes near the overlap concentration, respectively. Burchard et al.[18] observed that the bimodal behavior in the crossover regime suggested that the slow motion was very likely the consequence of the highly hindered motion of clusters in the inhomogeneous structures of semidilute polymer solutions. Thus, there is substantial disagreement on the interpretation of the slow mode.

With recent development of labeling techniques, such as forced Rayleigh scattering, pulsed field gradient spin-echo NMR and index-matching dynamic light scattering, more polymer probe-diffusion experiments have been reported.[19-25] The reptation concept in self-diffusion of polymer chains in semidilute solutions is being questioned by the phenomenological approach proposed by Phillies et al.[26-27] They suggested a stretched exponential equation for the concentration dependence of polymer self-diffusion

$$D_s = D_o \, exp(-\alpha \, C^{\nu}) \, , \tag{1}$$

where D_0 is the probe diffusion coefficient at infinite dilution; C is the concentration of background polymer, and α, ν are the scaling parameters which are molecular weight dependent. The equation was reasonably consistent with all experimental studies of polymer self-diffusion. Thus, the reptation concept was probably not an important issue for polymer self-diffusion in semidilute solutions. It should be noted that the same conclusion should not be extended to polymer melts.

Chu et al.[28] have studied the structure and dynamics of a polymer solution composed of PS(polystyrene), PMMA(poly(methyl methacrylate)), TOL(toluene), and CNA(α-chloronaphthalene). According to the photocount intensity-intensity time correlation function measurements of the PMMA probe in the PS/MS(mixed solvent) isorefractive matrix, one surprising feature was that at least two dominant characteristic modes appeared even at small scattering angles. While the slow mode could be identified with the translational motion of the center of mass of the PMMA probe chain, it remained unconfirmed that the fast mode might be related to a coupling of PMMA motion with the cooperative motion of the isorefractive PS/MS(matrix).

In order to understand the effects of polydispersity, chain configuration and entanglement mesh size on the characteristic decay times, a systematic study of PS/TOL binary solutions, PS1/PS2/TOL and PS1/PMMA2/TOL ternary solutions was initiated in the semidilute regime using dynamic light scattering. We denote the matrix polymer by "1" and the probe polymer by "2" in naming the ternary solutions. Firstly, we examined the binary solutions with different molecular weight distributions(*MWD*). For typical semidilute binary solutions of polystyrene in toluene, only one relaxation mode was found in those systems with high molecular weight and narrow molecular weight distribution, whereas two modes appeared in a broad *MWD* PS/TOL binary solution. Secondly, we chose two types of ternary model systems: one with two identical chemical species(PS) but different molecular weights, and the other with two similar molecular weights but different chemical species(PS and PMMA). In the first case, we observed two modes in a ternary system satisfying certain conditions. The effect of star configuration on the characteristic decay time(τ) was also examined. In the latter case, the probe PMMA was supposed to be "optically" invisible because of the refractive index matching with the solvent(TOL), but we could still observe two distinctly separated relaxation modes. We believe that it was the coupling of the probe and the matrix that had caused the slow mode.

The organization of this paper is as follows. In Section II, we describe dynamic light scattering and data analysis methods. In Section III, details are given about experimental materials and facilities. Results are presented in Section IV. We conclude with a discussion of several related works in Section V.

II. DYNAMIC LIGHT SCATTERING

Light-scattering spectroscopy is a common method for studying the dynamics of polymers in solution.[29] The measured self-beating photocount autocorrelation function has the form

$$G^{(2)}(t) = A(1 + b\,|g^{(1)}(t)|^2)\,, \tag{2}$$

where A is the background, which can be measured as well as computed; b is a spatial coherence factor depending upon various experimental conditions such as coherence and receiver area, and is usually taken as an unknown parameter in the data fitting procedure, and $g^{(1)}(t)$ is the first-order electric field correlation function. For a monodisperse solution, $g^{(1)}(t)$ is a single-exponential decay characterized by

$$|g^{(1)}(t)| = e^{-\Gamma t}, \tag{3}$$

where Γ is a characteristic linewidth. For monodisperse structureless particles,

$$\Gamma = DK^2, \tag{4}$$

where D is the translational diffusion coefficient, and $K = 4\pi \sin(\theta/2)/\lambda$, with λ and θ being the wave length in the scattering medium and the scattering angle, respectively. For a polydisperse solution, $g^{(1)}(t)$ becomes a sum or distribution of exponentials

$$|g^{(1)}(t)| = \sum_i w_i e^{-\Gamma_i t}, \tag{5}$$

where w_i ($\propto I_i$) is the normalized intensity weighting factor of each representative molecular weight fraction M_i. In the continuous limit,

$$|g^{(1)}(t)| = \int_0^\infty G(\Gamma) e^{-\Gamma t} d\Gamma, \tag{6}$$

where $G(\Gamma)$ is the normalized distribution of characteristic linewidths. The average linewidth $\overline{\Gamma}$ is defined by the following equation

$$\overline{\Gamma} = \int_0^\infty G(\Gamma)\Gamma d\Gamma. \tag{7}$$

The data analysis consists of inverting the Laplace transformation Eq. (6). However, the measured $g^{(1)}(t)$ has noise and is bandwidth limited (due to the limited channel number of the correlator) and the Laplace inversion of equation (6) is an ill-conditioned problem, which restricts the amount of information on $G(\Gamma)$ that we can retrieve from $g^{(1)}(t)$. Several approximate methods have been proposed.[30-32]

The cumulants method[31] is based on the formalism of the statistical cumulant generating function:

$$g^{(1)}(t) = e^{-\overline{\Gamma} t + \mu_2/2!t^2 - \mu_3/3!t^3 + (\mu_4 - 3\mu_2^2)/4!t^4 + \ldots}$$

$$= \sum_{i=1}^{\infty} K_i(\Gamma)(-t)^i/i!, \tag{8}$$

where the i-th cumulant has the form

$$K_i = \frac{d^i}{dt^i}(ln|g^{(1)}(t)|)|_{t=0}, \tag{9}$$

and the i-th moment of the decay rate has the form

$$\mu_i = \int_0^{\infty} (\Gamma - \overline{\Gamma})^i G(\Gamma) d\Gamma . \qquad (10)$$

The variance is identified by $\mu_2/\overline{\Gamma}^2$. The average linewidth and the width of $G(\Gamma)$ distribution can be obtained by this method. In practice, it is difficult to apply the cumulants method when the variance becomes large, say $\mu_2/\overline{\Gamma}^2 \geq 0.5$.

The CONTIN method[32] uses a regularization technique to seek smooth solutions, no matter whether the $G(\Gamma)$ distribution is unimodal, multimodal, or broad. So the CONTIN method is appropriate for photocount correlation profile analysis without an a priori assumption on the form of the $G(\Gamma)$ distribution. We used the CONTIN method, which was kindly provided by Dr. S.W. Provencher (European Molecular Biology Laboratory), mainly for correlation function profile analysis of unimodal and bimodal $G(\Gamma)$ distributions.

For a unimodal linewidth distribution, no matter how broad it is, $|g^{(1)}(t)|$ can be represented by the widely used Williams-Watts function[33]

$$|g^{(1)}(t)| = e^{-(\Gamma_o t)^{\beta}} , \qquad (11)$$

where Γ_o is some effective linewidth, and β ($0 < \beta \leq 1$) is a measure of the width of the specific distribution of decay times. The average linewidth is

$$\overline{\Gamma} = \Gamma_o \frac{\beta}{\Gamma(\beta^{-1})} , \qquad (12)$$

where $\Gamma(\beta^{-1})$ is the gamma function. By fitting the measured $g^{(1)}(t)$ to the Williams-Watts equation, we can distinguish a broad unimodal $G(\Gamma)$ distribution from a multimodal one. This is used to help verifying the bimodal distributions in our analysis of the characteristic linewidth distribution.

III. EXPERIMENT METHODS

In this section, we describe in detail experimental materials and facilities used.

III.1 Materials A series of linear PS, star PS and linear PMMA were used in our studies. The characteristics of these polymers are listed in Table 1, where "LPS" and "LPM" denote linear PS and linear PMMA, and "SPS4", "SPS12" denote the 4-arm and 12-arm star PS, respectively. The two star polystyrenes were previously characterized by light scattering in toluene at 25°C. Here we used the overlap concentration C^*[34] obeying the relation

$$C^* = \frac{M}{N_A R_g^3} \ (g/cm^3), \qquad (13)$$

where N_A is Avogadro's number in mole^{-1}, M is the polymer molecular weight in g/mole, R_g is the radius of gyration in cm. The overlap concentrations for all the polymers in TOL at 25°C were calculated by Eq. (13) and are also listed in Table 1. The solvent TOL (HPLC grade) from Fisher Scientific Inc. was used without further purification.

Table 1. Characteristics of Polymer Samples

Sample Code	M_w(g/mole)	$\frac{M_w}{M_n}$	C*(10^{-2}g/cm^3)	Source
LPS8	8.6×10^6	≤1.17	0.36	Toyo Soda Inc. (Japan)
LPS10	10×10^6	≤1.2	0.32	Polysciences Inc.
LPS23	23×10^6	≥2.0	0.17	IK3000 (Japan)
LPS2	2.3×10^5	~1.06	5.8	Pressure Chemical Co.
SPS4	3.0×10^5	<1.10	5.4	Prof. Q.W. Wang (China)
SPS12	1.6×10^6	<1.10	6.0	Dr. L.J. Fetters (Exxon)
LPS9	9.0×10^5	≤1.10	2.0	Pressure Chemical Co.
LPM8	8.4×10^5	≤1.12	3.2	Pressure Chemical Co.
LPM3	3.3×10^5	≤1.10	6.5	Pressure Chemical Co.

Table 2. Concentrations of Binary Solutions

Sample Code	LPS80	LPS100	LPS230
C(10^{-2}g/g)	1.20	1.15	0.57

Table 3. Concentrations of PS1/PS2/TOL Ternary solutions

Sample Code	$C \times 10^2$(g/g)					W_{LPS8}[a]
	Total	LPS8	LPS2	SPS4	SPS12	
LPS80	1.20	1.20				1.00
LPS82	1.17	0.881	0.287			0.75
LPS84	1.28	0.835		0.441		0.65
LPS812	1.22	0.809			0.410	0.66

[a]W_{LPS8} is the weight fraction of LPS8

Three higher molecular weight PS samples: LPS8 and LPS10 with small polydispersity ($M_w/M_n \leq 1.2$) and LPS23 with large polydispersity ($M_w/M_n \sim 2$), were chosen for preparing binary semidilute solutions with concentrations C being about $3C^*$. In PS1/PS2/TOL solutions, the higher molecular weight LPS8 was the matrix polymer in the semidilute regime, and

Table 4. Concentrations of PS1/PMMA2/TOL Ternary solutions

Sample Code	$C \times 10^2$(g/g)					W_{PS}[a]
	Total	LPS9	LPM8	SPS4	LPM3	
LPS90	3.45	3.45				1.00
LPS98	3.99	3.83	0.156			0.96
SPS40	8.60			8.60		1.00
SPS43	8.09			7.83	0.256	0.97

[a]W_{PS} is the weight fraction of PS

a relatively low molecular weight PS sample, i.e., LPS2, SPS4 or SPS12, was used as the probe polymer in its dilute regime if PS1 were absent. The total concentrations of the ternary solutions were also around 3C*. In PS1/PMMA2/TOL ternary solutions, LPS9 and SPS4 were used as matrix polymers and LPM8 and LPM3 were used as probe polymers. The total concentrations were about 1.5C*. The concentrations of binary and ternary solutions are listed in Tables 2,3 and 4, respectively. The last digit in the sample code denotes the type of polymers present as the third component, such as 0 for the absence of the third component and 2 for LPS2.

Semidilute solutions and dilute solutions were clarified for optical measurements by centrifugation and filtration, respectively. Semidilute solutions were centrifuged at gravity 2000 g for 8 to 10 hours. Under this condition, we examined that there was no concentration gradient in the centrifuge tube. Then they were transferred to dust-free light-scattering cells using dust-free pipets. For ternary solutions, dilute solutions of probe polymers were filtered by using a Millipore FG filter of 0.2 μm nominal pore-size diameter and by adding the filtered dilute solutions directly to the light scattering cells which contained known amounts of dust-free semidilute solutions of the matrix polymer. To insure homogenous mixing, the mixed ternary solutions were allowed to stand for ten to fifteen days before measurements. We remeasured those ternary solutions after another four months. The results were the same (error ≤ 2%), indicating that we achieved homogenous mixing and dust-free condition for our semidilute ternary solutions before light-scattering measurements.

III.2 Facilities We used a Lexel Model 95 Argon-ion laser operated at λ_o = 514.5 nm and ~200 mW as a light source, a computer controlled photon counting system to perform static light-scattering measurements and Brookhaven 136- and 72-channel digital correlators to perform dynamic light-scattering measurements. A Spectra-physics Model 125 Helium-Neon laser operated at λ_o = 632.8 nm was also used occasionally. The light-scattering spectrometer has been described elsewhere.[35] Figure 1 shows the schematics of our present automated data acquisition system for the light-scattering(LS) spectrometer.

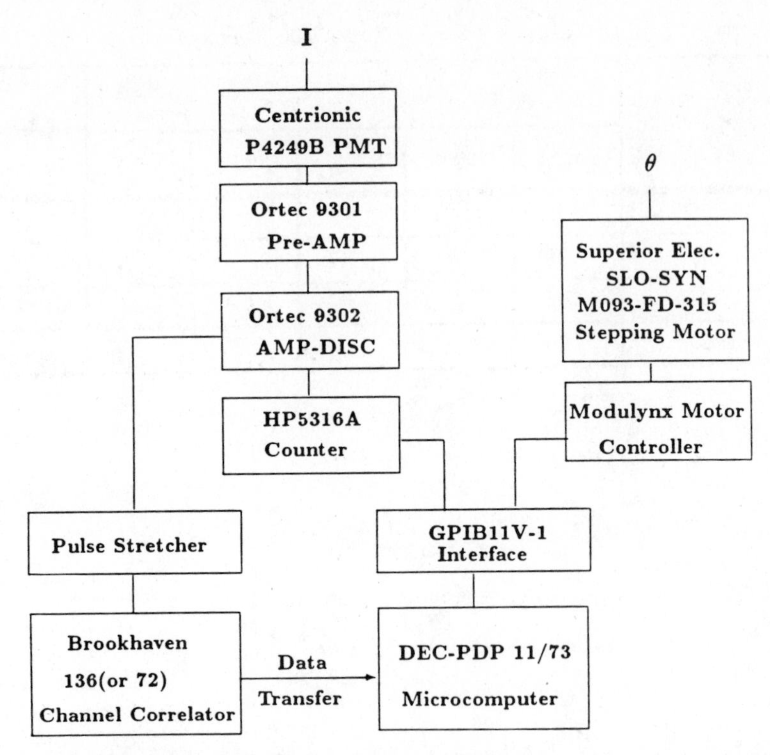

Fig. 1. Automated data acquisition system for the light-scattering (LS) spectrometer. Static LS data could be acquired by the DEC PDP-11/73 computer automatically. Dynamic LS data were transferred from the Brookhaven correlator to the computer for data analysis.

The viscosities of some semidilute solutions were measured by using a Brookfield cone/plate viscometer. All the measurements were performed at temperature 25°C.

IV. RESULTS

In light-scattering spectroscopy, the setting of experimental conditions, the theoretical model for solution dynamics and the method of data analysis are closely interrelated.[29] Before we proceed with our experiments, we have to know what we are searching for and to select the experimental conditions properly. Firstly, in order to avoid the complication of all kinds of internal motions, we should use only the small KR_g range for our light-scattering measurements. Thus, we chose a small scattering angle $\theta = 18°$ with $\lambda_o = 514.5$ nm for most of our measurements, and $\theta = 15°$ with $\lambda_o = 632.8$ nm for the ultrahigh molecular weight sample LPS230 in order to obtain an even smaller K value. At $\theta = 15°$, with $\lambda_o = 632.8$ nm, $KR_g \approx 1$ for LPS230 in toluene at 25°C. Secondly, we tried to avoid the crossover region ($C \sim C^*$) for the purpose of our present studies. Since the overlap concentration C^* is arbitrarily defined as the concentration at which polymer coils begin to overlap. There are several definitions based upon different viewpoints, such as: $C^* = 1/[\eta]$ with $[\eta]$ being the intrinsic viscosity of the polymer solution

$$C^* = \frac{M}{(2R_g)^3 N_A}\ ;\ C^* = \frac{M}{4/3\pi R_g^3 N_A}\ ;\ and\ C^* = \frac{M}{N_A R_g^3}$$

which is Eq. (13), and so on. As reported in (18), the crossover behavior was observed at concentrations $C \sim 3\ C^*$ with C^* being $1/[\eta]$. We used the largest C^* derived from Eq. (13) which for polystyrene in toluene corresponded to $3/[\eta]$. Thus, the actual concentrations used in our measurements were $9/[\eta]$, in order to ensure that most polymer chains were entangled even for a polydisperse system.

IV.1. PS/TOL Binary Solutions The measured, unnormalized, net intensity-intensity time correlation function and CONTIN fitting percent deviation (D%) of binary solutions: two narrow *MWD* samples LPS80 (M_w = 8.6 x 10^6, $M_w/M_n \le 1.17$), LPS100 (M_w = 10 x 10^6, $M_w/M_n \le 1.20$), and one broad *MWD* sample LPS230 (M_w = 23 x 10^6, $M_w/M_n \sim 2$), are shown in Figs. 2a, 3a and 4a respectively, where

$$D\% = \frac{b\,|g^{(1)}(t)|^2_{exp} - b\,|g^{(1)}(t)|^2_{cal}}{b\,|g^{(1)}(t)|^2_{exp}} \mathbf{x}\,100\,. \tag{14}$$

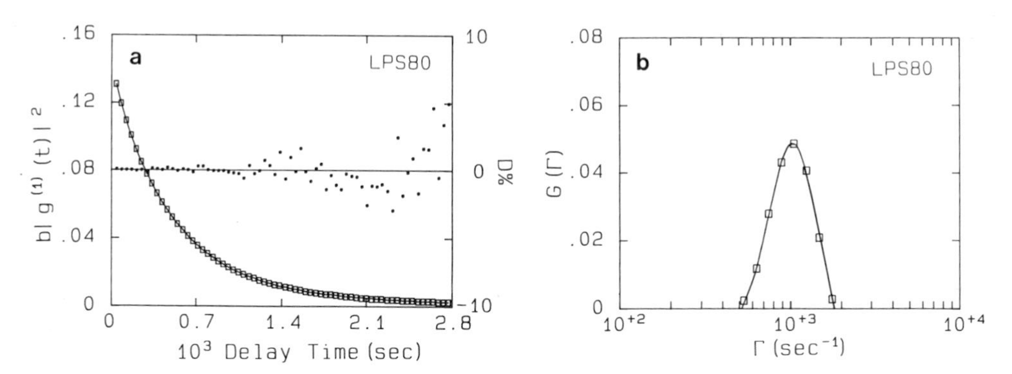

Fig. 2(a). Measured, unnormalized, net intensity-intensity time correlation function of binary polymer solution LPS80 (M_w = 8.6 x 10^6 g/mole) in TOL. C = 1.20 x 10^{-2} g/g at 25°C and θ = 18°. The solid curve represents a CONTIN fit with $\overline{\Gamma}$ = 1.10 x 10^3 sec^{-1}, $\mu_2/\overline{\Gamma}^2$ (VAR) = 0.05. Small squares denote a fitting deviation as defined by Eq. (14).

Fig. 2(b). Plot of the characteristic linewidth distribution $G(\Gamma)$ vs. Γ from the CONTIN analysis for LPS80. Only one narrow peak is obtained.

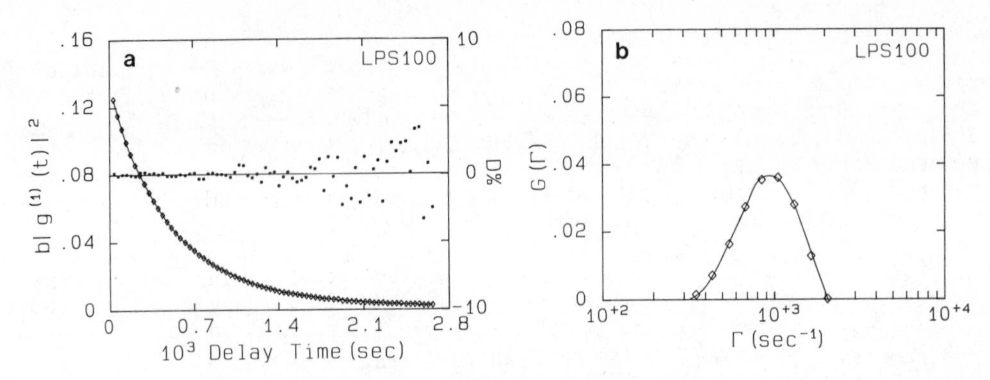

Fig. 3(a). Measured unnormalized net intensity-intensity time correlation function of binary polymer solution LPS100 (M_w = 10 x 10^6 g/mole) in TOL. C = 1.15 x 10^{-2}g/g at 25°C and θ = 18°. The solid curve represents a CONTIN fit with $\bar{\Gamma}$ = 1.07 x 10^3 sec^{-1}, VAR = 0.09. Small squares denote a fitting deviation as defined in Figure 2(a).

Fig. 3(b). Plot of characteristic linewidth distribution $G(\Gamma)$ vs. Γ from the CONTIN analysis for LPS100. Only one narrow peak is obtained.

By using the CONTIN analysis, we obtained one characteristic linewidth with a small variance for the two narrow *MWD* PS samples and two linewidths for the broad *MWD* PS sample. Figures 2b, 3b and 4b show the characteristic linewidth distribution of LPS80, LPS100 and LPS230 binary PS/TOL solutions, respectively. Numerical values of the data analysis of binary PS/TOL solutions are listed in Table 5. **A** is the integrated area of the observed peak for the characteristic linewidth distribution from the CONTIN method of analysis.

To ensure the bimodal behavior of LPS230 solution, we attempted to fit $|g^{(1)}(t)|$ to the Williams-Watts function (Eq. 11). The fitting results are shown in Figs. 4c and 4d. Obviously the behavior of a polydisperse LPS230 solution in the semidilute regime could not be fitted to a broad unimodal characteristic linewidth distribution even of the type exemplified by the Williams-Watts stretch exponential function which is assumed to fit a unimodal linewidth distribution. The Williams-Watts fitting was also applied to the LPS80 solution. $|g^{(1)}(t)|$ of LPS80 was well fitted and the linewidth parameter β was close to 1, which was in agreement with the CONTIN analysis results.

The two analysis techniques confirmed the unimodal behavior and the bimodal behavior, respectively, for our narrow and broad binary model solutions at semidilute concentrations. Furthermore, $\bar{\Gamma}/K^2$ values of LPS80 and LPS100 were independent of K, and two $\bar{\Gamma}_f/K^2$ (fast one) values (from θ = 15°, 17°) of LPS230 were almost the same. The fast mode could be considered as due to cooperative diffusion. While the slow mode appearing in the LPS230 binary solution, we believe, could be due to the polydispersity effect, involving translational motions from a fraction of the (smaller) polymer coils not yet entangled even though the overall concentration $C > C^*$.

IV.2. PS1/PS2/TOL (Ternary) Solutions In PS1/PS2/TOL ternary solutions, the concentration of the high molecular weight LPS8 was in the semidilute regime, which enabled LPS8 to form an entangled polymer pseudo network. The concentration of lower molecular weight component (LPS2, SPS4 or SPS12) would be in the dilute regime if LPS8 were absent. The total concentration of the ternary solution was almost the same as that of the binary LPS80 solution. The effect of polydispersity could be examined with respect to the reference binary solution (LPS80).

Figure 5a shows the measured net intensity-intensity time correlation function and the CONTIN fitting deviation of LPS82 ternary solution with the 8 million linear polystyrene LPS8 being the high molecular weight component in the semidilute regime and the 0.2 million linear polystyrene LPS2 being the low molecular weight component in the dilute regime. Two peaks appeared in the CONTIN analysis, as shown in Fig. 5b. The Williams-Watts stretch exponential fitting again failed to fit the measured net

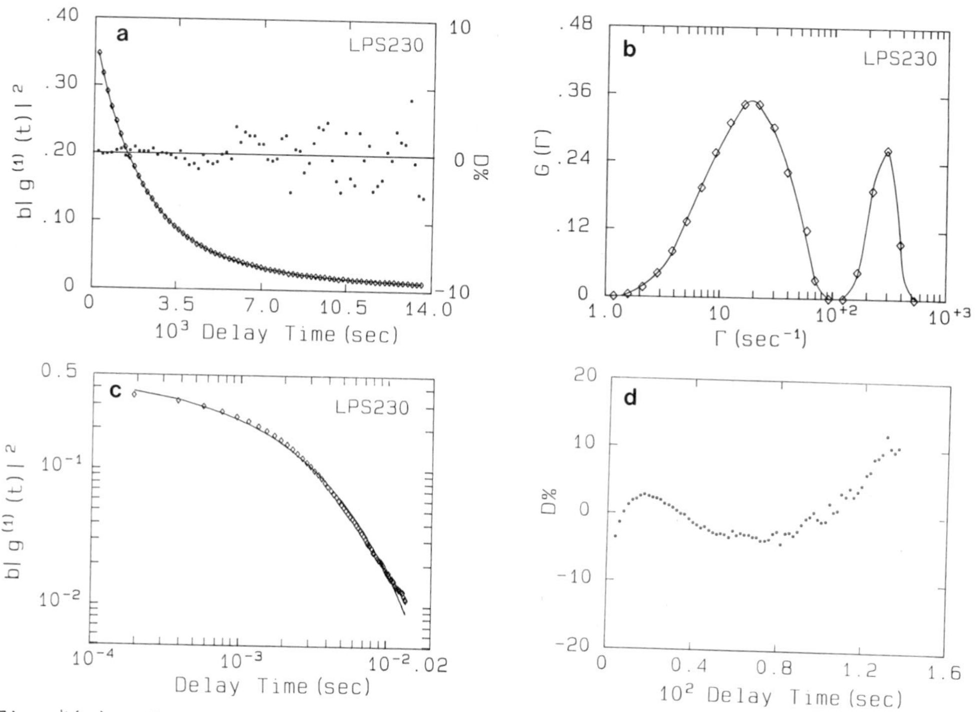

Fig. 4(a). Measured, unnormalized, net intensity-intensity time correlation function of binary polymer solution LPS230 (M_w = 23 x 10^6 g/mole) in TOL. C = 0.57 x 10^{-2}g/g at 25°C and θ = 15°. The solid curve represents a CONTIN fit with $\overline{\Gamma}_1$ = 297 sec^{-1}, VAR_1 = 0.06, A_1 = 0.77; Γ_2 = 29.6 sec^{-1}, VAR_2 = 0.3, A_2 = 0.23. Small squares denote a fitting deviation as defined in Figure 2(a).

Fig. 4(b). Plot of characteristic linewidth distribution $G(\Gamma)$ vs. Γ from the CONTIN analysis for LPS230. Two peaks have appeared in this polydisperse semidilute solution.

Fig. 4(c). Plot of log (b | $g^{(1)}$(t) |^2) vs. delay time for LPS230. Obviously, as compared with the Williams-Watts fitting result (denoted by solid line), bimodal behavior of LPS230 could not be fitted to the Williams-Watts function which is quite universal for unimodal distributions.

Fig. 4(d). Plot of deviation of Williams-Watts fitting for LPS230.

Table 5. CONTIN analysis results of PS/TOL binary solutions at 25°C

Sample	θ	$\overline{\Gamma}(sec^{-1})$	$\mu_2/\overline{\Gamma}^2$	A	$10^7\ \overline{\Gamma}/K^2$ (cm^2/sec)
LPS80	18	1.10×10^3	0.05	1.00	3.32
LPS100	18	1.07×10^3	0.09	1.00	3.24
LPS230	15	297 29.6	0.06 0.3	0.77 0.23	1.99 0.198

Table 6. CONTIN analysis results of PS1/PS2/TOL ternary solutions at 25°C

Sample	θ	$\overline{\Gamma}(sec^{-1})$	$\mu_2/\overline{\Gamma}^2$	A	$10^7\ \overline{\Gamma}/K^2$ (cm^2/sec)
LPS80	18	1.10×10^3	0.05	1.00	3.32
LPS82	18	1.16×10^3 89.8	0.06 0.3	0.86 0.14	3.51 0.273
LPS84	18	1.37×10^3	0.07	1.00	4.16
LPS812	18	1.04×10^3	0.11	1.00	3.16

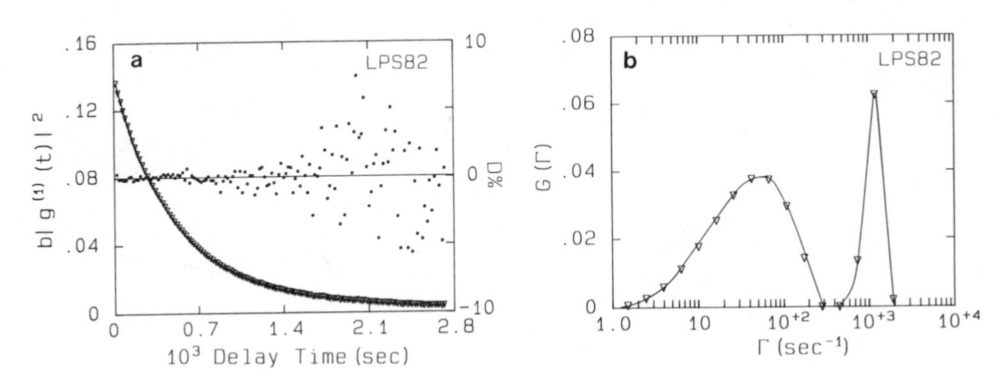

Fig. 5(a). Measured unnormalized net intensity-intensity time correlation function of LPS82 ternary solution with M_w(LPS8) = 8.6×10^6 g/mole in the semidilute concentration (C_{LPS8} = 0.88×10^{-2} g/g) and M_w(LPS2) = 2.3×10^5 g/mole in the dilute concentration (C_{LPS2} = 2.87×10^{-3} g/g) at 25°C and θ = 18°. The solid curve denotes a CONTIN fit with $\overline{\Gamma}_1$ = 1.16×10^3 sec^{-1}, VAR_1 = 0.06, A_1 = 0.86; $\overline{\Gamma}_2$ = 89.8 sec^{-1}, VAR_2 = 0.3, A_2 = 0.14. Small squares denote the deviation of CONTIN fittings.

Fig. 5(b). Plot of $G(\Gamma)$ vs. Γ from the CONTIN analysis for LPS82. There are two peaks in which the $\overline{\Gamma}$ for the fast mode is almost the same as that of the LPS80 binary solution. The magnitude of the slow mode A_2 is 14%.

intensity-intensity time correlation function. As compared with reference LPS80, the linewidth of the fast one of two modes in LPS82 had almost the same value as that of LPS80, and could be attributed to the cooperative motion. Intuitively, the slow one was caused by LPS2. However, the $D_s(=\overline{\Gamma}_s/K^2)$ was about ten times smaller than that of D_o (estimated at $\sim 3 \times 10^{-7} cm^2/sec$) with D_o being the diffusion coefficient of LPS2 in pure solvent toluene. After correction of the macroscopic viscosity (η), D_s became much larger than $D_o\eta_o/\eta$, where η_o is the viscosity of the solvent, and η is the viscosity $D_o$of the solution with $\eta_o/\eta \approx 8.5 \times 10^{-3}$. This difference implies that the slow motion is not only the translational motion of the lower molecular weight component LPS2 in the pseudo network, but also the result of coupling of the lower and higher molecular weight components in the mixture. In addition, we also observed that the variance (= 0.3) of the slow motion was large. The reason could be that some test chains LPS2 were attached to and/or detached from the entangled polymer coils from time to time.

The remaining two ternary solutions contained star polystyrenes as the low molecular weight components. Their behaviors were different from LPS82. The CONTIN analysis results of LPS84 with 4-arm star PS being the third component are listed in Table 6. Although only one peak was derived from the CONTIN analysis, the nonlinear semilogarithmic $|g^{(1)}(t)|$ curve suggested a non-single exponential decay, as shown in Fig. 6. Obviously, the average linewidth of LPS84 differed from those of LPS80 and LPS82. The faster cooperative motion was due to the smaller mesh size of entanglement. Thus the difference in the configuration of star and linear polymers should be responsible for this increase in $\overline{\Gamma}$.

In the LPS812 ternary solution, the molecular weight of SPS12 was close to that of the matrix polymer LPS8, in contrast to the condition of the previous two ternary solutions. The results of LPS812, as shown in Fig. 7 and Table 6, were similar to those of reference binary solution LPS80. The $|g^{(1)}(t)|$ of LPS812 can be fitted to a Williams-Watts function with a relatively large $\beta(=0.86)$. This behavior indicated that the two components LPS8 and SPS12 became entangled entirely.

IV.3. PS1/PMMA2/TOL Ternary Solutions Our second type of ternary solution contained two different polymers, PS and PMMA, which had comparable molecular weights. Since PS and PMMA are incompatible with each other, the compositions of solutions were chosen very carefully in order to avoid phase separation. There was only a small amount of PMMA acting as a probe polymer in PS1/PMMA2/TOL ternary solutions.

The behavior of reference binary solution LPS90 ($M_w = 9.0 \times 10^5$, $M_w/M_n \leq 1.10$, $C/C^* \sim 1.5$), was the same as those of LPS80 and LPS100. The unimodal behavior was obtained from both CONTIN analysis and Williams-Watts fitting. Figure 8(a,b) shows the logarithmic correlation function, the Williams-Watts fitting result, and the Williams-Watts fitting deviation of solution LPS90. A trace amount of LPM8 (linear PMMA, $M_w = 8.4 \times 10^5$, $M_w/M_n \leq 1.12$) was introduced and mixed with LPS9 to form the ternary solution LPS98. Multiple delays had to be used to obtain a reasonably complete correlation curve (Fig. 9). Two modes both with small variances were observed (see Table 7). The fast mode of LPS98 came from the cooperative motion. The source of the slow mode must be referred back to the addition of the probe LPM8, but LPM8 was isorefractive with toluene.

Another ternary solution SPS43 was a 4-arm star PS (M_w=3.0 x 10^5, $M_w/M_n \leq 1.10$) mixed with a similar molecular weight probe linear PMMA (M_w=3.3 x 10^5, $M_w/M_n \leq 1.10$). The results were very similar to those of LPS98. Figure 10 shows the measured correlation function and the CONTIN

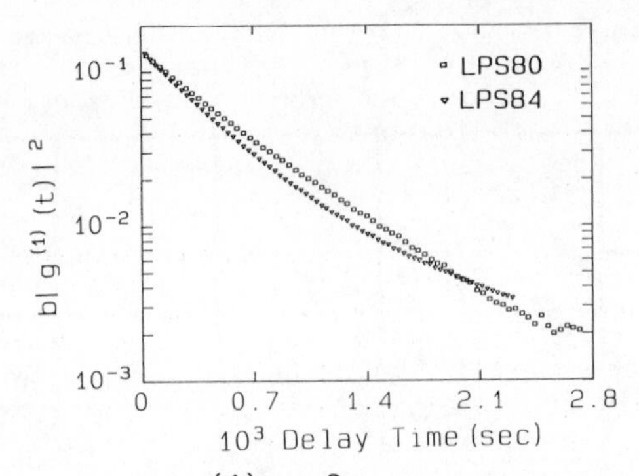

Fig. 6. Plot of log (b | $g^{(1)}(t)$ |^2) vs. delay time of a LPS84 ternary solution as compared with that of the LPS80 binary solution. Squares denote the result of LPS80 and inverse triangles denote the result of LPS84. | $g^{(1)}(t)$ | of LPS84 in a semi-logarithmic plot deviates from that of LPS80.

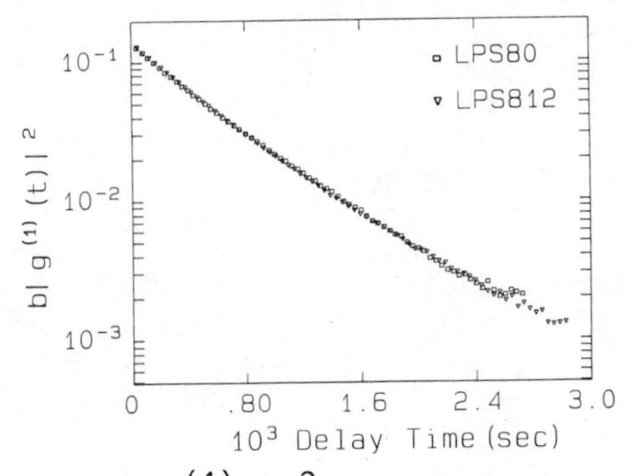

Fig. 7. Plot of log(b | $g^{(1)}(t)$ |^2) vs. delay time for a LPS812 ternary solution (denoted by inverse triangles) as compared with that of a LPS80 binary solution (denoted by squares). Most of the LPS812 curve overlaps with that of LPS80.

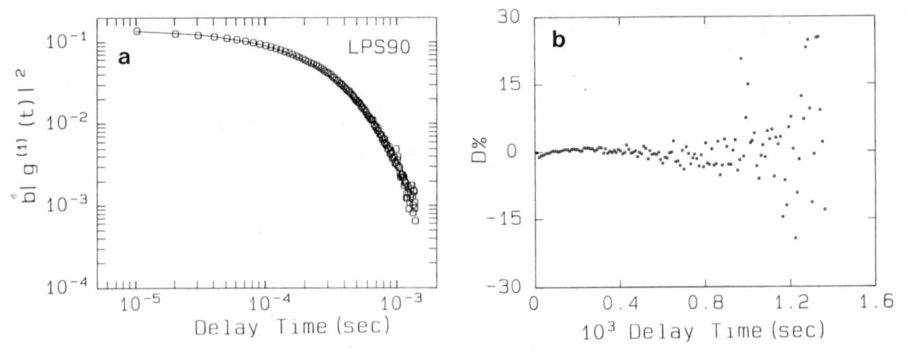

Fig. 8(a). Measured, unnormalized, net intensity-intensity time correlation function of binary polymer solution LPS90 (M_w = 9.0 x 10^5 g/mole) in TOL. C = 3.45 x 10^{-2} g/g at 25°C and θ = 18°. The solid curve represents a Williams-Watts fit with β = 0.92, Γ_o = 2.06 x 10^3 sec^{-1}.

Fig. 8(b) Plot of Williams-Watts fitting deviation for LPS90.

Table 7. CONTIN analysis results of PS1/PMMA2/TOL ternary solutions at 25°C

Sample	θ	$\bar{\Gamma}(\mathrm{sec}^{-1})$	$\mu_2/\bar{\Gamma}^2$	A	$10^7\bar{\Gamma}/K^2$ ($\mathrm{cm}^2/\mathrm{sec}$)
LPS90	18	2.09×10^3	0.08	1.00	6.36
LPS98	18	2.30×10^3 41.1	0.03 0.06	0.50 0.50	7.00 0.125
SPS40	18	3.68×10^3	0.09	1.00	11.2
SPS43	18	3.61×10^3 90.4	0.01 0.09	0.47 0.53	11.0 0.274

fitting curve of SPS43 ternary solution. All the CONTIN analysis data of PS1/PMMA2/TOL ternary solutions are listed in Table 7.

V. DISCUSSIONS

V.1. Effect of Polydispersity Our experimental results for all binary solutions, including three model solutions (LPS80, LPS100, LPS230) and two reference solutions (LPS90, SPS40), demonstrate that the behavior of typical monodisperse polymer/solvent binary solutions is unimodal in the

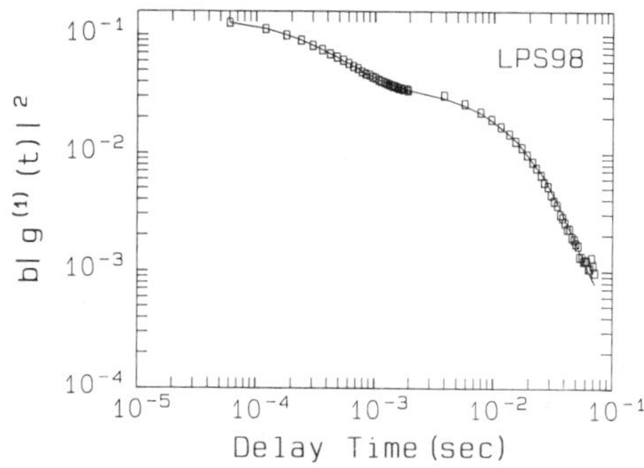

Fig. 9. Measured unnormalized net intensity-intensity time correlation function of a LPS98 ternary solution with LPS9 (M_w = 9.0 x 10^5 g/mole) as a matrix polymer (C_{LPS9} = 3.83 x 10^{-2} g/g); LPM8 (M_w = 8.4 x 10^5 g/mole) as a probe polymer (C_{LPM8} = 0.156 x 10^{-2} g/g) at 25°C and θ = 18°. Multiple delays were used in order to get a complete decay. The solid line denotes a CONTIN fit with $\bar{\Gamma}_1$ = 2.30 x 10^3 sec^{-1}, VAR_1 = 0.03, A_1 = 0.50, and $\bar{\Gamma}_2$ = 41.1 sec^{-1}, VAR_2 = 0.06, A_2 = 0.50.

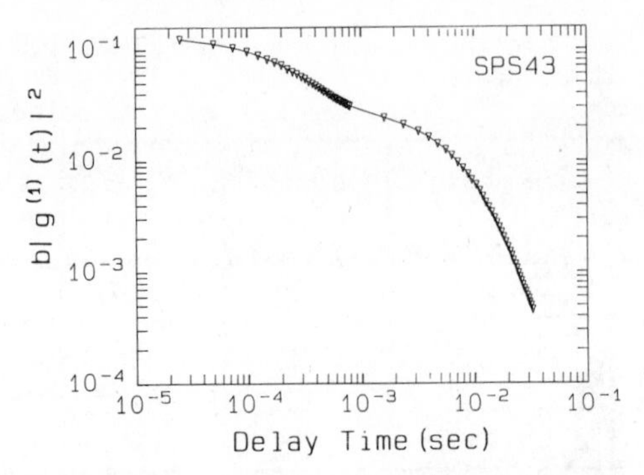

Fig. 10. Measured unnormalized net intensity-intensity time correlation function of a SPS43 ternary solution with SPS4($M_w = 3.0 \times 10^5$ g/mole) as a matrix polymer ($C_{SPS4} = 7.83 \times 10^{-2}$g/g); LPM3 ($M_w = 3.3 \times 10^5$ g/mole) as a probe polymer ($C_{LPM3} = 0.256 \times 10^{-2}$ g/g) at 25°C and $\theta = 18°$. Multiple delays were used in order to get a complete decay. The solid line denotes a CONTIN fit with $\overline{\Gamma}_1 = 3.61 \times 10^3$ sec^{-1}, VAR$_1$ = 0.01, A_1 = 0.47, and $\overline{\Gamma}_2$ = 90.4 sec^{-1}, VAR$_2$ = 0.09, A_2 = 0.53.

semidilute regime. The polydispersity effect could lead to bimodal behavior. For semidilute polymer solutions, polymer chains begin to overlap and entangle with each other. A transient network with a certain average mesh size ξ is assumed to be formed by entangled chains. In these kinds of homogeneous structures, it is evident that the hydrodynamic motions of entangled polymer chains, which are related to the cooperative diffusion coefficient D_c, exist and are observable. The motions of individual chains, called self-diffusion D_s, are effectively restricted by the surrounding chains in a pseudo network. So the magnitude of D_s is nonexistent if all the chains have been involved in the entanglement. In any case, D_c is much slower than D_c if some of the chains have not been entangled due to the polydispersity effect. De Gennes[(1)] proposed that there are three types of motions for self-diffusion of individual chains: reptation, tube renewal and Stokes-Einstein friction. Differences in chain lengths between individual chains and entangled chains determine the type of motion of individual chains.

For monodisperse polymers, all polymer chains are involved in the entanglement to almost the same extent because their lengths are comparable. The opportunity for chains to disentangle is small. Therefore, we have observed only one cooperative motion in monodisperse semidilute polymer solutions. For polydisperse polymers, a certain amount of relatively smaller polymer chains exists. When the concentration is larger than the overlap concentration C^* which is computed from an average molecular weight and an average geometric size, not all the chains are involved in the entanglement. It is reasonable to assume that some smaller chains may be left free in the transient network and are responsible for the slow mode, as shown by the bimodal behavior in a polydisperse LPS230 solution. Since there are no quantitative data about those small chains in the polydisperse polymer sample, it is difficult to account for the observed effect quantitatively. Nevertheless, it is clear that the effect of polydispersity should not be ignored in the investigation of semidilute polymer solutions. The bimodal behavior should disappear at $C >> C^*$ even for polydisperse binary solutions.

V.2. Simulating Polydispersity Effect To simulate the polydispersity effect, we intentionally incorporated with the matrix some small chains of the same chemical nature. A small amount of low molecular weight polystyrene LPS2 was introduced into a monodisperse semidilute LPS80 solution, while keeping the same total concentration. The slow motion appeared besides the predicated fast motion. The resulting two modes conformed with the BBA theory about ternary mixtures of two polymers and a solvent.

Previously, Nemoto et al.[5] also tried to find two modes in PS1/PS2/Benzene ternary solutions. They have tested eight sets of combinations as listed in Table 8. The two modes were observed in only one of the eight solutions. According to our analysis, two modes could be observed under the following conditions. First, the size of the test chain cannot be too close to that of the matrix polymer. Otherwise the test chain may also be involved in entanglement and there is no slow mode. Second, if the size or the number of the test chain is too small relative to the matrix polymer, its self-diffusion may be invisible because of the weak scattering signal even though the slower mode does exist. For example, in all solutions except PS7804 in the experiments performed in Kyoto,[5] either the molecular weight of the test chain and that of the matrix polymer were too close, e.g., PS7819, or the intensity ($\propto M_w C$ where M_w is the molecular weight and C is the concentration) of the test chain was too small for self-diffusion to be visible, e.g., PS8404, PS8419, PS8442. Therefore, except PS7804, only one mode was observed in those solutions.

In the present study, we selected the conditions for the purpose of observing two modes. The fast motion, whose average linewidth $\bar{\Gamma}_f$ is almost the same as that of reference LPS80, was identified as the cooperative

Table 8. Molecular weights and concentrations of polystyrenes in benzene as reported by Nemoto[5]

Sample code[a]	Component 1		Component 2		Ratio		No. of Modes
	M_w (g/mole)	$C \times 10^2$ (g/g)	M_w (g/mole)	$C \times 10^2$ (g/g)	$\frac{Mw1}{Mw2}$	$\frac{(M_w C)_2}{(M_w C)_1}$	
ps7804	7.75×10^5	4.18	4.28×10^4	2.58	18	0.034	2
ps7819	7.75×10^5	4.19	1.86×10^5	2.60	4	0.155	1
ps8404	8.42×10^6	1.75	4.28×10^4	1.45	197	0.002	1
ps8419	8.42×10^6	0.389	1.86×10^5	0.138	45	0.009	1
ps8419	8.42×10^6	0.718	1.86×10^5	0.254	45	0.009	1
ps8419	8.42×10^6	1.67	1.86×10^5	0.560	45	0.007	1
ps8442	8.42×10^6	1.68	4.22×10^5	0.229	20	0.007	1
ps8442	8.42×10^6	2.28	4.22×10^5	0.304	20	0.007	1

[a]Sample code can be denoted as follows: ps, linear polystyrene; first two digits, molecular weight in 10^5 g/mole of the first polymer; last two digits, molecular weights in 10^5 g/mole of the second polymer.

motion of entangled chains; the slow motion may result from the self-diffusion of those smaller LPS2 test chains which have not yet been entangled. The self-diffusion coefficient $D_S(=\Gamma_S/K^2)$ due to the slow mode after correction of the matrix viscosity was indeed larger than the overall self-diffusion coefficient of all the test chains. In addition to the translational motion of the short LPS2 chains in the entangled pseudo network, there could also exist a coupling of the shorter chains and the entangled chains. It could be a combination of these two effects that contributed to the slow mode. It is not straightforward to predict which binary or ternary solution may exhibit two modes. However, our results clearly suggest that the polydispersity effect could lead to two characteristic modes.

V.3 Effect of Coupling To further examine the effect of coupling, we added PMMA test chains to the polystyrene matrix based upon two reasons: 1) PMMA is isorefractive with solvent toluene. It would be invisible if coupling did not exist; 2) PMMA is incompatible with PS. Thus the effect of compatibility can be studied. In principle, if the self-diffusion of the test chain comes only from the translational motion, then this motion should be invisible when the refractive index of the test chain matches with that of the solvent. The appearance of the slow motion in ternary mixtures LPS98 and SPS43 further ascertained that the coupling between the test chain and the matrix polymer not only existed, but also played an important role in "self-diffusion" in semidilute solutions. On the other hand, since the test chain and the matrix polymer are incompatible, it is energetically unfavorable for the test chain to be attached to the matrix and be involved in network entanglement in the same way as among the homopolymers. So the variance of self-diffusion should be quite small. In our experiment, it was 0.06 for LPS98 and 0.09 for SPS43.

V.4. Effects of Configuration and Entanglement We also demonstrated the effect of configuration on the dynamic behavior of semidilute polymer solutions. Different results have been obtained by adding to the matrix a 4-arm star polystyrene SPS4 which was comparable with the linear test chain LPS2 in number and size. First, the self-diffusion (the slow mode) became invisible (disappeared). Second, the cooperative diffusion became faster.

In the case of star polymers as test chains, the more the number of arms, the smaller the possibility for stars to be detached from the matrix once they are attached to the matrix over much longer periods. Even though star test chains are smaller, their motions are still hindered by other entangled polymer coils. If most star test chains participate in the entanglement, the amount of "free" stars is too small to be detected. So the CONTIN analysis did not show two modes for LPS84, but we do not want to deny the possible existence of self-diffusion of star test chains because $|g^{(1)}(t)|$ of LPS84 evidently deviated from the single exponential decay.

On the other hand, as soon as many stars are in the matrix, the entanglement points of the matrix are increased due to additional permanent crosslink points (i.e., centers of star polymers). Thus the average mesh size of the matrix decreases. In consequence, the cooperative diffusion becomes faster. But in the case of 12-arm star polystyrene as the test chain, the effect of star centers was not so obvious. According to the model of star shaped polymers by Daoud and Cotton,[36] there is a region with size χ around the center of a star, inside which the chains of other polymers do not penetrate. The distance χ is a function of the number of arms, i.e., $\chi \propto f^{\frac{1}{2}}$, where f is the arm number of the star. So the unpenetrable distance χ of 12-arm star is larger than that of a 4-arm star, leading to an effective decrease in the number of entangled points.

Furthermore, we tested the effect of entanglement on the characteristic linewidth. When the molecular weight of the probe polymer is close to that of the matrix polymer, test chains will certainly participate in the entanglement as long as they are compatible. This effect will be even stronger when the test chains are stars. In our experiment, the test chain and the matrix polymer of LPS812 were both polystyrenes. Therefore, for the 12-arm SPS and LPS8 they mixed and entangled entirely with each other. The dynamic behavior should be similar to that of the pure matrix. This could be the reason that LPS812 and LPS80 obeyed almost the same single exponential decay. As LPS812 was a ternary mixture, it is reasonable that the variance of cooperative diffusion increased slightly.

VI. CONCLUSIONS

We noted as follows: 1) The polydispersity could have significant effect on the dynamic behavior of semidilute solutions and could be one source of the slow motion. 2) The theory on the ternary mixtures is applicable to a certain extent, depending upon experimental conditions. If two polymer components are compatible and have close molecular weights, they will entangle together and exhibit unimodal behavior. In the case of two modes, the slow mode was due not only to the translational motion of the lower *MW* fractions, but also to the coupling between the two polymer components, i.e., the unentangled coils and the pseudo network. 3) The configuration of polymers (stars vs linear) could also affect the characteristic linewidth of semidilute solutions.

ACKNOWLEDGEMENTS

We gratefully acknowledge support of this work by the U.S. Army Research Office (DAAL0387K0136) and the National Science Foundation, US-China Cooperative Program (INT 8619977).

NOMENCLATURE

A:	background, Eq.(2)
$\mathbf{A}$:	integrated area of characteristic linewidth distribution
b:	coherence factor, Eq.(2)
C:	concentration
C^*:	overlap concentration
D:	translational diffusion coefficient
D_c:	cooperative diffusion coefficient
D_o:	probe diffusion coefficient at infinite dilution
D_s:	self-diffusion coefficient
$D\%$:	% deviation, Eq. (14)
f:	arm number of the star
g:	gravity
$G(\Gamma)$:	normalized distribution of characteristic linewidths
$g^{(1)}(\tau)$:	normalized first-order electric field correlation function
$G^{(2)}(\tau)$:	normalized intensity-intensity time correlation function
I_i:	intensity of representative molecular weight fraction M_i
K:	scattering vector
K_i:	ith cumulant
M:	molecular weight
MS:	mixed solvent
MWD:	molecular weight distribution
M_n:	number-average molecular weight
M_w:	weight-average molecular weight
N_A:	Avogadro's number

R_g: radius of gyration
t: time
w_i: normalized intensity weighting factor of representative molecular weight fraction M_i
α: a scaling parameter, Eq.(1)
β: a scaling parameter in Williams-Watts function, Eq.(11)
$\underline{\Gamma}$: characteristic linewidth
$\overline{\Gamma}$: average characteristic linewidth
$\overline{\Gamma}_f$: fast average characteristic linewidth
$\overline{\Gamma}_s$: slow average characteristic linewidth
Γ_o: 1 some effective linewidth
$\Gamma(\beta^-)$: gamma function
η: viscosity of solution
η_o: viscosity of solvent
$[\eta]$: intrinsic viscosity
θ: scattering angle
λ: wavelength in the scattering medium
λ_o: wavelength in vacuo
μ_i: ith moment of decay rate
ν: a scaling parameter, Eq.(1)
ξ: mesh size
τ: decay time
χ: distance from the center of the star

REFERENCES

1. For example, see P.-G. de Gennes, Scaling Concepts in Polymer Physics, Cornell University: Ithaca, N.Y. (1979).
2. For example, see S.F. Edwards and K.F. Freed, "Theory of the Dynamical Viscosity of Polymer Solutions," J. Chem. Phys. 61, 1189-1202 (1974); K.F. Freed and S.F. Edwards, "Polymer Viscosity in Concentrated Solutions," J. Chem. Phys., 61, 3626-3633 (1974).
3. M. Benmouna, H. Benoit, M. Duval and Z. Akcasu, "Theory of Dynamic Scattering From Ternary Mixtures of Two Homopolymers and a Solvent," Macromolecules, 20, 1107-1112 (1987).
4. R. Borsali, M. Duval, H. Benoit and M. Benmouna, "Diffusion of Polymers in Semidilute Ternary Solutions. Investigation by Dynamic Light Scattering," Macromolecules, 20, 1112-1115 (1987).
5. N. Nemoto, Y. Makita, Y. Tsunashima and M. Kurata, "Dynamic Light Scattering of Polymer Solutions. 4. Semidilute Solutions of Polystyrenes and Their Binary Blends in Benzene," Macromolecules, 17, 2629-2633 (1984).
6. B. Chu and T. Nose, "Static and Dynamical Properties of Polystyrene in trans-Decalin. 4. Osmotic Compressibility, Characteristic Lengths and Internal and Pseudogel Motions in the Semidilute Regime," Macromolecules, 13, 122-132 (1980).
7. P. Mathiez, C. Mouttet and G. Weisbuch, "On the Nature of the Slow Modes Appearing in Quasi-Elastic Light Scattering by Semi-dilute Polymer Solutions," J. Phys. (Orsay, Fr.) 41, 519-523 (1980).
8. I. Nishio and A. Wada, "Quasi-Elastic Light Scattering Study of Concentration Dependence of Diffusion and Internal Motion in Chain Polymers," Polym. J., 12, 145-152 (1980).
9. T.L. Yu, H. Reihanian and A.M. Jamieson, "Diffusion of Linear and Branched Polystyrenes in Tetrahydrofuran in the Dilute and Semidilute Regimes," Macromolecules, 13, 1590-1594 (1980).
10. J.G. Southwick, A.M. Jamieson and J. Blackwell, "Quasi-Elastic Light Scattering Studies of Semidilute Xanthan Solutions," Macromolecules, 14, 1728-1732 (1981).
11. E.J. Amis and C.C. Han, "Cooperative and Self-Diffusion of Polymers in Semidilute Solutions by Dynamic Light Scattering," Polymer, 23, 1403-1406 (1982).

12. E.J. Amis, P.A. Janmey, J.D. Ferry and H. Yu, "Quasielastic Light Scattering Measurements of Self-Diffusion and Mutual Diffusion in Gelatin Solutions and Gels," Polym. Bull. (Berlin), 6, 13-19 (1981).
13. W. Brown, "Slow-Mode Diffusion in Semidilute Solutions Examined by Dynamic Light Scattering," Macromolecules, 17, 66-72 (1984).
14. M. Tirrell, "Polymer Self-Diffusion in Entangled Systems," Rubber Chem. Technol., 57, 523-556 (1984).
15. M. Eisele and W. Burchard, "Slow-Mode Diffusion of Poly(vinylpyrrolidone) in the Semidilute Regime," Macromolecules, 17, 1636-1638 (1984).
16. S. Balloge and M. Tirrell, "The QELS 'Slow Mode' Is a Sample-Dependent Phenomenon in Poly(methyl methacrylate) Solutions," Macromolecules, 18, 817-819 (1985).
17. B. Chu, "Static and Dynamic Properties of Polymer Solutions," in: Scattering Techniques Applied to Supramolecular and Nonequalibrium Systems (S.-H. Chen, B. Chu and R. Nossal, ed.), pp.231-264, Plenum Press, New York (1981).
18. K. Huber, S. Bantle, W. Burchard and L. Fetters, "Semidilute Solutions of Star Branched Polystyrene: A Light and Neutron Scattering Study," Macromolecules, 19, 1404-1411 (1986).
19. L. Leger, H. Hervet and F. Rondelez, "Reptation in Entangled Polymer Solutions by Forced Rayleigh Light Scattering," Macromolecules, 14, 1732-1738 (1981).
20. H. Kim, T. Chang, J.M. Yohanan, L. Wang and H. Yu, "Polymer Diffusion in Linear Matrices: Polystyrene in Toluene," Macromolecules, 19, 2737-2744 (1986).
21. P.T. Callaghan and D.N. Pinder, "Self-Diffusion of Random-Coil Polystyrene Determined by Pulsed Field Gradient Nuclear Magnetic Resonance: Dependence on Concentration and Molar Mass," Macromolecules, 14, 1334-1340 (1981).
22. E.D. von Meerwall, E.J.Amis and J.D. Ferry, "Self-Diffusion in Solutions of Polystyrene in Tetrahydrofuran: Comparison of Concentration Dependences of the Diffusion Coefficients of Polymer, Solvent and a Ternary Probe Component," Macromolecules, 18, 260-266 (1985).
23. D.B. Cotts, "Properties of Semidilute Polymer Solutions: Investigation of an Optically Labeled Three-Component System," J. Polym. Sci.: Polym. Phys. Ed., 21, 1381-1388 (1983).
24. T.P. Lodge, "Self-Diffusion of Polymers in Concentrated Ternary Solutions by Dynamic Light Scattering," Macromolecules, 16, 1393-1395 (1983).
25. J.E. Martin, "Polymer Self-Diffusion: Dynamic Light Scattering Studies of Isorefractive Ternary Solutions," Macromolecules, 17, 1279-1283 (1984); "Polymer Self-Diffusion in Bimodal Semidilute Solutions," Macromolecules, 19, 922-925 (1986).
26. G.D.J. Phillies, G.S. Ullmann, K. Ullmann and T.-H. Lin, "Phenomenological Scaling Laws for 'Semidilute' Macromolecule Solutions From Light Scattering by Optical Probe Particles," J. Chem. Phys., 82, 5242-5246 (1985).
27. G.D.J. Phillies, "Universal Scaling Equation for Self-Diffusion by Macromolecules in Solution," Macromolecules, 19, 2367-2376 (1986); "Dynamics of Polymers in Concentrated Solutions: The Universal Scaling Equation Derived," Macromolecules, 20, 558-564 (1987).
28. B. Chu and D.-Q. Wu, "Polymer Probe Dynamics," Macromolecules, 20, 1606-1619 (1987).
29. B. Chu, "Light-Scattering Studies of Polymer Solution Dynamics," J. Polym. Sci.: Polym. Symp., 73, 137-155 (1985).
30. J.R. Ford and B. Chu, "Correlation Function Profile Analysis in Laser Light Scattering. III. An Iterative Procedure," in Photon Correlation Techniques in Fluid Mechanics, (E.O. Schultz-Dubois, ed.), pp.303-314, Springer-Verlag: New York (1983).

31. D.E. Koppel, "Analysis of Macromolecular Polydispersity in Intensity Correlation Spectroscopy: The Method of Cumulants," J. Chem. Phys., 57, 4814-4820 (1972).
32. S.W. Provencher, "Inverse Problems in Polymer Characterization: Direct Analysis of Polydispersity With Photon Correlation Spectroscopy," Makromol. Chem., 180, 201-209 (1979).
33. G. Fytas, T. Dorfmuller and B. Chu, "Photon Correlation Spectra of Poly(phenylmethyl siloxane) Under High Pressures," J. Polym. Sci. Polym. Phys. Ed., 22, 1471-1481 (1984).
34. Q.-C. Ying and B. Chu, "Overlap Concentration of Macromolecules in Solution," Macromolecules, 20, 362-366 (1987).
35. B. Chu, M. Onclin and J.R. Ford, "Laser Light Scttering Characterization of Polyethylene in 1,2,4-Trichlorobenzene," J. Phys. Chem., 88, 6566-6575 (1984).
36. M. Daoud and J.P. Cotton, "Star Shaped Polymers: A Model for the Conformation and Its Concentration Dependence," J. Physique, 43, 531-538 (1982).

THE SCALING LAWS IN TERNARY SYSTEMS: POLYMER - POLYMER - GOOD SOLVENT. AN EXPERIMENTAL STUDY

L. Ould Kaddour* and Cl. Strazielle

Institut Charles Sadron (CRM-EAHP), CNRS-ULP
6 rue Boussingault
67083 Strasbourg-Cedex
France

ABSTRACT

The dependence of the interaction between two unlike polymers A and B (characterized by ΔA_2 which is proportional to the interaction parameter χ_{ab}) and of their demixing concentration c_k, on their molecular weight is investigated. The experiments were performed by light scattering on four polymer pairs in good solvent. The results are in agreement with the prediction of scaling laws ($\Delta A_2 \sim M^{-0.44}$ and $c_k \sim M^{-0.63}$) derived from renormalization group calculations.

I. INTRODUCTION

In the last few years, large efforts in studying ternary systems, i.e., two unlike polymers in a solvent, have been made from both theoretical[1,2] and experimental points of view.

Recently, renormalization group calculations[3,4] have been used to derive new scaling laws for the molecular weight dependence of the interaction parameter, χ_{ab}, between unlike polymers (A and B) in a good solvent and the critical demixing concentration. The purpose of this paper is to present some experimental results which verify this theory for several mixtures of two polymers: polystyrene (PS) - poly(dimethyl-siloxane) (PDMS), poly(methyl-methacrylate) (PMMA)-PDMS, PS-PMMA and PS-poly(vinylacetate) (PVAc).

II. THEORETICAL PART

II.1. <u>Molecular weight dependence of the interaction parameter χ_{ab} (or ΔA_2) in dilute solution</u> The aim of the new theories[3,4] has been to predict the dependence of χ_{ab}, which is proportional to the difference of

---

* Present address: University of Tlemcen, B.P. 119, Tlemcen, Algeria

the second virial coefficients: $\Delta A_2 = A_{2,ab}-(A_{2,a} + A_{2,b})/2$, on molecular parameters such as molecular weight. Here $A_{2,a}$ and $A_{2,b}$ are the second virial coefficients due to the interactions between solvent and chains of polymer A and polymer B, respectively; and $A_{2,ab}$ the coefficient of two different chains.

From the general relation of light scattering established for the ternary system[1,2], i.e., two unlike polymers A and B in solvent S, one can deduce an expression of scattering intensity ΔI, ($Intensity_{solution}$ - $Intensity_{solvent}$), which is a function of the second virial coefficients. This expression in the case of dilute solution and at wavevector $\mathbf{q}$ equal to zero, ($\mathbf{q} = (4\pi/\lambda)\sin\theta/2$, where λ is the wavelength of the incident beam in the medium and θ the scattering angle), is:

$$K'^{-1}\Delta I = \Delta n_a^2 M_a c_a + \Delta n_b^2 M_b c_b - 2\Delta n_a^2 M_a^2 c_a^2 A_{2,a}$$

$$-2\Delta n_b^2 M_b^2 c_b^2 A_{2,b} - 4\Delta n_a \Delta n_b M_a M_b c_a c_b A_{2,ab}, \qquad (1)$$

where c_i and M_i, ($i = a$ or b), are the concentration and the molecular weight of the polymer and Δn_i their refractive index increment, K' being the optical constant equal to $(2\pi^2/\lambda_o^4)n^2/N_a$, with λ_o the wavelength of the incident radiation in vacuum, n the refractive index of solution and N_a the Avogadro's number.

Now, according to the Flory-Huggins theory, the second virial coefficients are related to the interaction parameters as follows:

$$A_{2,i} = \frac{\overline{v}_i^2}{2V_o}(1 - 2\chi_i), \; with\, i = a\, or\, b$$

and (2)

$$A_{2,ab} = \frac{\overline{v}_a \overline{v}_b}{2V_o}(1 - \chi_a - \chi_b + \chi_{ab}),$$

where V_o is the molar volume for solvent, and $\overline{v}_i$ (a or b) the partial specific volume of the polymers.

The second virial coefficients can also be related to the interpenetration functions Ψ defined by:[5]

$$\Psi_i = A_{2,i} M_i^2/(4\pi^{3/2} R_i^3 N_a), \quad (with\, i = a\, or\, b)$$

and (3)

$$\Psi_{a,b} = A_{2,ab} M_a M_b/(4\pi^{3/2} R_a^{3/2} R_b^{3/2} N_a), \qquad (4)$$

where R_a and R_b are the radii of gyration of the chains A and B, respectively, and M_a and M_b their molecular weights. The function Ψ_{ab}, similar to Ψ_a and Ψ_b, tends to the asymptotic limit $\Psi^* = 0.24$ when the molecular weight increases.

When the polymers are in good solvents and for a symmetrical system: $M_a = M_b$, $R_a = R_b$, and therefore $\Psi_a = \Psi_b = \Psi$, it has been demonstrated[3,4] that the difference of the interpenetration functions: $\Delta\Psi = \Psi_{ab} - \Psi$, is related to the molecular weight by the scaling law:

$$\Delta\psi \simeq M^{-\chi} with \quad \chi = 0.22 \pm 0.02$$

On the other hand, the difference $\Delta\Psi = \Psi_{ab} - \Psi/2$ is proportional to ΔA_2 and thus to the Flory interaction parameter χ_{ab}, if we assume that $\overline{v_a} = \overline{v}_b = \overline{v}$, as follows:

$$\Delta A_2 = \frac{\overline{v}^2}{2V_o \chi_{ab}} = 4\pi^{3/2} \Delta\psi R^3 / M^2 . \tag{5}$$

If the system deviates slightly from symmetry, $\Psi = (\Psi_a + \Psi_b)/2$ is used.

From relation(5), ΔA_2 (or χ_{ab}) is related to molecular weight as follows:

$$\Delta A_2 = k M^{-a} , \tag{6}$$

where the exponent $a = 2 + \chi - 3\nu = 0.44 \pm 0.02$. The quantity ν is the exponent of the scaling law for the radius of gyration R; $R \simeq M^{-\nu}$ with $\nu = 0.588$[6] (renormalization group) or $\nu = 0.6$ (Flory's value). The prefactor k is related to the degree of incompatibility between polymer A and B.

II.2 Molecular weight dependence of critical concentration c_k (semidilute solution). When demixing of a symmetrical ternary system occurs, the spinodal curve, which corresponds to the limit of metastability of the homogeneous solution, and the binodal curve, which corresponds to the coexisting phase, are tangent at one point corresponding to the critical concentration c_k which is higher than the overlap concentration $c^* = 3M/(4\pi N_a R^3)$. Recently, it has been found[4,7] that the critical concentration in a semidilute solution is related to molecular weight by the following scaling law:

$$c_k = K M^{-b}, with\ b = \frac{3\nu - 1}{1 + \chi} = 0.63 , \tag{7}$$

and K the prefactor depending on the temperature. In dilute solution, b should be equal to 0.55.

III. EXPERIMENTAL RESULTS

III.1. Samples and method The PS samples were unfractionated polymers obtained by anionic polymerization using the classical procedure used in the laboratory. The PDMS, PMMA and PVAc samples were either commercial samples or prepared in the laboratory by cationic (PDMS) or free radical (PMMA) polymerization. These polymers were fractionated. The index of polydispersity of the fractions determinated by size exclusion chromatography (SEC) was in the range of 1.1 to 1.3. The average molecular weights measured by light scattering are given in the Tables 1 and 2.

All the experiments were performed at room temperature, using a FICA 50 light scattering apparatus with a vertically polarized incident beam at $\lambda_o = 546$nm. The clarification of dilute and semidilute solutions was obtained by centrifugation (15000 rpm during 2h). The solvents were redistilled twice before use.

III.2. Determination of the exponent a of the scaling law $\Delta A_2 \simeq M^{-a}$ (dilute solution) The χ_{ab} parameter (or ΔA_2) was determined by means of light scattering for the PS-PDMS-toluene, PMMA-PDMS-chloroform and PS-PVAc-styrene. The method used here was proposed by Inagaki,[8] and is

Table 1. Molecular weights, interaction parameter and second virial coefficient ΔA_2 for PS-PDMS, PMMA-PDMS and PS-PVAc mixtures at mass fraction $y = c_a/(c_a+c_b) = 0.5$. $M = (M_a + M_b)/2$.

PS-PDMS-toluene system				
$M_a.10^{-3}$	$M_b.10^{-3}$	$M.10^{-3}$	χ_{ab}	$\Delta A_2.10^4$ mol.ml.g-2
25.3	25.5	25.4	0.076	3.43
60.0	57.5	58.75	0.0575	2.60
85.0	79.2	82.1	0.0457	2.066
230.0	280.0	255.0	0.032	1.446
740.0	718.0	730.0	0.0183	0.827
1,400.0	1,220.0	1,310.0	0.0125	0.556

PMMA-PDMS-$CHCl_3$ system				
$M_a.10^{-3}$	$M_b.10^{-3}$	$M.10^{-3}$	χ_{ab}	$\Delta A_2.10^4$ mol.ml.g-2
28	25.5	26.75	0.055	3.01
66	57.5	61.75	0.0326	1.787
177	164.0	170.5	0.0278	1.52
305	280.0	292.0	0.017	0.932
506	605.0	580.0	0.015	0.822
1,030	1,220.0	1,120.0	0.0122	0.669

PS-PVAc-styrene system				
$M_a.10^{-3}$	$M_b.10^{-3}$	$M.10^{-3}$	χ_{ab}	$\Delta A_2.10^4$ mol.ml.g-2
120	120	120	0.0057	2.175
245	245	245	0.0043	1.64
396	412	404	0.0034	1.297

Table 2. Critical concentration c_K and overlap concentration c^* as dependence of molecular weights for polymer A-polymer B-good solvent at mass fraction y = 0.5

System	$M_a 10^{-3}$	$M_b 10^{-3}$	$c_k 10^2$ g.ml^{-1}	$c^* 10^2$ g.ml^{-1}
A:PS B:PDMS S:THF	98.0 230.0 740.0 1,400.0 3,000.0	99.5 280.0 718.0 1,320.0 4,200.0	4.88 3.20 1.82 1.06 0.62	2.82 1.36 0.61 0.39 0.18
A:PS B:PDMS S:toluene	25.3 98.0 230.0 740.0 3,000.0	25.6 99.5 280.0 718.0 4,200.0	9.7 4.35 2.70 1.52 0.475	7.84 2.82 1.36 0.61 0.18
A:PS B:PMMA S:benzene	135.0 460.0 1,400.0 3,000.0	128.0 458.0 1,300.0 3,000.0	16.2 7.5 3.82 2.36	2.0 0.75 0.30 0.14
A:PS B:PVAc S:styrene	120.0 245.0 396.0	120.0 245.0 412.0	6.8 5.3 3.38	0.64 0.37 0.24

called "optical theta conditions" or "nul average contrast," defined by the following relation:

$$y \Delta n_a M_a + (1-y) \Delta n_b M_b = 0 \ , \tag{8}$$

where y, ($y = c_a/(c_a + c_b)$), is the composition of polymer A in the mixture. In this condition, the solvent is chosen so that the refractive index increments Δn_a and Δn_b of the two polymers are opposite in sign, but not very much different in absolute value. Considering these absolute values, the composition y is determined in order to have relation(8). Under these conditions, Eq. (1) becomes:

$$\frac{K' c_T}{\Delta I(\mathbf{q}=0)} = A + BA^2 \chi_{ab} c_T \ , \tag{9}$$

with

$$A = \Delta n_a^2 y M_a + \Delta n_b^2 (1-y) M_b ; \tag{10}$$

and

$$B = 2 \Delta n_a \Delta n_b y (1-y) M_a M_b \frac{v_a v_b}{V_0} \ , \tag{11}$$

where c_T, ($c_T = c_a + c_b$), is the total concentration of the polymers.

The slope of the variation of $K' c_T / \Delta I(\mathbf{q} = o)$ as a function of c_T gives directly χ_{ab} with good precision in dilute solution, as is shown in Fig. 1 for some mixtures. Curve 1 corresponds to a mixture of PS (M_w = 240,000) and PVAc (M_w = 240,000) in styrene, and curve 2 for a mixture of PS (M_w = 230,000) and PDMS (M_w = 280,000) in toluene. All the results in dilute solution (χ_{ab} or ΔA_2) are listed in Table 1 for the systems PS-PDMS, PMMA-PDMS and PS-PVAc.

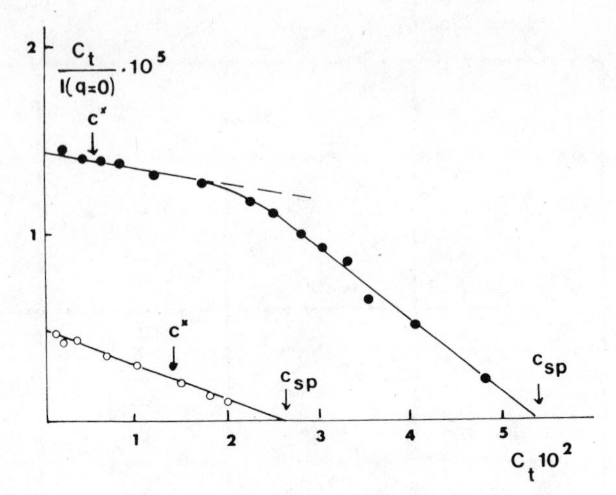

Fig. 1. Light scattering experiments (in conditions of "nul average contrast" see text) for two systems:
1. PS-PVAc-styrene (M = 245,000)
2. PS-PDMS-toluene (M = 255,000)

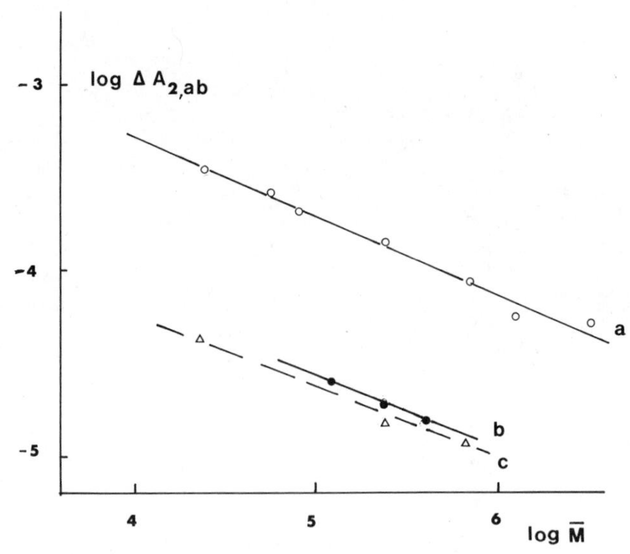

Fig. 2. Dependence of ΔA_2 on molecular weight M.
a : PS-PDMS-toluene
b : PS-PVAc-styrene
c : PS-PMMA-bromobenzene (ref. 8)

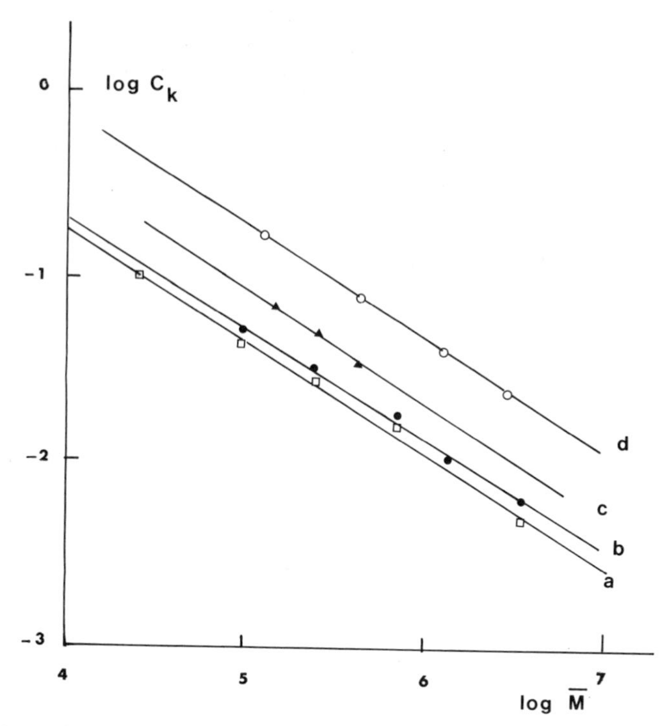

Fig. 3. Dependence of critical concentration c_K on molecular weight M
a : PS-PDMS-toluene
b : PS-PDMS-THF
c : PS-PVAc-styrene
d : PS-PMMA-toluene

Figure 2 shows the variation of ΔA_2 versus M (with $M = (M_a + M_b)/2$) for 3 systems; Curve C is obtained from literature data[8] for the system PS-PMMA. From these variations, the scaling laws are respectively:

$\Delta A_2 = 3.54 \quad (\pm 0.08).10^{-2} \quad M^{-0.45(\pm 0.02)}$ (PS-PDMS-toluene),

$\Delta A_2 = 1.58 \quad (\pm 0.08).10^{-2} \quad M^{-0.40(\pm 0.04)}$ ($PDMS-PMMA-CHCl_3$),

$\Delta A_2 = 3.45 \quad (\pm 0.08).10^{-3} \quad M^{-0.42(\pm 0.04)}$ (PS-PVAc-styrene),

$\Delta A_2 = 4.15 \quad 10^{-3} \quad M^{-0.45}$ (PS-PMMA-bromobenzene),

From the precision of the slope of the experimental curves giving the variations of the quantity $K'c_T/\Delta I(\mathbf{q} = o)$ as function of c_T, the average uncertainty of the exponent a has been determined equal to $\pm$ 0.03.

This exponent a is in good agreement with the theoretical value. However, the exponent a = 0.40 is slightly lower than expected for the mixture PDMS-PMMA. It is experimentally shown that ΔA_2 is not constant but decreases when M increases with a characteristic exponent a = 0.45, higher than that of scaling law for A_2 ; $A_2 \simeq M^{-0.24}$, $(3\nu - 2)$.

III.3. Determination of the exponent b of the scaling law: $c_K \simeq KM^{-b}$

The critical concentration of demixing was determined for 3 mixtures of polymers PS-PMMA, PS-PDMS and PS-PVAc at constant composition $y = c/c_T$ = 0.5. The solvents used for the PS-PDMS couple were toluene, in which the "optical theta" condition was verified: $\Delta n_{PS} = - \Delta n_{PDMS}$, and THF, in which the PDMS was matched : $\Delta n_{PDMS} = 0$. The system PS-PMMA was mixed in benzene where Δn_{PMMA} is 0.

The solvent for the system PS-PVAc was styrene "nul average contrast." For the symmetrical system, at y = 0.5, c_k is the concentration at the spinodal, c_{SP}, which corresponds to the concentration when the scattering intensity at $\mathbf{q}$ = 0 diverges : ΔI^{-1} ($\mathbf{q}$ = 0) = 0 (see Fig. 1).

Table 2 gives the critical concentration c_k of demixing compared to the overlap concentration c^*. The experimental result for the 4 systems are shown in Fig. 3.

From these plots, one can deduce for the exponent b the value of 0.61 + 0.02 in agreement with the theoretical value (though slightly lower for PS-PDMS mixture (b = 0.58) :

$$c_k = 44.6\ M^{-0.58(\pm 0.02)} \quad \text{(PS-PDMS-THF)}\ ,$$

$$c_k = 43.6\ M^{-0.59(\pm 0.03)} \quad \text{(PS-PDMS-toluene)}\ ,$$

$$c_k = 227\ M^{-0.61\ (\pm 0.02)} \quad \text{(PS-PMMA-benzene)}\ ,$$

and

$$c_k = 81.7\ M^{-0.62(\pm 0.03)} \quad \text{(PS-PVAc-styrene)}\ ,$$

IV. CONCLUSIONS

It may be noted that the concentration c_k for the PS-PMMA-benzene and for PS-PVAc-styrene characterizing a weak incompatibility, is much higher than the overlap concentration c^*, i.e., c_k/c^* 8 to 10, this ratio is much lower (2 to 4) than that for the PS-PDMS pair.

NOMENCLATURE

- a exponent
- A_2 second virial coefficient
- b constant
- c_i concentration
- c_k critical demixing concentration
- C^* overlap concentration
- C_{sq} concentration at which the scattering intensity at $\mathbf{q}$ = 0 diverges
- I scattering intensity
- θ scattering angle
- k' optical constant
- K prefactor
- λ_o wavelength of the incident team
- M molecular weight
- N_a Avagadro's number
- n refractive index of the solution
- Δn_i refractive index increment
- $\mathbf{q}$ wavevector
- R radius of gyration

Ψ interpenetration function
V_o molar volume of the solvent
$\overline{v}_i$ partial specific volume of the polymers
ν exponent of the scaling law for the radius of gyration
χ_{ab} interaction parameter
Y $c_a/(c_a+c_b)$

REFERENCES

1. W.H. Stockmayer, "Light Scattering in Multi-component Systems," J. Chem. Phys., 18, 58-61 (1950).
2. H. Benoît and M. Benmouna, "New Approach to the Problem of Elastic Scattering From a Mixture of Homopolymers in a Concentrated Solution", Macromolecules, 17, 535-540 (1984).
3. J.F. Joanny, L. Leibler and R. Ball, "Is Chemical Mismatch Important in Polymer Solutions?", J. Chem. Phys., 81, 4640-4656 (1984).
4. L. Schäfer and Ch. Kappeler, "A Renormalization Group Analysis of Ternary Polymer Solutions," J. Phys. (Paris), 46, 1853-1864 (1985).
5. H. Yamakawa, in "Modern Theory of Polymers Solutions," Chap. 7, Harper & Row Publ., New York (1971).
6. J.P. Cotton, "Polymer Excluded Volume Exponent ν : An Experimental Verification of the n Vector Model for n = 0," J. Phys. Lett. (Paris), 41, 231-234 (1980).
7. D. Broseta, "Demixation dans les mélanges de polymères en Solution," Thesis (Paris), (1987).
8. T. Fukuda, M. Nagota and H. Inagaki, "Light Scattering From Ternary Solutions. I. Dilute Solutions of Polystyrene and Poly(methylmethacrylate)," Macromolecules, 17, 548-553 (1984).

EXCIMER STUDIES OF POLYMER SOLUTIONS OVER THE ENTIRE CONCENTRATION RANGE

Renyuan Qian

Institute of Chemistry
Academia Sinica
Beijing 100080
People's Republic of China

ABSTRACT

Results of excimer fluorescence studies on solutions of polystyrene, polyethylene terephthalate and polybenzamide from dilute, semidilute and concentrated solutions and on the solid films have been discussed and summarized. For flexible chain polymers, a fractional power dependence of the excimer to monomer fluorescence intensity ratio on concentration in the semidilute region have been found to be a polymer effect resulting from a shrinkage of the coil dimension in solution with increasing concentration. This is true also in a θ-solution. In very concentrated solutions and in the amorphous solid film of polyethylene terephthalate pronounced excimer fluorescence observed indicates the presence of local stacking of terephthaloyl chromophores during the process of condensed phase formation and some local parallel chain segment aggregation. For the rigid chain polybenzamide evidence for the stacking of the planar chromophores in going from a dilute solution to a nematic solution has been found in concentrated sulfuric acid solution by excimer fluorescence.

I. INTRODUCTION

The study of polymer solutions has had a long history and played an important role in the development of the very concept of macromolecules. Most studies have been concerned with dilute solutions. Besides some technical interest like fiber spinning from solution and so on, the study of concentrated polymer solutions has been handicapped by the complexity of interchain interactions and lack of proper experimental tool and theoretical guide. The introduction of scaling theory by de Gennes[1] to polymers made a strong impetus to the study of polymer solutions over the entire concentration range, i.e., from dilute to semidilute, to concentrated solutions to solvent-plasticized film and eventually to a solid film. Macromolecules do not exist as a gaseous phase before decomposition. The molecular chain aggregation process involved from isolated chains to the formation of a condensed phase could only be followed by the changes in a properly chosen physical property which is sensitive to intermolecular interactions when going from a dilute solution to a condensed phase.

The application of scaling concept and renormalization theory to polymer solutions leads to several "phase" regions in the temperature-concentration diagram.[2] Since then, numerous attempts have been made to find experimentally the existence of any sharp changes in certain solution property or its slope with respect to concentration when the solution is going across the concentration boundaries of the solution "phase" regions. Unfortunately most experimental results seem to deny such an existence. The thermodynamic as well as transport properties of a polymer solution change rather gradually and continuously throughout the concentration regions without any sharp change of slope in going across these "phase" boundaries.[3-6] From a naive physical picture of a polymer solution based on the spatial segment density distribution and inter-segmental interactions[7] it is reasonable to expect such boundaries, although not in the sense of phase boundaries, to exist when going from dilute to semidilute and to concentrated regions of a polymer solution. In a dilute solution macromolecular chains exist as isolated coils when the chain is flexible. At and beyond a concentration defined as c_s the effect of inter-chain interactions begins to be felt by the chains in solution. On further increase of solution concentration, the coils will overlap each other and the spatial segment density distribution will become continuous but undulating. Further increase of concentration to a value c^+ the chains in solution will be densely overlapped so that the spatial segment density becomes statistically uniform everywhere throughout the solution, analogous to the case of a solid amorphous polymer but with a smaller average spatial segment density. The situation is schematically illustrated in (Fig. 1). The problem at hand is to find if there exists any solution property which will show the transitions between these solution concentration regimes when going from a dilute solution to a solid film.

For rigid chains in solution, our understanding of the molecular chain packing and molecular aggregation process involved in going from a dilute solution to a solid film is very superficial at present. For a solution

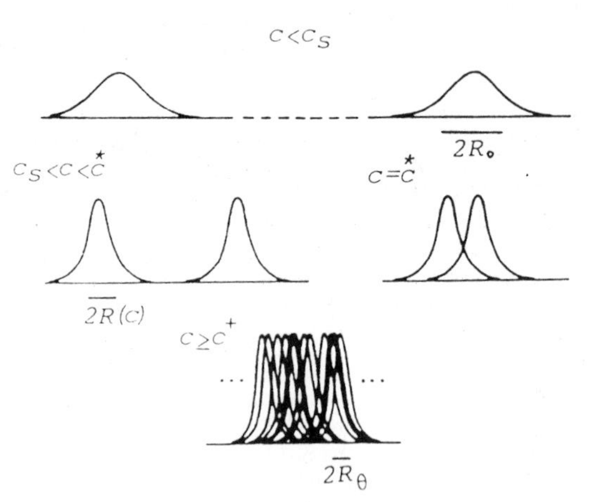

Fig. 1. Spatial segment density distribution of a flexible polymer coil in a good solvent from a dilute to a concentrated solution (schematic).

of rod-like chains there exists nematic liquid crystalline domains above certain critical concentration in the concentrated solution regime. The role played by the specific polymer-solvent interaction needed for the rigid chains to dissolve and by the anisotropic interchain interactions to the molecular chain aggregation process occurring in the transition from a dilute solution to a nematic phase and to a solid film is little understood.

In the following sections the application of excimer fluorescence as a probe for the study of the interchain interaction in polymer solutions over the entire concentration range will be discussed. The experimental results obtained in this laboratory over the past few years are encouraging, with some degree of success as well as arousing controversies on the problems of macromolecular chain condensation process in solution from a dilute solution to a solid film.

II. EXCIMER FLUORESCENCE OF POLYSTYRENE SOLUTION

The excimer fluorescence of polystyrene (PS) solution has been widely studied[8] since early sixties, predominately for intrachain excimer formation between adjacent pendant phenyl chromophores on the chain in dilute solutions, $c < 5.10^{-2}$ g/dl. As the excimer interaction is a very short range interaction between an excited state and a ground state chromophore at a parallel planar configuration of 0.33-0.35 nm apart, it should be useful to probe the molecular aggregation processes. For the present discussion only the concentration dependence of the intensity ratio $\zeta \equiv I_E/I_M$ of the excimer to monomer (isolated excited chromophore) fluorescence will be of interest. Typical fluorescence spectrum of PS in solution is shown in Fig. 2. The overlap between the monomer fluorescence peak (λ = 285 nm) and the excimer fluorescence peak (λ = 330 nm) is not severe so that simple peak intensity ratio is adequate for the evaluation of ζ within usual experimental errors. Results of first such investigation by Roots and Nyström[9] to show a sharp turn at the overlap concentration c^* ($c^*[\eta] \simeq 1$) were not confirmed.[10,11] Extensive studies carried out in the author's laboratory showed a peculiar concentration

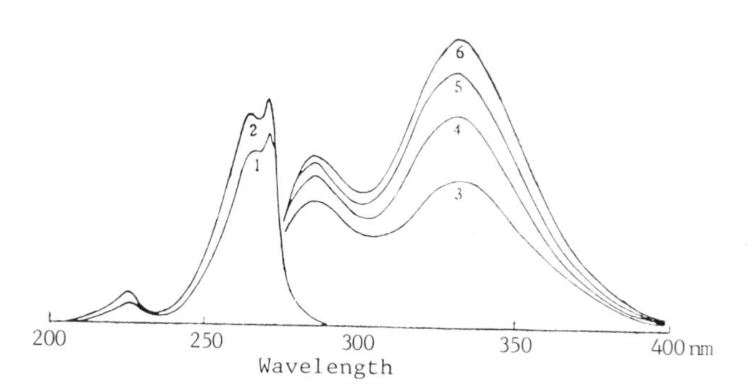

Fig. 2. Fluorescence excitation spectra (left) and emission spectra (right) of PS in DCE.

Excitation: 1 - λ_{em} 330 nm (excimer); 2 - λ_{em} 285 nm (monomer)

Emission: 3,4,5,6-solution concentration 0.011, 0.256, 3.82, 11.4g/dl dependence of $\zeta(c)$ in the semidilute region.[10] A typical example of PS in a good solvent, 1,2-dichloroethane (DCE), is shown in Fig. 3. Two critical concentration values could be identified clearly, designated as c_s and c^+. c_s is located at the transition from a concentration independent ζ at $c < c_s$ to a fractional power dependence at $c_s < c < c^+$, while c^+ is located at the transition from fractional power dependence of $\zeta(c)$ to a linear and super-linear dependence beyond c^+.[12] Such concentration dependence $\zeta(c)$ was also observed by Graley <u>et al</u>.[12] for the polymer PPDA (poly(1,6-cyclohexane di-oxyethylene glycol p-phenylene diacrylate).

$$-O-CO-CH=CH-C_6H_4-CH=CH-CO-O-CH_2CH_2-O-C_6H_{10}-O-CH_2CH_2-$$

in DCE although interpreted differently from the present discussion. Since then such behavior of $\zeta(c)$ have been found to be rather general as in the case of PS star polymers[13] in DCE and poly(polytetramethylene ether glycol naphthalene 2,6-dicarboxylate) (PTMN) $-OC-C_{10}H_6-CO-O-(CH_2CH_2CH_2CH_2-O)_m$, $m\sim 23$ in chloroform[14] as shown in Fig. 3. Whatever the aromatic ring chromophores are attached to the side chain or in the back-bone chain of good flexibility, similar concentration

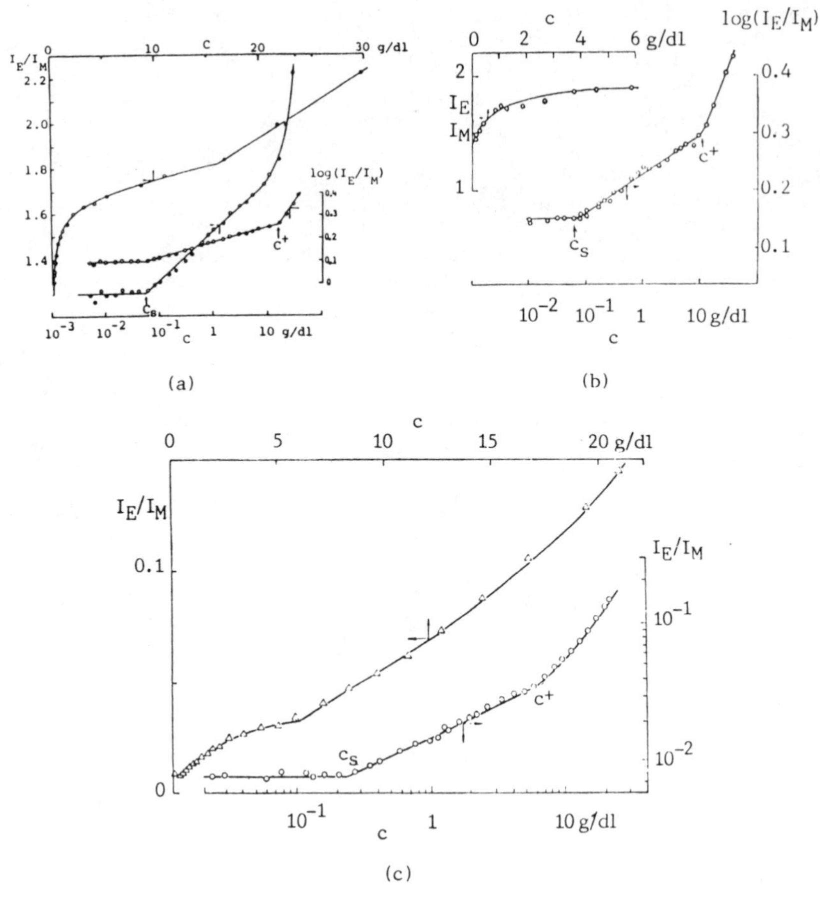

Fig. 3. $\zeta(c)$ for (a) PS of M_w = 250 kD in DCE; (b) 12-arm star PS of M_w = 800 kD in DCE; and (c) poly(polytetramethylene ether glycol naphthalene 2,6-dicarboxylate) of M_n = 18 kD in $CHCL_3$, all measured at 25°C.

dependences of $\zeta(c)$ in the semidilute region were observed. At concentrations above c^+ the values of ζ increase steadily. In the solid film of PS the excimer fluorescence is the sole emission observed presumably due to the migration of excitation energy[15] to the excimer sites lying within a sphere of action of ca. 2-3 nm[16] from the excited chromophore. The relative probability of excitation migration to excimer sites over monomer fluorescence emission will depend very much on the density of excimer sites prevailing. Consequently it plays a more important role in a condensed state than in a dilute solution.

The fractional power dependence

$$\zeta(c) \propto c^{\varepsilon}, \qquad \varepsilon = 0.07 - 0.5 \tag{1}$$

in semidilute solutions of flexible chain polymers has been interpreted as a manifestation of shrinkage of the polymer coil in solution due to repulsive interaction between segments of neighboring chains in solution.[7] The values of ε so far found for the polymer-solvent systems are listed in Table 1. The values of c_s were found to be of the order of 10^{-1} g/dl, except PPDA-DCE which turned out to be 2.10^{-2} g/dl. For PS the molar mass dependence of c_s was found to be weak, $c_s \propto M_w^{-0.1}$. The concentration region between c_s and c^+ covers two orders of magnitude. As the solution around the transition region of c_s is still rather dilute in the ordinary sense there is no reason to expect a sudden change of energy migration rate except an enhancement of energy migration reflected from a change of chain conformation leading to a shrinkage of the coil dimension in solution, thus an increase of spatial segment density. It is well known that energy migration is very sensitive to the chromophore density because of strong distance dependence.[17]

This fractional power dependence of $\zeta(c)$ has shown to be a unique flexible polymer effect by experiments with PS samples of low molar mass[18] and PS in solid miscible blends.[19] When the molar mass of PS is less than 2 kD the concentration dependence $\zeta(c)$ showed a linear dependence without a region of fractional power dependence as shown in (Fig. 4). For small molecules in solution $\zeta(c)$ has been found to be proportional to the concentration in all cases and the value of ζ starts from zero. For low molar mass polymer PS because of the intrachain excimers existing in solution $\zeta(c)$ extrapolates to some finite value at zero concentration. A demarkation of the behavior of $\zeta(c)$ for PS-DCE solutions of different molar mass samples has been found to lie between 1.7-4.2 kD.

Table 1. The values of ε.

Polymer-solvent system	ε	Ref.
PS-DCE	0.07	10
Star PS (3 and 12 arm)-DCE	0.07	13
PS-cyclohexane (35.4°C)	0.15	18
PPDA-DCE	0.21	12
PTMN-$CHCl_3$	0.47	14

PS and polyvinylmethylether (PVME) form miscible blends over all proportions at room temperature. When a series of solution-cast films of PS/PVME blends from 2.5 to 70 wt% PS were examined for excimer fluorescence, the $\zeta(c)$ curve obtained is shown in (Fig. 5). Below 5% PS the value of ζ remains constant and then increases with increasing concentration with a curvature opposite to that observed in semidilute solutions, showing only the regions of linear and super-linear dependence. In a solid PS/PVME miscible blend no coil shrinkage due to interaction of neighboring chain segments of PS would be possible because of the stronger attractive interaction between segments of PS and PVME that prevails, which is essential to rendering the system miscible over the entire concentration range.

From the arguments presented above we are led to consider with some confidence that the fractional power dependence of $\zeta(c)$ is a flexible polymer effect in semidilute solutions originated from the coil shrinkage due to interchain interactions. As the transition in the observed data are rather sharp the value of c_s could be located with good confidence. For the transition from a semidilute to a concentrated solution the value of c^+ has been found to be in the order of 10 g/dl for all polymers examined, being roughly independent of molar mass. As an amorphous polymer has the same density value irrespective of its molar mass it is reasonable to

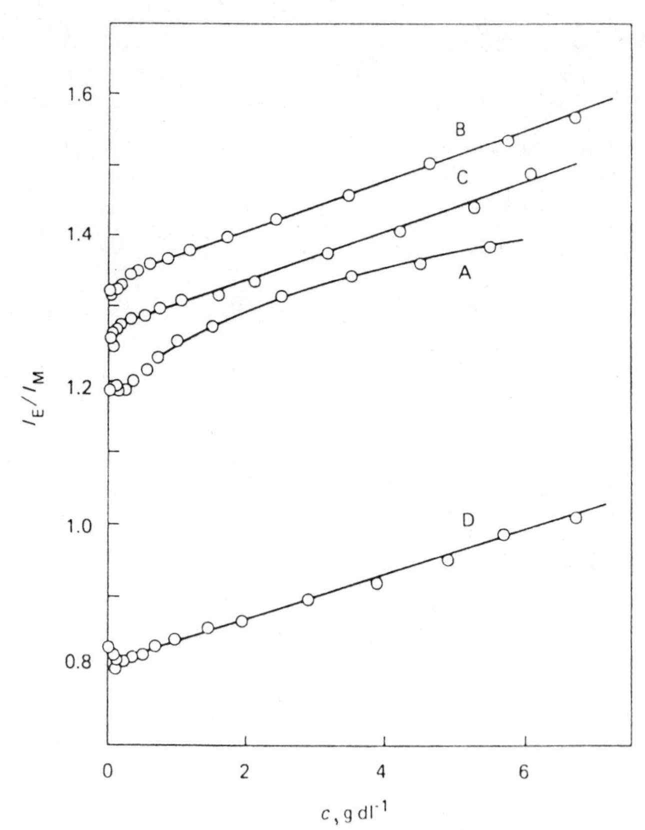

Fig. 4. ζ(c) for PS in DCE at 25°C. M_n of PS: A,B,C,D-4.2, 1.7, 1.4, 0.7 kD.

expect the same spatial segment density irrespective of molar mass when the spatial segment density distribution becomes uniform throughout the solution above the concentration c^+. The overlap concentration c^* as usually defined from the coil dimension in a dilute solution will lose its meaning because of shrinkage of coil dimension before the concentration reaching c^*. At present, the only sensible method of experimental estimation of c^* is the concentration dependence of the sedimentaion coefficient $s(c)$ which is molar mass dependent at $c < c^*$ and becomes molar mass independent at $c \geq c^*$.[20,21]

III. POLYSTYRENE IN θ-SOLVENT

A very interesting and unexpected experimental fact is the finding of the same fractional power dependence of $\zeta(c)$ for PS in a θ-solvent cyclohexane at 25.4°C as shown in Fig. 6[18] for a PS sample of $M_w = 6.1 \cdot 10^6$ as compared to the same sample in DCE, a good solvent. The c_s value was found to be higher in the θ-solvent (0.15 g/dl) than in DCE (0.44 g/dl). This is reasonable since the polymer coil dimension in a θ-solvent is significantly smaller than that in a good solvent. The value of ζ in dilute solution is higher in the θ-solvent than that in DCE, which also reflects a higher spatial segment density of the coil in a θ-solvent. The important issue here is that a flexible polymer coil will shrink its coil dimension even in a θ-solution when going from a dilute solution to a semidilute solution. In this concentration region, it is still quite dilute in the usual sense so that the third osmotic virial coefficient is not significant and binary intercoil encounters only will be important. Even if the long-range intersegmental repulsion vanishes in

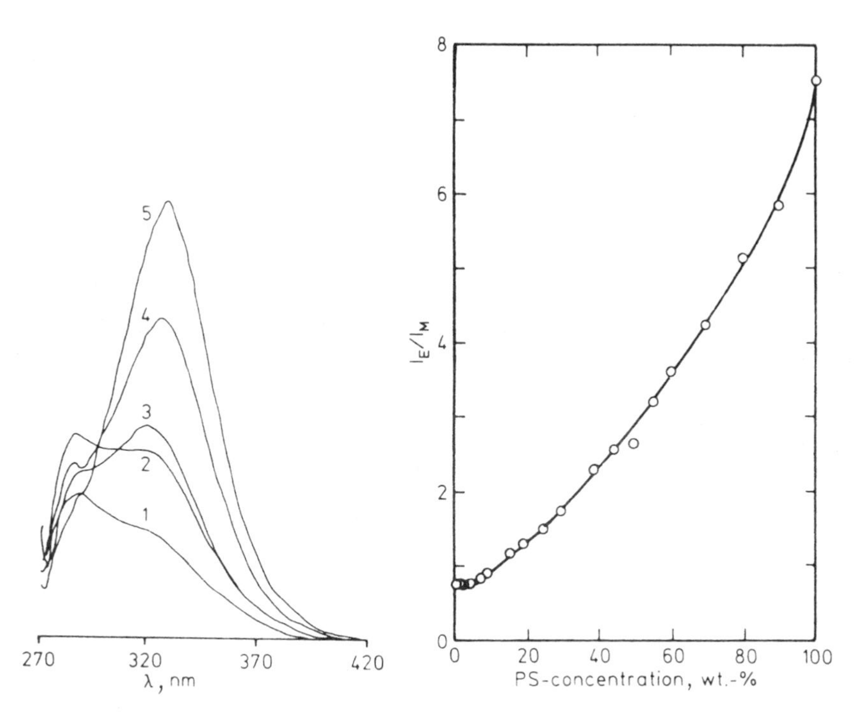

Fig. 5. Fluorescence spectra (left) of PS/PVME miscible blend films and $\zeta(c)$ for these blends (right).

1,2,3,4,5-2.5, 11.3, 19.1, 46.0, 70.0 wt% PS.

a θ-solution there will be a short-range repulsion, the hard-core volume effect, existing under all circumstances. Due to the flexibility of a polymer chain the coil conformation seems to be very sensitive to the intercoil interaction if the foregoing interpretation is correct.

A theoretical model has been proposed[22] to treat the concentration dependence of the flexible polymer coil dimension in semidilute solutions by a perfectly reflecting spherical wall around each segment of the coil, which influences the random walk conformation. In a θ-solution the intersegmental interaction is expressed by a potential

$$\phi(r) = \begin{cases} \infty, & r \le a \\ 0, & a < r < t\,\mathbf{h}_0 \\ \infty, & r \ge t\,\mathbf{h}_0 \end{cases} \tag{2}$$

where a is the hard-core radius of the segment, t a parameter characterizing the radius of the reflecting wall and $\mathbf{h}^2{}_0$ the random walk value of $< r^2 >$. Thus,

$$h_\theta^2(c) = <r^2> = \mathbf{h}_0^2 - N^2/2\int_0^a (r^2 - \mathbf{h}_0^2)P_N(r)\,4\pi r^2 dr - \int_{th_0}^{\infty} (r^2 - \mathbf{h}_0^2)P_N(r)\,4\pi r^2 dr\ ; \tag{3}$$

$$P_N(r) = (3/2\pi)^{3/2} h_0^3\, exp(-3r^2/2\mathbf{h}_0^2)$$

in which the pre-integral factor of the third term in the right-hand side of the above equation is unity because of the binary contact between the wall and one segment of the coil. Performing the integration we get

$$h_\theta^2(c) = h_\theta^2(0) - \mathbf{h}_0^2\,(3/2\pi)^{\frac{1}{2}} 2\,t^3 exp(-3t^2/2) = \mathbf{h}_0^2\{1 - 2\,(3/2\pi)^{\frac{1}{2}}\,t^3 exp(-3t^2/2\}\ . \tag{4}$$

The radius of the spherical reflecting wall around each segment of the coil in solution should be connected to the concentration of the solution. At least it is reasonable to assume a relation $c \propto t^{-3}$ in the semidilute region. The function $h^2{}_\theta(c)/\mathbf{h}^2{}_0$ as a function t^{-3} is shown in Fig. 7, which should simulate the concentration dependence of the flexible polymer coil dimension in a θ-solution. It is interesting to note that the normalized coil dimension, $h^2{}_\theta(c)/\mathbf{h}^2{}_0$ decreases with increasing concentration to a minimum value of 0.69 at t = 1 and then it will increase with further increase in concentration in a concentrated solution. The function $h^2{}_\theta(c)/\mathbf{h}^2{}_0$ given by (Eq. 4) approaches unity when $t \to 0$, i.e., $c \to \infty$, in accord with the experimental fact that in an amorphous solid film the coil dimension is the same as that in a dilute θ-solution.[23]

The light-scattering data of Nose and Chu[3] on a PS sample of M_w = 190kD in trans-decalin at 20.0°C) seem to fit fairly well to the curve of Fig. 7, if one fits the first data point (c = 0.21 g/dl) to the theoretical curve at $h^2{}_\theta(c)/\mathbf{h}^2{}_0$ = 0.91 to fix the concentration scale of the figure. However, another publication in literature, using a mixture of randomly substituted p-iodostyrene and a PS sample of M = 110 kD in trans-decalin at T_θ = 21.2°C, showed by SAXS that the polymer coil dimension remained constant independent of concentration up to a volume fraction of 0.4 within the experimental error of ±5%.[24] More experimental studies are certainly needed before a reliable conclusion could be drawn, especially when results obtained by independent methods are avail-

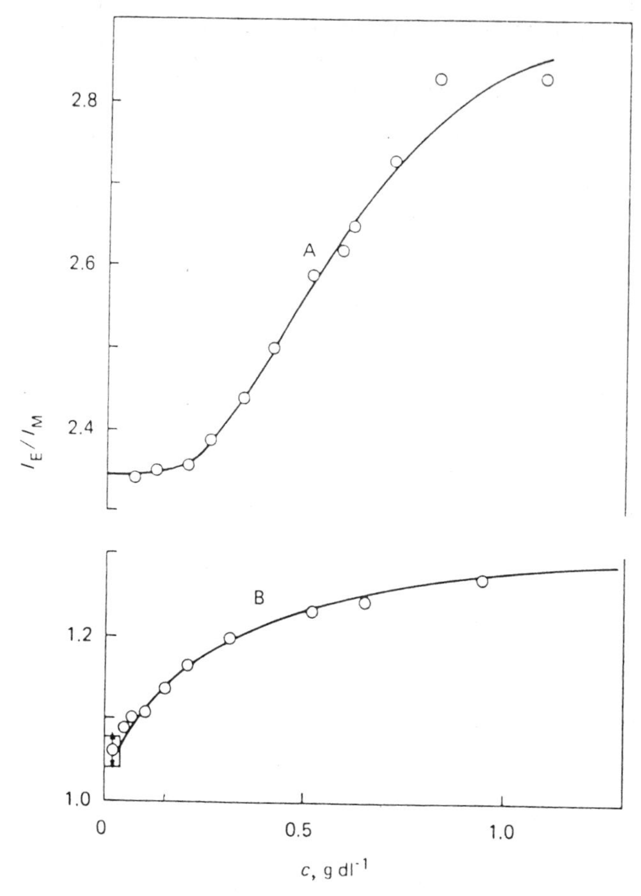

Fig. 6. $\zeta(c)$ for PS of M_n = 6.1 MD in solutions of theta solvent, cyclohexane at 35.4°C and B-good solvent, DCE at 25°C.

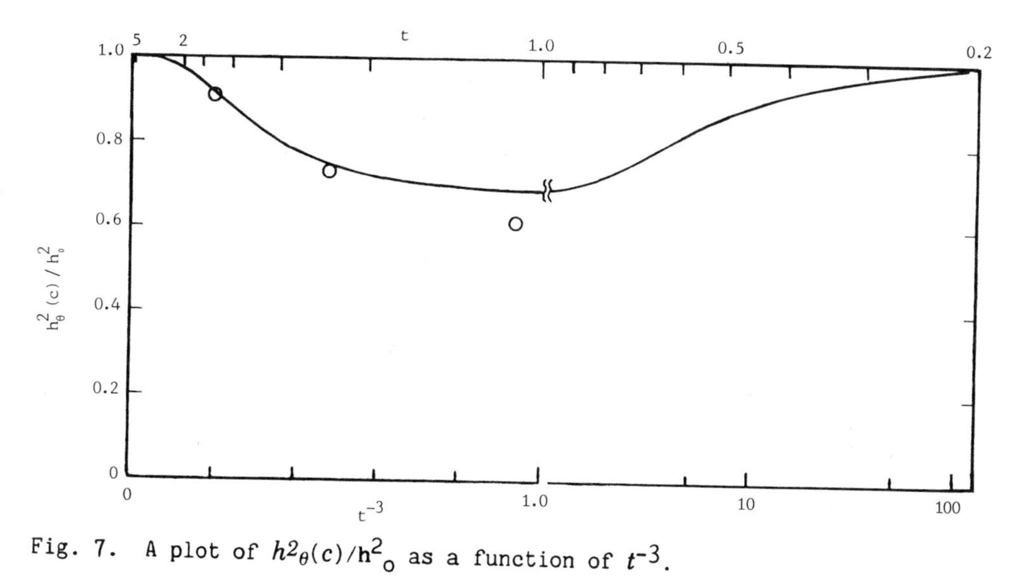

Fig. 7. A plot of $h^2_\theta(c)/h^2_0$ as a function of t^{-3}.

Circles: experimental data of PS of M_w = 190 kD in trans-decalin at 20.0°C.[3]

able. In this connection, it is interesting to mention that the decay of fluorescence polarization recently developed[25,26] may offer a promising tool for such investigations. Although a fluorescent-tagged polymer is needed in such experiments, owing to very high sensitivity of fluorescence the tagged macromolecule needed will be in much lower concentration than those for SAXS measurements.

IV. POLYETHYLENE TEREPHTHALATE

Polyethylene terephthalate (PET) has chromophores in the backbone chain and an intrachain excimer site from adjacent chromophores is impossible. Excimer fluorescence from dilute solutions of PET is not significant until concentrations reaching over 30 wt% in CF_3COOH.[27] As the chromophores are in the backbone chain the excimer fluorescence in very concentrated solutions can serve as a probe for the interchain aggregation process during the transition from a concentrated solution to a solid film. In amorphous PET solid film the excimer fluorescence is almost the sole emission observed even down to 77 K[7] which implies that the excimer sites are pre-formed since the backbone chain motions are frozen at this temperature (for PET, T_g = 351 K, β-transition at 240 K). The changes of the fluorescence spectra of PET in CF_3COOH from a concentrated solution to a solid film are shown in Fig. 8.[27] The steady increase of the excimer intensity with concentration is readily seen. In the solid film owing to fast excitation energy migration to the excimer sites fluorescence emission comes almost solely from excimers. The presence of excimer sites in PET condensed phase implies that there are local parallel alignment of PET chains to within 0.35 nm so as to form the excimer site. It is interesting to see (Fig. 9) that when the amorphous PET film was thermally crystallized at 200°C for 4 h the monomer fluorescence intensity was greatly enhanced at the expense of the excimer intensity.[28] In fact, a linear correlation exists between the IR crystalline/amorphous ratio and the I_{320}/I_{380} emission intensity ratio. The crystalline structure of PET[29] precludes the formation of an excimer site in the crystalline region. The morphology of partially crystalline PET is in the form of crystalline lamellae of some 10 nm thickness interposed by amorphous regions of similar thickness organized into a superstructure of spherulites. The greatly enhanced monomer emission in partially crystalline PET indicates limited range of energy migration. It must be much less than 10 nm. Thus the excited chromophores in the crystalline lamellae relax predominately through monomer fluorescence emission, while the chromophores excited in the amorphous regions relax predominately through energy migration to the pre-formed excimer sites lying within the effective range.

The experimental findings described above tell us the interesting fact that during the molecular chain condensation process from concentrated solution to solid film the stacking of terephthaloyl groups of PET chains takes place to some significant extent. In amorphous PET film there are local regions where the interchain distance may even be shorter than that in the crystalline lattice (0.35 nm for the excimer site and 0.43 nm in the crystalline region). Of course the mean density of amorphous phase must be smaller than that of the crystalline region.

When the ethylene glycol moiety of PET is replaced by a flexible polytetramethylene ether glycol chain of molar mass 2 kD, the fluorescence spectrum in concentrated solution appears very similar to that of PET[14] and it was found to be able to thermally crystallize at 50-114°C for days into spherulities with diameters up to 20 μm.[30] When the terephthaloyl group is replaced by naphthalene 2,6-dicarboxyloyl group, it has been possible to get TEM pictures showing crystalline lamellae of ca. 10 nm

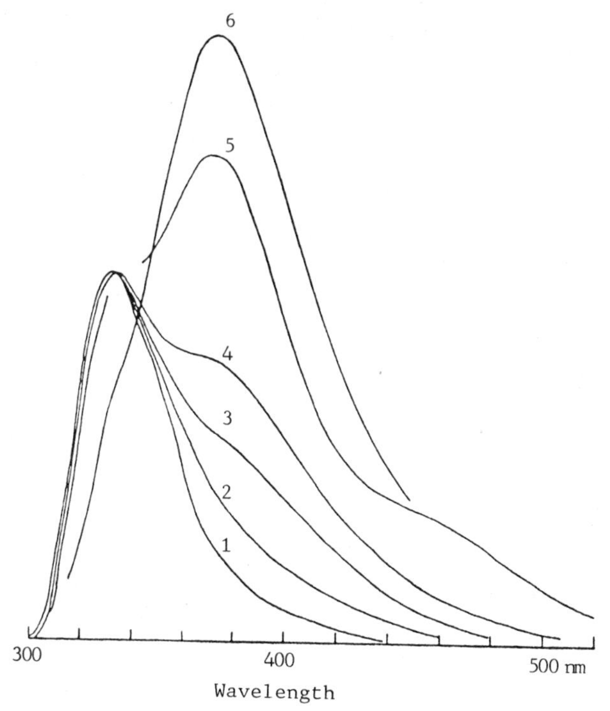

Fig. 8. Fluorescence spectra of PET in CF_3COOH. For curves 1,2,3,4,5-concentrations 9.34, 27.7, 45.0, 64.0, 93.1 wt%; curve 6-solution-cast film.

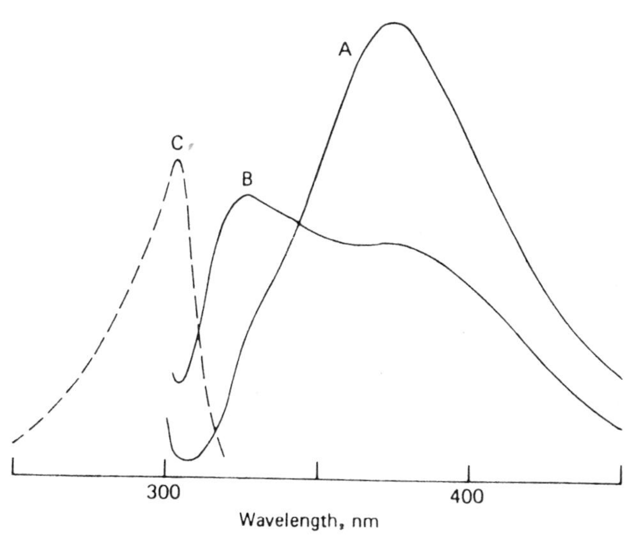

Fig. 9. Fluorescence excitation (broken curve) and emission (full curve) spectra of solution-cast PET films.

A-amorphous film; B-film A after thermal crystallization at 200°C for 4 h; C-excitation spectrum for the excimer emission.

thickness.[30] In these cases, crystallization is the stacking of planar terephthaloyl or naphthalene dicarboxyloyl groups.

V. RIGID CHAIN POLYBENZAMIDE

For a rigid chain polymer like polybenzamide (PBA) in solution it goes from a dilute solution to a concentrated one through a nematic liquid crystalline phase when the concentration is above certain critical concentration in the range of a few %w/v depending on the solvent used and molar mass of the polymer. The molecular chain condensation process involved during the transition from a dilute soluution to a nematic phase and to a solid film is of great interest and remains little understood. Whether the anisotropic intermolecular interaction as the stacking of the planar phenylene rings in the chain plays a role or not is by no means clear. Being a rigid chain polymer PBA is very difficult to dissolve except in solvents like dimethyl acetamide containing 3% LiCl (DMAC-LiCl) in which Li^+ plays an important role, or concentrated H_2SO_4 in which protonation plays an important role in rendering the rigid chain soluble.

PBA in DMAC-LiCl showed in its absorption spectrum and its fluorescence excitation spectrum (emission at λ = 320-500 nm having no difference) a peak at λ = 347-350 nm and a shoulder at λ = 332 nm. A solid film was prepared by spreading a 3-4% solution layer on a glass slide and then soaked in water so that the film was precipitated out and floated to the water surface. It was then washed and dried. The solid film showed the same excitation spectrum except much broader in shape, especially in the longer wavelength side of the peak as shown in Fig. 10a.[31] This longer wavelength part of the excitation spectrum might come from stacked ground-state chromophores present in the film. The excitation spectrum of PBA in H_2SO_4 (emission at λ = 430 nm) showed a red shifted single peak at λ = 400 nm, presumably due to protonation of the amide groups in the PBA chain.

In isotropic dilute solution of PBA in DMAC-LiCl, 0.2% w/v, the fluorescence spectrum showed a monomer band with fine structure peaked at λ = 355 nm and λ = 372 nm and a broad structureless band peaked around λ = 440 nm Fig. 10a. The peak separation of the monomer band corresponds to a very strong IR absorption band at 1236 cm^{-1}. The λ = 440 nm broad band emission decreased in intensity with decreasing LiCl concentration, becoming a tail below 1% LiCl at the same PBA concentration. A nematic solution of PBA in DMAC-LiDl, 6% w/v, showed almost the same fluorescence spectrum as in the isotropic dilute solution.

The fluorescence spectrum of an isotropic dilute solution of PBA in H_2SO_4, 0.15% w/v, showed only a single broad band peaked around λ = 435 nm as in Fig. 10b similar to the broad emission band observed in the dilute solution in DMAC-LiCl but without a monomer emission band. However a well-defined excimer band peaked at λ = 435 nm appeared in a nematic solution of 12% concentration with λ = 400 nm excitation.

A comparison of the fluorescence behavior PBA solutions in DMAC-LiCl and in H_2SO_4 indicates that probably the broad emission band around λ = 440 nm in isotropic dilute solutions in DMAC-LiCl or H_2SO_4 may be attributed to exciplex formation between Li^+ or H^+ in the two cases. The exciplex formation between aromatic rings and H^+ was first elucidated by Chen <u>et al</u>.[32] This interpretation, of course, needs further studies and confirmation. In DMAC-LiCl solution the presence of Li^+ apparently hinders the stacking of phenylene chromophores of PBA chain as no difference in the fluorescence spectra of a dilute and a nematic solution could be noted.

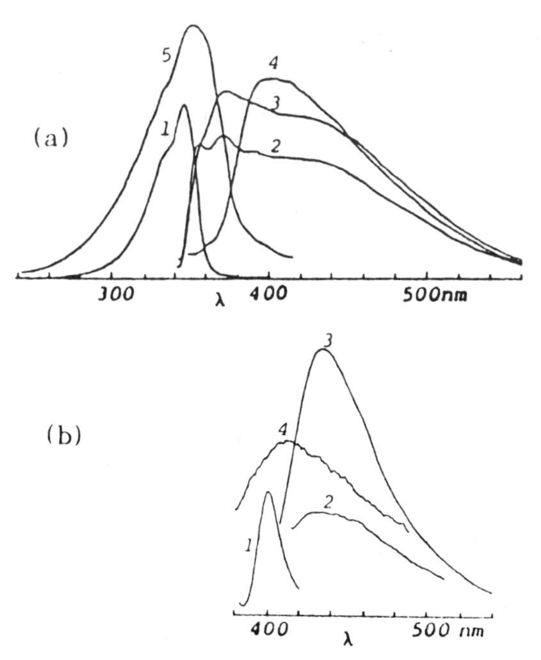

Fig. 10. Fluorescence excitation and emission spectra of PBA in (a) DMAC-LiCl; (b) H_2SO_4.

1 - excitation spectrum from solution; λ_{em}: 370 nm (a), 430 nm (b)

2 - emission from dilute solution (a-0.2%, b-0.15%; λ_{ex}: 325 nm (a), 400 nm (b)

3 - emission from nematic solution (a-6%, b-12%); λ_{ex}: 325 nm (a), 400 nm (b)

4 - emission from solid film; λ_{ex} 325 nm (a), 360 nm (b)

5 - excitation spectrum (film)

The solid films precipitated from both DMAC-LiCl and H_2SO_4 solutions showed similar excitation spectrum and a well-defined excimer band in emission peaked at λ = 400 nm for the film from DMAC-LiCl and at λ = 413 nm from the film from H_2SO_4. The blue shift of the excimer band in the solid film is compared to that in a nematic solution in H_2SO_4 indicates a loosening of the stacking distance or the degree of overlap of phenylene chromophores in the solid.

VI. CONCLUSIONS

1. Excimer fluorescence has been demonstrated to be useful for probing the process of interchain aggregation in the formation of a condensed phase from a dilute solution when the macromolecule bears a planar aromatic chromophore, especially in the main chain.

2. For solutions of flexible chain polymers two sharp transitions at concentration values c_s and c^+ could be easily located in the log-log plot of the excimer to monomer fluorescence intensity ratio ζ vs the concentration c,

$$\zeta(c) \quad \text{being independent of } c \text{ for} \quad c < c_s ;$$
$$\zeta(c) \quad \propto c^{\varepsilon}, \ \varepsilon < 1 \text{ for} \quad c_s \le c \le c^{+} ;$$
$$\zeta(c) \quad \propto c^{\varepsilon}, \ \varepsilon \ge 1 \text{ for} \quad c > c^{+} .$$

3. The fractional power dependence of ζ in the semidilute solution of flexible chain polymers is the manifestation of the polymer coil shrinkage with increasing concentration. This is true also for a flexible chain polymer in a θ-solution.

4. In very concentrated solutions and in the amorphous solid film of polyethylene terephthalate some local parallel aggregation of chain segments must have been present.

5. In the nematic solutions of polybenzamide in dimethylacetamide containing 3% LiCl and in concentrated sulfuric acid the chain aggregation state was somewhat different in the two solutions. In concentrated sulfuric acid solution a close stacking of the planar chromophores of the chain in the nematic phase led to pronounced excimer fluoresence.

ACKNOWLEDGEMENTS

The author is indebted to his colleagues and former students, particularly Dr. Ti Cao and Professor Fenglian Bai for most of the experimental work described in this paper. This work was supported by the National Natural Science Foundation, China.

NOMENCLATURE

a	hard-core radius of a segment
c	solution concentration
c_s	critical solution concentration for the initiation of coil shrinkage due to interaction with neighboring chains
c^+	critical solution concentration for a roughly uniform segment density throughout the solution
c^*	overlap concentration
D	Dalton
ε	power exponent in the concentration dependence of $\zeta(c)$
ζ	intensity ratio of excimer to monomer fluorescence
$[\eta]$	intrinsic viscosity
h_θ	end-to-end distance of a chain in a θ-solution
h	end-to-end distance of a chain
$\mathbf{h_o}$	end-to-end distance of a random walk chain
I_E	excimer fluorescence intensity
I_M	monomer fluorescence intensity
λ	wavelength
M_w	weight average molar mass
N	number of segments of a chain
P_N	radial distribution function of segments of a chain
r	distance between two segments
s	sedimentation coefficient
t	ratio of the radius of a spherical reflecting wall around a segment to the random walk end-to-end distance of the chain
T_θ	theta-temperature
T_g	glass transition temperature
$\phi(r)$	potential function for the binary inter-segmental interactions
w/v	solution concentration in weight per unit volume of solution

REFERENCES

1. P.G. de Gennes, Scaling Concepts in Polymer Physics, Cornell University Press, Ithaca (1979).
2. M. Daoud and G. Jannink, J. Phys. (Paris) 37, 973-979 (1976).
3. T. Nose and B. Chu, Macromolecules, 12, 590-599 (1979).
4. B. Nyström and J. Roots, J. Macromol. Sci., Rev. Macromol. Chem., C19, 35-82 (1980).
5. J.A. Wesson, I. Nole, T. Kitano and H. Yu, Macromolecules, 17, 782-792 (1984).
6. E.D. van Meerwall, E.J. Amis and J.D. Ferry, Macromolecules, 18, 260-266 (1985).
7. R. Qian, in Macromolecules, H. Benoît, P. Rempp, Eds., Pergamon Press, Oxford (1982), p. 139-154.
8. C.W. Frank and S.M. Semerak, Adv. Polym. Sci., 54, 31-85 (1984).
9. J. Roots and B. Nyström, Eur. Polym. J., 15, 1127-1131 (1979).
10. R. Qian, T. Cao, S. Chen and F. Bai, Zhongguo Kexue, B1983, 1080-1087.
11. J.M. Torkelson, S. Lipsky, M. Tirrell and D.A. Tirrell, Macromolecules, 16, 326-330 (1983).
12. M. Graley, A. Reiser, A.J. Roberts and D. Phillips, Macromolecules, 14, 1752-1757 (1981).
13. R. Qian, Q. Ying, T. Cao and L.J. Fetters, Kexue Tongbao, 1983, 354-356.
14. T. Cao and R. Qian, Acta Chim. Sin., 42, 51-62 (1984).
15. Th. Förster, Diss. Faraday Soc., 27, 7-17 (1959).
16. N.J. Turro, Modern Molecular Photochemistry, W.A. Benjamin Publishing Co., Menlo Park, California (1978), p.325-328.
17. C.W. Frank, G.H. Fredrickson and H.C. Anderson, in Photophysical and Photochemical Tools in Polymer Science, M.A. Winnik, Ed., D. Reidel Publishing Co., Dordrecht, Holland (1986), p. 495-522.
18. R. Qian and T. Cao, Polym. Commun., 27, 169-171 (1986).
19. R. Qian and X. Gu, Angew. Makromol. Chem., 141, 1-9 (1986).
20. P. Vidakovic, C. Allain and F. Rondelez, Macromolecules, 15, 1571-1580 (1982).
21. G. Meyerhoff and T. Huang, Makromol. Chem., 185, 2459-2465 (1984).
22. Y. Huang and R. Qian, Polym. Commun., 26, 242-243 (1985).
23. D.G.H. Ballard, G.D. Wignall and J. Schelten, Eur. Polym. J., 9, 965-969 (1973).
24. H. Hayashi, F. Hamada, A. Nakajima, Polymer, 18, 638-639 (1977).
25. K.A. Peterson, M.D. Fayer, J. Chem. Phys., 85, 4702-4711 (1986).
26. K.A. Peterson, M.B. Zimmst, S. Linse, R.P. Domingue, M.D. Fayer, Macromolecules, 20, 168-175 (1987).
27. R. Qian, F. Bai and S. Chen, Kexue Tongbao, (Eng. Ed.) 27, 725-729 (1982).
28. T. Cao, S.N. Magonov and R. Qian, Polymer Commun., 29, 41-44 (1988).
29. R. de P. Daubeny, C.W. Bunn and C.J. Brown, Proc. Roy. Soc., (London) A226, 531-542 (1954).
30. T. Cao and R. Qian, New Polym. mater., in print 1, 87-97 (1988).
31. R. Qian, T. Cao and S.N. Magonov, Preprints, IUPAC MACRO 88, Kyoto, 1988, p. 779.
32. S. Chen, F. Bai and Z. Wang, Zhongguo Kexue, B1986, 113-120.

POLYMERS AT INTERFACES: A STUDY BY NEUTRON REFLECTION

X. Sun, B. Farnoux, and G. Jannink

Laboratoire Léon Brillouin
(CEA-CNRS) CEN-Saclay
91191 Gif-sur-Yvette
Cedex, France

J. des Cloizeaux

Service de Physique
Théorique, CEN-Saclay
91191 Gif-sur-Yvette
Cedex France

ABSTRACT

The vicinity of total reflection of a polymer solution - air interfaces is investigated using a neutron radiation. Experiments and calculations reveal a significant singular behaviour, from which the exact nature of the polymer concentration profile can be determined.

I. INTRODUCTION

Since P.G. de Gennes[1] introduced the scaling method for the study of polymer conformations near an interface, many predictions were given to polymer concentration profiles. These are different from mean field results and depend in a subtle manner on surface interaction and bulk properties.[2]

In general, the concentration profiles derived from scaling arguments decrease more slowly than those given by mean field, and therefore such profiles have a wider extension in space.

Scaling is not able to predict exactly how the concentration profile merges respectively into surface and bulk concentrations. This can be considered as a simple crossover problem, but it may turn out to be a major issue.

For these reasons and many others, it is of interest to test the nature of the polymer concentration profile by performing an experiment and comparing the data with predicted values. There are already many tests using various techniques such as ellipsometry,[3] force-distance microscopy,[4] surface tension[5] and small-angle scattering.[6] It seems that the observed dependence with respect to polymer molecular weight agrees with scaling. Still, the more direct results yielding, for instance, the Fourier transform of the concentration profile are not yet convincing.

We discuss here the use of neutron reflectivity at a polymer solution-air interface, for the experimental determination of the polymer concentration profile. The method consists of analyzing the neutron reflectivity R, near total reflection, as a function of the incident inverse wavelength k. This method was developed at the LLB[7] in

Saclay. First predictions[8] and results[9] are indeed promising, but the determination of the polymer concentration profile is not yet made. We wish to discuss how close the analysis of the experimental data brings us to the true concentration profile.

Here we show that the neutron reflectivity method combined with contrast variation allows us, in principle, to determine as precise as we wanted. Practically, however, the collimation imperfection reduces the resolution to a level which is unsatisfactory, but can be improved. Perhaps, the major interest in the method is conceptual.[10] Using the one-dimensional Schrödinger equation to predict the reflectivity, we have to consider the concentration profile as a smooth interaction potential. This point of view is fruitful,[10] and we shall adopt it in this paper.

II. EXPERIMENT AND ITS INTERPRETATION

We examine here the interface between a polymer solution and air. Figure 1 gives a representation of this situation; the interface being materialized by the ideal line $z = 0$. The neutron beam comes from the left, at an angle θ with the interface $z = 0$. It is, in part, reflected at the surface and, in part, transmitted through the solution. The range over which the polymer concentration varies with z, i.e., z_{max}, is of the

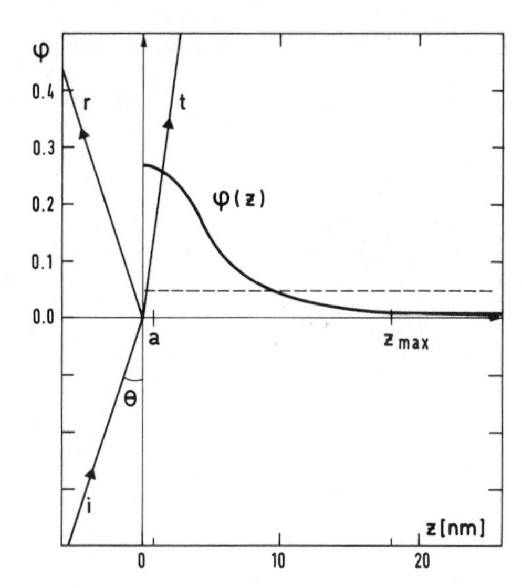

Fig. 1. Schematic representation of the reflectivity experiment. The horizontal interface between the polymer solution and air, is here shown in vertical. The incoming neutron beam (i) makes an angle $\theta \simeq$ 10' of arc with the interface; (r) reflected beam; (t) transmitted beam.

____ theoretical polymer concentration profile[1] in the strong adsorption case, drawn in the half space $z > 0$.

$\phi(z) = 1,\ z \leq a$ (monomer length)

$\phi(z) = a/z^{\mu},\ a < z \leq \xi = z_{max}$

$\phi(z) = a/\xi^{\mu} = \phi,\ z > \xi$

--- other theoretical profile extending to very large distances.

order of the polymer average radius of gyration (for example 50 nm). The polymer concentration is a continuous function of z. The formulation of the neutron reflectivity experiment can be reduced to a one-dimensional problem. The stationary state of a neutron with incident inverse wavelength $\boldsymbol{k}$ is represented by the wave function $\Psi(z)$, and the solution of the equation

$$\left(\frac{\partial^2}{\partial^2 z} + K^2 - V(z) \right) \psi(z) = 0 , \tag{1}$$

where

$$K = k \sin \theta \tag{2}$$

and where $V(z)$ is the smoothed neutron interaction potential with the atoms. In particular

$$V_- \equiv V(z), z < 0, and\ V_+ = \lim_{z \to +\infty} V(z) . \tag{3}$$

The reflectivity experiment is described by the solution $\Psi(z)$ of equation (1), satisfying the boundary conditions

$$\Psi(z) = e^{iz\sqrt{K^2 - V_-}} + A e^{-iz\sqrt{K^2 - V_-}}, \quad z \to -\infty \tag{4}$$

$$\Psi(z) = B e^{iz\sqrt{K^2 - V_+}} . \quad z \to -\infty \tag{5}$$

The reflectivity is the quantity

$$R = \left| \frac{\sqrt{K^2 - V_-} - \dfrac{\Psi'(0)}{i\Psi(0)}}{\sqrt{K^2 - V_-} + \dfrac{\Psi'(0)}{i\Psi(0)}} \right|^2 . \tag{6}$$

We consider R as a function of $K - \boldsymbol{k} \sin \theta$. In the experiment, the angle θ (about 10' of arc) is fixed and the inverse wavelength $\boldsymbol{k}$ is varied above the critical value

$$k_c = \frac{\sqrt{V_+ - V_-}}{\sin\theta} , \tag{7}$$

defining the onset of total reflection of the bulk solution. The method is described in Ref. 7; it allows us to determine the reflectivities for density distributed values of K.

The potential $V(z)$, $z > 0$, is directly related to the polymer concentration profile. The concentration is given in terms of the monomer volume fraction $\phi(z)$. By definition

$$V(z) = 4\pi\left(\frac{b_p}{v_p} \phi(z) + \frac{b_s}{v_s} (1 - \phi(z)) \right), \tag{8}$$

where b_i, v_i are respectively the collision length and the partial volume of the monomers of species i ($i = p$ for polymer, $i = s$ for solvent). The solvent molecule is considered as a monomer.

It will be convenient to rewrite Eq.(8) by introducing the concentration profile

$$\Delta\phi(z) = \phi(z) - \phi , \tag{9}$$

where

$$\phi = \lim_{z \to \infty} \phi(z) , \tag{10}$$

and the isotopic composition ratio

$$\gamma = \left(\frac{b_p}{V_p} \right) \left(\frac{v_s}{b_s} \right) - 1 . \tag{11}$$

The sign of $\Delta\phi(z)$ is positive in the adsorption case; it is negative in the depletion case.

The sign of γ is positive if the polymer is deuterated and the solvent is non-deuterated; it is negative for the opposite composition.

We get

$$V(z) = V_+ + \Delta V(z) , \tag{12}$$

where

$$\Delta V(z) = 4\pi\gamma \, \Delta\phi(z) \frac{b_s}{v_s} , \tag{13}$$

and

$$V_+ = 4\pi(1 + \gamma\phi) \frac{b_s}{v_s} . \tag{14}$$

Figure 2 shows schematic variations of $V(z)$ in the case of a given adsorption polymer profile, for the two types of isotopic compositions

a) $\gamma > 0$,

b) $\gamma < 0$.

The same figure holds qualitatively true in the case of a depletion layer, but with reversed isotopic compositions.

The problem to be discussed is the determination of the potential ($V/(z)$ and ultimately of the concentration profile $\Delta\phi(z)$, from a given set of experimental values of the reflectivity $R\{k\}$.

We shall make full use of the fact that, the concentration profile $\Delta\phi(z)$ and the isotopic composition coefficient γ are independent parameters in the interaction potential.

III. REFORMULATION OF THE PROBLEM, THE BARRIER AND THE WELL

We can represent, in a more significant manner, the differences displayed in Fig. 2, between the potentials associated with each type of isotopic composition ($\gamma > 0$, $\gamma < 0$). For this purpose, we redefine the energy K^2 (see Eq.(2)) introducing

$$\mathbb{K}^2 = K^2 - V_+ , \tag{15}$$

where V_+ is given by Eqs.(3) and (14).

Then the wave equation (1) becomes

$$\left(\frac{\partial^2}{\partial z^2} + \mathbb{K}^2 - (V_- - V_+) \right) \psi(z) = 0 , \quad z < 0 \tag{16}$$

$$\left(\frac{\partial^2}{\partial z^2} + \mathbb{K}^2 - \Delta V(z) \right) \psi(z) = 0 . \quad z < 0 \tag{17}$$

The reflectivity in Eq.(7) will be considered as a function of $\mathbb{K}$. We have, explicitly

$$R(\mathbb{K}) = \frac{\left| \sqrt{\mathbb{K}^2 + V_+ - V_-} - \dfrac{\psi'(0)}{i\psi(0)} \right|^2}{\left| \sqrt{\mathbb{K}^2 + V_+ - V_-} - \dfrac{\psi'(0)}{i\psi(0)} \right|^2} . \tag{18}$$

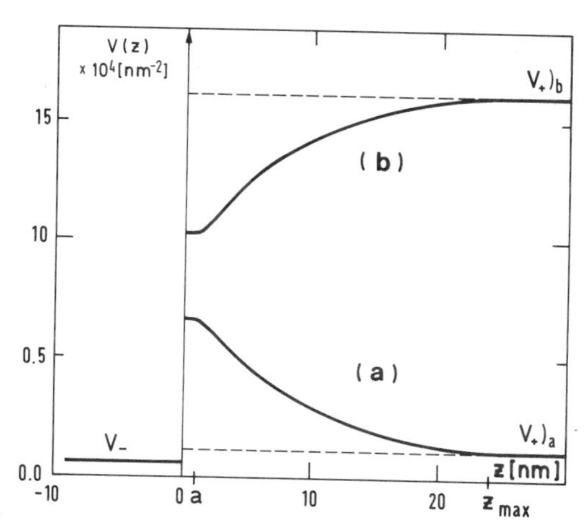

Fig. 2. Schematic representation of the potential seen by the neutron, in the case of strong polymer adsorption at the interface (see Fig. 1).
a) for an isotopic composition $\gamma = 4.321$ (below)
b) for an isotopic composition $\gamma = -0.98$ (above)

The potential V_- is the neutron interaction potential of the medium facing the polymer solution (for example, air)

Figure 3 displays the perturbation potentials $\Delta V(z)$ for the two types of isotopic compositions (a) and (b). With Eqs.(16) and (17), the problem raised by the polymer concentration profile becomes now similar to the classical problem of the reflectivity of a barrier (type a) and of a well (type b). The step $(V_+ - V_-)$ between bulk solution and air is accounted for in Eq.(18).

Finally, equations (16) and (17) can be given a dimensionless form, if we use the transformation

$$y = z\sqrt{V_+ - V_-} \,, \tag{19}$$

$$Y^2 = K^2/(V_+ - V_-) \,. \tag{20}$$

The result is

$$\left(\frac{\partial^2}{\partial y^2} + Y^2 + 1\right)\chi(y) = 0 \,, \qquad y < 0 \tag{21}$$

$$\left(\frac{\partial^2}{\partial y^2} + Y^2 - \frac{\Delta V(y)}{V_+ - V_-}\right)\chi(y) = 0 \,, \qquad y < 0 \tag{22}$$

The potential in Eq.(17) appears as a perturbation. In terms of the scattering length Eqs.(3) and (4) we get

$$\frac{\Delta V(y)}{V_+ - V_-} = \frac{\gamma\, \Delta\phi(y)}{1 + \gamma\phi - \delta} \,, \tag{23}$$

where

$$\delta = V_-(v_s/b_s) \,. \tag{24}$$

For dilute bulk solutions having an interface with air, $\gamma\phi << 1$ and $\delta = 0$. As a consequence

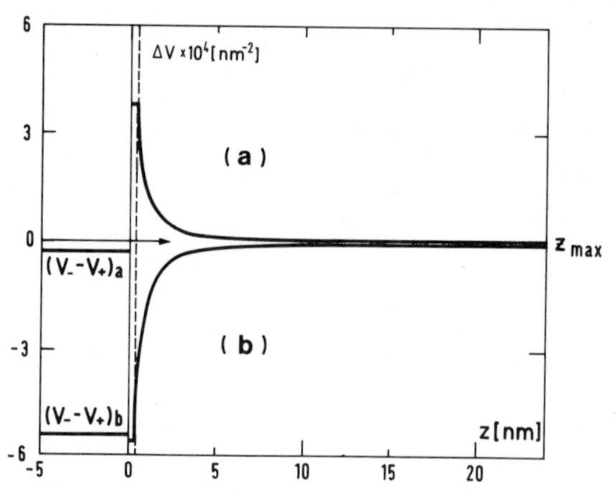

Fig. 3. Representation of the neutron interaction potential derived from Fig. 2, using transformation in Eq.(15)
a) γ = 4.321, the barrier
b) γ = -0.98, the well

$$\frac{\Delta V(y)}{V_+ - V_-} \simeq \gamma\, \Delta\phi(y)\ . \tag{25}$$

Figure 4 displays a theoretical representation of function in Eq. (23).

Notice that the interval $(0, y_{max})$ over which (23) is non-zero, depends on the isotopic composition of the bulk solution (see Eq. (19)).

Varying the factor $\sqrt{V_+ - V_-}$ by isotopic substitutions in solute and solvent, we have in principle, an infinite set of different reflectivity profiles $R(Y;\gamma,\delta)$, for a given polymer concentration profile. This is the essence of the method.[8] The reflectivity is now written.

$$R(Y,\gamma,\delta) = \left| \frac{\sqrt{Y^2+1} - \dfrac{\chi'(0)}{i\chi(0)}}{\sqrt{Y^2+1} + \dfrac{\chi'(0)}{i\chi(0)}} \right|^2 , \tag{26}$$

where by definitions in Eqs. (20), (15), (7) and (2)

$$Y^2 = \frac{k^2}{k_c^2} - 1 - \left[\frac{V_-}{V_+ - V_-} \right] .$$

In particular, if $V_- = 0$ (air)

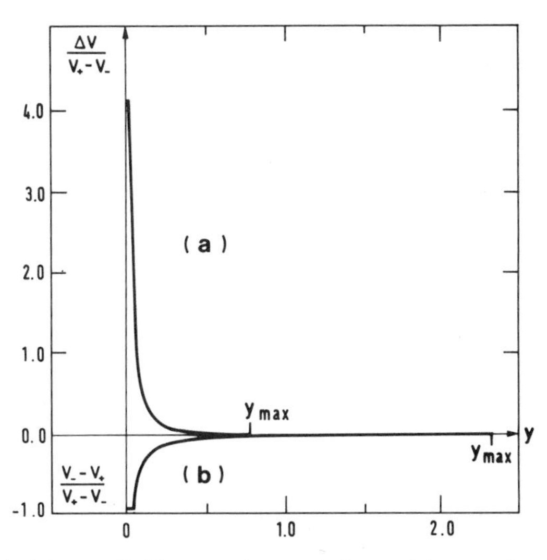

Fig. 4. Representation of the neutron interaction potential derived from Fig. 3, using dimensionless variables.

The effective range y_{max} and the effective intensity of the potential depend on the step $(V_+ - V_-$ (Eqs.(19) and (20)).

$$Y^2 = \frac{k^2}{k_c^2} - 1 . \tag{27}$$

The functions $R\{Y;\gamma\}$ have characteristic features related to the sign of γ. These are presented in the next section.

IV. CHARACTERISTIC FEATURES OF THE REFLECTIVITY FUNCTION

In section III, we showed that the study of the reflectivity function can be divided into two types of problems, according to the sign of the perturbation $\Delta V(y)$. For each case, predictions can be given about the reflectivity function, near total reflection.

Comparison with the Fresnel Function

In absence of interaction ($\Delta V(z) \equiv 0$), the reflectivity function in Eq.(21) reads

$$R_o\{Y\} = \left| \frac{\sqrt{Y^2+1} - Y}{\sqrt{Y^2+1} + Y} \right|^2 . \tag{28}$$

In the presence of a potential barrier ($\Delta Vy) > 0$), the reflectivity $R(Y)$ corresponding to a given reduced energy Y^2 is larger than $R_o(Y)$: the incident neutron beam interacts more strongly with the adsorbed deuterated polymer than with the bulk polymer solution, and as a result the reflectivity is greater than in the case of $(\Delta Vy) \equiv 0$. Consider the solution $X(y)$ of the wave equation. In the WKB approximation and for $Y^2 > \Delta V(0) / (V_+ - V_-)$, we can write

$$\frac{X'(0)}{i\,X(0)} \simeq \sqrt{Y^2 - \frac{\Delta V(0)}{V_+ - V_-}} < Y ,$$

and as a consequence $R(Y) > R_o(Y)$.

The same argument, applied to the case of a well ($\Delta V(y) < 0$), gives the opposite result:

$$R(Y) < R_o(Y) .$$

Reflectivity in the limit $Y \to 0$

Dietrich[10] and Schack studied the limit

$$\lim_{Y \to 0} R(Y) ,$$

and concerning polymer concentration profiles, they obtained interesting results which should be tested.

These authors predict the existence of singularities for $R(Y)$, when $(\Delta Vy) / (V_+ - V_-)$ decreases as a power law of the type $y^{-\mu}$, $y > 0$, where μ is a characteristic exponent

In the case of a potential barrier, with $\mu < 2$,

$$\lim_{Y\to 0} \left(1 - R(Y)\right) \propto \exp\{-2I(\mu)\, A^{1/\mu}/Y^{(2/\mu-1)}\} \quad , \tag{29}$$

where $I(\mu)$ is a universal function of μ and A is a constant related to the intensity of the potential. (for example, $A = \gamma(U_+ - V_-)^{\mu-2}\, a^{\mu}$ in the model of ref. 1, where a is the size of a monomer).

In the case of a potential well,[11,12] with $\mu = 2$:

$$\lim_{Y\to 0} R(Y) < 1 \quad .$$

For $\mu = 2$, the limit $Y = 0$ is an accumulation[10,12] point of oscillations of $R(Y)$.

Effect of the potential

Total reflection is caused by the step $(V_+ - V_-) > 0$ between bulk potentials. An increase of $(V_+ - V_-)$ extends the effective range of the potential created by the concentration profile (see Eq.(19)), on the other hand it decreases with the effective height of this potential (see Eq.(20)).

The singularities predicted by Dietrich and Schack only exist at finite k, if $(V_+ - V_-) > 0$. In general, the finite step magnifies the effects of the concentration profiles on the reflectivity curve. It is of practical interest to experiment in a situation where $(V_+ - V_-)$ is as large as possible.

V. EXPERIMENTAL RESULTS AND PRELIMINARY INTERPRETATION

The reflectivity of the solution-air interface was measured[9] for polydimethylsiloxane solutions in toluene. In such systems, the polymer is known to be attracted at the surface because of the negative increment of the surface tension (~8 mN/m). Two isotopic compositions were tested, and they are characterized respectively by $\gamma_a = 4.321$ and $\gamma_b = -0.98$ (see Eq.(11)). Ideally in the experiment, the polymer molecular weight $M_w)i$ and the monomer volume fraction ϕ_i should be independent of composition. This was however not exactly achieved:

$$M_w)_a = 7.83 \times 10^5 \quad ; \quad \Phi_a = 6.2 \times 10^{-3} \quad ,$$

and

$$M_w)_b = 8.98 \times 10^5 \quad ; \quad \Phi_b = 4.7 \times 10^{-3} \quad .$$

The overlap volume fraction is typically $\phi^* = 3.5 \times 10^{-3}$. The inverse wave numbers defining the onset of total reflection are respectively

$$k_c)_a = 5.77 \text{ nm}^{-1} \quad ; \quad k_c)_b = 6.73 \text{ nm}^{-1} \quad .$$

Measured reflectivities are given in Fig. 5, with the representation of $R(Y)$ discussed in the section on reformulation of the problem (Eq.(26)). One recognizes immediately the attractive nature of the interface with respect to the polymer chains: the $\gamma > 0$ reflectivity curve is higher than the Fresnel function and therefore the potential associated with this case acts as a barrier. Conversely, the $\gamma < 0$ reflectivity curve is below the Fresnel function.

In the case $\gamma > 0$ we notice a change of curvature of $R(Y)$ at about $Y = 0.45$; the negative curvature at $Y = 0$ is not in contradiction with the Dietrich and Schack prediction,[10] but the data reveal other features. The observed reflectivity curve can be roughly interpreted as a shift of the Fresnel curve:

$$R_a\{Y\} = R_o(Y - \Delta_a Y), \tag{30}$$

with $\Delta_a Y \simeq 0.4$.

In a similar manner, one can argue that the experimental curve for the case of the potential well ($\gamma = -0.98$) is also a translation of the Fresnel curve, but in the opposite direction:

$$R_b\{Y\} = R_o(Y - \Delta_b Y), \tag{31}$$

The shift appears, however, to be smaller than in the barrier case. We can test this idea quantitatively, since the two opposite shifts must be related to each other according to the isotopic compositions. For this, we suppose that the shift is caused by a nearly uniform concentration increase $\Delta\Phi$ with respect to the bulk.

$$\frac{\Delta V}{V_+ - V_-} \simeq \gamma \Delta \Phi .$$

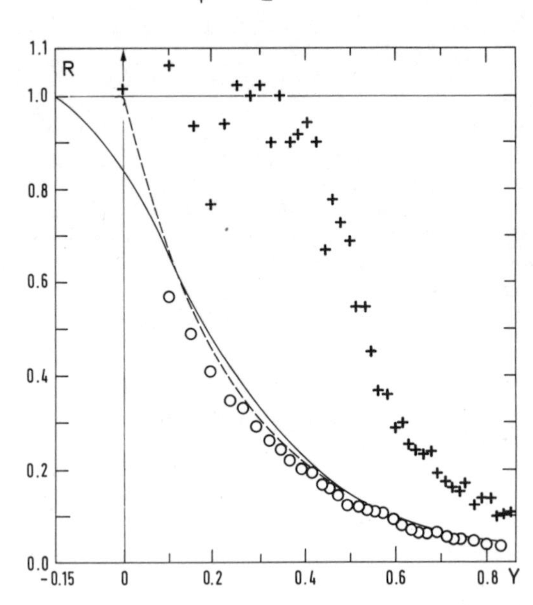

Fig. 5. Result of the reflectivity experiment on the solution-air interface for polydimethylsiloxane - toluene solutions

+ case a) : $\gamma = 4.321$,
o case b) : $\gamma = -0.98$.

The discontinuous curve is the Fresnel function. The continuous curve represents the Fresnel profile modified by resolution effects.

If we set, for the case of the barrier (a)

$$\gamma_a \Delta \phi = (\Delta_a Y)^2 = 0.16, \tag{32}$$

then

$$\gamma_b \Delta \phi = (\Delta_b \gamma)^2 = 0.16 \frac{\gamma_b}{\gamma_a} = 0.036 \; . \tag{33}$$

We note that Eqs.(32) and (33) imply $\Delta\phi$ = 0.037, which is an order of magnitude greater than the overlap volume fraction ϕ^*.

Figure 6 displays the functions $R_i(Y)$ (i = a, b), Eqs.(28), (32) and (33). Comparison with Fig. 5 shows that the curve $R_b(Y)$ is able to reproduce the overall variation of experimental points of case b; the discrepancy at $Y \to 0$ is probably due to the resolution effects. For the case of the barrier (a) the fit is not satisfactory. A larger value of the perturbation would be necessary to reproduce the overall variation of these experimental points, but the good fit in case b) would then be lost.

The more realistic model for $\Delta\phi(y)$ will have to comply with both experimental results.

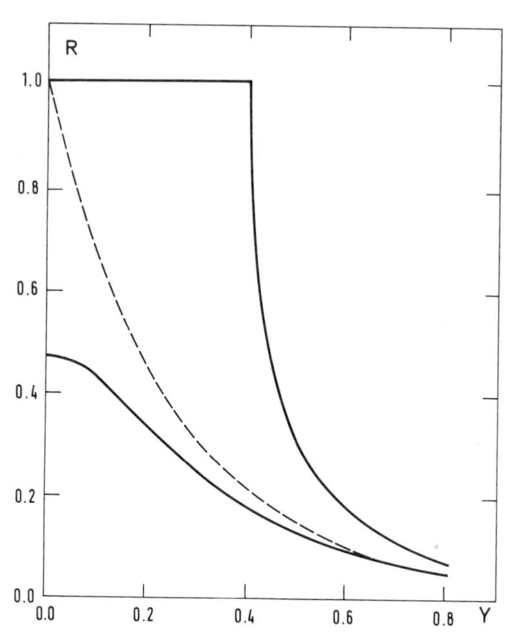

Fig. 6. Calculated reflectivity functions given by Eqs.(28), (32) and (33).

top: case a, $\Delta_a Y^2$ = 0.16,
bottom: case b, $\Delta_b Y^2$ = 0.036,
broken line : ΔY^2 = 0.

Comparison between Figs. 5 and 6 also suggests that the bulk concentrations ϕ introduced in the calculation of $k_c)_i$ could have been underestimated. However, even this interpretation cannot explain all observed features of the experimental result. Therefore, the data shown in Fig. 5 reveal information on the polymer concentration profile due to the polymer adsorption at the solution-air interface. This remains to be determined more precisely.

VI. CONCLUSIONS

The method used to derive polymer concentration profiles from neutron reflectivity experiments is based on the sensitivity of the reflectivity to the isotopic composition, when one approaches total reflection. It requires a very accurate knowledge of the critical inverse wavelength, k_c, associated with each isotopic bulk composition.

When this requirement is fulfilled, and when the resolution related to collimation of the beam is improved, the method is indeed powerful: eventually the singularities associated with the existence of a slowly decreasing perturbation potentials becomes observable.

A preliminary interpretation of the data suggests that adsorption at the solution-air interface creates a deep layer in which the polymer concentration is higher than that in the bulk and varied slowly. This is not predicted by the scaling approach.

ACKNOWLEDGEMENTS

The authors thank M. Daoud, S. Dietrich and A. Lapp for interesting discussions.

NOMENCLATURE

K	inverse wavelength of the incident neutron
θ	angle of reflection
$K^2 = k^2 \sin^2\Phi$	energy
K^2	redefined energy
Y	dimensionless energy
$\phi(z)$	monomer volume fraction of the polymer chains, at a distance z from the interface
$V(z)$	neutron interaction potential with the polymer solution, at a distance z from interface
V_+	$V(z) \to +\infty)$
V_-	$V(z) \to -\infty)$
$\Psi(z)$	solution to the Schrödinger equation describing the neutron reflectivity
R	reflectivity
y	distance to the interface expressed in dimensionless form.
γ	isotopic composition parameter

REFERENCES

1. P.G. de Gennes, Macromolecules 14, 1637 (1981).
2. A recent review is found in : E. Bouchaud, Thesis, Orsay (1988).
3. R. Varoqui and P. Dejardin, J. Chem. Phys. 66, 4395 (1977).
4. J. Klein Colloïdes et Interfaces, A.M. Cazabat and M. Veyssié, eds., Editions de Physique (1983) - Les Ulis, France p.289.
5. R. Ober, L. Paz, C. Taupin, P. Pincus and S. Boileau, Macromolecules 16, 50 (1983).
6. L. Auvray and J.P. Cotton, Macromolecules 20, 202 (1987).

7. B. Farnoux, Proceedings of the AIEA, Jülich Meeting, January (1985), p.205-208.
8. E. Bouchaud, B. Farnoux, X. Sun, M. Daoud and G. Jannink, Europhys. Lett. 2, 315 (1986).
9. X. Sun, E. Bouchaud, A. Lapp, B. Farnoux, M. Daoud and G. Jannink Europhys. Lett. 6, 207 (1988).
10. S. Dietrich and R. Schack, Phys. Rev. Lett. 58, 140 (1987).
11. A. Steyerl, W. Drexel, S.S. Malek and E. Gutsmiedl, Workshop on Neutron Interferometry, Wien (1987).
12. R. Schack, Thesis, Munich (1986).

THEORY OF DYNAMIC SCREENING IN MACROMOLECULAR SOLUTIONS

Andrzej R. Altenberger, John S. Dahler and Matthew Tirrell

Department of Chemical Engineering and Materials Science
University of Minnesota
Minneapolis, MN 55455

ABSTRACT

A semi-microscopic treatment is presented of hydrodynamic screening in a suspension of elastic-dumbbell, two-bead "polymers." The theory is cast in a form that serves to emphasize its similarities to an earlier theory due to de Gennes. It is concluded that screening is a transient phenomenon which vanishes for stationary flows unless the particles interact so strongly that they form a rigid network of immobilized obstacles.

I. INTRODUCTION

In the second of his two well-known papers (of 1976) devoted to the dynamics of entangled polymer solutions,[1] de Gennes used a "two-fluid model" to determine the effects of hydrodynamic interactions. Among the many results reported in this paper are two of particular interest to us here, viz., (1) that these interactions are not screened in the static, long wavelength limit and (2) that static screening does occur provided that the wavelength is less than the radius of the polymer chain.

These conclusions are of obvious and significant consequence to the dynamics of macromolecular solutions. The issues involved can be addressed in the following way: a relatively weak, sharply localized force $\mathbf{f}$ $(\mathbf{r},t)$ is applied to the solvent component of a macromolecular solution. The solvent velocity field that develops in response to this force may be expressed in the manner

$$\mathbf{w}_S(\mathbf{r},t) = \int_0^t dt' \int d\mathbf{r}' \mathbf{T}(\mathbf{r}-\mathbf{r}', t-t') \cdot \mathbf{f}(\mathbf{r}',t') , \tag{1}$$

or, in the equivalent Fourier representation,

$$\mathbf{w}_S(\mathbf{q},\omega) \equiv \int_{-\infty}^{+\infty} dt \int d\mathbf{r} e^{i(\omega t - \mathbf{q}\cdot\mathbf{r})} \mathbf{w}_S(\mathbf{r},t) = \mathbf{T}(\mathbf{q},\omega) \cdot \mathbf{f}(\mathbf{q},\omega) . \tag{2}$$

When the macromolecules are absent the response function (propagator, Green function) $\mathbf{T}$ $(\mathbf{q},\omega)$ is known to be the Oseen tensor

$$\mathbf{T}_S(\mathbf{q},\omega) = [\rho_S(i\omega + \nu_S q^2)]^{-1}(\mathbf{I} - \hat{\mathbf{q}}\hat{\mathbf{q}}). \tag{3}$$

Here, the symbols ρ_S and $\nu_S = \eta_S/\rho_S$ indicate the density and kinematic viscosity of the neat solvent (S) and $\hat{\mathbf{q}} = \mathbf{q}/q$. A second system for which the response function is known is a solvent littered with randomly distributed <u>immobile</u> objects, each acting as a point center of hydrodynamic scattering with a friction coefficient ξ characteristic of the strength of its (Stokes) interaction with the solvent. The response function specific to this second system[2] is given by the formula

$$\mathbf{T}_{SC}(\mathbf{q},\omega) = [\rho_S(i\omega + \nu_S\{q^2 + q_{SC}^2\})]^{-1}(\mathbf{I} - \hat{\mathbf{q}}\hat{\mathbf{q}}) , \tag{4}$$

with $q^2_{SC} = \xi c_B/\eta_S$ and where c_B denotes the concentration (number density) of the immobile scattering centers.

The spatial dependences of these two response functions can be extracted from the formula

$$\mathbf{T}(\mathbf{r},\omega) = (8\pi\eta_S r)^{-1}\{(\mathbf{I} + \hat{\mathbf{r}}\hat{\mathbf{r}})e^{-y} - (\mathbf{I} - 3\hat{\mathbf{r}}\hat{\mathbf{r}})\frac{2}{y^2}[1 - e^{-y}(1 + y + \tfrac{1}{2}y^2)]\} , \tag{5}$$

with $\hat{\mathbf{r}} = \mathbf{r}/r$, $y = Q(\omega)r$ and $Q^2(\omega) = q^2_{SC} + i\omega\,\nu_S^{-1}$, which is applicable to both. Specifically, in the low-frequency limit $\exp(-y)$ reduces to $\exp(-q_{SC}r)$ so that $q_{SC}^{-1} = (\xi c_B/\eta_S)^{-\frac{1}{2}}$ can be identified as a "screening length" that becomes infinite as ξc_B, the friction per unit volume, tends to zero. (In this limit $\mathbf{T}$ $(\mathbf{r}, 0) \rightarrow (8\pi\eta_S r)^{-1}$ $(\mathbf{I} + \hat{\mathbf{r}}\,\hat{\mathbf{r}})$.)

At issue is the question of whether screening of this sort occurs when the hydrodynamic scattering centers no longer are constrained to be immobile. We recently have studied this question[3] using an approach which, although apparently quite different, actually can be shown to share many features with that of de Gennes. However, in contrast to him we found the hydrodynamic screening to be a transient phenomenon that vanishes as the flow became stationary, regardless of wavelength. One possible source of this difference is that de Gennes' analysis of screening was couched in the language of chain-polymer solutions whereas ours was specific to "single-center macromolecules." To test this possibility we shall consider a solution of elastic-dumbbell, two-bead polymers. Also, our presentation here will differ somewhat from its predecessor in order to expose the assumptions and approximations upon which our theory rests and, at the same time, to emphasize its similarity to that of de Gennes.

II. SEMI-PHENOMENOLOGICAL THEORY OF SCREENING

The solute component of the solution consists of dumbbells labelled i = $1,2,\ldots,N_B$. Each of these is composed of two equi-mass "monomer units" or beads with position vectors $\mathbf{R}_{1i}$ and $\mathbf{R}_{2i}$. The solvent consists of N_S structureless particles with position vectors $\mathbf{r}_\alpha$; $\alpha = 1,2,\ldots,N_S$. The masses of the polymer beads and solvent particles are denoted by $M/2$ and m, respectively.

The microscopic mass densities of polymer beads and solvent particles are

$$\hat{\rho}_B(\mathbf{r},t) = \frac{M}{2}\hat{n}_B(\mathbf{r},t) \equiv \frac{M}{2}\sum_i \sum_p \hat{n}_{pi}(\mathbf{r},t) \; ; \; \hat{n}_{pi}(\mathbf{r},t) = \delta[\mathbf{r} - \mathbf{R}_{pi}(t)], \tag{6}$$

and

$$\hat{\rho}_S(\mathbf{r},t) = m\hat{n}_S(\mathbf{r},t) \equiv m\sum_\alpha \hat{n}_\alpha(\mathbf{r},t) \; ; \; \hat{n}_\alpha(\mathbf{r},t) = \delta[\mathbf{r} - \mathbf{r}_\alpha(t)]. \tag{7}$$

These are related to the corresponding momentum densities

$$\hat{\mathbf{p}}_B(\mathbf{r},t) = \sum_i \sum_p \frac{M}{2}\dot{\mathbf{R}}_{pi}(t)\;\hat{n}_{pi}(\mathbf{r},t) \; ; \tag{8}$$

$$\hat{\mathbf{p}}_S(\mathbf{r},t) = \sum_\alpha m\dot{\mathbf{r}}_\alpha(t)\hat{n}_\alpha(\mathbf{r},t), \tag{9}$$

through the equations of change $\partial_t\hat{\rho}_B + \nabla\cdot\hat{\mathbf{p}}_B = 0$ and $\partial_t\hat{\rho}_S + \nabla\cdot\hat{\mathbf{p}}_S = 0$.

The dynamics of the system is assumed to be governed by the equations

$$\frac{M}{2}\ddot{\mathbf{R}}_{pi}(t) = z_p \mathbf{E}(\mathbf{R}_{pi}, t) - \nabla_{R_{pi}} [U_B(|\mathbf{R}_{1i} - \mathbf{R}_{2i}|) + \sum_j{}' \sum_q U_{BB}(|\mathbf{R}_{pi} - \mathbf{R}_{qj}|)$$

$$+ \sum_\alpha U_{BS}(|\mathbf{R}_{pi} - \mathbf{r}_\alpha|)]; \tag{10}$$

$$m\ddot{\mathbf{r}}_\alpha(t) = \mathbf{F}(\mathbf{r}_\alpha, t) - \nabla_{r_\alpha} [\sum_\beta{}' U_{SS}(|\mathbf{r}_\alpha - \mathbf{r}_\beta|) + \sum_i \sum_p U_{BS}(|\mathbf{r}_\alpha - \mathbf{R}_{pi}|)]. \tag{11}$$

Here U_B is the potential of the "connector force" that binds together the two beads of a dumbbell. U_{BB} is the site-site interaction between beads belonging to different dumbbells, U_{SB} the energy of interaction between a bead and a solvent particle, and U_{SS} the interaction between two solvent particles. Finally, $\mathbf{F}$ is the external force on a solvent molecule and $z_p\mathbf{E}(\mathbf{R}_{pi},t)$ the external force on bead p of dumbbell i. From all of this we obtain for $\hat{\mathbf{p}}_S$ and $\hat{\mathbf{p}}_B$ the equations of change[4]

$$\partial_t \hat{\mathbf{p}}_B + \nabla \cdot (\hat{\sigma}_B + \hat{\sigma}_{BS}) = \hat{\mathbf{F}}_{BS} + \hat{z}_B \mathbf{E}; \tag{12}$$

$$\partial_t \hat{\mathbf{p}}_S + \nabla \cdot (\hat{\sigma}_S + \hat{\sigma}_{SB}) = \hat{\mathbf{F}}_{SB} + \hat{n}_S \mathbf{F}, \tag{13}$$

with

$$\hat{z}_B(\mathbf{r},t) = \sum_i \sum_p z_p \hat{n}_{pi}(\mathbf{r},t)$$

and where

$$\hat{\mathbf{F}}_{BS}(\mathbf{r},t) = -\frac{1}{2}\int d\mathbf{r}' \frac{\mathbf{r}-\mathbf{r}'}{|\mathbf{r}-\mathbf{r}'|} U'_{BS}(|\mathbf{r}-\mathbf{r}'|)[\hat{n}_B(\mathbf{r},t)\hat{n}_S(\mathbf{r}',t) - \hat{n}_B(\mathbf{r}',t)\hat{n}_S(\mathbf{r},t)], \tag{14}$$

denotes the density of the "frictional" force that the solvent exerts upon the polymer beads. $\hat{\mathbf{F}}_{SB} = -\hat{\mathbf{F}}_{BS}$ is the equal and opposite force experienced by the solvent.

The function

$$\hat{\sigma}_B(\mathbf{r},t) = -\int d\mathbf{p} \frac{\mathbf{p}\mathbf{p}}{M/2} \hat{f}_B(\mathbf{r},\mathbf{p},t) + \frac{1}{2}\int d\mathbf{x}\; \mathbf{x}\mathbf{x}\; x^{-1} U'_{BB}(x) \int_0^1 da\, \{\hat{n}_B(\mathbf{r}-a\mathbf{x},t)\hat{n}_B(\mathbf{r}+(1-a)\mathbf{x},t)\}$$

$$+ \frac{1}{2} \int d\mathbf{x}\ \mathbf{x}\mathbf{x}\ x^{-1} U'_B(x) \int_0^1 da \hat{P}_B(\mathbf{r} - a\mathbf{x}, \mathbf{r} + (1 - a)\mathbf{x}, t) , \tag{15}$$

is the contribution to the microscopic stress tensor due exclusively to the polymeric component of the solution. The first term of Eq. (15), with

$$\hat{f}_B(\mathbf{r}, \mathbf{p}, t) = \sum_i \sum_p \delta[\mathbf{r} - \mathbf{R}_{pi}(t)]\delta[\mathbf{p} - \frac{M}{2} \dot{\mathbf{R}}_{pi}(t)] , \tag{16}$$

is the portion of this stress associated with the translational kinetic motions of the beads. The second is the "collisional transfer"[5] rate of momentum exchange due to interactions between different pairs of dumbbells, and the third is the rate at which the connector force transfers momentum from one end of a dumbbell to the other. This third contribution is a functional of

$$\hat{P}_B(\mathbf{r}, \mathbf{r}', t) = \sum_i \{\hat{n}_{1i}(\mathbf{r}, t)\hat{n}_{2i}(\mathbf{r}', t) + \hat{n}_{1i}(\mathbf{r}', t)\hat{n}_{2i}(\mathbf{r}, t)\} , \tag{17}$$

the density of dumbbells with one bead located at $\mathbf{r}$ and the other at $\mathbf{r}'$. In many treatments of polymeric solutions the first and third terms of $\hat{\sigma}_B$ are taken to be the only polymeric contributions to the fluid stress tensor.

The functions $\hat{\sigma}_{BS}$ and $\hat{\sigma}_{SB} = \hat{\sigma}_{BS}$ appearing in Eqs. (12) and (13) are "collisional transfer" portions of the stress tensor, completely analogous to the second term of $\hat{\sigma}_B$ but arising from the solvent-bead interactions U_{BS}. Finally, $\hat{\sigma}_S$ is the solvent analogue of the first plus second terms of $\hat{\sigma}_B$. In the absence of the dumbbells this would be the microscopic stress tensor of the neat solvent.

We now indicate the ensemble averages of the quantities $\hat{\sigma}_B$, $\hat{\sigma}_S$, $\hat{\sigma}_{BS}$, $\hat{\mathbf{F}}_{BS}$, $\hat{z}_B\mathbf{E}$ and $\hat{n}_S\mathbf{F}$ by the symbols σ_B, σ_S, σ_{BS}, $\mathbf{k}$, $\mathbf{e}$ and $\mathbf{f}$, respectively. Additionally the solute (B) and solvent (S) velocity fields $\mathbf{w}_B$ $(\mathbf{r},t)$ and $\mathbf{w}_S$ $(\mathbf{r},t)$ are defined as follows:

$$\langle \hat{\mathbf{p}}_B(\mathbf{r}, t) \rangle = \rho_B \mathbf{w}_B(\mathbf{r}, t) ; \tag{18}$$

$$\langle \hat{\mathbf{p}}_B(\mathbf{r}, t) \rangle = \rho_S \mathbf{w}_S(\mathbf{r}, t) . \tag{19}$$

Here $\rho_B = (M/2)c_B$ and $\rho_S = mc_S$ with $c_B \equiv 2N_B/V$ and $c_S \equiv N_S/V$ denoting the mean concentrations of polymer beads and solvent molecules.

Next we introduce the first of several simplifying assumptions; namely, that the average stress tensors $\boldsymbol{\sigma}_B + \boldsymbol{\sigma}_{BS}$ and $\boldsymbol{\sigma}_S + \boldsymbol{\sigma}_{SB}$ can be replaced with the Newtonian approximations

$$\boldsymbol{\sigma}_S + \boldsymbol{\sigma}_{BS} \doteq -p_B\mathbf{I} + \eta_B[\boldsymbol{\nabla}\mathbf{w}_B + (\boldsymbol{\nabla}\mathbf{w}_B)^T - \frac{2}{3}\mathbf{I}\boldsymbol{\nabla}\cdot\mathbf{w}_B] \; ; \tag{20}$$

$$\boldsymbol{\sigma}_S + \boldsymbol{\sigma}_{SB} \doteq -p_S\mathbf{I} + \eta_S[\boldsymbol{\nabla}\mathbf{w}_S + (\boldsymbol{\nabla}\mathbf{w}_S)^T - \frac{2}{3}\mathbf{I}\boldsymbol{\nabla}\cdot\mathbf{w}_S] \, . \tag{21}$$

It is natural to identify p_S and η_S, respectively, with the pressure and shear viscosity coefficient of the solvent component, but one should remember that $\boldsymbol{\sigma}_{SB}$ directly involves solvent-solute interactions and that the presence of the solute will affect the solvent distribution function $<\hat{n}_S(\mathbf{r},t)\,\hat{n}_S(\mathbf{r}',t)>$ upon which $\boldsymbol{\sigma}_S$ depends. p_B and η_S are similarly defined properties (osmotic pressure and viscosity) associated principally but not exclusively with the solute component of the solution. Finally, it is convenient (but not essential) to assume that the densities $<\hat{n}_B(\mathbf{r},t)>$ and $<\hat{n}_S(\mathbf{r},t)>$ are constants so that

$$\boldsymbol{\nabla}\cdot\mathbf{w}_B = \boldsymbol{\nabla}\cdot\mathbf{w}_S = 0 \, . \tag{22}$$

With these definitions and the assumptions of Eqs. (20) and (21) the momentum balance Eqs. (12) and (13) become

$$\rho_B\partial_t\mathbf{w}_B - \eta_B\nabla^2\mathbf{w}_B + \boldsymbol{\nabla}p_B = \mathbf{e} + \mathbf{k} \; ; \tag{23}$$

$$\rho_S\partial_t\mathbf{w}_S - \eta_S\nabla^2\mathbf{w}_S + \boldsymbol{\nabla}p_S = \mathbf{f} - \mathbf{k} \, . \tag{24}$$

By invoking the conditions in Eq. (22) the Fourier transforms of these then may be written as

$$\mathbf{w}_B(\mathbf{q},\omega) = \mathbf{T}_B(\mathbf{q},\omega)\cdot(\mathbf{e}+\mathbf{k}) \, , \tag{25}$$

and

$$\mathbf{w}_S(\mathbf{q},\omega) = \mathbf{T}_S(\mathbf{q},\omega) \cdot (\mathbf{f} - \mathbf{k}) . \tag{26}$$

Here $\mathbf{T}_S$ is given by Eq. (3) and $\mathbf{T}_B$ is an analogously defined Green function specific to the solute component of the solution.

To proceed further a constitutive equation is needed for the friction force density $\mathbf{k}$. Our ultimate goal is to obtain this constitutive relationship from a systematic microscopic theory but here we make do with less. Thus, we abandon the formula Eq. (14), which relates $\mathbf{k} = < \hat{\mathbf{F}}_{BS} >$ to the solvent-bead interactional potential U_{BS}, in favor of a phenomenological model that treats each polymer bead as a point center of hydrodynamic frictional interaction immersed in a continuous medium. According to the Faxen theorem[6] the frictional force on a bead is given by the formula

$$\mathbf{K}(\mathbf{R}_{pi}, \dot{\mathbf{R}}_{pi}, t) = -\xi[\dot{\mathbf{R}}_{pi} - \mathbf{w}_S^*(\mathbf{R}_{pi}, t)] , \tag{27}$$

with ξ a phenomenological friction coefficient and $\mathbf{w}_S^*(\mathbf{R}_{pi}, t)$ the solvent velocity at $\mathbf{R}_{pi}$, calculated as if there were no bead at this location. Our assumption then can be identified as replacing $\mathbf{k} = < \hat{\mathbf{F}}_{BS} >$ in Eqs. (23) and (24) with the Stokes-Faxen friction force density

$$\mathbf{k}_f(\mathbf{r},t) \equiv < \sum_i \sum_p \mathbf{K}(\mathbf{R}_{pi}, \dot{\mathbf{R}}_{pi}, t) \hat{n}_{pi}(\mathbf{r},t) > = -\xi c_B(\mathbf{w}_B - \mathbf{w}_S^*) . \tag{28}$$

By substituting Eq. (28) into Eq. (23) we obtain an equation

$$\rho_S \partial_t \mathbf{w}_S - \eta_S \nabla^2 \mathbf{w}_S + \nabla p_S = c_B \xi(\mathbf{w}_B - \mathbf{w}_S^*) + \mathbf{f} , \tag{29}$$

which closely resembles Eq. (II.1) of de Gennes. Indeed, the only differences aside from notation are that the external force $\mathbf{f}$ is missing from de Gennes' equation and that the friction term of his equation contains $\mathbf{w}_S$ in place of $\mathbf{w}_S^*$.

A second consequence of the constitutive relation of Eq. (28) is obtained by inserting it into Eq. (25) and, with $z_1 = z_2 = 1$, replacing $\mathbf{e} = < \hat{n}_B > \mathbf{E}$ with $c_B \mathbf{E}$. This produces the connection

$$\mathbf{w}_B(\mathbf{q},\omega) = \mathbf{M}(\mathbf{q},\omega) \cdot [\xi^{-1}\mathbf{E} + \mathbf{w}_S^*] . \tag{30}$$

and

$$\mathbf{M}(\mathbf{q},\omega) = M(\mathbf{q},\omega)(\mathbf{I} - \hat{\mathbf{q}}\hat{\mathbf{q}}) , \tag{31}$$

and where $M(\mathbf{q},\omega) = [1+(\rho_B/\xi c_B)(i\omega + \nu_B q^2)]^{-1}$. Except for the appearance of $\mathbf{w}_S^*$ in place of $\mathbf{w}_S$, Eq. (30) is a precise analogue of de Gennes' Eq.(II.3). Furthermore, the function $\mathbf{M}(\mathbf{q},\omega)$ defined by Eq. (31) possesses the properties that de Gennes attributes to the analogous quantity that occurs in his paper; namely, that $\mathbf{M} \to 0$ in the short wavelength $q \to \infty$ limit and that $\mathbf{M} \to \mathbf{I}$ as $q \to 0$ and $\omega \to 0$. Finally, by inserting Eq. (30) into the Fourier transform of Eq. (24) we obtain an equation

$$\rho_S(i\omega + \nu_S q^2)\,\mathbf{w}_S = (\mathbf{I} - \hat{\mathbf{q}}\hat{\mathbf{q}})\cdot\ [\mathbf{f} + \xi c_B(1 - M)\mathbf{w}_S^*] , \tag{32}$$

that closely resembles de Gennes' (obviously misprinted) Eq. (II.7).

Now that these connections have been established between the present analysis and that of de Gennes, all that remains is to obtain the desired functional relationship Eq. (2) between the solvent flow field and the external force density **f**. This can be done in several essentially equivalent ways. For example, in our recent paper on this subject we did so by invoking a relationship

$$\mathbf{k}_f = -\xi c_B \mathbf{w}_B + \xi c_B \mathbf{T}_S\cdot\ (\mathbf{f} - \mathbf{k}_f) , \tag{33}$$

that can be derived [using the mean-field approximation $<\mathbf{k}_{pi}(\mathbf{r},t)\, n_{qj}(\mathbf{r}',t)> = <\mathbf{k}_{pi}(\mathbf{r},t)><n_{qj}(\mathbf{r}',t)>$] from the definition of the friction force density $\mathbf{k}_f \equiv <\Sigma_i\ \Sigma_p\ \mathbf{k}_{pi}>$. Since this method of derivation is equally applicable to the dumbbell model of the present paper, it could be used here as well. However, we shall use an alternative scheme that combines Eqs. (28) and (30) to obtain the formula

$$\mathbf{T}_S\cdot\ \mathbf{k}_f = \frac{\rho_B(i\omega + \nu_B q^2)}{\rho_S(i\omega + \nu_S q^2)}\mathbf{M}\cdot\ \mathbf{w}_S^* .$$

Then, by inserting this result into Eq. (26) and ignoring the presumably slight difference between $\mathbf{w}^*_S$ and $\mathbf{w}_S$, it is found that

$$\mathbf{w}_S(\mathbf{q},\omega) = \chi\mathbf{T}_{SC}\cdot\ (\mathbf{f} + M\mathbf{e}) \tag{34}$$

Here $\mathbf{T}_{SC}$ is the screened Oseen tensor of Eq. (4), $\mathbf{e} \doteq c_B\mathbf{E}$ is the external force density on the dumbbells, and

$$\chi(q,\omega)=[(\rho_B/\xi c_B)(i\omega+\nu_B q^2)+1][(\rho_B/\xi c_B)(i\omega+\nu_B q^2) + (i\omega+\nu_S q^2)(i\omega+\nu_S q^2+\xi c_B/\rho_S)^{-1}]^{-1} . \tag{35}$$

These are precisely the same results that we previously obtained and so it would appear that the mean-field approximation used to obtain Eq. (33) is "equivalent" to neglecting the difference between the actual and virtual flow fields $\mathbf{w}_S$ and $\mathbf{w}^*_S$.

III. CONCLUSIONS AND REMARKS

The conclusions that can be drawn from the formula Eq. (34) already have been considered in our earlier paper, cf. Ref. 3. These may be summarized as follows:

1.) The low frequency ($\omega \to 0$) limit of the Green function

$$\mathbf{T}^{EFF}(\mathbf{q},\omega) \equiv \chi\mathbf{T}_{SC} = [\rho_S(i\omega+\nu_S\{q^2+q_{SC}^2\})]^{-1}(\mathbf{I}-\hat{\mathbf{q}}\hat{\mathbf{q}}) , \tag{36}$$

is given by the formula $[q^2\eta^{EFF}(q)]^{-1}(\mathbf{I}-\hat{\mathbf{q}}\hat{\mathbf{q}})$ with

$$\eta^{EFF}(q) = \eta_S[1+(\rho_B/\xi c_B)\nu_B q^2]^{-1}[1+(\rho_B/\xi c_B)\nu_B(q^2+q_{SC}^2)]$$
$$= \eta_S[1+(\xi c_B/\eta_S q^2)(1+\xi c_B/\eta_B q^2)^{-1}] . \tag{37}$$

When ν_B is set equal to zero, the effective viscosity $\eta^{EFF}(q)$ reduces to η_S and $\mathbf{T}^{EFF}(\mathbf{q}, 0)$ becomes the familiar, <u>unscreened</u> Oseen tensor.

2.) The static unscreened Oseen tensor also will be obtained even when $\nu_B \neq 0$, provided that the wavenumber is so large that $(\rho_B\xi/c_B)\,\nu_B q^2 >> 1 + \eta_B/\eta_S$. Thus, the effective viscosity becomes equal to η_S in the immediate vicinity of the source of the perturbing force. The reason for this is that there are too few solute particles to interfere with the solvent flow in the space between the source and a nearby field point.

3.) At large distances from the source (or equivalently, for small values of q) the effective viscosity reduces to the sum of the solvent and solute viscosity coefficients, η_S and η_B. Thus, static screening is absent in this limit as well.

4.) Indeed, it can be seen from the second line of Eq. (37) that static screening occurs only if both of the conditions $\eta_B q^2/\xi c_B >> 1$ and $\eta_S q^2/\xi c_B \approx 1$ are satisfied. The corresponding physical requirements are that the dumbbells be extremely massive (ρ_B/c_B very large) and/or that their interactions be so strong that $\eta_B/\eta_S >> 1$. These are conditions that prevail in gel-like suspensions of immobilized molecules. In this limit $\eta^{EFF}(q) \to \eta_S + \xi c_B/q^2$ and

$$\mathbf{T}^{EFF}(\mathbf{q},0) \to \mathbf{T}_{SC}(\mathbf{q},0) = [\eta_S(q^2 + q^2{}_{SC})]^{-1}(\mathbf{I} - \hat{\mathbf{q}}\hat{\mathbf{q}}) . \tag{38}$$

Since we have demonstrated several remarkable similarities between our theory and that of de Gennes, why should we and he have drawn such different conclusions about screening in the static, low-frequency limit? The origin of this disparity can be traced to an incorrect assessment by de Gennes of the large wavenumber limit; he missed the contribution of χ to the Green function $\mathbf{T}^{EFF} = \chi\,\mathbf{T}_{SC}$ by calculating the asymptotic behavior of $\mathbf{T}^{EFF}(\mathbf{q},\omega = 0)$ using the improper scheme

$$\mathbf{T}^{EFF}(\mathbf{q},\omega = 0) = \mathbf{T}_{SC}(\mathbf{q},\omega = 0) \lim_{\mathbf{q} \to \infty} \chi(\mathbf{q},\omega = 0) .$$

Results rather similar to those of de Gennes have been reported in one recent study by Muthukumar and Edwards[7] and another by Odijk.[8] However, it does not seem to us that either of these studies provides a truly compelling mathematical demonstration that screening occurs in stationary flows for high or moderately high values of q. In another relevant study Muthukumar[9] identified several deficiencies of the Freed-Edward theory[10] of polymer solution dynamics and then used their multiple scattering technique to conclude that there is no hydrodynamic screening in infinitely dilute solutions. This same conclusion was reached by Freed and Perico,[11] based on an analysis of the model of phantom polymer chains. Thus, these two independent studies, although limited in applicability to very dilute solutions, lead to conclusions that agree with ours.

There is a phenomenological argument that, without recourse to calculation, discredits the notion of screening in steady fluid flows. It begins with the observation that screening is evidence of (or a consequence of) a process that weakens the fluid flow by an irreversible dissipation of its energy. In our model the mechanism for this dissipative process is the interaction between the fluid and the point centers of friction. The flow exerts forces on these mobile sites which, unless constrained, accelerate until the drag vanishes and along with it the attendant hydrodynamic screening. Consequently, no hydrodynamic screening is to be expected when the flow is steady and the solute particles are mobile.

A second very fundamental objection to the existence of static screening is especially convincing when applied to the steady flows of homogeneous solutions. It relies upon the principle of Galilean invariance, according to which the form of the linear equations of motion of the medium (composite of solvent and solute) should be unchanged when the velocity is everywhere increased by an additive constant. This principle would be violated if these equations included a uniformly distributed frictional force. A variation on this theme was offered by an (unidentified) referee of our earlier paper, Ref. 3, who stated that "... I am somewhat surprised by the confusion in the literature. Specifically, when all components are mobile, the total momentum density (i.e., that of the fluid and particles) is conserved; hence, a mode whose lifetime diverges as $q \to 0$ is to be expected."

None of this argues that screening need be absent under non-stationary conditions and, indeed, for finite frequencies of the perturbing force there are two situations where screening obviously will occur. The first of these is encountered when the frequency of $\mathbf{f}(\mathbf{q},\omega)$ exceeds all of the characteristic frequencies of the fluid. [These consist of $\omega_B = \xi c_B/\rho_B$ (the inverse of the momentum relaxation time of the dumbbells), $\omega_B^{VIS} = \nu_B/\ell^2_B = \nu_B c_B^{2/3}$ (the inverse of the solute component shear-wave relaxation time), $\omega_S^{VIS} = \nu_S/\ell^2_B = \nu_S c_B^{2/3}$ and $\omega_{SC} = \rho_S/c_S\xi$]. Then, provided that $\ell_B q$ remains finite, $\chi(q,\omega) \to 1$ as $\omega \to \infty$ and $\mathbf{T}^{EFF} \to \mathbf{T}_{SC}$. In this case the screening arises because the solute particles are unable to follow the rapidly changing direction of the solvent flow. A similar situation prevails when $\omega_B \to 0$, for then the dumbbells are so massive that their motions are unaffected by the solvent flow. In both cases, the screening is identical in form to that produced by an array of immobile hydrodynamic scattering centers.

By way of concluding remarks, let us reiterate the limitations of the theory presented here. The ensemble averages of the microscopic stress tensors appearing in the equations of motion Eqs. (12) and (13) were replaced with the Newtonian expressions Eqs. (22) and (23). Although the subsequent analysis would have been no more difficult had these been replaced with the corresponding linear viscoelastic approximations [η_S and η_B then would have been replaced with frequency dependent coefficients], the low order of the spatial gradients included in these expressions does severely limit the reliability of the theory at large wave numbers. The second critical approximation was our use of the Stokes-Faxen ansatz of Eq. (7) for the "frictional" interaction between the solvent and solute components of the solution. However reasonable this may seem from a continuum-mechanical point of view, it is a step for which a more rigorous microscopic justification (or replacement) is sorely needed. There surely are conditions for which this constitutive relationship can be used with confidence but there are others where it cannot. Finally, we came to our end results by ignoring the distinction between the solvent velocity $\mathbf{w}_S$ and the virtual flow field $\mathbf{w}^*_S$. Some evidence exists that this is equivalent to a mean-field approximation for the friction force density. In any case, it is an approximation of uncertain validity that deserves more careful scrutiny than has been accorded to it here.

It is no defense of the approximations listed above to point out that they are common to all of the previous theories referenced here and to several others as well. What we do feel confident about is the correctness of the conclusions that we have drawn from them. To move beyond this point the state of the microscopic theory must be advanced.

It has long been argued that screening of the hydrodynamic interactions is responsible for the shift from Zimm-like to Rouse-like dynamics that accompanies an increase of polymer concentration. The obvious question to ask is whether the conclusions presented here contradict this conventional wisdom. The short answer is yes. The long answer is a bit more equivocal. What we have shown is that a theory based upon a very specific (but familiar and widely accepted) model of polymeric solutions predicts that screening will not occur in a steady-state flows <u>unless</u> the parameters of the system are such that it is "gel-like". The conditions to which we refer are those which will immobilize the polymer molecules, e.g., conditions that favor strong entanglements among the polymer threads as well as strong interactions between the network and the boundaries of the system. It is important to recognize that interactions of these types may have significant effects upon the solvent dynamics at concentrations well below those at which the system begins to exhibit gel-like <u>mechanical</u> properties. There is a definite possibility that it is this immobilization which causes the Zimm-to-Rouse transition. But if it is not, then something about the model must be changed to make it capable of accounting for the Zimm-to-Rouse shift. Whatever this change might be, it would have to involve one or more of the three constitutive relationships upon which the present theory is based. These consist of the quasi-Newtonian partial stress tensors, Eqs. (20) and (21), and the Stokes-Faxen friction force density Eq. (28). At high solute concentrations these relationships may fail to faithfully describe the rates of momentum transfer which they are intended to represent. Finally, it should be recognized that the model of solute Brownian particles immersed in a viscous solvent must itself fail when the solvent concentration becomes so low that it no longer can act as a dispersive medium.

ACKNOWLEDGEMENTS

This research has been supported in part by grants from the National Science Foundation.

NOMENCLATURE

$\mathbf{r}_\alpha, \dot{\mathbf{r}}_\alpha$	position and velocity of solvent particle α
$\mathbf{R}_{pi}, \dot{\mathbf{R}}_{pi}$	position and velocity of pth bead of dumbbell i
$t, \mathbf{r}$	time and position of field point
$\omega, \mathbf{q}$	frequency (angular velocity) and wave vector
B,S	indices labelling solute (Brownian particles) and solvent
c_B, c_S	stoichiometric concentrations
$\nu_B, \nu_S, \eta_B, \eta_S$	kinematic and shear viscosities
$\mathbf{E}, \mathbf{F}$	external forces on solute and solvent, respectively
$\mathbf{e}, \mathbf{f}$	ensemble average densities of solute and solvent external forces
$U_B, U_{BB}, U_{BS}, U_{SS}$	interaction energies
$\hat{\rho}_j, \hat{\mathbf{p}}_j$	microscopic mass and momentum densities of species j (B or S)
$\hat{n}_\alpha, \hat{n}_{pi}$	microscopic particle densities
$\hat{\sigma}_B, \hat{\sigma}_S, \hat{\sigma}_{SB}, \hat{\sigma}_{BS}$	microscopic stress tensor contributions

$\hat{\mathbf{F}}_{BS}, \hat{\mathbf{F}}_{SB}$	microscopic densities of solute-solvent interspecies drag ("friction") forces
$\mathbf{k}, \mathbf{k}_f$	average of $\mathbf{F}_{BS}$ and of a phenomenological analogue thereof
$\mathbf{w}_B; \mathbf{w}_S, \mathbf{w}^*_S$	solute and solvent velocity fields
$\mathbf{T}_B, \mathbf{T}_S$	Oseen-like Green functions
$\mathbf{K}, \xi$	bead friction force and friction coefficient
$\mathbf{M}, M, \chi$	response functions defined in text
$\mathbf{T}_{SC}, \mathbf{T}^{EFF}, \eta^{EFF}$	generalized Green functions and viscosity coefficient defined in text
q^{-1}_{sc}	hydrodynamic screening length

REFERENCES

1. P.G. de Gennes, Macromolecules, 9, 587 (1976); ibid. 9, 594 (1976).
2. See for example, A.R. Altenberger, M. Tirrell and J.S. Dahler, J. Chem. Phys., 84, 5122 (1986).
3. A.R. Altenberger, J.S. Dahler and M. Tirrell, Macromolecules, 21, 464 (1988).
4. To obtain these equations use has been made of the identity (W. Noll, J. Rat. Mech. Anal., 4, 625 (1955)).

$$\int d\mathbf{y} f(\mathbf{x},\mathbf{y}) = \int d\mathbf{y} f^s(\mathbf{x},\mathbf{y}) + \int d\mathbf{y} f^a(\mathbf{x},\mathbf{y})$$

$$= \int d\mathbf{y} f^s(\mathbf{x},\mathbf{y}) + \nabla_x \cdot \frac{1}{2} \int d\mathbf{z}\,\mathbf{z} \int_0^1 da\, f(\mathbf{x} - a\mathbf{z}, \mathbf{x} + (1-a)\mathbf{z})$$

wherein f^s and f^a denote the symmetric (s) and antisymmetric (a) parts of the function $f(\mathbf{x},\mathbf{y})$.

5. J.H. Irving and J.G. Kirkwood, J. Chem. Phys., 18, 817 (1950).
6. P. Mazur and D. Bedeaux, Physica, 76, 235 (1975); see also J.J. Erpenbeck and J.G. Kirkwood, J. Chem. Phys., 29, 909 (1958) and Chapter 6 of Modern Theory of Polymer Solutions, H. Yamakawa, Harper and Row (1971).
7. M. Muthukumar and S.F. Edwards, Macromolecules, 16, 1475 (1983).
8. T. Odijk, Macromolecules, 19, 2073 (1986).

9. M. Muthukumar, J. Phys. A; Math. Gen., 14, 2129 (1981).

10. K.F. Freed in Chapter 8 of Progress in Liquid Physics, Ed. C.A. Croxton, Wiley (New York, 1978).

11. K.F. Freed and A. Perico, Macromolecules, 14, 1290 (1981).

TRACER DIFFUSION IN POLYMER-LIKE NETWORKS

R. Messager, D. Chatenay, W. Urbach and D. Langevin

Laboratoire de Physique de l'Ecole Normale Supérieure
24, rue Lhomond - 75231
Paris Cedex 05
France

ABSTRACT

We have studied self-diffusion in polymer-like networks: self-diffusion of transient chains (long entangled micelles) and tracer diffusion in the solvent (silica gels). Our results illustrate the role of reversible chain breaking and the scale-dependence of diffusion in the solvent.

I. INTRODUCTION

Polymeric networks are interesting structures which have complex dynamic properties. Self-diffusion of the polymeric chains can be studied in polymer melts or in semidilute solutions. For long linear polymers, this process, as well as other properties like viscoelasticity, are governed by the reptation mechanism.[1] In crosslinked polymer solutions, the self-diffusion of the polymer chains is locked. But one can study the obstruction effects created by the gel structure by analyzing the diffusive motion of a tracer molecule in the surrounding solvent.

We will present here experimental results dealing with self-diffusion in polymer-like systems. We have first investigated chain self-diffusion in semidilute solutions of long cylindrical micelles which can break and reform: "living polymers". We have also studied self-diffusion of a tracer molecule inside porous silica gels.

II. SELF-DIFFUSION IN LIVING POLYMERS

Surfactants, when dissolved in water, can form aggregates called micelles.[2] The micelles are generally spherical at small surfactant concentration and become rod-like at higher concentrations. The evolution is faster when salt is added to an aqueous solution of an ionic surfactant: the salt screens the electrostatic interactions between the polar parts of the surfactant molecules, and lowers the aggregates curvature. In some cases, very long aggregates can form.[3,4] Because of thermal fluctuations, they adopt, like polymers, worm-like structures. These structures have also been observed in microemulsions (oil-water-surfactant mixtures),[5] where their occurrence was predicted earlier theoretically.[6] The polymer analogy was recently demonstrated by light scattering experiments in the semidilute regime where the worm-like micelles are entangled.[7]

There is, however, a fundamental difference between micelles and polymers: the micelles are transient aggregates and their dynamic properties are expected to be different from those of polymer solutions.[8] This problem was addressed theoretically by Cates[8] who called the transient chains "living polymers." He considered the meantime τ for such a chain to break into pieces. If the time of the measurements is shorter than τ, the analogy with polymers is conserved. This was probably the case in the quasi-elastic light-scattering experiments of Candau et al.,[7] where the mutual diffusion coefficient was found to obey power laws with semidilute polymer solutions exponents.[1] If the time of the measurements is longer than τ, like in the stress relaxation experiments of Candau et al.,[9] the situation is less well understood. Cates introduced the reptation time of the chain τ_r. If $\tau_r<\tau$, the dominant stress-relaxation mechanism is simply reptation. If $\tau_r>\tau$, stress relaxation should be characterized by a new intermediate time scale $\sqrt{\tau\ \tau_r}$ associated with a process whereby the chain breaks at a point close enough to a given segment of tube for reptative relaxation of that segment to occur before the new chain end is lost by recombination.

A further difference with polymers arises from the fact that the chain length ℓ of the worm-like micelles depends on the surfactant concentration because aggregation is an equilibrium process. The polymer scaling laws which deal both with polymer chain length and concentration can be tested here only for surfactant concentration c. Moreover, the relation between ℓ and c is not well known: according to different models, ℓ scales with c or $c^{\frac{1}{2}}$.[10] In the Cates theory $\ell \sim c^{\frac{1}{2}}$.

In order to clarify the problem, we have undertaken a study of the self-diffusion coefficient of the micelles as a function of surfactant concentration. The corresponding theoretical variations have been predicted by Cates:

$$D \sim \ell^{-2} c^{-7/4} \sim c^{-2.75}\ ; \ for\ \tau_r < \tau \tag{1}$$

$$D \sim \ell^{-2/3} c^{-5/4} \sim c^{-1.58}.\ for\ \tau_r > \tau \tag{2}$$

II.1. Experimental Procedure. The micelles self-diffusion coefficient has been determined by fluorescence recovery after fringe pattern photobleaching (FRAP).[11] A fluorescent probe was incorporated in the micelles: it was checked that its finite residence time in the micelle did not affect the diffusion process.

The surfactant used was cetyltrimethyl ammonium bromide (CTAB). Micellar growth was enhanced with potassium bromide. For 0.1 M KBr at 30°C, the light-scattering measurements of Candau et al. demonstrated that the polymer analogy was valid. We studied other salinities: 0.05 M, 0.25 M at 30°C and 35°C.

II.2 Discussion The curves in Fig. 1 show two regimes: a dilute regime at small surfactant concentration where the recovery curves were observed to be non-exponential (as a consequence of the large micelle size polydispersity) and a semidilute regime in which D exhibits a power law dependence on concentration. In this regime, the recovery curves are single exponentials.[11] If τ_r were less than τ, D being strongly dependent on ℓ and the micelle size polydispersity being large, the recovery curves would be strongly non-exponential. This proves that our experiments, where the time scales were typically 1-100 sec, were performed in the regime $\tau_r > \tau$. A similar conclusion was obtained by Candau et al., who showed that the stress relaxation is mono-exponential in these systems, with relaxation times in the range 0.01-1 sec. The time scale for quasi-elastic light-scattering experiments being 10^{-6}-10^{-3} sec, the micelle lifetime lies probably in the range 10^{-3}-10^{-2} sec.

The data of the figure also show an unexpected behaviour. The exponent of the power law dependence $D(c)$ is salinity and (slightly) temperature-dependent: it varies between -4.6 and -1.4 (± 0.1). For 0.1 M salt (system studied by Candau et al.), its value is -2 ± 0.1. The theoretical value -1.6 is in agreement with the data for 0.25 M solution. The origin of the discrepancy for the other salinities is not known. Although the limiting behaviour for time shorter than τ for 0.1 M KBr solution follows a polymer behaviour, Candau et al. also reported that, in the "living polymer" regime, the stress relaxations properties did not vary with concentration, as predicted by Cates' model.[9] The micellar growth mechanism is likely to be more complex than predicted and in particular might be salinity-dependent.

In conclusion, our experiments strongly support the analogy between giant worm-like micelles and "living polymers." In the semidilute regime, the micellar self-diffusion mechanism is a reptation process truncated by a micellar breaking-recombination process. Further work is however needed to clarify the dependence of the dynamic properties on surfactant concentration.

III. SELF-DIFFUSION IN POROUS SILICA GELS

Recent studies of the structure of several silica gels have shown the relevance of new concepts developed in disordered matter physics: fractal objects, percolation structures,...[12] It must be pointed out, however, that, if the solid matrix is fractal, the solvent subspace is not neces-

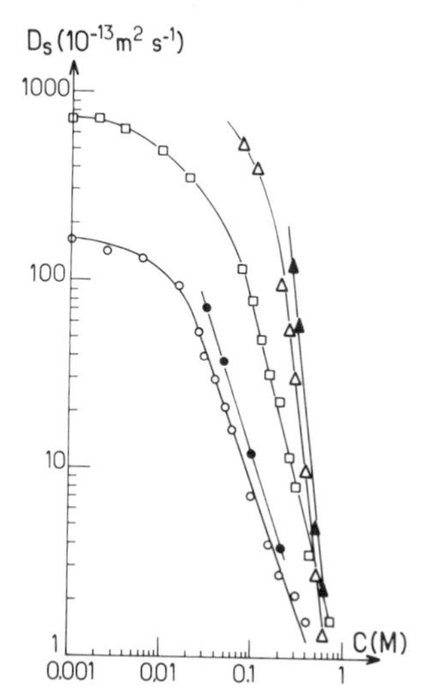

Fig. 1. Self-diffusion coefficients of aqueous CTAB solutions for 0.05 M(▲); 0.1 M(□); and 0.25 M(o) KBr at 30°C-0.05 M (Δ) and 0.25 M (●)KBr at 35°C.

sarily also fractal. Diffusion in the solvent subspace is nevertheless expected to be scale-dependent: For scales much smaller than the smallest pore size, the diffusion coefficient will be the same as in the pure solvent D_0. For scales much larger than the largest pore size, the diffusive motion will be averaged over the different paths and the effective diffusion coefficient becomes scale-independent:

$$D_\infty = D_o / T \tag{3}$$

where T is the tortuosity of the medium.[13]

If the solvent subspace is fractal in the scale range $\ell_1 < L < \ell_2$, the diffusive behaviour is similar: $D = D_0$ for $L << \ell_1$ and $D = D_\infty < D_0$ for $L >> \ell_2$. This arises because in the range $\ell_1 < L < \ell_2$, diffusion is anomalous: the average square distance from the origin grows more slowly than linearly with time[14] ("ant in a labyrinth" problem) : $L^2 \sim t$ with $\alpha < 1$ (α is the ratio of the spectral and fractal dimensions of the medium).[15] This leads to a scale-dependent diffusion coefficient $D \sim L^{2(\alpha-1)/\alpha}$ and from crossover arguments to $D_\infty = D_0 \, (\ell_1/\ell_2)^{2(1-\alpha)/\alpha}$.

III.1 Experimental Procedure The samples were made of variable amounts of hydrophobic colloidal silica (Aerosol R972 Degussa) mixed with an index matching solvent (ethanol + ethanolamine) containing the tracer molecule (fluorescein isothiocyanate, Molecular Probes). We have checked that tracer adsorption on the silica was negligible. Further details on the sample preparation can be found in reference [16].

Diffusion coefficients of the tracer molecules have been measured also by fluorescence recovery after fringe pattern photobleaching recovery. The fluorescence recovery curves are pure exponentials but the diffusion

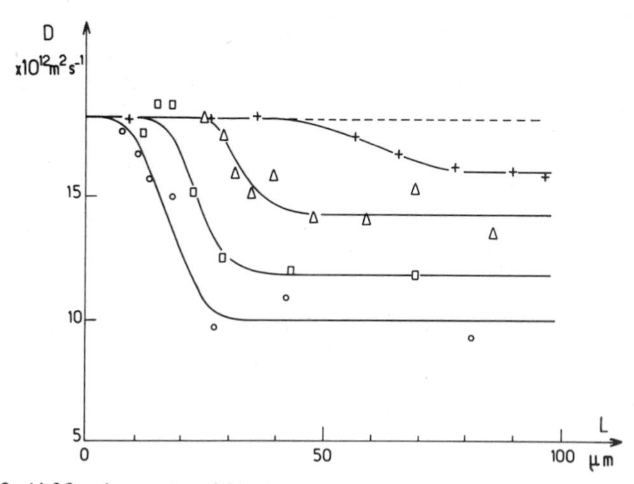

Fig. 2. Self-diffusion coefficient versus the fringe spacing L at different silica concentrations: (+) : Φ = 1.4%; (Δ) : Φ = 2.8%; ($\square$) : Φ = 4.1%; (O): Φ = 5.5%. The dashed line is the D value in the pure solvent.

coefficient depends on the fringe spacing L. The results are shown in Fig. 2. We observe clearly two regimes where D is constant. At small scales L, D is the same for all the samples and is also identical to the diffusion coefficient of the tracer in the pure solvent D_o. At large scale $L, D = D_\infty$ is a decreasing function of the porosity $1 - \Phi$, Φ being the silica volume fraction. In an intermediate range of scales, which also depends on Φ, we observe a variation of D, although the recovery curves remain exponential.

III.2 Discussion. We have defined a characteristic length ξ as the fringe spacing for which $D(\xi) = (D_o + D_\infty)/2$. We find that ξ (μm) = $1.2\ \Phi^{-0.93 \pm 0.05}$. ξ is expected to scale with the mesh size of the silica gel. By analogy with semidilute polymer solutions, we will assume that the gel is made of a random-walk association of elementary objects of size a:

$$\xi = N^{\nu} a \ , \tag{4}$$

where N is the number of associated elements along the distance ξ. From $\Phi = Na^3/\xi^3$, we get

$$\xi = a\,\Phi^{\nu/(1-3\nu)} \ , \quad and \tag{5}$$

$$\nu = 0.52 \pm 0.03\,. \tag{6}$$

The value $\nu = 0.5$ is consistent with a random-walk structure for the gel, as well as any other structure of fractal dimension 2. This result was frequently found in colloidal silica aggregates[17] and was confirmed by X-rays experiments performed on dry silica.[18] Structures of fractal dimension 2 include random-trees (lattice animals) with excluded volume. The random-tree would seem a likely suggestion for the short-scale structure of a silica gel.

Figure 3 shows the behaviour of D_∞ versus Φ with a slope equal to -9. Using Eq. 1 with a tortuosity of packed glass beads $T = (1 - \Phi)^{-\frac{1}{2}}$,[19] one gets for small Φ: $D_\infty = D_o(1-0.5\Phi)$, i.e., a slope smaller than observed. A possible explanation of the discrepancy comes from the fact that the medium is made of random-walk association of larger objects than the elementary silica spheres: a (~1 μ) is much larger than the silica sphere diameter a_o (~18 nm). The larger objects are themselves silica sphere aggregates formed during the hydrophobic treatment.[20] If we assume that the large scale value of D is D_a inside the aggregates, and that these aggregates occupy a volume fraction Φ_a, we get from an effective medium argument

$$D_\infty = D(1 - k\,\Phi), and\, k = \frac{3(D_o - D_a)}{3D_o + D_a}\,\frac{\Phi}{\Phi_a} \ , \tag{7}$$

where $\Phi' = \Phi a/\Phi$ is the silica volume fraction in the aggregates. The data leads to $k = 9$, in agreement with rough estimations: $D_a{\sim}4D_o$ and $\Phi'{\sim}0.1$. A better knowledge of the structure of the aggregates would be needed to estimate separately D_a and Φ_a.

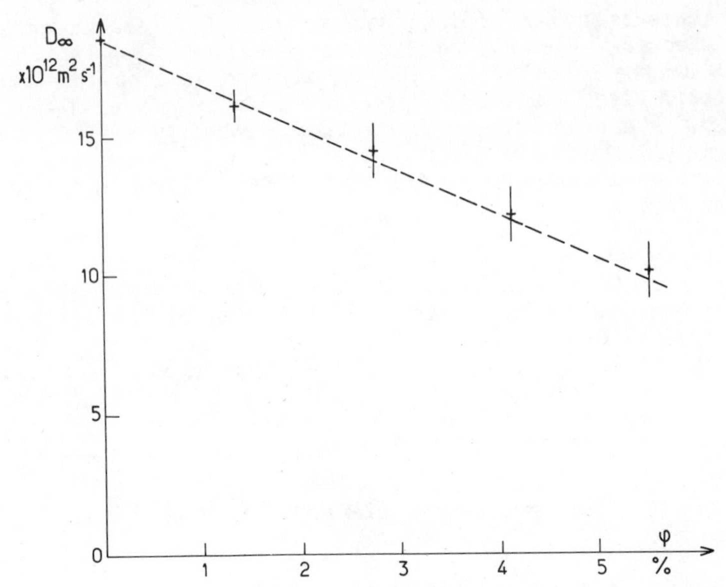

Fig. 3. Large-scale diffusion coefficient versus silica volume fraction. With $D_\infty = D_0(1 + 9\Phi)$, the dashed line a fit with $D_\infty = D_0(1 - 9\Phi)$.

In brief, we have observed the scale-dependent diffusion of a tracer particle in a porous silica gel, although the explored medium is probably not fractal. At small scales, the diffusion is free; at large scales it is slowed due to the tortuous paths of the porous structure. The crossover region corresponds to sizes comparable with the mesh size of the silica gels. The data support a structural description in terms of double porosity.

IV. GENERAL CONCLUSION

We have studied several original aspects of self-diffusion in polymer-like networks: the role of the reversible chain breaking, and the role of the tortuous character of the space in which diffusion takes place. These studies illustrate the complexity of the problem and emphasize its connection with other fields of physical chemistry; e.g., surfactant association, heterogeneous media.

ACKNOWLEDGEMENTS

The contributions of A. Ott for the experiments on micelles and of J.P. Bouchaud for the interpretation of the silica gel data are gratefuly acknowledged. We also thank a referee for his constructive remarks.

NOMENCLATURE

a size of the elementary structural objects
c surfactant concentration
D Self-diffusion coefficient
D_0 self-diffusion coefficient in the pure solvent
D_∞ large-scale self-diffusion coefficient

ℓ chain length
L length scale of the measurement
ξ mesh size
τ chain lifetime
τ_r reptation time
T tortuosity of the porous medium
Φ silica volume fraction

REFERENCES

1. P.G. de Gennes, Scaling Concepts in Polymer Physics, Cornell University Press (1979).
2. V. Degiorgio and M. Corti, editors, Physics of Amphiphiles: Micelles, Vesicles and Microemulsions, North Holland, (1985).
3. G. Porte, J. Appell and Y. Poggi, J. Phys. Chem., 84, 3105 (1980); J. Colloid and Interface Sci., 87, 492 (1982).
   G. Porte and J. Appell, J. Phys. Chem., 85, 2511 (1981).
4. H. Hoffmann, M. Lobl and H. Rehage in Ref. 2, p.237; H. Hoffmann, G. Platz, H. Rehage and W. Schorr, Ber. Bunsenges Phys. Chem., 25, 877 (1981); Adv. Colloid and Interface Sci., 17, 275 (1982).
5. B.W. Ninham, I.S. Barnes, S.T. Hyde, P.J. Derian and T.N. Zemb, Europhys. Lett., 4, 561 (1987).
6. S.A. Safran, L.A. Turkevich and P. Pincus, J. Phys. Lett., 45, L-69 (1984).
7. S.J. Candau, E. Hirsch and R. Zana
   a) J.Colloid and Interface Sci., 105, 521 (1985).
   b) J.Phys. (Paris), 45, 1263 (1984).
   c) in Physics of Complex and Supermolecular Fluids, eds. S.A. Safran and N.A. Clark, Wiley, New York (1987).
8. M.E. Cates, Macromolecules, 20, 2289 (1987).
9. S.J. Candau, E. Hirsch, R. Zana and M. Adam, in Physics of Amphiphilic Layers, eds. J. Meunier, D. Langevin and N. Boccara, Springer, New York p.268 (1987).
10. See Ref. 7c for a discussion.
11. R. Messager, A. Ott, D. Chatenay and D. Langevin, Phys. Rev. Lett., 60, 1410 (1988).
12. N. Boccara and M. Daoud editors, Physics of Finely Divided Matter, Springer, New York (1985).
13. J.M. Smith, Chemical Engineering Kinetics, McGraw Hill, New York, (1981).
14. P.G. de Gennes, La Recherche, 72, p.919 (1976).
15. J. Vannimenus in Ref. 12, p. 317.
16. R. Messager, D. Chatenay, W. Urbach, J.P. Bouchaud and D. Langevin, submitted for publication.
17. D.W. Schaefer, J.E. Martin, A.J. Hurd and K.D. Keefer in Ref. 12, p.31.
18. R. Ober and D. Chatenay, unpublished data.
19. D.L. Johnson, T.J. Plona, C. Scala, F. Pasierb and H. Kojuma, Phys. Rev. Lett., 49, 1840 (1982).
20. Silica Degussa, technical bulletin.

DYNAMICAL PROPERTIES OF SEMIDILUTE POLYMER SOLUTIONS AT THE THETA TEMPERATURE

M. Adam and M. Delsanti

Service de Physique du Solide
et de Résonance Magnétique
CEA-CEN, 91191 Gif-sur-Yvette Cedex
France

ABSTRACT

Experimental results obtained on semidilute polymer solutions are reviewed. The static and dynamic properties observed show that the correlation length of concentration fluctuations and the distance between entanglements are not proportional at the θ-temperature whereas they are in good solvent conditions.

I. INTRODUCTION

Contrary to the earlier beliefs, dynamical properties of semidilute solutions at the theta temperature are much more difficult to comprehend than properties of good-solvent solutions. Here, without going either into the details of the experimental determination or theoretical considerations, we will review different experimental results obtained on polystyrene solutions. Our aim is to show that, at the theta temperature, the mesh size of the transient gel is not proportional to the correlation length of the concentration fluctuations. The coexistence of these two characteristic lengths explains why some dynamical properties do not obey scaling laws at the theta temperature.

II. STATIC PROPERTIES

A polymer solution is semidilute[1] when the monomer concentration C is much smaller than the solvent concentration but larger than the overlap concentration: $C^* \simeq (M/R^3)$, M and R are the mass and the radius of gyration of the isolated polymer respectively. Both quantities, M and R, are linked by $R \sim M^{\nu}$ thus $C^* \sim M^{1-3\nu}$ with $\nu = 1/2$ at the theta temperature and $\nu = 3/5$ in a good solvent.[1,2] For polystyrene-cyclohexane at the theta temperature, ($\theta = 35°$ C), we used : $C^* = 40/\sqrt{M_w}$ and $C < 10^{-1}$g/cm^3, M_w represents the molecular weight of the polymer.

In a semidilute solution, the pair correlation function $g(r)$ between monomers has the form:

$$g(r) = \frac{kT}{4\pi\xi^2\left(\frac{\partial\pi}{\partial c}\right)} \times \frac{e^{-r/\xi}}{r} , \qquad (1)$$

where ξ is the correlation length or the screening length of concentration fluctuations[3] and π the osmotic pressure. ξ was measured by neutron scattering,[3] it is mass-independent and inversely proportional to the concentration $C(g/cm^3)$:

$$\xi(cm) = 5.5 \times 10^{-8}/C . \qquad (2)$$

The osmotic bulk modulus $K = C \ \partial\pi/\partial C$ was measured by static light-scattering;[4] it was found that K is mass-independent and increases with concentration following:

$$K\,(dynes/cm^2) = 2.93 \times 10^7 \times C^3 . \qquad (3)$$

In dilute ($C \rightarrow 0$) and semidilute solutions, the osmotic bulk modulus (or osmotic pressure) is proportional to the number of non-interacting particles.[1] In a dilute solution $\pi = K = kT\,C/M$; whereas in a semidilute solution $K \simeq \pi \simeq kT\,C/g$, where g is the monomer mass inside a volume ξ^3: $g \simeq C\,\xi^3$. This implies, the relationship between osmotic bulk modulus and the correlation length: $K \simeq kT/\xi^3$ with $\xi \sim g^{\nu}$ and $\nu = 1/2$ at θ. Using the experimental results (Eqs. 2 and 3), we obtain $K = 0.114 \times kT/\xi^3$.

From neutron- and light-scattering measurements[3,4] it has been shown that the reduced quantities ξ/R and $K/(kT \times C/M)$ are only a function of the reduced concentration C/C^*. Those observations are in agreement with theory[1] and similar to the results obtained in good solvent,[5] if ν is switched from 3/5 to 1/2.

III. SEDIMENTATION COEFFICIENT MEASUREMENTS

The sedimentation coefficient is proportional to the ratio of the mass to the friction of the entity which sedimentes. In a semidilute solution the entity which sediments is the blob of mass g[1]

$$S \simeq \frac{g}{6\pi\eta_o\xi_H}(1 - \bar{v}\rho) \qquad (4)$$

$\bar{v}$ is the partial specific volume of polystyrene (0.92 cm^3/g) and ρ is the density of cyclohexane at 35°C (0.764 g/cm^3). ξ_H represents the hydrodynamic screening length, i.e., the hydrodynamic radius of the blob, and η_o the viscosity of the solvent (= 0.758 Cp). It was found that:[6]

$$S\,(s) = 1.15 \times 10^{-14} \times C^{-0.96 \pm 0.04} , \qquad (5)$$

and that S/S_o is a function only of the reduced concentration C/C^*, where S_o is the sedimentation coefficient of the isolated polymer chain.[6] Combining relations 2, 4, 5 this implies that $\xi_H(cm) \simeq 3 \times 10^{-8}/C^{1.04 \pm 0.04}$. Thus in a semidilute θ-solution, hydrodynamic and thermodynamic screening lengths can be described with one characteristic length ξ as is the case in good solvent.

IV. VISCOELASTIC MEASUREMENTS

Due to the presence of entanglements, a semidilute solution has an elastic behavior over short times and a viscous behavior over long times. The characteristic time which separates these two behaviors is the longest viscoelastic relaxation time T_R.

In good-solvent semidilute solutions[7] (polystyrene-benzene), the viscosity η is proportional to that of the solvent η_o and is a function of C/C^*. The relative viscosity $\eta_r = \eta/\eta_o$ can be described by a power law at $C > 4C^*$:

$$\eta_r = 0.35 \times \left(\frac{C}{C^*}\right)^{4.07 \pm 0.06} \tag{6}$$

at a concentration smaller than 0.1 g/cm^3 and for a range of molecular weight between 10^5 and 2×10^7 (see Fig. 1). The longest relaxation time (see Fig. 2) also obeys a scaling law:

$$T_R / T_1 = 1.34 \times \left(\frac{C}{C^*}\right)^{2.05 \pm 0.10} \tag{7}$$

where T_1 is the characteristic time of the first Zimm-mode of the isolated polymer chain. T_1 was calculated from the intrinsic viscosity $[\eta]$:

$$T_1 \simeq \frac{M[\eta]}{kT}\eta_o \, .$$

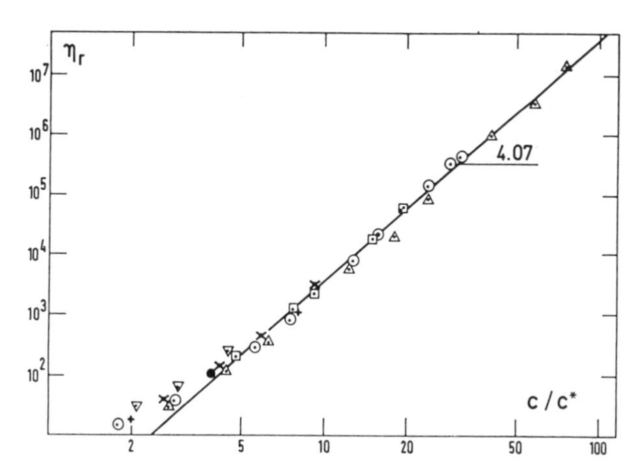

Fig. 1. Relative viscosity as a function of the reduced concentration measured in polystyrene-benzene solutions (log-log scale). The following symbols are used for the different molecular weights: M_w = 20.6 x 10^6 Δ, 6.77 x 10^6 ⊙, 3.84 x 10^6 ., 2.89 x 10^6 ⊡, 1.26 x 10^6 +, 4.22 x 10^5 x, 1.71 x 10^5 ▽.

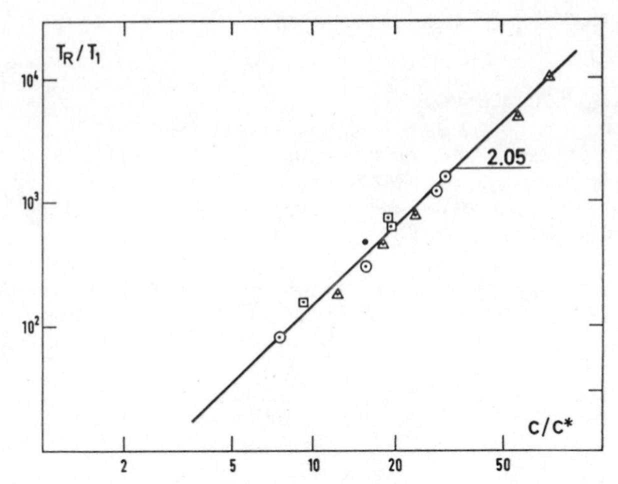

Fig. 2. Log-log plot of the reduced relaxation time T_R/T_1, as a function of c/c^*. The symbols and solutions studied are the same as in Fig. 1.

In contrast, at the theta temperature[8] C/C^* is neither a reduced variable of the relative viscosity η_r nor of the reduced time T_R/T_1. It was found that:

$$\eta_r = A \, \mathbf{x} \left(\frac{C}{C^*} \right)^{5.14 \pm 0.16} ;$$

$$\frac{T_R}{T_1} = B \, \mathbf{x} \left(\frac{C}{C^*} \right)^{2.8 \pm 0.05} , \tag{8}$$

the prefactors A and B increase with the molecular mass. At a monomer concentration of 5% the viscosity and longest relaxation time increase strongly with the molecular mass (see Fig. 3)

$$\eta \sim T_R \sim M_w^x ; x = 3.75 \pm 0.15 \, at \theta = 35°C . \tag{9}$$

The mass exponent value x decreases as the temperature increases to reach a value of 3.45 at 55°C when cyclohexane can be considered as a good solvent for polystyrene. One has to note that the exponent x is larger than the predicted value $x = 3$. Self-diffusion coefficient measurements[9-13] lead to $D_S \sim M^{-2}$. Assuming that T_R is the time required for the polymer chain to diffuse on a distance of order of its radius of gyration $R \sim M^{1/2}$ ($R^2 = T_R D_S$) we should have $T_R \sim M^3$. Present results for relaxation time therefore disagree with self-diffusion coefficient measurements. From viscosity and relaxation time measurements, we can deduce the shear elastic modulus at short times: $G = \eta/T_R$. We find that G is mass-independent and increases with concentration:[8]

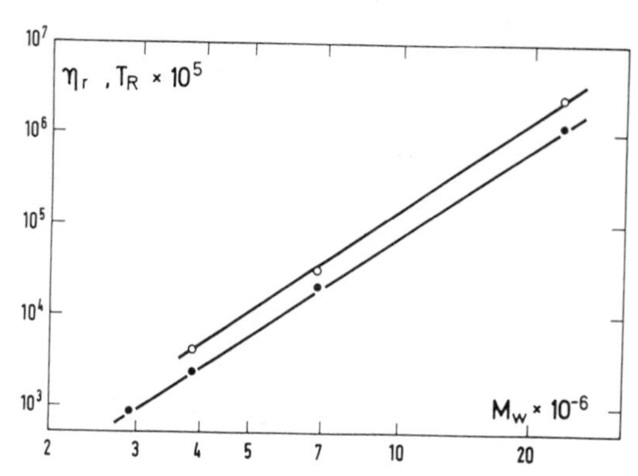

Fig. 3. Relative viscosity (•) and longest relaxation time (o) in seconds as a function of molecular weight at a concentration of 5% in polystyrene cyclohexane semidilute solutions at 35°C.

$$G\,(dynes/cm^2) = 1.34 \times 10^6 \times C^{2.5 \pm 0.2} \; at \; \theta = 35°C \; . \tag{10}$$

It is important to note that at the θ-temperature the shear elastic modulus G does not have the same concentration-dependence as the osmotic bulk modulus K, but both quantities are of the same order of magnitude:

$$K/G = 2.9 \; at \; C = 0.02 \; g/cm^3 \; ;$$

$$K/G = 5.9 \; at \; C = 0.1 \; g/cm^3 \; . \tag{11}$$

On the other hand, in a good solvent (polystyrene-benzene), G and K are proportional but they differ by two orders of magnitude:

$$K/G = 145 \; . \tag{12}$$

V. QUASI-ELASTIC LIGHT-SCATTERING MEASUREMENTS[14]

Quasi-elastic light-scattering experiments allow the time-dependence of the dynamical structure factor $S(q,t)$ to be determined

$$S(q,t)/S(q,o) = \frac{< \delta C_q(t) \delta C_{-q}(o) >}{< |\delta C_q|^2 >} \; , \tag{13}$$

where $\delta C_q(t)$ is the concentration fluctuation of wave vector $\mathbf{q}$. In dilute solutions, light-scattering experiments, performed at a momentum transfer $\mathbf{q}$, reflect macroscopic properties if $\mathbf{q}^{-1}$ is larger than the size of the polymer ($\mathbf{q}^{-1} > R$) and local properties if $\mathbf{q}^{-1} < R$. In a semidilute θ-solution, one will observe macroscopic properties if $\mathbf{q}^{-1} > \xi$ but one has to deal also with a time scale. If the time needed to relax a concentration by cooperative motions τ_s is longer than the longest viscoelastic relaxation time T_R, then the dynamical structure factor reflects the properties of a liquid. In contrast, if τ_s is smaller than T_R, $S(q,t)$ is sensitive to the elastic properties of the transient gel.

We will describe mainly the macroscopic properties ($q^{-1} > \xi$). Figure 4, a schematic representation of the typical variation of τ_s^{-1} as a function of q^2, is separated into two domains depending on the value of τ_s compared to the longest viscoelastic relaxation time T_R: a liquid regime for $\tau_s > T_R$, a gel regime for $\tau_s < T_R$. $q^\dagger$ corresponds to the transfer vector where τ_s is equal to the longest relaxation time. As T_R is strongly mass-dependent and τ_s is not, at a given concentration depending on the molecular weight ($q^\dagger \sim M_w^{-1.9}$) $q^\dagger$ may or may not be in the q range ($1.4 \times 10^4 < q$ (cm^{-1}) $\leq 3.6 \times 10^5$) investigated by the light-scattering experiment.

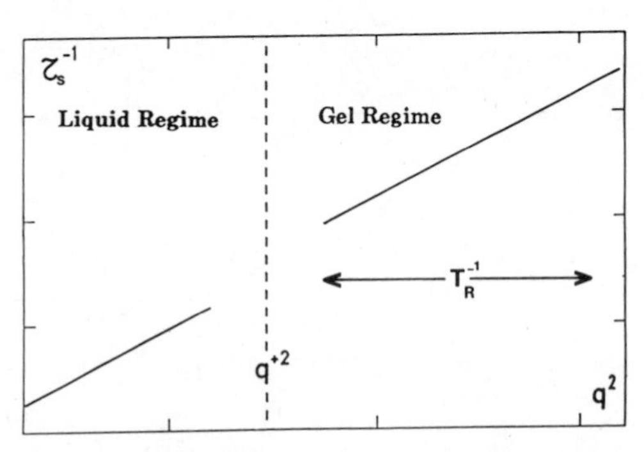

Fig. 4. Schematic representation of the two different domains observed by quasielastic light scattering experiments: $\tau_s > T_R$ corresponds to the liquid regime and $\tau_s < T_R$ corresponds to the gel regime.

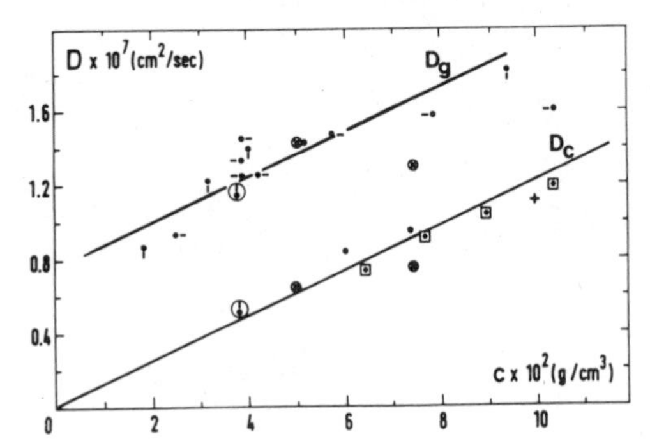

Fig. 5. Linear plot of the diffusion coefficients as a function of concentration. The full lines D_c and D_g represent the behaviors of the diffusion coefficients in the liquid (L) and gel (G) regimes of Fig. 4. The symbols inside circles are obtained in the transition regime from L to G (see paragraph 4.3.). For the meaning of the symbols, see Table 1. (+) ($M_w = 3.9 \times 10^5$) and (■) ($M_w = 8.6 \times 10^5$) correspond to the diffusion coefficient measured by the gradient concentration technique.[15]

Table 1. Sample characteristics and corresponding symbols used in the figures for Q.E.L.S. measurements

$M_w \times 10^{-6}$	$10^2 c^*$ g/cm^3	symbol	T_R(sec)
0.422	6.16	•, no pip	$8.7 \times 10^{-2} c^{2.8}$
1.26	3.56	x	$5.5 c^{2.8}$
2.89	2.35	•, pip up	$1.3 \times 10^{2} c^{2.8}$
3.84	2.04	•, pip left	$3.29 \times 10^{2} c^{2.75}$
6.77	1.54	•, pip right	$3.30 \times 10^{3} c^{2.85}$
20.6	0.88	•, pip down	$1.95 \times 10^{5} C^{2.75}$

M_w and C^* represent the molecular weight and the overlap concentration respectively. The shear vicoelastic relaxation time T_R (expressed in sec) is obtained from mechanical measurements.[8]

V.1. LIQUID REGIME $\tau_s > T_R$ ($q < q^\dagger$)

In this regime the time dependence of the dynamical structure factor has an exponential profile. The characteristic time τ_s is q^{-2}-dependent, the corresponding diffusion coefficient $D_c = 1/\tau_s q^2$ is mass-independent and proportional to the concentration (see Fig. 5):

$$D_C (cm^2/s) = (1.25 \pm 0.1) \times 10^{-6} \times C . \tag{14}$$

As usual, the osmotic diffusion coefficient is linked to the bulk osmotic modulus and to the effective mobility per monomer μ : $D_c = \mu \times K/C$. The quantity μ can be measured through the sedimentation coefficient $S = \mu(1- \bar{v} \rho)$. Combining the experimental results[4,6] for K and S, we obtain $D_C = 1.13 \times 10^{-6} \times C^{1.04}$ in agreement with the result presented above. This result confirms that the hydrodynamic screening length ξ_H is proportional to that of the concentration fluctuations. Actually the scaling law is well verified (see Fig. 6): D_C/D_o, where D_o is the diffusion coefficient of the isolated polymer chain, is only a function of C/C^*.

V.2. GEL REGIME $\tau_S < T_R$ ($q > q^\dagger$)

In this regime the time-dependence of the dynamical structure factor is strongly non-exponential; it is a smoothly decreasing function of time. Our intention is not to describe the entire profile but only to extract the shortest τ_S and the largest relaxation time τ_L. By shortest relaxation time, we mean the initial slope of the dynamical structure factor and by longest relaxation time we mean the time needed

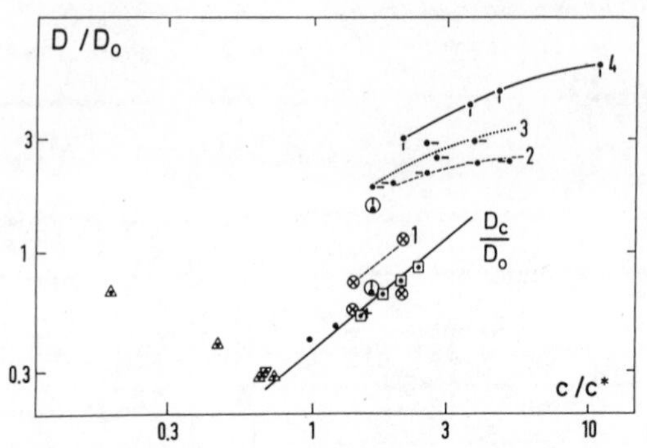

Fig. 6. Log-log plot of the variation of the reduced variable D/D_o as a function of C/C^*, where D_o is the diffusion of a single chain ($D_o(\mathrm{cm}^2/S)=1.3 \times 10^{-4}/M^{1/2}$ from Ref. 17). The straight, solid line represents $D_c/D_o \simeq C/C^*$ and the lines 1,2,3 and 4 qualitatively represent the variations of D_g/D_o. Triangles represent values obtained from Ref. 16 (▲) M_w = 1.3 x 10^6; (▼) M_w = 2 x 10^6.

for $S(q,t)$ to reach its zero value (for experimental details see Ref. 14). The longest relaxation time τ_L is q-dependent, we observe how the stress due to concentration fluctuations relaxes: this is a structural relaxation. τ_L is strongly molecular mass- and concentration-dependent, $\tau_L \times M_w^{-3.8} \simeq C^{3.07 \pm 0.15}$. In fact, it is identical to the longest viscoelastic relaxation time T_R measured by viscoelastic experiments. This is ilustrated in Fig. 7 where $\tau_L \times M_w^{-3.8}$ is plotted as a function of the concentration. The dotted line corresponds to the best fit of T_R measured by viscoelastic experiments.

The shortest relaxation time τ_s is q^{-2}-dependent and the corresponding diffusion coefficient $D_g = 1/\tau_s q^2$ is an increasing function of the concentration. D_g is larger than the diffusion coefficient measured in the liquid regime (see Fig. 5). In contrast to the behavior of D_c/D_o, D_g/D_o is not simply just a function of C/C^* (see Fig. 6). This indicates that D_g is sensitive to a characteristic length which is not proportional to ξ. We observe a gel behavior[14] and the motion is controlled by a restoring force composed of an osmotic force plus an elastic force, then

$$D_g = \frac{\mu}{C}(K + M_g), \tag{15}$$

where M_g is the longitudinal elastic modulus of the transient gel in which entanglements play a major role. Here we assume implicitly that the mobility is independent of the time scale. From the ratio of the two diffusion coefficients measured in the gel (D_g) and liquid (D_c) regime we can deduce the ratio of the two moduli

$$\frac{D_g - D_c}{D_c} = M_g/K, \tag{16}$$

and we find (see Fig. 8) that:

$$\frac{M_g}{K} = 5.22 \times 10^{-2} \times C^{-1.1 \pm 0.2} \tag{17}$$

The concentration dependence of the bulk osmotic modulus ($K \sim C^3$) implies that M_g increases as C^2. Within the experimental precision, longitudinal and shear elastic moduli are proportional: $M_g/G = 6.5$.

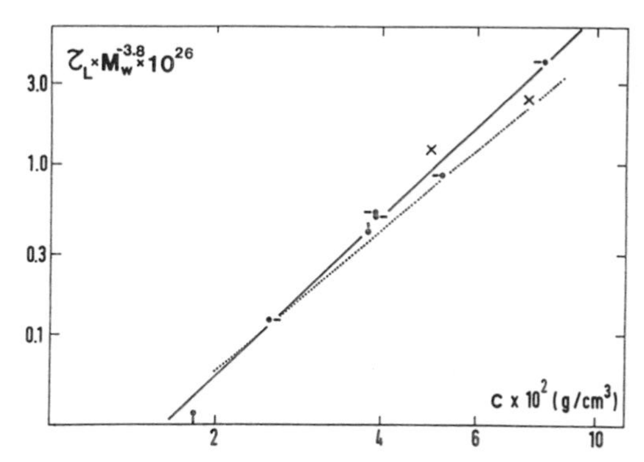

Fig. 7. Variation, as a function of the concentration, of the longest decay time $\tau_L(s)$ divided by the molecular weight-dependence ($M_w^{3.8}$). The solid line represents the concentration-dependence of $\tau_L \times M_w^{-3.8} \sim C^{3.07}$. For the symbols see Table 1. For comparison we report (dotted line) the concentration-dependence of $T_R/M_w^{3.8}$, T_R is the shear-viscoelastic relaxation time measured by mechanical measurements.

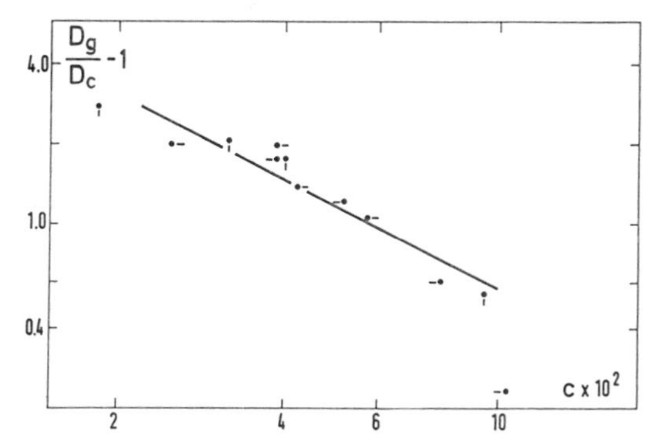

Fig. 8. Log-log plots of the relative variation of the diffusion coefficients in the liquid and gel regimes (D_g/D_c-1) as function of concentration. For the meaning of the symbols see Table 1; the straight line corresponds to $(D_g-D_c)/D_c = M_g/K \sim C^{-1.1}$.

The elastic modulus M_g is proportional to the density of entanglements : kTC/p where p is the monomer mass between two successive entanglements ($p \sim 1/C$). As the polymer is Gaussian over all scales, we have for the distance between entanglements: $\ell \sim \sqrt{p}$, and $\ell \sim 1/\sqrt{C}$. The distance ℓ corresponds to the distance between two binary contact points. At the theta temperature, this distance does not play any thermodynamic role but it does affect the viscoelastic properties.[18,19]

Increasing the quality of solvent the elastic modulus decreases slightly whereas the osmotic bulk modulus increases strongly. Typically at $C = 4 \times 10^{-2}$ g/cm^3:[4,20,7,8]

$$G \text{ (good solvent)} \approx 0.9\, G\ (\theta \text{ solvent}) ;$$

$$K \text{ (good solvent)} \approx 32\, K\ \ (\theta \text{ solvent}) . \tag{18}$$

Such a dramatic change is also reflected in the time-dependence of the dynamical structure factor. The line shape, non-exponential in the gel regime at the θ-temperature, becomes more and more exponential as we increase the temperature[14] from 35°C to 65°C. A single exponential line is observed in polystyrene-benzene system.[21] This is due to the fact that the relative intensity I_L/I_S of the slow mode (τ_L) to the fast mode (τ_S) is proportional to the ratio of the elastic modulus to the osmotic bulk modulus

$$\frac{I_L}{I_S} \simeq \frac{D_g}{D_c} - 1 \simeq \frac{M_g}{K} . \tag{19}$$

Physically, this means that by increasing the quality of solvent, the screening length ξ of the monomer-monomer interaction tends toward the distance ℓ between two binary contact points or entanglements.

V.3 CROSS-OVER REGIMES

The situation schematically described in Fig. 4, can be realized experimentally by choosing a sample having a viscoelastic relaxation time such that $\mathbf{q}^\dagger$ is in the range of the $\mathbf{q}$ investigated. The results are reported in Fig. 9. For $\mathbf{q} < \mathbf{q}^\dagger$ the profile of the dynamical structure factor is exponential, the characteristic decay time varies as $\mathbf{q}^{-2}$. For $\mathbf{q} > \mathbf{q}^\dagger$ the profile is not exponential and extends from τ_s ($\mathbf{q}^{-2}$-dependent)

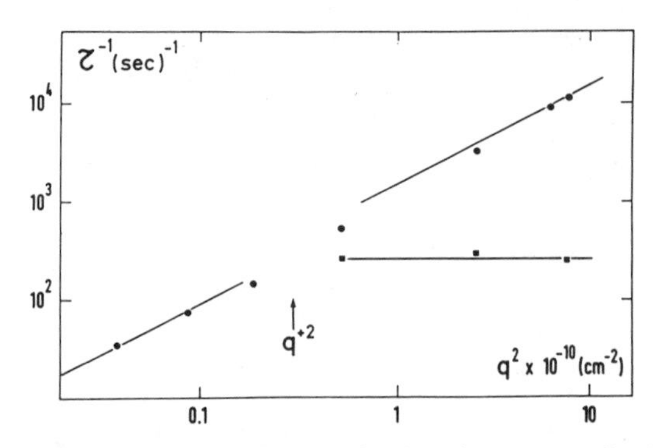

Fig. 9. Log-log plots of variations of the inverse of the decay times in the transition from liquid to gel regime as a function of $\mathbf{q}^2$. On a sample having a molecular weight of 1.26×10^6 and a concentration of 7.42×10^{-2} g/cm^3. The dots correspond to the shortest time τ_s and the squares to the longest relaxation time which is measured only in the gel regime $\mathbf{q} > \mathbf{q}^\dagger$.

to τ_L (q-independent). Thus, liquid behavior or gel behavior can be observed only by changing the momentum transfer.

The influence of the two lengths ξ and ℓ at the θ-temperature can also be seen if experiments are performed at $\mathbf{q}\,\xi > 1$, i.e., on a scale length smaller than ξ. In good-solvent semidilute solutions, it has been shown that the reduced quantity $1/\tau_S D_g \mathbf{q}^2$ is only $\mathbf{q}\,\xi$ dependent and proportional to $\mathbf{q}\,\xi$ for large $\mathbf{q}\,\xi$ value.[22] If the same representation is used (see Fig. 10) in a θ-solvent, it is found that there is no universal curve and one can show[14] that for $\mathbf{q}\,\xi \gg 1$

$$\frac{1}{\tau_s D_g \mathbf{q}^2} \simeq \frac{1}{\mathbf{q}\,\xi} \left[\frac{M_g/K + \mathbf{q}^2 \xi^2}{M_g/K + 1} \right] . \tag{20}$$

M_g/K is not a small constant, as it is in a good solvent; but it is of the order of one and it is concentration-dependent. Thus $1/\tau_S D_g \mathbf{q}^2$ remains concentration-dependent even for $\mathbf{q}\,\xi \gg 1$.

VI. CONCLUSIONS

All the experimental results obtained in the semidilute θ-solutions are consistent with the idea of de Gennes and Brochard[18,19] that at the theta temperature, where the second virial coefficient of the osmotic pressure vanishes, the mesh size of the transient gel ℓ is not proportional to the correlation length of concentration fluctuations ξ, as it is in the case of a good solvent.[23,24]

All the thermodynamic and hydrodynamic quantities, which are sensitive to ξ, such as osmotic bulk modulus, sedimentation coefficient, and the osmotic diffusion coefficient obey the scaling law in C/C^*.

In contrast, the dynamic properties such as viscosity, viscoelastic relaxation time, gel diffusion coefficient, which depend on the entanglements and thus on the length ℓ, do not scale with C/C^*.

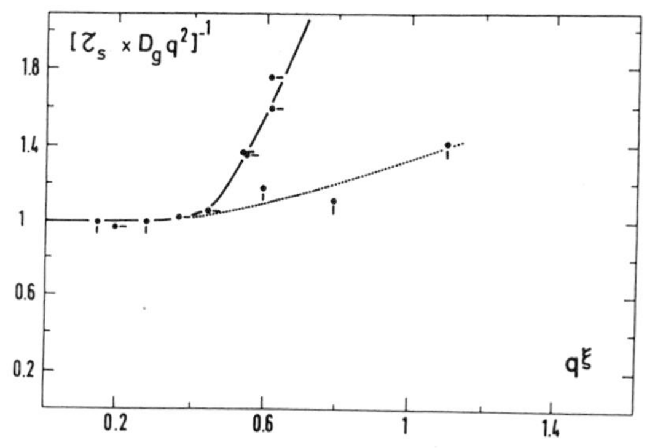

Fig. 10. Variation of the reduced effective diffusion coefficient, $1/(\tau_s D_g q^2)$, as a function of the dimensionless quantity $q\xi$. Solid lines are only guides for the eye. (•, pip down) $M_w = 20.6 \times 10^6$, $C = 1.82 \times 10^{-2}$ g/cm^3; (•, pip right) $M_w = 6.77 \times 10^6$, $C = 2.48 \times 10^{-2}$ g/cm^3.

The puzzling quantity is the longest viscoelastic relaxation time T_R. If the friction involved in the relaxation is purely hydrodynamic, it has been shown[1] that $T_R \sim L^3$ where L is the length of the tube in which the chain has to reptate: $L \simeq (M/g)\xi \sim MC$. Experimentally we find $T_R \sim C^3$,thus it seems that T_R is mainly influenced by the hydrodynamic friction. But as T_R/T_1 does not scale with C/C^*, this means that entanglements play a role in the friction involved in the disentanglements of one chain. This additional friction is a function of the number of entanglements per chain M/p. In a good solvent, $M/p \sim M/g$ is a function of C/C^*, whereas in the theta solvent it is not. This explains the breakdown of scaling laws at theta temperature. This additional friction could be the reason for increase of the concentration-dependence of the viscosity in the good solvent and of the mass-dependence whatever the quality of solvent. ($T_R \sim \eta \sim M^x$ with $x > 3$).

ACKNOWLEDGEMENTS

We thank P.G. de Gennes, F. Brochard and J.D. Ferry for stimulating discussions. We also wish to thank J. Roots, B. Nyström, J.P. Munch and S. Candau for kindly communicating the numberical values of Ref. 16.

NOMENCLATURE

$A \simeq B$	corresponds to A = constant x B, the constant is dimensionless
$A \sim B$	A is proportional to B
$A \approx B$	means that A is of the order of B
C	monomer concentration
C^*	overlap concentration
D_c	osmotic diffusion coefficient
D_g	gel diffusion coefficient
D_o	diffusion coefficient of a single polymer chain at infinite dilution
G	shear modulus of the transient gel
$g(r)$	pair correlation function between monomers
g	monomer mass inside a blob of volume ξ^3
I_L/I_s	relative intensity of the slow-mode to the fast-mode of the dynamical structure factor in the gel regime
K	osmotic bulk modulus $K = C\ \partial\pi/\ \partial C$
k	Boltzmann's constant
L	length of the tube in which the chain reptates
ℓ	distance between two successive entanglements
M	mass of the polymer chain
M_g	longitudinal elastic modulus of the transient gel
M_w	molecular weight (weight average)
p	monomer mass between two successive entanglements
$\mathbf{q}$	momentum transfer
$\mathbf{q}^\dagger$	momentum transfer which separates the two dynamical regimes: liquid regime and gel regime
R	radius of gyration of a single polymer chain at infinite dilution
S	sedimentation coefficient
S_o	sedimentation coefficient of a single chain at infinite dilution
$S(q,t)$	dynamical structure factor
T	temperature
T_R	longest viscoelastic relaxation time
T_1	characteristic time of the first Zimm-mode
t	time
$\bar{v}$	partial specific volume of polystyrene
δC_q	concentration fluctuation
η	viscosity of the polymer solution
η_o	viscosity of the solvent

η_r relative viscosity $\eta_r = \eta/\eta_s$
$[\eta]$ intrinsic viscosity
θ theta temperature
μ effective mobility per monomer
ν exponent which links radius of gyration and mass
ξ screening length of concentration fluctuation
ξ_H hydrodynamic screening length
π osmotic pressure
ρ density of solvent
τ_s characteristic relaxation time of the dynamical structure factor in the liquid regime or shortest relaxation time in the gel regime
τ_L longest relaxation time of the dynamical structure factor in the gel regime

REFERENCES

1. P.G. de Gennes, Scaling Concepts in Polymer Physics, Cornell University Press, Ithaca, NY, (1979).
2. P.J. Flory, Principles of Polymer Chemistry, Cornell University Press, Ithaca, NY, (1979).
3. J.P. Cotton, M. Nierlich, F. Boue, M. Daoud, B. Farnoux, G. Jannink, R. Duplessix and C. Picot, J. Chem. Phys. 65, 1101 (1976).
4. P. Stepanek, R. Perzynski, M. Delsanti and M. Adam, Macromolecules, 17, 2340 (1984).
5. M. Daoud, J.P. Cotton, B. Farnoux, G. Jannink, G. Sarma, H. Benoit, R. Duplessix, C. Picot and P.G. de Gennes, Macromolecules, 8, 804 (1975).
6. P. Vidakovic, C. Allain and F. Rondelez, J. Phys. (Paris, Fr.) 42,. L323 (1981).
7. M. Adam and M. Delsanti, J. Phys., (Paris, Fr.) 44, 1185 (1983).
8. M. Adam and M. Delsanti, J. Phys., (Paris, Fr.) 45, 1513 (1984).
9. L. Leger, H. Hervet and F. Rondelez, Macromolecules, 14, 1732 (1981).
10. J.A. Wesson, I. Noh, T. Kitano and Y. Yu, Macromolecules, 17, 782 (1984).
11. P.T. Callaghan and D.N. Pinder, Macromolecules, 14, 1334 (1981).
12. E.D. von Meerwall, E.J. Amis and J.D. Ferry, Macromolecules, 18, 260 (1985).
13. H. Deschamps and L. Leger, Macromolecules, 19, 2760 (1986).
14. M. Adam and M. Delsanti, Macromolecules, 18, 1760 (1985).
15. J. Roots and B. Nyström, Macromolecules, 13, 1595 (1980).
16. J.P. Munch, G. Hild and S.J. Candau, Macromolecules, 16, 71 (1983).
17. M. Schmidt and W. Burchard, Macromolecules, 14, 210 (1981).
18. F. Brochard and P.G. de Gennes, Macromolecules, 10, 1157 (1977).
19. F. Brochard, J. Phys. (Paris, Fr.), 44, 39 (1983).
20. I. Noda, N. Kato, T. Kitano and M. Nagasawa, Macromolecules, 14, 668 (1981).
21. M. Adam and M. Delsanti, Macromolecules, 10, 1229 (1977).
22. P. Wiltzius and D.S. Cannell, Phys. Rev. Lett., 56, 61, (1986).
23. P.G. de Gennes, Macromolecules, 9, 587 (1976).
24. P.G. de Gennes, Macromolecules, 9, 594 (1976).

Discussion

On the Paper by L. Ould Kaddor and Cl. Strazièlle

E.F. Casassa (Carnegie Mellon University): In the instances you discuss, ΔA_2 is positive. Is this a necessary condition? Are there any constraints on the relative magnitudes of $A_{2,a}$, $A_{2,b}$ and $A_{2,ab}$?

Cl. Strazièlle (Institut C. Sadron (CRM, Strasbourg)): In ternary systems: two polymers in good solvent, the second virial coefficients $A_{2,a}$, $A_{2,b}$ and A_{2ab} in dilute solution are of the same order of magnitude for symmetrical systems. But the difference $\Delta A_2 = A_{2ab} - (A_{2a} + A_{2b})/2 \sim \chi_{ab}$ can be positive (incompatible system) or negative (compatible system).

On the Paper by B. Farnoux, J. des Cloizeaux and G. Jannink

T.P. Russel (IBM Almaden Research Center): One means by which the critical angle can be measured is to examine the off-specular scattering. Here, the scattering exhibits a maximum at the critical angle. Have you examined this for the solution reflection case?

G. Jannink (Laboratoire Léon Brilloun, CEN-Saclay, France): The critical angle is calculated from the bulk solution concentration and scattering length. The calculated value is checked experimentally with the help of an equivalent binary mixture of deuterated and non-deuterated toluene.../.. having an average collision length equal to that of the bulk of the polymer solution. The experimental difficulty arises because of angular resolution effects near total reflection. We shall try to improve the determination following your suggestion.

On the Paper by R. Messager, D. Chatenay, W. Urbach and D. Langevin

E.F. Casassa (Carnegie-Mellon University, Pittsburg): I would urge that these most interesting micelles not be called "living polymers." Polymer chemists universally use this term to mean something very different.

P.T. Callaghan (Massey University): I wish to follow on from Professor de Gennes' point about the motion of the probe. It strikes me that measuring the diffusion of the surfactant by P&SE nmr would be extremely interesting. You will recall that we obtained an interesting result when we worked on the same microemulsion system which you had studied by light scattering. In the present case, the time scale of P&SE is ideally suited to the 1 ms to 10^3 ms range which you suggested as being a possible time scale for chain breaking.

D. Langevin (Ecole Normal Superieure, Paris): I perfectly agree.

A. Silberberg (Weizmann Institute): There should be two persistent lengths characterizing a surface. Are these really independent or would the pressure of a given curvature in one direction not lie in the envelop bending tending in the other direction.

D. Langevin: In the de Gennes-Taupin model, there is a single persistence length because the surface is assumed to be isotropic and only energy variations due to differences between surface mean curvature and spontaneous curvature are taken into account. The coupling between local curvatures in two opposite directions appears in the Gaussian curvature term of the surface bending energy and might lead to the definition of a second persistence length. But this calculation has not yet been made at least from this point of view.

K.J. Mysels (La Jolla, CA): CTAB is a typical association colloid which should not be called a "living polymer," a term reserved for Dr. Michael Szwarc's products. The dynamic equilibrium of monomers entering and leaving the long micellar saussage is probably as a microsecond time scale as in other micelles and is likely to be a more important mechanism of equilibration than the breaking and reforming of the saussages.

D. Langevin: The term "living polymer" was introduced for worm-like micelles by M. Cates, and is now currently used for this purpose; but we agree that it could be confusing for polymer scientists.

The dynamic equilibrium of micelles involves two characteristic times: a fast one related to monomer exchange and a slow one related to micelle breakage. The first one is unlikely to affect the micelles self-diffusion process that we studied.

PART FOUR:

DIFFUSION AND INTERDIFFUSION OF POLYMERS

SELF-CONSISTENT FIELDS AND CRITICAL EXPONENTS FOR POLYMERS

Isaac C. Sanchez*

Aluminum Company of America
Alcoa Technical Center
Alcoa Center, Pennsylvania 15069

ABSTRACT

A self-consistent field (SCF) is constructed for a polymer chain with excluded volume modeled as a self-avoiding random walk (SAW) of N steps ($N \rightarrow \infty$). The SCF requires the introduction of a second exponent $\overline{\theta}$ in addition to the usual ν exponent that characterizes the size of a SAW. The SCF equals N times the probability $\overline{P}_N$ of an interaction of the chain end with a distant part of the chain. In d-dimensional space $\overline{P}_N$ scales as $N^{-\nu(d+\overline{\theta})}$ ($d \leq 4$); self-consistency of the field yields the relations $\theta = (4 - d)/3$ and $\nu(d + \overline{\theta}) = 2$. It is shown that $\theta_0 < \overline{\theta} < \theta_1$ where θ_0 is the exponent associated with the probability of a SAW returning to its origin and θ_1 is the exponent associated with the probability of a SAW forming a large loop with a long tail. A SCF (Φ) is also determined for a semidilute solution of polymer volume fraction ϕ. The number of binary interactions scales as $\Phi\phi \sim \phi^{9/4}$.

I. INTRODUCTION

It is well-known that the Flory mean-field theory of the polymer self-excluded volume problem yields excellent results.[1] If R is some scalar measure of the polymer chain size, such as the radius of gyration, and N is the number of monomer units in the chain, then R scales with N as

$$R \sim N^{\nu}, \tag{1}$$

where the Flory exponent ν is given in d-dimensional space by

$$\nu = 3/(2 + d), \quad 1 \leq d \leq 4. \tag{2}$$

---

*Current address: Chemical Engineering Department and Center for Polymer Research, University of Texas, Austin, Texas 78712

In the derivation of this result the mean number of repulsive binary contacts is taken to be proportional to $N\rho_N(R)$ where $\rho_N(R)$ is the mean monomer density:

$$\rho_N(R) \sim N/R^d \, . \tag{3}$$

Minimization of the sum of the elastic energy ($\sim R^2/N$) and the repulsive energy $[\sim N\rho_N(R)]$ yields the famous Flory result. Fisher is usually given credit for generalizing the Flory argument to d-dimensions.[2]

In 1972 de Gennes, in a seminal paper, showed the analogy between self-avoiding random walks (SAWs), a magnetic system (the zero-component spin vector model) and critical phenomena.[3] The Flory ν exponent is related to the critical exponent for the correlation length in the magnetic analogy. Later des Cloizeaux, using field-theoretic methods, showed that the exponent θ_0 associated with the probability of a SAW returning to its origin is related to the critical exponent γ associated with the magnetic susceptibility.[4] In particular he showed that

$$\theta_0 = (\gamma - 1)/\nu \, . \tag{4}$$

In critical phenomena there are always at least two independent critical exponents. The Flory mean-field theory yields only one exponent and raises the question: can mean-field or self-consistent field arguments yield the missing second exponent? The answer to this question is the main subject of this paper.

II. CONSTRUCTION OF THE SCF

The classical model for a polymer chain with self-excluded volume is the SAW. Let Φ_N be the appropriate self-consistent field (SCF) for a SAW of N steps but of any size R. By analogy with the Flory argument, we are tempted to assume that Φ_N for a SAW is proportional to the mean density of occupied sites, $\Phi_N \sim \bar{\rho}_N \sim N^{1-d\nu}$, where the volume R^d in $\rho_N(R)$ has been replaced by its mean value $N^{\nu d}$ because we are considering the total ensemble of N-step SAWs and not just the subset of size R. The SCF for a SAW must satisfy certain properties. As will be shown below, this choice for Φ_N is not self-consistent for $d < 4$.

Consider a random walk on a lattice in which the fraction of occupied sites is Φ_N (the field). This random walker is supposed to be a SAW. At every step of the walk the probability p_n that the random walker will encounter an occupied site (i.e., encounters itself) at the nth step is given by the following intuitive mean-field expression (n-1 successful steps followed by an encounter):

$$p_n = (1 - \Phi_N)^{n-1}\Phi_N \, . \tag{5}$$

The average number of steps $\bar{n}$, that the walker executes before it encounters an occupied site ("dies") is easily calculated to be

$$\bar{n} = \sum_{n=1}^{\infty} n p_n = \Phi_N^{-1} \, . \tag{6}$$

Using the Flory value for ν, Eq. (2), and the aforementioned guess that $\Phi_N \sim N^{1-\nu d}$ yields

$$\bar{n} \sim N^{\mu} \ , \tag{7}$$

where

$$\mu = 2(d-1)/(2+d) \ . \tag{8}$$

For $d = 4$, $\mu = 1$ and for $d < 4$, $\mu < 1$. To be self-consistent the random walker should, on average, execute N steps before it dies because the SCF is supposed to represent successful N-step SAWs. The statistical properties of the random walker in this field should be the same as the SAWs that the SCF is supposed to represent. This result suggests that the ν exponent alone is not sufficient to describe this SCF. The situation is rescued by introducing a new exponent $\bar{\theta}$; i.e., let's assume

$$\Phi_N \sim \bar{\rho}_N N^{-\nu\bar{\theta}} \sim N^{1-\nu(d+\bar{\theta})} \ . \tag{9}$$

Now the self-consistent requirement that $\bar{n} \sim N$ translates to $\nu\bar{\theta} + \mu = 1$ or

$$\bar{\theta} = (4-d)/3 \ , \tag{10}$$

where the Flory value of ν has been used. As the random walker proceeds through the field Φ_N, the average number of times it will encounter an occupied site is $N\Phi_N$ (or $N\Phi_N/2$ self-encounters). Thus, the self-consistency requirement also translates to

$$\lim_{N \to \infty} N\Phi_N = constant, \tag{11}$$

or $\Phi_N \sim N^{-1}$. In more popular jargon, Eq. (11) implies that the appropriate SCF is one in which two-body interactions are marginal. Thus, we are led to the following definition of Φ_N:

> The appropriate SCF (Φ_N) for an ensemble of N-step SAWs of any size R is one in which a random walker on average does not intersect itself until the Nth step; this requires that $\Phi_N \sim N^{-1}$.

The physical significance of the $\bar{\theta}$ exponent becomes clearer if we construct the thSCF in a different way: Let $P(n \mid N)$ be the probability that at the N step of a SAW, a loop of size n is formed. In the ordered sequence of steps 1,2, ..., N of a N-step random walk, a loop of size n is formed when the Nth step comes arbitrarily close to the occupied site at $N - n$. Therefore, by virtue of the SCF definition adopted above, Φ_N must also be given by

$$\Phi_N \sim \sum_{n=N_c}^{N} P(n \mid N) \ , \tag{12}$$

where N_c is a cutoff for the smallest loop size that is relevant in determining the asymptotic, statistical properties of a SAW. The asymptotic properties are controlled by "long-range interactions" (large loops) which implies that $N_c >> 1$ for large N. The self-similar nature of random walks also implies that $N_c \sim N$. These properties of the cutoff allow Eq. (12) to be expressed as

$$\Phi_N \sim N\overline{P}_N , \tag{13}$$

where $\overline{P}_N$ is an average probability of a long range interaction of the chain-end with a distant part of the chain:

$$\overline{P}_N \equiv \left(\frac{1}{N - N_c} \right) \sum_{n = N_c}^{N} P(n \mid N) \sim \frac{1}{N} \sum_{n = N_c}^{N} P(n \mid N) . \tag{14}$$

Note that $\overline{P}_N$ must equal zero for a SAW in one-dimension. Furthermore, Eqs. (9) and (13) for the SCF imply that

$$\overline{P}_N \sim N^{-\nu(d + \overline{\theta})} , \tag{15}$$

and we see that the $\overline{\theta}$ exponent required to define the SCF is associated with the probability of making a large loop in a SAW. Equations (11) and (13) also imply that $\overline{P}_N \sim N^{-2}$ or by Eq. (15) and for $1 < d \leq 4$

$$\nu\,(d + \overline{\theta}) = 2 . \tag{16}$$

This exponent relation is more general than Eq. (10) because the Flory value of ν, Eq. (2), has not been invoked. Substitution of Eq. (2) into Eq. (16) yields Eq. (10). Although Eq. (16) is also derivable from Eqs. (9) and (11), it does not give us any clue as to the physical significance of $\overline{\theta}$; Eq. (15) makes the connection to loop exponents clear.

III. LOOP EXPONENTS

Now des Cloizeaux has shown that $P(N \mid N)$, the probability for a SAW to return to its origin (a loop of size N), has the following asymptotic form ($1 < d \leq 4$):[4,5]

$$P(N \mid N) \sim N^{-\nu(d + \theta_0)} . \tag{17}$$

For a loop of size N_c, where N_c is the cutoff defined in Eq. (12), des Cloizeaux also showed that

$$P(N_c \mid N) \sim N_c^{-\nu(d + \theta_1)} , \tag{18a}$$

where both N_c and N are very large. As was mentioned previously, the self-similar character of random walks implies that N_c is proportional to N; thus, Eq. (18a) can also be expressed as

$$P(N_c \mid N) \sim N^{-\nu(d+\theta_1)} . \tag{18b}$$

The renormalization group calculation of the exponents θ_0 and θ_1 in three-dimensions yields:[5,6]

$$\theta_0 = 0.275; \quad \theta_1 = 0.46, \qquad d = 3 \tag{19}$$

and in two-dimensions:

$$\theta_0 = 11/24 = 0.458; \quad \theta_1 = 5/6 = 0.833, \qquad d = 2 \tag{20}$$

whereas the values of $\bar{\theta}$ from Eq. (10) are

$$\bar{\theta} = \begin{cases} 1/3; & d = 3 \\ 2/3. & d = 2 \end{cases} \tag{21}$$

If Eq. (16) is used to calculate $\bar{\theta}$ in 3-d using the widely accepted value[7] of $\nu = 0.588$, then instead of 1/3 one obtains $\bar{\theta} = 0.401$. The ε expansions ($\varepsilon \equiv 4 - d$) are also known[5,6] for θ_0 and θ_1 :

$$\theta_0 = \varepsilon/4 + 9\varepsilon^2/128 + \dots;$$

$$\bar{\theta} = \varepsilon/3 ;$$

$$\theta_1 = \varepsilon/2 - 3\varepsilon^2/64 + \dots . \tag{22}$$

Thus, $\theta_0 < \bar{\theta} < \theta_1$ in both 2-d and 3-d. Notice that the relation for the SCF given in Eq. (12) depends on loop probabilities and the corresponding exponents for all loop sizes between N_c and N. As the loop size increases from N_c to N, the exponent must crossover from θ_1 to θ_0. In this sense $\bar{\theta}$ plays the role of a crossover exponent with a value intermediate between θ_0 and θ_1.

It is also interesting to make the comparisons with Eq. (16): Using $\nu = 0.588$ for $d = 3$ and $\nu = 3/4$ for $d = 2$, we find

$$\nu(d+\theta_0) = \begin{cases} 1.92, & d=3 \\ 59/32 = 1.84; & d=2 \end{cases}$$

$$\nu(d+\theta_1) = \begin{cases} 2.03, & d=3 \\ 17/8 = 2.125. & d=2 \end{cases} \tag{23}$$

A previous approximation[8] has suggested that $\nu(d+\theta_0)$ equals 2.

IV. SEMIDILUTE SOLUTIONS

A polymer solution of volume fraction ϕ is semidilute (SD) when

$$\phi^* << \phi << 1 , \tag{24}$$

where ϕ^* is the crossover concentration from dilute to SD solutions.[1] In a good solvent

$$\phi^* \sim \overline{\rho}_N \sim N^{1-\nu d} \sim N^{-4/5} . \tag{25}$$

In a SD solution, binary interactions dominate. A given polymer chain interacts with many other polymer chains and will encounter a monomer belonging to another polymer chain on average every $\overline{n}$ steps. If Φ is the SCF for a SAW in the presence of other SAWs, then by Eq. (6) $\overline{n} = \Phi^{-1}$. As occurred earlier with the isolated SAW, our intuitive guess ($\Phi = \phi$) will not be self-consistent.

In between encounters, the polymer chain is self-avoiding. Thus, the distance between encounters, the correlation length ξ, is given by

$$\xi \sim \overline{n}^{\nu} \sim \Phi^{-\nu} . \tag{26}$$

The SD solution of concentration ϕ appears homogeneous down to size scales of order ξ; thus, we must also have

$$\phi \sim \overline{n}/\xi^3 \sim \Phi^{3\nu-1} \sim \Phi^{4/5} , \tag{27}$$

or

$$\Phi \sim \phi^{5/4} , \tag{28}$$

and by Eq. (26)

$$\xi \sim \phi^{-3/4} . \tag{29}$$

Note that at ϕ^*, $\Phi \sim N^{-1}$ and $\xi \sim N^{\nu} \sim N^{3/5}$ as self-consistency requires.

As before, the average number of binary encounters (now intermolecular monomer-monomer interactions) per chain is given by $N\Phi = N/n$. Therefore, the number of interactions per unit volume of solution scales as $\Phi\phi \sim \phi^{9/4}$; the latter is a well-known result due to des Cloizeaux.[9]

The intermolecular interactions screen the self-excluded volume interaction[10] and, as a result, each correlated sequence of $\overline{n}$ monomers acts independently of the others. Thus, the mean square end-to-end of the chain will be given by

$$R^2 = (N/\overline{n})\xi^2 \sim N\Phi^{1-2\nu} \sim N\phi^{-1/4}, \quad (30)$$

which, along with Eq. (29), are well-known scaling results.[1]

V. SUMMARY

The main idea here is that the appropriate SCF for an ensemble of N-step SAWs, but of any size R, is one which two-body interactions are marginal, or equivalently, one in which a random walker on average does not intersect itself until the Nth step; this requires that the SCF vary as N^{-1}. Construction of this SCF requires the introduction of a second critical exponent $\overline{\theta}$ and the self-consistency requirement yields the exponent relations Eqs. (10) and (16); Eq. (16) is the more general of the two relations. The exponent $\overline{\theta}$ lies intermediate in value to the des Cloizeaux loop exponents θ_0 and θ_1 and appears to play the role of a crossover exponent.

A SCF can also be determined for a semidilute polymer solution, but here the $\overline{\theta}$ exponent plays no role; only ν is important. All of the known scaling results for semidilute solutions are recovered by the SCF approach which supports its efficacy. The SCF approach can also be applied systematically to concentrated solutions where ternary and higher order interactions dominate, but has not been presented here.

ACKNOWLEDGEMENTS

The author is pleased to acknowledge M. Muthukumar, P.D. Gujrati and J. des Cloizeaux for their helpful comments on an earlier version of this paper.

DEDICATION

This paper is dedicated to Prof. de Gennes in honor of his receiving the ACS Polymer Chemistry Award of 1988.

NOMENCLATURE

SCF	self-consistent field
SAW	self-avoiding random walk
R	scalar measure of polymer chain size
N	number of monomer units in a polymer chain
ν	critical exponent associated with chain dimensions or correlation length
d	dimensionality of Euclidean space
$\rho_N(R)$	mean monomer density of a chain of N monomer units and size R
θ_0	scaling exponent associated with a SAW returning to the origin
γ	critical exponent associated with magnetic susceptibility
Φ_N	SCF for a self-avoiding random walk of N steps
$\overline{\rho}_N$	mean monomer density of a chain of N units and of mean size N^{ν}.

p_n	probability that a random walker will encounter an occupied site.
$\overline{n}$	mean number of steps a random walker executes before encountering an occupied site.
μ	scaling exponent associated with n
$\overline{\theta}$	new exponent associated with self-consistent field for a self-avoiding random walk of N steps
$P(n \mid N)$	probability that at the Nth step of a SAW, a loop of size n is formed
N_c	smallest loop size that is relevant in determining the asymptotic statistical properties of a SAW
P_N	average probability of a long range interaction of the chain end with a distant part of the chain
θ_1	scaling exponent associated with a SAW of N steps forming a loop of size N_c.
ε	epsilon expansion parameter equal to $4-d$
ϕ	polymer volume fraction
ϕ^*	crossover concentration
Φ	SCF for a semi-dilute solution
ξ	correlation length

REFERENCES

1. P.-G. de Gennes, Scaling Concepts in Polymer Physics, Cornell University Press, Ithaca, N.Y., 1979.
2. M.E. Fisher, J. Phys. Soc. (Japan) (suppl), 26, 44 (1969).
3. P.-G. de Gennes, Phys. Lett. A, 38, 339 (1972).
4. J. des Cloizeaux, Phys. Rev. A, 10, 1655 (1974).
5. J. des Cloizeaux, J. Phys. (Paris), 41, 223 (1980).
6. J. des Cloizeaux and G. Jannink, Les Polymeres en Solution: leur Modelisation et leur Structure, Z.I. de Courtavoeuf, France, 1988, pp. 538-545.
7. J. C. Le Guillou and J. Zinn-Justin, Phys. Rev., B, 21, 3976 (1980).
8. L. Pietronero and L. Peliti, Phys. Rev. Lett. 55, 1479 (1985).
9. J. des Cloizeaux, J. Phys. (Paris), 36, 281 (1975).
10. S.F. Edwards, Proc. Phys. Soc 88, 265 (1966).

DIFFUSION IN BLOCK COPOLYMERS AND ISOTOPIC POLYMER MIXTURES

Peter F. Green

Sandia National Laboratories
Albuquerque, New Mexico 87185-5800

ABSTRACT

We present here a forward recoil spectrometry (FRES) study of "thermodynamic slowing down" of mutual diffusion in isotopic polymer mixtures and of the diffusion of homopolymers into symmetric diblock copolymer structures. The measurements of "thermodynamic slowing down" were performed on binary mixtures of normal and deuterated polystyrene (PS). Both the Flory interaction parameter, χ, and the upper critical solution temperature (UCST) of the d-PS/PS system were obtained from these measurements. In addition, measurements of the temperature and molecular weight dependence of the diffusion of deuterated polystyrene (d-PS) and of deuterated polymethylmethacrylate (d-PMMA) homopolymer chains into a series of symmetric diblock copolymers of PS and PMMA were compared with the diffusion of d-PS and d-PMMA chains into their respective homopolymer analogues.

I. INTRODUCTION

The current understanding of the translational diffusion of a single chain comprised of N monomer segments in a highly entangled melt of chains, each comprised of P monomer segments, is based on the concept of reptation which was introduced by de Gennes in 1971.[1] In the melt, the topological constraints on the N-mer chain provided by the neighbouring P-mer chains are such that the motion of the chain is restricted to a tube-like region.[2,3] The large scale translational motion of the chain is believed to occur as a result of the propogation of loops or "kinks" along the contour of the chain. If the environment of the N-mer chain remains relatively immobile over a time scale characterized by $\tau_d(N) \sim N^3$ then the diffusion coefficient of the N-mer chain is $D^* = D_{REP} = D_0 N^{-2}$, which is independent of the length of the P-mer chains. The constant D_0 is a function of the Rouse mobility of the chain, the molecular weight between entanglements and the temperature.[2,3,4] One of the consequences of reptation is that the ratio D^*/T should have essentially the same temperature dependence as that of the inverse of the zero shear rate viscosity, η_0, and consequently be described by the WLF equation using the same constants that describe the temperature dependence of η_0.[3,5]

When the P-mer chains are sufficiently short the surroundings of the N-mer chain become altered on time scales comparable to $\tau_d(N)$. Consequently, the "tube" may undergo a series of displacements. The foregoing description of the motion of the N-mer chain occurring in a fixed "tube" is therefore inadequate. Theories[6-9] have shown that under these conditions the D^* of the N-mer chain may be modified by the addition of a term, D_{CR}, which describes the effects of the environment on its motion; $D^* = D_{REP} + D_{CR}$. The exact dependence of D_{CR} on the molecular weight of the P-mer chain is still an open question. Theories due to Graessley[6] to Watanabe and Tirrell[7] and to Daoud and de Gennes[8] predict that the dependence of D_{CR} on the host molecular weight varies as P^{-3}; Klein[9] predicts a dependence of $P^{-2.5}$ and Hess predicts a P^{-1} dependence.[10] The prediction of Hess should, however, in a strict sense, be applicable only to the semidilute regime. A number of diffusion and viscoelastic experiments have been done to investigate this problem.[11-17] The experimental results suggest that $D_{CR} \sim P^{-\alpha}$ where α is between 2 and 3.[11-13,15,16]

This paper is divided into four sections. Section II is concerned with a brief discussion of an ion-beam analysis experimental technique called forward recoil spectrometry (FRES) used for the diffusion measurements. In section III we will discuss the diffusion of linear deuterated polystyrene (d-PS) chains into PS hosts and of deuterated polymethylmethacrylate (d-PMMA) chains into PMMA hosts. Here the molecular weight and the temperature dependence of a tracer chain diffusing into a highly entangled melt is addressed. This will provide a basis for the subsequent discussions in section IV of mutual diffusion in isotopic polymer mixtures and of the diffusion of homopolymers in diblock copolymer structures in section V. The Flory interaction parameter, χ,[18,19] in isotopic polymer mixtures is believed to be slightly positive. This enables one to study "thermodynamic slowing down" of the mutual coefficient $D(\Phi)$, where Φ is the average composition of the mixture. This experiment provides an independent way by which one may experimentally determine χ instead of using small-angle neutron scattering (SANS). Furthermore, from the temperature dependence of $D(\Phi)$ the upper critical solution temperature (UCST) of the mixture may be determined. A brief discussion of theories that describe the origins of χ will also be presented. The final section will be concerned with the diffusion of d-PS and d-PMMA homopolymers into nonequilibrium symmetric diblock copolymer structures of PS and PMMA. The temperature and molecular weight dependence of the tracer diffusion coefficient of the homopolymer chains is addressed.

II. EXPERIMENTAL

A number of papers have appeared giving a complete description of this technique which is sometimes known in the literature as elastic recoil detection (ERD).[20-23] The discussion presented here will therefore be brief. FRES is used to determine the volume fraction versus depth profile of a species labelled by deuteration which was allowed to diffuse into an unlabelled host. The diffusion coefficient is extracted from this profile using a solution to the diffusion equation.

Shown in Fig. 1 is schematic of the FRES experimental arrangement. Here a monoenergetic beam of helium ions of energy $E_0 = 3.0$ MeV impinges on the target at an angle of α with respect to the target normal. Some of these projectiles kinematically recoil some of the nuclei in the target. A nucleus which recoils from the surface of the target at an angle of θ with respect to the direction of the incident beam has an energy of E_1 which is given by

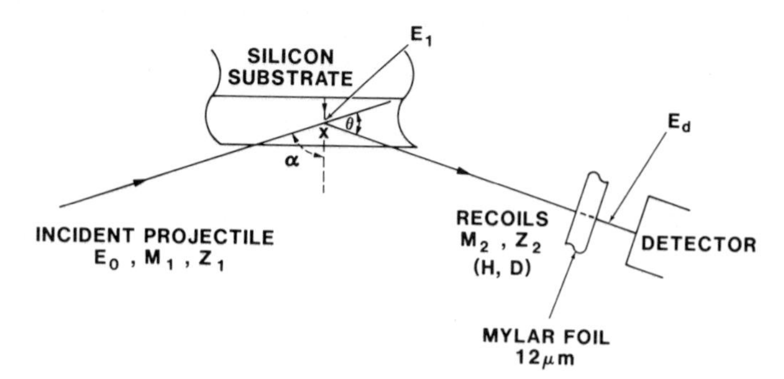

Fig. 1. Schematic of the FRES scattering experiment.

$$E_1 = KE_0, \qquad (1)$$

where $K=[4M_{He}M_r/(M_{He}+M_r)^2]\cos^2\theta$ is the kinematic factor for elastic scattering. Here M_{He} and M_r are the masses of the incident and recoiling ions, respectively. It is clear from Eq. 1 that through measurement of E_1 the FRES technique is able to distinguish between the protons and deuterons based on their masses. One may consider the situation where the nuclei recoil from a collision at a depth x beneath the target surface. The energy E_d with which a recoiling nucleus is detected is

$$E_d = KE_0 - [S]x - \delta E . \qquad (2)$$

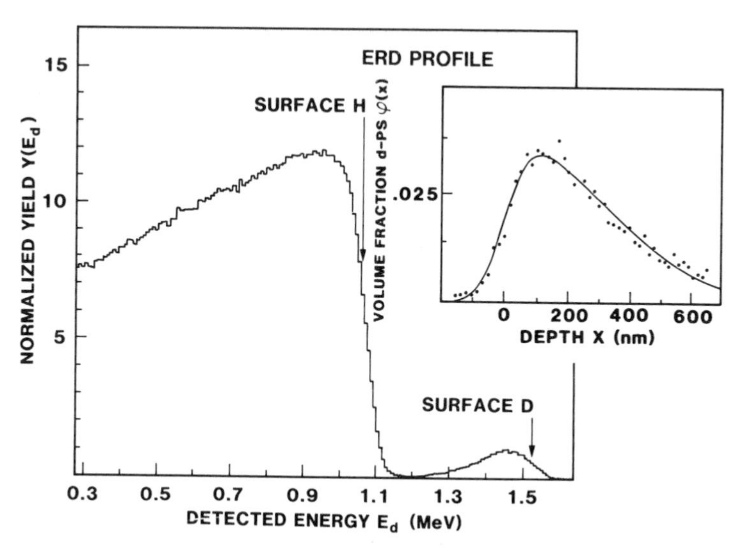

Fig. 2. FRES profile of d-PS of N = 260 monomer segments which diffused into the copolymer of N_c = 830 monomer segments at 162°C. The insert is the volume fraction versus depth profile of the homopolymer in the copolymer; D^* = 3.8x10^{-13}cm^2/s.

The effect of the energy loss of the projectile and of the recoiling nucleus upon traversing a depth x is included in $[S]$; δE is the energy which the nucleus loses upon traversing a 12μm thick mylar foil, shown in Fig. 1, placed there to prevent the most energetic forward scattered particles from being detected. $[S]$ is defined in terms of the stopping powers S_1 of the projectile as it passes through the sample, and S_2 of the recoiling nucleus as it travels out of the target. If they are both assumed to be constant then

$$[S] = \frac{KS_1}{cos\alpha} + \frac{S_2}{cos(\alpha + \theta)} . \tag{3}$$

This experiment provides a spectrum of the particle (recoils) yield, $Y(E_d)$ (number of recoils detected with an energy between E_d-$\Delta E/_2$ and E_d+$\Delta E/_2$ where ΔE is the energy width of the multichannel analyzer), versus E_d. The depth resolution of this technique is ~80 nm and the labelled diffusant can be followed to depths ~800 nm.

Illustrated in Fig. 2 is a FRES profile of $Y(E_d)$ versus E_d for d-PS chains of N_H = 260 monomer segments which were allowed to diffuse into a nonequilibrium symmetric diblock copolymer structure, of N_C = 830 monomer segments, for 21 min at 162°C. Seen at higher energies is the deuterium profile of the deuterated polymer in the copolymer and at lower energies (below 1.1 MeV) is the hydrogen profile from the copolymer host. The insert in this figure is the profile of volume fraction versus depth of d-PS in the copolymer. This was obtained from the $Y(E_d)$ versus E_d profile using a deuterated standard and knowledge of the energy loss rates of the projectile and recoils in the sample and in the range foil. The line drawn through the data was computed using a solution to the diffusion equation assuming a concentration-independent diffusion coefficient which should be approximately valid since the concentrations of homopolymer considered here is extremely low. A tracer diffusion coefficient of D^*=3.8x10^{-13} cm^2/s was obtained. This profile is typical of the ones found for the diffusion of homopolymers into homopolymers. The procedure for obtaining D^* in those cases is identical.[20-22]

III. TRACER DIFFUSION IN HOMOPOLYMERS

In this section experiments on the diffusion of d-PS and d-PMMA chains in PS and PMMA hosts, respectively, are discussed in light of the Doi-Edwards model.[2,3] In this model the motion of a labelled chain in an entangled melt is imagined to occur along an average trajectory called the primitive path which is defined as the center line of the "tube". The dynamical modes of this chain in its "tube" are assumed to be described by the Rouse model; therefore the diffusion coefficient of the chain along the primitive path is denoted by[3]

$$D_t = \frac{k_B T}{n\zeta} , \tag{4}$$

where ζ is the monomeric friction coefficient, k_B is the Boltzmann constant and T is the temperature; $N = M/M_0$, where M_0 is the molecular weight of a monomer in the chain. The ends of the chain that emerge from the "tube" choose random directions in space in which to move. As the chain diffuses, portions of a new "tube" are created ahead while portions of the old are abandoned. At long-time scales defined by τ_d the chain loses complete memory of the old tube. τ_d = $L^2/\pi^2 D_t$ where L is the length of the primitive path. The center of mass diffusion coefficient of a chain diffusing by reptation alone is given by[2-4]

$$D^* = D_0 M^{-2}, \tag{5}$$

where

$$D_0 = \frac{4}{15} \frac{M_0 M_e k_B T}{\zeta},$$

M_e is the molecular weight between entanglement.

The temperature dependence of D^* may be obtained from the zero shear-rate viscosity of the polymer in an indirect way. Both D^* and the zero shear-rate viscosity, η_0, are related through τ_d and hence the monomeric friction coefficient. One can obtain η_0 from the stress relaxation modulus, G(t)[3,4]

$$\eta_0 = \int_a^a G(t) dt = \int_0^a G_N^0 F(t)\, dt = \frac{\pi^2}{12} G_N^0 \tau_d, \tag{6}$$

where $F(t)$ is the fraction of the tube that was occupied at time $t = 0$ that is still occupied at time t

$$F(t) = \frac{8}{\pi^2} \sum_{p\,odd} \frac{1}{p^2} exp(-tp^2/\tau_d), \tag{7}$$

and $G_N0 = \rho RT/M_e$. Graessley[4] has shown that one may easily express D^* in terms of η_0

$$D^* = \frac{G_N^0}{135} M_e^2 K \frac{M_c}{\eta_0(M_c) M^2}. \tag{8}$$

In this equation M_c is the critical molecular weight for viscous flow and $K = \langle R^2 \rangle / M$ where $\langle R^2 \rangle$ is the root mean square end-to-end vector. The temperature dependence of D^* may be obtained by recognizing that Equation 8 may be separated into temperature and molecular weight independent terms

$$log\, D^*/T = C(M) - \ell n \eta_0, \tag{9}$$

where $C(M)$ is a function of the molecular weight. $C(M)$ does, however, vary slowly with temperature due to ρ and K; it may change by a few percent over 100 degrees whereas η_0 and D^* will vary by a few orders of magnitude over this temperature range in most polymers.[5,25] It is well known that the temperature dependence of η_0 is accurately described by the Vogel-Fulcher equation, or equivalently by the WLF equation[5]

$$log\, \eta_o = A + B/(T - T_o), \tag{10}$$

where T_0 and B are Vogel constants. The temperature dependence of D^*/T is easily shown to be

$$log\, [(D^*/T)/(D^*_{ref}/T_{ref})] = B/(T_{ref} - T_0) - B/(T - T_0). \tag{11}$$

In the above equation, T_{ref} is a reference temperature at which D^*_{ref} is determined.

In what follows we discuss the molecular weight and temperature dependence of the translational diffusion of d-PS and d-PMMA chains in PS and PMMA hosts, respectively. The results of a FRES analysis of the M dependence of the diffusion of d-PS into PS at 171°C are represented by the triangles in Fig. 3. The squares represent the diffusion of PS into d-PS. Both sets of data are identical. For a comparison, the results of a Rutherford backscattering (RBS) marker experiment of the diffusion of PS into PS are represented by the filled circles.[26] Independent measurements of the diffusion of fluorescent labelled PS into PS by a holographic grating technique are represented by the open circles.[27] The data obtained from nuclear magnetic resonance spectroscopy (---)[28] are somewhat larger in magnitude than all the other sets of data; this discrepancy is believed to be due to polydispersity effects.[11] It is clear, however, that an independent comparison of studies performed using different techniques yield results that are in excellent agreement with each other. All the data, except that from the NMR experiment, may adequately be described by the equation

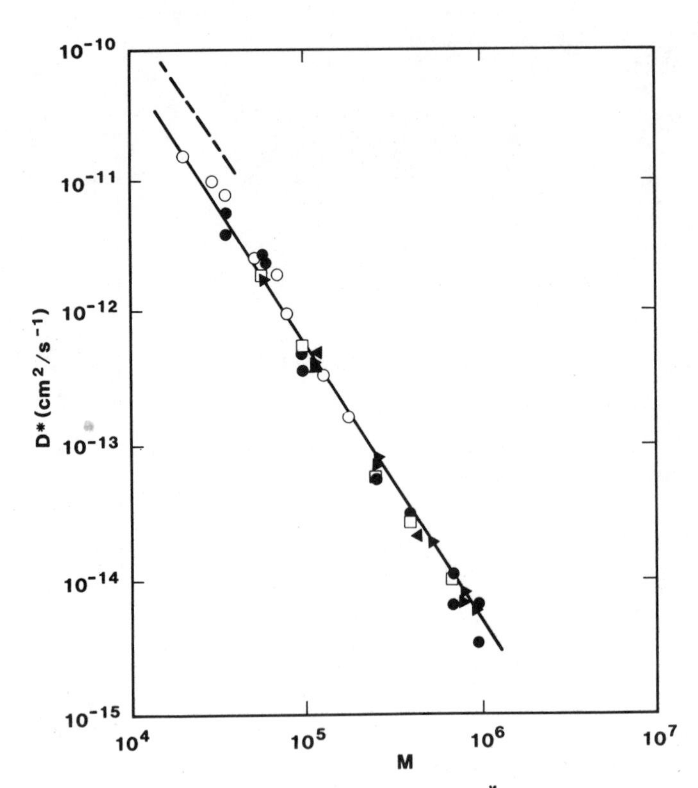

Fig. 3. Molecular weight dependence of the D^* of PS as a function of M from various techniques: ERD of PS into d-PS(□); ERD of d-PS into PS(▲); Holographic grating technique(O); NMR(---) and the RBS marker experiment (●) (reproduced from ref. 20).

$$D^* = 0.005M^{-2}, \tag{12}$$

which is in agreement with the reptation prediction. The diffusion of d-PMMA into PMMA exhibits similar behavior. Shown in Fig. 4 is a plot of D^* versus M for the diffusion of d-PMMA into PMMA at 185°C. These results may be described by the following equation

$$D^* = 0.00009M^{-2}. \tag{13}$$

This power-law dependence has been observed in a number of other systems,[24,29] including polymer blends.[30,32] It may be pointed out that equation 8 has been used to accurately predict D_0, of Eq. 5, in a number of polymer systems.[24,26,29]

A comparison of the temperature dependence of D^* and η_0^{-1} in the PS and the PMMA systems is made below. Recall that the ratio D^*/T and η_0^{-1} should exhibit nearly the same temperature dependence. Plotted in Fig. 5 is the ratio D^*/T as a function of $(T-T_0)^{-1}$ for a d-PS chain of M = 1.1×10^5 (Δ) which diffused into a PS host. The open triangles represent

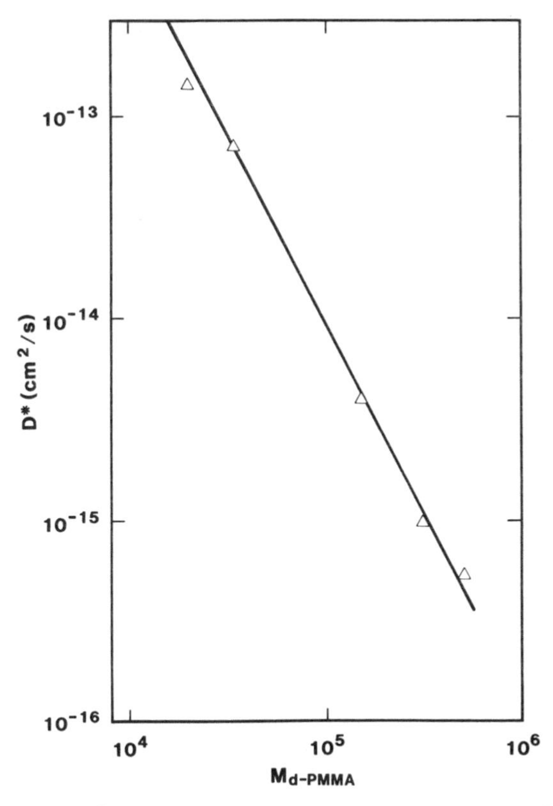

Fig. 4. Data of the diffusion of d-PMMA into PMMA at 185°C.

data for the diffusion of d-PS of $M = 4.3 \times 10^5$ into a PS host that was normalized by a factor of $(430/110)^2$ so that it could be superimposed on the other set of d-PS data. The line drawn through the data was computed with Equation 11 using constants $B = 710$ and $T_0 = 49°C$. These constants are the identical ones used to fit the η_0 data of PS by Roovers and Graessley.[31]

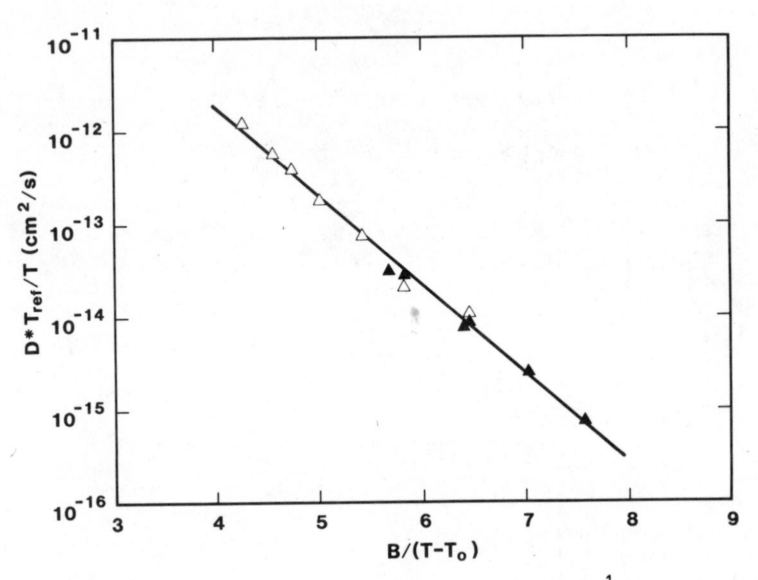

Fig. 5. Plot of D^*/T as a function of $(T-T_0)^{-1}$ for polystyrene; $M = 1 \times 10^5$, $T_0 = 49°C$ and $B = 710$.

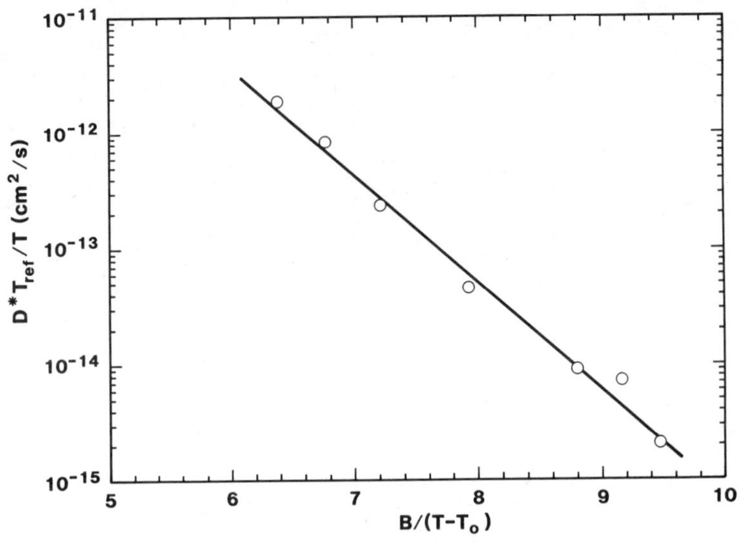

Fig. 6. Plot of D^*/T as a function of $(T-T_0)^{-1}$ for PMMA; $M = 2 \times 10^4$, $T_0 = 35°$ and $B = 1118$.

The temperature dependent data of PMMA is shown in Fig. 6 where D^*/T_0 is plotted as a function of $(T-T_0)^{-1}$. The line drawn through the data was computed using Eq. 11. The constant B = 1118 and T_0 = 35°C were obtained from fits to the viscosity data of PMMA.[5] These data show conclusively that the temperature dependences of D^*/T in PS and PMMA are different and that one should be able to predict D^*/T based on knowledge of the temperature dependence of η_0^{-1}. Excellent agreement between the temperature dependences of D^*/T and η_0^{-1} have also been observed in the polyisoprene[33] and polypropylene oxide systems.[34]

The results shown here indicate that the Doi-Edwards model correctly predicts the molecular weight and temperature dependence of chains diffusing in entangled melts. It is important to point out, however, that extensive modifications have been made to the model to describe viscoelastic properties such as the stress relaxation modulus $G(t)$ in bimodal systems and the correct power-law dependence of the viscosity in a homopolymer melt.[7,35-39] The agreement between theory and experiment is generally quite good in many cases.

The ideas discussed in this section will be used in subsequent sections.

IV. "THERMODYNAMIC SLOWING DOWN" OF MUTUAL DIFFUSION IN ISOTOPIC POLYMER MIXTURES

There is great current interest in the thermodynamics of isotopic polymer mixtures.[40-50] Recent small-angle neutron scattering (SANS) measurements strongly suggest that mixtures of normal polymers and their deuterated analogues are characterized by a slightly positive Flory-Huggins interaction parameter, χ, and exhibit an upper critical solution temperature (UCST).[40-45,50] If this is true then measurements of the mutual diffusion coefficient $D(\Phi)$ at finite compositions Φ in a blend of a polymer and its deuterated analogue should reveal that $D(\Phi)$ undergoes a "thermodynamic slowing down" in the vicinity of a critical composition, Φ_C. In this section we investigate "thermodynamic slowing down" effects in blends of normal and deuterated polystyrene. Using the composition dependence of $D(\Phi)$ we determine the χ parameter. The UCST is determined from the temperature dependence of χ. This procedure[32] for determining χ differs from the conventional method of using SANS measurements where the experimental scattering curves are fit by de Gennes' random phase approximation (RPA) of the static structure factor using χ in an adjustable parameter. At the end of this section comparisons are made with the results of the SANS measurements and with theoretical predictions of χ.[42]

In systems where the value of the χ parameter is positive but the mixture is still in the range of single phase stability, defined by

$$\chi_s(\Phi) = \{[N_D\Phi]^{-1} + [N_H(1-\Phi)]^{-1}\}/2 \; . \tag{14}$$

$D(\Phi)$ should experience a minimum or a "thermodynamic slowing down" in the vicinity of the critical composition Φ_C where

$$\Phi_c = \frac{N_H^{\frac{1}{2}}}{[N_H^{\frac{1}{2}} + N_D^{\frac{1}{2}}]} \; . \tag{15}$$

In the above equations N_D and N_H are the number of monomer segments that comprise the deuterated and undeuterated polymer chains, respectively. As χ approaches χ_s, or equivalently, as T approaches the UCST, the system experiences large fluctuations in composition. The extent of the "thermodynamic slowing down" should therefore be enhanced. In cases where $\chi > \chi_s$ the mixture is unstable and undergoes phase separation. "Thermodynamic slowing down" effects should not, however, be observable when $\chi = 0$.

In section III we discussed how to determine tracer diffusion coefficients. We will now briefly discuss the method used to determine the mutual diffusion coefficient. The details may be found in separate publications.[44,45] High molecular weight mixtures of deuterated and undeuterated polystyrenes ($N_D = 10^4$, $N_H = 8.8\text{x}10^3$) were used in this study. Diffusion couples were produced where on either side of the interface is a blend of D-PS and H-PS; one of composition Φ_0 and the other of $\Phi_0 + \Delta\Phi$; $\Delta\Phi$ is chosen to be about 10%. The samples were prepared by coating a polished silicon wafer with ~1µm film of a blend of a given composition. A second film (~0.45µm) was produced separately by spinning a solution, of a composition that differed from the previous film by ~10%, on a glass slide. The film was then floated onto a bath of distilled water where it was transferred onto the surface of the coated silicon wafer. Fig. 7 shows the FRES profile of a film of average composition $\Phi_{ave} = 0.55$ of D-PS; $\Delta\Phi = 0.05$. The line drawn through the data represents the instrumental broadening which is Gaussian with a full width at half maximum of 100nm. This line is not a fit to the data. The resolution of this system is fixed. The initially steplike concentration profile in this sample was allowed to broaden by interdiffusion at elevated temperatures. Shown in Fig. 8 is the FRES profile of the sample that was allowed to diffuse at 174°C for 15.5 hr. The mutual diffusion coefficient was then extracted by fitting it with a solution to the diffusion equation. The mutual diffusion coefficient extracted from the data at this composition is $D(0.55) = 2.0\text{x}10^{-15}\text{cm}^2/\text{s}$. Mutual diffusion coefficients were extracted at different temperatures using couples at varying compositions. The procedure just described has been used to determine χ in compatible polymer mixtures.[32]

$D(\Phi)$ is plotted as a function of the average blend composition, Φ_{ave}, in Fig. 9 at temperatures of 166, 174, 190 and 205°C. The data at each temperature exhibits a pronounced minimum or "thermodynamic slowing down"

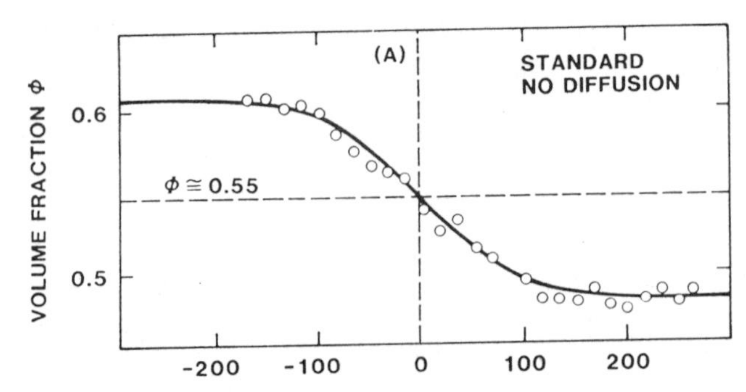

Fig. 7. Volume fraction versus depth profile for a standard. The average volume fraction is 0.55.

in the vicinity of Φ_{crit} = 0.5. The magnitude of this effect increases with decreasing temperature (χ approaches χ_s), which one should anticipate if the system exhibits a UCST. Note that the correlation length, ξ, of the concentration fluctuations increase,

$$\xi^{-2} \sim [\Phi(1-\Phi)(\chi_s(\Phi)-\chi], \tag{16}$$

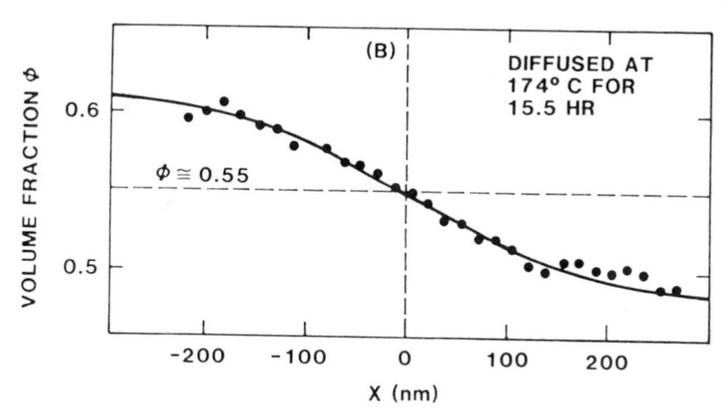

Fig. 8. Sample diffused for 15.5 hr at 174°C; N_H = 8.7x10^3 and N_D = 9.8x10^3 (reproduced from ref. 44).

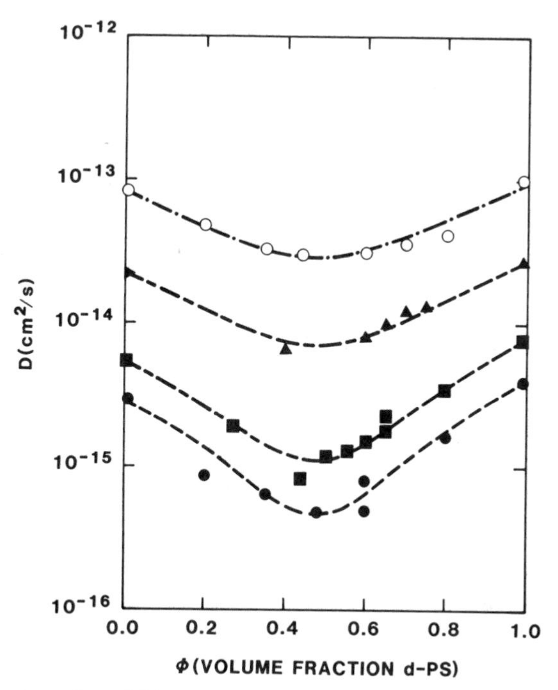

Fig. 9. A plot of $D(\Phi)$ versus Φ at 205(o), 190(▲), 174(■) and 166(●)°C (reproduced from ref 45).

as χ approaches χ_s which suggests that $D(\Phi)$ should decrease. The mean field prediction for the compositional dependence of the mutual diffusion coefficient may be expressed as[50-55]

$$D(\Phi) = 2\Phi(1-\Phi)D_T(\chi_s(\Phi) - \chi), \tag{17}$$

where

$$D_T = D^*_{d-PS}N_D(1-\Phi) + D^*_{h-PS}N_H\Phi.$$

D_T is the transport coefficient which is expressed in terms of, $D^*_{d\text{-PS}}$, the tracer diffusion coefficients of the deuterated chains diffusing into the undeuterated host and, $D^*_{h\text{-PS}}$, the undeuterated chains diffusing into a deuterated host. The above equation may be expressed in terms of the correlation length $D(\Phi) \sim \xi^{-2}$, which shows that when the correlation length of the concentration fluctuations in the system increase $D(\Phi)$ decreases.

Since $D(\Phi)$ is sensitive to small changes in χ, especially near the stability limit, knowledge of $D(\Phi)$ is a useful way to determine χ. The lines drawn through each set of data were computed using equation 17 in which χ is the only adjustable parameter. Its value is adjusted to yield the best fit to the data in the middle of the concentration regime. The data of the FRES experiments may be fit by the following equation

$$\chi = \frac{0.22(+0.06)}{T} - 3.2(+1.2)x10^{-4}. \tag{18}$$

A plot of this data is shown in Fig. 10. The stability limit of this mixture was calculated, using Eq. 14, to be $\chi_s(\Phi_{crit}) = 2.1 \times 10^{-4}$. Using Eq. 18 one predicts the UCST to be about 130°C. The SANS data of Bates and Wignall[41] may be expressed as

$$\chi = \frac{0.02(+0.01)}{T} - 2.9(+0.4)x10^{-4}. \tag{19}$$

The agreement between both sets of experimental data is excellent. The thermodynamic slowing down" effects are difficult to observe far from the stability limit as shown in Fig. 11. The stability limit of this PS/d-PS mixture of chains of $N_D = N_H = 2.3 \times 10^3$ was calculated to be $\chi_s(0.5) = 8.7 \times 10^{-4}$ using equation 14. The solid line was computed using $\chi = 1.9 \times 10^{-4}$, a value which one calculates at a temperature of 160°C using equation 18. The broken line represents a smaller value of 1.4×10^{-4}. This fit is not very different from that obtained using $\chi = 0$. It is clear from this result that $D(\Phi)$ is not very sensitive to χ in this molecular weight regime. The use of high molecular weight polymers is necessary in order to see these effects.

One should easily anticipate these effects for the following reason. The combinatorial entropy of mixing varies as $1/N$ which approaches zero in the limit of high N which suggests, within the framework of the Flory-Huggins free energy, that the enthalpic contributions, embodied by χ, dominate the free energy change upon mixing. Since $\chi > 0$, completely random mixing is unfavoured, therefore the system should show a tendency to demix.

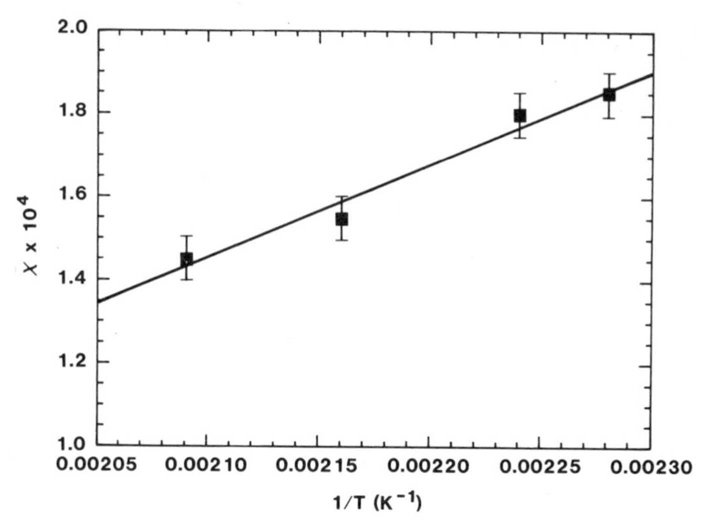

Fig. 10. ERD data of the dependence of χ on $1/T$. (reproduced from ref. 45).

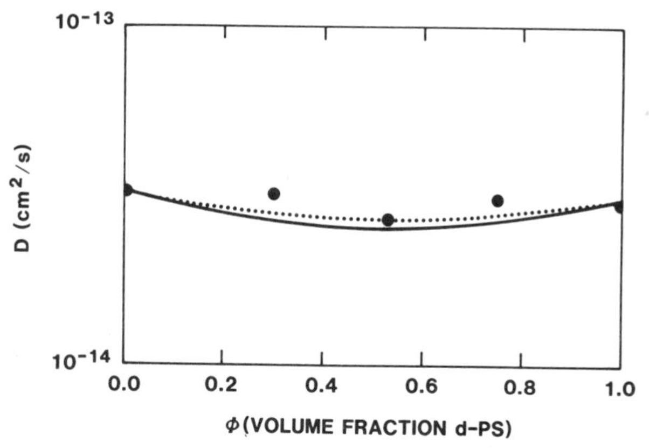

Fig. 11. $D(\Phi)$ versus Φ of chains of $N_D = N_H$ = 2×10^3 which diffused at 160°C.

Earlier studies of critical phenomena in the PS cyclohexane system by Strazielle and Benoit[58] showed that the θ-temperature was markedly influenced by whether the PS and or the cyclohexane was deuterated. They estimated a difference in the cohesive energy densities between the deuterated and undeuterated polystyrenes to be $\delta_{h\text{-}PS}-\delta_{d\text{-}PS} = \Delta\delta = 0.038$ $(cal/cm^3)^{1/2}$. The χ parameter may be approximately expressed as

$$\chi = \frac{V_0}{RT} (\Delta\delta)^2, \tag{20}$$

where V_0 is the molar segment volume and R is the universal gas constant. This estimate gives values of the same order as that determined by the FRES and SANS measurements.[41] The most noteworthy point about this study is that one could anticipate the existence of a UCST in this mixture based on knowledge of $\Delta\delta$ prior to the SANS of FRES measurements.

The current understanding of these effects appear to be somewhat incomplete; and, to a certain degree, controversial.[40,43,48-50] It is well documented that deuteropolymers have slightly lower segmental volumes than normal polymers. This arises from the fact that by replacing a proton with a deuteron on a C-H bond results in a reduction of the bond length; hence a smaller segmental volume. This reduction in volume induces a reduction in the polarizability.[42] Despite the fact these changes are less than a few percent, they appear to have a major impact on the thermodynamics of these systems (i.e., χ is positive) because the combinatorial entropy of mixing is so small. The χ parameter described by the Flory-Huggins expression for the free energy change that arises from mixing two polymer species is assumed to be controlled by dispersive forces in nonpolar isotopic mixtures and as such should vary as T^{-1}. Bates et al.[42,50] made rough estimates of χ based on the foregoing ideas. These results are in good agreement with the measured values.[42,43,45]

There have been sophisticated calculations of χ in these systems. In the Flory-Huggins model χ is independent of molecular weight and composition. In reality, this parameter has been shown, by scattering experiments and by theoretical calculations to be a function of M, T and Φ.[43,48,49] Theoretical attempts, which are beyond the scope of the mean field predictions of the Flory-Huggins approach which applies strictly to incompressible systems, have been made to address these questions.[43,48,49] The theories of Bates and Muthukumar[43] and of Schweizer and Curro[48,49] both have predictions which may be written in the following form

$$\chi_{eff} = \chi + \beta(N, T, \Phi) . \tag{21}$$

Here χ_{eff} is an effective interaction parameter which depends on the Flory-Huggins χ, described above, and on β which is a function of N, T and Φ. Both theories predict an approximately parabolic compositional dependence of χ, where in the middle of the composition regime the curve is relatively flat and toward the ends it increases. This is in agreement with measurements of χ versus Φ data of poly(vinylethylene) and poly(ethylethylene) systems.[43] There are, however, many important differences between both predictions. This is however not an appropriate place for such a discussion but a brief comment may be made with regard to both theoretical approaches. The approach of Bates et al.[49] is based on a correction to the Flory-Huggins free energy expression which accounts for monomer density fluctuations. The approach of Schweizer and Curro, unlike that of Flory-Huggins, is an off-lattice approach using integral equations. In this theory the function β is expressed in terms of direct correlation functions, which can be thought of as effective or renormalized pair potentials that describe the correlations between the particles. A complete discussion of the merits of both predictions may be found elsewhere.[48,49]

As a final note, it may be pointed out that the current approach using the FRES technique to determine χ is most sensitive at the middle of the composition range where χ is essentially independent of composition.

V. DIFFUSION OF HOMOPOLYMERS INTO DIBLOCK COPOLYMER STRUCTURES

In section III we discussed the diffusion of a homopolymer chain comprised of N monomer segments into a highly entangled melt where it was shown that its tracer diffusion coefficient, D^*, varied as N^{-2} and that the temperature dependence of D^*/T was virtually the same as that of the inverse of the zero shear-rate viscosity of the homopolymer. Here we discuss the diffusion of homopolymers into microphase-separated diblock copolymer structures.

The bulk properties of microphase-separated diblock copolymers are well understood. Studies show that the relative volume fraction of each component of a microphase separated copolymer essentially determines the structure that it exhibits.[60-63,67,68] If the average volume fraction of one of the blocks of the copolymer chain is between 0.4 and 0.6 then the equilibrium structure is comprised of alternating lamellae of each component. When the volume fraction of the minor component is less than 0.2 it forms spherical microdomains, exhibiting a body-centered cubic structure, in a matrix of the major component. The minor component forms cylindrical microdomains which exhibit a hexagonal close-packed (hcp) structure in the major component of its volume fraction is between 0.2 and 0.4. It has recently been shown that these cylindrically shaped microdomains form a structure which exhibits a double diamond-like symmetry when the volume fraction of the minor component lies in the narrow regime between those which predict the formation of the lamellar and the cylindrical microdomains arranged in a hcp structure.[64] The size of the microphase separated domains is determined by the total length of the copolymer chain.

Before discussing the diffusion of the homopolymer chains into the copolymer structure it is important to discuss the bulk and surface properties of the copolymers. The bulk characterization was accomplished using small-angle X-ray scattering (SAXS). Ideally, electron microscopy (EM) measurements should be done but preliminary studies were unsuccessful because of considerable electron beam damage to the PMMA. The surface characterization was done using X-ray photoelectron spectroscopy (XPS). This technique is more usefull for determining surface compositions than EM.

Small-Angle X-ray Scattering Characterization of the Bulk At equilibrium the lamellae of microphase-separated symmetric diblock copolymers should, in general, be aligned parallel to the copolymer/air interface. The method used to prepare copolymer films has a marked influence on the details of the microstructure. The PS/PMMA copolymer samples used in this study were all produced by solution casting using toluene which does not preferentially solvate either of the components. Since the preparation of these samples involved rapidly evaporating the solvent from a concentrated copolymer solution, the resulting copolymer structure may not be at equilibrium. Small-angle X-ray scattering (SAXS) which was used to characterize the bulk morphology of these films indicate that this is true. These studies show that the lamellae are isotropically arranged in space and are not aligned parallel to the copolymer/air interface as would be expected under equilibrium conditions. In what follows we present a summary of the results; the details of these measurements may be found elsewhere.[65,66,69]

Shown in Fig. 12 is a typical SAXS profile of a PS/PMMA copolymer where the molecular weight, M_{PS}, of the PS block is 42,000 and that of the PMMA block is M_{PMMA} = 42000. Note that the volume fraction of PS is $f_{PS} = (M_{ps}/M_0)/[(M_{ps}/M_0) + (M_{PMMA}/M_0')] = 0.49$; M_0 is the molecular weight of a PS monomer and M_0' is the molecular weight of a PMMA monomer. Henceforth, this copolymer will be known as SMI. In this Figure, the intensity is plotted as a function of scattering vector $\mathbf{q} = (4\pi/\lambda)\sin(\varepsilon/2)$, where λ is the wavelength and ε is the scattering angle. This data was taken at 175°C. There is an intense first-order maximum at low scattering vectors. The insert at higher scattering vector is a magnification, by a factor of 20, of a third-order maximum at $\mathbf{q}$ = 0.04 (Angstroms)$^{-1}$; the ratio of the Bragg spacing corresponding to the peak position is 2.96 which clearly shows this peak to be a third-order maximum. The presence of the higher-order maximum demonstrates that the copolymers are indeed microphase-separated in contrast to previous measurements.[71] Even-order maxima should not be observed because of symmetry conditions. The observation that the third-order maximum is very weak suggests that the interface between the two domains is diffuse and not very sharp.

An evaluation of the long period, the center-to-center distance between like phases, of this copolymer is 442 Angstroms and the width of the interface is about 50 Angstroms. Measurement of the long period of symmetric PS/PMMA copolymers, each comprised of N_C monomer segments, shows that $L = 13.5N_C^{1/2}$, suggesting that the size of the domains varies as the radius of gyration of the chains. These results are plotted in Figure 13. Liebler's[67] theoretical treatment of diblock copolymers in the weakly segregated limit, which assumes that the order parameter, a measure of the local deviation of the composition from its equilibrium value, varies smoothly from one domain to another (i.e., broad interface), predicts a power-law exponent of 1/2. Other theories that consider copolymers in the strongly segregated limit (narrow interphase boundary)[68] predict power-law exponents of 2/3. Based on the molecular weight dependence of L and the breadth of the interphase region, E, one may conclude that these copolymers are in the weakly segregated limit. Temperature dependent measurements of L and E indicated that this system exhibited no tendency to undergo a disordering transition even at temperatures in excess of 200°C.

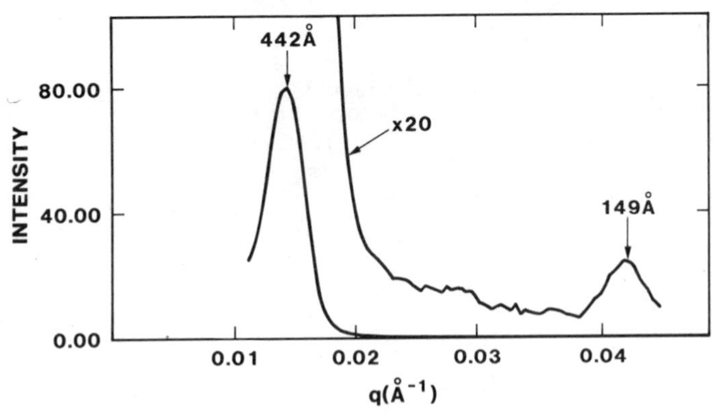

Fig. 12. SAXS profile of the SMI copolymer at 175°C (reproduced from ref. 65).

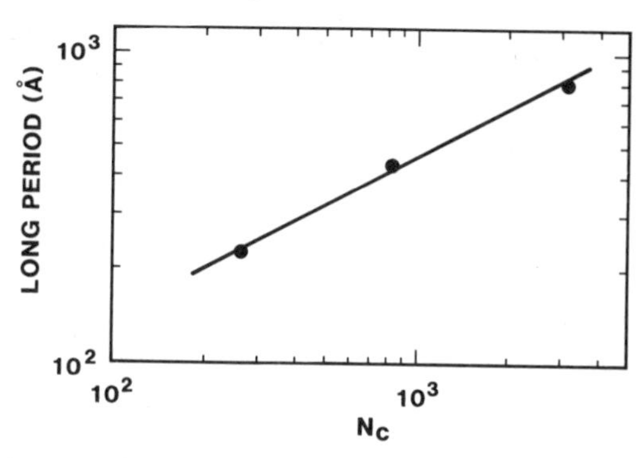

Fig. 13. Dependence of the long period, L, on the total copolymer chain length; $L \sim {N_C}^{1/2}$.

SAXS studies of mixtures comprised of SMI and varying compositions of the PS and the PMMA homopolymers were also done.[66,67] The long period was observed to increase linearly with the volume fraction of homopolymer for homopolymer chains of $N_H < N_C$. This is to be expected if the homopolymer chains were being incorporated in the domains of the copolymer. This observation was made for volume fractions in excess of 20% of the homopolymer for homopolymers of $N_H < N_C/2$. When N_H was comparable to N_C deviations from linearity occurred at 10% of the homopolymer content. The long period showed no change when the homopolymer chain lengths were sufficiently greater than that of the copolymer. This suggests that the homopolymer was excluded from the domains of the copolymer.

Surface Characterization Using X-ray Photoelectron Spectroscopy The presence of a free surface alters the effective interactions between the unlike segments in a copolymer chain. Furthermore, because of the difference in surface free energies between both components the component of lower surface energy shows a preferential affinity for the free surface. For this reason, it is important to know whether the surface is completely covered by one component or both components exist at the surface. This will determine what fraction of the surface is accessible to the homopolymers through which they may freely bypass in order to get to the bulk. In the PS/PMMA copolymer system, XPS studies showed that both PS and PMMA coexist at the surface. In all cases studied the composition of PS, Φ_{PS}, was higher than that of PMMA. Our XPS measurements show that Φ_{PS} is in fact a monotonically increasing function of N; at large N, however, Φ_{PS} approached a constant. This in contrast to studies of solvent cast symmetric microphase-separated styrene-isoprene[58] and styrene-butadiene[59] copolymers which show that the surface of these copolymers is completely covered by the lower surface energy component regardless of the orientation of the lamellae (i.e., whether or not equilibrium structures are formed). This is not an appropriate place for a discussion of block copolymer surfaces; one may refer to reference 70 for a discussion of factors that have been found to influence the composition at block copolymer surfaces.

VI. RESULTS AND DISCUSSION

The experimental procedure used to obtain the tracer diffusion coefficients described in section II was also used here. A thin layer (~15 nm) of labelled homopolymer was placed at the surface of the copolymer and allowed to diffuse into the bulk. The volume fraction versus depth profiles obtained for homopolymer chains of $N_H < N_C/2$ in cases where Φ < 0.06 are typical of the one shown in Figure 2. All such profiles could be accurately described by a solution to the diffusion equation assuming that D was independent of Φ at these compositions. In some experiments we prepared the thin layer such that it was comprised of a mixture of the labelled homopolymer and an unlabelled high molecular weight homopolymer of the same chemical structure. The molecular weights were chosen such that the shorter chain homopolymer would diffuse much more rapidly into the bulk of the host while the other species would remain relatively immobile. By doing this, one could achieve more dilute volume fractions of the labelled diffusant in the copolymer host. The diffusion coefficients extracted from these profiles were virtually the same as those obtained using the other procedure provided that N_H was less than N_C.

When N_H is comparable to N_C the diffusion profiles assumed a bimodal shape indicating that only a small portion of the thin layer initially on the surface was able to diffuse and the other portion remained on the surface of the copolymer. In these cases, the diffusion coefficient was

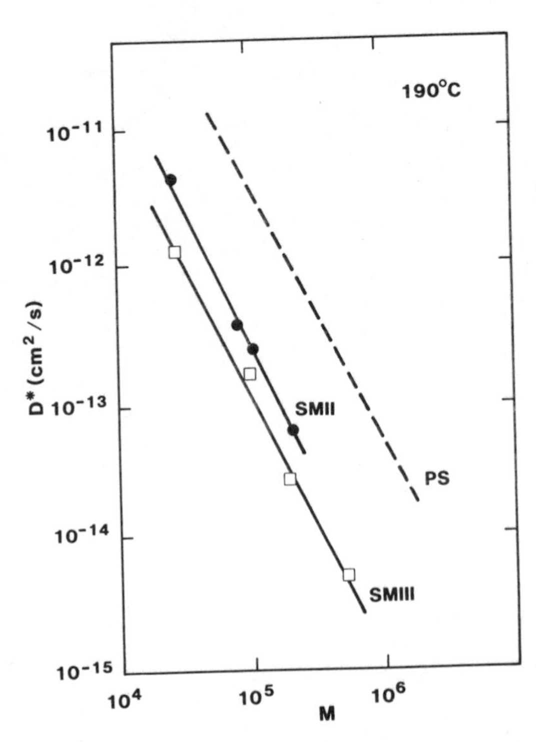

Fig. 14. Plot of D^* versus N_H for the diffusion of d-PS into SMII(●) and SMIII(□) ($N_H < N_C$).

extracted by fitting the portion of the profile that diffused into the copolymer. As suggested by the SAXS measurements[66] and by theory,[72] when $N_H \sim N_C$ the solubility of the homopolymer in the copolymer structure is low which explains the occurrence of the bimodal shaped profiles.

The dependence of the tracer diffusion coefficient of PS homopolymer chains of $N_H < N_C$ on N_H diffusing into two copolymer hosts, SMII and SMIII, is shown in Figure 14. This data was taken at 190°C. The SMII copolymer is composed of PS of M_{PS} = 157000 and of PMMA of M_{PMMA} = 162000 (N_C = 3130); the SMIII copolymer of M_{PS} = 275000 and M_{PMMA} = 260000 (N_C = 5244). The broken line represents the diffusion of d-PS into PS (slope = -2). The molecular weight dependence of d-PS into the copolymer is

$$D_{HC} \sim N_H^{-2}, \tag{22}$$

which is identical to that which one finds for the diffusion of homopolymers into homopolymers. It is clear from this figure that the homopolymer chains diffuse into the copolymer structure an order of magnitude slower than into its homopolymer analogue. The diffusion rates into the copolymer structures, SMII and SMIII, are also slightly different which reflects a slight difference in the details of the microstructure of the copolymers, as discussed below.

When N_H is comparable to N_C and larger one finds somewhat different behavior than that described by Eq. 22. The results in Figure 15 exhibit limiting power-law dependence with an exponent that is greater than 2. The power-law dependence should increase since one eventually finds a situation where the diffusion ceases since at this point the homopolymer is not soluble.

Figure 16 is a plot of the N_H dependence of the diffusion of d-PMMA into the copolymer host. The squares represent the diffusion of d-PMMA into SMII and the circles that of d-PMMA into SMI. The homopolymers are all of $N_H < N_C/2$. The broken line (slope = -2) represents the diffusion of d-PMMA into PMMA. Though the data is limited, the results are similar to that found for the diffusion of d-PS, $D_{HC} \sim N_H^{-2}$. In addition, the diffusion into the copolymers occurs considerably slower than into PMMA. For reasons discussed below the N_H^{-2} dependence may not be observed over a molecular weight regime as wide as that observed for the d-PS diffusion.

Figure 17 is a plot of log D_{HC}/T versus $(T\text{-}T_0)^{-1}$ for the diffusion of a d-PMMA chain of N_H = 210 into the SMI copolymer ($N_H/N_C \sim 1/4$). The line drawn through the data was computed with Eq. 11 using the same constants (B = 1118 and T_0 = 35°C) used to describe the d-PMMA/PMMA diffusion. Similar comparisons may be made between the diffusion of d-PS into PS and d-PS into the copolymers. The dependence of log D_{HC}/T on $(T\text{-}T_0)^{-1}$ for the diffusion of d-PS into the SMI copolymer is shown in Figure 18. The constants B = 710 and T_0 = 49°C were the same ones used to fit the PS diffusion data in section III.

The temperature dependent measurements corroborate the SAXS findings that the copolymers are microphase separated. This data also strongly suggests that the d-PS chains diffuse within the PS domains and the d-PMMA chains diffuse within the PMMA domains. The phases of PS and PMMA are continuous, at least on length scales on the order of 800nm, the distance probed by the FRES experiment. It appears that the difference between the magnitudes of D_{HC} and D^* for the diffusion of the homopolymer chains is related to the structure of the copolymer and not due to the

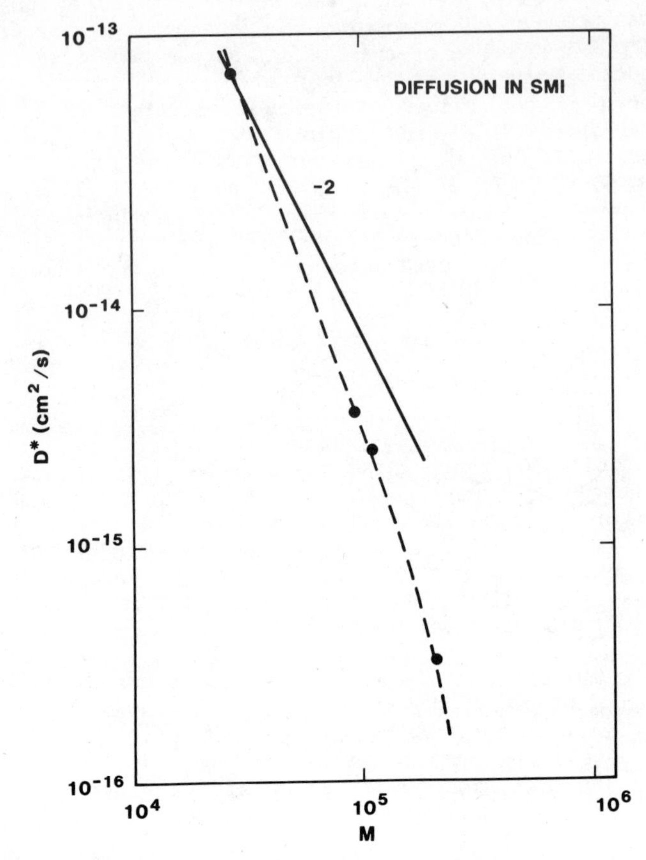

Fig. 15. N_H ($N_H > N_C$) dependence of the diffusion coefficient of d-PS which diffused into SMI.

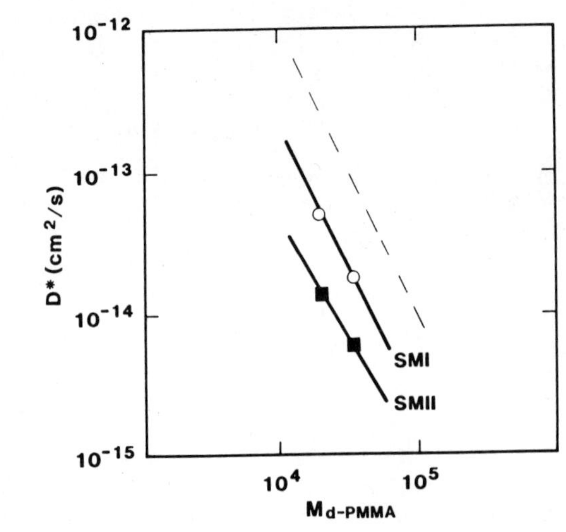

Fig. 16. D^* vs. M for the diffusion of d-PMMA into SMI(O) and SMII(■).

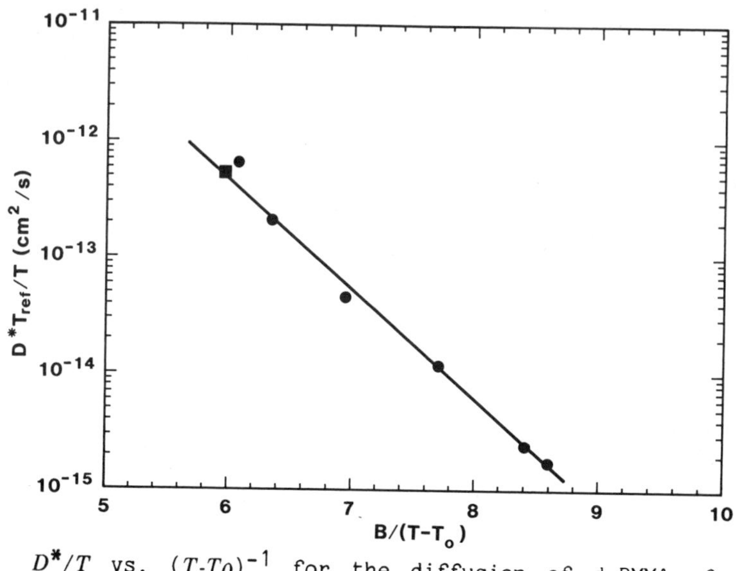

Fig. 17. D^*/T vs. $(T\text{-}T_0)^{-1}$ for the diffusion of d-PMMA of N_H = 210 ($N_H \sim N_C/4$) into SMI.

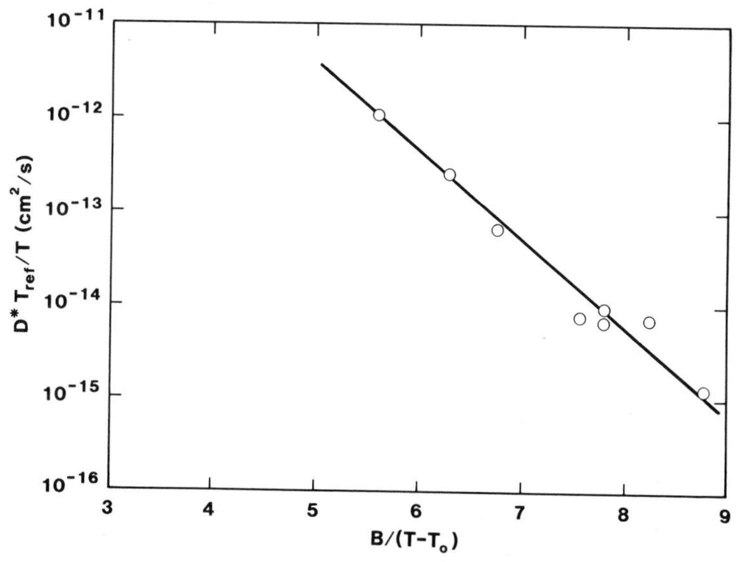

Fig. 18. The temperature dependence of d-PS of N_H = 260 ($N_H \sim N_C/4$) which diffused into SMI.

fact that a diffusing homopolymer chain may penetrate both phases in the copolymer. The difference is related to a number of factors, one of which is a partitioning effect since half of the total volume is accessible. The accessible domains have random orientations and are evidently tortuous. This is to be expected if the structure was not at equilibrium. The SAXS measurements suggest that the structures are not at equilibrium. The differences in microstructure of the SMII and SMIII should account for the difference in the D_{HC} observed for the diffusion of d-PS into both hosts. The same should also be true of the diffusion of d-PMMA into SMI and SMII.

In view of the fact that the copolymer structure changes as the diffusion process commences and the homopolymer chains appear to diffuse within the domains of structure that is identical to their own, it does, at first glance, appear surprising that we observe the mentioned N_H and T dependence of D_{HC}. This observation simply reflects that the diffusion times are small in comparison to the times that characterize the ordering of the copolymer structure from a non-equilibrium to an equilibrium structure. While $D_{HC} \sim N_H^{-2}$ $(N_H < N_C)$ for the diffusion of d-PS chains, it may not be true for the diffusion of d-PMMA chains over the same N_H range of d-PMMA chains. This is because the d-PMMA chains diffuse over an order of magnitude slower than the d-PS chains of comparable length. This process may be sufficiently slow for the d-PMMA chains that the copolymer structure may undergo significant changes. Our preliminary results suggest that this may be true.

In order to investigate the effect of structure on D_{HC} the copolymers were subjected to different preparation procedures. The SMI copolymers were heated for 3 hours at 170°C before the diffusion of the homopolymers. Diffusion experiments conducted at 160°C showed that D_{HC} increased by a factor of 2. The same effect was observed when it was exposed to toluene vapor for the same period of time. However when they were treated (annealed at 170°C or exposed to vapor) for 24 hours D_{HC} decreased by a factor of ~3. The increase in D_{HC} may be related to an increase in the purity and long-range order of the structures which one would expect to occur at early times if the structures are not well developed after the solution-casting process. However, over long-time periods one expects appreciable changes in structure since the copolymer structure should approach an equilibrium structure where the lamellae become alinged parallel to the free surface. This may explain the decrease in D_{HC}.

VIII. CONCLUDING REMARKS

The homopolymers diffuse into the copolymer hosts at a rate that is an order of magnitude lower than that which they diffuse into their analogues. Based on the temperature dependent measurements and the SAXS measurements, it is reasonable to conclude that the diffusion of a homopolymer chain is restricted to the domains whose structure is identical to its own. The difference in magnitude may be related partially to a partitioning effect since only half of the total volume is accessible to the diffusant. The other important factors are the tortuosity and the orientation of the microdomains in the structure. One should not, in general, observe the same chain length and temperature dependence of D_{HC} in homopolymer/copolymer system if the copolymer structure changes appreciably during the diffusion process since the structure affects D_{HC}.

ACKNOWLEDGEMENTS

The work described here has benefited from very important collaborations with Prof. E.J. Kramer, Dr. T.P. Russell and Dr. B.L. Doyle over the years. This work was performed at Sandia National Laboratories and was supported by the U.S. Department of Energy under contract number DE-AC04-76DP00789.

NOMENCLATURE

M - molecular weight of a chain
N - the number of monomers on a polymer chain
τ_d - the longest relaxation time of a chain in a polymer melt
D^* - the tracer diffusion coefficient of a chain of N monomer segments diffusing into a medium
η_0 - the zero shear rate viscosity
D_{REP} - diffusion coefficient of a chain diffusing by reptation
D_{CR} - the constraint release contribution to the tracer diffusion coefficient of a chain diffusing by reptation
χ - Flory interaction parameter
Φ - average composition of a blend
E_0 - energy of the incident helium beam
E_d - energy with which a recoil is detected
$[S]$ - energy loss factor
$Y(E_d)$ - recoil yield
D_t - diffusion coefficient of a chain along the "tube"
M_0 - molecular weight of a monomer
M_e - molecular weight between entanglement
M_C - critical molecular weight
G_N^0 - plateau modulus
$G(t)$ - stress relaxation modulus
$\langle R^2 \rangle$ - root mean square end-to-end vector
B - Vogel constant
T_0 - Vogel constant
$\chi_s(\Phi)$ - value of χ at the spinodal
ξ - correlation length of the concentration fluctuations
D_T - Onsager transport coefficient
$\Delta\delta$ - cohesive energy density difference between d-PS and PS
$\beta(N,T,\Phi)$ - function that describes the N,T and Φ dependence of χ
Φ_{PS} - volume fraction of PS on the copolymer chain
$\mathbf{q}$ - scattering vector
D_{HC} - diffusion coefficient of a homopolymer chain of N_H segments that diffused into a copolymer of N_C segments

REFERENCES

1. P.G. de Gennes, J. Chem. Phys. 55, 572 (1971).
2. M. Doi and S.F. Edwards, J. Chem. Soc., Faraday Trans. 2, 8, 1798 1809 1978 (1978).
3. M. Doi and S.F. Edwards, The Theory of Polymer Dynamics, Oxford University Press, Oxford, UK, (1986).
4. W.W. Graessley, J. Polymer Sci., Poly. Phys. Ed., 18, 27 (1980).
5. G.C. Berry and T.G. Fox, Adv. Polym. Sci., 5, 261 (1968).
6. W.W. Graessley, Adv. Polym. Sci., 47, 67 (1982).
7. H. Watanabe and M. Tirrell, Macromolecules (submitted) (1988).
8. M. Daoud and P.G. de Gennes, J. Polym. Sci., Polym. Phys. ed., 17, 1971 (1979).
9. J. Klein, Macromolecules, 19, 105 (1986).
10. W. Hess, Macromolecules, 21, 2587 (1988).
11. G. Fleischer, Colloid and Polymer Sci., 265, 89 (1987).
12. P.F. Green, P.J. Mills, C.J. Palmström, J.W. Mayer and E.J. Kramer, Phys. Rev. Lett. 53, 2145 (1984).
13. P.F. Green and E.J. Kramer, Macromolecules, 19, 1108 (1986).
14. B.A. Smith, E.T. Samulski, L.P. Yu and M.A. Winnik, Phys. Rev. Lett., 52, 45, (1984).

15. M. Antonietti, J. Coutandin and H. Sillescu, Macromolecules, 19, 793 (1986).
16. J.P. Monfort, G. Marinand P. Monge, Macromolecules, 17, 1551 (1984).
17. R.H. Colby, L.J. Fetters and W.W. Graessley, Macromolecules, 20, 2226 (1987).
18. P.J. Flory, Principles of Polymer Chemistry, Cornell University Press, Ithaca, N.Y. (1953).
19. P.G. de Gennes, Scaling Concepts in Polymer Physics, (Cornell University Press, Ithaca, N.Y. 1979).
20. P.F. Green, P.J. Mills and E.J. Kramer, Polymer, 27, 1063 (1986).
21. P.J. Mills, P.F. Green, C.J. Palmström, J.W. Mayer and E.J. Kramer, Appl. Phys. Lett., 45, 9 (1984).
22. P.F. Green and B.L. Doyle, Nuclear Instruments and Methods in Physics Res. B18, 64, (1986).
23. B.L. Doyle and P.S. Peercy, Appl. Phys. Lett., 34, 811, 1979).
24. W.W. Graessley, J. Polym. Sci., Polym. Phys. Ed., 18, 27 (1980).
25. T.G. Fox and V.R. Allen, J. Chem. Phys., 41, 244, (1964).
26. P.F. Green, C.J. Palmstrom, J.W. Mayer and E.J. Kramer, Macromolecules, 18, 501, (1985).
27. M. Antonietti, J. Coutandin and H. Sillescu, Macromol. Chem. Rapid Commun., 5, 525 (1984).
28. G. Fleischer, Polymer Bull., 9, 152 (1983).
29. M. Tirrell, Rubber Chemistry and Technology, 57, 523 (1984).
30. P.F. Green, unpublished results.
31. J. Roovers and W.W. Graessley, Macromolecules, 12, 1959 (1979).
32. R.J. Composto, Ph.D. Thesis, Cornell University, Ithaca, N.Y. 1987.
33. N. Nemoto, M.R. Landry, I. Noh and H. Yu, Polym. Commun., 25, 141, (1984).
34. S.J. Mumby, B.A. Smith, E.T. Samulski, L.-P. Yu and M.A. Winnik, Polymer, 27, 1826 (1986).
35. M. Doi. J. Polym. Sci., Polym. Phys. Ed., 19, 265 (1981).
36. M. Rubinstein, E. Helfand and D.S. Pearson, Macromolecules, 20, 822 (1987).
37. M. Rubinstein and R.H. Colby, J. Chem. Phys., (accepted).
38. M. Doi, E. Helfand, W.W. Graessley and D.S. Pearson, Macromolecules, 20, 1900, (1987).
39. W.W. Graessley and M.J. Struglinski, Macromolecules, 19, 1754 (1986).
40. F.S. Bates, G.D. Wignall and W.C. Koehler, Phys. Rev. Lett, 55, 2425 (1985).
41. F.S. Bates and G.D. Wignall, Macromolecules, 19, 934 (1986).
42. F.S. Bates and G.D. Wignall, Phys. Rev. Lett., 57, 1425 (1986).
43. F.S. Bates, M. Muthukumar, G.D. Wignall and L.J. Fetters, J. Chem. Phys., (accepted).
44. P.F. Green and B.L. Doyle, Phys. Rev. Lett., 57, 2407 (1986).
45. P.F. Green and B.L. Doyle, Macromolecules, 20, 2471 (1987).
46. H. Yang, R.S. Stein, C.C. Han, B.J. Bauer and E.J. Kramer, Polymer Commun., 27, 132 (1986).
47. A. Lapp, C. Picot and H. Benoît, Macromolecules, 18, 2437 (1985).
48. K.S. Schweizer and J.G. Curro, Phys. Rev. Lett., 60, 809, (1988); J. Chem. Phys. (in preparation).
49. J.G. Curro and K.S. Schweizer, J. Chem. Phys. (accepted).
50. F.S. Bates, Macromolecules, 20, 2221 (1987).
51. E.J. Kramer, P.F. Green and J.W. Mayer, Polymer, 25, 473 (1984).
52. H. Sillescu, Macromol. Chem., Rapid, Commun., 5, 519 (1984).
53. H. Sillescu, Macromol. Chem., Rapid, Commun., 8, 393 (1987).
54. G. Foley and C. Cohen, J. Polym. Sci. Polym. Phys. ed., 25, 2027 (1987).
55. There is a second prediction for $D(\Phi)$ (Ref. 56,57) which differs from Eq. 17 such that $D_T^{-1} = [D^*_{d\text{-PS}} N_D]^{-1} (1-\Phi) + [D^*_{h\text{-PS}} N_H]^{-1}\Phi$.

Since we have chosen $N_H = N_D$, both predictions yield the same result.
56. F. Brochard, J. Jouffroy and P. Levinston, Macromolecules, 17, 2925 (1984).
57. K. Binder, J. Chem. Phys, 79, 6387 (1983); J. Colloid Polym. Sci., 265, 273 (1987).
58. C. Strazielle and H. Benoît, Macromolecules, 8, 203 (1975).
59. H. Hasegawa and T. Hashimoto, Macromolecules, 18, 590 (1985).
60. C.S. Henkee, E.L. Thomas and L.J. Fetters, J.Mater. Res., (1988) (in press).
61. T. Hashimoto, M. Fujimura and H. Kawai, Macromolecules, 13, 1660 (1980).
62. G.E. Molau, Block Copolymers, S.L. Aggarwal ed. Plenum Press (1970).
63. D.J. Meir, Block and Graft Copolymers, J.J. Burke and V. Weidss eds. Syracuse University Press, Syracuse, N.Y. (1973).
64. E.L. Thomas, D.B. Alward, D.J. Kinning, D.C. Martin, D.L. Handlin and L.J. Fetters, Macromolecules, 19, 2197 (1986).
65. P.F. Green, T.P. Russell, M. Granville and R. Jerome, Macromolecules (in press).
66. P.F. Green, T.P. Russell, M. Granville and R. Jerome, Macromolecules (accepted).
67. L. Leibler, Macromolecules, 13, 1602 (1980).
68. T. Ohta and K. Kawasaki, Macromolecules, 19, 2621 (1986).
69. T.P. Russell and R. Jerome, (in preparation).
70. P.F. Green, T.M. Christensen, T.P. Russell, R. Jerome, Macromolecules, (submitted).
71. H. Benoît, W. Wu, M. Benmouna, B. Mozer, B. Bauer and A. Lapp, Macromolecules, 18, 986 (1985).
72. M.D. Whitmore and J. Noolandi, Macromolecules, 18, 2486 (1985).

MUTUAL DIFFUSION IN THE MISCIBLE POLYMER BLEND: POLYSTYRENE:POLY (XYLENYL ETHER)

R.J. Composto* and
E.J. Kramer

Dept. of Mater. Sci. & Eng.
Cornell University
Ithaca, New York 14853

Dwain M. White

General Electric Corp.
Research and Development
Schenectady, New York 12301

ABSTRACT

We have used forward recoil spectrometry to measure the mutual diffusion and tracer diffusion coefficients, D and D^*, in the miscible polymer blend of deuterated polystyrene (d-PS):poly(xylenyl ether) (PXE). Using the "fast theory" of mutual diffusion, D is related to the D^*, degree of polymerization N, and volume fraction ϕ of the individual blend components by,

$$D = 2\phi(1-\phi)[D^*_{PS}N_{PS} + \phi D^*_{PXE}N_{PXE}](\chi_s - \chi),$$

where χ and χ_S are the Flory interaction parameter of the blend and its value at the spinodal. From the measured values of the D^*'s and D at a volume fraction ϕ=0.55 of d-PS, the interaction parameter χ=0.112-62/T was estimated as a function of temperature $T(K)$. At low T, D was much larger than D^*'s due to the large negative value of χ whereas at high T, D becomes less than the D^*'s as χ becomes positive (thermodynamic slowing down). Similar measurements show that χ is not markedly composition-dependent in the d-PS:PXE blends.

I. INTRODUCTION

In the early 1970's de Gennes[1] proposed that in an entangled system the tracer or self-diffusion coefficient, D^*, of a long polymer chain of degree of polymerization N should decrease as N^{2}. This prediction sparked a wealth of experiments which are still going on today.[2] About ten years later, de Gennes[3] was the first to note that because the combinatorial entropy of mixing of polymers is so small, scaling as N^{-1}, the mutual diffusion of chemically dissimilar polymers will be dominated by the excess enthalpy and entropy of segment-segment mixing. Although this second prediction has many practical applications, such as in the

*Current Address: Polymer Science and Engineering, University of Massachusetts, Amherst, MA 01003

welding of miscible polymers and the control of phase separation in immiscible polymers, few experiments have rigorously tested the relationship between the thermodynamics of such blends and their diffusion behavior.[4-12] In this paper we summarize experiments in which we examine the temperature- and composition[4]-dependence of the mutual diffusion coefficient in a miscible polymer blend.

Using the Flory-Huggins[13] theory the mutual diffusion coefficient D can be written as[14-18]

$$D = 2(\chi_s - \chi)\phi_A\phi_B D_T, \tag{1}$$

where $\phi_A\phi_B D_T$ is an Onsager transport coefficient, ϕ_A and ϕ_B are the volume fractions of the components A and B in the binary blend, χ is the Flory segment-segment interaction parameter and χ_s is its value at the spinodal

$$\chi_s = \tfrac{1}{2}\left[\frac{1}{\phi_A N_A} + \frac{1}{\phi_B N_B}\right], \tag{2}$$

where N_A and N_B are the degrees of polymerization of the polymer components represented by the subscripts. In order to relate D_T of the binary blend to the D^*, N and ϕ of the individual components, several theories have been proposed. While the "slow" theory,[14,15] which proposed that D_T was controlled by the slower diffusing component in the blend, received early experimental support,[9,11] more recent theory and experiments favor the "fast" theory. The "fast" theory assumes that the diffusion fluxes of A and B, rather than cancelling as in the "slow" theory, are compensated for by a bulk flow in the diffusion couple. Unlike the "slow" theory, the "fast" theory[16-18] assumes that the pressure gradients in the pair vanish so that

$$D_T = \phi_B D^*_A N_A + \phi_A D^*_B N_B. \tag{3}$$

Note that D is controlled by the diffusion of the faster moving component in the polymer blend.

Using irreversible thermodynamics, a more recent theory of Brochard-Wyart[19] also arrives at Eq. 3. This paper revises the original "slow" theory[14] and thus resolves the "fast" versus "slow" controversy from a theoretical standpoint for entangled polymers. Both our earlier experimental results,[5,21] on d-PS:PXE blends with ϕ=0.55 where we measured the dependence of D^* and D on d-PS molecular weight and those of Jordan et al.[12] on mutual diffusion in entangled polyethylenes, produced by deuterating or protonating narrow molecular weight distribution polybutadienes, provide strong experimental support for the "fast" theory.

In a miscible polymer blend, mutual diffusion measurements provide a very sensitive method for probing the thermodynamics of interaction between two chemically different segments. From the relationship between the transport coefficient D_T and the D^*'s of the polymer components one can extract the Flory parameter χ from the measured values of D and the D^*'s. Upon rewriting Eq. 1, χ is given by

$$\chi = \chi_s - \frac{D}{2\phi_A\phi_B D_T}. \tag{4}$$

In order for high N polymers to be miscible, χ normally must be negative and consequently the second term should dominate Eq. 4. In some cases, blends which have a small positive χ can be compatible.[20] However, for a blend which exhibits a lower critical solution temperature, χ should become less negative with increasing temperature until, finally, χ becomes greater than χ_S. Since χ strongly influences the diffusion in concentrated pairs, one should witness a corresponding decrease in the value of D, compared with the D^*'s of the individual components of the blend, as temperature increases. These predictions will be tested quantitatively.

In this paper, we summarize measurements of the temperature- and composition-dependence of the mutual diffusion of deuterated polystyrene (d-PS) chains and protonated poly(xylenyl ether) (PXE) chains which have appeared in more detail elsewhere.[4,21] By measuring D and D^* as a function of temperature, one can extract the temperature dependence of $\chi(T)$. Because the PS:PXE system exhibits a lower critical solution temperature,[22] a large thermodynamic "speeding up" of diffusion will be observed at low temperatures ($\chi<0$) while a thermodynamic "slowing down" will occur at higher temperatures ($\chi>0$). Using these values for $\chi(T)$, we will examine the composition-dependence of the mutual diffusion coefficient $D(\phi)$ for d-PS:PXE blends ranging from pure d-PS to pure PXE.

II. EXPERIMENTAL METHODS

For the miscible blend PS:PXE, the D^*'s of the individual components were determined from the concentration versus depth profiles, measured by forward recoil spectrometry (FRES)[23] of trace amounts of the deuterated polymer d-PS and d-PXE that were diffused into a blend of the protonated polymers which had a volume fraction ϕ of PS. The degree of polymerization characteristics of the polymers are given in Table 1. Bilayer films of a very thin layer of the pure deuterated polymer (~20 nm thick) on top of a thick layer of the blend were diffused in vacuum ($<10^{-6}$ torr) at temperature T and the FRES spectra were analysed by methods described previously.[23] Figure 1 shows the depth profile for d-PS(a) and d-PXE(b) tracer films diffused into protonated PS:PXE blends which contained 0.55 volume fraction of PS. The d-PS or d-PXE from the thin layer becomes rapidly diluted so that the combinatorial entropy of mixing is the only driving force for diffusion over most of the diffusion time and thus a true tracer diffusion coefficient is measured. The solid lines correspond to the solution to the diffusion equation[24,25] convoluted with the instrumental resolution function, a Gaussian with a full width at half maximum of 80 nm. The best fits to the data in Figs. 1(a) and 1(b) require $D^*_{PS} = 1.2 \times 10^{-14}$ cm^2/sec and $D^*_{PXE} = 2.2 \times 10^{-15}$ cm^2/sec, respectively.

The mutual diffusion couples consisted of two films of deuterated polystyrene (D-PS):protonated PXE blends which have a small difference in the volume fraction ϕ of d-PS (~10%), from ϕ_2 in the ~350 nm-thick top film to ϕ_1 in the 2 μm-thick bottom film.[4] The initial step function of the d-PS concentration profile was broadened by mutual diffusion at T in vacuum ($<10^{-6}$ torr). The volume fraction versus depth profile of d-PS was measured by FRES and a mutual diffusion coefficient D was extracted by fitting these profiles to a standard error-function solution;[4] the D measured corresponds to the D at the average composition of the diffusion couple, $(\phi_1 + \phi_2)/2$. Figure 2 shows the volume fraction profile of d-PS versus depth x for a mutual diffusion couple before (a) and after (b) diffusion. The initial ϕ_1 of the bottom film is 0.60 while that of the top film is 0.51. The solid line in Fig. 2(a) corresponds to a step

Table 1. Weight-average degrees of polymerization and polydispersity indices of the various polystyrene and poly(xylenyl ether) polymers together with their sources.

Deuterated PS and PXE

N_{PS}	N_{PXE}	Polydispersity index	Source
529		1.06	Polymer Laboratories
1 058		1.1	" "
2 452		1.1	" "
5 000		1.1	" "
7 885		1.3	" "
	225	1.9	General Electric
	383	2.4	General Electric

Protonated PS and PXE

N_{PS}	N_{PXE}	Polydispersity index	Source
529		1.06	Pressure Chemical
865		1.06	" "
2 452		1.06	" "
3 750		1.06	" "
19 230		1.2	" "
	292	2.3	General Electric

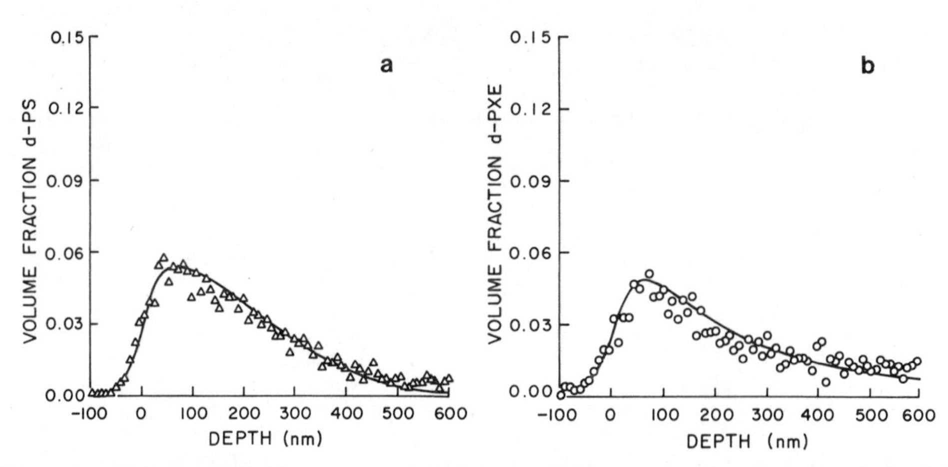

Fig. 1. Volume fraction versus depth profiles corresponding to (a) d-PS (N_{PS} = 2,452) and (b) d-PXE (N_{PXE} = 383) diffused in the same PS:PXE blend at 206°C. The solid lines were theoretical fits using $D_{PS}* = 1.2 \times 10^{-14}$ and $D_{PXE}* = 2.2 \times 10^{-15} cm^2/sec$ for d-PXE, respectively.

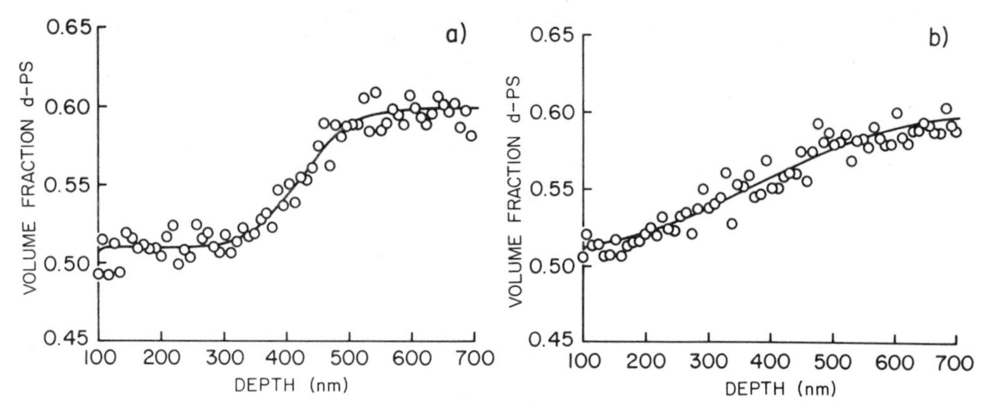

Fig. 2. Volume fraction of deuterated polystyrene in a d-PS(N_{PS}=2,452): PXE(N_{PXE} = 292) mutual diffusion couple: (a) as deposited and (b) after diffusion for 1800 sec at 206°C. The solid lines represent a best fit of the standard error function solution to the data using an instrumental resolution of 80 nm in (a) and a D of 1.1 x 10^{-13}cm^2/sec in (b).

function convoluted with the instrumental resolution function. The solid line in Fig. 2(b) is the solution to the diffusion equation using D = 1.1 x 10^{-13}cm^2/sec.

III. RESULTS AND DISCUSSION

Temperature Dependence The temperature dependence of the tracer diffusion coefficients D^* and mutual diffusion coefficient D was measured for PS:PXE blends containing 0.55 volume fraction PS. In the tracer experiments, the weight average degrees of polymerization N for d-PS and d-PXE were 2, 452 and 292, respectively, while the N's for the matrix chains were large enough to ensure that the D^*'s were dominated by reptation. We measured D^*_{PXE} for N_{PXE} = 225 and 384 and then scaled these D^*'s (i.e., from reptation $D^* \propto N^{-2}$) so as to find the D^*_{PXE} for N_{PXE} = 292. For temperatures not too much greater than the glass transition temperature, the temperature dependence of D^*/T has been shown to be described quite well by a form of the Vogel-Fulcher equation,[26]

$$log\left(\frac{D^*T_{ref}}{T}\right) = A' - \frac{B}{(T-T_\infty)}, \tag{5}$$

where A', B and T_∞ are empirical parameters and where T_{ref} is a reference temperature. As shown in Fig. 3, the tracer diffusion coefficients for d-PS (circles) and d-PXE (squares) increased rapidly as the diffusion temperatures increased from 180° to 310°C. The dashed lines are nonlinear least square fits of Eq. 5 to the experimental D^*'s where the empirical parameters B_{PS} = 601, B_{PXE} = 555 and T_∞ = 361K were obtained using a reference temperature of 500K. Except at the lowest T, the temperature dependence of the D^*'s for d-PS and d-PXE diffusing in a ϕ = 0.55 blend are in good agreement with Eq. 5. Note that both D^*_{PS} and D^*_{PXE} increase at about the same rate with increasing temperature (i.e., both D^*_{PS} and D^*_{PXE} can be described by using B's that are nearly the same).

Also shown in Fig. 3 are the values of the mutual diffusion coefficients D (diamonds) for d-PS:PXE blends measured for a 0.55 volume

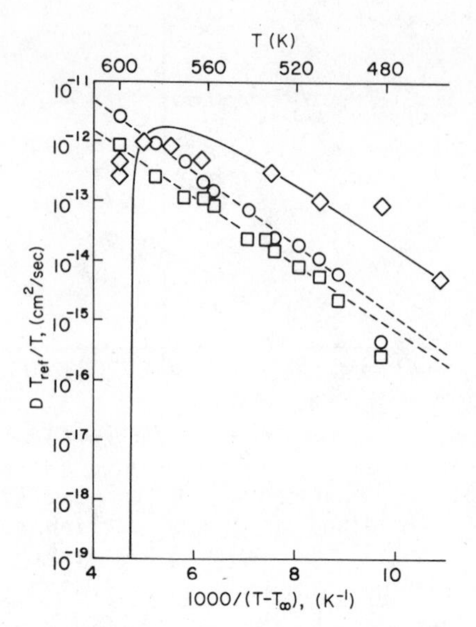

Fig. 3. The temperature dependence of the tracer diffusion coefficients of d-PS (circles) and d-PXE (squares) of weight average degree of polymerizations 2,452 and 292, respectively, and the mutual diffusion coefficients (diamonds) of d-PS:PXE blends of weight average degree of polymerizations 2,452 and 292, respectively. The values used for T_{ref} and T_∞ were 500K and 361K, respectively. The matrix films for the tracer experiments were a PS:PXE blend containing 0.55 volume fraction PS while the average volume fraction of d-PS in the mutual diffusion pair was also 0.55.

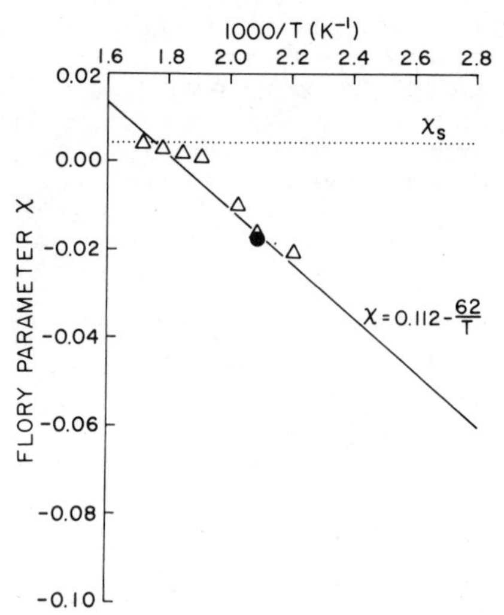

Fig. 4. Temperature dependence of the segment-segment (Flory) interaction parameter, χ, in a d-PS:PXE blend containing 0.55 volume fraction of d-PS. The dotted line represents $\chi_S = 0.0039$ and the solid line $\chi = 0.112 - [62/T(K)]$. The open triangles represent the χ's from D and D^* in Fig. 3 while the solid circle represents the average value of $\chi = -0.0178$ as computed from the $D(N)$ and $D^*(N)$ data given in reference 5.

fraction d-PS blend. The weight-average degrees of polymerizations of the d-PS (N_{PS} = 2,452) and PXE (N_{PXE} = 292) were the same as the tracer molecules used to measure the D^*'s. Due to the attractive interaction between PS and PXE at low temperatures, D is almost an order of magnitude greater than the D^*'s of the components of the blend. Note also that the temperature dependence of D deviates from that of the D^*'s such that the difference between D and the D^*'s gradually diminishes with increasing temperature until finally D becomes somewhat less than the D^*'s at 310°C.

This decrease in D derives from the temperature dependence of the interaction parameter χ. Using Eq. 4 and the measured values of $D(T)$ and $D^*(T)$, the values of the interaction parameter at various temperatures were calculated and are plotted in Fig. 4 for a ϕ = 0.55 blend. The open triangles correspond to χ's computed from the diffusion coefficients in Fig. 3 while the closed circle represents the average χ calculated from previous molecular weight studies.[5,21] Our results show that χ becomes less negative (i.e., there is a less attractive segment interaction between d-PS and PXE) as T increases and that χ scales roughly as 1/T. A least squares fit to the data yields χ = 0.112 -62/T (solid line). Also shown in Fig. 4 is the interaction parameter at the spinodal, χ_s = 0.0039, which was calculated by Eq. 2 using N_{PS} = 2,452 and N_{PXE} = 292. Our values of χ are in good agreement with the small-angle neutron-scattering measurements of Jelenic et al.[28,29] Thus measurements of the mutual diffusion coefficient provide a very sensitive means for determining χ.

Substituting $\chi(T)$ = 0.112 - 62/T and the D^*'s from Fig. 3 into Eqs. 1 and 3, the temperature dependence of the mutual diffusion coefficient D corresponding to a ϕ = 0.55 blend can be predicted. This result, represented by a solid line in Fig. 3, shows that at temperatures less than 280°C, the predicted values of D agree quite well with the experimental values of D. Note that the retardation of D observed experimentally can be explained quite nicely by the temperature dependence of χ. At low T or, equivalently, negative χ, a thermodynamic "speeding up" of mutual diffusion is observed because the excess enthalpy and entropy of mixing strongly favors mixing. However, as T increases or as χ approaches χ_s, a thermodynamic "slowing down" occurs because the excess term is retarding mutual diffusion. Green and Doyle[27] observe a similar slowing down of D in the PS:d-PS system due to a small positive χ.

Composition Dependence The tracer diffusion coefficients D^* and mutual diffusion coefficient D were also measured as a function of the volume fraction ϕ of PS in PS:PXE blends ranging from ϕ = 1.0 (pure PS) to ϕ = 0.0 (pure PXE). The values of D^* are plotted in Fig. 5 for d-PS (N_{PS} = 2,452) and d-PXE (N_{PXE} = 292). As before, we measured D^*_{PXE} for N_{PXE} = 225 and 384 and then scaled these D^*'s so as to find the D^*_{PXE} for N_{PXE} = 292. To partially compensate for the decreasing glass transition temperature T_g of the blend with increasing ϕ, the diffusion temperature T was adjusted so that T-T_g was held constant at 66°C. The weight-average degrees of polymerization of the matrix PS and PXE components were 3,750 and 292, respectively. As shown in Fig. 5, the D^*'s depended differently on blend composition, with D^*_{PS} surprisingly exhibiting a strong minimum with ϕ, whereas D^*_{PXE} was at first constant and then increased monotonically as ϕ approached 1.0. Note also that D^* for 292 PXE component was about a factor of 80 lower than D^* for the 2,452 d-PS component at the lower ϕ's.

Values of the mutual diffusion coefficient are also plotted as a function of composition in Fig. 5. The weight-average degree of polymerization for the components in the d-PS:PXE blend were 2,452 and 292 for d-PS and PXE, respectively. Note that at intermediate compositions D is almost an order of magnitude higher than either of the D^*'s of the components of the blend, again providing strong qualitative evidence for the thermodynamic "speeding up" of the mutual diffusion coefficient.

The mutual diffusion coefficient $D(\phi)$ was predicted using Eqs. 1 and 3 for the entire composition range ($0.0 \leq \phi \leq 1.0$) from the χ determined for the $\phi = 0.55$ blend. The predicted D's are shown as the solid line in Fig. 5, and are in good agreement with the experimental data. Since these values of χ appear to predict the magnitude of $D(\phi)$ quite well, it appears that χ in the d-PS:PXE system is not a strong function of composition. The dashed line in Fig. 5 represents the predicted D's expected if the mixing were ideal, i.e., $\chi = 0$. Therefore, at inter-mediate values of ϕ, where the effects of χ on diffusion are strongest, the mutual diffusion coefficient is strongly enhanced compared to the ideal mixing prediction ($\chi = 0$).

IV. CONCLUSIONS

We have measured the temperature and composition dependence of the mutual diffusion coefficient in a miscible blend of d-PS:PXE. Previously we have shown that in a blend containing 0.55 volume fraction of PS, the mutual diffusion coefficient D increases as N_{PS}^{-1} with decreasing N_{PS}, where N_{PS} is the degree of polymerization of the faster diffusing PS molecules. Because the species (PS) with the largest D^*N product dominates D, these results are in excellent agreement with the "fast" theory of mutual diffusion. From the measured values of $D(T)$ and $D^*(T)$ at

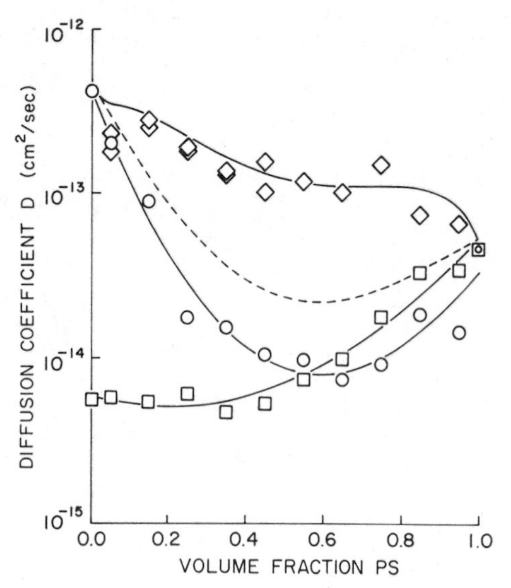

Fig. 5. Diffusion coefficients versus composition in the PS:PXE system. The squares and circles represent the tracer diffusion coefficients D^* of d-PXE ($N_{PXE} = 292$) and d-PS ($N_{PS} = 2,452$) in blends of PS:PXE. The diamonds represent the measured mutual-diffusion coefficients in the d-PS:PXE diffusion pairs. The lower solid lines represent smoothed values of $D^*_{d\text{-}PXE}$ and $D^*_{d\text{-}PS}$ used in the calculation for D. The upper solid line and dashed line represent the D predicted by Eqs. 1 and. 3 with, and without, segment-segment-interaction enhancement, i.e., $\chi = 0.112 - [62/T(K)]$ and $\chi = 0.0$, respectively.

ϕ = 0.55, the Flory-Huggins interaction parameter χ is found to be χ = 0.112 - 62/T. For χ < 0, $D(T)$ is enhanced relative to the values of D^*; however as χ approaches χ_S a thermodynamic "slowing down" of $D(T)$ is observed. Thus diffusion measurements are sensitive probes for determining the thermodynamics of mixing in miscible polymer blends. Also, an enhanced mutual-diffusion coefficient $D(\phi)$ was observed at intermediate compositions of d-PS:PXE. Using calculated values of $\chi(T)$ for the ϕ = 0.55 blend and the measured values of $D^*(\phi)$ for d-PS and d-PXE we can successfully predict $D(\phi)$ across the entire composition range suggesting that the composition-dependence of χ is weak in the d-PS:PXE system.

ACKNOWLEDGEMENTS

This work was supported by the Division of Materials Research, NSF Polymers program, and we benefitted from the use of the facilities of Cornell Materials Science Center which is funded by the Division of Materials Research-Materials Research Laboratory programme of the NSF. We especially thank J.W. Mayer for his encouragement and appreciate useful discussions and correspondence with H. Sillescu, K. Binder, J.H. Wendorff, G. Fytas, P.G. de Gennes, F. Brochard-Wyart, J. Klein, R.A.L. Jones and A.M. Donald.

NOMENCLATURE

D	mutual diffusion coefficient
D^*	tracer diffusion coefficient
N	degree of polymerization
T_{ref}	reference temperature
T_∞	empirical parameter
ϕ	Onsager transport coefficient
χ	interaction parameter

REFERENCES

1. P.G. de Gennes, J. Chem Phys., 55, 572 (1971).
2. For a review of tracer diffusion see M. Tirrell, Rubber Chem. Technol., 57, 523 (1984).
3. P.G. de Gennes, J. Chem. Phys., 72, 4756 (1980).
4. R.J. Composto, J.W. Mayer, E.J. Kramer and D.M. White, Phys. Rev. Lett., 57, 1312 (1986).
5. R.J. Composto, E.J. Kramer and D.M. White, Nature, 328, 1980 (1987).
6. R.A.L. Jones, J. Klein and A.M. Donald, Nature, 321, 161 (1986).
7. R.A.L. Jones, S.A. Rocca, A.M. Donald and J. Klein, to be published.
8. P.T. Gilmore, R. Falabella and R.L. Laurence, Macromolecules, 13, 880 (1980).
9. (a) R.W. Garabella and J.H. Wendorff, Makromol. Chem. Rapid Commun., 7, 591 (1986).
   (b) R.W. Garabella and J.H. Wendorff, submitted to Macromolecules.
10. S. Wu, H.K. Chuang and C.D. Han, J. Polym. Sci.;Polym. Phys. Ed., 24, 143 (1986).
11. (a) U. Murshall, E.W. Fisher and C. Herkt-Maetzky and G. Fytas, J. Polym. Sci.;Polym. Lett., 24, 191 (1986).
    (b) M.G. Brereton, E.W. Fischer, G. Fytas and U. Murschall, submitted to J. Chem. Phys.
12. E.A. Jordan, R.C. Ball, A.M. Donald, L.J. Fetters, R.A. Jones and J. Klein, Macromolecules, 21, 235 (1988).
13. P.J. Flory, Principles of Polymer Chemistry; Cornell University Press: Ithaca, N.Y., 1953, p.507.
14. F. Brochard, J. Jouffroy and P. Levinson, Macromolecules, 16, 1638 (1983).

15. K. Binder, J. Chem. Phys., 79, 6387 (1983).
16. E.J. Kramer, P.F. Green and C.J. Palmström, Polymer, 25, 473 (1984).
17. H. Sillescu, Macromol. Chem. Rapid Commun., 5, 519 (1984).
18. H. Sillescu, Macromol. Chem. Rapid Commun., 8, 393 (1987).
19. F.C. Brochard-Wyart, R. Acad. Sci. (Paris), t.305 Serie II, 657 (1987).
20. (a) F.S. Bates, G.D. Wignall and W.C. Koehler, Phys. Rev. Lett., 55, 2425 (1985).
(b) F.S. Bates and G.D. Wignall Macromolecules, 19, 934 (1986).
21. R.J. Composto, E.J. Kramer and D.M. White, Macromolecules, in press.
22. A. Maconnachie, R.P. Kambour, D.M. White, S. Rostami and D.J. Walsh, Macromolecules, 17, 129 (1984).
23. P.J. Mills, P.F. Green, C.J. Palmström, J.W. Mayer and E.J. Kramer, Appl. Phys. Lett., 45, 957 (1984).
24. P.G. Shewmon, Diffusion in Solids, McGraw-Hill: New York, 1963.
25. J. Crank, The Mathematics of Diffusion, 2nd ed. Oxford Univ. Press: Oxford, 1975.
26. P.F. Green and E.J. Kramer, J. Mater. Res., 1, 202 (1986).
27. P.F. Green and B.L. Doyle, Phys. Rev. Lett., 57, 2407 (1986).
28. J. Jelenic, R.G. Kirste, R.C. Oberthur, S. Schmitt-Strecker and B.J. Schmitt, Macromol. Chemie, 185, 129 (1984).
29. E.J. Kramer and H. Sillescu, to be published.

SELF-DIFFUSION OF POLY(DIMETHYLSILOXANE) CHAINS

Leoncio Garrido* and
Jerome L. Ackerman

Dept. of Radiology, NMR facility
Baker-2
Massachusetts General Hospital
Boston, Massachusetts 02114

James E. Mark

Dept. of Chemistry and the
Polymer Research Center
University of Cincinnati
Cincinnati, Ohio 45221

ABSTRACT

In this review we have described the basic concepts involved in the diffusional behavior of polymers and the NMR technique used to measure diffusion of poly(dimethylsiloxane) (PDMS) in melts and in model networks. We have found that the final value calculated for the exponent α in the scaling law $D_s \propto M^{\alpha}$ for diffusion constant D with molecular weight M depends strongly on the fitting procedures used in data analysis. Only for very low molecular weights the exponent is affected by free volume effects. The self-diffusion of PDMS chains in melts obeys the predicted dependence on the molecular weight while their diffusion in PDMS model networks remains unclear, in part because of the short range of M studied. For melts, low molecular weight polymers behave in accordance with the Rouse model while their hydrodynamic behavior for $M_n \geq 4200$ g-mol^{-1} might be explained in terms of the reptation model. The value found for the exponent α is -2.02, in excellent agreement with the theoretical prediction.

I. INTRODUCTION

The study of molecular dynamics in polymers is of crucial importance in the understanding of processes in which diffusion is involved. Examples of such processes are diffusion-controlled chemical reactions, sorption-extraction, dissolution, interactions across polymer-polymer interfaces, etc.[1] This has been a very active area of research, especially during the past few years, due to new theoretical approaches and the development and refinement of experimental techniques.[1,2] Different polymer systems have been used to test the theoretical predictions based on scaling methods. One of these systems is poly(dimethylsiloxane) (PDMS).[3] Other factors have also contributed to the interest in studying the molecular dynamics of PDMS.[3-9] Siloxane polymers have widespread commercial applications,

*Author to whom correspondence should be directed.

especially elastomers synthesized from them.[10,11] In rubber elasticity, "model" networks prepared from PDMS chains with reactive end groups have been widely used to get a better understanding of the properties of elastomeric materials at molecular level.[12-19] Although the chemical reactions used to obtain those networks are highly specific and allow independent knowledge of the network structure (crucial in studying rubberlike elasticity), there are different interpretations of the experimental results. In some cases,[14] the conclusions are based on the extraction of unreacted material (soluble fraction) from the final network to calculate the extent of the crosslinking reaction. The knowledge of PDMS hydrodynamic behavior in sorption and extraction of free polymer chains in polymer networks has provided important information for evaluating the efficiency of removal of the soluble fraction in those networks.[7-9] However, due to factors such as differences in experimental setup, with the corresponding systematic errors or inadequate data analysis, among others, it is often very difficult to reach definitive conclusions on the molecular dynamics for polymers based on the experimental data available. This is particularly true in the case of PDMS, for which exponents have been reported varying between -1 and -2 independently of the range of molecular weights studied.[3-9]

In this work, we discuss briefly some aspects of the molecular theory for diffusion in entangled polymer systems proposed by de Gennes,[20] and we review the studies on diffusion of PDMS chains in melts and in networks, not only considering the influence of free volume or polydispersity effects but also focusing attention on data analysis procedures. Since most of the data available for self-diffusion of PDMS has been obtained using NMR techniques, some considerations of the NMR methods and models used for analysis of the experimental results are included as well.

II. THEORY

Molecular motion in entangled systems has been analyzed by several authors[20,21] and very recently has been reviewed by Tirrell.[1] Although it is not the scope of this paper to review those theories, we will focus attention on some aspects of the "reptation" theory developed by de Gennes.[20] This theory imagines the polymer chain trapped in a "tube" formed by transient topological constraints: the so-called "transient network". On a short-time scale (i.e., less than the tube-renewal time) the polymer chain is confined within the tube. The chains may diffuse along the tube length via worm-like motions (reptation). To analyze the possible motions of a molecule in this system, it is assumed that the length of the chain is very long compared to the distance between crosslinks in the network and its motion is associated with the propagation of kink-type "defects" along the chain.

Reptation theory predicts that the self-diffusion coefficient D_s varies with the molecular weight M according to

$$D_s(M) \propto M^{-2} . \tag{1}$$

If the polymer chains are relatively short it is not possible to have an entangled regime. The process is controlled by the Brownian motion of the molecules diffusing in a Rouse-like fashion.[22] In this case,

$$D_s(M) \propto M^{-1} . \tag{2}$$

Considerable work has been done to test the validity of Eq. (1).[1] Although it has been confirmed in some systems[23-27] there are other systems in which this relationship is not satisfied.[4-6,28-32] The observed exponents were found to be between -1 and -2. These discrepancies have been attributed to several factors, among others, relating to non-entangled behavior, free volume and polydispersity effects.[2]

Some studies done on the onset of entangled behavior suggest that[33,34]

$$D_s(M,H) \propto M^{-2}H^0, \quad (H \gg M_{cr}) \tag{3}$$

and

$$D_s(M,H) \propto M^{-1}H^{-1}, \quad (H \leq M_{cr}) \tag{4}$$

where H represents the molecular weight of the host (matrix) in which the molecules of molecular weight M diffuse and M_{cr} is the molecular weight for the onset of entangled behavior. In Table 1 is shown the reported values for the molecular weight between entanglements, $M_e \simeq M_{cr}/2$, of some polymers of interest.[35] In cases of self-diffusion $H = M$ and Eq. (1) is obtained. Experimental results on polystyrene[24] satisfy this equation above and below M_{cr}. However, exponents between -1.7 and -1.5 have been reported in PDMS[5,6] over a wide range of molecular weights and in polyethylene[28] with molecular weights below 6,000 g-mol^{-1}. Also, these results are not in agreement with the expected behavior for a non-entangled regime described by Eqs. (2) and (4).

Free volume effects have been considered in studies done on polyisoprene[29] and polybutadiene.[31] It has been shown that the inclusion of corrections for changes in free volume with molecular weight in calculating diffusion coefficients brings closer agreement between theory and experiments.

The change in monomeric friction for a polymer chain in a melt will arise from differences in the fractional free folume, f. The fraction of free volume f in a polymer melt of molecular weight M at a temperature T varies approximately according to[35,36]

Table 1. Molecular Weights Between Entanglements, M_e

Polymer	M_e(g-mol^{-1})[a]	Temperature (K)
Polyethylene	1750[b]	----
Polystyrene	18100	433
Polydimethylsiloxane	8100	298
1,4-Polybutadiene (96% cis)	2950	298
Polyisobutylene	8900	298

[a] from reference 35 [b] from reference 32

$$f(M,T) = f_g + \Delta\alpha(T - T_{g\infty}) + (V_e \rho/M), \ (T > T_g) \tag{5}$$

where f_g is the free volume fraction at the glass transition temperature, T_g. $T_{g,\infty}$ is the value of T_g for $M = \infty$, $\Delta\alpha$ is the thermal expansivity of the free volume (estimated from the difference in measured bulk volume expansivities above and below T_g), ρ is the density of the melt, and V_e represents the free volume contribution of the ends of the polymer molecule. An increase in the molecular weight of the diffusant or the temperature at which the experiment is carried out will reduce the variation of free volume fraction with the molecular weight.[26]

The third factor cited to explain deviations from Eq. (1) is the polydispersity of the sample. Different models have been proposed to analyze the influence of polydispersity on measured diffusion coefficients.[37-41] One such model for the NMR experiments is considered in the next section.

III. EXPERIMENTAL

Steady (SFG) and pulsed (PFG) field gradient NMR techniques have been widely used to study the self-diffusion of polymers in melts, dilute and semidilute solutions.[2] The NMR self-diffusion measurements are based on local differences in Larmor precession frequencies when applying a spatially varying magnetic field across a given volume. When diffusion is present, the mean-square displacement $<r^2>$ of a molecule in some direction, r, during a time t is determined by the macroscopic self-diffusion coefficient D_s of the chains, and is given by

$$<r^2> = 2D_s t , \tag{6}$$

resulting in an attenuation of the spin-echo amplitude, in addition to that from intrinsic nuclear magnetic relaxation (T_2).

In earlier studies on diffusion of PDMS,[4,5] the Carr-Purcell technique[42] was used with magnetic field gradients present at all times (SFG NMR). The measurement of small diffusion coefficients (e.g., in polymers) requires strong gradients which means that as gradient strength is increased wider resonance lines and narrower echo envelopes will be produced. This represents a decrease in the information available from the echo, an increase in the bandwidth of the detection system to observe the narrower echoes and a corresponding decrease in signal-to-noise. Additionally, B_1 may have to be increased in order to satisfy the requirement that $\gamma B_1/2\pi \gg \nu_e$, ν_e being the width of the resonance at half-height.

The PFG NMR technique was first described by Stejskal and Tanner,[43] and it offers the advantage of allowing the measurement of diffusion coefficients over a wider range of values and more precise definition of the time period over which diffusion is being measured. Assuming the diffusion to be time-independent, the echo attenuation for a single spectral component is given by

$$A(g)/A(0) = exp[-\gamma^2 D_s g^2 \delta^2(\Delta - \delta/3)] , \tag{7}$$

where $A(g)$ and $A(0)$ are the echo amplitudes with and without the field gradient pulses, respectively. The magnitude of the gradient is g, γ is the magnetogyric ratio of the nuclei being observed, δ and Δ are the lengths of the pulses and spacing between them as specified in Figure 1 respectively, and D_s is the self-diffusion coefficient.

In general, polymer samples are polydisperse and are expected to exhibit a distribution of diffusion coefficients. Previous studies on polydisperse melts have shown the influence of the molecular weight distribution on the echo attenuation.[30-32,37,41] In the model proposed by von Meerwall,[37] Eq. (7) is extended to a multicomponent system considering additivity of the NMR signals of the N components:

$$A(g)/A(0) = \sum_{i=1}^{N} h_i \, exp[-\gamma^2 D_s(M_i) g^2 \delta^2 (\Delta - \delta/3)] \; , \tag{8}$$

where $D_s(M_i)$ is the self-diffusion coefficient for the i-th component and

$$h_i = n w_i (M_i) exp[-2\tau / T_2(M_i)] \; , \tag{9}$$

with n being a normalization factor, $w_i(M_i)$ the fraction of nuclei at resonance contributed by the i-th component and $T_2(M_i)$ the spin-spin relaxation time of the same component. If it is possible to choose a value for the time between RF pulses in the spin-echo sequence $\tau \ll T_2(M_i)$, the distinction between h_i and w_i vanishes.

In order to apply this model, it is necessary to know the molecular weight distribution and the form of the molecular weight dependence of the individual diffusion coefficients. If for any reason τ can not be chosen sufficiently short, it would be a requirement to measure the molecular weight dependence of the individual transverse relaxation times, $T_2(M_i)$. Generally, the molecular weight distribution can be easily determined by gel permeation chromatography (GPC) or be numerically generated using standard models (e.g., log-normal distribution). Typically, ten equispaced values of weight fraction w_i are enough to characterize the molecular weight distribution.[37]

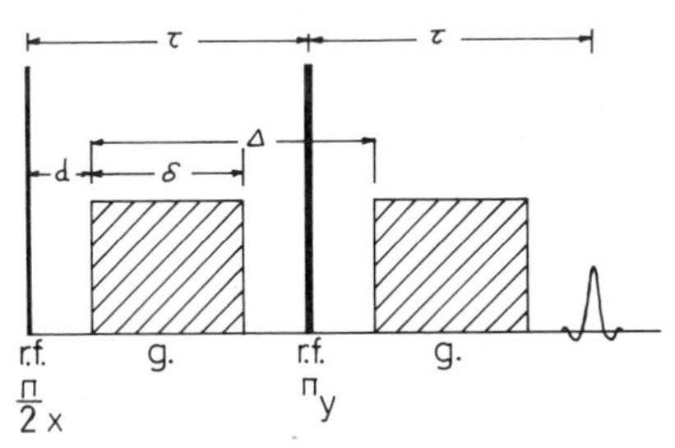

Fig. 1. The RF and magnetic field gradient pulse sequence used in references 3 and 6.

For the relaxation times a simple model has been applied[30,31,44]

$$T_2(M) \propto M^{-\beta}\ , \tag{10}$$

with β depending upon the polymer. Accepting the validity of this model and assuming that $D_s(M_i)$ obeys the predicted scaling law, the echo attenuation can be fitted to the experimental data allowing the exponent of molecular weight α (as in Eqs. (3) and (4)) to be a variable parameter (not restricted to either -1 or -2) in the power law

$$D_s(M_i) = D_s(M_{Ref})(M_i / M_{Ref})^{\alpha}\ . \tag{11}$$

Then, the diffusion coefficient can be obtained at some reference molecular weight M_{Ref} (typically M_n). An alternative fitting procedure has been described in previous work.[3]

Although there is a difference between the exponent in Eq. 11 used to fit the spin-echo attenuation data and the exponent obtained from plots of D_s vs. M_n, for consistency we used the same symbol α regardless of the procedure employed to determine its value. In the limit of a monodisperse sample with $M \geq M_{cr}$ both exponents should have the same value and obey the scaling law (Eq. (1)).

VI. DIFFUSION OF PDMS IN MELTS

Measurements of the self-diffusion coefficients of PDMS in melts have been reported covering a wide range of molecular weights,[3-6] from oligomers starting with an M_n = 88 g-mol^{-1} up to M_n of 10^5 g-mol^{-1}. The results are summarized in Table 2. The third column shows how different are the values obtained for the exponent in the power law that relates D_s and M making it very difficult to reach a definitive conclusion. However, our analysis (shown below) of the literature data reveals the source of these differences.

Table 2. Values for α in PDMS Melts

Reference	M_n Range (g-mol^{-1})	Reported	Linear Fit	Exponential Fit
5	180-32000	-1.55	----	----
5 (low M_n)	180-790	----	-1.44	-1.18
5 (high M_n)	15000; 32000	----	----	-1.90
6	150-105000	-1.70	----	----
6 (high M_n)	4200-105000	----	-1.59	-1.70
3	4200-12000	-1.96	-1.86	-1.96
3 and 6	4200-105000	----	-1.66	-2.02

At this point, several considerations have to be taken into account. First, two techniques, SFG and PFG NMR, are used to get those measurements. Both use the same principle but McCall, et al.[5] did some of their experiments at the lower limit of measurable diffusivity using SFG NMR with increased uncertainties in their measurements. Second, in the case of PDMS melts, the effect of polydispersity upon the echo attenuation and its interpretation is repeatedly pointed out but only Garrido, et al.[3] explicitly included its effect in the numerical analysis of the data. In this case, no significant effect of polydispersity was detected, at least within the range of echo attenuations observed. Third, as mentioned earlier, free volume effects cannot be ignored in explaining diffusion in polymer melts when their M is smaller than M_{cr}. However, for PDMS chains with $M_n \geq 3,000$ g-mol^{-1} the contribution of the third term in Eq. (5) to free volume fraction would be relatively small (less than 1%) at the experimental conditions at which the diffusion measurements have been performed (150 K above its T_g).[35,36]

Fourth, there is a significant influence of the type of function chosen to fit D_s vs. M upon the value for the exponent α. Fitting the data to the exponential form of this equation gives values higher in magnitude than fitting to the linear form. This is due to the variation of the statistical contribution of each experimental point to the fit from one type of function or another. Considering the expected behavior of D_s vs. M, it seems obvious to choose a fit to an exponential function. Table 2 shows in columns 4 and 5 the calculated values for α using a linear fit $y = a + bx$ and an exponential fit $y = ax^b$, respectively.

Our exponential fit of the results of McCall, et al.[5] shows close agreement with the expected exponents for both the low (α = -1) and the high (α = -2) molecular weight regimes. The value for the low molecular weight exponent is slightly higher than that expected for Rouse-like behavior, α = - 1.0. A better agreement might be obtained if free volume effects are considered because of their importance at low molecular weight. For the high molecular weights it is not possible to perform a statistical analysis due to the small number of experimental points. However, with the two points available we find by direct calculation a value for α of -1.9 in relatively good agreement with the theoretical predictions for diffusion in entangled systems.

In the case of Tanner, et al.,[6] the exponential fit of their data improves the quality of agreement with the theory, obtaining α = -1.83. When the low molecular weight data is excluded from the fit, the value obtained for the exponent is larger than expected but only three points are left. Again, the exponential fit brings the results closer to the expected value. However, an excellent agreement is obtained when data from reference 3 is included in the fit, giving a value of -2.02 for α. Both data sets have been obtained using the same technique (PFG NMR). The Tanner, et al. results were extrapolated to 299K (experimental temperature in reference 3) using an Arrhenius type equation with an activation energy of 4.5 kcal mol^{-1}.[5]

Nevertheless, a value of α = -2 does not guarantee the presence of an entangled regime when $H = M$ as pointed out in previous section. Studying chain diffusion in a crosslinked matrix helps avoid the tube-renewal factor, the presence of permanent topological constraints is assured, and an independent measure of the diffusion coefficient for the free chain can be obtained but mechanisms in addition to reptation could still appear, i.e., fluctuations of network junctions.

VI. DIFFUSION OF PDMS IN MODEL NETWORKS

The use of "model" networks provides an independent control of the structure of the medium in which the diffusion will occur. Additionally, on long-time scales, the permanent constraints of the real network more closely resemble the assumed constraints of reptation theory than do the transient constraints in an entangled melt. Also, the use of a model network, as opposed to a randomly crosslinked network or the transient network of a melt, assures that properties such as crosslink densities are reasonably homogeneous. In our previous work[3] done on diffusion of PDMS chains in PDMS model networks a value of -1.3 was found for α. This value could indicate that the system network-diffusant has a behavior intermediate between free and entangled diffusion. However, the limited range of molecular weights for the diffusants (M_n between 3,100 and 6,200 g-mol^{-1}) and for the networks (molecular weight between crosslinks of 3,700 and 7,400 g-mol^{-1}) used does not allow us to reach more definitive conclusions.

It should be mentioned that free volume effects will reduce the value of α (be more negative) and unentangled diffusion ($\alpha = -1$) might appear as an intermediate behavior as well.

VI. CONCLUSIONS

The self-diffusion of PDMS chains in melts satisfied Eq. (1) according to the theoretical predictions of reptation theory to a degree of accuracy much higher than that suggested by a cursory examination of the literature. Very good agreement between theory and experiment may be obtained by properly separating low and high molecular weight results and using more robust numerical analysis. In PDMS, the scaling law exponent -2, predicted to occur when reptation is operative, appears in material with $M_n \sim 4200$ g-mol^{-1} (about one fourth the value expected for the onset of entangled behavior). The understanding of diffusional behavior of PDMS chains in model networks requires additional experimental work using polymer chains with M_n above M_{cr}. The use of cyclic PDMS chains might help to elucidate the mechanisms of molecular dynamics in those networks when compared with diffusion of linear chains in them.

NOMENCLATURE

$A(g)$	echo amplitude with field gradient on
$A(0)$	echo amplitude with field gradient off
α	exponent of M in the scaling law
$\Delta\alpha$	thermal expansivity of free volume
B_1	oscillating magnetic field
β	molecular weight exponent for T_2 dependency with M
γ	magnetogyric ratio
D_s	self-diffusion coefficient
$D_s(M_i)$	self-diffusion coeffieient for the i-th component
$D_s(M_{Ref})$	self-diffusion coefficient for the polymer molecule with molecular weight M_{Ref}
Δ	spacing between leading edges of field gradient pulses
δ	length of field gradient
f	free volume fraction
f_g	free volume fraction at T_g
g	magnitude of the field gradient
H	molecular weight of the matrix
M	molecular weight of the diffusant
M_n	molecular weight number average
M_{cr}	molecular weight for the onset of entanglements
M_e	molecular weight between entanglements

M_{Ref} reference molecular weight
n_2 normalization factor
$< r >$ mean-square displacement
ν_e width of resonance at half height
ρ density of the melt
T absolute temperature
T_1 spin-lattice relaxation time
$T_2(M_i)$ spin-spin relaxation time of the i-th component
T_g glass transition temperature
$T_{g\infty}$ glass transition temperature for $M = \infty$
t time
τ time between the RF pulses
V_e molecular end free volume
$w_i(M_i)$ weight fraction of the i-th component

REFERENCES

1. M. Tirrell, "Polymer Self-diffusion in Entangled Systems," Rubber Chem. Technol., 57, 523-556 (1984).
2. E.D. von Meerwall, "Pulsed and Steady Field Gradient NMR Diffusion Measurements in Polymers," Rubber Chem. Technol., 58, 527-560 (1984).
3. L. Garrido, J.E. Mark, J.L. Ackerman and R.A. Kinsey, "Studies of Self-diffusion of Poly(dimethylsiloxane) Chains in PDMS Model Networks," J. Polym. Sci., Polym. Phys. Ed., 26, 2367 (1988).
4. D.W. McCall, E.W. Anderson and C.M. Higgins, "Self-diffusion in Linear Dimethylsiloxanes," J. Chem. Phys., 34, 804-808 (1961).
5. D.W. McCall and C.M. Higgins, "Self-diffusion in Linear Polydimethyl Siloxane Liquids," Appl. Phys. Lett., 7, 153-154 (1965).
6. J.E. Tanner, "Diffusion in a Polymer Matrix," Macromolecules, 4, 748-750 (1971).
7. A.N. Gent and R.H. Tobias, "Diffusion and Equilibrium Swelling of Macromolecular Networks by Their Linear Homologs," J. Polym. Sci, Polym. Phys. Ed., 20, 2317-2327 (1982).
8. J.E. Mark and Z.M. Zhang, "The Use of Model Networks With Reptating Chains to Study Extraction Efficiencies," J. Polym. Sci., Polym. Phys. Ed., 21, 1971-1979 (1983).
9. L. Garrido and J.E. Mark, "Extraction and Sorption Studies Using Linear Polymer Chains and Model Networks," J. Polym. Sci., Polym. Phys. Ed., 23, 1933-1940 (1985).
10. W Noll, The Chemistry and Technology of Silicones, Academic Press, New York (1968).
11. W.J. Bobear, in: Rubber Technology (M. Morton, ed.), Van Nostrand Reinhold, New York (1973).
12. P.J. Flory and B. Erman, "Theory of Elasticity of Polymer Networks. 3," Macromolecules, 15, 800-806 (1982), and "Relationship Between Stress, Strain and Molecular Constitution of Polymer Networks. Comparison of Theory With Experiments," Macromolecules, 15, 806-811 (1982).
13. J.E. Mark, "The Use of Model Polymer Networks to Elucidate Molecular Aspects of Rubberlike Elasticity," Adv. Polym. Sci., 44, 1-26 (1982).
14. M. Gottlieb, C.W. Macosko, G.S. Benjamin, K.O. Meyers and E.W. Merrill, "Equilibrium Modulus of Model Poly(dimethylsiloxane) Networks," Macromolecules, 14, 1039-1046 (1981).
15. J.E. Mark, in: Elastomers and Rubber Elasticity, (J.E. Mark and J. Lal, eds.), Am. Chem. Soc., Washington, D.C. (1982).
16. J.P. Queslel and J.E. Mark, "Elasticity", in: Encyclopedia of Polymer Science and Engineering, 2nd ed., Vol. 5, pp. 365-407, Wiley, New York (1986).
17. M.Y. Tang, L. Garrido and J.E. Mark, "The Effect of Crosslink Functionality on the Elastomeric Properties of Bimodal Networks," Polym. Commun., 25, 347-350 (1984).

18. C.Y. Jiang, L. Garrido and J.E. Mark, "Dependence of Elastomeric Properties on Network Junction Functionality," J. Polym. Sci., Polym. Phys. Ed., 22, 2281-2284 (1984).
19. L. Garrido, J.E. Mark, S.J. Clarson and J.A. Semlyen, "Studies of Cyclic and Linear Poly(dimethylsiloxanes):17. Elastomeric Properties of Model PDMS Networks Containing Either Cyclics or Linear Chains as Diluents," Polym. Commun., 26, 55-57 (1985).
20. P.G. de Gennes, Scaling Concepts in Polymer Physics, Cornell University Press, Ithaca, New York (1979).
21. M. Doi and S.F. Edwards, "Dynamics of Concentrated Polymer Systems. I, II and III," J. Chem. Soc. Faraday Trans. 2, 74, 1789-1801, 1802-1817, 1818-1832 (1978).
22. P.E. Rouse, "A Theory of Linear Viscoelastic Properties of Dilute Solutions of Coiling Polymers," J. Chem. Phys., 21, 1272-1280 (1953).
23. J. Klein, "Evidence for Reptation in an Entangled Polymer Melt," Nature, 271, 143-145 (1978).
24. G. Fleischer, "Self-diffusion in Melts of Polystyrene and Polyethylene Measured by Pulsed Field Gradient NMR," Polym. Bull, 9, 152-158 (1983).
25. I. Zupancic, G. Lahajnar, R. Blinc, D.H. Reneker and D.L. VanderHat, "NMR Self-diffusion Study of Polyethylene and Paraffin Melts," J. Polym. Sci., Polym. Phys. Ed., 23, 387-404 (1985).
26. R. Bachus and R. Kimmich, "Molecular Weight and Temperature Dependence of Self-diffusion Coefficients in Polyethylene and Polystyrene Melts Investigated Using Modified n.m.r. Field Gradient Technique," Polymer, 24, 964-990 (1983).
27. P.J. Mills, P.F. Green, C.J. Palmstrøm, J.W. Mayer and E.J. Kramer, "Polydispersity Effects on Diffusion in Polymers: Concentration Profiles of d-polystyrene Measured by Forward Recoil Spectrometry," J. Polym. Sci., Polym. Phys. Ed., 24, 1-9 (1986).
28. D.W. McCall, D.C. Douglass and E.W. Anderson, "Diffusion in Ethylene Polymers. IV," J. Chem. Phys., 30, 771-773 (1959).
29. E.D. von Meerwall, J. Grigsby, D. Tomich and R. van Antwerp, "Effect of Chain-end Free Volume on the Diffusion of Oligomers," J. Polym. Sci., Polym. Phys. Ed., 20, 1037-1053 (1982).
30. E. von Meerwall and K.R. Bruno, "Pulsed-gradient Spin-echo Diffusion Study of Polydisperse Paraffin Mixtures," J. Magn. Reson., 62, 417-427 (1985).
31. E. von Meerwall and P. Palunas, "The Effects of Polydispersity, Entanglements and Crosslinks on Pulsed-gradient NMR Diffusion Experiments in Polymer Melts," J. Polym. Sci., Polym. Phys. Ed., 25, 1439-1457 (1987).
32. G. Fleischer, "The Chain Length Dependence of Self-diffusion in Melts of Polyethylene," Colloid Polym. Sci., 265, 89-95 (1987).
33. J. Klein, "The Onset of Entangled Behavior in Semidilute and Concentrated Polymer Solutions," Macromolecules, 11, 852-858 (1978).
34. J. Klein, "Effect of Matrix Molecular Weight on Diffusion of a Labeled Molecule in a Polymer Melt," Macromolecules, 14, 460-461 (1981).
35. J.D. Ferry, Viscoelastic Properties of Polymers, 2nd ed., Wiley, New York (1970).
36. F. Bueche, in: Physical Properties of Polymers, Interscience, New York (1962).
37. E.D. von Meerwall, "Interpreting Pulsed-gradient Spin-echo Diffusion Experiments in Polydisperse Specimens," J. Mag. Reson., 50, 409-416 (1982).
38. D.A. Bernard and J. Noolandi, "Polydispersity Variation and the Polymer Self-diffusion Exponent," Macromolecules, 16, 548-555 (1983).
39. P.T. Callaghan and D.N. Pinder, "A Pulsed Field Gradient NMR Study of Self-diffusion in a Polydisperse Polymer System: Dextran in Water," Macromolecules, 16, 968-973 (1983).

40. P.T. Callaghan and D.N. Pinder, "Influence of Polydispersity on Polymer Self-diffusion Measurements by Pulsed Field Gradient Nuclear Magnetic Resonance," Macromolecules, 18, 373-379 (1985).
41. G. Fleischer, D. Geschke, J. Kärger and W. Heink, "Peculiarities of Self-diffusion Studies on Polymer Systems by the NMR Pulsed Field Gradient Technique," J. Magn. Reson., 65, 429-443 (1985).
42. H.Y. Carr and E.M. Purcell, "Effects of Diffusion on Free Precession in Nuclear Magnetic Resonance Experiments," Phys. Rev., Ser. 2, 94, 630-638 (1954).
43. E.O. Stejskal and J.E. Tanner, "Spin-diffusion Measurements Spin Echos in the Presence of a Time-dependent Field Gradient," J. Chem. Phys., 42, 288-292 (1965).
44. D.W. McCall, D.C. Douglass and E.W. Anderson, "Nuclear Magnetic Relaxation in Polymer Melts and Solutions," J. Polym. Sci., 59, 301-315 (1962).

CHAIN SEGMENT ORDERING IN STRAINED RUBBERS

B. Deloche and P. Sotta

Labor. de Physique des Solides
(CRNS-LA2)
Université de Paris Sud
91405 - Orsay, France

J. Herz

Inst. Charles Sadron
(CRM-EAHP)
67083 Strasbourg, France

ABSTRACT

The orientational order generated in a polymer network by an uniaxial stress is probed with deuterium NMR. The experiments are performed either on polydimethylsiloxane (PDMS) network chains or on PDMS probe chains dissolved in the network. The stress-induced orientation observed on both kind of chains is explained by short-range, orientational correlations between chain segments.

I. INTRODUCTION

Recently great effort has been devoted to developing spectroscopic techniques (infrared dichroism, polarized fluorescence, nuclear magnetic resonance,...) sensitive to the microscopic chain behaviour in dense amorphous polymer systems. Indeed, most of the physical properties of rubber networks are determined by the configurations accessible to the chains. Macroscopically aligned systems, such as uniaxially deformed networks are helpful to investigate the interactions which determine the chain configuration on a local scale. In particular, it is essential to establish clearly whether the applied constraint transmitted through the crosslink junctions are the only mechanism for the stress-induced chain segment orientation. To clarify this point, we have used the deuteron nuclear magnetic resonance (^{2}H-NMR) for monitoring the orientational behaviour of both network chains[1] and linear chains dissolved inside the network;[2] the dissolved chains are chemically identical to the network chains and short enough to be considered as "free", i.e., devoid of entanglements with the network structure as schematized in Fig. 1.

II. EXPERIMENTAL

II.1 Samples Experiments were performed either on polydimethylsiloxane (PDMS) networks with a known fraction of deuterated crosslinked chains (O-Si$(CD_3)_2$)$_n$ or on perdeuterated linear PDMS chains dissolved

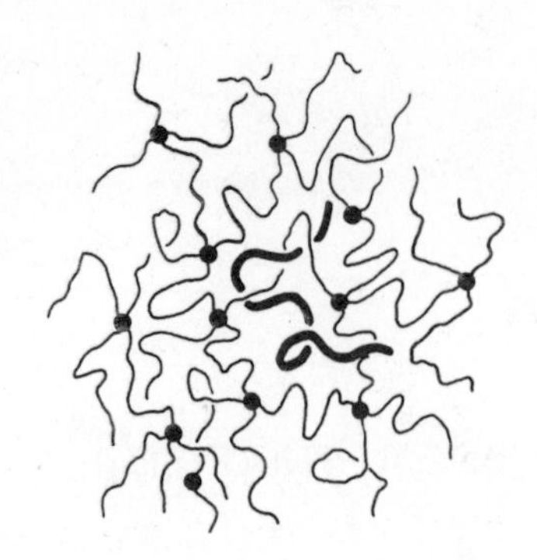

Fig. 1. Schematic diagram of a probe chain dissolved into a rubber matrix.

inside an unlabelled PDMS network. Tetrafunctional networks, prepared by an end-linking reaction in the bulk, were used. The labeling procedure and the network synthesis are outlined in refs. 3 and 4. The linear chains, used as NMR probes, have either the same molecular weight as the network chains (M_n = 10500) or a lower molecular weight (M_n = 6000; 3000; 450). In each case the molecular weight distribution M_w/M_n is about 1.6. Let us note that the molecular weight of the linear chains is lower than the average molecular weight between entanglements (M_c) in the melt of such materials (M_c~27000 for PDMS).

II.2 Swelling process A drop of liquid linear PDMS chains (Tg=-120°C) was spread on the surface of a dry network film at room temperature. The PDMS molten chains were short enough to be perfectly compatible with the host matrix,[5] and free to diffuse through the network, with a diffusion coefficient of about $10^{-12}m^2s^{-1}$.[6] After several days the chains had totally penetrated into the matrix and the surface of the film appeared dry. The weight fraction of free chains introduced into the network was about 10%. Additional neutron-scattering experiments performed on a relaxed sample show that no effect of demixing or inhomogeneities occurred in such homopolymer solutions for that degree of swelling.[7] Let us note also that the deuterated probe chains can be removed from the matrix: no ^{2}H-NMR signal was detected after an immersion of the film in a good solvent for 48 hours. This indicates that no permanent links occurred between the probe chains and the matrix.

II.3 Stretching device Sample elongation was performed as described earlier.[1] Both ends of the sample were gripped by jaws. One of these jaws was fixed while a calibrated screw moved the other along the NMR tube. The sample elongation was monitored before and after each NMR experiment with a micrometer on a microscope stage. In this fashion, the elongation ratio $\lambda = L/L_0$ (where L and L_0 were the lengths of the (swollen) network elongated and relaxed respectively) was measured with an accuracy of about 0.1%.

II.4 NMR conditions A CXP-90 BRUKER spectrometer was used, operating at 12 MHz with a conventional electromagnet (1.8 Tesla), locked by an external proton probe. The magnetic field was normal to the uniaxially applied constraint. The temperature was regulated at 293 K (± 1 K). The spectra were obtained by Fourier transforming averaged free induction decays.

II.5 ^{2}H-NMR background Fast, anisotropic reorientations of a C-D bond lead to a quadrupolar interaction which is no longer averaged to zero.[8] When motions are uniaxial around a macroscopic symmetry axis, such a residual interaction splits the liquid-like NMR line into a doublet whose spacing is in frequency units:

$$\Delta \nu = \frac{3}{2} \nu_q \left| P_2\left(\cos \Omega\right)\right| \quad <P_2\left(\cos \theta\ (t)\right)> , \qquad (1)$$

where ν_q denotes the static quadrupolar coupling constant (ν_q~200 kHZ). The angles in the second Legendre polynominals characterize the experimental geometry and the molecular dynamics: Ω is the angle between the spectrometer magnetic field and the sample symmetry axis; namely, the axis of the applied constraint ($\Omega = 90°$ in Eq. [1], so that $| P_2(\cos\Omega) | = \frac{1}{2}$); $\theta(t)$ is the instantaneous angle between the C-D bond and the constraint axis. The brackets indicate an average over the motions faster than the characteristic ^{2}H-NMR time, ν_q^{-1}. Then $<P_2(\cos\theta)>$ is the orientational order parameter of the C-D bond with respect to the symmetry axis of the sample.

III. RESULTS

III.1 Induced Orientational Order

III.1.1 Network chains Figure 2 shows the change in the ^{2}H-NMR spectra of the network chains on deforming the samples. Single doublets appear when the sample is either uniaxially compressed ($\lambda < 1$) or elongated

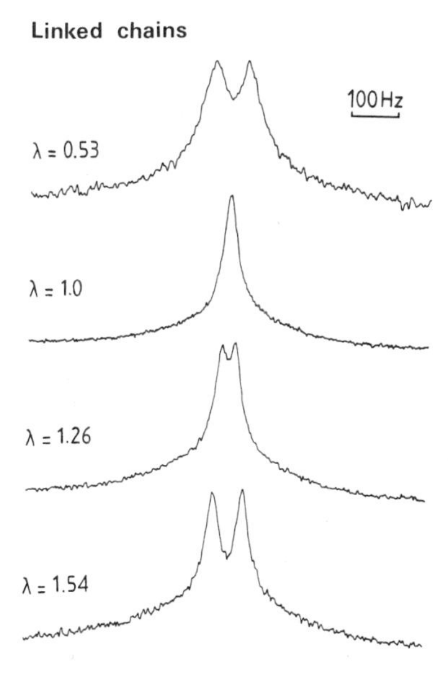

Fig. 2. ^{2}H-NMR spectra (13 MHz) of perdeuterated crosslinked chains of a dry PDMS network, uniaxially compressed ($\lambda < 1$) or elongated ($\lambda > 1$) along a direction perpendicular to the spectrometer magnetic field.

($\lambda > 1$). Additional experiments performed at various angles Ω for a fixed λ show that the angular dependence of the observed splitting $\Delta\nu(\Omega)$ reproduces exactly the $|P_2(\cos\Omega)|$-curve given by relation [1], as indicated in Fig. 3. These results prove that the observed quadrupolar interactions are averaged along the stress axis by all molecular motions(methyl group rotation, chain isomerization and larger-scale reorientation) that are fast on the ^{2}H-NMR time scale ($\tau\nu_q<1$). In other words, the stress axis is a symmetry axis for the reorientational motions of the chain segments which give rise to the doublet structure. The corresponding orientational order parameter $S = <P_2(\cos\theta)>$ can be easily deduced from Eq. [1] exactly as it has been done in reference (9): for instance, at $\lambda = 1.30$ the order parameter of the methyl symmetry axis (i.e., the Si-CD_3 bond) is 1.3×10^{-3} and that of the segment connecting two adjacent oxygen atoms along the PDMS chain skeleton is 2.6×10^{-3}.

III.1.2 Free Chains Figure 4 shows that the spectrum in the relaxed state ($\lambda = 1$) is a single line characteristic of a liquid: fast isotropic motions totally average the quadrupolar interactions. Surprisingly, just as in the case of crosslinked chains discussed above, well-defined doublets appear when the sample is uniaxially stretched. Moreover, for a given deformation rate $\lambda \neq 1$, the splitting $\Delta\nu$ remains constant over a time span larger than a few 10^5 sec, so that we may consider the system to be in equilibrium state. Lastly, the splitting variation $\Delta\nu(\Omega)$ reproduces hereto the $|P_2(\cos\Omega)|$-curve. This means that as the probe chain diffuses through the deformed network its segments undergo uniaxial reorientations around the stress direction on the NMR-time scale and acquire an orientational order. This induced order is practically independent of the free chain length ($400<M_n<10500$) for a given deformation of the matrix.(2)

III.1.3 Crosslinked and uncrosslinked chain segment orders A relevant comparison of the segmental order induced on the two kinds of chain required to use two identical PDMS networks (labelled and unlabelled) swollen at the same degree of swelling ($\Phi\sim10\%$) with unlabelled and labelled free PDMS chains respectively. Figure 5 shows the splittings of the crosslinked ($\Delta\nu_c$) and uncrosslinked ($\Delta\nu_u$) chains (measured at maximum height) versus $\lambda^2-\lambda^{-1}$. Both the host matrix and the guest probe exhibit the same strain dependence and reflect the same degree of order: at the same λ the ratio $\Delta\nu_c/\Delta\nu_u$ is close to one.

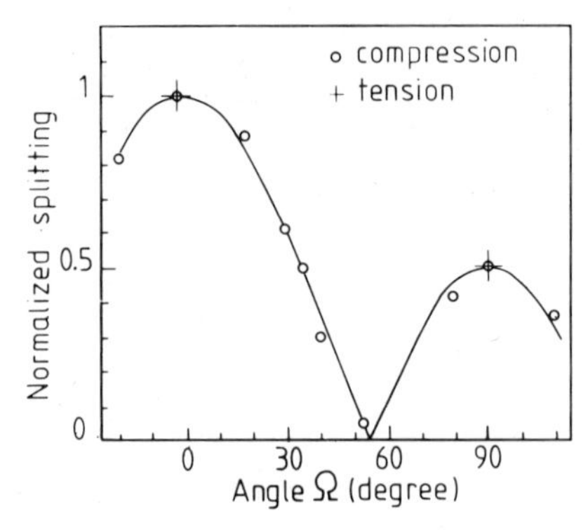

Fig. 3. Variation of the normalized splitting as a function of the angle Ω between the applied uniaxial constraint and the spectrometer magnetic field. The continuous line represents the function $|P_2(\cos\Omega)|$.

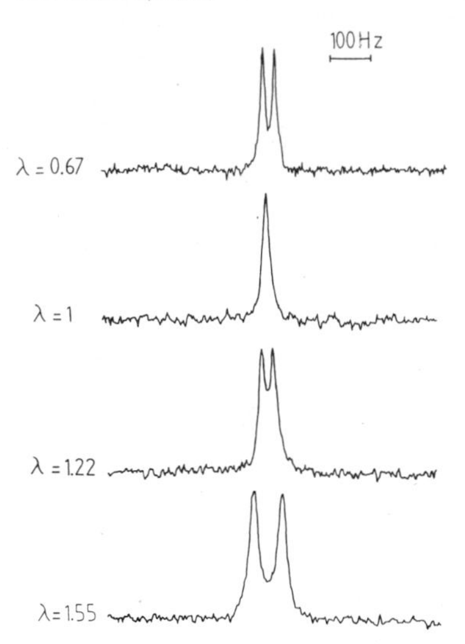

Fig. 4. 13 MHz ^{2}H-NMR spectra of perdeuterated free chains of polydimethylsiloxane (PDMS) dispersed in a PDMS network which is uniaxially compressed ($\lambda < 1$) or elongated ($\lambda > 1$) along a direction perpendicular to the spectrometer magnetic field ($\Omega = 90°$ in Eq. [1] of the text). Both probe chains and crosslinked chains have the same size (M_n = 10500). The volume fraction of free chains is 8%.

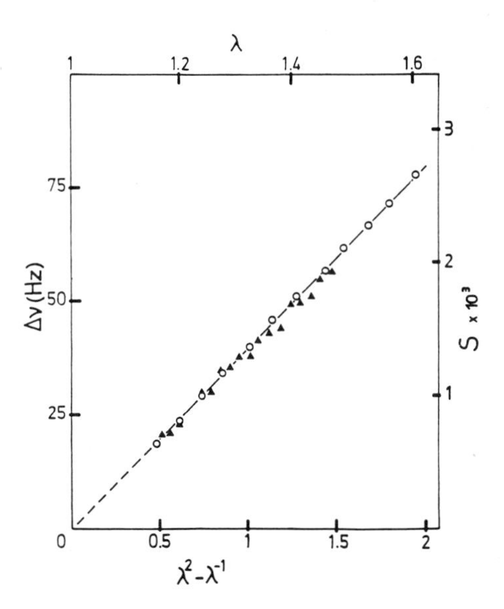

Fig. 5. Probe chain (circles O) and network chain (triangles Δ) quadrupolar splittings as a function of ($\lambda^2-\lambda^{-1}$). The free chain volume fraction is 8%. On the right vertical ordinate is the calculated order parameter S for the Si-CD_3 bonds of the PDMS chains.

Despite these similarities, some differences between the two kinds of chains are also noticeable. By contrasting the spectra of Figs. 2 and 4, it is apparent that those corresponding to the network chains are broadened and exhibit spectral wings (which do not disappear totally at the magic angle $\Omega = 55°$ for $\lambda \neq 1$). These wings are attributed to deuterons whose dynamical behavior is different from that of deuterons associated with the doublet structure (in particular, a non-uniaxial behavior relative to the stress axis on the ^{2}H-NMR time scale) and obviously are related to the presence of crosslinking junctions.

IV. DISCUSSION

The fact that the segments of probe chains acquire a permanent orientational order although the extremities of the chains are not linked to the network is striking. This result attenuates somewhat the role of the crosslink junctions in the stress-induced orientation process. Obviously the network junctions are essential to establish a permanent deformation of the rubber matrix; however, the induced uniaxial order observed on free chains necessarily involves cooperative orientational correlations between chain segments. Such short-range correlations are strong enough to prevent the free chains from relaxing to an isotropic equilibrium configuration in the deformed network. Anisotropic excluded volume effects due to the local confinement of polymer chains probably dominate these correlations; in any case they have to be distinguished from the purely intramolecular interactions among nearest-neighbour chain segments dictated by dihedral angle energetics.[10]

These orientational correlations also play a major role in the orientation process of crosslinked chains. According to the results reported in Fig. 5, it appears clear that most of the orientational anisotropy as manifested by the doublet structure in the ^{2}H-NMR line is attributed to interchain effects rather than the conventional end-to-end stretching of isolated chains. Indeed, as it seems reasonable to assume that both contributions are operative in elastic network chains we would have anticipated much larger splittings for such chains; but this is not borne out by the experiments. From that point of view both kinds of chains, crosslinked and uncrosslinked, appear to be locally indistinguishable.

The cooperative orientational correlations between chain segments tend to align the chains in the direction of the applied force, as suggested in Fig. 6, and then are the source of the dynamics uniaxiality observed on the segments of crosslinked chains (Fig. 3). Such a property which is typical of an uniaxial fluid is contrary to various classical descriptions of rubber elasticity wherein the segmental orientation is presumed to arise solely from independent chains which are stretched by their extremities. Indeed, these descriptions lead to a segmental anisotropy along the end-to-end vector of each chain which is fixed to the network and in that case the macroscopic uniaxality arises from an uniaxial static distribution of these end-to-end vectors relative to the strain direction,[11] as schematized in Fig. 6.

A system exhibiting short-range orientational correlations among interacting species requires to modify the usual pair interaction parameters χ_{ij} (i = network chain segment, j = free chain segment or solvent molecule). An expansion to the second-order correction leads

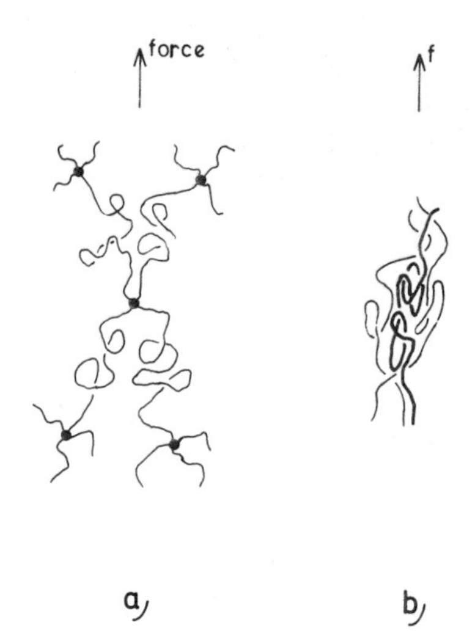

Fig. 6. Schematic diagram of:
a) independent chains in a uniaxially strained network.
b) one chain interacting with the uniaxial environment created by the neighbouring chains.

to effective interaction parameters, $\chi_{ij} + \chi'_{ij}S_iS_j$, which are dependent on the relative orientation of the species under consideration (the parameter S is defined in Section III). Therefore the Flory interaction parameter χ_0, a combination of the χ_{ij}, is anisotropic in presence of orientational correlations.[13] Including such corrections into the free energy of the network[14] and introducing a coupling between the orientation S and the deformation λ, as it has been proposed by de Gennes in case of network composed of nematogenic polymers,[15] allow one to discuss our results[2][12][13] and to analyse the effects of orientational correlations on rubber elasticity. In particular, it appears clear that the presence of orientational correlations which favour the chain alignment along the strain direction reduce the chain stiffness;[16] consequently the applied stress is attenuated relative to that derived from the classical description of rubber elasticity.

V. CONCLUSIONS

The above results show that the stress-induced chain orientation in dense polymer networks is not a single-chain process but is dominated by a cooperative effect of orientational couplings between chain segments. The physical origin of such couplings is probably related to anisotropic repulsive interactions between linear chains. Then the role of network chains in rubber elasticity would not be restricted to their sole actions on the crosslink junctions, the chains between junctions being devoid of any material proberties. In fact, effects arising from the physical volume occupied by chains have been considered in rubber elasticity recently.[17]

NOMENCLATURE

A instantaneous angle between the C-D bond and the constraint axis

i network chain segment
j free chain segment
λ elongation ratio
L length of (swollen) network elongated
L_o length of (swollen) network relaxed
M_n number-average molecular weight
q static quadrupolar coupling constant
S orientational order parameter
Φ degree of swelling
χ interaction parameter
Ω the angle between the spectrometer magnetic field and the sample symmetry axis

REFERENCES

1. B. Deloche, M. Beltzung and J. Herz, J. Phys. Lett., 43, 763 (1982).
2. P. Sotta, B. Deloche, J. Herz, A. Lapp, D. Durand and J.C. Rabadeux, Macromoleclues, 20, 2769 (1987).
3. J. Herz, A. Belkebir and P. Rempp, Eur. Polym. J., 9, 1165 (1973).
4. M. Beltzung, C. Picot, P. Rempp and J. Herz, Macromolecules, 15, 1594 (1982).
5. A. Lapp, C. Picot, H. Benoît, Macromolecules, 18, 2437 (1985).
6. L. Garrido, J.E. Mark, S.J. Clarson and J.A. Semlyen, Polymer, 25, 218 (1984).
7. F. Boué, B. Farnoux, J. Bastide, A. Lapp, J. Herz and C. Picot, Europhys. Lett., 1, 637 (1986).
8. M.H. Cohen and F. Reiff, Solid State Phys., 5, 321 (1975).
9. H. Toriumi, B. Deloche, E.T. Samulski and J. Herz, Macromolecules, 18, 305 (1985).
10. P.J. Flory, Statistical Mechanics of Chain Molecules, Interscience, New York (1969).
11. R.J. Roe and W.R. Krigbaum, J. Appl. Phys., 35, 2215 (1964).
12. B. Deloche and E.T. Samulski, Macromolecules, 14, 575, (1981).
13. P. Sotta, B. Deloche and J. Herz, Polymer, 29, 1171 (1988).
14. P.G. de Gennes, Scaling Concepts in Polymer Physics, Cornell University Press, Ithaca, N.Y. (1979).
15. P.G. de Gennes, C.R. Acad. Sci., (Paris) B281, 101 (1975).
16. B. Deloche and E.T. Samulski, to be published in Macromolecules (1988).
17. R.J. Gaylord and J.F. Douglas, Polymer Bull., 18, 347 (1987).

VISCOELASTICITY AND SELF-DIFFUSION IN MISCIBLE HETEROPOLYMER BLENDS

Christos Tsenoglou

Department of Chemistry and Chemical Engineering
Stevens Institute of Technology
Hoboken, New Jersey 07030, USA

ABSTRACT

A theory has been developed for the viscoelasticity and diffusion of miscible fluid blends of entangled linear polymers with dissimilar chemical structure. The architecture and molecular dynamics of the temporary network formed by the intermeshing species are analyzed by assuming that the coupling frequency between two macromolecular components along a single chain is proportional to the fractional participation of these components in the blend as a whole. The population densities and lifetimes of the various types of entanglements among similar and dissimilar polymers, and the frictional properties of the interacting chains are related to the composition and the corresponding properties of the unmixed polymeric precursors. This information is used to derive mixing laws for the time-dependent linear relaxation modulus, the rubbery plateau modulus, the zero-shear viscosity, the recoverable compliance, and the self-diffusion coefficients in a binary heteropoloymer blend.

I. INTRODUCTION

In an effort to develop novel materials for technological applications, blending already available polymers is often a preferable alternative to chemically synthesizing new ones.[1] Optimal combination of the properties of the individual components is usually accomplished when heteropolymer mixing occurs on a molecular level.[2] There is an increasing number of exceptions to the rule that dissimilar polymers are immiscible. It mostly comprises of pairs of macromolecular species able to form athermal or exothermic blends due to the presence of attractive interactions between chemical groups they possess.[3-5] The present work attempts to develop methods of predicting the mechanical and diffusional properties of these homogeneous fluids. An appropriate way to introduce this subject is by reviewing the current understanding on the dynamics of one-component polymer systems.

The rheological behavior of dense macromolecular fluids depends on the structure of their polymeric network and the mobility of the individual chains in this network.[6,7] The material rigidity, for example, is proportional to the entanglement density, and the fluidity is affected by the rate of renewal of the molecular configurations. The fundamental property describing the viscoelasticity of flexible polymers under small deformations is the time-dependent, stress relaxation modulus, $G(t)$. It is defined as the ratio of stress over strain under a constant deformation. In order to predict $G_i(t)$ for a one-component monodisperse melt of a polymer (i) in the entanglement regime, the classical theory of rubber elasticity has been adopted for the case of a temporary network.[8-10] It is assumed that the stress value at time t after a sudden strain imposition (at t = 0) depends on the number density of the original primitive steps (i.e., sub-segments lying between two successive entanglements) surviving at this time.

$$G_i(t) = kT\nu_i N_i(t) = (RT\rho_i/M_{ei})F_i(t) \ , \tag{1}$$

where ν_i is the number of i chains per unit volume,

$$\nu_i = \rho_i v_i N_A/M_i \ , \tag{2}$$

k is the Boltzmann constant, T the temperature (in K), $N_i(t)$ the number of surviving steps per chain at the time t ($N_i(0) = M_i/M_{ei}$), M_i the molecular weight of the polymer, M_{ei} the molecular weight between entanglements, ρ_i the density, v_i the polymer volume fraction (v_i = 1 for a melt), R the gas constant, N_A the Avogadro number, and $F_i(t)$ the relaxation function of an individual molecule in the melt ($F_i(t) \equiv N_i(t)/N_i(0) \leq 1$). For linear chains the de Gennes-Doi-Edwards theory[8-10] predicts:

$$F_i(t) = F(t/\tau_i) \simeq exp(-t/\tau_i) \ , \tag{3}$$

where τ_i is the characteristic relaxation time,

$$\tau_i = \frac{<R_i^2>}{36D_i} \simeq \frac{N_i(0)<R_i^2>M_i\zeta_i}{12kTm_i} \ , \tag{4}$$

where $<R_i^2>$ is the end-to-end distance of the chain ($<R_i^2> \sim M_i/m_i$), D_i the self-diffusion coefficient, ζ_i the monomeric friction coefficient and m_i the molecular weight of the monomeric unit. It should be reminded that the ratio of ζ_i at a given temperature T over the corresponding value, ζ_{gi}, at the glass transition temperature (T_{gi}) is a decreasing function of the fractional free volume, f_i:[6]

$$ln(\zeta_i T_{gi}/\zeta_{gi} T) = ln(a_i(T)) = 1/f_i - 1/f_{gi} \ , \tag{5}$$

where f_{gi} is the free volume fraction at T_{gi}:

$$f_i = f_{gi} + \alpha_i(T - T_{gi}) \ , \tag{6}$$

α_i is the free-volume thermal expansion coefficient and a_i the necessary shift factor for accomplishing a time-temperature superposition of non-isothermal G_i (t) responses onto T_g curve.

II. NETWORK STRUCTURE AND MOLECULAR DYNAMICS IN BLENDS

In a miscible heteropolymer blend, the lifetimes of the various entanglements depend on the combined length and mobility of the two chains from which they are formed. Consider a composite network formed by two dissimilar species (i,j = 1,2) each of which participates with a volume fraction v_i $(v_1 + v_2 = 1)$. Let the index ij signify a segment lying on a i polymer chain and defined by two successive entanglements with j chains; N_{ij} is the number of such steps, and N_{Bi} the total number of steps per i chain in the blend (Fig. 1):

$$N_{Bi}(t) = N_{ii}(t) + N_{ij}(t) \ . \tag{7}$$

The fraction of entanglements along an i chain caused by associations with j chains should be equal to the fractional participation of the j steps in the total step population:

$$N_{ij}(t)/N_{Bi}(t) = \nu_j N_{Bj}(t)/[\nu_i N_{Bi}(t) + \nu_j N_{Bj}(t)] \ . \tag{8}$$

Provided that the presence of a dissimilar species does not perturb the ability of a polymer to entangle, it is reasonable to stipulate that the number of associations between similar chains decreases linearly with the volume fraction:

$$N_{ii}(t) = v_i(M_i/M_{ei})F(t/\tau_{Bi}) \ . \tag{9}$$

where τ_{Bi} is the relaxation time of an i chain in the mixture. From Eqs. 1,2 and 7-9 it can be shown that:

$$N_{ij}(t) = v_j M_i \left(\frac{\rho_j/\rho_i}{M_{ei} M_{ej}} \right)^{\frac{1}{2}} \left[F(t/\tau_{Bi}) F(t/\tau_{Bj}) \right]^{\frac{1}{2}} \ . \tag{10}$$

In a homopolymer blend, where the i and j species have identical chemical structure but dissimilar size, and where reptation is the prevailing mode of macromolecular relaxation, it is to be expected that τ_{Bi} and τ_i are equal. This is not the case in heteropolymer blends due to modifications of (a) the entanglement topology and (b) the frictional resistance encountered by an i chain after mixing.

According to Eq. 4:

$$\frac{\tau_{Bi}}{\tau_i} \equiv \delta_i = \frac{N_{Bi}(0)}{N_i(0)} \ \frac{\zeta_{Bi}}{\zeta_i} \ . \tag{11}$$

The first of these ratios can readily be evaluated from Eqs. (7-10). At the glass transition temperature of the blend ($T_{gB} \neq T_{gi}$) the monomeric friction coefficient of a blended i chain, ζ_{Bi}, is presumed equal to the weighted average of the individual ζ_{gi}; at higher temperatures it decreases due to an increasing free volume fraction, f_B, as demonstrated in Eq. 5:

$$\frac{\zeta_{Bi}}{\zeta_i} = \frac{v_i\zeta_{gi} + v_j\sqrt{\zeta_{gi}\zeta_{gj}}}{\zeta_{gi}}\left(\frac{T_{gi}}{T_{gB}}\right)exp\left(\frac{1}{f_B} - \frac{1}{f_i}\right). \tag{12}$$

In deriving Eq. 12, it is also assumed that f_g is equivalent for both species. This assumption is somewhat simplified because glass transition is known to be non-isofree volume. Though linear free-volume additively is not generally true because miscibility requires specific interactions which, in turn, can lead to the change of f_B. If we stipulate linear free-volume additivity between components ($f_B = v_i f_i + v_j f_j$)[6,11], and combine Eqs. (11) and (12), we can show that:

$$\frac{\tau_{Bi}}{\tau_i} \equiv \delta_i = \left[v_i + v_j\left(\frac{\rho_j M_{ei}}{\rho_i M_{ej}}\right)^{\frac{1}{2}}\right]\left[v_i + v_j\left(\frac{\zeta_{gj}}{\zeta_{gi}}\right)^{\frac{1}{2}}\right]\left[\frac{T_{gi}(v_i a_i + v_j a_j)}{v_i a_i T_{gi} + v_j a_j T_{gj}}\right]\left(\frac{a_B}{a_i}\right), \tag{13a}$$

where $a_B(T)$, the time-temperature superposition factor for the blend, can be obtained from the following expression:

$$ln\left(\frac{a_B}{a_i}\right) = \frac{v_j(f_i - f_j)}{f_i(v_i f_i + v_j f_j)} = \frac{(1 - v_i)(1 + f_g\, ln a_i)}{1 + f_g(v_i\, ln a_j + v_j\, ln a_i)}\, ln\left(\frac{a_j}{a_i}\right). \tag{13b}$$

It is clear that, due to its exponential nature, this last ratio is the most significant factor of the δ_i variation and thus, in a first approximation, $\delta_i \simeq a_B/a_i$.

The shear stress relaxation modulus of the blend, $G_B(t)$, may then be evaluated following the same reasoning which leads to Eq. (1):

$$G_B(t) = kT(\nu_1 N_{B1}(t) + \nu_2 N_{B2}(t)) \Rightarrow$$

$$\sqrt{G_B(t)} = v_1\sqrt{G_1}\left(\frac{t}{\tau_1\delta_1}\right) + v_2\sqrt{G_2}\left(\frac{t}{\tau_2\delta_2}\right). \tag{14}$$

The preceding analysis can also be used to estimate D_{Bi}, the self-diffusion coefficient of a chain of the i species in the blend. Provided that reptation is the dominant mechanism of molecular motion, Eqs. 4 and 11 suggest that:

$$D_{Bi} = \frac{D_i}{\delta_i} \simeq \frac{kTM_{ei}m_i}{3M_i^2\zeta_i\delta_i}, \tag{15}$$

where D_i is the self-diffusion coefficient in the one-component melt, and δ_i represents the shift of the relaxation spectrum of the i species due to the modified chain topology and frictional resistance after blending (Eq. 13).

III. VISCOELASTIC PROPERTIES

The knowledge on the stress relaxation progression may be utilized for the prediction of a series of rheological properties of the blend. The plateau modulus, G^o, an index of material rigidity, is obtained from (Eq. 14), at the onset of the terminal relaxation region (t = 0):

$$(G_B^o)^{\frac{1}{2}} = v_1 (G_1^o)^{\frac{1}{2}} + v_2 (G_2^0)^{\frac{1}{2}} \quad , \tag{16}$$

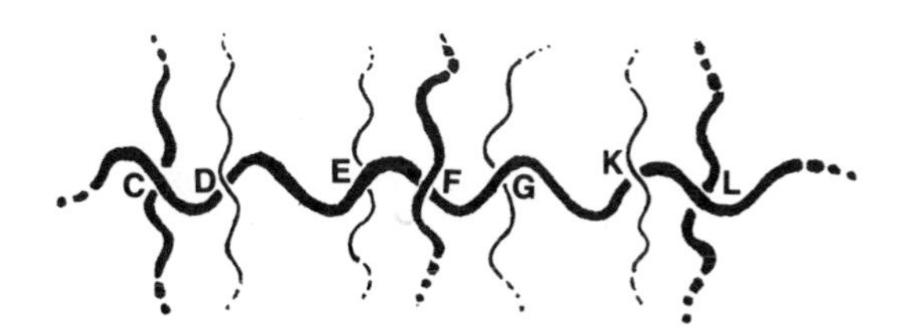

Fig. 1. Entanglements and molecular segmentation in a miscible blend. Lines of variable thickness signify dissimilar polymeric species. Enganglements are represented by (lettered) line interceptions and network continuation by dots. Segments CF and FL are (11) steps while DE, EG and GK are (12) steps.

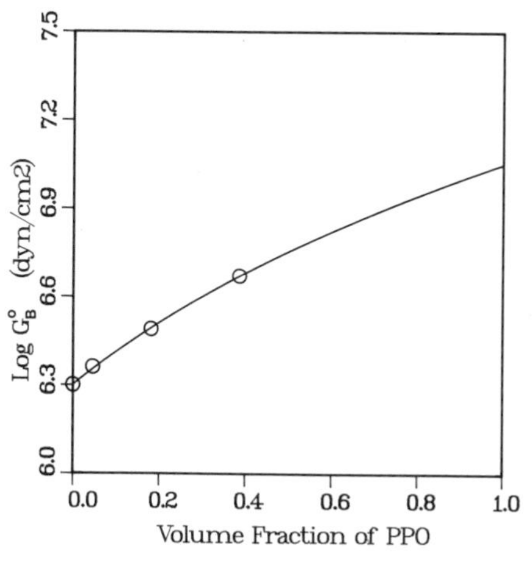

Fig. 2. The plateau modulus (in dyn/cm^2) of PPO/PS blends. Circles correspond to the experimental results by Prest and Porter.[12] Continuous curve represents the predictions of Eq. (16) (G^o is a temperature invariant property).

where G^o_{ei} is the corresponding modulus of the unmixed melt of the i species ($G^o_{ei} = \rho_i R T/M_{ei}$). The zero-shear viscosity (η_B) and recoverable shear compliance (J_B^o) are measures of the fluidity and elasticity of the blend:[6]

$$\eta_B = \int_0^\infty G_B(t)\, dt \, ; \tag{17a}$$

$$J_B^o \eta_B^2 = \int_0^\infty t G_B(t)\, dt \, . \tag{18a}$$

Provided that the $G_i(t)$ of each of the individual components can be approximated by a single exponential term (Eqs.1-4), these properties may be obtained by the following expressions:

$$\eta_B = \upsilon_1^2 \delta_1 \eta_1 + 4\upsilon_1 \upsilon_2 \left[\frac{(G_1^o/G_2^o)^{\frac{1}{2}}}{\delta_1 \eta_1} + \frac{(G_2^o/G_1^o)^{\frac{1}{2}}}{\delta_2 \eta_2} \right]^{-1} + \upsilon_2^2 \delta_2 \eta_2 \, ; \tag{17b}$$

$$J_B^o \eta_B^2 = \upsilon_1^2 \delta_1^2 J_1^o \eta_1^2 + 8\upsilon_1 \upsilon_2 \left[\frac{(G_1^o/G_2^o)^{\frac{1}{4}}}{\delta_1 \eta_1 \sqrt{J_1^o}} + \frac{(G_2^o/G_1^o)^{\frac{1}{4}}}{\delta_2 \eta_2 \sqrt{J_2^o}} \right]^{-2} + \upsilon_2^2 \delta_2^2 J_2^o \eta_2^2 \, , \tag{18b}$$

where η_i and J_i^o are the corresponding properties of the unmixed polymeric precursors.

Prest and Porter[12] measured the viscoelastic properties of several poly(2,6-dimethylphenylene oxide)/polystyrene (PPO/PS) blends at $T = 220°C$.

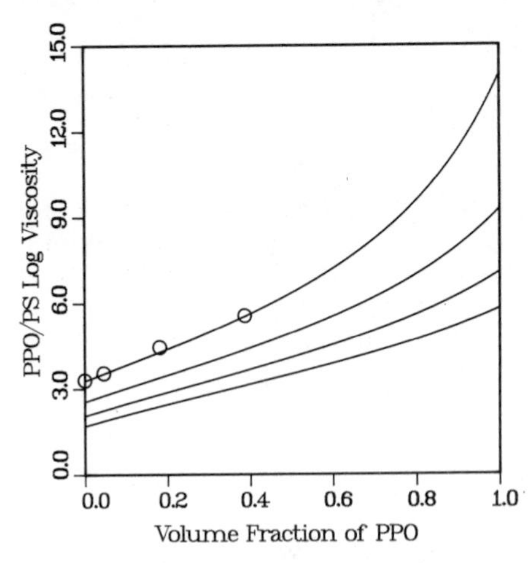

Fig. 3. The zero-shear viscosity (in poise) for the PPO/PS blends reported in Table 1. Circles correspond to the experimental results from Ref. 12. Curves represent the predictions of Eq. (17b) for (from top to bottom) T = 220°, 260°, 300° and 340°C.

Table. 1. The properties of PPO and PS used in the rheological study on blends (T = 220°C) by Prest and Porter.[12] These same values were utilized in Eqs. (5,6,13,16 and 17) for the calculation of the viscoelastic properties presented in Figs. 2 and 3. (Numbers in parenthesis correspond to extrapolated estimates).

	M_w g/mol	M_w/M_n	T_g °C	f_{gi}	$\alpha_i 10^4$ 1/°C	ρ_i g/cc	$G^o_i/10^6$ dyn/cm^2	M_{ei} g/mol	η_i poise
PPO	69000	2.10	219	0.033	4.8	0.953	(11.3)	3459	(10^{14})
PS	97200	1.04	101	0.033	6.9	1.025	2.0	21021	2000

Their data on this system of commercial importance has been selected for comparison with the results of the present theory. The experimentally determined and theoretically predicted values for $G_B{}^o$ (Eq. 16) and η_B (Eqs. 17b, 13, 5 and 6) are displayed in Figs. 2 and 3. The properties of the pure components are listed in Table 1. It was also reported that for the materials involved, $\zeta_{gPPO}/\zeta_{gPS} \approx 1$. Since measurements at high PPO concentrations were prevented by the sensitivity of this polymer to thermal degradation, the tabulated values for $G^o{}_{PPO}$ and η_{PPO} correspond to extrapolated estimates from Eqs. (16) and (17b). Predicting the viscosity of the blend at higher temperatures (Fig. 3) was possible by utilizing the fact that:

$$\eta_i(T_1)/a_i(T_1) = \eta_i(T_2)/a_i(T_2) = \eta_i(T_{gi}) . \qquad (19)$$

A similar examination of the validity of Eq. (18) was not possible due to the difficulty in determining the recoverable shear compliance in polydisperse samples.[12]

IV. DISCUSSION

Based on first principles, a relatively simple method has been developed for predicting the rheological properties of miscible polymer blends from the corresponding properties of their pure constituents. The preceding analysis is expected to apply primarily for athermal blends, where it is safe to assume that mixing does not affect the ability of the individual species to entangle and that the principle of linear free volume additivity is not defied. There is, however, an important class of heteropolymer blends where miscibility is achieved not due to entropic reasons but due to the negative enthalpy of mixing. In highly exothermic blends the above mentioned assumptions do not apply. This is apparent from Wu's recent experiments on a series of poly(methyl methacrylate) blends[13,14] which demonstrate a dependence of the rheological properties on the (negative) Flory-Huggins thermodynamic parameter (χ). Successful semi-empirical methods for estimating the behavior of these systems have already been presented[13] and extension of this theory to accommodate the presence of thermodynamic interactions is currently under way.[15]

On the other hand, after appropriate adaptation, the results of the present study (Eqs. 14-18) may also be useful in describing the properties of the following systems:

(a) Polydisperse homopolymer blends, i.e., mixtures of polymers of identical chemistry but unequal molecular weights.[16] For the binary case: $G^o_1 = G^o_2$, $J^o_1 = J^o_2$, $\zeta_1 = \zeta_2$ and $\delta_1 = \delta_2 = 1$.

(b) Concentrated solutions of a polymer[1] in a low molecular weight solvent[2] in which, $1/M_{e2} = G^0_2 = \tau_2 = 0$, $\zeta_{g2}/\zeta_{g1} \simeq 0$, $\eta_2/\eta_1 \simeq 0$ and yet, $f_1 \neq 0 \neq f_2$.

(c) Molecularly interpenetrating polymeric elastomers.[15,17]

(d) Imperfect rubber networks which may be treated as "blends" of fixed, crosslinked strands with mobile, tethered chains.[18]

(e) Copolymer fluids in the disordered state, i.e., above the microphase separation transition temperature.

It is therefore, hopeful that the present study will constitute a reasonable first step towards the development of a unified theory for the interpretation of the rheological behavior of homogeneous, multicomponent polymeric materials in the entangled regime.

NOMENCLATURE

a_i	shift factor
D_i	self-diffusion constant
f_i	fractional free volume
$F_i(t)$	relaxation function of an individual molecule in the melt
ζ_i	monomeric friction coefficient
$G_B(t)$	shear stress relaxation modulus of the blend
η	viscosity
δ_i	shift of relaxation spectrum
M_{e_i}	molecular weight between entanglements
M_i	molecular weight of the polymer
k	Boltzmann constant
N_A	Avogadro number
$N_i(t)$	number of surviving steps per chain at time t
N_{ij}	number of each step
N_i	number of i chains per unit
R	gas constant
$<R_i^2>$	end-to-end distance of the chain
ρ_i	density
T	temperature
T_g	glass temperature
τ_i	characteristic relaxation time
v_i	polymer volume fraction

REFERENCES

1. B.J. Feder, The New York Times, p. D2, Oct. 16, 1986.
2. O. Olabisi, L.M. Robeson and M.T. Shaw, Polymer-Polymer Miscibility, Chapter 2, Academic Press, New York (1979).
3. P.G. de Gennes, Scaling Concepts in Polymer Physics, Chapter 4, Cornell University Press, Ithaca, New York (1979).
4. S. Krause in Polymer Blends, (D.R. Paul and S. Newman, eds.), Chapter 2, Academic ress, New York (1978).
5. D.R. Paul and J.W. Barlow, J. Macromol. Sci.-Rev. Macromol. Chem., 18, 109 (1980).

6. J.D. Ferry, Viscoelastic Properties of Polymers, 3rd ed., Wiley, New York (1980).
7. W.W. Graessley, Adv. Polym. Sci., 16, 1 (1974).
8. P.G. de Gennes, J. Chem. Phys., 55, 572 (1971).
9. M. Doi and S.F. Edwards, J. Chem. Soc. Faraday Trans. II, 74, 1782, 1802 (1978).
10. W.W. Graessley, Adv. Polym. Sci., 47, 67 (1982).
11. F. Beuche, Physical Properties of Polymers, Chapter 4 & 5, Wiley, New York (1965).
12. W.M. Prest and R.S. Porter, J. Polym. Sci., Part A-2, 10, 1639 (1972).
13. S. Wu, J. Polym. Sci. Polym. Phys. Ed., 25, 557, 2511 (1987).
14. S. Wu, Polymer, 28, 1144 (1987).
15. C. Tsenoglou, J. Polym. Sci. Polym. Phys. Ed., 26, 2329 (1988).
16. C. Tsenoglou, ACS Polymer Preprints, 28(2), 185 (1987).
17. C. Tsenoglou, in Crosslinked Polymers, (R.A. Dickie, S.S. Labana and R.S. Bauer Eds.), Chapter 4, ACS Symposium Series 367 (1988).
18. C. Tsenoglou, Macromolecules, to be published (1988).

Discussion

On the Paper by Isaac Sanchez

P.G. de Gennes (College de France): Your ideas might be particularly useful for problems of <u>depletion layers</u> (with semidilute solutions near a hard wall). Here we would like to know what is the probability of contact between wall and polymer. I have only a very rough argument for this probability. Have you thought about depletion layers?

I. **Sanchez** (University of Texas, Austin): No, I have not given any thought to the problem of depletion layers. If the solution is dilute near the wall or in a crossover region between dilute and semidilute regimes, this approach might prove useful.

On the Paper by P.F. Green

Jeff Koberstein (University of Connecticut, Storrs):

In block copolymer/homopolymer blends there is some evidence for lamellar systems (Quan and Koberstein, <u>Macromolecules</u>, 1987) that the homopolymer is localized in the center of the microdomain. This effect would lead to the kind of diffusion resistance that you have observed, that is, a greater resistance than that caused simply by dilution by the second component. Secondly, the solubility of the homopolymer in the microdomains falls off rapidly when its *MW* is about 1/2 that of the like copolymer sequence. In your case, however, you still get diffusion up to the case of equal *MW*'s. This may be related to your conclusion that your system behaves as if it is in the weak segregation limit.

On the Paper by R.J. Composto, E.J. Kramer and Dwain M. White

M. Rubinstein (Eastman Kodak Company): There are two types of parameters in tube models: friction coefficient and molecular weight between entanglements. In order to get one, you need to assure something about the other. Is it inconsistent to assume two friction coefficients but only one mean molecular weight between entanglements for the blend?

We used Prest and Porter values of the molecular weight between entanglements for the blend.

R. Composto (University of Massachusetts): In the tube models the entanglement molecular weight M_e is a topological parameter (related to the tube diameter) which can be obtained from the rubbery plateau storage modulus without assumptions about the friction coefficient(s) of the molecules. While each chemically different polymer molecule in the blend could have its own tube diameter with the experimental M_e corresponding to some appropriate average over these, we believe that our assumption that the tube diameter for each molecule in the blend is the same is more consistent with current theoretical models. In the Graessley and Edwards (Polymer 22 1329 (1981)) theory of M_e for example, M_e depends principally on the total contour length of polymer per unit volume. For a blend this leads to the following expression for M_e if the persistence lengths for the two polymers are not very different:

$$M_e^{-1/2} = \Phi_A M_{e,A}^{-1/2} + \Phi_B M_{e,B}^{-1/2}$$

[Reference: W.W. Graessley, private communication; R.J. Composto Ph.D. Thesis, Cornell University (1987)]

This expression fits the existing data for M_e of blends, both PS/PXE and PS/polytetramethylbisphenol-A carbonate, reasonably well. It is also worth emphasizing that even if we make the opposite (extreme) assumption that PS and PXE have the same tube diameters (and M_e's) in the blend as in the pure polymer, the qualitative result that the monomeric friction coefficient of PXE is much larger than that of PS is not changed.

PART FIVE:

ENTANGLEMENT AND REPTATION OF POLYMER MELTS AND NETWORKS

MOLECULAR WEIGHT DEPENDENCE OF THE VISCOSITY OF A HOMOGENEOUS POLYMER MELT

A. Silberberg

Polymer Research Department
Weizmann Institute of Science
Rehovot 76100
Israel

ABSTRACT

In the polymer melt conformational rearrangements are associated with the motions of the macromolecule (reptation) curvilinearly along its contour in an environmental tube to which the molecule is confined. The tube can only be altered, topologically, by contractions and re-elongations of the chain as its ends diffuse, eliminate and redefine, new tube sections. The time for the renewal of the entire tube, by this process, is proportional to the cube of the molecular weight. The rotation of the macromolecule in simple shear, as well, requires up and down motions along the tube. Hence, if the viscosity of the melt is governed by the same friction factor as in diffusion it should scale as M^3. Experimentally, however, it is found that viscosity is proportional to $M^{7/2}$. This discrepancy is here analysed and shown to be due to the shear-imposed, active pumping of the chains up and down their tubes.

I. INTRODUCTION

Two rather remarkable features characterize the polymer melt. In the melt, firstly, the macromolecules are randomly coiled and their radius of gyration is distributed in Gaussian fashion over the accessible conformations, and in the melt, secondly, the viscous response is Newtonian, over a broad range of rates of shear, and increases with the 3.5 power of the molecular weight, if the chains exceed a certain critical size. Below that size the viscosity increase is linear with molecular weight.

The macromolecules are Kuhn coils because the excluded volume effect between segments in the same macromolecule, which in dilute solution produces coil expansion, is overwhelmed in the melt by very strong shielding by segments from other chains. Flow behavior is Newtonian because conformational reorganization occurs fast enough so that the Gaussian conformation distribution remains essentially undisturbed. Though any particle, even a long chain macromolecule in a melt, must, in simple shear, rotate with an angular velocity proportional to the rate of shear q, the conformational reorganization has a finite time interval, of length $1/q$, available for it to take place. This length of time can be adequate, even in cases when the chain is constrained to move, by its environment, along a path essentially outlined by its own contour (reptation).[1] The time associated with a conformation change, i.e., the time required for the macromolecule to build itself an entirely new environment by curvilinear diffusion, the time for tube renewal, scales as M^3, where M is the molecular weight. It has thus become something of a puzzle why, experimentally, the viscosity is found to scale as $M^{7/2}$. An M^3-dependence would have been expected on a model which relates the viscous relaxation time to the time of conformation change.

It is the aim of this paper to show that in the simple shear of a melt of chain molecules undergoing reptation an additional effect accounts for the higher power.

II. THE VISCOSITY OF A SUSPENSION OF RIGID SPHERES

Using the equations of hydrodynamics for an incompressible fluid in Stokes approximation, Einstein[2] solved the problem of the velocity and pressure fields created by a sphere of radius R when at a distance far from the sphere a simple shear field (velocity gradient q) is imposed on the fluid. He showed that if there are c spheres in the system per unit volume and the effects of these are linearly superimposed, there is an effective increase in the viscosity from a value η, when there are no spheres, to a value η' when they are present. Einstein found that

$$\eta' = \eta(1 + 2.5\phi) , \tag{1}$$

where

$$\phi = 4\pi c R^3/3 << 1 , \tag{2}$$

is the volume fraction of the spheres and is assumed to be very low.

The points to note about this result is that η' does not depend upon the size of the sphere and is independent of q, i.e., is still a Newtonian viscosity. The effectively increased viscosity is due to the increased energy dissipation. The presence of the spheres locally raises the rate of shear above the nominal externally applied value q. The medium, in reality, is of unchanged viscosity η.

When no spheres are present, q is uniform and the energy dissipated per unit volume and unit time is ηq^2 everywhere. The same rate of shear is present in all volume elements and the total dissipation in a volume V is

$$dE_V/dt = \int \eta q^2 dV . \tag{3}$$

In the presence of the spheres, on the other hand, this becomes

$$dE'_V/dt = \int_V \eta(q')^2 dV = \int_{V(1-cV_S)} \eta(q')^2 dV , \tag{4}$$

where

$$V_S = 4\pi R^3/3 \tag{5}$$

is the volume of the sphere. The rate of shear, q', is the real rate of shear as it is locally modified by the presence of the spheres. It is clearly zero inside the spheres and this explains the third expression in Eq. (4). It is also clear, by definition, that q' becomes q on the outer periphery of the volume V. It follows also, given the Einstein result, that

$$\eta'/\eta = (dE'_V/dt)/(dE_V/dt) = 1 + 2.5\,\phi\ . \tag{6}$$

The volume integrals in Eqs. (3) and (4) are now converted into integrals over the surface. Since the rate of shear is q over the outer periphery in the case of both Eqs. (3) and (4), these two surface integrals have to be equal and their ratio accounts for the first term, 1, on the right-hand side of Eq. (6). It thus follows that the second term is the ratio of the surface integral over the surface of the spherical inclusions divided by the integral over the periphery of V. When this calculation is performed the result, 2.5ϕ, is obtained.[3]

We see that the viscosity increase in suspensions can be calculated from energy dissipation effects related to the rotations of the dispersed particles and the tractions which arise over their surfaces. The surface velocity of the sphere is

$$v_S \sim qR\ , \tag{7}$$

and the surface traction is

$$F_S \sim \eta R v_S\ . \tag{8}$$

The energy dissipated per particle in unit time is thus

$$F_S v_S \sim \eta v_S^3/q\ , \tag{9}$$

so that

$$\eta'/\eta = 1 + const.\,(c V F_S v_S)/V\eta q^2 = 1 + const.\,(c v_S^3/q^3)\ . \tag{10}$$

Substituting Eq. (7) into Eq. (10) thus gives

$$\eta'/\eta = 1 + const.\,\phi \ , \tag{11}$$

which reproduces the essence of Eq. (1).

III. THE VISCOSITY OF A SUSPENSION OF COILING MACROMOLECULES

A macromolecule comprised of N segments of length a forms a coil in dilute solution whose average radius R_F is given by

$$R_F = a N^{\nu} \ . \tag{12}$$

The exponent ν is 0.5 in the case of a θ-solvent and becomes 0.6 in the case of an ideally good, athermal solvent medium.[4] In either instance, of course, the contour length L is given by

$$L = a N \ . \tag{13}$$

To the extent that each coil and the solvent it contains can be regarded as an inclusion within the solvent medium which rotates with an angular veolcity proportional to q, while, within its domain, i.e., within the radius R_F, the rate of shear is essentially zero, the current situation can be regarded as quite analogous to the case of the sphere suspension just discussed. In place of Eq. (7) we thus write

$$v_S \sim q R_F \sim q a N^{\nu} \ , \tag{14}$$

and, substituting Eq. (14) into Eq. (10), find that

$$\eta'/\eta = 1 + const.\,\phi\,(N^{3\nu - 1}) \ , \tag{15}$$

where the volume fraction ϕ in this case is given by

$$\phi = c a^3 N \ .$$

If it is recalled that the so-called intrinsic viscosity, defined as

$$[\eta] = \lim_{\phi = 0} \ [(\eta'/\eta - 1)/\phi] \ , \tag{16}$$

we see that intrinsic viscosity is 2.5 in the sphere case and is given by the Flory's relation

$$[\eta] \sim N^{3\nu - 1} , \tag{17}$$

in the present instance.

IV. THE VISCOSITY OF A MELT OF COILING MACROMOLECULES

We are now ready to apply the principles we have developed to an analysis of simple shear in the polymer melt. Here too, as was already pointed out, the macromolecules are coiled, rotate with an angular velocity proportional to q and retain their average, ideal (in this case), random coil shape as they rotate.

It is no longer correct to say, however, that there is no flow within the domain of the coil. In fact the major, the only motions which occur are movements of the chain, along its contour, as it reptates and as its environment extends and contracts during the rotation. The ends of the chain are, in turn, pulled in and then pulled out thereby allowing the macromolecule to restructure its confining tube and rotate. Energy is dissipated due to friction generated by these imposed curvilinear flows. The velocity v_{Ln} within the rotating tube which envelops the rotating chain depends on the order number n, along the chain, of the segment in question. Numbering is from the central segment of the chain outwards. At a radial position r from the center of rotation the average tangential velocity in that region of the coil is given by

$$v \sim rq , \tag{18}$$

where the radial location r is related to the length of chain, n, enclosed within by

$$n \sim (r/a)^2 . \tag{19}$$

If the number of segments at the internal coil surface at r, is dn and these are convected away, tangentially, at a rate v, given by Eq. (18), an increase in the curvilinear velocity, v_{Ln}, that supplies segments to the layer surface at r, must occur which is given by

$$dv_{Ln} \sim v\,dn \sim aqn^{\frac{1}{2}}dn . \tag{20}$$

It follows that the effective curvilinear surface velocity of the chain at the backbone position n along the chain is given by

$$v_{Ln} = \int_o^n dv_L \sim aq \int_o^n n^{1/2} dn \sim aqn^{3/2} , \tag{21}$$

where it was assumed that the curvilinear velocity in the center of the tube is zero.

At n along the chain, therefore, the additional length of chain which is being speeded up by the flow is dv_{Ln}/q and the additional force on this portion of chain is given by

$$dF_n = \eta v_{Ln} dv_{Ln}/q \ .$$

Hence the incremental energy dissipated per unit time by the acceleration effect at n is

$$dE_n/dt = v_{Ln} dF_n = \eta v_{Ln}^2 dv_{Ln}/q \ ,$$

which, after integration becomes proportional to $\eta v_{LN}{}^3/q$ for the entire chain of volume Na^3. Per unit volume, therefore, after substituting from Eq. (21), we have

$$\eta_{melt} q^2 \sim \eta q^2 a^3 N^{9/2}/Na^3 \ , \tag{22}$$

which gives

$$\eta_{melt} \sim N^{9/2-1} = N^{7/2} \ . \tag{23}$$

Thus, the melt viscosity scales with the 3.5 power of the molecular weight, as observed.

V. DISCUSSION

In order that the coiled chain can rotate in the melt at the angular velocity imposed and can maintain its average conformation intact there must be very active convection of chain sections towards the center of rotation and then away from it as the coil alternately experiences the elongational and then the compressive forces created by the symmetric part of the simple shear field. Of all segments, those at the chain ends move fastest and are alternately decelerated and accelerated as they are pushed in and out. These nonuniform curvilinear movements, which destroy and create new tubes as the chain rotates, have nothing to do with diffusion. They are due to externally applied forces tending to distort the conformations and are powered by the entropy-driven tendency of the chain to restore the optimum distribution. As long, therefore, as these restoring forces are large enough to overcome friction and drive the chains up and down their tubes at an adequate rate, i.e., within the time $1/q$ available, the system will respond with a viscosity which is Newtonian.

It is also clear that Eq. (23) will apply only as long as the chain is required to reptate. If its molecular weight drops below the critical size M_c essentially stiff rotations of the coil will occur in an environment which has a viscosity η, but is free of topological constraints. In such a case, the mean velocity of the N segments composing the macromolecule is qR_F. The resistance they experience is $\eta(aN)(qR_F)$ and the energy dissipated per particle in unit time becomes $\eta(aN)(qR_F)^2$. Per unit volume, the melt in this case dissipates

$$\eta_{melt} q^2 = \eta(aN)(qR_F)^2/Na^3$$

energy per unit time and we find

$$\eta_{melt} \sim N; \qquad M < M_c, \tag{24}$$

a result found experimentally in the low molecular weight regime.

Equation (24) represents a molecular weight-dependence characteristic of the well-drained random coil. While the coil remains effectively stiff the increases induced locally in the rate of shear by the presence of the chain are contributed in linear, additive fashion by a set of effective beads per chain proportional in number to the molecular weight.

In the reptation region, this picture changes drastically. The coil still rotates and retains its mean shape, but in order to do so the ends of the coils have to execute elongational and compressive curvilinear motions which in each half cycle of rotation create a new, angularly rotated, environmental tube.

NOMENCLATURE

a	segment length of macromolecular coil.
c	number of spheres per unit volume.
E_V	energy associated with a volume V of the system, no spheres.
E'_V	energy associated with a volume V of the system, with spheres.
E_n	energy associated with the curvilinear motion of the macromolecular chain at position n.
F_n	additional force at position n along the chain due to local curvilinear velocity changes.
F_S	surface traction on sphere.
L	contour length of macromolecule
n	ordinal number of segment along the chain.
N	total number of segments of length a per chain.
q	apparent, uniform rate of shear.
q'	actual, local rate of shear.
R	radius of sphere.
R_F	Flory radius of macromolecular coil.
t	time
v_S	surface velocity of sphere.
v_{Ln}	effective, curvilinear velocity of chain locally at n.
V	volume of system.
V_S	volume of each sphere.
η	actual viscosity.
η'	effective viscosity.
ν	exponent of N in expression for R_F.
ϕ	volume fraction of spheres.

REFERENCES

1. P.-G. de Gennes, Scaling Concepts in Polymer Physics, Cornell University Press, Ithaca, 1979, Chapter VIII.
2. A. Einstein, Ann. d. Phys. 19, 289-306 (1906); 24, 591-592 (1911).
3. L.D. Landau and E.M. Lifshitz, Fluid Mechanics, Pergamon Press, Oxford (1959); pp.76-79.
4. P.-G. de Gennes, Scaling Concepts in Polymer Physics, Cornell University Press, Ithaca, 1979, Chapter I.

Note (added in proof)

The observation that the melt viscosity of polymers scales as $M^{3.4}$, once M is larger than some critical molecular weight (M_c), has been in the literature for over thirty years and may be regarded as a very well documented result experimentally. The reason is quite simple. The result was a challenge; in the early years because the high exponent was not understood and then because it failed to correspond to a simple reptation result. Several theoretical explanations have been proposed. These include some where the reptation result is said to become correct only at molecular weights very much larger than M_c. Some recent experimental results of R.H. Colby, L.J. Fetters and W.W. Graessley Macromolecules 20, 2226 (1987) are quoted in support. Close examination of these data, however, shows that they follow the usual pattern and that the reduced slope at high molecular weight depends on only one data point, out of 24, and this is the point for the highest molecular weight sample.

DYNAMICS OF ENTANGLED POLYMERS : REPTATION

L. Léger, H. Hervet,
M.F. Marmonier

*Collège de France
Laboratoire de Physique de la
Matiére Condensée, U.A.CNRS 792
11 Place Marcelin-Berthelot
75231 Paris Cedex 05
France

J.L. Viovy

Ecole Supérieure de Physique
et Chimie Industrielles
(E.S.P.C.I.)
Laboratoire de Physicochimie
Macromoléculaire
10 rue Vauquelin
75231 Paris Cedex 05, France

ABSTRACT

Polymer liquids have fascinating dynamical properties: they flow like conventional viscous liquids when subjected to perturbations slowly varying with time, but behave like elastic solids at higher frequencies. These mechanical properties strongly depend on the molecular structure (linear, branched, flexible, rigid...) but unifying concepts have emerged in the past two or three decades, especially for the easiest case of linear flexible chains, pointing out the crucial role of chain entanglements. The reptation model has provided a fruitful and simple framework to analyze the wide range of available dynamical experiments. The reptation model qualitatively well describes how the dynamic properties depend on the molecular weight of the chains, and yields simple and good estimations of the different time scales involved. Small quantitative discrepancies remain, however, which are at the origin of a vigorous debate on the validity of the reptation hypothesis, and of a growing number of proposed refinements.

In this review, we shall present the reptation ideas in their simple version, see how they apply to self-diffusion, viscoelasticity and more local processes, critically analyze the underlying hypothesis, and sketch the additional effects which have to be included to get a realistic description of the dynamic properties of entangled polymers.

I. INTRODUCTION

Polymeric materials play a growing role in our everyday life. A wide range of their applications relies on their unique and spectacular viscoelastic behavior. In the melt state, a polymer sample can be strongly deformed (deformation ratio of several hundred of percent) without breakage, and then, recover almost completely its original shape if the stress is relaxed after a short enough time. This memory is completely lost at long times, when the same material flows like any ordinary viscous liquid. A number of experimental methods have been developed to study this behaviour (these methods are presented in detail in the book by Ferry)[1] and try to relate the time after which the system behaves like a liquid to its molecular characteristics. A particularly illustrative example is provided by a step-strain experiment in which a polymer sample, initially at rest, is instantaneously deformed at time zero, and then kept at constant strain. The reactive, time-dependent modulus exerted by the sample, or relaxation modulus, $G(t)$, presents unique features, characteristic of long-chain polymer material, as shown in Fig. 1:

i) After a decrease at short times, $G(t)$ levels off at intermediate times, and displays a "plateau" with an amplitude essentially independent of the molecular weight of the polymer, M_w. In this time range, the polymer melt behaves like a rubber. This plateau zone is more pronounced for larger molecular weights, and totally disappears for short enough molecules.

ii) At long times, in the "terminal zone", $G(t)$ decays to zero, as must be the case for a liquid, in a roughly exponential manner, with a characteristic time increasing rapidly with the molecular weight, and scaling approximately as $M_w^{3.4}$.

The Brownian diffusion of polymer chains in melts or in concentrated solutions also presents unique features: it is much slower than in ordinary liquids and strongly molecular weight-dependent: the self-diffusion coefficient scales as M_w^{-2}.

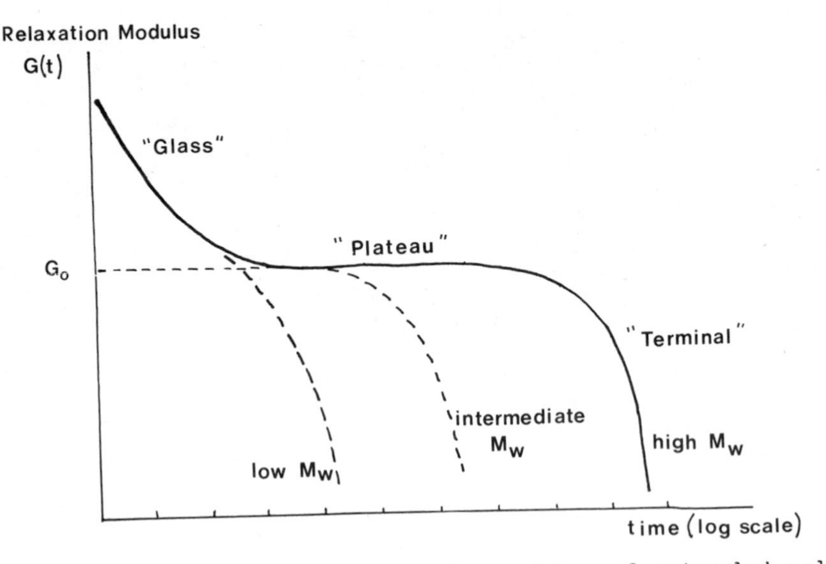

Fig. 1. Time evolution of the relaxation modulus of entangled polymers for three different molecular weights. The plateau zone, where the system behaves like a transient rubber, is more pronounced for larger molecular weights.

Of course, viscoelasticity (response to an external mechanical excitation) and diffusion (response to thermal fluctuations) must have their origin in the same dynamical process at the molecular level, and any model proposed to explain one aspect must also consistently account for the other (as a consequence of the fluctuation dissipation theorem). The elaboration of such models has remained a challenge to polymer scientists for many years, and is indeed a complicated many-body problem.

It has been suggested rather early that the unique dynamics of dense, long-chain polymer materials was due to molecular "entanglements",[2] a rather loose concept which expresses the strong restrictions that uncrossable and highly unidimensional objects must exert on the motion of their neighbours. The reptation model[3] has played an essential role in amening this concept to mathematical formulation, and it still is the privileged tool for understanding the dynamics of entangled polymers.

In the first section of this review we shall present the reptation ideas in a simple manner, concentrating on the best understood case of linear flexible chains, with a large number (N) of monomers per chain, leaving aside all mathematical developments or refinements, but pointing out and analyzing critically the underlying assumptions of the model.

In the second section we shall briefly review more recent trends in molecular rheology, including attempts to deepen the understanding of entanglements on a molecular level, and the use of the reptation ideas to describe various practical problems such as polymer/polymer welding, phase separation of polymer mixture, flow behaviour of a polymer melt close to a solid wall...

II. DYNAMICS OF LINEAR FLEXIBLE CHAINS

II.1 Motion of one chain among fixed obstacles: reptation A quantitative description of the Brownian diffusion of one long polymer chain among fixed obstacles has been proposed by de Gennes in 1971.[3] As schematically represented in Fig. 2a, the chain cannot cross the obstacles. At any time, its conformation depends on the actual monomer-obstacles interactions; it is confined by the obstacles. The only way for the chain to change its conformation is to choose its way among the obstacles. The process takes place essentially at the extremities (the probability for the chain to form a loop and to have a leading motion from any other part of the chain (Fig. 2b) is very small as it implies a strong entropy loss). Due to small conformation fluctuations, the extremities can engage between new obstacles, and progressively pull all the chain in a new environment (Fig. 2c), a process named reptation by de Gennes. As a consequence, the central part of the chain feels during a rather long time the same topological constraints.

An average way of taking into account those constraints is to assume that the chain is trapped into a virtual tube, envelope of all the obstacles which directly surround it (dotted line of (Fig. 2c)), a picture first proposed by Edwards.[4] The chain can move freely along the curvilinear axis of the tube, but it cannot escape laterally. At any time, by fluctuations of its local kinks, the chain leaves some parts of the initial tube, and the chain extremities define new portions of the tube. The detailed statistical description of de Gennes leads to definite predictions for the molecular weight dependences of both the relaxation times and the diffusion coefficient of the chain, which we shall present in a simplified version.

An ideal flexible chain made of N monomers of size a obeys a Gaussian statistics and is characterized by a radius of gyration $R_G \sim N^{1/2}a$. The following discussion can easily be extended to real chains, by renormalizing the statistical unit and replacing the monomer length by the Kuhn's length.[5] In any case, the curvilinear length of the tube is proportional to N.

The chain is not rigid, and it permanently samples all its possible conformations.

For an isolated chain, the dynamics of these fluctuations has been worked out by Rouse[6,7] in terms of modes, each having a characteristic wavelength λ_p and a characteristic time $\tau_R(p)$ with $\tau_R(p) \sim \tau_1 N_p^2$, where τ_1 is a monomeric characteristic time and N_p is such that $\lambda_p \simeq N_p^{1/2}a$. This means that small local fluctuations are sampled much more rapidly than fluctuations on long length scales.

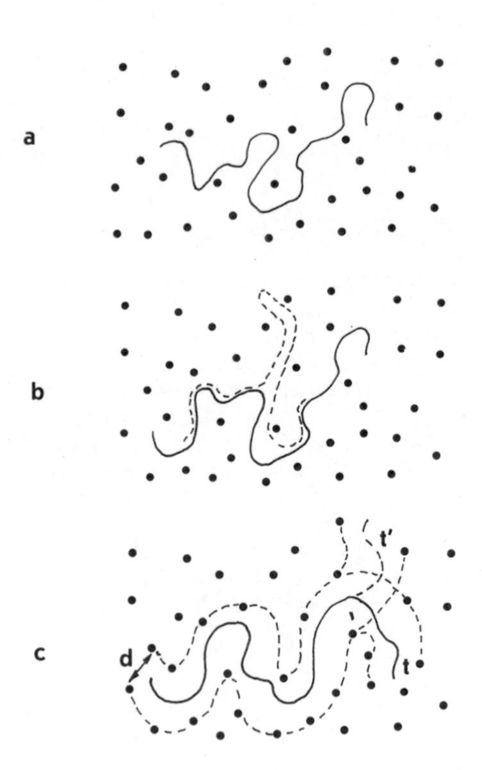

Fig. 2. Schematic representation of one chain among obstacles:

(1a) The chain is constrained by the obstacles.

(1b) By local fluctuations, the chain changes its conformation. The probability of forming a loop is very small (strong entropy loss), and the role of the extremities is dominant.

(1c) The chain reptates like a snake in the virtual tube (dotted line) envelope of all the topological constraints exerted on it by the obstacles. The tube is progressively redefined from the extremities.

For the chain among fixed obstacles, it is natural to assume that fluctuations taking place over distances smaller than the tube diameter, d, are not affected by the tube and are that of a free chain. The overall chain inside its tube can be considered as a necklace of beads of size d (N_d monomers per beads, with $N_d^{1/2}a \sim d$). The longest relaxation time of one bead is $\tau_R(d) \sim \tau_1 N_d^2$, and the associated diffusion coefficient $D_d \sim D_1/N_d$, with D_1 a monomeric diffusion coefficient and τ_1, D_1, on one hand, $\tau_R(d), D_d$ on the other hand, related by Einstein's diffusion laws.

$$\tau_1 D_1 \sim a^2 ,$$

$$\tau_R(d) D_d \sim d^2 . \tag{1}$$

The mobility of the whole chain, free to move along the curvilinear axis of its tube is N/N_d smaller than that of one bead (the friction on the full necklace is the friction on one bead times the number of beads), leading to a curvilinear diffusion coefficient

$$D_t \sim D_1/N . \tag{2}$$

The reptation time τ_{Rept}, or time necessary for a complete renewal of the chain configuration, is related to D_t through a relation analogous to (1) : $D\tau_{Rept} \sim L_t$, where $L_t = d(N/N_d)$ is the total length of the tube. Thus:

$$\tau_{Rept} \sim \tau_R(d)\left(\frac{N}{N_d}\right)^3 \sim \tau_1 \frac{N^3}{N_d} . \tag{3}$$

τ_{Rept} is the longest characteristic time of the chain constrained by the obstacles. It is much longer than the longest Rouse time of the free chain $\tau_R(N) \sim \tau_1 N^2$. One also has to notice that the measurable self-diffusion coefficient D_s of the chain is not D_t : the tube is contorted, and when the chain travels a distance L_t along the tube axis, it only travels R_G in a given direction of the real space, so that $\tau_{Rept} \sim R_G^2 \sim N a^2$, i.e.,

$$D_s \sim D_1/N^2 . \tag{4}$$

Again, this is a much slower diffusion (by a factor $1/N$) than for a Rouse-type free chain.

Obviously, if the chain is short enough and $R_G < d$, it does not feel the tube and one recovers a Rouse-like dynamics.

II.2 Extension to polymer melts or solutions The above model should correctly describe the dynamics of isolated chains trapped into a porous medium or a crosslinked gel. In polymer melts or concentrated solutions, the chains are interpenetrated. Since they cannot cross each other, they are strongly constrained, in a way somewhat similar to the above situation. One can again define a tube around each chain, envelope of all the topological constraints exerted on it by the surrounding chains. However,

i) all the chains in the system move, and obstacles are not permanent. In order to apply reptation ideas to polymer melts or entangled solutions, it is necessary to assume that the evolution of the tube, due to the motion of the surrounding chains, is much slower than the reptation.

ii) the tube diameter is no longer an external parameter. It represents the distance perpendicular to the local chain direction that one monomer can travel, due to local chains flexibility, before being blocked by the surrounding chains. An ab initio determination requires a complete description of the actual monomer-monomer interactions. In a polymer melt, the excluded-volume interactions between monomers are totally screened out, and the chains still obey a Gaussian statistics.[8,9] This, however, does not yield a sufficiently precise description of the dynamical interactions on a local scale to allow for an a priori calculation of the tube diameter d. Up to now, d has been introduced through the phenomenological parameter N_e, number of monomers necessary to get one efficient topological constraint or "entanglement", with $d \sim N_e^{1/2} a$. τ_{Rept} and D_S can then be obtained from Eqs. (3) and (4), replacing N_d by the parameter N_e to be determined experimentally. In entangled polymer solutions the situation appears a priori more complicated : due to excluded-volume interactions, the chains no longer obey a Gaussian statistics, and are partly swollen by the solvent. Recent successful developments of the scaling description of polymer solutions have led to clear prediction for the tube diameter, d, which has to be identified with the monomer-monomer correlation length ξ[7] (Chapters III, VII and VIII), and decreases with increasing monomer concentration c. Here, ξ is the average distance at which chains meet each other, and is thus the natural quantity to be identified with d. It is less obvious that ξ is also the length to be used to evaluate chain mobility, i.e., that it is identical to the scale ξ_H over which the monomers are hydrodynamically correlated. Scaling arguments show that ξ_H and ξ become proportional in the asymptotic large-N limit,[7] but it is now well established that ξ_H converges towards its asymptotic limit much more slowly than does ξ.[73,74] The modifications that this slow convergence introduces into the concentration exponents remain small, however, and of the same order as the experimental error in the determination of the self-diffusion versus concentration exponent. In all the following, we shall assume that $\xi \sim \xi_H$. The concentration dependencies of D_S and τ_{Rept} can thus readily be obtained from the $\xi(c)$ scaling law and, for a good solvent where monomers prefer to be surrounded by solvent molecules rather than by other monomers, one obtains:

$$D_s \sim N^{-2} c^{-1.75} ,$$

$$\tau_{Rept} \sim N^3 c^{1.5} . \tag{5}$$

An important prediction of the reptation model is the fact that the normalized diffusion coefficient, D_s/D_0 (with D_0 the diffusion coefficient at infinite dilution), should be a universal function of the number of entanglements per chain.

A growing experimental effort has developed in the past decade to test these predictions, essentially through self-diffusion measurements. The process is very slow (typically, for $N \sim 10^3$, $a \sim 3$ Å, $\tau_1 \sim 10^{-10}$ sec, the D_s range is of the order of 10^{-11} cm^2/sec) and techniques in which the diffusion length can be decreased down to the micron range are necessary to avoid prohibitively long experimental times. Another difficulty stems from the necessity of labeling a few chains in the system. Because mixing chemically different polymers is difficult (a small repulsion between monomers results in an N-times larger repulsion between chains), only minor chemical modifications are allowed (deuteration, or change of a few monomers among N), and one must watch for possible dependences on the concentration of labeled chains. Finally, in the case of solutions, the model is valid only for low enough concentrations ($c < 10\%$), otherwise, delicate concentration-dependent friction corrections have to be taken into account.[10,11] Very long chains must be used if one wants a wide concentration range in which chains are entangled, still keeping $c < 10\%$.

Rather quickly, the N^{-2} law has become well established for sufficiently long chains, both in melts and in solutions,[12-15] as noted in the review by Tirrell.[16] An example of a universal D_s/D_0 curve for polystyrene/benzene solutions is presented in Fig. 3a. A careful examination of the experimental data, however, indicated deviations from a simple power law, especially when a wide range of molecular weights was investigated.[17]

II.3 <u>Viscoelasticity. The Doi-Edwards model</u> The reptation model is also an essential tool to understand the response of polymer materials to external solicitations. Its extension to viscoelasticity has been performed by Doi and Edwards[18] on the basis of two main assumptions:

i) <u>Affine deformation</u> : during the macroscopic deformation, the "tube" of a chain affinely follows the strain of the sample, i.e., the shape of the tube is deformed in the same way as the sample.

ii) <u>Fixed tube</u> : this affine deformation is the only modification of the tube.

Under such hypothesis, for the step-strain experiment of Fig. 1, the evolution of one chain in the sample is schematically presented in Fig. 5 : immediately after the deformation the chain is elongated in its deformed tube. In the first stage, it re-equilibrates inside its deformed tube in a time scale comparable, at most, to the Rouse time of the free chain $\tau_R(N) \sim \tau_1 N^2 \sim \tau_R(d)\ (N/N_e)^2$. In the second stage, it progressively relaxes its remaining stress by reptating out of its deformed tube, a process which takes a time $\tau_{Rept} \sim \tau_R(d)\,(N/N_e)^3$. During the time interval between $\tau_R(N)$ and τ_{Rept}, the stress is proportional to the "tube memory", i.e., to the fraction $M(t)$ of the chain which at time t is still in the deformed tube. $M(t)$ decays exponentially[3] at large t like exp $-t/\tau_{Rept}$, and the Doi-Edwards terminal relaxation time, proportional to τ_{Rept}, i.e., to N^3, is in qualitative agreement with experiments. Several viscoelastic quantities can be calculated from the memory function. For example, the plateau modulus G^o is simply related to N_e ($G^o \sim kT/N_e$) and the zero-shear viscosity, which for $N > N_e$, is expected to scale like $\eta = G^o\ \tau_{Rept} \sim N^3$. This is again in good qualitative agreement with experiments, but quantitative difficulties exist: the N_e values deduced from elastic plateau modulus data are not identical to those deduced from viscosity data; the experimental exponent for the viscosity is slightly larger than predicted (3.3 or 3.4 instead of 3).[1,18] These discrepancies have stimulated a critical re-examination of the reptation hypothesis, both theoretically and experimentally, and are at the origin of the growing number of papers published on the subject in the past three or four years.

II.4 <u>Self-consistency : removing the hypothesis of fixed obstacles</u> Obviously the hypothesis of fixed obstacles, crucial for the application of reptation ideas to melts or entangled polymer solutions, and which makes it possible to reduce the real many-body problem to a one-chain problem, might not be valid and has to be carefully tested experimentally. Diffusion measurement techniques in which labelled chains are used are unique tools to do so, as labelled and unlabelled chains of different polymerization index (respectively N and P) can be used systematically, to sort out the effect of the environment on the motion of the labelled chains. Such systematic experiments have recently been performed both in solutions and in melts.[19,21] In both cases, for $P >> N$, a clear reptational regime is observed, i.e., a regime in which the tracer diffusion coefficients is independent of P and scales as N^{-2}. Indeed in that case the matrix can be considered as immobile during the reptation time of the labelled N chain. Typical results, in the case of entangled solutions (polystyrene in benzene) are reported in Fig. 3. The open symbols correspond to a frozen matrix, and to labelled chain molecular weights in the range 7.8 x 10^4 to 7.5 x 10^5. In the reduced units D_s/D_0 as a function of c/c_N^* (c_N^* is the first overlap concentration of the labelled chains), a universal power law is well observed, and the exponent -1.8±0.1 is in very good agreement with the reptation prediction (Eq. 5).

The data for $N = P$ are also reported (filled symbols) in the same units. A surprising result is that they also form a universal curve, which, in the logarithmic scales of Fig. 3 is, over one decade in concentration, a straight line, parallel to that of the $P >> N$ limiting curve. This experimental fact calls for several comments.

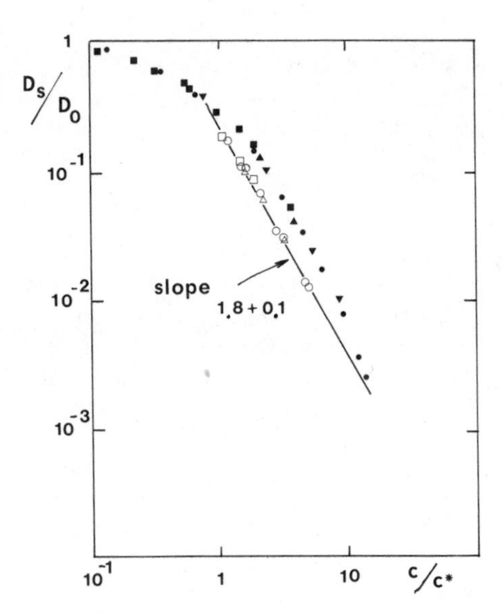

Fig. 3. Universal behaviour of the normalized self-diffusion coefficient D_s/D_0 (D_0 is the diffusion coefficient at infinite dilution) as a function of the reduced concentration c/c^*, where c^* is the concentration at which the labelled chains start to overlap, for polystyrene in benzene solutions. The different symbols correspond to different molecular weights in the range 78000-750000. Full symbols correspond to labelled and unlabelled chains having the same molecular weight, while open symbols correspond to a frozen matrix ($P>>N$) for which a pure reptational behaviour is observed (from Ref. (19)).

1) It is clearly not sufficient to observe the predicted power law behaviour in a finite concentration range (data for $N = P$) to ensure that reptation is observed, even if the test of the universal curve is a quite constraining one, checking at the same time both the concentration and the molecular weight dependencies.

2) The universal curve obtained for $N = P$ in the reduced units of Fig. 3 means that the additional mechanism which accelerates the chain diffusion with respect to pure reptation is only a function of the number of entanglements along the test chain, $N/g = (c/c_N^*)^{4/5}$ with g the number of monomers in a subunit of size ξ.

3) For large concentrations ($c/c_N^* > 10$) the curve for $P = N$ bends down, possibly joining the pure reptation curve. Experiments to completely establish that point are very difficult to perform: due to the labelling technique, the molecular weight of the test chains cannot be very large ($N < 10^4$), and reaching $c/c_N^* > 20$ means working with too high concentrations to ensure that friction corrections remain weak.

If P is further decreased, a transition towards another dynamical regime, in which the big N chain, partly deswollen by the small P chains, diffuses as a rigid sphere (Stokes diffusion) is predicted[20] and observed,[21,22] as illustrated in Fig. 4. Raw data for two different N values are reported in Fig. 4a, while in Fig. 4b, the product ηD_s (with η the bulk viscosity of the solutions) is reported as a function of the overall polymer concentration. A break is observed in the curves when c crosses c_P^*, the first overlap concentration of the matrix P chains. Above c_P^*, ηD_s increases and seems to follow a scaling law $\eta D_s \sim c^{1/4}$ (data are available over a limited concentration range). This is exactly what is expected for the inverse of the radius of the big N chains, which are

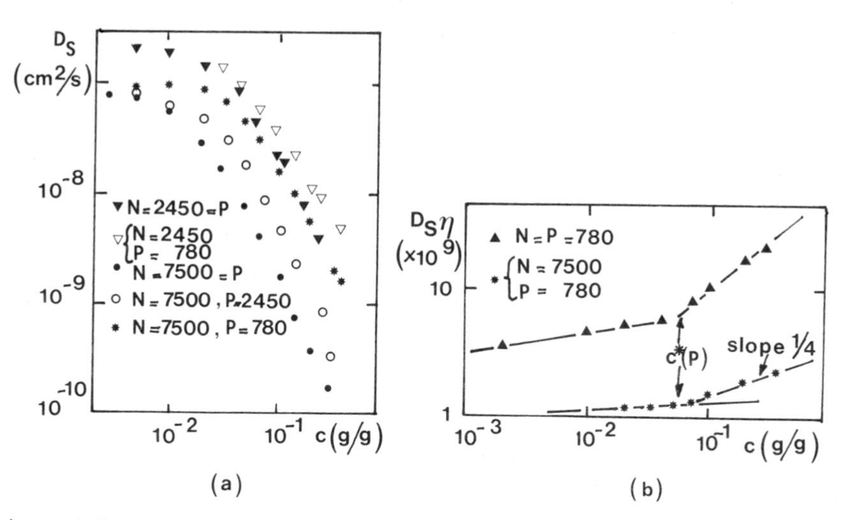

Fig. 4. (a) Evolution of the diffusion coefficient of the labelled chains for different molecular weights of the unlabelled chains ($P<N$).

(b) The product ηD_S with η the bulk viscosity of the solution shows that the big N chains diffuse as a Stoke's sphere swollen by the small P chains (from Ref. (22)).

more efficiently deswollen by the short P chain than by an ordinary solvent.[23]

The mixture of long and short chains also has striking effects on the viscoelastic properties. Thorough experimental studies of polymer melts with sharply peaked bimodal distribution of polymerization indexes,[24,25,26,51b] have shown qualitative discrepancies with the blending law which can be expected from the original Doi-Edwards model:

$$G(t) = \Phi_L G_L(t) + \Phi_S G_S(t) \,, \tag{6}$$

Φ represents the volume fraction and G the "monodisperse" relaxation modulus for the long (L) and short (S) chains respectively. Equation (6) grossly overestimates the stress, an effect now rather unanimously attributed to a failure of the "fixed tube" approximation.

II.5 Removing the hypothesis of fixed obstacles : the models In the context of the reptation model, the finite lifetime of the tube constraints leads to an extra relaxation process called tube-renewal or constraint-release.

The first attempts to model the process consisted of assuming the tube itself to be a Gaussian flexible chain made of N/N_e subunits of size d (the tube diameter), and thus obeying Rouse-type dynamics. The characteristic time for tube-renewal τ_{ren} can then be identified with the longest relaxation time of this chain :

$$\tau_{ren} \sim w_d^{-1} \left(\frac{N}{N_e} \right)^2 \,, \tag{7}$$

with w_d a jump frequency of the subunits, characteristic of the tube-renewal process, and to be determined.

II.5.1 Decorrelated subunits In the first approach, de Gennes[27] suggested that w_d was just the jump frequency of a free subunit of size d, $1/\tau_R(d)$, times the probability that an extremity of a matrix chain was located within the distance d from the test chain (a necessary condition for the corresponding constraint to be released), i.e., $w_d \sim 1/\tau_R(d)\,(2N_e/P)$ if P is the polymerization index of the matrix chains. This makes it possible to evaluate $\tau_{ren} \sim \tau_R(d)\; P/2N_e (N/N_e)^2 \sim \tau_{Rept}\, P/N$. In this description, for the particular case P = N, constraint release gives a contribution comparable to that of pure reptation (for independent processes, the inverses of the characteristic times add) and has the same scaling behaviour with molecular weight and concentration.

II.5.2 Correlated subunits A few years later, however, J. Klein noticed that, if the presence of an extremity of a matrix chain close (within d) to the test chain was a necessary condition for constraint-release, it was not a sufficient condition. If the matrix chain extremity remains in the vicinity of the test chain during a time comparable to or longer than the reptation time of the test chain, the constraint remains and the tube-renewal process vanishes. The diffusion dynamics of the matrix chain is thus a leading parameter. Following this line, Klein,[28] Daoud and de Gennes[20] and Graessley[30] have modeled the tube-renewal process using three slightly different approaches. They all identify w_d with the inverse of the reptation time of the matrix chains $1/\tau_{Rept}(P)$. On the average,

$\tau_{Rept}(P)$ is the time necessary for a chain forming the tube to bring its extremity into the immediate vicinity of the test chain. The characteristic time of the process is then:

$$\tau_{ren}(N) \sim \tau_{Rept}(P)\left(\frac{N}{N_e}\right)^2 \sim \tau_{Rept}(N)\left(\frac{P^3}{NN_e^2}\right). \tag{8}$$

For $P \geq N$, tube renewal becomes negligible, but more rapidly than what is observed experimentally (see the experimental results presented in section II.4).

II.5.3 Self-consistency The above description is only a perturbative one, in which tube-renewal is introduced as a correction to pure reptation, correction evaluated by assuming that all the chains in the medium move by reptation only. An obvious improvement of such a mean-field treatment is to introduce self-consistency. This amounts to write the parameters defining the tube as function of the evolution of the different types of chains in the medium, to reinject these (yet unknown) parameters in the dynamic equations for the chains in their tube, and to solve self-consistently the corresponding set of coupled differential equations. Decisive progress has been achieved recently:

For diffusion, the regime in which the constraint-release process remains a small correction to pure reptation is well described,[10,28] and appears in good agreement with experiments.[10,19,22] This is not at all the case for the regime where the number of entanglements along the test chain is smaller than 10.[19]

For viscoelasticity, only bimodal distributions of M_w have been considered in detail. Several recent approaches agree on a general factorization of the blending law for the relaxation modulus :

$$G(t) = \Phi_S C_{LS}(t, \Phi_L, \Phi_S) G_S(t) + \Phi_L C_{SL}(t, \Phi_L, \Phi_S) G_L(t) , \tag{9}$$

but the assumptions underlying the calculations of the intermolecular contributions C_{LS} and C_{SL} differ from one model to the other.[26,31,32,51b] In Eq. (9), $G_S(t)$ and $G_L(t)$ are the relaxation modulus evaluated from the Doi-Edwards approach, i.e., proportional to the memory function of the one-chain problem, $M(t)$, introduced in II.3.

A more original approach has been proposed by Marrucci[76] and Viovy[77] and, more recently, put into a concise sketch by des Cloizeaux.[33] In the framework of the reptation model, noting that a topological constraint between two chains, A and B, can be relaxed either (and independently) by the motion of A or B, des Cloizeaux suggests that the relaxation modulus should in fact be proportional to $M^2(t)$, with $M(t)$ obeying normal blending laws $M(t) = \Phi_L P_L(t) + \Phi_S P_S(t)$. $P(t)$ is the fraction of unrelaxed chains (long and short, respectively) at time t. This approach seems to lead to an interesting improvement of the agreement between model and experiments and calls for more detailed experimental tests.

In any case, whatever the model, since tube-renewal is more important (compared to reptation) and accelerates the motion more efficiently for short chains than for long chains, it introduces additional molecular weight dependences of the dynamical quantities, and certainly contributes to the experimental deviation of the viscosity/molecular weight exponent from the reptation value 3. All treatments, including tube renewal, exhibit such deviations which vanish for asymptotically long chains. Detailed quantitative tests are, however, very difficult to perform: when tube-renewal is taken into account, polydispersity becomes an essential parameter (the shortest mechanism, reptation and tube-renewal, dominates the relaxation process). No complete set of experiments, either for diffusion or for viscoelasticity, with constant polydispersity at all molecular weights, are presently available.

II.6 Higher order modes of relaxation In paragraph II.3, when describing the Doi-Edwards model, we have introduced three time scales : $\tau_R(d)$, $\tau_R(N)$ and τ_{Rept}. For time scales $t < \tau_R(d)$, the monomers move over distances smaller than d, and no entangled behavior is observable. On length scales larger than R_G, in contrast, the motions are dominated by reptation, which, in a mode language, represents the zero-order mode of the longitudinal fluctuations of the chain. The original reptation theory does not consider higher order longitudinal modes which imply local compressions or extensions of the chain in regard to its average curvilinear length. These modes have characteristic relaxation times ranging from $\tau_R(d)$ to $\tau_R(N)$, the longest Rouse relaxation time of the chain or time necessary to relax all its internal modes. The existence of these modes which can be activated either thermally or by a deformation of the system has important consequences on the viscoelasticity of polymer melts. In the step-strain experiment of Fig. 1, the sudden deformation of the entangled polymer medium locally changes the state of extension of the chains.[18] The internal modes re-equilibrate this extension and lead to a lowering of $G(t)$ in the plateau region. This contribution contains two terms : a linear one which "smoothens" $G(t)$ in the plateau region[34] and a strain-dependent term[35] which accounts for the observed decrease of the plateau modulus at large strain.[36,37]

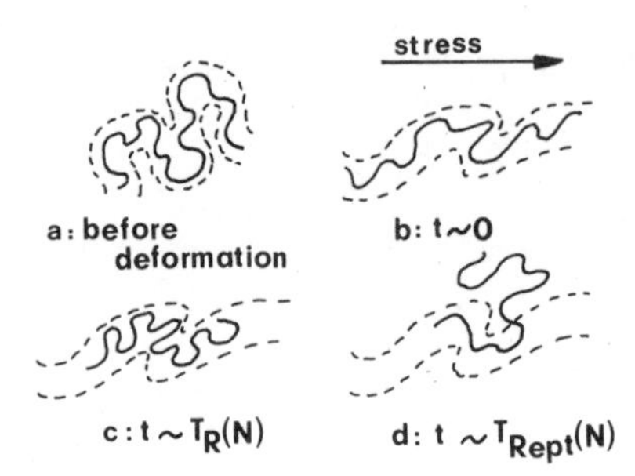

Fig. 5. Schematic representation of the different stages of the relaxation of a deformed sample, in the framework of the Doi-Edwards description:

(a) For times smaller than $\tau_R(N)$, the chain relaxes inside its deformed tube;

(b) For $\tau_R(N)<t<\tau_{Rept}(N)$, the chain progressively returns to isotropy, escaping from its deformed tube by reptation. The stress at time t is directly related to the fraction $M(t)$ of monomers still in the deformed tube at time t.

The thermal activation of the internal modes is at the origin of thermal fluctuations of the curvilinear length of the chain L_t.[38] The amplitude of these fluctuations is of the order of $L_t(N_e/N)^{1/2}$, and their longest characteristic time is $\tau_R(N)$. Therefore, during one reptation time of the chain, $\tau_{Rept} >> \tau_R(N)$, the chain has time to sample these fluctuations, and its configuration is renewed when it has reptated over a distance $L_t((1-2/\sqrt{N})$, rather than L_t for pure reptation. The fluctuations of the chain inside its tube thus introduce a negative correction term to the reptation prediction for the viscosity. This correction, similar to that of tube-renewal, has a relative importance which increases for low N values. This leads to an increased apparent molecular weight exponent for the viscosity, and a shift of the cross-over towards entangled behavior extracted from viscosity data ($N_c > N_e$). Inclusion of length fluctuation effects strongly improves the agreement between theory and experiment.[39] Recently, M. Rubinstein[51a] has proposed an elegant discretized model of the dynamics of a single chain which makes it possible to include chain fluctuations efficiently.

The most spectacular effect of non-zero order modes, however, should occur for branched chains. In this case, typically represented by "stars polymers" in which several identical chains (the arms) are chemically attached together by one extremity (the core), the reptation mode is strongly hindered by branching, and higher orders modes (essentially the breathing mode of order one) are responsible for the chain relaxation. Several theoretical treatments have been proposed[27,40,41] essentially relying upon the same physical idea : to renew its conformation, an arm must retract completely inside its tube (Fig. 6) and the rate of such events is given by the Boltzmann weight of such a retracted conformation, leading to a dominant term for the viscosity of a melt of stars

$$\eta \sim A\, exp(-\alpha N_{arm}/N_e) , \qquad (10)$$

with N_{arm} the number of monomers per arm. The way A and α vary with the number and the length of the arms, however, depends on the authors.

Present experiments agree with an exponential dependence[42] but are still inconclusive in this latter issue. Anyhow, constraint-release and polydispersity may be particularly important there, and a complete treatment of the dynamics of a melt of stars including all these aspects still has to be done.

III. RECENT TRENDS

III.1. Semi-local scale spectroscopy Viscoelastic properties, which convey informations on motions at all length scales have been studied first, owing to their practical importance, and there is a wide range of well-established data, reviewed in the book of J.D. Ferry.[1] An increasing number of diffusion measurements have come in the last few years, which probe motions on a macroscopic scale, and can thus be analyzed in the framework of the reptation model. These approaches are presently supplemented by spectroscopic studies probing the dynamics on the scale of the radius of gyration of the chains or smaller. Infrared dichroism,[43,44] fluorescence polarization[45] and N.M.R.[46,47] are sensitive to the orientation of the monomer segments, and yield an orientation correlation time. The experiments can be performed at rest or after the application of a macroscopic strain to the sample. Generally, the agreement with the reptation description is reasonable, but these techniques reveal a coupling between the orientation of the segments of one chain on a local scale, and its environment. This coupling, which seems to have only a weak effect on the rheological properties, is not accounted for by the reptation theory. These studies are just starting but appear to be a promising tool in the understanding of polymer dynamics. In particular, as they are only sensitive to local scales, subtle effects such as the evolution of the dynamics with the location of the probing segment along the chain, can be investigated.

Another quite powerful tool for discussing reptation on a molecular basis is time-resolved small-angle neutron scattering : the evolution with time of the structure factor after the step-strain or during an oscillatory deformation is recorded in all directions and in a wide wave vector range, in order to characterize the relaxation of the deformation on length scales ranging from the monomer size (a few Å) to the overall chain size (a few 1000 Å). Typical results obtained by F. Boué[48] on polystyrene, after a step uniaxial elongation, are reported in Fig. 7. The polystyrene sample (M_w = 780,000) is pulled rapidly in the fluid state, reaching elongation ratio of order 3. It is then allowed to relax during a time θ (10 sec < θ < 30 min) at a fixed temperature slightly above the glass transition, and quenched at room temperature. A certain fraction of polymer chains have been deuterated, and by suitable difference procedures, the diffraction pattern $S_\theta(\mathbf{q})$ of one chain in the partly elongated state can be extracted from the neutron diffraction data; $\mathbf{q}$ is the scattering wave vector and can lie either parallel to the stretching direction ($\mathbf{q}_{\parallel}$) or perpendicular ($\mathbf{q}_{\perp}$). Before the stretching $S(\mathbf{q})$ is isotropic ($S(\mathbf{q}) = S_i(\mathbf{q})$). After the stretching $S(\mathbf{q})$ becomes anisotropic ($S(\mathbf{q}) = S_a(\mathbf{q})$). After the time θ, a part of the chain has returned isotropy, and $S_\theta(\mathbf{q}) = M(\theta)\ S_a(\mathbf{q}) + (1-M(\theta))\ S_i(\mathbf{q})$ with $M(\theta)$ the memory function of the chain at time θ. These experiments are complex, and their interpretation is still incomplete, but striking qualitative features are already well established : for large scales ($\mathbf{q}R_G << 1$) the data agree with the reptation description, while the chain appears more relaxed than predicted at intermediate scales ($1/\mathbf{q} \sim$ a few tube diameter) and less relaxed than predicted for $1/\mathbf{q} \sim d$. This suggests that the reptation theory which includes the constraints in an average fashion through the notion of tube, does not take correctly into account all the degree of freedom available to the chain for minimizing the stress.

III.2 <u>The fundamental assumptions revisited</u> A renewed theoretical questioning of the founding assumptions of the reptation model has accompanied these more recent studies at molecular scales. The concept of entanglement, even if very useful, does not rely on molecular basis, and the question "What <u>is</u> an entanglement in a melt" is still relevant.

Extensive (and expensive) computer simulations of dense polymer systems have been performed[49,50,75] but they raised new questions without supporting unambiguously the reptation model. In particular, these simulations, which are restricted to moderately entangled systems in order to avoid prohibitively long computing times, do not show clear evidence of the unidimensional motion characteristic of the reptation.

Computer simulations, neutron-scattering data and the small enduring discrepancies between viscoelasticity data and theory suggest that the transition from "Rouse" to entangled behaviour has to be better understood. Very early, attempts to describe entanglements on a rigorous mathematical basis were made,[52] but the extreme difficulty of the problem did not permit really useful achievements. A rather opposite point of view was put forward by Curtis and Bird,[53] generalizing hydrodynamic theory without considering the topological aspect of entanglements. They

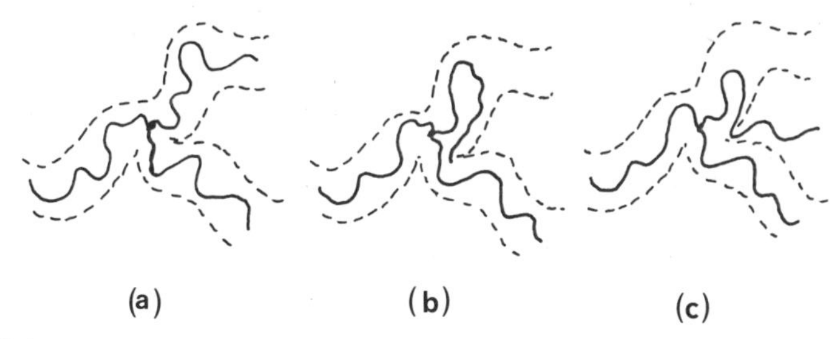

Fig. 6. Schematic representation of the relaxation process in a melt of star molecules.

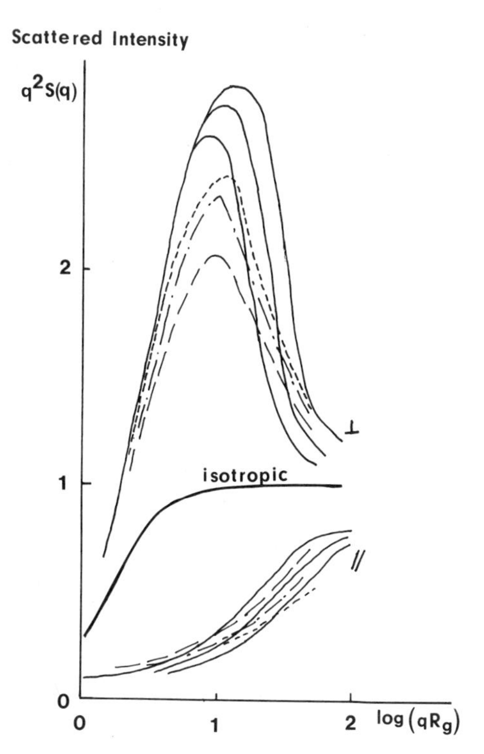

Fig. 7. Tentative fit of the structure factors predicted by the reptation theory (full lines) to neutrons small-angle scattering experiments performed after various relaxation times:

short dashes 15 sec.

dash-dot 40 sec.

long dashes 70 sec.

The sample is polystyrene, M_w=780,000, T=117°C (from ref. (48)).

introduced a non-isotropic friction coefficient into the phase-space kinetic theory, and were able to account for some features of entangled polymers, but apparently failed to provide a unifying frame consistent with the wide variety of available experimental observations. More recently, W. Hess[54] has applied the Zwanzig-Mori theory of generalized hydrodynamics to entangled polymer solutions : each chain interacts with its neighbour through excluded volume interactions, invariant upon curvilinear displacements (the chains freely "slip" along each other). By projecting the generalized equations of motion onto the chain axis, a "reptation transition" is revealed for sufficiently long chains. This theory which is still under development provides interesting predictions for the crossover between Rouse and reptation behaviour, which seem to well agree with self-diffusion data,[55] but the underlying physics is not as easy to extract as in the classical reptation approach. A more conservative, but also more comprehensive improvement of the reptation theory has been proposed very recently by Kavassalis and Noolandi.[56] This model, in which the number of segments between entanglements is derived by an enumeration of neighboring chains, seems to account for the difference between the critical molecular weight for interpenetration deduced from viscosity measurements M_c or from the plateau modulus M_e, and recovers the classical de Gennes' theory in the long-chain limit.

Progress must still be made before clear answers can be provided to the questions: "What is an entanglement?" and "How is N_e related to the nature of the polymer?"

III.3 Connection with other problems A number of practical situations are conditioned by dynamics on a semilocal scale. We shall draw up a list (of course incomplete) of such situations and briefly discuss their present state of understanding.

III.3.1 Dynamics of polymer mixtures Most polymers do not mix. However, certain exceptional polymer pairs A + B are compatible in the melt, at least in a certain temperature range. For some applications, one may want to produce a segregation of A and B chains, for example by a temperature jump. As for usual small molecules binary mixtures, several regimes of segregation exist, depending on how far into the two-phase zones of the phase diagram one penetrates by temperature quenching. If one stays in the "Spinodal decomposition zone", a fluctuation of the relative concentration A/B, with a well-defined wavelength λ^* becomes amplified.[57] For binary mixtures of small molecules, λ^* is comparable to the thickness of the interface between A_{rich} and B_{rich} domains and is thus very sensitive to the distance in temperature from the critical temperature. For polymer alloys, this is no longer the case, and $\lambda^* \sim R_G$,[58] much larger than the interface thickness in most cases.

The kinetics of phase separation is controlled by A/B interdiffusion. Polymer/polymer interdiffusion is an important field which has remained poorly understood for many years, and is indeed rather complicated. For the above binary mixture system, three distinct diffusion coefficients can be defined. The self-diffusion coefficients of A or B, and the mutual diffusion coefficient D_M, which all strongly depend on the respective A and B volume fractions.

The possible molecular weight and volume fraction dependence of D_M expected in the framework of the reptation model for all possible cases (attractive or repulsive interactions between A and B, different or identical molecular weights, entangled or non-entangled chains) have been computed recently by F. Brochard[59] and compared with the available experimental data. An important feature is that, when the polymer mixture is compatible, the Flory interaction parameter between the two species is negative (attraction) and large. There is then a chemical driving force for the mixing. The reptation under such a driving force corresponds to a diffusion coefficient inversely proportional to the molecular weight, and not to the square of the molecular weight.[60]

III.3.2 Polymer/polymer welding This is a particularly important situation in which semi-local dynamics plays an important role. Systematic experiments have been performed by Kausch and his coworkers,[61] measuring the fracture energy for two blocks of the same polymer put into close contact during a certain healing time t, above the glass temperature. The fracture energy G_{IC} increases like $t^{1/2}$ at small time interval.

This result has been interpreted by de Gennes in the framework of the reptation model[62] and by Prager and Tirrell.[63] When the two blocks are put into contact, the chains begin to migrate by diffusion across the interface. The healing process is essentially terminated for $t > \tau_{Rept}$. For $t < \tau_{Rept}$, the interdigitated region has a thickness smaller than R_G. The controlling parameter for the fracture energy is the number of bridges per unit area which can be established between the two blocks during the time t, $n(t)$. The reptation prediction yields $n(t) = n_{max}(t/\tau_{Rept})^{1/2}$ and thus seems to indicate that $G_{IC}(t)$ is just proportional to $n(t)$.

III.3.3 Flow of a polymer liquid close to a solid wall This is again a situation quite often encountered in practical situations, for example when a polymer liquid is extruded or laminated. It is quite important to correctly model the flow in such situations, as large velocity gradients may modify the conformation of the polymer chains and condition the final properties of the product.

Several experimental evidences[64] seem to indicate that the boundary condition for the flow of an entangled polymer close to a smooth solid wall is a slipping boundary condition (i.e., a velocity which remains finite at the wall). The model proposed by de Gennes[65] indeed indicates a slipping boundary condition, with an extrapolation of the parabolic velocity profile decaying to zero well below the solid interface, and charaterized by an extrapolation length $b \sim \eta/\eta_0$ with η the viscosity of the entangled polymer melt, η_0 that of a liquid of monomers, i.e., very rapidly increasing with the molecular weight. These predictions have still to be tested experimentally.

III.3.4 Rigid polymers The tube concept has also been applied to long, rigid macromolecules. In the description proposed by Doi,[66] the chain rotation is hindered by an elongated "cage" of entangled neighbouring chains. The predicted rotational diffusion coefficient scales line $c^{-2}L^{-9}$, and the translational diffusion coefficient should increase with concentration, the trajectory of the rod being less contorted because of the cage.[67] This theory has raised controversies[68] which settle rather slowly for several reasons : i) the idealized situation of completely rigid, "infinitely thin" polymers is not as easy to approach experimentally as the long flexible chain limit, ii) such systems generally present strong static correlations responsible for a nematic-like order,[69] which complicate the interpretation, iii) self-consistency is not easily introduced into the theory. Recent molecular dynamics simulations seem to confirm the essential features of the model.[67] Finally, it is worth mentioning recent progress in the reptation theory of semi-rigid polymers,[70] which should ultimately bridge the gap between the "flexible" and "rigid" limits considered so far.

III.3.5 Nonlinear viscoelasticity A most important issue in the dynamics of entangled polymers is the nonlinear viscoelasticity. For weak nonlinear situations in which the deformation rate has the same order of magnitude as $1/\tau_{Rept}$, the application of the reptation model has been successfully explored recently. Striking nonlinear effects such as the positive Weissenberg effect (responsible for climbing of a polymer melt along a rod rotating in it), or the anomalous sign of the stress in some double step-strain experiments have received a consistent molecular interpretation in the frame of generalized Doi-Edwards theories. Such theories are probably able to provide useful qualitative guidelines for numerous problems in polymer processing, such as drawing of fibers or films, molding or extrusion. However, getting a general and quantitative constitutive equation for complicated nonlinear deformations is still a controversial hope. In the limit of high shear and/or molecular weights, unexplained departures are observed. The chains are strongly deformed and aligned, even on a local scale, and the tube concept may become irrelevant. One major feature of nonlinear viscoelasticity is a progressive decrease of the viscosity beyond some range of deformation rate. Rather accurate predictions were obtained using a simple model due to Graessley.[71] In this model, the viscosity is obtained by a balance between the loss of entanglements carried out by the flow, and their replacement, which occurs on a time scale of the order of the relaxation time of the chain at rest, τ. Using the reptation time, τ_{Rept}, as an imput for τ leads to a good agreement with typical viscosity-shear rate reduced master curves.

IV. CONCLUSION

The reptation model has provided an intuitive unifying frame which qualitatively describes well a large variety of specific dynamic properties of polymers.[72] It has also raised a lot of controversies, and it still does to some extent. This is because the models developed by de Gennes, Doi-Edwards, and all the improvements based on this original approach, use the tube as an ad hoc constraint imposed on the chains. Therefore, accounting for a new situation or effect most often implies further assumptions (and controversies) about the tube. Very extensive experimental and theoretical studies have been developed in the last few years, and the scenery for a rather deep renewal of our understanding of entangled polymers in the following years seems to be settled.

NOMENCLATURE

a	size of monomer for an infinitely flexible chain (size of the Kuhn's length for a real chain)
c	monomer concentration
$c^*_{N(P)}$	first overlap concentration of chains made of N,(P) monomers
C_{LS}, C_{SL}	intermolecular contributions in the blending law of the elastic modulus of bimolecular samples
d	width of the tube
D_0	center-of-mass self-diffusion coefficient at infinite dilution
D_1	diffusion coefficient of a monomer (or of a Kuhn's unit)
D_d	diffusion coefficient of a chain with N_d monomers
D_M	mutual diffusion coefficient
D_t	curvilinear diffusion coefficient of a chain along its tube
D_s	center-of-mass self-diffusion coefficient
g	number of monomers in a subunit of size ξ
G^0	plateau elastic modulus

$G(t)$	elastic relaxation modulus
$G_L(t)$,$G_s(t)$	relaxation modulus for long (L) or short (S) chains
G_{IC}	fracture energy
η	bulk zero-shear viscosity
η_o	zero-shear viscosity of a liquid of monomers
k	Boltzmann constant
L	length of a rigid rod polymer
L_t	curvilinear length of the tube
$M(t)$	memory function of one chain at time t, i.e. fraction of monomers still inside the tube defined at $t = 0$
M_w	weight-average molecular weight
$n(t)$	number of bridges established per unit area, at time t, between two blocks put into close contact at time zero
N	polymerization index of a linear infinitely flexible chain
N_{arm}	number of monomers per arm of a star molecule
N_d	number of monomers (Kuhn units) in a portion of chain having a radius of gyration d
N_e	average number of monomers between entanglements (in a melt)
ξ	static monomer-monomer correlation length of a semi-dilute solution
ξ_H	hydrodynamic correlation length of a semidilute solution
$P(t)$	fraction of unrelaxed chains at time t
$\mathbf{q}$	wave vector transfer in a scattering experiment, $\mathbf{q}_{\parallel}$ ($\mathbf{q}_{\perp}$)parallel (perpendicular) to the stretching direction
R_G	radius of gyration of a chain
$S(\mathbf{q})$	scattering intensity function at the wave vector $\mathbf{q}$
t	time
T	temperature
τ_1	characteristic time of a monomer (or of a Kuhn's unit)
τ_{Rept}	reptation time of a chain
τ_{ren}	characteristic time of the constraint release process
$\tau_R(p)$ or $\tau_R(\lambda p)$	Rouse time of a chain with Np monomers, or characteristic time of the Rouse mode of wavelength $\lambda p \sim Np^{1/2}$ a

REFERENCES

1. J.D. Ferry, Viscoelasticity of Polymers, 3rd Edition, Wiley and Sons, New York (1980).
2. W.W. Graessley, Adv. Polym. Sc., 16, (1974).
3. P.G. de Gennes, J. Chem. Phys., 55, 572 (1971).
4. S.F. Edwards, Proc. Phys. Soc., 92, 9 (1967).
5. P.J. Flory, Statistics of Chain Molecules, Interscience Publisher, New York (1969).
6. P.E. Rouse, J. Chem. Phys., 21, 1273 (1953).
7. P.G. de Gennes, Scaling Concepts in Polymer Physics, Cornell University Press, Ithaca (1979).
8. P.J. Flory, J. Chem. Phys, 17, 303 (1949).
9. J.P. Cotton, D. Decker, B. Farnoux, G. Jannink, R. Ober and C. Picot, Phys. Rev. Lett., 32, 109 (1974).
10. M.F. Marmonier, Ph.D. Thesis, Université Paris VI (1985).
11. J.A. Wesson, I. Noh, T. Kitano and H. Yu, Macromolecules, 17, 782 (1984).
12. J. Klein and B.J. Briscoe, Proc. Roy. Soc. Lond., A365, 53 (1979).
13. L. Léger, H. Hervet and F. Rondelez, Macromolecules, 14, 1732 (1981).

14. C.R. Bartell, B. Crist and W.W. Graessley, Macromolecules, 17, 2702 (1984), and C.R. Bartel, B. Crist, L.J. Fetters and W.W. Graessley, Macromolecules, 19, 785 (1986).
15. P.F. Green, C.J. Palmstrom, J.W. Mayer and E.J. Kramer, Macromolecules, 18, 501 (1985).
16. M. Tirrell, in Rubber Chem. and Technol., 57, 523 (1984).
17. N. Nemoto, M.R. Landry, I. Noh, and H. Yu, Polymer Commun., 25, 141 (1984).
18. M.Doi and S.F. Edwards, J. Chem. Soc. Faraday Trans. II, 74, 1789 (1978); M. Doi and S.F. Edwards, Theory of Polymer Dynamics, Oxford (1986).
19. M.F. Marmonier and L. Léger, Phys. Rev. Lett., 55, 1078 (1985).
20. M. Daoud and P.G. de Gennes, J. Polymer Sci., Polymer Phys. Ed., 17, 1971 (1979).
21. P.F. Green and E.J. Kramer, Macromolecules, 19, 1108 (1986) and P.F. Green and E.J. Kramer, Phys. Rev. Lett., 53, 2145 (1985).
22. L. Léger, H. Hervet and M.F. Marmonier, to appear.
23. J.F. Joanny, P. Grant, L.A. Turkevich and P. Pincus, J. Phys. (Paris), 42, 1045 (1981).
24. H. Watanabé, T. Sakamoto and T. Kotaka, Macromolecules, 18, 1436 (1985).
25. J.P. Montfort, G. Marin and P. Monge, Macromolecules, 19, 393 (1986) and 19, 1979 (1986).
26. M.J. Struglinski and W.W. Graessley, Macromolecules, 18, 2630 (1985) and 19, 1754 (1986).
27. P.G. de Gennes, J. Phys. (Paris), 36, 1199 (1975).
28. J. Klein, Macromolecules, 11, 852 (1978).
29. P.G. de Gennes and L. Léger, Ann. Rev. Phys. Chem., 33, 49 (1982).
30. W.W. Graessley, Adv. Polym. Sci., 47, 67 (1982).
31. M. Rubinstein, E. Helfand and D.S. Pearson, Macromolecules, 20, 822 (1987).
32. M. Doi, W.W. Graessley, E. Helfand and D.S. Pearson, Macromolecules, 20, 1900 (1987).
33. J. des Cloizeaux, Europhys. Lett., 5, 437 (1988).
34. J.L. Viovy, J. Polym. Sci., Polym. Phys. Ed., 23, 2423 (1985).
35. M. Doi, J. Polym. Sci., Polym. Phys. Ed., 18, 1005, 1891, 2055 (1980).
36. K. Osaki and M. Doi, Polym. Eng. Rev., 4, 35 (1984).
37. W.W. Graessley, Polymers in Solution, W.C. Forsman ed., Plenum Press (1986).
38. M. Doi, J. Polym. Sci., Polym. Phys. Ed., 21, 667 (1983). J. Polym. Sci., Polym. Phys. Ed., 19, 265 (1981).
39. R.H. Colby, L.J. Fetters and W.W. Graessley, Macromolecules, 20, 2226 (1987).
40. M. Doi and N.Y. Kuzuu, J. Polym. Sci., Polym. Lett. Ed., 18, 775 (1980).
41. D.S. Pearson and E. Helfand, Macromolecules, 17, 888 (1984).
42. J. Klein, Macromolecules, 19, 105 (1986).
43. D. Lefebvre, B. Jasse and L. Monnerie, Polymer, 22, 1616 (1981); D. Lefebvre, B. Jasse and L. Monnerie, Polymer, 23, 701 (1982); D. Lefebvre, B. Jasse and L. Monnerie, Polymer, 25, 318 (1984) and D. Lefebvre, B. Jasse and L. Monnerie, Polymer, 26, 879 (1985).
44. J.P. Faivre, Z. Xu, J.L. Halary, B. Jasse and L. Monnerie, Polymer, 28, 1881 (1987).
45. F. Pinaud, J.P. Jarry, P. Sergot and L. Monnerie, Polymer., 23, 1575 (1982).
46. B. Deloche, A. Dubault, J. Hertz and A. Lapp, Europhys. Lett., 1, 629 (1986).
47. J.P. Cohen-Addad and R. Dupeyre, Macromolecules, 18, 1101 (1985).

48. F. Boué, Thesis, University Paris XI, (1982), and F. Boué, K. Osaki, R.C. Ball, J. Polymer Sci., Polymer Phys. Ed., 23, 833 (1985).
49. A. Baumgartner, K. Kremer and K. Binder, Faraday Symp. Chem. Soc., 18, 37 (1983).
50. J.M. Deutsch, Phys. Rev. Lett., 54, 56 (1983).
51 (a). M. Rubinstein, Phys. Rev. Lett., 59, 1946 (1987) and (b). M. Rubinstein and R.H. Colby, J. Chem. Phys., to appear.
52. S.F. Edwards and J.W.V. Grant, J. Phys., A 6, 1169 (1973).
53. C.F. Curtiss and R.B. Bird, J. Chem. Phys., 74, 2016 (1981).
54. W. Hess, Macromolecules., 19, 1395 (1986).
55. Y. Shiwa, Phys. Rev. Lett., 58, 2102 (1987).
56. T.A. Kavassalis and J. Noolandi, Phys. Rev. Lett., 59, 2674 (1987).
57. J.W. Cahn and J.E. Hilliard, J. Chem. Phys., 28, 258 (1958).
58. P. Pincus, J. Chem. Phys., 75, 1996 (1981).
59. F. Brochard-Wyart, in Fundamentals of Adhesion, Ed. L.H. Lee, Plenum Press, to appear in 1990.
60. P.G. De Gennes, J. Chem. Phys., 72, 4756 (1980).
61. K. Jud, H.H. Kausch and J.G. Williams, J. Mater. Sci., 16, 204 (1981).
62. P.G. de Gennes, C.R. Acad. Sci. (Paris), B291, 219 (1980).
63. S. Prager and M. Tirrell, J. Chem. Phys., 75, 5194 (1981).
64. R.H. Burton, M.J. Folkes, K.A. Narh and A. Keller, J. Mater. Sci., 18, 315 (1983).
65. P.G. de Gennes, C.R. Acad. Sci. (Paris), 288B, 219 (1979).
66. M. Doi. J. Polym. Sci., Polym. Phys. Ed., 19, 229 (1981).
67. J.J. Magda, H.T. Davis and M. Tirrell, J. Chem. Phys., 88, 1207 (1988).
68. M. Fixman, Phys. Rev. Lett., 55, 2429 (1985).
69. S. Venkatraman, G.C. Berry and Y. Einaga, J. Polym. Sci., Polym. Phys. Ed., 23, 1275 (1985).
70. A.N. Semenov, J. Chem. Soc. Faraday Trans., 2 82, 317 (1986).
71. W.W. Graessley, J. Chem. Phys., 47, 1942 (1967).
72. D.S. Pearson, to appear in Rubber Chemistry and Technology.
73. G. Weill, J. des Cloizeaux, J. Phys. (Paris), 40, 99 (1979).
74. P. Wiltzius and D.S. Cannell, Phys. Rev. Lett., 56, 61 (1986).
75. A. Kolinski, J. Scolnick and R. Yaris, J. Chem. Phys., 86, 1567, 7164 (1987).
76. G.G. Marrucci, J. Polym. Sci., Polym. Phys. Ed., 83, 159 (1985).
77. J.L. Viovy, J. Phys. (Paris), 46, 847 (1985).

THE ENTANGLEMENT CONCEPT IN POLYMER PHYSICS REVISITED

Tom A. Kavassalis and Jaan Noolandi

Xerox Research Centre of Canada
2660 Speakman Drive
Mississauga, Ontario L5K 2L1 Canada

ABSTRACT

In this paper we discuss some recent developments on entanglements in dense polymer systems. The model employed here describes tube constraints in polymer melts and solutions by an averaged topological parameter $\bar{N}$. Conjecturing that N is universal, we are able to predict the mean entanglement spacing, N_e, in terms of the individual polymer chain chemical and structural parameters. Good correlations with experimental data support the universality conjecture. A simple entanglement criterion is introduced in the model which predicts a mean-field like transition from entangled chains to unentangled chains at a specific degree of polymerization N_c. In the transition region, the model predicts that polymer chains are weakly entangled as compared to longer chains. In this region, the effective tube diameter assumes a chain length dependence which leads to a significant correction to the molecular weight scaling law for the zero-shear-rate viscosity. Similar scaling laws are developed for entanglements in polymer solutions and several universal relations are predicted.

A methodology is outlined for applying a Rouse description of polymer dynamics to entangled melts. A tube axis is defined by coarse-graining the polymer conformations using the mean entanglement spacing as a filtering length scale. By imposing a wavenumber-dependent friction coefficient we are able to incorporate entanglement effects in a manner which yields reptational motion for the collective modes of the polymer conformation. The shorter length scale modes remain Rouse like.

I. INTRODUCTION

The tube model[1] and reptation concept[2] have become the cornerstones of most of our understanding of the mechanical and transport properties of dense polymer systems. In 1967, S.F. Edwards[1] originated the tube model while studying rubber elasticity. The concept of a tube was introduced in order to describe the topological constraints imposed on a polymer chain by the neighbouring polymer strands in a lightly crosslinked rubber. The tube model successfully explained the dynamical properties of these systems.

In 1971, P.G. de Gennes[2] wrote a seminal paper on the stochastic diffusion of a long polymer molecule in a strongly crosslinked gel. In this system, the gel network imposes topological constraints on the motion of the mobile polymer chain and can be viewed as forming a tube surrounding the chain. De Gennes postulated that if the polymer chain is very long compared to the distance between the crosslinked strands of the gel, then the allowed motions could be described as the migration of noninteracting "defects" along the chain which will tend to create and destroy the tube at the ends. Two distinct relaxation times appear in this problem. The first relaxation time, which resembled a Rouse relaxation time, corresponds to the relaxation of the density of defects. This time scales with chain molecular weight as $\tau_d \propto M^2$. The second relaxation time corresponds to the time required for the polymer to disengage completely from its original tube into a new tube. This process requires a much longer time and scales differently with molecular weight $\tau_r \propto M^3$. The diffusion coefficient, for this model system, was shown to scale with molecular weight according to $D \propto M^{-2}$.

De Gennes recognized the similarity of these results with the scaling relations that were (experimentally) derived for polymer melts. He conjectured that even though all chains are mobile in a melt, a similar mechanism for motion may be involved in this system. The topological constraints in this case arise from chain entanglements and not crosslinks.

In 1978, Doi and Edwards[3] adopted this view and developed the tube theory for polymer melts. The main objective of this work was to derive the constitutive equations for polymer melts from a molecular level model. It was in this work that the tube concept and reptational motion were combined for a concise mathematical description of polymer melt dynamics.

Doi and Edwards derived theoretical expressions for various quantities including the relaxation spectrum; the stress relaxation modulus $G(t)$ and the in-phase and out-of-phase tranforms $G'(\omega)$ and $G''(\omega)$; the plateau modulus G_N; the steady state compliance J_e^o; the self-diffusion coefficient D_G; the complex viscosity $\eta(\omega)$; and the zero-frequency viscosity η_o. For example, the self-diffusion coefficient and the zero-frequency viscosity, according to the tube theory of polymer melts are

$$D_G = (a^2 k_B T) / (3 \zeta N^2 b^2) \quad , \tag{1}$$

$$\eta_o = (c \zeta N^3 b^6) / (12 a^2) \quad , \tag{2}$$

respectively. In the previous equation ζ is the monomeric friction coefficient, N the chain degree of polymerization, b the effective bond length, c the concentration, a the tube diameter and $k_B T$ the Boltzmann temperature. The two principal parameters in this model are therefore the monomeric friction coefficient ζ and the tube diameter a.

The tube model predictions for polymer melts have been extensively compared against experimental data. While there appears to be qualitative agreement between theory and experiment, the agreement is not quantitative and several discrepencies between theory and experiment exist. For example, the theoretical predictions of the zero-shear viscosity, η_o, are about 15 times larger than experimental values. Also, the molecular weight dependence of η_o is typically stronger than the $\eta_o \propto M^3$ predicted by the tube model. Most polymers show a molecular weight dependence of $\eta_0 \propto M^a$ where a is in the range 3.0 to 3.7 with 3.4 being the most typical value of a. Several researchers have proposed theoretical explanations for the discrepency in the viscosity power law. For example, Wendel and Noolandi[4] suggested that the reptational motion relaxation times would be greatly altered by the presence of a small number of long lived knots. Graessley[5] has proposed that the static tube picture is inadequate for finite length chains and the 3.4 power law is due to a broad cross-over regime between Rouse behaviour and true reptation behaviour. Curtiss and Bird[6] proposed that for finite length chains, the monomeric friction coefficient may possess a molecular weight dependence. Doi[7] has proposed a refinement of the tube model which includes tube contour length fluctuations that can account for departures from pure reptation behaviour. The present authors[8-10] have suggested that the tube diameter has a molecular weight dependence which translates into departures from the $\eta_0 \propto M^3$ scaling law. From the diversity of ideas proposed, it is very clear that the debate has not been settled yet.

In the original tube theory for polymer melts[3], and the various modifications and refinements[5,7], the monomeric friction coefficient and tube diameter are phenomenological parameters. It is important therefore to have a molecular level understanding of these parameters. In the present paper, we will concentrate on the tube diameter in polymer melts and solutions. Our main aim is to provide a quantitative description of the tube diameter in terms of the polymer properties.

In section II we review[8-10] a mean-field model for molecular entanglements. In this section we discuss the basic assumptions of our entanglement theory and introduce an entanglement criterion and a topological parameter with which we describe the tube. We demonstrate that a very simple entanglement criterion leads immediately to a molecular weight dependence of the tube diameter. In section III we apply the entanglement model to polymer melts and formulate scaling laws that describe the tube diameter and the molecular weight threshold for entanglements M_c, in terms of the chemical and structural properties of the polymer. These predictions are shown to be in good agreement with experimental data. In section IV we apply the entanglement model to polymer solutions in the semidilute regime. Here we predict scaling laws for the tube properties in terms of the polymer density and molecular weight. A "phase" diagram is discussed that describes the entangled and overlapped regimes in terms of the density and chain molecular weight. In section V we digress from the main subject and discuss polymer dynamics. In this section we outline a prescription for applying a Rouse-like model to entangled polymers. A polymer tube axis is defined by a coarse-graining operation using the entanglement spacing as a filtering length scale. We then calculate the dynamics of the tube using a wavenumber dependent friction coefficient to account for entanglements. The long wavelength modes of this tube axis correspond to reptational motion.

II. ENTANGLEMENT MODEL

In this section, we review a model for describing entanglements that we have recently proposed.[8-10] The purpose of this model is to relate the tube diameter to the relevant properties of the polymeric system. More specifically, we will concentrate on a quantity N_e defined as the mean number of skeletal or backbone bonds between entanglements. We will then take the definition of the tube diameter a in a melt as

$$a = C_\infty^{1/2} N_e^{1/2} \ell \quad , \tag{3}$$

where ℓ is the mean skeletal bond length and C_∞ is the polymer's characteristic ratio. In polymer solutions the tube diameter will be defined more generally.

In Fig. 1 we schematically depict the density of entanglements as a function of the degree of polymerization for linear chain polymers. The density of entanglements may be defined as

$$\rho_e = \frac{\rho_m}{N_e} \quad , \tag{4}$$

where ρ_m is the number density of skeletal bonds. For short chains, the probability of chain entanglement is less likely and ρ_e should therefore vanish as depicted in Fig. 1. For very long chains, the conventional view is that N_e and hence ρ_e should be independent of chain length. In this case the value of N_e is very specific to the polymer under investigation. For chains of intermediate length we presumably have a transition region as depicted in Fig. 1. The value of N_c, that is the critical chain degree of polymerization for entanglements, presumably lies somewhere in this transition region.

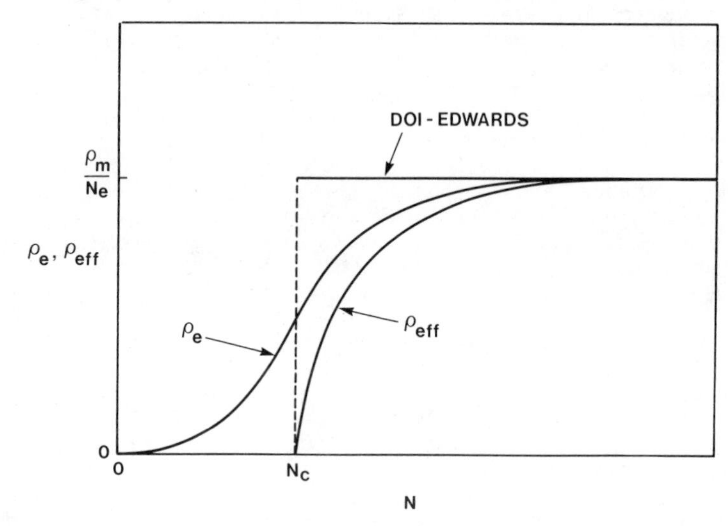

Fig. 1. The density of entanglements, ρ_e, versus degree of polymerization, N, depicted schematically for polymer melts. The variable ρ_e can be defined operationally as $\rho_e = \rho_m/N_e$ where ρ_m is the density of skeletal bonds and N_e is the mean number of skeletal bonds between entanglements. For short chains, ρ_e should vanish since short chains are less likely to be entangled. For long chains, the density of entanglements is independent of N which corresponds to a constant mean entanglement spacing N_e. For intermediate length chains, a transition region should exist where the chains are weakly entangled relative to the long-chain limit. The curve labelled ρ_e may correspond to the actual density of entanglements. This function may be difficult to define without a criterion for describing entanglements. The curve labelled ρ_{eff} may correspond to the density of those entanglements that persist longer than an appropriate microscopic time. Of particular interest would be the entanglements that are long lived compared to the time for rearrangement of chain segments shorter than N_e. The solid horizontal curve corresponds to the conventional assumption that N_e or ρ_e is independent of N for $N \geq N_c$.

The curve labelled ρ_e is difficult to define uniquely. This stems from the difficulty of defining an entanglement itself. For example if we define entanglements as those conformations that include at least on winding loop between a pair of chains, then we would obtain one specific curve. Another definition of an entanglement would yield another curve. The problem is made more difficult by the fact that entanglements in these systems are not static, that is they are continually being formed and undone. Also, many dynamic entanglement interactions may be totally ineffective in forming a "tube" around a given polymer chain. For example, a short-tail segment of a polymer strand that is entangled with another polymer may become undone faster than the characteristic Rouse time for the motion of a segment with N_e bonds. In this situation, the second polymer may evolve its conformation without experiencing a strong influence by the other. We will use this point later in developing a specific criterion for describing dynamic entanglements.

As we have already indicated, molecular entanglements are often described as topological interactions ranging from simple loops to perhaps even knots. The tube dimensions are then associated with the mean separation between such conformations. This description, apart from being difficult to quantify mathematically, ignores a common and perhaps more likely group of conformations that can also be used to describe a tube. This description of the tube is best illustrated by a series of thought experiments.

Consider a system of very long and completely rigid rods of length L and inscribe one labelled rod in a sphere of diameter D where $D \ll L$. Now begin passing other rods through the sphere, in a random fashion (Fig. 2), ignoring for the moment the finite thickness of the rods. As each new rod is added to the sphere, the probability that the labelled rod is trapped, or cannot be removed through lateral movements, increases. Eventually the probability of entrapment will level off to a value of 1 and the addition of more rods will only restrict the transverse mobility further. This thought experiment can be repeated to obtain an ensemble average for the mean number of rods required to constrain the mobility of the labelled rod.

Alternatively we can repeat the above experiment by first constructing an array of rods up to a desired rod density and then computing the sphere diameter that describes the lateral mobility of a labelled rod. An appropriate ensemble average here would yield a "tube" diameter for the system of rods.

The same approach can be used to describe the tube dimensions of a system of flexible chains in both solutions and melts. The system of interest here consist of monodisperse polymers with degree of polymerization N. Letting D_e represent the tube diameter and ρ_m the density of skeletal bonds in the system, we want to relate D_e and ρ_m to the mean number of polymers that contribute to the formation of the tube. Since the chains here are flexible, we expect that the average tube diameter is determined by both the presence of specific loop and knot like entities, involving pairs of chains, and entrapment geometries involving larger numbers of chains. The approach we use is a mean-field approach where we do not distinguish between different types of entanglement geometries.

We begin by inscribing a subsection of a labelled polymer, with N_e bonds, in a sphere of diameter D_e. The sphere diameter D_e is choosen to correspond to the mean tube diameter. N_e is the number of skeletal bonds whose mean end-to-end distance is D_e. In a polymer melt N_e and D_e are related according to

$$D_e = C_\infty^{1/2} N_e^{1/2} \ell \quad . \tag{5}$$

In general however D_e, may be a complicated function and we will chose the form

$$D_e = C(\rho_m, T) N_e^a \ell \quad , \tag{6}$$

which will allow us to discuss melts and solutions with a single model.

The contents of the sphere can be decomposed into two contributions as follows:

$$\pi \rho_m D_e^3 / 6 = N_e + \sum_{m=1}^{N} N(m)\, m \quad . \tag{7}$$

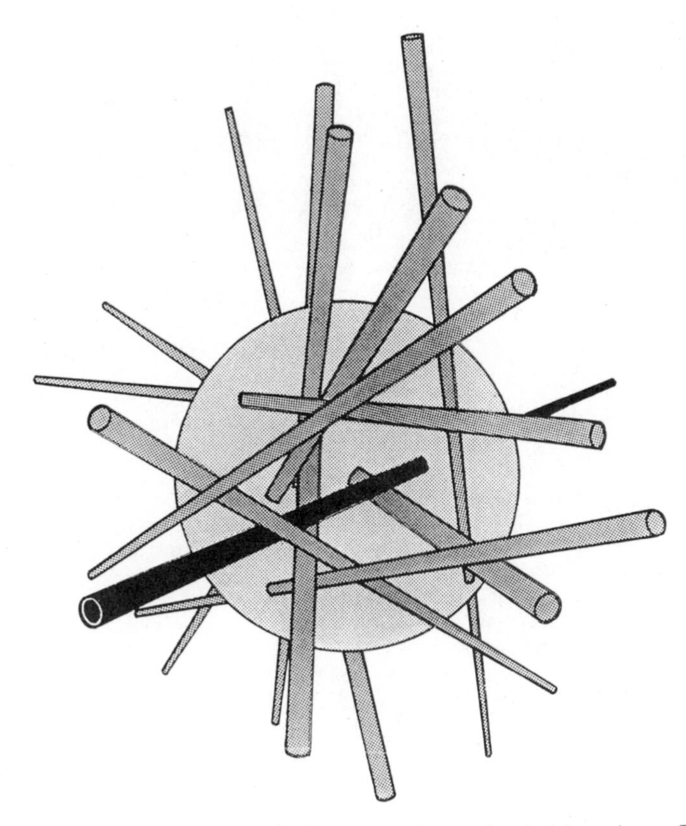

Fig. 2. Illustration of the "tube" for a system of rigid rods. The shaded rod corresponds to a labelled rod which is surrounded by a random array of other rods. The sphere is drawn here to represent the extent to which the rod is confined in space by the neighbouring rods.

The left-hand side of this equation represents the total number of skeletal bonds in the sphere. On the right-hand side we have expressed the sum of skeletal bonds as arising from N_e bonds of the labelled chain and a number of terms due to the other chains. The summation on the right-hand side is over the distribution function $N(m)$ which is defined as the mean number of polymer segments in the sphere that are exactly m-bonds long.

The development thus far has a distinct mean-field flavour. First, the N_e segments of the labelled chain were assumed to span a mean distance D_e, and therefore we are neglecting fluctuations in the length scale D_e. The total number of monomers in the sphere was specified by the left-hand side of Eq. (7) which is a representative mean value. Also, the contents of the sphere are described by the mean number of other polymer segments of length m. Preaveraging all of these quantities results in a classical mean-field model which we will elaborate on further in a later part of this section.

The distribution $N(m)$ is a statistical quantity and can therefore be computed from the conformational statistics of the chains. We can simplify this task by noting that the polymer segments can be classified into two contributions

$$N(m) = N_{tails}(m) + N_{nontails}(m) \quad . \tag{8}$$

In the last equation $N_{tails}(m)$ refers to the distribution function of the polymer segments that terminate in the sphere and $N_{nontails}(m)$ refers to segments that don't terminate in the sphere. Since the chains have a finite length, there is a nonzero probability that the sphere will contain some polymer end or tail-segments. For example, the density of tail-segments is $\rho_{tails} = 2\ \rho_m/N$ and therefore the mean number of tails found in a sphere with diameter D_e is $(\pi D_e^3 \rho_m)/(3N)$ which increases as the chains are shortened.

With the definitions in Eq.(8) we can rewrite our mean-field book-keeping equation as

$$\pi \rho_m D_e^3/6 = N_e + N_e \tilde{N} + \sum_{m=1}^{N} N_{tails}(m)\, m \quad , \tag{9}$$

where we have defined a new variable $\tilde{N}$ as

$$\tilde{N} = N_e^{-1} \sum_{m=1}^{N} N_{nontails}(m)\, m \quad . \tag{10}$$

The parameter $\tilde{N}$, which we call a coordination number, is a measure of the mean number of nontail segments that traverse the sphere. In reference (9) we discuss the interpretation of $\tilde{N}$ in more detail and the distributions $N_{tails}(m)$ and $N_{nontails}(m)$ are computed using a Monte-Carlo method. Here we will only give a brief argument that illustrates the interpretation of $\tilde{N}$.

Consider a change in the diameter of sphere from D_e to D_e' where $D_e' > D_e$. An enlarged sphere will enclose, on average, a longer portion of the labelled chain according to the scaling law in Eq. (6). Similarly, the enlarged sphere will enclose larger segments of the surrounding chains.

Since the conformational statistics of the surrounding chains are indentical to the labelled chain statistics, the average length of the nontail-segments will scale in the same manner as the labelled chain segment. So we may write

$$<nontail> \sim N_e \ , \tag{11}$$

where we have excluded a proportionality constant of order unity. The average nontail length can also be expressed as

$$<nontail> = \sum_{m=1}^{N} N_{nontails}(m)\, m \,/ \sum_{m=1}^{N} N_{nontails}(m) \sim N_e \ . \tag{12}$$

Noting that the summation in the denominator is the mean number of nontails in the sphere, $\bar{N}$, we arrive at Eq. (10). Therefore we see that $\bar{N}$, as defined by Eq.(10), is a measure of the number of neighbouring nontails.

The tail contribution in Eq. (9) can be computed in a similar manner. First we note that the tail segments are uniformly distributed in an amorphous system and therefore the mean number of tails in the sphere is

$$\sum_{m=1}^{N} N_{tails}(m) = (\pi \rho_m D_e^3)/(3N) \ . \tag{13}$$

Here we have defined both ends of the polymer to be tails. The average tail, segment length in the sphere is therefore

$$<tail> = \sum_{m=1}^{N} N_{tails}(m)\, m \,/ \sum_{m=1}^{N} N_{tails}(m) \ . \tag{14}$$

$$= \sum_{m=1}^{N} N_{tails}(m)\, m \,/ \left[(\pi \rho_m D_e^3)/(3N)\right] \ , \tag{15}$$

or

$$\sum_{m=1}^{N} N_{tails}(m)\, m \approx \pi \rho_m N_e D_e^3 / (6N) \ . \tag{16}$$

In going from Eq. (15) to Eq. (16) we have approximated the average tail length as

$$<tail> \approx N_e / 2 \ . \tag{17}$$

This relation follows from the uniform spatial distribution of tail-segments but is only approximate since both ends of the polymer cannot be independently uniformly distributed. The correlations between the two ends of a polymer chain, which are often of Gaussian form, would make Eq.(17) a rough approximation when D_e was comparable to the total size of a polymer chain.

Combining Eqs. (9), (12) and (16) and rearranging slightly we obtain

$$\tilde{N} + 1 = \frac{\pi \rho_m D_e^3}{6 N_e} (1 - N_e / N) \ , \tag{18}$$

which relates the variables ρ_m, N_e, N, D_e to the mean number of nontail-segments in the sphere.

At this point we introduce our strongest assumption in the entanglement model. We assume that the tube like constraints imposed on the labelled chain are due totally to the presence of the $\tilde{N}$ neighbouring nontail segments. Specifically, we have excluded the tail segments, that may be in the vicinity of the test chain, from contributing to the tube. While the tail-segments may be interacting with and topologically constraining the labelled chain, this sort of constraint will be short lived compared to the constraints formed by nontail segments.

Equation (18) is our fundamental relation between the tube dimensions or equivalently N_e and the polymer properties. To see this more readily we can substitute Eq. (6) into Eq. (18) and write

$$\tilde{N} + 1 = \pi \rho_m C(\rho_m, T)^3 N_e^{3a-1} \ell^3 (1 - N_e / N) / 6 \ . \tag{19}$$

On the right-hand side of this equation we have a number of material specific parameters such as ρ_m, N_e, N, ℓ and the function $C(\rho_m, T)$. According to this relation, this particular combination of material properties yields a pure number, $\tilde{N}$ + 1. We have conjectured[8,9] that this pure number is not material specific and is therefore universal. This hypothesis can be tested by comparing experimental data for polymer melts against Eq. (19).[10] These results will be reviewed in the next section.

Before discussing specific systems such as polymer melts and solutions, we can derive some general properties from Eq. (19). The scaling properties of the entanglement spacing, N_e, with density cannot be specified directly without a model for the function $C(\rho, T)$ and the exponent a. We can however consider some special limiting cases. For example, in the long chain limit, the bracketed quantity on the right-hand side of Eq. (19) goes to one and we have

$$\tilde{N} + 1 = \pi \rho_m C(\rho, T)^3 N_e^{3a-1}(\infty) \ell^3 / 6 \ , \tag{20}$$

where $N_e(\infty)$ refers to the mean number of skeletal bonds between entanglements in the infinite chain limit. Rearranging this equation to isolate the variable $N_e(\infty)$ we obtain

$$N_e(\infty) = \left[\frac{6(\tilde{N} + 1)}{\pi \rho_m C(\rho_m, T)^3 \ell^3} \right]^{\frac{1}{3a-1}} \tag{21}$$

$$\propto \rho_m^{-\left(\frac{1}{3a-1}\right)} C(\rho_m, T)^{-\left(\frac{3}{3a-1}\right)} \ . \tag{22}$$

For polymer melts and semidilute solutions, the functions $C(\rho_m, T)$ are known and in these cases we obtain[10]

$$N_e \propto \rho_m^{-2} \; ; \tag{23}$$

$$N_e \propto \rho_m^{-5/4} \; , \tag{24}$$

respectively. These examples are discussed in more detail in the next two sections.

From the structure of Eq. (19) we see that for any finite value of N, the corresponding value of N_e that satisfies the equation is larger than $N_e(\infty)$. This conclusion follows from the bracketed quantity on the right-hand side of Eq. (19) which is less than 1 for any value of N, thus requiring a larger value of N_e to satisfy the relation. However, it is clear that when $N_e/N \ll 1$, then N_e is not a strong function of N. As the variable N is decreased sufficiently the ratio of N_e/N can increase more rapidly than one would first expect since the value of N_e increases as N is decreased. This relationship and its properties can qualitatively explain the curve labelled ρ_{eff} in Fig. 1. For very long chains, N_e, and hence ρ_e, are insensitive to the chain degree of polymerization. For shorter chains the degree of entanglement is lower, corresponding to a larger mean spacing between entanglements. The dilution of entanglements here is due to the increase in density of polymer end segments which according to our entanglement criterion are not effective in forming long lived entanglements.

The present model also predicts a lower critical value of chain degree of polymerization, N_c, below which the polymer chains are not entangled according to our definition of the tube. In references (8,9) we showed that the lower critical value satisfies

$$N_c = 3\,N_e(N_c) \tag{25}$$

Any system of chains with $N < N_c$ will not satisfy the entanglement criterion. In other words, for a system with $N < N_c$, any sphere drawn around the entire chain with diameter $D_e = C(\rho_m, T)\, N^{\alpha} \ell$ will contain fewer than N nontail-segments from other chains.

III. POLYMER MELTS

In this section we apply the entanglement model, reviewed in section II, to monodisperse polymer melts. We make specific predictions for the mean entanglement spacing, N_e, as a function of the polymer melt properties. The expressions derived here relate N_e to the structural parameters of the polymer chain and predict correlations between the mechanical properties of melts, such as the plateau modulus, and the microscopic properties of the individual chains. We also discuss the predicted chain length dependence of N_e and the implications on the rheological properties such as the plateau modulus and zero-shear viscosity. Where possible, we compare our theoretical predictions against experimental data. A more general approach for polydisperse systems can be found in reference (9).

General Considerations The application of the entanglement model to polymer melts proceeds as follows: In a polymer melt, chain conformations are known to be ideal, that is the large scale properties obey Gaussian statistics. This result, which was surprising at first, has been explained using several different arguments[11,12] and is known today as Flory's theorem.[12] Without going into the details of this theorem, we will simply say that the ideality stems from the strong screening of excluded volume interactions in a melt, and proceed to use this result in our development.

The mean spacing between entanglements, N_e, is typically on the order of 10^2 in polymer melts, and therefore the chain conformations can adequately be described by Gaussian statistics on this length scale. Therefore the sphere of tube diameter D_e appropriate to polymer melts is given by

$$D_e = C_\infty^{1/2} N_e^{1/2} \ell \tag{26}$$

where C_∞ is the characteristic ratio and ℓ is mean skeletal bond length. Substituting this into Eq. (18), our basic equation describing entanglements in polymer melts is

$$\bar{N} + 1 = \frac{\pi \Psi}{6} N_e^{1/2} (1 - N_e / N) \ , \tag{27}$$

where we have introduced a dimensionless constant Ψ, that contains various parameters. In the last equation Ψ is given by

$$\Psi = \frac{\rho\, C_\infty^{3/2} \ell^3}{\mu_m} \ , \tag{28}$$

where ℓ is the mean skeletal bond length, ρ the mass density and μ_m the monomer mass per skeletal bond.

With equation (27) we can compute the value of N_e as a function of the degree of polymerization N. The differences between different polymeric systems enter through their respective values of Ψ. In Fig. 3 we have computed several curves of N_e vs N for values of Ψ ranging from 1 to 4. Each of these curves exhibits the properties discussed in section II. For example, for large N, the value of N_e is independent of N. The plateau value of N_e is determined entirely through the value of Ψ. From Eq. (27) we can see that the plateau value of N_e is

$$N_e(\infty) = \left[\frac{6\,(\bar{N} + 1)}{\pi \Psi} \right]^2 \tag{29}$$

$$= \left[\frac{6\,(\bar{N} + 1)\,\mu_m}{\pi\, \rho\, C_\infty^{3/2} \ell^3} \right]^2 \ . \tag{30}$$

For any finite value of N, the corresponding value of N_e is larger, but the difference betwee $N_e(N) - N_e(\infty)$ is only significant as we approach N_c.

In Fig. 3 we can see that the value of N_c also varies with the value of Ψ. From Eq. (27) it follows[9] that N_c satisfies the relation

$$N_c = 3\, N_e(N_c) \ . \tag{31}$$

However since N_e varies with N, it is more convenient to express N_c in terms of $N_e(\infty)$. Doing so we obtain[10]

$$N_c = 27/4\ N_e(\infty) \quad . \tag{32}$$

The exact solution of Eq. (27) can be obtained from straightforward algebra and is

$$N_e = (4/3)\, N \cos^2 \left| (1/3)\, \mathrm{arcos}(N_c / N)^{1/2} + 4\,\pi/3 \right| . \tag{33}$$

In section II, we labelled the present model as a mean-field model. We now explore the analogy to other mean-field models further. Squaring both sides of Eq. (27) and rearranging terms we obtain

$$N_e^3 - (2N)\, N_e^2 + N^2 N_e - \left[\frac{6\,(\tilde{N} + 1)}{\pi \Psi} \right]^2 N^2 = 0 \, . \tag{34}$$

This equation has the same form as the familiar van der Waals equation of state for a gas[13]

$$V^3 - \left(b + \frac{RT}{p} \right) V^2 + \frac{a}{p}\, V - \frac{ab}{p} = 0 \quad , \tag{35}$$

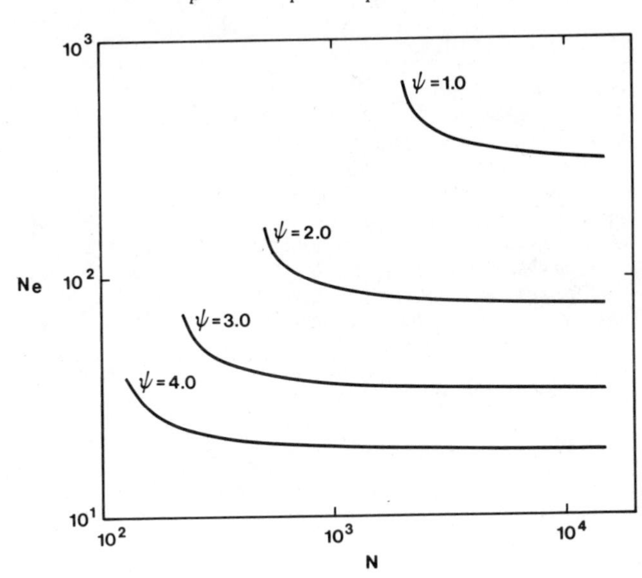

Fig. 3. Computed values of the mean numbers of skeletal bonds between entanglements, N_e, versus degree of polymerization, N, for a melt. Here we have used a value of $\tilde{N}$ = 8.1 and several values of $\Psi = \rho\, C_\infty^{3/2} \ell^3 / \mu_m$. The dimensionless constant Ψ characterizes the polymer system completely in this model. The asymptotic values of N_e are $N_e(\infty) = [6\ (\tilde{N} + 1)/\pi\Psi]^2$ and N_c is given by $N_c = 3^5\ [(\tilde{N} + 1)/\pi\Psi]^2$. An entanglement is defined here as consisting of $\tilde{N}$ neighbouring chain segments excluding tail segments. The value of $\tilde{N}$ we used was derived from an average of the experimental data in Table 1.

where a and b are positive, material specific constants. It is not surprising therefore that the critical properties of the entanglement model are similar to those of the van der Waals equation. We can see this directly by expanding the exact solution, Eq. (33) , around the critical point at N_c

$$\frac{N_e(N) - N_e(N_c)}{2N_c} = -\frac{1}{\sqrt{3}}\left(\frac{N - N_c}{N_c}\right)^{1/2}. \qquad (36)$$

The critical exponent of 1/2 appears in the expansion of the volume in the van der Waals equation around the liquid gas critical point as well as in other mean-field models.

The mean-field nature of our entanglement model is perhaps its weakest point. It has long been recognized that mean-field theories are rather rough approximations near the critical points for systems with $d < 4$ dimensions.[13] For example, the experimentally measured exponent for the liquid-gas critical point is $\delta = 0.33$ and not $\delta = 0.5$. Mean-field theories generally predict more abrupt transitions than are actually observed. The discrepency between mean-field theories and experiments stems from the absence of fluctuation phenomena which are often more important near the transition.[13]

The Long Chain Limit In this part we consider a melt of polymer chains far away from the transition, where our mean-field approximation should be of little consequence. In the long chain limit our equation relating the mean entanglement spacing to the polymers properties of the polymer is

$$\tilde{N} + 1 = \left[\frac{\pi \rho C_\infty^{3/2} \ell^3}{6\mu_m}\right] N_e^{1/2} . \qquad (37)$$

In section II, we conjected that this particular combination of parameters should yield a universal number. A critical test of this hypothesis is therefore to determine the accuracy of this relation.

In Table 1, we have reproduced[14] a compilation of experimental values of C_∞ and N_e for a variety of polymers along with respective values of ρ and ℓ. The values of N_e are derived from the experimental values of the plateau modulus, G_N, which is related to N_e according to

$$G_N^o = \frac{\rho_m k_B T}{N_e} . \qquad (38)$$

In column 6 we report the value of $\tilde{N} + 1$ computed from Eq. (37). In column 7, we compute an additional quantity K, which follows from a "universal" relation proposed by Graessley and Edwards[15] using dimensional analysis arguments. The expression for K, used here, is

$$K = \left[\frac{\rho C_\infty \ell^3}{\mu_m}\right]^{1/2} N_e^{1/2} , \qquad (39)$$

Table 1. Polymer Structural and Entanglement Properties Compared Against Entanglement Models (a)

	ρ g /cm^3	$\ell \times 10^8$ cm	μm Na g	C_∞	N_e	$\tilde{N}+1$	K
Polystyrene	1.007	1.54	52	9.4	259	10.3	10.2
Poly(α-methyl-styrene)	1.04	1.54	59	10.1	173	8.59	8.23
Polybutadiene (high cis)	0.9	1.47	13.5	4.9	174	9.53	10.4
Polyisobutylene	0.89	1.54	28	6.2	252	8.95	10.5
Polyisoprene (high cis)	0.9	1.47	17	5.3	238	9.95	11.3
Hydrogenated polyisoprene	0.854	1.54	17.5	6.8	84	9.16	7.83
Poly(vinyl acetate)	1.14	1.54	43	9	162	10.5	9.22
Polybutadiene, $\varkappa_{12}$=0.08	0.896	1.47	14.1	5.1	101	7.59	7.91
Polybutadiene, $\varkappa_{12}$=0.43	0.9	1.47	17.2	6.1	108	8.48	8.12
Polybutadiene, $\varkappa_{12}$=0.99	0.883	1.54	26.7	7	160	8.90	9.03
Polyethylene	0.802	1.54	14	7.4	52.6	9.27	6.91
Hydrogenated polybutadiene, $\varkappa_{12}$=0.43	0.832	1.54	17.8	6.4	95.5	8.53	7.93
Hydrogenated polybutadiene, $\varkappa_{12}$=0.99	0.819	1.54	27.7	5.5	386	8.64	11.8

(a) Data taken from reference (14). $\tilde{N}+1$ computed from present work (Eq. (37)), K is from Graessley and Edwards[15] (Eq. (39)).

which is essentially the square root and reciprocal of the original expression proposed by Graessley and Edwards[15]. This particular form of their expression was chosen in order to do a quantitative comparison of the two relationships. In the experimental data, in Table I, the variable with the greatest uncertaintly is N_e, hence it is desirable that the two relations be compared with the same power of the variable N_e.

The mean value of $\tilde{N}$ + 1 and K are 9.1 ± .8 and 9.2 ± 1.5 respectively. The error reported here is the standard deviation and appears to be significantly smaller in the $\tilde{N}$ + 1 value than the K value for the data set considered. Both relations do remarkably well given that the N_e values vary by about 700%, the C_∞ values vary by 100% and the μ_m values vary by 400%. The standard deviation, as a percentage of the mean value, for $\tilde{N}$ + 1 is 8% and for K is 16%. The computed values of $\tilde{N}$ are also within physical reason. Too large ($\tilde{N} \sim 100$) or too small (~ 1) a number would have been suspect. The fact that the mean K value is close to the mean $\tilde{N}$ values is probably a coincidence, since no obvious physical interpretation can be ascribed to K.

Several other researchers have attempted to correlate mechanical properties such as G_N (or N_e) to molecular properties and have met with varying degrees of success. For example, Privalko[16] had deduced an equation similar to Eq.(37) by using dimensional analysis. Still others have used statistical approaches to establish correlations. For example Aharoni[17] inferred a correlation between the characteristic ratio and the critical chain length for entanglements of the form $N_c \propto C_\infty^2$. These correlations were rather weak since the monomer mass, bond length and polymer density were omitted from the relation. Recently Lin[14] has independently proposed a universal relation between N_e and the polymer parameters that is identical to our result in the long chain apart from the factor of $\pi/6$.

The Transition Regime Now we study the transition regime where the chain degree of polymerization is above N_c but less than about 30 N_c. In this range of N values, we have predicted by Eq. (33), and illustrated in Fig. 3, that polymer chains are weakly entangled as compared to longer chains and therefore the tube is dilated. We will show here that the predicted variation of N_e with N translates into departures from reptation theory scaling laws in the range $N_c \leq N \leq 30\ N_c$.

The maximum variation of N_e, that is the value of N_e in the long chain limit to the value of N_e for chains with $N = N_c$ can be computed from Eqs. (31) and (32)

$$N_e(N_c)/N_e(\infty) = 9/4 \quad . \tag{40}$$

Note that this ratio is independent of the value of N and the polymer properties ρ, ℓ, C_∞ and μ_m and is therefore predicted to be universal. Since the tube diameter a is proportional to $N_e^{1/2}$, the predicted variation in the tube diameter is

$$a(N_c)/a(\infty) = 3/2 \quad , \tag{41}$$

where $a(N_c)$ is the effective tube diameter for a system of chains with degree of polymerization N_c and $a(\infty)$is the tube diameter of long chains. The variation in tube diameter that the model predicts is therefore small. We will show later that the effect on properties such as the zero-shear viscosity is significantly larger.

The ratio $N_e(N_c)/N_e(\infty) = 9/4$ translates into a similar variation of the plateau modulus

$$G_N^o(N_c)/G_N^o(\infty) = 4/9 \quad , \tag{42}$$

with the lowest value of the plateau modulus occuring at N_c. A variation of the plateau modulus in the regime $N_c \leq N \leq 30N_c$ has not been reported in experimental studies. However, the plateau region is seldom resolved for $N \sim N_c$ because of it's narrow width and the strong influence of polydispersity. Although the effect is not documented, a visual examination of stress relaxation master curves[18] does show that the curves generated for $N \approx N_c$ are somewhat separated from the higher molecular weight curves in time domain where the plateau region is expected. These curves are usually drawn on a logarithmic scale where it is difficult to resolve differences of 4/9. A confirmation of the effect awaits further experimental investigation.

The variation of N_e with N has greater consequences for the zero shear viscosity where the entanglement spacing appears as a squared quantity. According to the reptation/tube theory[1,3] of polymer melts, the zero-shear viscosity scales with degree of polymerization and entanglement spacing according to

$$\eta_o \propto N^3/N_e^2 \; . \tag{43}$$

Since N_e is predicted to be larger, by a factor of 9/4, at N_c, the zero-shear viscosity is therefore lower by a factor of about 5 than would be predicted by reptation theory using a constant value of $N_e = N_e(\infty)$.

In our entanglement model, N_e is given by the rather complicated function given in Eq. (33). If we substitute this equation into Eq. (43) we do not derive a simple scaling correction to the 3.0 power law. If we however insist on representing Eq. (43) as a scaling law, the appropriate expression is

$$\eta_o \propto N^{3+\delta(N)} \quad , \tag{44}$$

where the scaling correction $\delta(N)$ is a continuous function of N for $N > N_c$ and has a value of about 1/2 for $N \approx N_c$ and tends to zero for higher N. The deviations predicted from the 3.0 power law, persist for about one and a half decades of molecular weight and disappear for higher molecular weight.

Equations (43) and (33) therefore predict that the entanglement variation contributes to the deviation from reptation theory in a manner similar to Doi's length fluctuation model.[7] Unlike the fluctuations approach, we have not introduced any parameters to adjust the magnitude of the effect. We might say the present approach reduces the number of parameters in the problem by replacing N_e by a relation which contains the polymer properties (ρ, μ_m, C_∞, ℓ, N) and a constant N which we demonstrated to be universal. We are not however suggesting that the entanglement model corresponds to the length fluctuations that Doi has discussed. These length fluctuations are an important factor and should be included in the description of entanglements.

The addition of fluctuations would probably smoothen and extend the width of the transition as in the liquid-gas problem.

We can test our predictions for the scaling correction $\delta(N)$ to the zero-shear viscosity by comparing them against experimental data. In Fig. 4 we have plotted experimental values of the zero-shear viscosity of polybutadiene versus molecular weight as reported by Colby, Fetters and Graessley.[19] The continuous curve in this figure was computed from Eqs.(43) and (33) using the polymer data in Table 1. The viscosity values were divided by M^3 to remove the pure reptation component. Therefore any non-zero slope in Fig. 4 represents a departure from reptation theory which corresponds to the function $\delta(N)$ in Eq. (44).

The first twelve points in Fig. 4 from left to right correspond to molecular weight $M < M_c$, hence the negative slope of about -2. The curve we have computed of η_0/M^3 varies sufficiently to account for the magnitude of

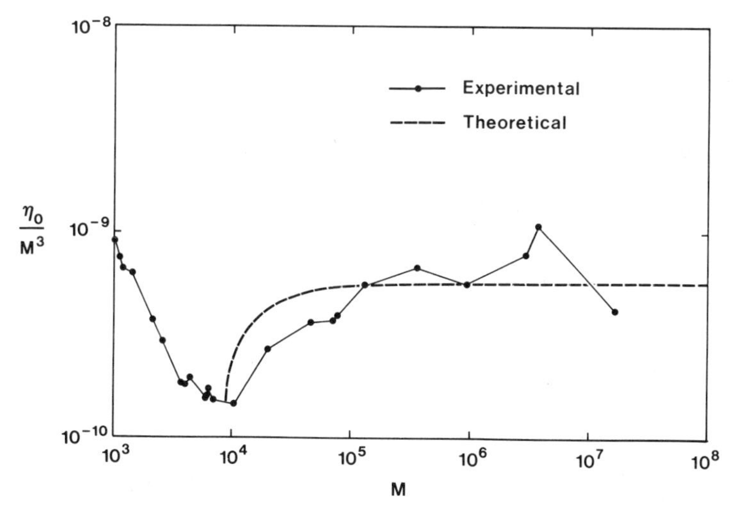

Fig. 4. The zero shear viscosity, η_0, as a function of molecular weight, M, of polybutadiene as reported by Colby, Fetters and Grasseley.[19] The data in this plot has not been corrected to the iso-free-volume state. The viscosities are reduced by M in order to remove any pure reptation component. Therefore, any non-zero slope represents a departure from reptation theory. The negative slope of about -2 for the first dozen points corresponds to the Rouse regime ($M<M_c$) where the reptation theory is not applicable. Beyond M_c a positive slope of approximately 0.4 extends for about one and a half decades and then levels off to zero slope although the data is somewhat oscillatory here. The theoretical curve expresses the departure from reptation theory computed from our entanglement model where N_e has the molecular weight dependence of Eq. (33). This curve was computed using the mean value of $\bar{N}$= 8.1 and the polymer properties from Table 1. The magnitude of the effect predict by the theoretical curve is in agreement with the data. The fit in the transition region is somewhat poor. In this region the theoretical curve varies more rapidly than the experimental data which like other mean-field theories is due to the absence of fluctuation phenomena.[13]

the deviation from pure reptation theory, but the fit is somewhat poor. The data varies more slowly than the computed curve. The lack of quantitative agreement we feel is a consequence of the mean-field nature of the entanglement model.

The Critical Degree of Polymerization at N_C The presence of entanglement effects can be probed in a number of different ways. One common method is to plot the zero-shear viscosity as a function of molecular weight and determine the point at which a change in slope from $\eta_0 \propto M$ occurs. The value of M_c determined in this manner is typically in the range of $2M_e$ to $3M_e$ although some polymers such as the methacrylates yield larger values of about $6M_e$.[20] Perhaps a more accurate estimate of M_c would come from a plot of η_0/M^3 versus M, as we illustrated in Fig. 4 for polybutadiene. In this type of plot, the values of η_0/M^3 only span one decade for a four decade span in molecular weight. Also, the change in slope expected at M_c is large, changing from a value of -2 below M_c to a small positive number of about 0.4 above M_c. However after reducing the viscosity values by M^3 we see significantly more scatter in the data relative to plot of η_0 versus M.

The determination of M_c in this manner is not as simple as we have stated. For example, the viscosity measurements for samples spanning four decades in molecular weight are frequently made at different temperatures and are often made using different methods in the high and low molecular weight limits. One then must correct the data to a common value of the monomeric friction coefficient. This in turn requires knowing the temperature dependence of the viscosity for each molecular weight, which is often imprecise and also subject to an empirical analysis. Once these corrections have been applied to the data, the value of M_c is determined by extrapolating between limiting models such as the Rouse and reptation models. The value of M_c may therefore be biased depending on the correctness of the dynamic models. Finally, the molecular weight and viscosity determinations are also subject to errors.

The onset of entanglement effects can also be observed by measuring the molecular weight dependence of the creep compliance.[20] Above a certain molecular weight, the creep compliance exhibits a rubbery plateau value, J_e^o, which is associated with a transient network of entanglements. Using this method one observes that the onset of entanglements occurs at a higher molecular weight M_c than the viscosity method. Values of M_c are typically about $7M_e$. A major drawback of this method is the high sensitivity of the value of J_e^o to polydispersity.[20] The physical origin of the difference between M_c values and M_e values has been discussed in the literature[20] but is still not fully resolved.

In Table 2 we list a few values of M_e, M_c and M'_c from experiments on polymer melts.[20] We also have tabulated the ratios M_c/M_e and M'_c/M_e which have average values of 2.4 and 7.4 respectively. One readily notices that there is less scatter in the ratio M'_c/M_e. This may be due to the uncertainty in the methods used to determine M_c. The entanglement model that we have developed predicts a single transition which according to Eq. (32) occurs at

$$N_c = 6.75\, N_e \,, \qquad (45)$$

Table 2. Entanglement Parameters[a]

	M_e	M_c	M_c/M_e	M_c'	M_c'/M_e
Polystyrene	18,100	31,200	1.7	130,000	7.2
Poly(α-methyl-styrene)	13,500	28,000	2.1	104,000	7.7
1,4-Polybutadiene	1,900	5,900	3.1	13,800	7.3
Poly(vinyl acetate)	12,000	24,500	2.0	86,000	7.2
Poly(dimethyl siloxane)	8,100	24,400	3.0	61,000	7.5

(a) Data from reference (20). M_c values are extrapolated from viscosity measurements, M_c' values are from steady-state compliance measurements and M_e values are from plateau modulus measurements.

or

$$M_c / M_e = 6.75 \quad . \tag{46}$$

This ratio is close to the ratios that are typically reported from steady state compliance measurements. The present model however offers no explanation for the difference between M_c and M'_c.

IV. APPLICATION TO POLYMER SOLUTIONS

In this section we apply our entanglement model to solutions of long linear chain polymers. Here we are interested in describing the entanglement parameter N_e as a function of the concentration of solvent. We are also interested in characterizing the transition from entangled chains to overlapped chains that occurs with the addition of solvent. Several of our results parallel earlier developments by P. G. de Gennes[11] on overlap of polymer chains in solution. The remainder of this section deals with semidilute solutions. An analogous discussion of concentrated θ solutions can be found in reference (10).

In semidilute solutions, the polymer volume fraction is low compared to a melt but the polymer chains are overlapped. We are particularly interested in solution concentrations where the chains are entangled and not simply overlapped. We therefore require that the chain degree of polymerization be larger than the critical chain length for entanglements in a melt of pure polymer. It is obvious that the concentration range over which a solution is entangled will depend on chain length and that longer chains will be entangled at lower concentrations as compared to shorter chains.

For our entanglement model, an entanglement volume is defined by the volume that contains an average $\tilde{N}+1$ nontail polymer segments. We have demonstrated that $\tilde{N}+1$ is approximately 9 by analyzing polymer melt data. Assuming that entanglements in solution can be defined using the same entanglement criterion, we conclude that the excluded volume extraction is important on the length scale of entanglements. Therefore we will assume that the Flory theorem holds in this situation and that the large-length scale conformational properties of the chains are Gaussian.

Neutron-scattering studies[21] on polystyrene solutions have shown that while short-length scale correlations are perturbed, in the same manner as single chains in good solvent, the large-length scale properties are indeed Gaussian. Daoud[21] predicted that the mean squared end-to-end distance of the chains is swollen relative to a pure melt but can be described simply by a swollen characteristic ratio $C_\infty(\phi)$

$$\langle R^2 \rangle = C_\infty(\phi)\, N \ell^2 \ , \tag{47}$$

where

$$C_\infty(\phi) = C_\infty(1)\, \phi^{-1/4} \ , \tag{48}$$

where $C_\infty(1)$ refers to the melts and ϕ is the polymer volume fraction. He verified this result by measuring the radius of gyration of polystyrene molecules in solution. Good agreement was found between Eq. (47) and the experimental data from solutions ranging in concentration from 10^{-2} to 1.0 g/cm^3.

With Eq. (47), the application of our entanglement model to semidilute solutions is straightforward. The "tube" diameter D_e or sphere diameter that contains $\tilde{N}$ nontail segments is related to the mean number of skeletal bonds between entanglements N_e according to

$$D_e = C_\infty^{1/2}(\phi)\, N_e^{1/2}\, \ell \ , \tag{49}$$

where we have used the Gaussian properties of the chain conformations on the entanglement length scale. Subsitituting this relation into Eq. (18) we obtain

$$\tilde{N} + 1 = \left(\frac{\pi \rho\, C_\infty(1)^{3/2}\, \ell^3}{6\, \mu_m\, \phi^{3/8}} \right) N_e^{1/2} (1 - N_e / N) \ , \tag{50}$$

which relates N_e to $\tilde{N}$, the polymer concentration and the structural parameters $C_\infty(1)$, μ_m and ℓ. In the long-chain limit ($N_e/N \ll 1$) the entanglement spacing is therefore

$$N_e = \left[\frac{6(\bar{N}+1)\mu_m}{\pi C_\infty^{3/2}(1)\ell^3} \right]^2 \frac{\phi^{3/4}}{\rho^2} \quad . \tag{51}$$

Since $\phi \propto \rho$, this equation leads to scaling relation between N and ρ of the form

$$N_e \propto \rho^{-5/4} \quad . \tag{52}$$

The mean entanglement spacing N_e is experimentally derived from the plateau modulus which is related to N_e according to Eq. (38). Substituting Eq. (52) into this expression, we predict that the plateau modulus of semidilute solutions will scale with polymer density according to

$$G_N^o \propto \rho^{9/4} \quad . \tag{53}$$

This prediction is in agreement with experimental data[20] where the scaling exponent is usually in the range 2.0 to 2.3.

The limit $N_e/N \ll 1$ is a high concentration and long-chain limit simultaneously. We have already predicted that N_e scales with concentration according to Eq. (52). Therefore as the concentration is decreased, N_e will increase in the fashion predicted by Eq. (51). However the bracketed quantity on the right-hand side of Eq. (50) can no longer be ignored. This bracketed quantity leads to an entangled/unentangled transition[10] that is analogous to the discussion on polymer melts in section III. Here we will review the model predictions for this transition. The details can be found in reference (10).

The critical chain length for entanglements in semidilute solution from Eqs. (32) and (51) scales with density according to[10]

$$N_c \propto (\bar{N}+1)^2 \rho^{-5/4} \quad . \tag{54}$$

This equation expresses the manner in which the minimum chain size, N_c, for the presence of entanglements, varies with the polymer concentration. We can invert this expression to obtain

$$\rho_c \propto (\bar{N}+1)^{8/5} N^{-4/5} \quad , \tag{55}$$

which expresses the minimum concentration, ρ_c, for which entanglements will exist, for a given chain degree of polymerization N.

The scaling of ρ_c for entanglements in semidilute solutions parallels the results by de Gennes[11] for the overlap density $\rho_{o.c}$.

$$\rho_{o.c.} \propto N^{-4/5} \ . \tag{56}$$

The fact that the two scaling laws are similar is not too surprising, since we could have choosen a low value of $\bar{N}$ (of say $\bar{N}$ = 1) in Eq. (55) to describe an overlap criterion rather than a high value of $\bar{N}$ ($\bar{N}$ = 8.1) to describe an entanglement criterion. Accepting this argument we therefore predict that ρ_c and $\rho_{o.c.}$ are related by a universal number

$$\rho_c / \rho_{o.c.} \approx \left(\frac{\bar{N} + 1}{2} \right)^{8/5} \ . \tag{57}$$

The value of $\bar{N}$ from Table 1 is approximately 8 and therefore the prediction is that

$$\rho_c / \rho_{o.c.} \approx 10 \ . \tag{58}$$

This implies that there is a large solution regime between entangled chains and separated chains where the chains are overlapped but insufficiently overlapped to have many effective entanglements. A diagram showing the entangled and overlapped regimes as a function of concentration and degree of polymerization is shown in Fig. 5.

V. POLYMER DYNAMICS IN ENTANGLED MELTS

In this section we discuss the influence of lateral constraints, of the form used to define the tube diameter in sections II, III and IV, on polymer dynamics. Here we offer an alternative derivation of the reptation theory

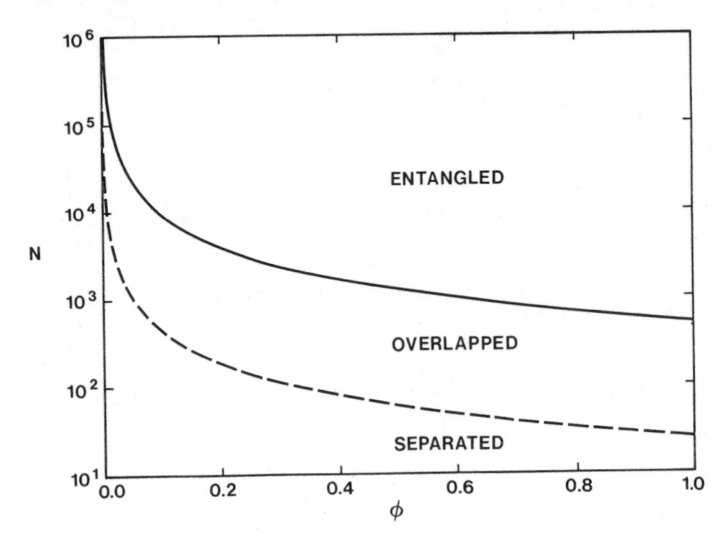

Fig. 5. "Phase" diagram for polymer solutions. N is the degree of polymerization and ϕ is the polymer volume fraction. The solid line separates the entangled/unentangled regimes. The dotted line denotes the overlap threshold. The curves were computed from the relation $\phi_c = (\bar{N} + 1)^{8/5} N_c^{-4/5}$ (see Section IV) where a constant of order unity is omitted. A value of $\bar{N}$ = 8.1 was used as an entanglement criterion and $\bar{N}$= 1 was used as an overlap criterion.

starting from a more fundamental description of the polymer dynamics via the Rouse model and imposing a tube constraint. The approach used here introduces a definition of the tube axis that is natural in the context of Rouse models and can also be related to the primitive path that Doi and Edwards[3] defined. Specifically we employed a version of the Rouse model, that has been called a generalized Rouse model (GRM), and was developed originally by Allegra[22] to describe polymer dynamics in semidilute polymer solutions. The major assumptions of the GRM are discussed in the Appendix where we review the derivation of the model equations. The application of the GRM to polymer melts[8,10,23] is described below.

Like the original Rouse model, the GRM involves a bead-spring friction coefficient. The three-dimensional conformation of a polymer chain is decomposed into Fourier modes which relax according to spectrum of relaxation times. The relaxation time corresponding to the Fourier mode with index p is

$$\tau(p) = [\zeta(p)\,\tilde{\alpha}^2(p)\,\tilde{C}(p)\,N^2\,\ell^2]\;/\;[3\,\pi^2 k_B T\,p^2]\;, \tag{59}$$

where $p = 1,2,3,\ldots N$ and $\alpha^2(p)$ is the square of the strain ratio

$$\tilde{\alpha}^2(p) = \langle|\tilde{\mathbf{r}}(p)|^2\rangle\,/\,\langle|\tilde{\mathbf{r}}(p)|^2\rangle_0\;, \tag{60}$$

$C(p)$ is a generalized characteristic ratio

$$\tilde{C}(p) = \langle|\tilde{\mathbf{r}}(p)|^2\rangle_0\,/\,N\ell^2\;, \tag{61}$$

$\zeta(p)$ is a wavenumber-dependent friction coefficient and ℓ is mean skeletal bond length. In Eqs. (60) and (61), $\langle\;\rangle$ represents an average over the chain configuration and $\langle\;\rangle_0$ represents an average over unperturbed Gaussian conformations.

From the Flory[12] theorem it follows that polymer conformations in a melt are Gaussian and hence $\tilde{\alpha}^2(p) = 1$. This also holds for entangled chains as well. However, the physical constraints imposed by surrounding polymers must force the chains to evolve in an anisotropic manner. To see this more clearly, imagine a system of entangled chains. In order to displace a particular chain through a finite distance along the contour of the chain conformation, one will encounter a certain amount of resistance due to the presence and contacts with other chains. A displacement of polymer segment perpendicular to the chain axis will meet with even greater resistance, particularly if the size of the displaced segment is larger than the mean spacing between the entanglements. The presence of entanglements therefore destroys the 3-dimensional isotropy of a polymer diffusional motion and can be thought of as producing non-equivalent longitudinal and transverse modes of motion. In our model, we have represented the effect of entanglements by defining wavevector-dependent longitudinal and transverse friction coefficients, $\zeta_\ell(p)$ and $\zeta_t(p)$ respectively, whose properties are discussed below.

In order to define the longitudinal and transverse directions, we have proposed[8] the following definition of a tube axis. The tube axis is defined by a set of vectors $\hat{\mathbf{r}}(n)$, given by the truncated Fourier series

$$\hat{\mathbf{r}}(n) = \sum_{p=1}^{\bar{p}} \tilde{\mathbf{r}}(p) \sin(\pi p n/N) \quad , \tag{62}$$

where $\tilde{\mathbf{r}}(p)$ are the Fourier transforms of the bond vectors and $\bar{p} = N/N_e$. The effect of the cut-off wavenumber in Eq. (62) is to compute a weighted average over N/p, which in this case is N_e, bonds around bond n. The $\hat{\mathbf{r}}(n)$ vectors are therefore a Fourier smoothened image of the actual chain conformation using the variable N_e to determine the extent of smoothening. In Fig. 6 we plot both a 2-dimensional polymer conformation with 500 randomly oriented bonds and a Fourier smoothened image of the chain using $p = 5$ thus smoothening over about 100 bond vectors for each bond vector.

The Fourier smoothened image of the chain in Fig. 5 can be seen to have essentially the same overall conformation as the original chain. It does

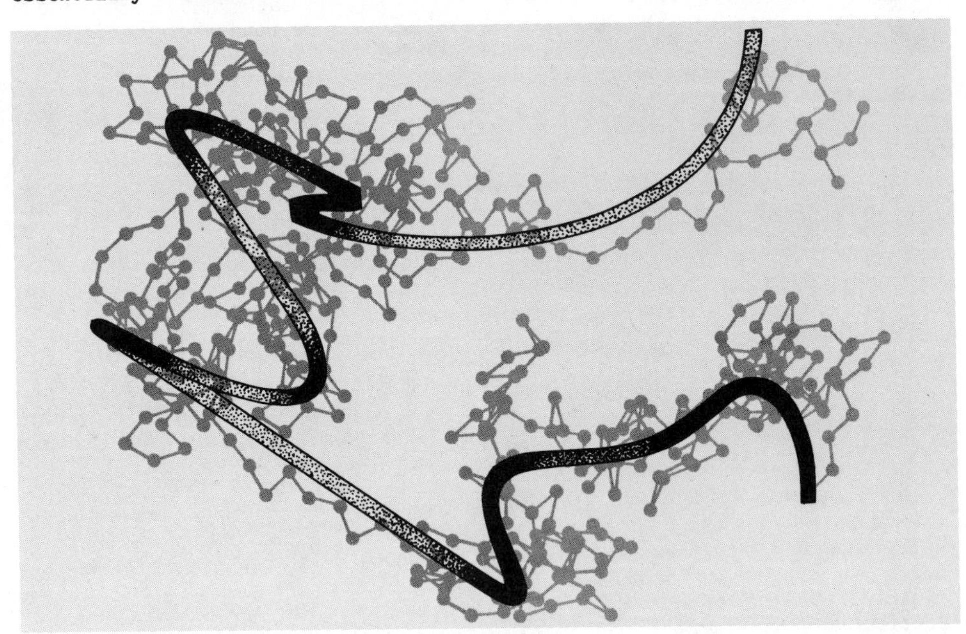

Fig. 6. A two-dimensional polymer conformation with $N = 500$ bonds generated by a random flight ($C_\infty = 1$). The smooth curve represents the "tube" axis defined by the truncated Fourier smoothening procedure in section V. This curve was generated using a cut-off wavenumber $\bar{p} = 5$, thus averaging each bond over the 100 neighbouring bonds. The tube axis has the same overall shape as the original random flight, but lacks the small scale details. The local tangent to the tube axis defines the direction along which unconstrained (longitudinal) diffusion can occur. Transverse diffusion is limited to a tube with diameter $(N/\bar{p})^{1/2} = 10$ units whose center is defined by the axis.

however lack the small-scale detailed structure of the chain. This detailed structural information we feel is not necessary in describing the dynamics of the chain since most properties are determined by the time scale of large-scale changes in conformation. Since our definition of a tube axis is related to the large-scale properties of the chain conformation, it should be sufficient to study the evolution of the tube axis in order to study the overall chain dynamics. This argument was used previously by Doi and Edwards[3], when they introduced the notion of a primitive path and the stochastic earthworm equations. We note here that a tube with diameter $C_\infty^{\frac{1}{2}}(N/\bar{p})^{\frac{1}{2}}\ \ell$, drawn along the tube axis will contain the Doi-Edwards primitive path.

The tube axis that we have defined in Eq. (62) is a smoothened image of a static conformation. However, since we used the mean spacing between entanglements N_e as a smoothening length, then one could argue that the tube drawn along this axis also represents a short-time average or blurred image of the chain. In other words, as the chain conformation undergoes small-length scale changes the tube axis that we have defined is not altered appreciably.

Now we decompose the Langevin equation for each Rouse bond into longitudinal and transverse motions by projecting the bond vectors along the tube axis. This will illustrate that the cut-off $\bar{p}$ serves to separate the time scales for conformational changes into two groups. The longitudinal components of the bond vectors are defined by

$$\lambda(n) = \mathbf{r}(n) \cdot \hat{\mathbf{r}}(n) / <\hat{\mathbf{r}}(n) \cdot \hat{\mathbf{r}}(n)>^{\frac{1}{2}} \quad , \tag{63}$$

where the diameter is a normalization factor equal to[9]

$$<\hat{\mathbf{r}}(n)\,\hat{\mathbf{r}}(n)> = \ell^2/N_e - (\ell^2/N)\left[\sin(2\pi n/N_e)/(2\tan(\pi n/N) - \sin^2(\pi n/N_e)\right] \quad . \tag{64}$$

Apart from the Rouse bonds near the chain ends, this factor can be well approximated by

$$<\hat{\mathbf{r}}(n) \cdot \hat{\mathbf{r}}(n)> \;\approx\; \ell^2/N_e \quad . \tag{65}$$

Next we define a set of longitudinal or tube Fourier modes, $\lambda(p)$

$$\tilde{\lambda}(p) = 2^{1/2} \sum_{n=1}^{N} \lambda(n) \sin\,(\pi pn/N) \quad . \tag{66}$$

which are the analogs of the chains spatial Fourier modes in the Appendix. The relaxation times of these tube modes are equivalent to the polymers longitudinal diffusion modes and are given by

$$\tau_\ell(p) = [<\tilde{\lambda}^2(p)> N\, \zeta_\ell(p)] \,/\, [3\pi^2 k_B T\; p^2] \quad , \tag{67}$$

where $<\tilde{\lambda}^2(p)>/N\ell^2$ is the square of the strain ratio computed along the tube axis and $\zeta_\ell(p)$ is the longitudinal friction coefficient. In reference (9) we computed $<\tilde{\lambda}^2(p)>$ and showed that to leading order in N (i.e., to order N^2) can be approximated as

$$<\tilde{\lambda}^2(p)> \;\approx \begin{cases} 2\ell^2/N_e \; cotan^2(\pi p/2N) & ; \; odd\ p \\ 0 & ; \; even\ p \end{cases} \tag{68}$$

which for small odd values of p yields the scaling law

$$<\tilde{\lambda}^2(p)> \propto N^2/(p^2 N_e) \quad . \tag{69}$$

The longitudinal mode relaxation times are then obtained by substituting Eq.(68) into Eq. (67) to yield

$$\tau_\ell(p) = \begin{cases} 2\ell^2 \zeta_\ell(p) N/(3\pi^2 p^2 N_e k_B T) \; cotan^2(\pi p/2N) & ; \; odd\ p \\ 0 & . \; even\ p \end{cases} \tag{70}$$

For small odd values of p we obtain the scaling law

$$\tau_\ell(p) \propto (\zeta_\ell(p) N^3/p^4 N_e) \quad . \tag{71}$$

The transverse relaxation terms can be obtained in a similar fashion. For wave-numbers $p \geq \bar{p}$, the diffusional motion is not hindered by entanglements and therefore for motions on these length scales we have

$$\tau_t(p) = [\zeta_t(p) N^2 \ell^2] \,/\, [3\pi^2 k_B T\; p^2] \quad . \qquad p \geq \bar{p} \tag{72}$$

For wave numbers $p < \bar{p}$ the transverse translational diffusion of a chain is hindered by the entangling chains. In the framework of the Rouse theory the interactions with other chains are purely frictional, since there is no elastic restoring force associated with interchain interactions. We will represent the effect of hindered lateral diffusion by a modification of the friction coefficient

$$\zeta_\ell(p) \;=\; \zeta_0 \;\; ; \; p = 1, 2, .. N$$

$$\zeta_t(p) = \begin{cases} \zeta_0 \; ; & p = \bar{p},\ \bar{p}+1, \ldots N \\ \infty \; . & p = 1, 2, \ldots \bar{p}-1 \end{cases} \tag{73}$$

Here, the effect of a very large, or in the present model infinite, transverse friction coefficient is to retard the lateral diffusion over length scales larger than the mean spacing between entanglements. Additional non-reptative processes, such as the release of constraints[5], can lead to transverse diffusion for $p < \bar{p}$. These effects can be included in this model by modifying Eq. (73) appropriately.

To combine Eqs. (70) and (72) and to summarize this section, the relaxation spectrum for entangled chains is

$$\tau(p) = \begin{cases} [2\ell^2 \zeta_0 N] / [3\pi^2 N_e(N)\, k_B T]\, cotan^2(\pi p/2N) \; , & N \geq N_c,\ p \leq \bar{p} \\ [\zeta_0 \ell^2 N^2] / [3\pi^2 k_B T p^2] \; , & N \geq N_c,\ p > \bar{p} \end{cases} \tag{74}$$

where $N_e(N)$ is the function given in Eq. (33). The first part of Eq. (74) describes diffusional relaxation over length scales larger than the mean spacing between entanglements. Motions on this length scale occur along the curvilinear coordinate of the tube axis because of the high friction coefficient we have imposed for transverse motion. The second part of Eq. (74) describes diffusional relaxation over shorter length scales which can be either transverse of longitudinal with respect to the tube axis. The relaxation times for longitudinal motion have the same scaling form as predicted by the original reptation/tube theories. The essential difference here is that we derived these results from the Rouse model relaxation times using an alternative definition of the tube and introducing an asymmetric friction coefficient to represent the tube constraints.

VI. SUMMARY

The purpose of this paper was to review some recent developments on entanglements in dense polymer systems. The paper is organized into two parts. The first part (Sections II, III, IV) addressed the entanglement structure of polymer melts and polymer solutions. The concept of a tube was redefined using a specific definition or criterion for entanglements. We defined an entanglement or tube element as consisting of certain numbers of neighbouring nontail polymer segments. This number, $\tilde{N}$, which we call the entanglement coordination number, represents the average number of neighbouring chains that restrict the lateral degrees of freedom of a given chain. The introduction of $\tilde{N}$ allowed us to relate the mean number of skeletal bonds N_e between entanglements to the properties of a given polymer according to the relation

$$N_e = \left[\frac{6(\tilde{N}+1)\mu_m}{\pi \rho_m C_\infty^{3/2} \ell^3} \right]^2 . \tag{75}$$

The correlations predicted by this expression are in agreement with current experimental data. The experimental data yield an average value for $\tilde{N}$ of $\tilde{N}$ = 8.1± 8%. The small standard deviation of values of $\tilde{N}$ supports our conjecture that $\tilde{N}$ is universal.

In section III we argued that tail segments do not contribute to the formation of long lived lateral constraints. Incorporating this idea into our mean-field description of entanglements leads to a prediction of a minimum degree of polymerization N_c for the presence of entanglements. A transition regime between strongly entangled and unentangled chains followed from this model and we showed that the model transition is analogous to other mean-field models such as the van der Waals theory of the liquid-gas critical point. In the transition regime, the model predicts that the mean entanglement spacing N_e varies with chain degree of polymerization in a manner which can, in part, account for departures from the $\eta_o \propto M^3$ scaling law that is frequently observed. Other polymer properties, such as the plateau modulus and self-diffusion coefficient, are also influenced by the swelling of the tube diameter. However, since these properties scale with a lower power of N_e the effect of tube enlargement is not as great.

In section IV we applied this entanglement model to polymer solutions in the semidilute regime. Here we derived scaling laws for the entanglement spacing as a function of polymer density and found for example that the mean entanglement spacing and plateau modulus scale with density according to

$$N_e \propto \rho^{-5/4} \quad ; \tag{76}$$

$$G_N^o \propto \rho^{9/4} \quad , \tag{77}$$

which is in agreement with experimental data.[20] We also showed that the approach used to describe entanglements in solution can also be used to describe chain overlap in the semidilute regime.

To summarize the first half of this paper, a relatively simple description of chain entanglement has provided considerable insight into an otherwise intractable topology problem. Our model is based on a series of conjectures regarding the nature of the tube constraints. In particular we conjectured the universality of chain entanglement via a coordination number $\tilde{N}$. At present, support for this conjecture has come from analyzing rheological data for monodisperse polymer melts. A similar analysis of polymer solution data would yield more insight into tube constraints. A theoretical or computational derivation of the value of $\tilde{N}$, would also be desirable.

In the last part of this paper (section V), we addressed the subject of polymer dynamics. Here we outlined a methodology for applying a Rouse description of polymer dynamics in a system with chain entanglement. A tube axis was defined by coarse-graining the polymer conformation using a smoothening length scale determined by the mean entanglement spacing. We then calculated the dynamics of the Fourier smoothened image by imposing a wavenumber dependent friction coefficient to account for the presence of entanglements. Here we recovered the reptation theory relaxation times for the most collective modes of motion. The small scale motions remain Rouse like.

NOMENCLATURE

a	tube diameter in Doi-Edwards theory
$\tilde{\alpha}^2(p)$	end-to-end vector of polymer
b	Kuhn's length
C_∞	characteristic ratio
$\tilde{C}(p)$	generalized characteristic ratio
D, D_G	diffusion coefficients
D_e	entanglement blob diameter in present theory
E	intramolecular potentia
ζ	monomer friction coefficient
$\zeta_{\ell'}$, ζ_t	longitudinol and transverse friction coefficient
$\eta(\omega), \eta_o$	complex and steady state viscosity
$G(t), G'(\omega), G''(\omega)$	stress relaxation, storage and loss moduli
G_N^o	plateau modulus
$g_\upsilon(n,t)$	Brownian force on nth bond of υth chain
J_e^o	steady-state compliance
K	universal parameter in Graessley-Edwards model
k_B	Boltzmann's constant
ℓ	mean skeletal bond length
$\lambda(n), \lambda(p)$	bond axis projections on tube axis and tube Fourier modes
M	molecular weight
M_e	mean molecular weight between entanglements
M_c	critical molecular weight for entanglements from viscosity
M_c'	critical molecular weight for entanglements from compliance
μ_m	monomer mass per skeletal bond
N	number of chain skeletal bonds
N_e	mean number of skeletal bonds between entanglements
N_c	critical value of N for entanglements
$\check{N}$	topological parameter defining entanglements also called "coordination number"
$N(m)$	distribution of segment sizes in mean entanglement
$N_{tails}(m)$	tail segment component of $N(m)$
$N_{nontails}(m)$	non-tail segment component of $N(m)$
ρ	Fourier wavenumber cut-off for defining tube axis
$\mathbf{r}(n)$, $\tilde{\mathbf{r}}(p)$	skeletal bond vectors and spatial transform
$\hat{\mathbf{r}}(n)$	tube axis vectors
$\mathbf{R}$	generalized strain ratio
ρ	polymer mass density
ρm	skeletal bond density
ρ_e ρ_{eff}	entanglement densities
ρ_{tails}	density of polymer ends

ρ_c	critical concentration for entanglements
$\rho_{o.c.}$	critical concentration for overlap
T	absolute temperature
τ_d	relaxation time for defects in reptation theory
τ_r	tube disengagement time
$\tau(p)$	relaxation times for Rouse modes
$\tau_\ell(p)$, $\tau_t(p)$	longitudinal and trasverse relaxation times
ϕ	polymer volume fraction
Φ	two and three bond interactions between chains
$W(R)$	distribution of end-to-end vectors ensemble size
$X_\nu(n,t)$	intramolecular force on nth bond of νth chain
Ψ	dimensionless constant

REFERENCES

1. S. F. Edwards, Proc. Phys. Soc., 92, 9 (1967).
2. P. G. deGennes, J. Chem. Phys., 55, 572 (1971).
3. M. Doi and S. F. Edwards, J. Chem. Soc. Faraday Trans., 2, 74, 1789,1802, 1818 (1978).
4. H. Wendel and J. Noolandi, Macromolecules, 15, 1318 (1982).
5. W. W. Graessley, Adv. Polym. Sci., 47, 67 (1982).
6. C. F. Curtiss and R. B. Bird, J. Chem Phys., 74, 2016, 2026 (1981).
7. M. Doi, J. Polym Sci. Lett., 19, 265 (1981).
8. T. A. Kavassalis and J. Noolandi, Phys. Rev. Lett., 59, 2674 (1987).
9. T. A. Kavassalis and J. Noolandi, Macromolecules 21, 2869 (1988).
10. T. A. Kavassalis and J. Noolandi, Macromolecules (in press) 1989.
11. P. G. deGennes, Scaling Concepts in Polymer Physics, Cornell University Press,Ithaca (1979).
12. P. J. Flory, J. Chem. Phys., 17, 303 (1949).
13. H. E. Stanley, Introduction to Phase Transitions and Critical Phenomena, Oxford University Press, Oxford (1971) and references contained within.
14. Y. -H. Lin, Macromolecules, 20, 3080 (1987).
15. W. W. Graessley and S. F. Edwards, Polymer, 22, 1329 (1981).
16. V. P. Privalko, Macromolecules, 13, 370 (1980).
17. S. H. Aharoni, Macromolecules, 16, 1722 (1983), 19, 426 (1986).
18. S. Onogi, J. Masuda, and K. Kitagawa, Macromolecules, 3, 109 (1970).
19. R. H. Colby, L. J. Fetters and W. W. Graessley, Macromolecules, 20, 2226 (1987).
20. W. W. Graessley, Adv. Polym. Sci., 16, 1 (1974).
21. M. Daoud, Macromolecules, 8, 804 (1975).
22. G. Allegra and F. Ganazzoli, J. Chem. Phys., 83, 397 (1985).
23. J. Noolandi, G. W. Slater and G. Allegra, Makromol. Chem. Rapid Commun., 8, 51 (1987).

APPENDIX

General Rouse Model

The purpose of this Appendix is to review the derivation of the relaxation spectrum (Eq. (15)) for the GRM. We begin by considering a polymer configuration under ideal or theta conditions. Letting $\{\mathbf{r}(n)\}$ represent the set of N skeletal bond vectors, we write the end-to-end vector, $\mathbf{R}$, of the chain as

$$\mathbf{R} = \sum_{n=1}^{N} \mathbf{r}(n) \quad . \tag{A.1}$$

Provided that the chain is not totally rigid, the probability distribution, $W(\mathbf{R})$, for the end-to-end vector, of a long chain, can be approximated by the Gaussian distribution

$$W(\mathbf{R}) = (3/2\pi <\mathbf{R}^2>_0)^{3/2} exp\,(-\,3\mathbf{R}^2/2<\mathbf{R}^2>_0) \ , \tag{A.2}$$

where $<\mathbf{R}^2>_0$ represents the unperturbed configuration average. This form of the distribution function is a consequence of the Central Limit Theorem.

Allegra[22] has shown that the probability distribution for more general linear combinations of the skeletal bond vectors of an unperturbed chain is also Gaussianly distributed. In particular, if we decompose the conformation into the Fourier modes, $\tilde{\mathbf{r}}\,(p)$, defined by

$$\tilde{\mathbf{r}}(p) = \sum_{n=1}^{N} \mathbf{r}\,(n)\, sin\,(\pi\, pn/N) \tag{A.3}$$

where the wavenumber p can have values

$$p = 1,\ 2,\ 3, \ldots N \ , \tag{A.4}$$

we find that the distribution function for the mode $\tilde{\mathbf{r}}(p)$ is

$$W\left(\tilde{\mathbf{r}}(p)\right) = \left(3/2\pi <\tilde{\mathbf{r}}(p)\cdot\ \tilde{\mathbf{r}}(p)>_o\right)^{3/2} exp\left(-3\tilde{\mathbf{r}}(p)\cdot\ \tilde{\mathbf{r}}(p)/2<\tilde{\mathbf{r}}(p)\cdot\ \tilde{\mathbf{r}}(p)>_o\right) \ . \tag{A.5}$$

Within the Gaussian approximation, it follows that the set of vectors $\{\mathbf{r}(p)\}$ constitutes an orthogonal set, in the sense that

$$<\tilde{\mathbf{r}}(p)\cdot\ \tilde{\mathbf{r}}(p')> = 0 , \quad p \neq p' \tag{A.6}$$

and therefore we can write the chain configuration distribution function $W\{\mathbf{r}(n)\}$ as a product of contributions from independent modes

$$W\left\{\mathbf{r}(n)\right\} = W\left\{\tilde{\mathbf{r}}(p)\right\} = \prod_{p=1}^{N} W\left(\tilde{\mathbf{r}}(p)\right) \quad . \tag{A.7}$$

The intramolecular potential, E, can therefore be seen to have the usual quadratic form

$$E = -k_B T \, \ell n \, W\left\{\tilde{\mathbf{r}}(p)\right\} \tag{A.8}$$

$$= E_o + (3k_B T/N\ell^2) \sum_{p=1}^{N} \tilde{\mathbf{r}}(p) \cdot \tilde{\mathbf{r}}(p)/\tilde{C}(p) \ , \tag{A.9}$$

where $\tilde{C}(p)$ is a generalized characteristic ratio and is defined by

$$\tilde{C}(p) = \langle \tilde{\mathbf{r}}(p) \cdot \tilde{\mathbf{r}}(p) \rangle_0 / N\ell^2 \quad . \tag{A.10}$$

The Gaussian approximation can therefore be seen to be equivalent to a quadratic potential or linear elastic restoring force. Deviations from the Gaussian distribution will correspondingly yield nonlinear force terms in the dynamics. The Gaussian approximation should therefore be an appropriate simplification for describing systems close to equilibrium or at most linearly displaced from the equilibrium state.

Allegra further generalized these considerations for polymers in various solvent media[22], and showed that the Boltzmann statistical weight for an ensemble of identical chains of length N, within the Gaussian approximation is

$$W\left\{\tilde{\mathbf{r}}(p)\right\} = exp\left(-\mathcal{N}\left\{\tilde{\alpha}^2(p)\right\}\right) \prod_{p=1}^{N} \left\{\left(3/2\pi\tilde{\alpha}^2(p)N\tilde{C}(p)\ell^2\right)^{3\mathcal{N}/2}\right.$$

$$\left. \times \, exp\left(-\sum_{\nu=1}\left(3\tilde{\mathbf{r}}_\nu(p) \cdot \tilde{\mathbf{r}}_\nu(p)\right)/\left(2\tilde{\alpha}^2(p)N\tilde{C}(p)\ell^2\right)\right)\right\} \ , \tag{A.11}$$

where the index $\nu = 1, 2, \ldots$ labels the chains, $A\{\tilde{\alpha}^2(p)\}$ is the excess free energy per chain and is given by

$$A\ \{\tilde{a}^2(p)\} = \frac{A - A_o}{k_B T} = \sum_{p=1}^{N} (3/2)\left[\tilde{a}^2(p) - 1 - \ln\tilde{a}^2(p)\right] + \Phi\ . \tag{A.12}$$

A_o is the free energy per chain in the unperturbed state, $\tilde{\alpha}^2(q)$ is a generalized strain ratio given by

$$\tilde{\alpha}^2(p) = \langle\tilde{\mathbf{r}}(p)\cdot\tilde{\mathbf{r}}(p)\rangle / \langle\tilde{\mathbf{r}}(p)\cdot\tilde{\mathbf{r}}(p)\rangle_o \tag{A.13}$$

and Φ is a sum of two and three body potential interactions between chain segments.

The quantity $\tilde{\alpha}^2(p)$ is a measure of the deviation from the unperturbed state, where $\tilde{\alpha}^2(p) = 1$. The first term in the RHS of Eq. (A.12) represents the elastic energy stored in the perturbed state, and vanishes when $\tilde{\alpha}^2(p) = 1$.

The x-component of the intramolecular forces acting on the nth atom of the νth chain can be calculated by analogy to Kirkwood's generalized thermodynamic force

$$f_{x\nu}(n,t) = k_B T\ \partial \ln W / \partial x_\nu(n,t)$$

$$= -k_B T \sum_{p=1}^{N} \left\{ \frac{\partial A}{\partial\tilde{\alpha}^2(p,t)} \frac{\partial\tilde{\alpha}^2(p,t)}{\partial r_\nu(p,t)} + \frac{3\tilde{r}_\nu(p,t)}{\tilde{\alpha}^2(p,t) NC(p)\ell^2} \right\} \frac{\partial\tilde{r}_\nu(p,t)}{\partial x_\nu(n,t)}\ , \tag{A.14}$$

where $x_\nu\ (n,\ t)$, is the x-component of $\mathbf{R}_\nu(n,\ t)$, the position of atom n on chain ν. In the equilibrium state, the excess free energy A is by definition a minimum with respect to the coordinate $\tilde{\alpha}^2$ and therefore

$$\frac{\partial A}{\partial\tilde{\alpha}^2(p,t)} = 0\ . \tag{A.15}$$

In addition, the quantity $\tilde{\alpha}^2(p,t)$ is independent of the time t. Multiplying Eq. (A.14) by $\sin(\pi pn/N)$, summing over n, and using Eq. (A.15), we can write

$$\tilde{f}_x(p,t) \equiv \sum_{n=1}^{N} f_{x\nu}(n,t) \sin(\pi pn/N) \tag{A.16}$$

$$= \left[6\, k_B T(1 - \cos(\pi pn/N))/\left(\tilde{C}(p)\ell^2\tilde{\alpha}^2(p,t)\right)\right]\tilde{x}(p,t)\ , \tag{A.17}$$

where we have dropped the subscript ν labelling the chain.

In addition to the intramolecular forces, the νth chain is subjected to a Brownian force $g(n,t)$ due to random collisions with other chains, and an intermolecular frictional force, $-\zeta\ \dot{x}(n,t)$ where ζ is an atomic friction coefficient. Defining the following Fourier transforms

$$\tilde{x}(p,t) = \sum_{n=1}^{N} x(n,t) \sin(\pi pn/N) \ ; \tag{A.18}$$

$$\tilde{g}(p,t) = \sum_{n=1}^{N} g(n,t) \sin(\pi pn/N) \ , \tag{A.19}$$

we can write the equation of motion for $\tilde{x}\ (p,t)$ as

$$\zeta\dot{\tilde{x}}(p,t) + \left[6\,k_B T \left(1-\cos(\pi pn/N)\right) / \left(\tilde{C}(p)\,\ell^2 \tilde{\alpha}^2(p)\right)\right] \tilde{x}(p,t) = \tilde{g}(p,t) \ , \tag{A.20}$$

The evolution of the general Fourier mode, $\tilde{x}\ (p,t)$ is therefore determined by the intramolecular elastic forces and a stochastic contribution, due to other chains or solvent, whose statistical properties must be specified. Adopting the traditional Langevin specification of the stochastic force we will say that $\tilde{g}\ (p,\ t)$ is Gaussian of zero mean, white and stationary

$$<\tilde{g}(p,t)> = 0 \ ; \tag{A.21a}$$

$$<\tilde{g}(p,t+\tau)\tilde{g}(p',\tau)> = <\tilde{g}(p,\tau)\tilde{g}(p',0)> \sim \delta(\tau)\,\delta_{p,p'} \ ; \tag{A.21b}$$

$$<\tilde{g}(p,t)^{2n+1}> = 0 \ ; \tag{A.21c}$$

$$\begin{aligned} &<\tilde{g}(p,t_1)\tilde{g}(p,t_2)\tilde{g}(p,t_3)\tilde{g}(p,t_4)> \\ &\quad = <\tilde{g}(p,t_1)\tilde{g}(p,t_2)><\tilde{g}(p,t_3)\tilde{g}(p,t_4)> \\ &\qquad + <\tilde{g}(p,t_1)\tilde{g}(p,t_3)><\tilde{g}(p,t_2)\tilde{g}(p,t_4)> \\ &\qquad + <\tilde{g}(p,t_1)\tilde{g}(p,t_4)><\tilde{g}(p,t_2)\tilde{g}(p,t_3)> . \end{aligned} \tag{A.21d}$$

In addition, the stochastic force is uncorrelated to $\tilde{x}\ (p,\ t)$ so that

$$\langle \tilde{g}(p',t')\,\tilde{x}(p,t)\rangle = 0 \ . \qquad (A.22)$$

Finally, the friction coefficient, ζ, is related to the random force $g(p,\ t)$ by a fluctuation-dissipation theorem

$$\langle \tilde{g}(p,t)\tilde{g}(p',t')\rangle = 2Nk_BT\,\zeta\,\delta(t-t')\,\delta_{p,p'} \, . \qquad (A.23)$$

The formal solution of Eq.(A.20) can be written as

$$\tilde{x}(p,t) = e^{-t/\tau(p)}\,\tilde{x}(p,o) + \int_0^t e^{-(t-t')/\tau(p)}\,\tilde{g}(p,t')\,dt' \ , \qquad (A.24)$$

where the relaxation time $\tau\ (p)$ is defined as

$$\tau(p) = [\zeta\tilde{a}^2(p)\,\ell^{\,2}\tilde{C}(p)]\,/\,[6\,k_BT(1-\cos\,(\pi p/N))] \ . \qquad (A.25a)$$

$$\approx [\zeta\tilde{a}^2(p)\,\tilde{C}(p)N^2\ell^{\,2}]\,/\,[3\pi^2k_BT\,p^2\ ] \ , \qquad (A.25b)$$

which is Eq. (59).

REPTON MODEL OF ENTANGLED POLYMERS

Michael Rubinstein

Corporate Research Laboratories
Eastman Kodak Company
Rochester, New York 14650-2115

ABSTRACT

A discretized version of the reptation model is proposed. The tube is modeled by a one-dimensional lattice, and the polymer is modeled by a cluster of walkers, called reptons, on this lattice. Each repton represents a part of the chain. Reptons are allowed to hop between neighboring sites in such a way that the cluster always remains connected. The fluctuations in tube length correspond to cluster length fluctuations in the repton model and are not pre-averaged. In the experimentally accessible range of molecular weights M, the repton model predicts the diffusion coefficient $D_{3D} \sim M^{-2} + O(M^{-3})$ and viscosity $\eta_0 \sim M^{3.4 \pm 0.1}$.

I. INTRODUCTION

This paper is dedicated to Professor P.G. de Gennes, the inventor of the reptation model.[1] This beautiful and simple idea significantly advanced our understanding of the dynamics of entangled polymers.[2,3]

The reptation model is based on the concept of a confining tube,[4] proposed by Professor S.F. Edwards. The polymer is confined to a tube-like region by neighboring chains (assumed immobile in the original reptation model). Unentangled loops are modeled by noninteracting defects diffusing along the contour of the tube and collectively causing the displacement of the whole chain. In order to obtain a simple analytical solution, the fluctuations in the density of randomly diffusing defects and the resulting fluctuations in the tube length were pre-averaged in the conventional treatments of the reptation model.[1,2,5] This solution of the reptation model explained the observed dependence[6] of the self-diffusion coefficient D_{3D} on polymer molecular weight, M

$$D_{3D} \sim M^{-2} \tag{1}$$

For the longest relaxation time τ_r, the prediction of the reptation model (with the pre-averaging approximation) is

$$\tau_r \sim M^3, \tag{2}$$

while the experimentally measured exponent[7] for the related viscosity η_0 is 3.4 ± 0.1. This difference between the predicted and observed exponent raised questions about the validity of the reptation model. A number of alternative (but more complicated) models have been proposed.[8-11]

Wendel and Noolandi[8] introduced a waiting time distribution for the hops of defects so that the mean square displacement of a defect along the contour of the tube is $<\ell^2> \sim t^{1-\gamma}$ with $0 < \gamma < 1$ instead of the ordinary diffusion ($\gamma = 0$). This leads to a longest relaxation time $\tau_r \sim M^{(3-\gamma)/(1-\gamma)}$ with an exponent equal to 3.4 for $\gamma = 1/6$. The universality of this choice of γ still remains unclear. The only reason for it seems to be that it produces the experimentally observed exponent. Another problem with this model is that it leads to a diffusion coefficient $D_{3D} \sim M^{-2/(1-\gamma)}$ with an exponent equal to 2.4 for $\gamma = 1/6$, while the experimentally observed exponent[6] is closer to 2 (Eq. 1). Thus one needs different values of γ to reproduce the experimental exponents of τ_r and D_{3D}.

A more universal way of obtaining a fractional exponent was proposed by Scher and Shlesinger[9] and later made more mathematically precise by Weiss, Bendler and Shlesinger.[10] They have demonstrated that the time τ_e taken by **all** the chains from a sphere of the size $R_G \sim M^{1/2}$ to escape completely from this sphere for the first time is $\tau_e \sim M^{10/3}/(\ell n M)^{2/3}$. Each chain was assumed to move by reptation with diffusion coefficient $D_{3D} \sim M^{-2}$. The only problem with this very interesting mathematical result is that the relevance of the escape time τ_e to the physics of polymer melts is not very clear. During this time, a chain diffuses a distance r that is much larger than the size of the sphere R_G ($<r^2> \sim D_{3D}\tau_e \sim M^{4/3} >> M$). Thus, to equate τ_e with the relaxation time τ_r one has to assume that if the polymer returns back to the sphere, it is unrelaxed (if some of other original chains are still there), even though this polymer traveled a distance many times its own size while the strain was removed. Thus, the correlations assumed by the model in its application to polymer melts are hard to explain.

An original physical idea was suggested recently by Deutsch,[11] who attempted to explain the 3.4 exponent by a different many-chain effect. He proposed that the reptation of polymers in a melt is hindered by excluded volume effects. Each chain is exponentially localized to its correlation hole,[2] so that the relaxation time is $\tau_r \sim M^3 \exp(\alpha M^{2/3})$. Deutsch argued[11] that this function could be indistinguishable from $M^{3.4}$ over several decades of molecular weight M. The problem with this explanation is that it implies that the diffusion coefficient should be exponentially small, $D_{3D} \sim M^{-2}\exp(-\alpha M^{2/3})$, or at least proportional to $M^{-2.4}$ over the same range of molecular weight M where relaxation time τ_r is proportional to $M^{3.4}$.

Doi suggested that the discrepancy between the exponents predicted by the reptation model and those observed experimentally is due to fluctuations in tube length.[12,13] He proposed an expression for the viscosity $\eta_0 \sim M^3[1-\nu(M_e/M)^{1/2}]^3$, where M_e is the molecular weight between entanglements and ν was estimated from the variational principle[13] to be around 1.47. The proposed function is approximately proportional to $M^{3.4}$ over a limited range of molecular weight M. But fluctuations in tube length do not affect the dependence of the diffusion coefficient on molecular weight (Eq. 1). This explanation was challenged analytically by des Cloizeaux[14] and numerically by Needs.[15] They argued that the effect of fluctuations in tube length on the longest relaxation time τ_r and the viscosity η_0 should be smaller than the predictions by Doi.

In the present paper we demonstrate that Doi's explanation[12,13] is qualitatively correct. We introduce the repton model,[16] which is a discretized model of the dynamics of a single entangled chain. This model follows in detail the diffusion of unentangled loops along the contour of the confining tube. The motion of defects was the basis of de Gennes' original reptation theory.[1] Thus both the reptation and repton models are built on the same physical ideas. In the repton model we do not pre-average the density of defects, and we address the basic question, whether the disagreement between the predictions of the reptation model and experiments is due to an approximation or whether there is some fundamental physics missing (e.g., many-chain effects). We conclude that if the fluctuations in tube length are not neglected, the experimentally observed exponents are obtained, as suggested earlier by Doi.[12,13] Many-chain effects, such as constraint release, may modify[17] the molecular-weight dependence, but the origin of the nontrivial crossover exponents is in the single-chain dynamics.

The subject of this paper is limited to the dynamics of a single entangled polymer chain, just as in the original de Gennes paper on reptation.[1] In a melt or concentrated solution, the dynamics of any chain would be affected by the motion of surrounding polymers (by constraint release), and this effect has to be self-consistently taken into account.[17] In order to do that, one has to start from a reliable model of the single-chain dynamics, such as the reptation model,[1] Doi's fluctuation theory,[12,13] or the repton model[16] described in the present paper.

In the following section we review the concepts of the confining tube, the primitive path, and unentangled loops. The repton model is introduced and discussed in detail in the third section. Relations between the repton model and other models are outlined in the fourth section. In the last section we mention some generalizations and possible applications of the repton model to various problems.

II. CONFINING TUBE, PRIMITIVE PATH, AND UNENTANGLED LOOPS

As stated in the Introduction, the reptation model is based on the idea of the confining tube proposed by Edwards.[2] This idea was first developed for crosslinked networks and later extended to polymer melts and concentrated solutions.[5] If all chains are connected by chemical bonds at both ends to a network, the topological configuration is permanently frozen at the time of the formation of the network. Each chain is permanently confined by its neighbors to a tube-like region. The central line of this confining tube is called the primitive path and it determines the average position of a constraint a given chain imposes on its neighbors. The set of primitive paths of all chains defines an entanglement net.[18]

These ideas can be extended to mobile polymers. The confining tube represents a temporary reduction of the configurational space a given polymer can explore due to the presence of the surrounding chains. If chains restricting the motion of a given polymer are mobile, there are several possible regimes of confinement.[19] In the present paper we will not discuss these phenomena but will concentrate on the dynamics of a single polymer in an array of fixed obstacles, such as a crosslinked network.

The width of the confining tube of a chain is determined by the spacing of the entanglement net, the distance between the primitive paths of the neighboring chains. This width corresponds to the average size of an unentangled loop. Unentangled loops of much larger molecular weight M_{loop} are exponentially unlikely[18,20,21]

$$p(M_{loop}) \sim exp(-\alpha M_{loop}/M_e), \tag{3}$$

where the prefactor α in the exponential is a function of the coordination number z of the entanglement net.[20] The (purely entropic) free energy of an unentangled loop is therefore proportional to its molecular weight M_{loop}. Thus, unentangled loops can combine and separate without any change in the total free energy. Note that in Eq. 3 we neglected an algebraic prefactor $M^{-3/2}$ in front of the exponential,[18,20] leading to additional logarithmic terms in the free energy and inducing weak interaction between unentangled loops. In the present paper we will neglect these logarithmic terms and keep the original de Gennes' assumption[1] of noninteracting unentangled loops.

The physical significance of separating the unentangled loops and the primitive path is that the modes related to them have very different time scales. The relaxation times of unentangled loops are very short and are independent of the chain molecular weight M because the motion of these loops is not severely restricted by topology. The modes, related to the primitive paths, relax much slower and control the important long-time properties, such as viscosity and diffusion coefficient.

In the following section we calculate these long-time properties from the simplest assumptions about the short-time dynamics of unentangled loops. These assumptions are based on the definition of unentangled loops given below, that provides a unique assignment of the loops to the steps of the primitive path.

There are several possible ways of defining and counting unentangled loops. The one we choose in this paper is similar to the B- and C-loops defined and used in references 18 and 20.

The entanglement net (set of primitive paths) of a crosslinked network divides space into cells. A coordination number z of a cell is the number of faces (called gates) it shares with its nearest neighboring cells.

The general procedure for assigning unentangled loops to the cells is then as follows (see Fig. 1a):

i) Consider only the cells that form the confining tube of the chain of interest - cells traversed by its primitive path (cells 1-4 in Fig. 1a).

ii) Choose a direction along the chain (say from the left-end A to the right-end E).

iii) This direction uniquely defines a face through which the primitive path enters a given cell, called the forbidden face for the cell. (There is no forbidden face for the first cell because the primitive path never enters it.)

iv) The entangled loop assigned to a given cell is the set of all unentangled loops that originate and end in it and do not leave this cell through the forbidden face (if it has one), either for the first time or after any subsequent unentangled returns to the cell.

For the chain in Fig. 1a the assignment is AB, BC, CD and DE. The reason for defining the forbidden face is to make the definition of unentangled loops unique and to avoid double counting. For example, loop A_1B could be assigned either to cell 1 or to cell 2, but according to the definition above, the face between cells 1 and 2 is the forbidden one for cell 2. Therefore, loop A_1B is assigned to cell 1. Similarly, we assign large loop B_1C to cell 2 rather than 3, because the corresponding face is forbidden for cell 3.

This definition divides the chain into a set of unentangled loops along the confining tube. In the next section we use this to analyze the motion of the polymer along this tube.

III. REPTON MODEL

The repton model[16] is a discretized version of the reptation model.[1] In the previous section the confining tube was defined as the set of cells of the entanglement net traversed by the primitive path of the test chain. Let us represent this set of cells by a one-dimensional lattice (Fig. 1b). The unentangled loops, assigned to each cell by the definition given in the previous section, are modeled by objects called "reptons" on the sites of this lattice. The number of reptons on each site is roughly proportional to the length of the unentangled loop assigned to the corresponding cell, e.g., the entangled chain in Fig. 1a is mapped onto a connected cluster of seven reptons in Fig. 1b. The coordinates $\{x_i\}$ of N reptons along the one-dimensional lattice represent the assignment of corresponding sections of the chain to the cells in the chosen direction along the tube and therefore always obey $x_1 \leq x_2 \leq \ldots \leq x_N$.

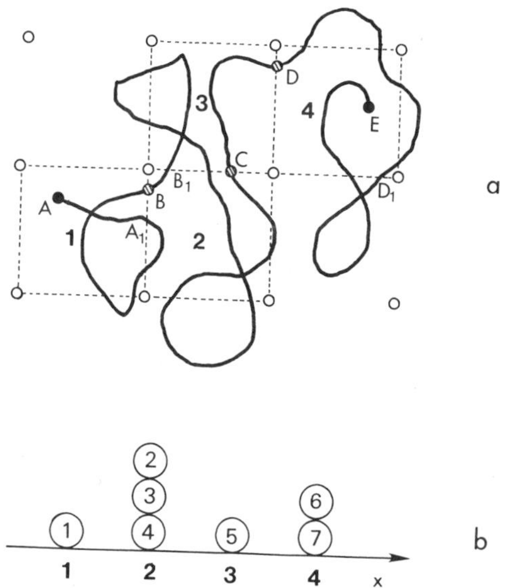

Fig. 1 (a) A polymer AE in a two-dimensional entanglement net, represented by open circles is confined to a tube consisting of four cells, numbered 1-4 and represented by dashed lines. Unentangled loops assigned to each cell are AB, BC, CD and DE, respectively.

(b) Repton model representation of the configuration of Fig. 1(a). Cells of the entanglement net along the confining tube are represented by a one-dimensional lattice. Unentangled loops corresponding to these cells are modeled by walkers (reptons) on the lattice sites.

In a discretization step (representing a section of a polymer by a single repton) we discard short-time modes and concentrate on the long-time behavior of the chain. The repton dynamics is defined by two axioms. The first one states that we are dealing with a connected object (polymer) and we do not change the assignment rules for the unentangled loops along the primitive path.

Axiom 1. Never vacate a site in the middle of the cluster of N reptons and always preserve their order: $x_{i+1} - x_i$ is either 0 or 1 for i = 1, 2,....,N-1, where x_i denotes the coordinate of the i-th repton.

The second axiom defines the hopping rates of a repton between neighboring sites. It is related to the fact that we are mapping a higher dimensional problem onto a one-dimensional repton model. Assume that each cell of the entanglement net has z faces (gates). The average coordination number of a three-dimensional random packing of spheres[22] suggests that in a random system $z \simeq 6$, but one could, in principle, synthesize networks with a different number of faces per cell. For an end section of a chain (e.g., section D_1E) there are z-1 possible gates to enter an empty cell and only one gate to go into an already occupied one. Since in the repton model all of these z-1 empty cells are mapped onto one empty site, we have to assume that the probability of a repton hopping onto this empty site is z-1 times higher than for hopping onto an already occupied one.

Axiom 2. If allowed by the Axiom 1, the probability of a repton hopping between neighboring occupied sites is $1/z$:

$$prob(x_i \rightarrow x_i + 1) = \frac{1}{z}\, \delta_{x_{i+1}, x_{i-1}+1} \delta_{x_i, x_{i-1}}, \tag{4a}$$

$$prob(x_i \rightarrow x_i - 1) = \frac{1}{z}\, \delta_{x_{i+1}, x_{i-1}+1} \delta_{x_i, x_{i+1}}, \tag{4b}$$

$$prob(x_1 \rightarrow x_1 + 1) = \frac{1}{z}\, \delta_{x_1, x_2-1}, \tag{4c}$$

$$prob(x_N \rightarrow x_N - 1) = \frac{1}{z}\, \delta_{x_N, x_{N-1}+1}, \tag{4d}$$

where the probability of hopping onto an empty site is $(z-1)/z$:

$$prob(x_1 \rightarrow x_1 - 1) = \frac{z-1}{z}\, \delta_{x_1, x_2}, \tag{5a}$$

$$prob(x_N \rightarrow x_N + 1) = \frac{z-1}{z}\, \delta_{x_N, x_{N-1}}, \tag{5b}$$

where $\delta_{i,j}$ is a Kronicker delta function (1 for i=j and 0 otherwise). These axioms completely define the repton model where each state of a system corresponds to a set of occupancy numbers of the sites (....,n_{-2}, n_{-1}, n_0 n_1, n_2....), with $n_i=0$ for $i \leq j$; $n_i>0$ for $j<i \leq j+K$; $n_i=0$ for $i>j+K$; and $\Sigma n_i=N$, where K is the number of occupied sites and j+1 is the coordinate of the left end of the cluster. Each state is a discretized representation of a polymer configuration in a network. The relative probabilities of different states are defined through the transition probabilities between them (Axiom 2).

Consider the subset of all allowed states of the system that corresponds to K occupied sites ($K = 1, 2, ..., N$). The detailed balance of the transitions between different states (Eqs. 4 and 5) implies that states within each subset have the same probability p_K, while the probabilities of states from different subsets are related by $p_L/p_K = (z-1)^{L-K}$.

A state is uniquely represented by a set of coordinates $\{x_i\}$ of N reptons. Let us define a set of N-1 "repton bonds", $\sigma_i = x_{i+1} - x_i$. From Axiom 1 we conclude that σ_i can only be equal to 0 or 1. It is very easy to prove that for each connected cluster of N reptons there is a unique set of N-1 zeros and ones $\{\sigma_i\}$, and vice versa. Ones correspond to the spacings between occupied lattice points (forbidden faces between the cells along the tube), while zeroes correspond to multiple reptons on a site. For example, the cluster of reptons in Fig. 1b can be represented by the set (100110). Note that the mode corresponding to the translation of the cluster as a whole (the center of mass coordinate) is not included in the "bond" representation.

The number of ones in a given representation is equal to the number of forbidden faces (K-1) between the cells along the tube. Therefore, the total number of states of N reptons in a subset with K occupied sites is equal to the number of combinations of K-1 ones and N-K zeroes:

$$S(K,N) = \binom{N-1}{K-1} = \frac{(N-1)!}{(K-1)!\,(N-K)!}\,. \tag{6}$$

States from different subsets have different probabilities; therefore, the probability of N reptons to occupy K sites is

$$P(K,N) = \frac{(z-1)^{K-1}(N-1)!}{z^{N-1}(K-1)!\,(N-K)!}\,. \tag{7}$$

The first moment of this probability distribution is the average size of the cluster,

$$<K> = \sum_{K=1}^{N} K\,P(K,N) = N - \frac{(N-1)}{z}\,. \tag{8}$$

As expected, the length of the tube grows linearly with the polymer molecular weight (the number N of reptons in a cluster). Notice that the entropic "tension" and the average cluster size $<K>$ increase with the coordination number z.

The fluctuations in the cluster size K around the average $<K>$ can be asymptotically represented by the Gaussian

$$P(K,N) \simeq \frac{z}{[2\pi(z-1)N]^{\frac{1}{2}}}\,exp\{-\frac{z^2}{2(z-1)N}[K - \frac{z-1}{z}N]^2\}, \tag{9}$$

which is very similar to the fluctuations in tube length described in earlier works.[18,20,21]

The motion of unentangled loops between the cells along the confining tube is represented in the discretized model by the hopping of reptons from site to site. In the repton bond representation $\{\sigma_i\}$,[23] this motion corresponds to a random interchange of nearest neighbors:[23]

$$prob(\{\sigma_1,\sigma_2,\ldots,\sigma_i,\sigma_{i+1},\ldots,\sigma_{N-1}\}\rightarrow$$

$$\{\sigma_1,\sigma_2,\ldots,\sigma_{i+1},\sigma_i,\ldots,\sigma_{N-1}\}) = 1/z. \tag{4e}$$

For example, the jump of a repton 6 in Fig. 1b from site 4 to site 3, corresponding to the motion of loop *DD1* through the gate between cells 4 and 3 in Fig. 1a, is described by the interchange of the last pair in the repton bond representation (100110)→(100101). If $\sigma_i = \sigma_{i+1}$, their interchange does not lead to a new state and in the repton language implies that this transition is forbidden (e.g., jump of repton 3 in Fig. 1b in either direction).

Random exchanges of nearest neighbors $\sigma_i \leftrightarrow \sigma_{i+1}$ corresponds to transitions within a subset of states with constant K because they conserve the number of zeroes and the number of ones. If the probability of each exchange is the same ($1/z$), then each state within a subset has the same probability, as discussed above.

Each subset of states corresponds to the fixed length of the confining tube (tube of K cells). The transitions between different subsets correspond to fluctuations in tube length. In the repton bond representation, these transitions can be modeled by random interchanges of the end variables of the set (σ_1 or σ_{N-1}) with those from an infinite set containing $(z-1)/z$ number fraction of ones and $1/z$ number fraction of zeroes. Operationally this corresponds to randomly assigning these end variables (σ_1 and σ_{N-1}) the value of 1 with probability $(z-1)/z$ and the value of 0 with probability $1/z$. If the value of, say, σ_1 after the assignment is the same as before, no motion of the end repton results from this step.

Thus, the axioms of repton dynamics are equivalent to very simple local rules[23] in the bond representation $\{\sigma_i\}$, with σ_i=0 or 1.

Rule 1. The probability of interchange of nearest neighbors in the set is $1/z$ (Eq. 4e).

Rule 2. The end variables of the set (σ_1 and σ_{N-1}) are set equal to 1 with probability $(z-1)/z$ and to 0 with probability $1/z$.

The probability of an attempted repton jump leading to a new state is calculated from rules 1 and 2 to be

$$J = \frac{2(z-1)}{z^3}\left[1 + \frac{2(z-1)}{N}\right]. \tag{10}$$

In the bond representation each state of the cluster of N reptons is represented by N-1 values of σ. The translation of the cluster as a whole without changing its internal configuration leads to the same set $\{\sigma_i\}$. In order to follow the motion of the cluster, one extra degree of freedom is needed, i.e., the coordinate of the center of mass or of any repton (say, x_1).

The diffusion coefficient of the center of mass of the cluster of N reptons is

$$D = \frac{(z-1)\,a^2}{z^3 N \tau_m} + O(N^{-2}), \tag{11}$$

where a is the lattice spacing (average cell size), and τ_m is the average microscopic time between attempted hops. The relative correction from the second term in Eq. 11 can be smaller than that in Eq. 10 because of correlations between steps. The one-dimensional diffusion coefficient of the repton model (Eq. 11) corresponds to the curvilinear diffusion coefficient of a polymer in a tube. It implies that the time it takes the center of mass of the chain to diffuse an average length of the tube $<aK>$ along the contour of the tube is $\tau_D = <aK>^2/(2D)$. If we assume a Gaussian random walk configuration of a chain, this curvilinear displacement corresponds to the three-dimensional displacement $R = a<K>^{1/2}$, leading to the three-dimensional diffusion coefficient $D_{3D} = R^2/(6\tau_D) = D/(3<K>) \simeq a^2/(z^2N^2\tau_m)$. Thus, the experimentally observed result (Eq. 1) is asymptotically recovered.

The numerically calculated diffusion coefficients of the repton model for $z = 2$, 6 and 12 are compared with the asymptotic expression $D(N,z) = (z-1)a^2/(z^3N\tau_m)$ in Fig. 2. This asymptotic behavior is reached relatively quickly (between $N = 30$ and 50 reptons). Equation 11 suggests a molecular weight dependence stronger than M^{-2} for the diffusion coefficient $D_{3D} \sim M^{-2-O(1/M)}$. Unentangled chains in a melt are expected to have a weaker molecular-weight dependence of the diffusion coefficient ($D_{3D} \sim M^{-1}$). Thus, this exponent as a function of molecular weight should have a minimum with a value less than -2. The repton model in its present form is unable to

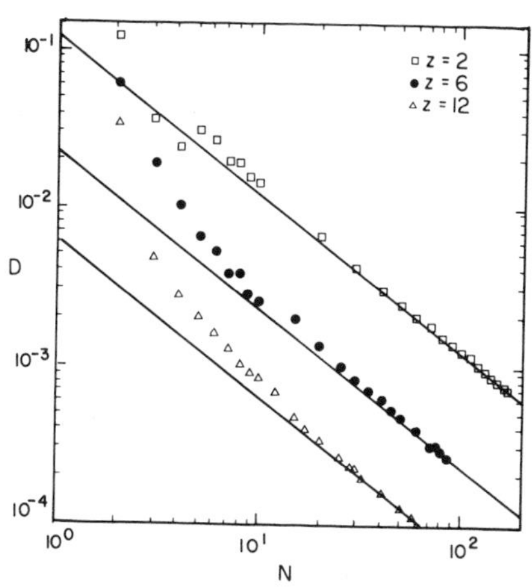

Fig. 2. Diffusion coefficient of the center of mass of the cluster of N reptons along the contour of the tube as a function of N for three different average coordination numbers of the entanglement net $z = 2$, 6 and 12. Straight lines correspond to the asymptotic behavior $D(N,z)=(z-1)a^2/(z^3N\tau_m)$.

accurately predict this minimum because it does not include the high frequency modes that are important in the dynamics of shorter chains. The crossover from entangled to disentangled polymers is beyond the scope of the present paper. A more appropriate model to study this crossover for a single chain in an array of fixed obstacles is the Evans - Edwards cage model.[24] An anomalous dependence of the diffusion coefficient D_{3D} on the polymer molecular weight M, suggested by the repton model, with a minimum value of the effective exponent less than -2 has already been observed for the Evans - Edwards model.[25]

If a step strain is imposed on a system of an unattached chain in an array of fixed obstacles, such as an ideal network, the resulting stress decays as the chain reptates into an undeformed configuration. The cells of the network that sustain stress are those from the original tube that has never been vacated by the chain since the time of the step strain.[1-3]

In the repton model, this relaxation process is described by the rates at which the cluster of reptons vacates initially occupied sites. Consider a step strain imposed at time $t = 0$. In Fig. 3, the coordinates of the cluster ends $x_1(t)$ and $x_N(t)$ are plotted as the functions of time.[12] Denote by $x_L(t)$ the rightmost propagation of the left end of the cluster between initial time and t: $x_L(t) = \max \{x_1(t')$ for $0<t'<t\}$. Similarly, let $x_R(t)$ be the leftmost propagation of the right end of the cluster $x_R(t) = \min \{x_N(t')$ for $0 <t'< t\}$. In Figure 3 $x_R(t)$ and $x_L(t)$ are represented by dashed lines.

Those and only those sites x of the lattice that satisfy $x_L(t) \leq x \leq x_R(t)$ have been continuously occupied between 0 and t. The stress is proportional to the number of these sites,

$$\mu(t) = \langle x_R(t) - x_L(t) + 1 \rangle / \langle K \rangle , \quad \text{for } x_R(t) \geq x_L(t), \tag{12}$$

where the angular brackets denote an ensemble average, and the normalization factor $1/\langle K\rangle$ is chosen so that $\mu(0) = 1$. (See Eq. 8 for the value of the average cluster size $\langle K\rangle$.)

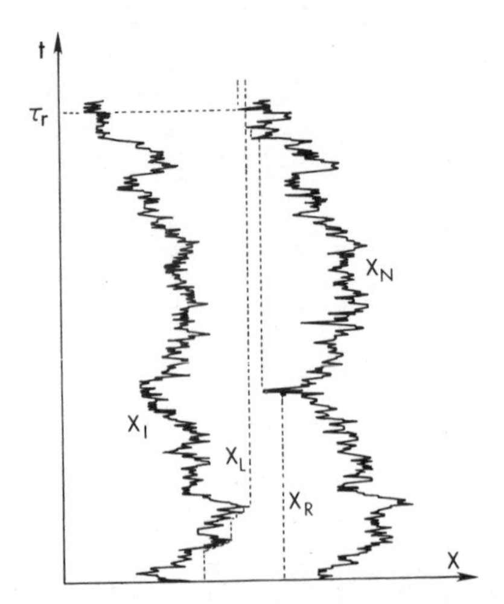

Fig. 3. Time dependence of the ends of the cluster $x_1(t)$ and $x_N(t)$ (solid lines) and their outmost propagations towards the center of the cluster $x_L(t)$ and $x_R(t)$ (dashed lines).

Zero-shear-rate viscosity is equal to the integral of the stress relaxation function:[7]

$$\eta_0 = \int_0^\infty \mu(t)\,dt\,, \tag{13}$$

where units are chosen to set plateau modulus $G_0=1$. An analytical expression for $\mu(t)$ for $N>2$ is still an open question. The numerically calculated values of η_0 for z = 2, 6 and 12 are presented in Fig. 4. The straight lines are the best fits to the viscosities for larger N. The slopes of these lines on the doubly logarithmic scale are 3.22 for z = 2 3.36 for z = 6 and 3.52 for z = 12. As mentioned above, the average number z of gates per entanglement net cell in most experimental systems is close to 6, and therefore the repton model predicts an experimentally observed[7] effective exponent $\eta_0 \sim M^{3.4\pm0.1}$, while $D_{3D} \sim M^{-2}+O(M^{-3})$ for those systems.

This demonstrates that a simple model of single-chain dynamics can exhibit nontrivial crossover behavior in the experimentally relevant range of molecular weights. Many-chain effects, such as constraint release, may modify[17] the molecular-weight dependence, but the origin of the phenomena is in the single-chain dynamics.

The results of the repton model qualitatively agree with the conclusions of Doi,[12,13] but represent a more accurate treatment of tube-length fluctuations and are in better agreement with experiments[26] at intermediate molecular weights. The two models probably coincide in the high-molecular-weight region where $M^{-1/2}$ expansion used by Doi is accurate, but this region is currently inaccessible both experimentally and numerically.

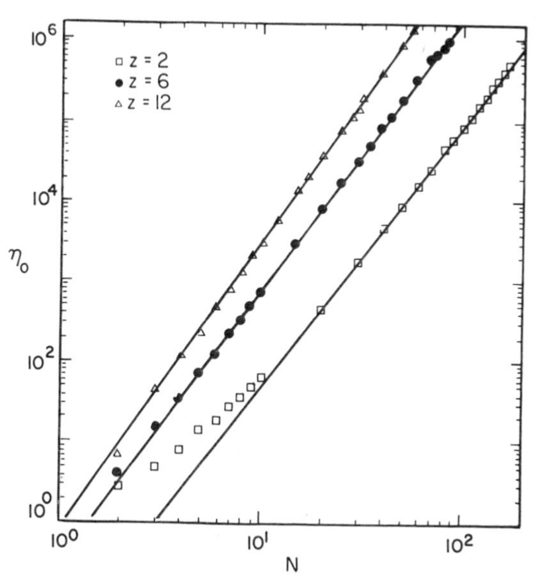

Fig. 4. Repton model predictions of the viscosity η_0 as a function of cluster size N. Straight lines are best fits of the higher N parts of the plot. Their slopes are 3.22 for coordination number z = 2, 3.36 for z = 6 and 3.52 for z = 12.

IV. RELATION TO OTHER MODELS

It is common to find various physical models related to each other. These relations often provide a deeper insight into the behavior of the systems in question. The repton bond representation $\{\sigma_i\}$ (Fig. 5a) is identical to the one-dimensional Ising spin representation $\{s_i\}$ of a state ($s_i = 2\sigma_i - 1$), where $s_i = \pm 1$ for $i = 1, \ldots, N-1$. Thus, the repton dynamics corresponds to random exchange of neighboring noninteracting spins (Fig. 5b). These models are also equivalent to the directed random walk (or interface) dynamics

$$t_i = \sum_{j=1}^{i} s_j . \tag{14}$$

A jump of a repton corresponds to a local flip of a section of an interface (Fig. 5c).

It is interesting to generalize these models to the case where each variable (σ_i or s_i) can assume more than two values. Let a configuration of a model be represented by a set $\{\sigma_i\}$ with $i = 1, \ldots, N-1$, but each σ_i can take any of q possible values. The dynamics of the model is defined by a random interchange of nearest neighbors (see Eq. 4e) and random reassignment of the boundary variables (σ_1 and σ_{N-1}). An example for $q = 2d$ is a Verdier-Stockmayer[27] model of an N-mer on a d-dimensional cubic lattice. An exchange of neighboring bond vectors σ_i and σ_{i+1} corresponds to the jump of the monomer $i + 1$ by the vector $\sigma_{i+1} - \sigma_i$ (see Fig. 5d for a two-dimensional example).

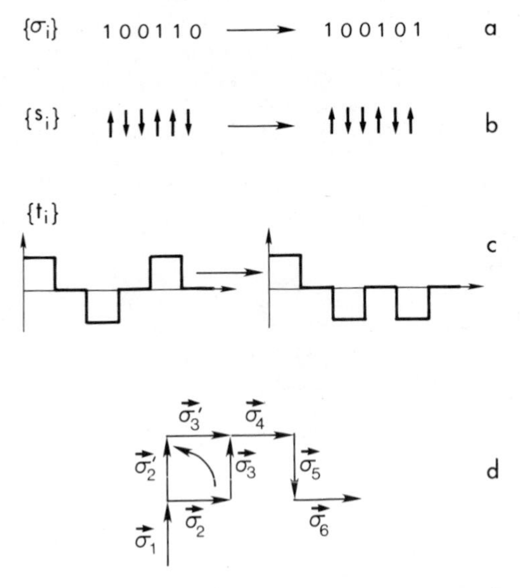

Fig. 5(a) Repton bond representation of the jump of repton #6 in Fig. 1(b) from site 4 to site 3, (b) Ising spin representation of the same jump, (c) Interface representation of the same jump, (d) Two-dimensional Verdier-Stockmayer model for N=7 monomers. The interchange of vectors σ_2 and σ_3 leads to a jump of the monomer #3.

Thus, the repton model ($q = 2$) could be thought of as a one-dimensional Verdier-Stockmayer model. In order to understand the physics of this connection let us place a Verdier-Stockmayer chain in a regular entanglement net with a spacing between obstacles equal to the bond length $|\sigma|$. Then the only configuration capable of diffusing is the one with all bonds parallel or antiparallel to each other, the case identical to the repton model with $z = 2$. In order to adapt the Verdier-Stockmayer model to the motion in the entanglement net, Evans and Edwards[24] had to introduce new jumps in addition to conventional Verdier-Stockmayer neighbor exchange.

V. BEYOND THE SIMPLE REPTON MODEL

By modifying Axiom 2 we can discuss a wide class of problems. In the model described above, there are no interactions between reptons occupying the same site. The probability of an unentangled loop decreases exponentially with the size of this loop[18,20,21] (Eq. 3). As mentioned in the second section, the algebraic prefactor of this dependence introduces a weak interaction between reptons occupying the same site and could be included in a more elaborate version of the model.

Another approximation was made by assuming uniform hopping probabilities. For a random entanglement net, the number of gates varies from cell to cell, as does the size of the cells, so the hopping probabilities fluctuate around their average values. It is important to estimate the effect of these fluctuations on the measurable quantities.

Random hopping probabilities are related to another important problem, gel electrophoresis. Reptation ideas were recently applied to this problem,[28] but tube-length fluctuations were neglected in these treatments. If an external electric field is applied to a system of a charged polymer in a neutral gel, the probability for a section of stored length (which is charged) to cross a face of a cell of an entanglement net depends on the relative position of this face with respect to the electric field. Since the initial chain conformation is random, a constant electric field induces additional random hopping probabilities. When the chain gets out of this initial isotropic conformation, it is stretched by the field, and the hopping probabilities acquire a bias in the direction of the field, leading to a finite mobility in addition to the random hopping component. The behavior of the repton model with such hopping probabilities will be discussed in future publications.

ACKNOWLEDGEMENTS

I thank E. Helfand for introducing me to the problem; C.L. Henley for a number of suggestions that led to the analytical solution of the model; M.E. Cates, I.S. Chang, R.H. Colby, M. Doi, J.M. Deutsch, P.G. de Gennes, W.W. Graessley, K. Kremer, D.S. Pearson, L.A. Ray and T.A. Witten for useful discussions and C. Lim for advice on the manuscript.

NOMENCLATURE

a	average cell size of an entanglement net
l	root mean square displacement of a defect
p	probability
q	number of possible states variable σ_i can be in
r	distance
s_i	Ising spin on the i-th site
t	time
t_i	coordinate of the i-th section of the interface
x_i	coordinate of the i-th repton along the contour of the tube

z	average coordination number of an entanglement net
D	curvalinear diffusion coefficient
D_{3D}	three-dimensional diffusion coefficient
G_0	plateau modulus
J	success probability of an attempted repton jump
K	number of occupied sites (size of the cluster)
M	polymer molecular weight
M_e	molecular weight between entanglements
N	number of reptons in a cluster
$P(K,N)$	probability of N reptons to occupy K sites
R_g	polymer radius of gyration
$S(K,N)$	number of states of N reptons on K sites
X_L	rightmost propagation of the left end of the cluster
X_r	leftmost propagation of the right end of the cluster
α	prefactor in the exponential
δ_{ij}	Kronicker delta function
η_0	zero-shear-rate viscosity
$\mu(t)$	stress relaxation function
ν	tube length fluctuation prefactor
σ_i	repton bond
τ_e	escape time
τ_m	average time between attempted repton jumps
τ_r	longest relaxation time
τ_D	time it takes a cluster to diffuse its size (one-dimensional diffusion)

REFERENCES

1. P.G. de Gennes, J. Chem. Phys., 55, 572 (1971).
2. P.G. de Gennes, Scaling Concepts in Polymer Physics, Cornell University Press, Ithaca and London (1979).
3. M. Doi and S.F. Edwards, Theory of Polymer Dynamics, Clarendon Press, Oxford (1986).
4. S.F. Edwards, Proc. Phys. Soc., (London) 92, 9 (1967).
5. M. Doi and S.F. Edwards, J. Chem. Soc., Faraday Trans., II, 74, 1789, 1802, 1818 (1978).
6. M. Tirrell, Rubber Chem. Technol., 57, 523 (1984).
7. J.D. Ferry, Viscoelastic Properties of Polymers, (Wiley, New York), 3rd ed. (1980).
8. H. Wendel and J. Noolandi, Macromolecules, 15, 1313 (1982).
9. H. Scher and M.F. Shlesinger, J. Chem. Phys., 84, 5922 (1986).
10. G.H. Weiss, J.T. Bendler and M.F. Schlesinger, Macromolecules, 21, 521 (1988).
11. J.M. Deutsch, Phys. Rev. Lett., 54, 56 (1985); J. Phys. (Paris), 48, 141 (1987).
12. M. Doi, J. Polym. Sci., Polym. Lett. Ed., 19, 265 (1981).
13. M. Doi., J. Polym. Sci., Polym. Phys. Ed., 21, 667 (1983).
14. J. des Cloizeaux, J. Phys. (Paris). Lett., 45, L17 (1984).
15. R.J. Needs, Macromolecules, 17, 437 (1984).
16. M. Rubinstein, Phys. Rev. Lett., 59, 1946 (1987).
17. M. Rubinstein and R.H. Colby, J. Chem. Phys., in press.
18. M. Rubinstein and E. Helfand, J. Chem. Phys., 82, 2477 (1985).
19. M. Doi, W.W. Graessley, E. Helfand and D.S. Pearson.Macromolecules, 20, 1900 (1987).
20. E. Helfand and D.S. Pearson, J. Chem. Phys., 79, 2054 (1983).
21. M. Doi and N.Y. Kuzuu, J. Polym. Sci., Polym. Lett. Ed., 18, 775 (1980).
22. K. Gotoh and J.L. Finney, Nature (London), 252, 202 (1974).
23. C.L. Henley, private communications.
24. K.E. Evans and S.F. Edwards, J. Chem. Soc., Faraday Trans. II, 77, 1891, 1913, 1929 (1981).

25. J.M. Deutsch, private communications.
26. R.H. Colby, L.J. Fetters and W.W. Graessley, Macromolecules, 20, 2226 (1987).
27. P.H. Verdier and W.H. Stockmayer, J. Chem. Phys., 36, 227 (1962); P.H. Verdier, J. Chem. Phys., 45, 2118, 2122 (1966); 52, 5512 (1970); R.A. Orwoll and W.H. Stockmayer, Adv. Chem. Phys., 15, 305 (1969).
28. O.J. Lumpkin, P. Dejardin, B.H. Zimm, Biopolymers, 24, 1573 (1985) and references therein.

ON THE EFFECT OF POSITIONAL ORDER ON THE SCATTERING BY COPOLYMERS AND NETWORKS WITH LABELED CROSSLINKS

Henri C. Benoît, Claude Picot

Institut C. Sadron (CRM-EAHP)
(CNRS-ULP) 6, rue Boussingault
67083 Strasbourg Cedex
France

Sonja Krause

Rensselaer Polytechnic Inst.
Troy, New York 12180-3590

ABSTRACT

In the frame of a "mean-field" approximation the intensity scattered by copolymers is evaluated taking into account the existence of correlations between the centers of the copolymer molecules. The results are discussed and extended to the case of networks with labeled crosslinks.

I. INTRODUCTION

A few years ago, R. Duplessix et al.[1-2] prepared networks with deuterium-labeled crosslinks in order to obtain information on their structure with the newly available small-angle neutron scattering technique. These networks were prepared by the following synthetic method. One first prepares a classical living polymer by anionic polymerization using a difunctional initiator (the tetramer of α-methylstyrene) and a mixture of benzene and tetrahydrofuran as solvent. This leads to a living polymer having two reactive ends. Immediately after, deuterated styrene monomer is added and one obtains a three-block deuterated polystyrene-hydrogenated polystyrene-deuterated polystyrene copolymer. This copolymer of low polydispersity is crosslinked by addition of divinylbenzene. The crosslinked network has no well-defined functionality, since one does not know the exact number of chains attached to the divinylbenzene nodules. Moreover, the labeling is not exactly on the crosslinks. Nevertheless, it is not unreasonable to assume, as a first approximation, that the behavior of these networks in neutron scattering experiments will be similar to that of better-defined systems. The results (obtained on these samples in bulk and in the swollen state) were surprising. The scattering intensity (Fig. 1) starts from a low value, increases with the modulus of the scattering vector $\mathbf{q}$ ($\mathbf{q} = 4\,\pi/\lambda) \sin(\theta/2)$, λ being the wavelength of the neutron beam and θ the observation angle), goes through a maximum and further decreases following approximately a $\mathbf{q}^{-2}$ law. The position of the maximum corresponds roughly to $\mathbf{q}\,R \approx 1$, calling R the radius of gyration of an elastic chain. Similar results have been obtained by Belkebir-Mrani

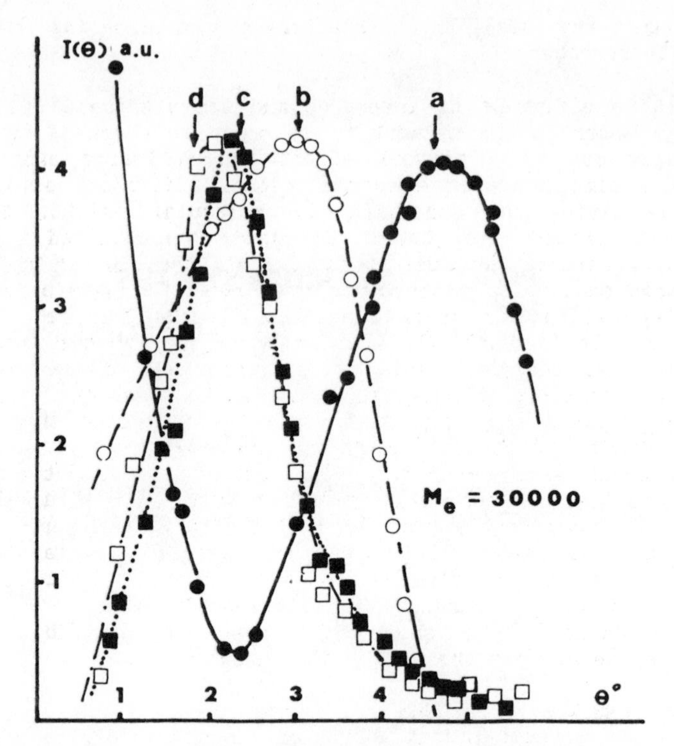

Fig. 1. Scattering intensity (in arbitrary units) of a polystyrene network with labeled crosslinks (from ref.(2)). M_e is the molecular weight of the elastic chains, (a) corresponds to experiments made in the bulk, (b) to the network swollen at equilibrium in cyclohexane, (c) in carbon disulfide and (d) in benzene.

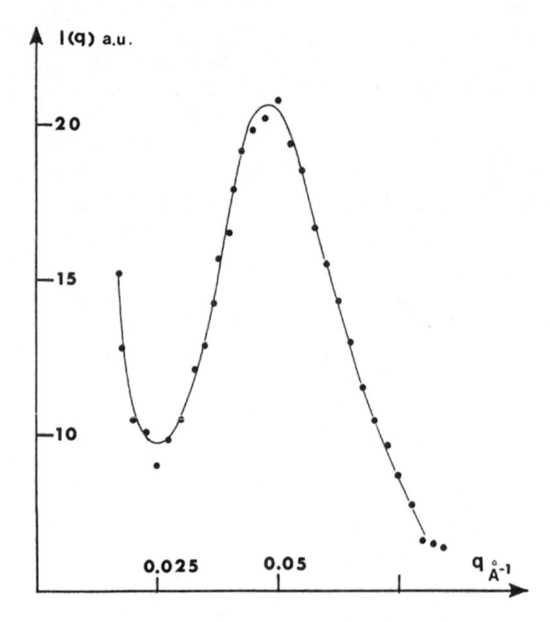

Fig. 2. Scattering intensity of stars swollen in carbon disulfide (The length of the arms of the stars is $M_e/2$).

et al.[3] using X-ray small-angle scattering and labeling the network crosslinks with ferrocene.

This result is difficult to interpret and was attributed at that time to short-range order in the network.[1] In order to check if the observed peak was really due to a network effect the following experiment was performed: this time one used a monofunctional initiator building chains with only one living end and half the molecular weight. Afterwards, exactly the same procedure as before was used; one obtained a collection of stars. The result is the same as cutting all the elastic chains in the middle. Neutron scattering experiments made under the same conditions (in bulk and in the swollen state) gave practically identical results. There is a maximum and its position is also the same as for the network (Fig. 2). This eliminates the hypothesis of the existence of an order in the network since, especially in bulk, there can be no order for stars as well as for linear polymers (In fact three-block copolymers of the type D-H-D gave similar results). At that time, the interpretation of these experiments was not possible since no theory was able to explain the properties of copolymers in concentrated solution and in bulk. The situation has now changed. The extension of the de Gennes[5] theory by Leibler[4] as well as other equivalent methods[6] allows the theoretical calculation of the intensity scattered by copolymers in bulk and in solution at any concentration. It is the purpose of the present paper to extend these theories to networks and to show how a nonrandom distribution of the crosslinks in space affects the scattering intensity.

II. BASIC EQUATIONS FOR COPOLYMERS

Let us consider a bulk specimen made of identical copolymrs having each n_a units A and n_b units of type B. These units are supposed to occupy the same volume. It has been shown that the intensity $i(q)$ scattered per unit can be written as:

$$\frac{(a-b)^2}{i(q)} = \left[\frac{1}{n u^2 v^2}\right] \frac{P_t(q)}{P_a(q) P_b(q) - P^2_{ab}(q)} - 2\chi . \quad (1)$$

In this expression, a and b are the coherent scattering lengths of the units A and B. The quantities n, u and v are defined as:

$$n = n_a + n_b ; \quad u = \frac{n_a}{n} ; \quad v = 1 - u = \frac{n_b}{n} . \quad (2)$$

The quantity χ is the classical Flory interaction parameter between units A and B. The P's are structure factors normalized to unity for $\mathbf{q} = 0$. $P_a(q)$ and $P_b(q)$ are the structure factors of the part A and the part B of the copolymer. For instance:

$$P_a(q) = \frac{1}{n_a^2} \sum_{i_a j_a} < exp(-i \mathbf{q}\, \mathbf{r}_{i_a j_a}) > . \quad (3)$$

$P_{ab}(q)$ is the cross-term characterizing the interferences between A and B parts of the copolymer:

$$P_{ab}(q) = \frac{1}{n_a n_b} \sum_{i_a j_b} < exp(-i\mathbf{q}\,\mathbf{r}_{i_a j_b}) > . \tag{4}$$

$P_t(q)$ is the structure factor of the whole molecule defined as:

$$P_t(q) = u^2 P_a(q) + v^2 P_b(q) + 2\,uv P_{ab}(q) \tag{5}$$

This formula predicts for a copolymer in bulk a curve presenting a maximum. When χ increases the height of this maximum increases, but its position does not change. For a critical value of χ depending on the copolymer architecture, $i(q)$ diverges and one reaches a mesophase.

Recently,[7] it has been shown that if, instead of having diblock copolymers in the system, one makes, without changing the size of the blocks, a multiblock copolymer $(A\text{-}B)_N$, the scattering intensity is not very sensitive to N and reaches a limiting value as soon as N is greater than 10. This means that there is no substantial difference between the scattering by one giant molecule built with all the blocks of the sample and a melt of randomly distributed diblock copolymers. This result is quite general and has been used to describe the scattering by networks.

III. THE CASE OF COPOLYMERS WHICH ARE NOT RANDOMLY DISTRIBUTED

In the classical Random Phase Approximation technique one assumes that the molecules are randomly distributed. It will be shown in this paragraph how the spatial distribution of the copolymers can be taken into account. For this purpose, a point O_p is attached to each copolymer molecule. This point is arbitrary but it is preferable to take a central point. If one deals with a diblock copolymer O could be the junction between the blocks. If it is a star the natural choice is the center of the star. (In more complex cases, it could be advantageous to use the center of mass as O.)

The distance between the monomer i of molecule p and the monomer j of molecule r can be written as:

$$M_{i_p} M_{j_r} = M_i O_p + O_p O_r + O_r M_r . \tag{6}$$

Assuming no correlations between the orientations of these three vectors one obtains:

$$< exp(-i\mathbf{q} M_{i_p} M_{j_r}) > = < exp(-i\mathbf{q} M_i O_p) > < exp(-i\mathbf{q}\, O_p O_r) > < exp(-i\mathbf{q}\, O_r M_j) > . \tag{7}$$

One considers now the sample as a unique molecule and one has to evaluate the form factors of this molecule which will be called Π. For the form factor Π_a of the part A, one writes:

$$\prod_a(q) = \frac{1}{N^2 n_a^2} \sum_{ijpr} < exp(-i\mathbf{q} M_{i_p} M_{j_r}) >$$

$$= \frac{1}{N} P_a(q) + \frac{A^2(q)}{N^2} \sum_{p \neq r} < exp(-i\mathbf{q} O_p O_r) > , \qquad (8)$$

calling N the total number of molecules and defining $A(q)$ as the amplitude scattered relatively to the center O:

$$A(q) = \frac{1}{n_a} \sum_{i_a} < exp(-i\mathbf{q} O M_i) > . \qquad (9)$$

The last term in Eq.(8) defines the correlations between the centers O of the molecules. It leads us to introduce a function $S(q)$ such as:

$$S(q) = \frac{1}{N} \sum_{p \neq r} < exp(-i\mathbf{q} O_p O_r) > . \qquad (10)$$

This quantity $S(q)$ is zero when the points O are randomly distributed. If in the system we had only the points O as scatterers, they would scatter an intensity $J(q)$ proportional to $1 + S(q)$. One can also say that $S(q)$ is the Fourier transform of the radial distribution function of the points O.

With these definitions one obtains for the factors Π:

$$N \prod_a(q) = P_a(q) + S(q) A^2(q) ;$$

$$N \prod_b(q) = P_b(q) + S(q) B^2(q) ; \qquad (11)$$

$$N \prod_{ab}(q) = P_{ab}(q) + S(q) A(q) B(q) .$$

This leads to a generalization of equation (1):

$$\frac{(a-b)^2}{i(q)} = \left[\frac{1}{n u^2 v^2} \right] \frac{P_t + S(q)(Au + Bv)^2}{P_a P_b - P_{ab}^2 + S(q)(A^2 P_b + B^2 P_a - 2ABP_{ab})} - 2\chi , \qquad (12)$$

not writing explicitly the dependence on $\mathbf{q}$.

This equation characterizes the scattering intensity of block copolymers in bulk when one assumes that there are correlations in position and no correlations in orientation. The use of the hypothesis which allows to write Eq.(7) explains why the point O has to be chosen, if possible, as a center of symmetry. We shall now discuss the applications of Eq.(12).

IV. APPLICATION TO COPOLYMERS

VI.1 The Case of Symmetric Copolymers At first sight, Eq.(12) seems to be difficult to use without any knowledge of the values of the function $S(q)$. This is not exactly so and we shall see that, even without a precise definition of $S(q)$, useful conclusions can be obtained. It is well known that, when one uses a symmetrical copolymer, i.e., a copolymer for which:

$$P_a(q) = P_b(q) \; ; \qquad n_a = n_b \; ,$$

the classical Eq.(1) of Leibler[4] is simplified and can be written as:

$$\frac{(a-b)^2}{i(q)} = \frac{8}{n} \, \frac{1}{P_a(q) - P_{ab}(q)} - 2\chi \, . \tag{13}$$

If we introduce the same hypothesis in Eq.(12) and if we also assume that the point O is such that $A(q) = B(q)$ we obtain exactly the same result: The scattering intensity is independent of $S(q)$. This result is very surprising; it can be explained qualitatively by saying that everywhere in the sample the composition is the average composition such that $u = v = .5$. Therefore, the points O have no contrast with respect to the background and cannot be seen. In some respect, this is an advantage because it makes the scattering intensity insensitive to heterogeneities. This symmetry condition is not as severe as one could believe. For instance, if a star copolymer is made of p identical branches A and q branches B of the same length and volume it can be shown that $i(q)$ is independent of $S(q)$.

IV.2 The General Case In order to visualize the effect of $S(q)$ on the scattering intensity it is convenient to study its effect on $i(o)$, on the initial slope of $i(q)$ as function of q^2 and on the behavior at large q. Equation (12) shows that regardless of the value of $S(o)$, $i(o) = O$. The curves always start from the origin (if the system is monodisperse). If one expands $i(q)$ as a function of q^2, one finds for the first term:

$$i(q) = n u^2 v^2 \frac{q^2}{3} \left[2R_{ab}^2 - R_a^2 - R_b^2 \right] . \tag{14}$$

In this expression R_a^2, R_b^2 and R_{ab}^2 are the partial radii of gyration of copolymer (they are, in fact, the coefficients of $q^2/3$ in the expansion of the P's as a function of q^2). Again we see that $i(q)$ is independent of $S(q)$ in the small q range. Regardless of the values of $S(q)$, the curve $i(q)$ starts with the same initial slope. In order to discuss the behavior at large q one has to make an assumption about the nature of the polymer chains. In order to simplify the discussion we shall assume that the chains are Gaussian. If one calls L the length of the Kuhn's segment, one has at large q:

$$P_a(q) = \frac{12}{q^2 L^2 n u} \; ; \qquad P_b(q) = \frac{12}{q^2 L^2 n v} \; ; \qquad P_{ab}(q) = \frac{K}{q^4 L^4} \; .$$

Similarly, $A(q)$ and $B(q)$ vary as q^{-2}. Regardless of the nature of the interactions between chains, the quantity $S(q)$ has to be a decreasing function of q which can be written as q^{-n} where n is positive. Putting these values in Eq.(12) one finds:

$$i(q) \approx (a - b)^2 \frac{12 u v}{q^2 L^2} \; , \tag{15}$$

a result again independent of $S(q)$. Since $S(q)$ has no influence on the initial part of the curve $i(q)$ versus q as well as on its last part, one is tempted to conclude that, unless $S(q)$ has strong maxima there will only be a very small effect of $S(q)$ on the scattering by a melt of copolymers. In order to illustrate this point, we have evaluated $S(q)$ in the following case; we consider a star copolymer made of four equal branches; each branch being a two-block copolymer A-B. The center of the star is the junction point of the A blocks (see Fig. 3):

It has been assumed that $u = 0.1$ in order to be far from the symmetrical regime. This means that the labeled chains merging at the crosslink are 9 times shorter than the external chains. One can first assume that the quantity $S(q)$ is constant; this means that there are short-range order interactions between the copolymer molecules as in classical liquids; this is not reasonable and has practically the same effect as changing the value of the χ parameter. Therefore, we have to introduce a q dependence of $J(q)$. It can be assumed, as a simple approximation, that $J(q)$ obeys an Ornstein Zernicke type of law and has the form:

$$J(q) = 1 + S(q) = \frac{2}{q^2 R^2 K^2 + 1} \; , \tag{16}$$

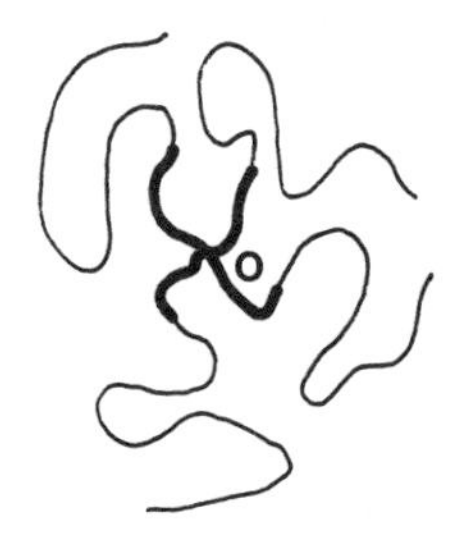

Fig. 3. The Star Model.

where K is the ratio of the correlation length ξ to the radius of gyration R of one branch of the star. Assuming the copolymer to be Gaussian and the interaction parameter χ to be zero, one obtains, for different values of K, the curves of Fig.(4). One sees that, as predicted by the theory, both the initial and the asymptotic parts are independent of $S(q)$. One sees also that for low values of the parameter K there is only a small displacement of the position of the maximum and a change in its height. On the contrary, for large values of K one obtains two maxima. The position of the maximum which comes from $S(q)$ can be changed as well as its height relatively to the second maximum. This kind of curve is similar to the results obtained by Wu and Bauer on epoxy resins.[8] For such systems it is practically impossible to measure the function $S(q)$, unless ξ and R are very different. The only thing which can be said is that $S(q)$ introduces changes in the shape of the curve $i(q)$, changes which can be easily misinterpreted.

IV.3 Very Asymmetrical Copolymers From the preceding results it is evident that, in order to separate $S(q)$ from the copolymer scattering one has to use very asymmetrical copolymers. It will therefore be assumed that in the example studied above the central part of the copolymer is small compared to q^{-1} and can be considered as one scattering point. In the frame of this hypothesis one can write:

$$u = 1/n \; ; \; v = 1 \; ; \qquad P_a(q) = A(q) = 1 \; ; \; P_b(q) = P_t(q) \; ; \; P_{ab}(q) = B(q) \; .$$

The classical Eq.(1) is then:

$$\frac{(a-b)^2}{i(q)} = n \left\{ \frac{1}{1 - \frac{B^2(q)}{P_t(q)}} \right\} - 2\chi \; . \tag{17}$$

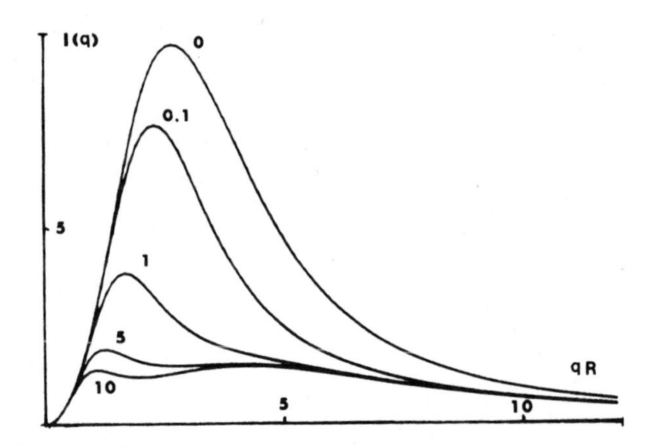

Fig. 4. Scattering intensity of four-arm star copolymers of composition $u = 0.1$. The position of their centers obeys the distribution of Eq.(16) with values of $K = \xi/R$ given in the figure.

If one introduces the intensity scattered per chain $i_o(q) = n_i(q)$ one sees that the contribution of the χ parameter can be neglected. This formula shows, as already pointed out by de Gennes,[5] that the intensity starts from zero at $\mathbf{q} = o$ and reaches a plateau when $\mathbf{q}\ R > 1$ since the quantity $B^2(q)/P_t(q)$ vanishes for large $\mathbf{q}$.

We can now, using the same model, take into account the correlations of the positions of the different copolymers, characterized by $S(q)$. Introducing the same modifications in Eq.(12) one obtains:

$$i_o(q) = (a-b)^2[1+S(q)]\ \frac{P_t(q)-B^2(q)}{P_t(q)+S(q)B^2(q)}\ ,$$

or:

$$\frac{(a-b)^2}{i_o(q)} = \frac{1}{J(q)} + \frac{B^2(q)}{P_t(q)-B^2(q)}\ . \tag{18}$$

This intensity is made of two terms: one is the scattering by the centers O of the copolymers and the other the contribution of the structure of the copolymer. The separation of these terms is not difficult to perform at large $\mathbf{q}$ ($\mathbf{q}\ R > 1$) since the copolymer term vanishes rapidly when $\mathbf{q}$ increases. As an example, we have considered in Fig.(5) the case of a four-arms star, plotting $i_o(q)$ as function of $\mathbf{q}^2$. The curve 1 corresponds to $S(q) = 0$; it is the contribution of the copolymer term and reaches a plateau. Curve 2 corresponds to an Ornstein Zernicke scattering function for $J(q)$. One sees clearly the $\mathbf{q}^{-2}$ tail. This shows that, if there is a fractal exponent, it is theoretically possible to measure it.

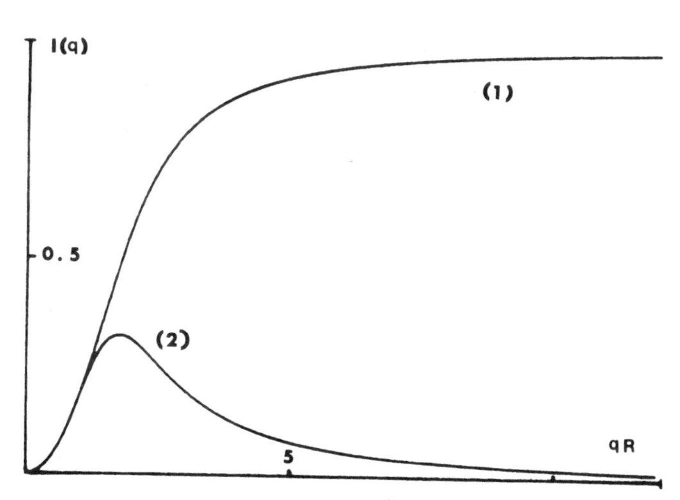

Fig. 5. Intensity scattered by stars with labeled centers as function of qR: Curve 1: the centers are randomly distributed. Curve 2: they obey an Ornstein-Zernicke type distribution.

V. COPOLYMERS IN SOLUTION

The determination of $S(q)$ is probably more interesting in solution than in bulk. It is therefore important to extend the preceding results to the case of solutions. In the general case it leads to cumbersome formulas, difficult to exploit: therefore, the discussion will be limited to the case where the copolymer is very dissymmetric and immersed in a solvent which has the same contrast factor as the sequences of type A (one sees only the centers O). It is also assumed that the quality of the solvent is the same for both sequences and that the interaction coefficient between the blocks is zero:

$$v_{as} = v_{bs} = v_{ab} = v, \qquad or \qquad \chi_{as} = \chi_{bs}; \quad \chi_{ab} = 0$$

Using a formula already published[6] and introducing the previous simplifications one obtains, calling ϕ the volume concentration, v the excluded volume parameter and $i_0(q)$ the intensity scattered per chain:

$$i_o(q) = (b-a)^2 J(q) \left[\frac{1 + v\phi n[P_t(q) - B^2(q)]}{1 + v\phi n[P_t(q) + S(q)B^2(q)]} \right] . \tag{19a}$$

or

$$\frac{(a-b)^2}{i_o(q)} = \frac{1}{J(q)} + \frac{v\phi n B^2(q)}{1 + v\phi n[P_t(q) - B^2(q)]} . \tag{19b}$$

One sees that, if $v\ \phi\ n$ which is equivalent, in the experimentalist notations, to A_2cM, (A_2 second virial coefficient, c concentration and M molecular weight) is large, one recovers Eq.(18). If one works in dilute solution, the scattering intensity depends only on $J(q)$ and the copolymer structure has no effect. One has to realize that in bulk as well as in solution, it will be difficult from an experimental point of view to use this last approximation for the determination of $J(q)$. If the labeled central part of the copolymer is small compared to the wavelength and to the other part of the copolymer, the contrast will be low and the experiment difficult, but, with the accuracy of the present neutron scattering equipment, it does not seem to be impossible to measure $J(q)$ on such systems.

VI. APPLICATION TO NETWORKS

Until now it has been shown that, if block copolymers, although being in homogeneous phase, are not distributed randomly in space one has to introduce a correction factor. This correction factor is small when the copolymer is nearly symmetrical but can become important for highly dissymetric labeling. It will be shown here how this calculation can be applied to networks in bulk or in the swollen state. In order to evaluate the intensity scattered by a network made of copolymers, the simplest idea is to calculate directly the total partial structure factors $\Pi_a(q)$, $\Pi_b(q)$ and $\Pi_{ab}(q)$ and to use Eq.(1) replacing the P's by the Π's. This is very difficult because the chains in a network are connected by many loops which depend strongly on the exact topology of the network, topology which is not known. Another approach, developed by S. Krause,[9] assumes that the network can be simulated by a very long chain to which all the building units are attached. This shows that the scattering intensity by the network is similar to that by a melt of the units. The third method derives directly from what has been discussed in this paper. One also divides the network into N identical units and assumes that the effect of the network structure is the same as the effect of a potential energy obliging the crosslinks to be distributed nonrandomly but to occupy relative positions characterized by the radial distribution function, Fourier transform of $S(q)$. This hypothesis does not take into account the exact effect of the elastic chains but assumes that the network can be considered as a collection of stars maintained together by a potential energy producing a nonuniform distribution. In the frame of this hypothesis, everything which has been said about the intensity scattered by copolymers with positional order can be applied to networks. If the copolymers forming the network are symmetric, with a ratio of labeled to unlabeled part of the order of 0.5, the scattering is practically that by the melt of the units. This explains the experimental results quoted at the beginning of this paper. The distribution of the crosslinks in space, in bulk as in solution, which is an important parameter for the discussion of the network properties has, until now, never been measured.

V. CONCLUSION

The method, which has been presented here, gives a possibility for extracting from experiments the quantity $S(q)$ which is the Fourier transform of the radial distribution function of the crosslinks. This could be done either at rest or in an extended state and would provide information on the mechanism of rubber elasticity. It is not easy because the conditions for distinguishing between the scattering by the copolymer units and the crosslinks are difficult to meet. Nevertheless, since we do not have any other method at our disposal, it is worth to be explored.

ACKNOWLEDGEMENT

The authors would like to thank Dr. G. Hadziioannou for fruitful discussions at the beginning of this work.

NOMENCLATURE

A and B	types of monomeric units in the copolymer.
a,b	coherent scattering length of the monomers A and B.
$A(q)$	amplitude scattered by part A of the copolymer.
$B(q)$	amplitude scattered by part B of the copolymer, (see Eq. 12).
$i(q)$:	intensity scattered per monomeric unit (total intensity divided by the number of monomers in the sample).
$i_o(q)$	the same quantity but in the case of solutions; the solvent is not taken into account.
$J(q)$	intensity scattered by the points 0 alone.
L	length of the statistical element of the chain (The same for the A and B parts).
n, n_a, n_b	number of monomers in a molecule: n total number, n_a number of monomers in part A, n_b in part B ($n_a + n_b = n$).
N	total number of molecules in the sample.
0	center of molecule.
$P(q)$	form factor of a molecule or a part of a molecule (see Eq. 3). $P_t(q)$ for the whole molecule, $P_a(q)$ and $P_b(q)$ for its parts A and B, respectively. $P_{ab}(q)$ is the cross term obtained from the interferences between a monomer A and a monomer B.

$\Pi(q)$ the $\Pi(q)$'s are similar to the $P(q)$'s but this time for the whole system considered as a giant molecule.

q scattering vector; its modulus is:

$q = 4\pi/\lambda \sin(\theta/2)$ where λ is the wavelength of the incident beam and θ the scattering angle.

s index characterizing the solvent.

$S(q)$ Fourier transform of the pair correlation function between the points 0.

$u = (n_a/n)$ $v = (n_b/n)$; $u + v = 1$

v_{ij} excluded volume parameter between species i and j.

ϕ^{ij} volume fraction occupied by the copolymer in the case of solutions.

χ Flory's interaction parameter between the two species present in the system; if there are more than two species one writes χ_{ij} where i and j characterize the interacting species.

REFERENCES

1. R. Duplessix, Thèse Doctorat d'Etat, Etude structurale de réseaux polymériques tridimensionnels par diffusion cohérente des neutrons et par diffusion inélastique de la lumière. Université Louis Pasteur Strasbourg (1975).
2. H. Benoît, D. Decker, R. Duplessix, C. Picot and P. Rempp, "Characterization of Polystyrene Networks by Small Angle Neutron Scattering", J. Polym. Sci., Polym. Phys. ed. 14, 2119-2128, (1976).
3. A. Belkebir-Mrani, G. Beinert, J. Herz and A. Mathis, "Experimental Evidence for an Isotropic Distribution of Cross-Linkages in Polystyrene Model-Networks," Europ. Polym. J., 12, 243-246 (1976).
4. L. Leibler, "Theory of Microphase Separation in Block Copolymers," Macromolecules, 13, 1602-1617 (1980).
5. P.G. de Gennes, "Theory of X-ray Scattering by Liquid Macromolecules With Heavy Atoms Labels," J. Phys. (Paris), 31, 235-238 (1970).
6. H. Benoît, W. Wu, M. Benmouna, B. Mozer, B. Bauer and A. Lapp, "Elastic Coherent Scattering From Multicomponent Systems- Application to Homopolymer Mixtures and Copolymers," Macromolecules, 18, 986-993, (1985).
7. H. Benoît and G. Hadziioannou, "Scattering Theory and Properties of Block Copolymers With Various Architectures in the Homogeneous Bulk State." Macromolecules, 21, 1449-1464 (1988).
8. W. Wu and B.J. Bauer, "Epoxy Network Structure: 3- Neutron Scattering Study of Epoxies Containing Monomers of Different Molecular Weights," Macromolecules, 19, 1613-1618 (1986).
9. S. Krause, C. Picot and H. Benoît, This book.

CALCULATION OF ELASTIC SMALL-ANGLE NEUTRON SCATTERING FROM LABELED GELS

Sonja Krause

Department of Chemistry
Rensselaer Polytechnic Institute
Troy, New York 12180

Henri C. Benoît and Claude Picot

Institut Charles Sadron (CRM)
6 Rue Boussingault, 67083
Strasbourg, France

ABSTRACT

Crosslinked network polymers which have been labeled around the crosslinks are essentially triblock copolymers connected in groups of three or four at the ends. Although such networks are infinite in extent, it has been possible to calculate elastic small-angle neutron scattering (SANS) patterns for them after making reasonable assumptions about correlations in the networks. It is shown that the SANS from such crosslinked labeled networks is indistinguishable from the SANS of long-branched polymers made up of the same triblock copolymers. In fact, little difference is calculated between the SANS of the networks and the SANS of the constituent triblock copolymers. The calculated SANS of three different-long branched polymers are compared with the SANS of the triblock copolymer. It is shown that SANS of gels in which all of the crosslinks are labeled depends only on the extent of labeling and on the radius of gyration of the constituent block copolymer.

I. INTRODUCTION

Up to this time, it has not been possible to compare the results of experimental studies of elastic small angle neutron scattering (SANS) from polymer networks with labeled crosslinks with theoretical treatments of this scattering. This problem existed because polymer networks are essentially infinite in extent and an infinite sample is expected to exhibit infinite scattering at very low scattering angles. In this work, we show that the SANS from crosslink labeled networks depends almost exclusively on the scattering from neighboring repeat groups in the network. That is, the scattering peak observed by various workers[1-3] who studied SANS by labeled networks is the result of a correlation hole as stated by Wu and Bauer.[2] Earlier work, like the work on polystyrene by Benoît et al.[1] has often been misinterpreted.

SANS studies on labeled networks generally result in plots of scattering intensity versus scattering angle, Θ, or **Q**, which have a maximum at a small value of Θ or **Q**. The parameter $\mathbf{Q} = (4\pi/\lambda)\sin^2(\Theta/2)$, where λ is the wavelength of the neutrons being scattered. The maximum in these plots was tentatively attributed to interference peaks from an ordered lattice of crosslink sites separated by Gaussian connecting chains by Benoît et al.,[1] who worked with crosslinked labeled networks. This interpretation could not, however, be correct because of the actual density of the distribution of crosslink sites in the network. That is, if one takes a particular crosslink site as a reference, there is a large number of other crosslink sites that are physically closer to this crosslink site than those connected by the shortest path along the network chains. It is thus more reasonable to interpret the observed maximum as a correlation hole effect. Wu and Bauer[2] made an attempt to do this which did not, however, take the connectivity of the network into account.

It is difficult, however, to deal with the almost infinite connectivity of a polymer network. In scattering calculations. such infinite connectivity usually leads to infinite scattering at zero scattering angle. This is not seen experimentally. In this work, the connectivity of the polymer network is approximated in a different way than that used in the preceding paper.[4] In that paper, some approximations that allow the calculation of correlations between scattering units in the labeled network were first justified and then used to calculate the expected scattering. Various labeled structures around each crosslink may be postulated, and the scattering from a large array of such structures connected by a force field but no physical links, is calculated. The method used in the present paper considers the gel as a connected network, but the calculations are made possible by means of some simplifications which are examined below. In spite of the major conceptual differences between the two methods of calculation, almost the same scattering pattern is predicted, not much different from that calculated and observed for simple diblock and triblock copolymers. That is, both theoretical treatments give results equivalent to the correlation hole observed in simple block copolymers.

In the present method of calculation, one first notes that any two scattering centers in a crosslinked gel are connected to each other through the network by a very large number of different paths, most of which are so interconnected that the statistics are forbidding to contemplate. Therefore, the first assumption in this method states that only the shortest path between two scattering centers need be considered in the calculation of correlations. The second assumption, that the shortest path between any two scattering centers in the network is a Gaussian chain, is closely related to the first assumption.

Both assumptions come from a consideration of the mean-square distance between any two scattering centers in a flexible ring polymer. Strictly speaking, whenever two points in a network or in a circular polymer can be reached by different paths, neither path has the characteristics of a linear Gaussian chain. In the case of a circular polymer made up of M monomer units, however, it can be shown that the exponent in the probability distribution of distances between monomer units that are less than $(M/8)$ monomer units apart is within 10% of that in the probability distribution of linear Gausian chains of the same length. For this calculation, one may use the probability distribution derived for flexible ring polymers by Casassa.[5] This calculation gives an approximate 10% error in the probability distribution when compared with that of the end-to-end distances of a linear Gaussian chain with the same number, $M/8$, monomer units. Thus, in a circular polymer containing M monomer units, the probability distribution of distances between monomer units that are less than $M/8$ monomer units apart can be considered as if it were that of a linear Gaussian chain containing the same number of monomer units. That is, the existence of the longer path, that of $7M/8$ or more monomer units, between monomer units can be ignored. We have assumed that the same considerations apply in a polymer network. Only a single, shortest path between any two monomer units in the network is therefore considered in this work. Thus, the present calculation will probably be in error whenever the network contains many small closed loops, when knots of closely connected network points exist in the network. The calculation will work best for a randomly connected network.

A third assumption was necessary in this work. Since a gel may become very large, the shortest path between some of the monomer units in the gel will also become very large. In order to avoid calculating infinite scattering at small values of **Q**, it is necessary to assume that the number of these long paths that must be considered does not increase rapidly with the length of the path. If one draws a Bethe lattice, which is not connected back on itself like a gel:

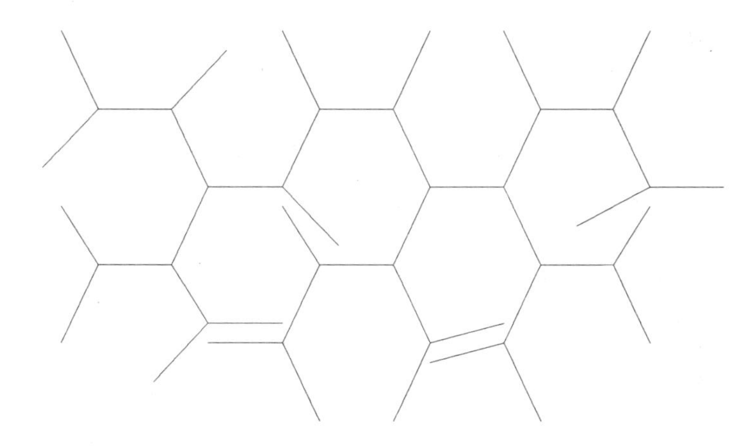

it is easy to see that, the longer the path between points on this lattice, the more such paths there are to be considered. The number of such paths increases without bound as the distance between two points increases. If one draws a gel, the situation becomes complicated and the number of shortest paths of a particular length in the gel depend upon the way in which the gel is interconnected. For this work, some completely interconnected trifunctional gels containing different numbers of the constituent triblock copolymers, were drawn. If very short loops, which also cause trouble with the first two assumptions, were avoided, it was found that the number of shortest paths of any particular type between distant monomer units remained constant, no matter how long these shortest paths became.

These three assumptions were used to calculate the expected neutron scattering from trifunctional gels labeled around the crosslinks by assuming that the gels were made up of trifunctionally connected triblock copolymers. For comparison, the expected scattering from the following infinite branched polymers, all made up of identical triblock copolymers, was also calculated. Each triblock copolymer was assumed to have deuterated ends of equal length.

Triblock Copolymer

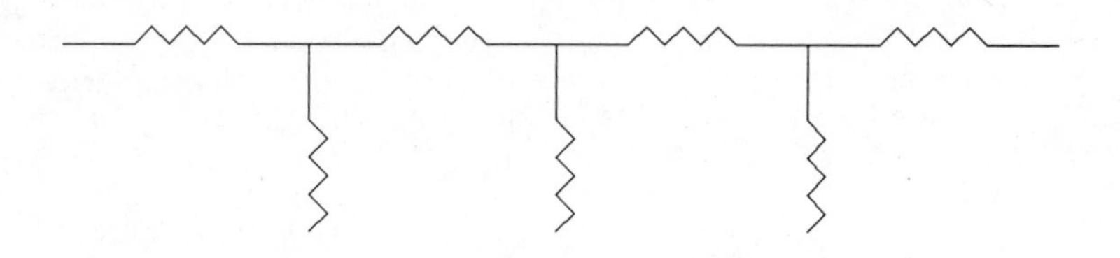

Infinite Branched Polymer 1

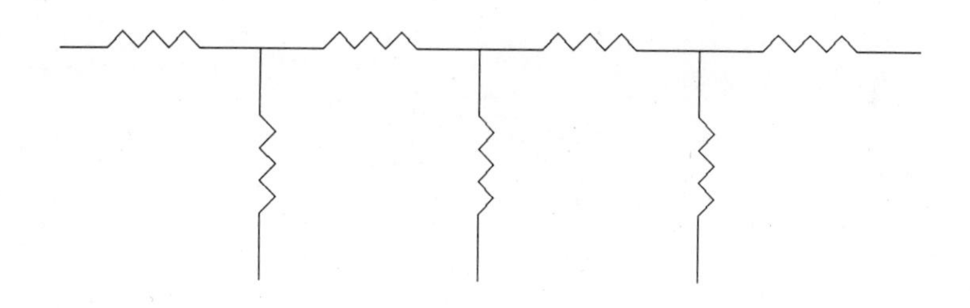

Infinite Branched Polymer 2

Infinite Branched Polymer 3

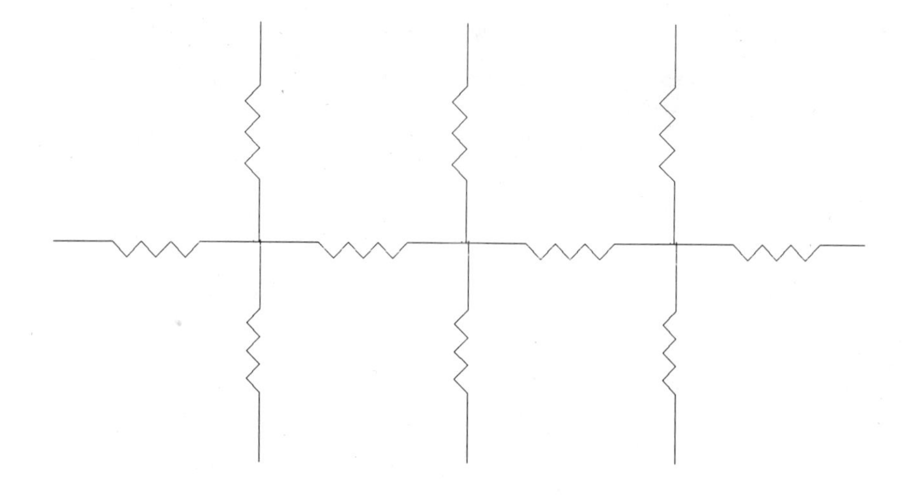

Because networks may be interconnected in different ways, that is, because the shortest path between points in the network depends on whether there are many small loops in the network, calculations had to be done for networks with different probabilities for small loops. An example of a small trifunctionally connected network, consisting of just twelve triblock copolymers, that can be interconnected in different ways, is shown below:

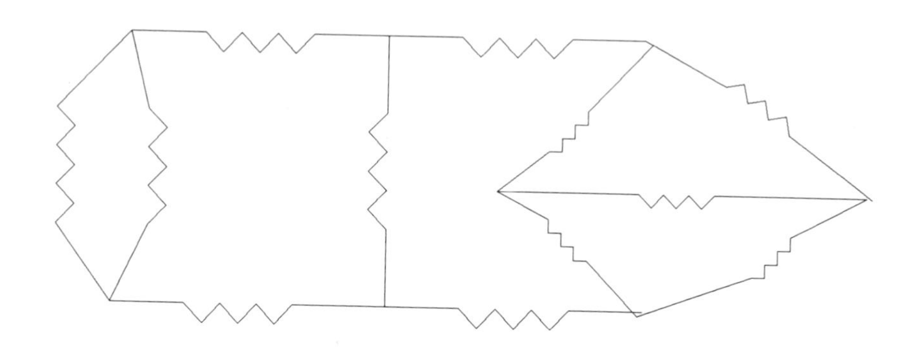

Network with Small Loops

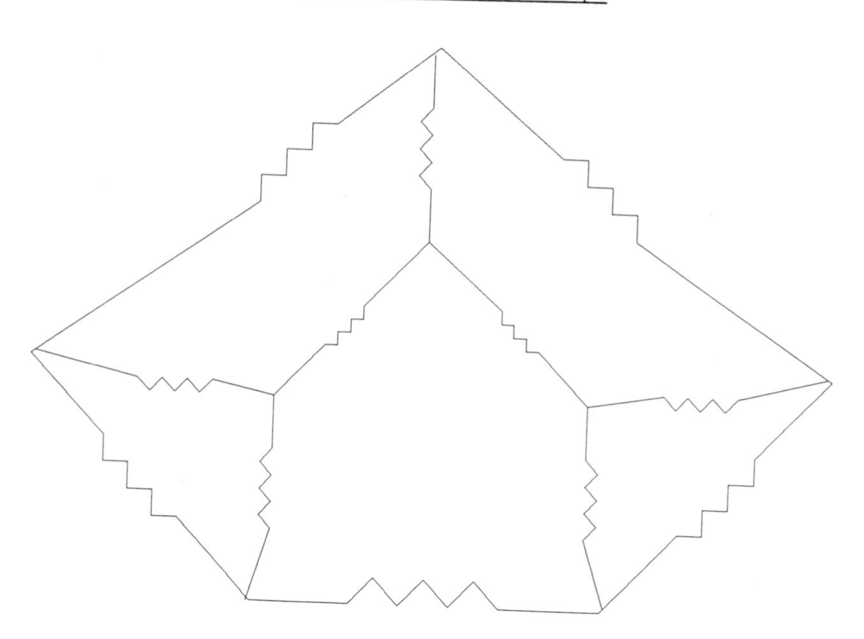

Tightly Interconnected Network

One may note that the network with small loops can be cut into two separate networks using just two cuts. This is not possible with the tightly connected network. The tightly connected network also, as expected, has a slightly larger average smallest loop size than the network with small loops. In the scattering calculations, although the exact connectivity of an infinite trifunctional network changed the mathematical expression for scattering as a function of **Q**, it had little effect on the graphical representation of the results. Because of the assumptions used for the calculations, however, networks with many small loops were not calculated.

II. Calculations and Results

The method of Benmouna and Benoît,[6] which consists of calculating density correlations for each type of monomer in the block copolymer or network and cross-correlations for the different types of monomer was used for the calculations. In the case in which the different types of monomer are hydrogenated and deuterated versions of the same monomer and the Flory interaction parameter between the different monomer units is assumed zero,[6] the intensity of scattering $I(\mathbf{Q})$ is proportional to:[6]

$$I(Q) \propto N u^2 v^2 (P_D(Q) P_H(Q) - [P_{DH}(Q)]^2 / P(Q)) , \tag{1}$$

where

$$P(Q) = u^2 P_D(Q) + v^2 P_H(Q) + 2 u v P_{DH}(Q) , \tag{2}$$

where N is the number of triblock repeat units per unit volume, u and v are the fraction of deuterated and hydrogenated monomers, respectively, $P(Q)$ is the total structure factor of the polymer or gel, $P_D(Q)$ and $P_H(Q)$ are the structure factors of the deuterated and hydrogenated portions of the polymer or gel, respectively, and $P_{DH}(Q)$ is the cross-term in the structure factor.

Results of the calculations of the structure factors of the various infinite branched polymers and gels are given below, together with the structure factor of the triblock copolymer given in the same terms. In the equations below, $\Lambda = (\mathbf{Q}\, R_g)^2$ where R_g is the root-mean-square radius of gyration of the triblock copolymer, $x_a = \exp(-\Lambda u/2)$ except in the case of infinite branched polymer 1 where $x_a = \exp(-\Lambda u/3)$, $x_b = \exp(-\Lambda v)$ except in the case of infinite branched polymer 1 where $x_b = \exp(-\Lambda v/2)$, $x_{2a} = x_a^2$, $x_{2b} = x_b^2$, $x_a x_b$ refers to the product of x_a and x_b, and so on.

Triblock Copolymer

$$N\ P_D(Q) = 2\{\Lambda u - 2 + 2 x_a + x_b (1 - x_a)^2\} / (\Lambda u)^2 ; \tag{3}$$

$$N\ P_H(Q) = 2\{\Lambda v - 1 + x_b) / (\Lambda v)^2 ; \tag{4}$$

$$N\ P_{DH}(Q) = 2 (1 - x_a)(1 - x_b) / \Lambda^2 u v ; \tag{5}$$

$$N P(Q) = 2 (\Lambda - 1 + x_{2a} x_b) / \Lambda^2 . \tag{6}$$

Infinite Branched Polymer 1

$$N\ P_D(Q) = 2\{\Lambda u\,(1 - x_{2a}x_b) - 3x_a + x_b + 3x_{2a} = 2x_a x_b$$

$$-3x_{2a}x_b - x_{3a}x_b + x_{4a}x_b\}/\Lambda^2 u^2(1 - x_{2a}x_b)\ ; \qquad (7)$$

$$N\ P_H(Q) = 2\{\Lambda v\,(1 - x_{2a}x_b - 2 + 2x_b + 3x_{2a} - 4x_{2a}x_b$$

$$+ x_{2a}x_{2b} + x_{4a}x_b - 2x_{4a}x_{2b} + x_{4a}x_{3b}\}/\Lambda^2 v^2(1 - x_{2a}x_b)\ ; \qquad (8)$$

$$N\ P_{DH}(Q) = \{3 + x_a - 3x_b - 6x_{2a} - 3x_a x_b + 7x_{2a}x_b$$

$$- x_{2a}x_{2b} + x_{3a}x_b - x_{3a}x_{2b} \qquad (9)$$

$$- 2x_{4a}x_b + x_{4a}x_{2b}\}/\Lambda^2 uv\,(1 - x_{2a}x_b)\ ;$$

$$N\ P(Q) = 2\{\Lambda + (1 - x_a x_b)(1 - x_{3a}x_{2b})/(1 - x_{2a}x_b)\}/\Lambda^2\ . \qquad (10)$$

Infinite Branched Polymer 2

$$2N\ P_D(Q) = \{2\Lambda u\,(1 - x_{2a}x_b) - 1 - 2x_a + 2x_b + 3x_b + 3x_{2a} + 2x_a x_b$$

$$- 5x_{2a}x_b + x_{2a}x_{2b} + x_{4a}x_b - 3x_{4a}x_{2b} + x_{4a}x_{3b}$$

$$+ 2x_{5a}x_{2b} - 2x_{5a}x_{3b} + x_{6a}x_{3b}\}/\Lambda^2(1 - x_{2a}x_b); \qquad (11)$$

$$2N\ P_H(Q) = \{2\Lambda v(1 - x_{2a}x_b) - 2 + 2x_b + 3x_{2a} - 4x_{3a}x_b + x_{2a}x_{2b}$$

$$+ x_{4a}x_b - 2x_{4a}x_{2b} + x_{4a}x_{3b}\}/\Lambda^2 v^2(1 - x_{2a}x_b)\ ; \qquad (12)$$

$$2N\ P_{DH}(Q) = \{2 + x_a - 2x_b - 3x_{2a} - x_a x_b + 4x_{2a}x_b - x_{2a}x_{2b} - x_{4a}x_b$$

$$+ 2x_{4a}x_{2b} - x_{4a}x_{3b} - x_{5a}x_{2b} + x_{5a}x_{3b}\}/\Lambda^2 uv(1 - x_{2a}x_b)\ ; \qquad (13)$$

$$2N\ P(Q) = \{2\Lambda + 1 - x_{4a}x_{2b}\}/\Lambda^2\ . \qquad (14)$$

Infinite Branched Polymer 3

$$3N\ P_D = 2\ \{\Lambda u(1 - x_{2a}x_b) - 2x_a + x_b + 2x_{2a} + 2x_a x_b - 4x_{2a}x_b + x_{2a}x_{2b}$$

$$+ x_{4a}x_b - 3x_{4a}x_{2b} + x_{4a}x_{3b} + 2x_{5a}x_{2b} - 2x_{5a}x_{3b} \qquad (15)$$

$$+ x_{6a}x_{3b}\}/\Lambda^2 u^2(1 - x_{2a}x_b)\ ;$$

$$3N\ P_H(Q) = 2\{\Lambda v(1 - x_{2a}x_b) - 1 + x_b + 2x_{2a} - 3x_{2a}x_b + x_{2a}x_{2b}$$

$$+ x_{4a}x_b - 2x_{4a}x_{2b} + x_{4a}x_{3b}\}/\Lambda^2 v^2(1 - x_{2a}x_b)\ ; \qquad (16)$$

$$3N\ P_{DH}(Q) = 2\{1 + x_a - x_b - 2x_{2a} - x_a x_b + 3x_{2a}x_b - x_{2a}x_{2b}$$

$$- x_{4a}x_b + 2x_{4a}x_{2b} - x_{4a}x_{3b} - x_{5a}x_{2b} \qquad (17)$$

$$+ x_{5a}x_{3b}\}/\Lambda^2 uv(1 - x_{2a}x_b)\ ;$$

$$3N\ P(Q) = 2\{\Lambda + 1 - x_{4a}x_{2b}\}/\Lambda^2\ . \qquad (18)$$

The six approximations for the trifunctional gel given below are for gels in which the loops essentially become larger and larger. In spite of the differences in the structure factors for the different approximations, calculations show that the final $I(\mathbf{Q})$ versus $\mathbf{Q}$ plots obtained for all the approximations are virtually identical.

Trifunctional Gel Approximation 1

$$N\,P_D(Q) = 2\{\Lambda u(1 - x_{2a}x_b) - 2x_a + 2x_{2a} + x_b + 2x_a x_b - 3x_{2a}x_b$$

$$+ x_{2a}x_{2b} - 2x_{3a}x_b - 2x_{3a}x_{2b} + 2x_{4a}x_b$$

$$+ x_{4a}x_{2b}\}/\Lambda^2 u^2(1 - x_{2a}x_b)\ ;$$

$$N\,P_H(Q) = 2\{\Lambda v(1 - x_{2a}x_b) - 1 + x_b + 2x_{2a} \qquad (19)$$

$$- 3x_{2a}x_b + x_{2a}x_{2b}\}/\Lambda^2 v^2(1 - x_{2a}x_b)\ ; \qquad (20)$$

$$N\,P_{DH}(Q) = 2\{1 + x_a - x_b - 2x_{2a} - x_a x_b + 3x_{2a}x_b - x_{2a}x_{2b}$$

$$+ x_{3a}x_b - x_{3a}x_{2b} - 2x_{4a}x_b + 2x_{4a}x_{2b}\}/\Lambda^2 uv(1 - x_{2a}x_b)\ ; \qquad (21)$$

$$NP(Q) = 2\{\Lambda(1 - x_{2a}x_b) + 1 - 4x_{3a}x_{2b} - 2x_{4a}x_b$$

$$+ 5x_{4a}x_{2b}\}/\Lambda^2(1 - x_{2a}x_b)\ . \qquad (22)$$

Trifunctional Gel Approximation 2

$$NP_D(Q) = \textit{all terms as in approximation } 1\ ;$$

$$NP_H(Q) = 2\{\textit{terms as in approximation } 1 + 2x_{4a}x_b - x_{2b}$$

$$+ 2x_{4a}x_{3b}\}/\Lambda^2 v^2(1 - x_{2a}x_b)\ ; \qquad (23)$$

$$N\ P_{DH}(Q) = \textit{all terms as in approximation } 1\ ;$$

$$N\ P(Q) = 2\{\Lambda(1 - x_{2a}x_b) + 1 - 4x_{3a}x_{2b} + x_{4a}x_{2b}$$

$$+ 2x_{4a}x_{3b}\}/\Lambda^2(1 - x_{2a}x_b)\ . \tag{24}$$

Trifunctional Gel Approximation 3

$$N\ P_D(Q) = 2\{\textit{terms as approximation } 1 + 4x_{3a}x_{2b} - 8x_{4a}x_{2b}$$

$$+ 4x_{5a}x_{2b}\}/\Lambda^2 u^2(1 - x_{2a}x_b)\ ; \tag{25}$$

$$N\ P_H(Q) = \textit{all terms as in approximation } 2\ ;$$

$$N\ P_{DH}(Q) = \textit{all terms as in approximation } 1\ ;$$

$$NP(Q) = 2\{\Lambda(1 - x_{2a}x_b) + 1 - 7x_{4a}x_{2b} + 2x_{4a}x_{3b}$$

$$+ 4x_{5a}x_{2b}\}/\Lambda^2(1 - x_{2a}x_b)\ . \tag{26}$$

Trifunctional Gel Approximation 4

$$N\ P_D(Q) = \textit{all terms as in approximation } 3\ ;$$

$$N\ PH(Q) = \textit{all terms as in approximation } 2\ ;$$

$$N\ P_{DH}(Q) = 2\{\textit{terms as in approximation } 1 + 2x_{4a}x_{2b}$$

$$- 2x_{4a}x_{3b} - 2x_{5a}x_{2b} + 2x_{5a}x_{3b}\}/\Lambda^2 uv(1 - x_{2a}x_b)\ ; \tag{27}$$

$$N\ P(Q) = 2\{\Lambda(1 - x_{2a}x_b) + 1 - 3x_{4a}x_{2b} - 2x_{4a}x_{3b}$$

$$+ 4x_{5a}x_{3b}\}/\Lambda^2(1 - x_{2a}x_b)\ . \tag{28}$$

Trifunctional Gel Approximation 5

$$N\ P_D(Q) = 2\{\textit{all terms as in approximation } 3 + 4x_{4a}x_{2b} - 8x_{5a}x_{2b}$$

$$+ 4x_{6a}x_{2b}\}/\Lambda^2u^2(1 - x_{2a}x_b)\ ; \tag{29}$$

$$N\ P_H(Q) = \textit{all terms as in approximation } 2\ ;$$

$$N\ P_{DH}(Q) = \textit{all terms as in approximation } 4\ ;$$

$$N\ P(Q) = 2\{\Lambda(1 - x_{2a}x_b) + 1 + x_{4a}x_{2b} - 8x_{5a}x_{2b} + 4x_{5a}x_{3b}$$

$$+ 4x_{6a}x_{2b}\}/\Lambda^2(1 - x_{2a}x_b)\ . \tag{30}$$

Trifunctional Gel Approximation 6

$$N\ P_D(Q) = 2\{\textit{terms as in approximation } 5 + 2x_{4a}x_{3b} - 4x_{5a}x_{3b}$$

$$+ 2x_{6a}x_{3b}\}/\Lambda^2u^2(1 - x_{2a}x_b)\ ; \tag{31}$$

$$N\ P_H(Q) = \textit{all terms as in approximation } 2\ ;$$

$$N\ P_{DH}(Q) = \textit{all terms as in approximation } 4\ ;$$

$$N\ P(Q) = 2\{\Lambda(1 - x_{2a}x_b) + 1 + x_{4a}x_{2b} - 8x_{5a}x_{2b} + 4x_{6a}x_{2b}$$

$$+ 2x_{6a}x_{3b}\}/\Lambda^2(1 - x_{2a}x_b) \tag{32}$$

Figure 1 shows the right-hand side of Eq. 1 versus $(\mathbf{Q}\ R_g)^2$, where R_g is the root-mean-square radius of gyration of the triblock copolymer subunit, for a 10% deuterium-labeled triblock copolymer, and of both branched polymer 2 and branched polymer 3. The branched polymers in Fig. 1 are made up of the triblock copolymer also shown in Fig. 1. As it happens, all the gel approximations give virtually the same curve that is shown for branched polymer 3 in Fig. 1. The infinite branched polymers shown in Fig. 1 have a more pronounced maximum in their scattering intensity curves than their constituent triblock copolymer, but the position of the maximum with respect to **Q** is hardly shifted at all. Figure 2 shows the same comparisons as Fig. 1, but now the triblock copolymers are 50% deuterated.

Figure 3 shows the right-hand side of Eq. 1 versus $(\mathbf{Q}\ R_g)^2$ for branched polymer 1 as compared with the triblock copolymer and branched polymer 2. Branched polymer 1 is the only one of the calculated polymers that is not composed strictly of triblock copolymers. Branched polymer 1 may be considered as an infinite polymer made up of tetrablock copolymers or of triblock copolymers in the backbone with diblock copolymers as branches. The maximum in the scattering curve for branched polymer 1 is at larger values of $(\mathbf{Q}\ R_g)^2$ than that of the other polymers, partly because the R_g of this polymer pertains to the tetrablock copolymer subunit. Branched polymer 1 was included in the calculation because synthetic methods for this polymer are easier to find than for the other branched polymers.

It is obvious from the Figures that the elastic neutron scattering to be expected from infinite polymers and gels with labeled branch points depends only on the percent labeling and the radius of gyration of the constituent repeat units. This radius of gyration of the repeat units can thus be obtained from the data.

Extension of the work presented here to gels which are composed of triblock copolymers in which one must postulate an interaction parameter between blocks that is not equal to zero is possible using the ideas of Benoît et al.[4,7,8]

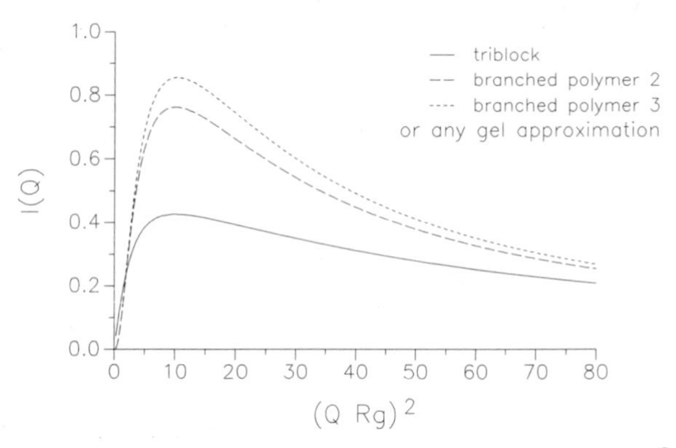

Fig. 1. The right-hand side of equation 1 versus $(\mathbf{Q}\ R_g)^2$ for two 10% deuterium-labeled infinite branched copolymers and the trifunctional gel compared with their constituent triblock copolymer.

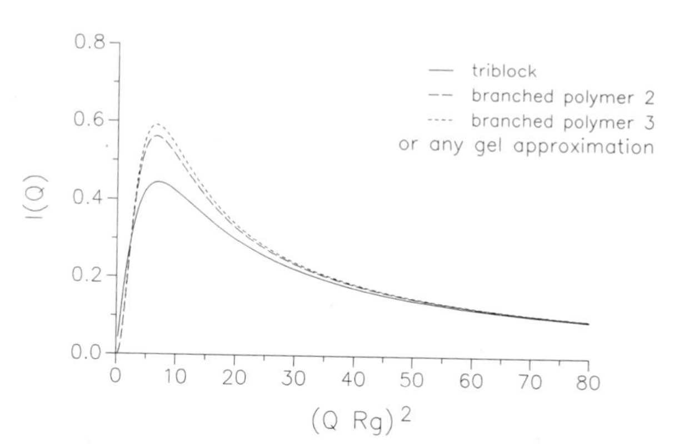

Fig. 2. The right-hand side of equation 1 versus $(\mathbf{Q}\ R_g)^2$ for two 50% deuterium-labeled infinite branched copolymers and the trifunctional gel compared with their constituent triblock copolymer.

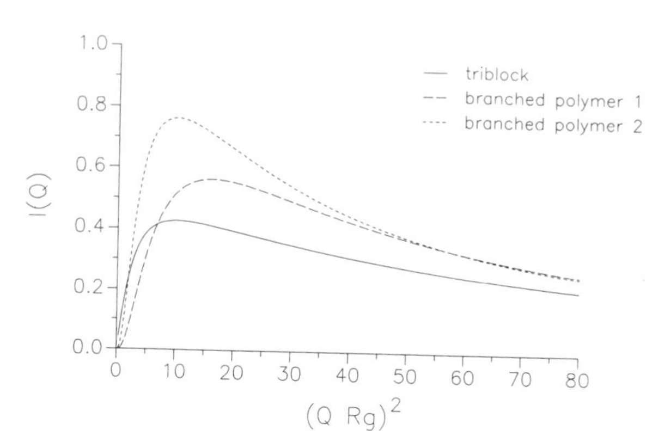

Fig. 3. The right-hand side of equation 1 versus $(\mathbf{Q}\ R_g)^2$ for two 10% deuterium-labeled infinite branched copolymers and a triblock copolymer.

ACKNOWLEDGEMENT

This work was done at the Institut Charles Sadron while one of us (SK) was on sabbatical leave there. The hospitality of the workers there to SK is gratefully acknowledged.

NOMENCLATURE

$I(Q)$	scattering intensity
M	number of monomer units in a circular polymer
N	number of triblock repeat units in unit volume
$P(Q)$	total structure factor of a polymer
$P_D(Q)$	structure factor of the deuterated portion of a polymer
$P_{DH}(Q)$	cross-term in structure factor of a partly deuterated polymer
$P_H(Q)$	structure factor of the hydrogenated portion of a polymer
Q	$(4\pi/\lambda) \sin^2 (\Theta/2)$
u	fraction of deuterated monomers in a polymer
v	fraction of hydrogenated monomers in a polymer
x_a	exp (Λu_2) in most cases (see text)
x_b	exp (Λv) in most cases (see text)
x_{ia}	x_a^i
x_{i_b}	x_b^i
$x_{i_b}x_{j_b}$	product of x_{ia} and x_{jb}
λ	wavelength of neutrons
Θ	scattering angle
Λ	$(Q\ Rg)^2$

REFERENCES

1. H. Benoît, D. Decker, R. Duplessix, C. Picot, P. Rempp, J.P. Cotton, B. Farnoux, G. Jannink and R. Ober, J. Polym. Sci. Polym. Phys. Ed., 14, 2119-2128 (1976).
2. W. Wu and B.J. Bauer Polymer, 27, 169-180 (1986).
3. W. Wu and B.J. Bauer, Macromolecules, 21, 457-464 (1988).
4. H. Benoît, C. Picot and S. Krause, This Volume, (1988) paper preceding this one.
5. E.F. Casassa, J. Polym. Sci., A3, 605-614 (1965).
6. M. Benmouna and H. Benoit, J. Polym. Sci. Polym. Phys. Ed., 21, 1227-1242 (1983).
7. H. Benoît, W. Wu, M. Benmouna, B. Mozer, B. Bauer and A. Lapp, Macromolecules, 18, 986-993 (1985).
8. H. Benoît and G. Hadziioannou, Macromolecules, 21, 1449-1464 (1988).

Discussion

On the Paper by A. Silberberg

Jim Mays (University of Alabama, Birmingham): Commenting on the power law exponent for the molecular dependence of viscosity:

Recent work by Graessley and coworkers with very high *MW* polybutadiene have indicated an asymptotic approach to a 3.0 exponent at the highest *MW*.

P.G. de Gennes (Collège de France): I understand your estimate for the curvilinear velocity $v_{cl} \sim qn^{3/2}$, but not the later estimate of the dissipation $T\dot{S}$. I would take $T\dot{S}/\mathrm{cm}^3$ to be proportional to $<v_{cl}^2> \sim <n^3> \sim N^3$, and this would bring me back to the standard result. Can you clarify this?

A. Silberberg (Weizmann Institute): I agree that

$$< v_{CL}{}^2 > = (1/N) \int_o^N v_{CL}{}^2 dn \sim N^3 .$$

It is my contention, however, that at n the flow injects a chain length (dv_{cl}/q) due to the rotation of the molecule in its conformation averaged mean state. This requires an increase of the traction by $dF_n \sim \eta_{eff}(dv_{cl}/q)v_{cl}$ and an increase in dissipation of $d\dot{E} \sim dF_n \cdot v_{cl} \sim (\eta_{eff}/q)\, v_{cl}^2 dv_{cl}$. In fact, the dissipation per chain of volume Na^3 is given by

$$\int_o^N d\dot{E} \sim (\eta_{eff}/q) v_{CL,N}^3 \sim (\eta_{eff}/q) a^3 q^3 N^{9/2} .$$

Hence the dissipation per unit volume becomes

$$\eta_{melt} q^2 = \eta_{eff} q^2 N^{7/2} .$$

Below the molecular weight range where the chains reptate the energy dissipation is given by

$$\dot{E} \sim \eta (Na)(N^{1/2} aq)^2 / Na^3$$

per unit volume, where $(N^{1/2}aq)$ and the effective velocity and (Na) is the length of chain being moved. In the reptation regime this becomes

$$\dot{E} \sim \eta (N^{3/2} a)(N^{3/2} a q)^2 / N a^3 .$$

The effective veolcity is $(N^{3/2}aq)$ instead of $(N^{1/2}aq)$ and the effective length of chain, due to pumping, is $(N^{3/2}a)$ instead of (Na). Hence the dissipation increases by a factor depending on the 2.5 power of N rather than only on 2, once the chains are obliged to rotate and thus reptate to and from in order to maintain an unmodified mean conformation.

On the Paper by L. Lèger, H. Hervet, M.F. Marmonier and J.F. Viovy

P.T. Callaghan (Massey University, Palmerston North, New Zealand): My question concerns the existence of entanglements in semidilute solution. From the beautiful experiments of Kramer and Green we know that in the melt, $M_c \sim 10^3$ to 10^4 daltons. This suggests that at 10% solution, M_c may be $\sim 10^5$ daltons. How many entanglements exist in your systems? Certainly in our own P&SE experiments we do not see N^{-2} scaling at fixed concentration except at concentrations much higher than those used in your experiments. We think that reptation does occur provided that the matrix molar mass is sufficiently larger than that of the labeled chain.

L. Lèger (Collège de France): Of course I am not able to see directly if the chains are entangled or not. My experimental answer is the following:

1) When the molecular weight of the labeled chains is identical to that of the unlabeled one, we can scale the data D_S/D_O on universal curve as a function of c/c^*, but this scaling does not correspond to a simple N^{-2} molecular weight dependence, especially for c/c^* of order 5 to 15. This is in total agreement with your experiments.

2) When the molecular weight of the unlabeled chains is large enough compared to that of the labeled chains ($P > 5N$), the scaling law of D_S/D_O as a function of c/c^* is compatible with an N^{-2} dependence of D_S down to $c/c^* \sim 2$. There are evidences of a change of dynamical regime for a number of entanglements per chain varying between a few units and more than 20. As, in this fixed matrix regime, D_S is P independent we think that this regime can be interpreted in terms of reptation, all the way down to $c/c^* \sim 2$. But this is a special case of frozen matrix. I fully agree with you on the fact that when matrix motions are present one has to go above $c/c^* = 10$ to observe a reptative behaviour.

On the Paper by T. Kavassalis and J. Noolandi

L.H. Lee (Xerox Webster Research Center):

1) Do we agree on Professor de Gennes' reptation theory?

2) What are the significant points of your paper in addition to the original theory.

T. Kavassalis (Xerox Research Center of Canada): Answer to (1) and (2) of the above question.

In the present work, we have focused our attention on the mean entanglement spacing N_e. This is a fundamental parameter in discussing melts and solutions. We have not addressed the dynamics explicitly. What we have contributed is a relationship between N_e and the polymer structural and chemical properties.

Liliane Lèger (Collège de France): Comment to P.G. de Gennes' question on this paper:

The self-diffusion measurements on semidilute solutions do show that for molecular weight of the labeled chains identical to that of the unlabeled chains one recovers the reptative behaviour observed in the case of a frozen matrix above $\phi/\phi^* = 10$. This could be an indication that an overlapping but not entangled regime exists for $1 < \phi/\phi^* < 10$, however, the self-diffusion coefficient in this regime is not of the Rouse type.

M. Rubinstein (Eastman Kodak Company): How is your theory modified for ring polymers and for miscible blends?

T. Kavassalis (Xerox Research Center of Canada): For ring polymers, you would not have a chain-end effect, but there should be a finite-length effect. We have not, however, addressed the question of rings. The model has been generalized to bimodal distributions of homopolymers and general polydispersity. It is also possible to discuss compatible blends of different polymers with the same model provided that there is no ordering. However, ordering is usually present.

On the Paper by M. Rubinstein

P.G. de Gennes (College de France): What do you think about the Des Cloizeaux picture where the stress relaxation function $\mu(E)$ goes like the square of the memory function.

M. Rubinstein (Eastman Kodak Company): In the des Cloizeaux double reptation model which is similar to the Marrucci Viovy model, it is assumed that a release of constraint chain A imposes on chain B when chain A reptates away completely relaxes the stress in that region for both chains. This would imply that for a homopolymer binary blend of long and short chains would be completely relaxed after each of these K entanglements is released only once. But if an entanglement is released, another one is formed nearby. I believe that to completely relax this section; one needs K^2 disentanglement events and that the Verdier-Stockmayer flip-bond model or the Rouse model is needed to describe the motion and relaxation of the primitive path due to the constraint release process, as was proposed by Prof. de Gennes, J. Klein, Daoud, G. de Bennes and Graessley and used recently by many other scientists. The fact that double reptation is an oversimplification of the constraint release process has been confirmed by experiments.

J. Noolandi (Xerox Research Center of Canada): What is the importance of the long-wavelength modes in your model, and could you clarify the relationship of your model to magnetic spin models?

M. Rubinstein: The long wavelength modes enter as a result of short wavelength jumps of reptons in the same way as long wavelength Rouse modes appear in Verdier-Stockmayer model as a result of local bond flips. Repton model is mapped onto a system of non-interacting spins with random nearest neighbor spin interchange.

G.R. Freeman (University of Alberta): Does the model take into account the possibility that the end of a chain might diffuse around a grid post and become entangled with it?

M. Rubinstein: Yes, it does. In fact, this is one of the main features of the repton model that it includes this type of fluctuation of tube length neglected in the conventional reptation model.

On the Paper by H. Benoît, C. Picot and S. Krause

R.P. Wool (University of Illinois, Urbana): What is the current status of affine/nonaffine deformation studies of gel points? Are there consequences for deformation of entanglements in dense linear polymers?

Henri Benoît (CNRS, Strasbourg): The problem of the deformation of networks is still controversial. When a network is stretched the scattered intensity shows reproducible anomalies. In some cases the deformation is nonaffine at the distance of the crosslinks (J. Bastide, F. Boué, Physica, 140 A, (1986) p.251) (R. Oeser, thèse Mainz Strasbourg (1981)).

No quantitative explanation of these results is yet available.

P.G. de Gennes (College de France): Let us assume that a small fraction of the nodes carries a strong scattering center (e.g., incoherent scattering by photons) and that we select the strictly elastic scattering. This measures $< \rho_g >^2$ where ρ_g is the F_T of the spatial distribution function for one mode. Have people measured or compared this $< \rho_g >^2$?

Henri Benoît: In actual polymer networks the size of the crosslinks is usually of the order of magnitude of one monomeric unit and will not be measurable in an elastic neutron-scattering experiment. The problem of inelastic scattering is more interesting since experiments on the quasielastic scattering by crosslinks can give informations on the Brownian motion of these units. Recent experiments made by Oeser et al. (Phys. Rev. Lett. 60, 1041 1988) suggests that this Brownian motion is much more limited than in the case of free chains.

PART SIX:

PHASE TRANSITIONS AND GEL ELECTROPHORESIS

PHASE TRANSITIONS IN POLYMER SOLUTIONS

J.F. Joanny
E.N.S. 46 allée d'Italie
69364 Lyon Cedex 07
France

L. Leibler
Laboratoire de Physico-Chimie
Macromoléculaire
E.S.P.C.I. 10 rue Vauquelin
75231 Paris Cedex 05, France

ABSTRACT

We review some recent results on phase separation of polymer mixtures and on mesophase formation of diblock copolymers in solution. Mixtures of two chemically different polymers most often segregate in solution. Their phase diagram and the related critical properties are studied in a solvent that can be either a good solvent for both polymers or a θ-solvent for both polymers or a selective solvent (good for one polymer, θ for the other). Diblock copolymers in dilute or semidilute solutions form several kinds of mesophases with different symmetries. The onset of formation of some of these phases is discussed in a highly selective solvent good for one block and poor for the other, and in a common good solvent.

I. INTRODUCTION

Systems containing macromolecules exhibit extremely strong segregation effects which have fascinated polymer scientists and engineers for many years. For example, liquid-liquid phase separation occurs in polymer solutions below the Flory compensation temperature θ and is also a very general phenomenon in polymer blends.[1-4] In molten block copolymers, the incompatibility between different monomer sequences may lead to the formation of various mesophases (lamellar, cylindrical, spherical, or recently discovered ordered co-continuous microphase structures) whose morphology and symmetry depends on the architecture of the copolymer molecules.[5-7] In the past decade, there has been considerable progress in the understanding of both phase separation in polymer blends[2,8] and mesophase formation in molten copolymers.[9-14] Our aim here is to review some more recent results obtained for solutions of incompatible blends and solutions of block copolymers. The addition of a solvent is very often necessary for practical reasons: it improves the compatibility and makes the mixture less viscous thus facilitating the processing; it also confers a new richness to the problem of phase behavior and arouses some fundamental issues concerning the chain conformations and interactions, the critical properties of demixing in the case of blend solutions and the intermolecular organization in the case of copolymer solutions. The recent increase of interest in these systems has been motivated by their numerous industrial applications. For example, phase separation in ternary polymer A-polymer B-solvent systems is used in biotechnological industries for enzyme and protein purification. Copolymers are used in membrane technology, adhesives and in high-impact-resistant plastics. Besides, many experimental difficulties encountered in studies of molten polymers are absent in solutions and it is important to know how information obtained by studying dilute and semidilute solutions may be used to predict the properties of more concentrated systems.

On the other hand, the addition of solvent gives rise to some important conceptual difficulties that can best be illustrated for the ternary polymer A-polymer B-good solvent S systems. Classical interpretations of properties of such systems have generally been based on a mean-field Flory-Huggins lattice model in which the free energy density (per site) is written as[16]

$$\left.\frac{F}{T}\right|_{site} = \frac{\Phi_A}{N_A}\,\ell og\,\Phi_A + \frac{\Phi_B}{N_B}\,\ell og\,\Phi_B + \Phi_S \ell og \Phi_S + \chi_{AS}\Phi_A\Phi_S + \chi_{BS}\Phi_B\Phi_S + \chi_{AB}\Phi_A\Phi_B \,, \tag{1}$$

where Φ_A, Φ_B and Φ_S denote the monomers A and B and solvent volume fractions respectively; N_A and N_B are the polymerization indices, χ_{iS} ($i = A, B$) are the monomer-solvent interaction parameters and χ_{AB} the interaction parameter between A and B monomers. The first three terms represent the translational entropies of the species. The model clearly shows that the entropy of mixing of the polymers is small and that an effective repulsion between unlike monomers leads to a phase separation into two phases with essentially one polymer in each phase. The phase diagrams are easily calculated from Eq. (1) for different molecular parameters. For completely symmetric systems ($\chi_{AS} = \chi_{BS} < 1/2$, $N_A = N_B$), the shape of the coexistence curve and other critical properties are characterized by classical exponents.[17] An important conclusion is that for a non-selective good solvent demixing should occur already in the dilute regime since the critical concentration scales as[16]

$$\Phi_k = \Phi_{Ak} + \Phi_{Bk} \sim 2/N\,\chi_{AB}\,. \tag{2}$$

However, Eq. (1) is based on the assumption of freely interpenetrating chains which cannot be valid in the dilute regime[1-3] and there seems to be an important inconsistency in the classical picture. The situation has been even more confused since experiments have shown that for long chains demixing occurs well in the semidilute regime[18-20] and the critical behavior seems to be characterized by nonclassical exponents.[21] Actually, the assumption that the chains interpenetrate freely and that the concentration fluctuations can be neglected, which is the basis of the Flory free energy Eq.(1) cannot be made to describe correctly the behavior of both dilute and semidilute solutions since the excluded volume effects introduce important correlations that must be included in the description of chain conformations and of thermodynamic properties. For a dilute solution, de Gennes[2] has pointed out that A and B coils interact as hard spheres which cannot interpenetrate and thus that the chemical differences between A and B monomers are not seen. As a consequence, demixing cannot occur below the overlap concentration ϕ^*. This picture has been confirmed by detailed renormalization group studies of Joanny et al.[22] (cf. also Ref. 23, 24) who have shown that in dilute systems the interactions between unlike monomers play only a marginal role and manifest themselves only through corrections to scaling laws. Recent precise light-scattering experiments that use the elegant technique of "optical-Θ solvent"[25] confirm well these conclusions. Also in semidilute solutions in a good solvent, excluded volume interactions play a crucial role in spatial scales smaller than the correlation length ξ.[8] The nonclassical excluded volume effects modify substantially the probability of contacts between both like and unlike species and govern the phase separation.[22] Everything seems to happen as if the interaction parameter χ_{ij} were "renormalized."[22] As a result, there seems to be no more confusion: the Flory-Huggins model does not apply and Scott's result for the critical point is modified[24,28] in the sense that for long chains the demixing occurs well into the semidilute regime. The conclusions of the renormalization theory are well confirmed by recent studies of critical properties of demixing and of semidilute solutions[20] as well as by Monte-Carlo simulations of Sariban and Binder.[9] We discuss the interactions in various dilute systems in Section II with an emphasis on a simple blob picture and a more detailed renormalization group analysis which follows Ref. 22. We consider various possible situations of nonselective and preferential solvents both above and in θ-conditions. We also treat a rather new and interesting case of systems containing rod-like and flexible molecules.[30]

Then in section III, we analyze the liquid-liquid phase separation of polymer blends in solvents of various quality. Particular emphasis is put on the case of a common good solvent and on the discussion of the critical properties of demixing which are very unusual as the critical behavior is not of the mean-field type (except for very long chains and low incompatibility degrees) and is also very different from that of low molecular weight ternary mixtures.[28] We also focus on well- demixed systems and consider the interfacial properties following the work of Broseta et al.[31]

The final section deals with the problem of intermolecular organization in block copolymer solutions. For dilute solutions in a selective solvent, we discuss micelle formation. For nonselective solvents, we analyze the formation of various organized mesophases. Also in these systems excluded volume interactions lead to nontrivial subtle effects.[31-33]

II. INTERACTIONS BETWEEN DIFFERENT MACROMOLECULES IN DILUTE SOLUTION

The principal question arising in investigating the thermodynamic behavior of systems containing chemically different polymers or copolymers concerns the role of interactions between unlike monomers. In this respect, the consideration of dilute solutions provides useful information as it elucidates the importance of chain conformations and monomer concentration fluctuations. In a dilute solution of two polymers A and B the osmotic pressure may be approximated by a virial expansion limited to the second order:

$$\frac{\Pi}{T} = C_A + C_B + \frac{1}{2} \sum_{ij} G_{ij} C_i C_j , \tag{3}$$

where C_i $(i = A, B)$ denotes the chain concentration and G_{ij} represent the chain virial coefficients. It is interesting to remark that a virial expansion of the free energy has often been used to describe the phase separation in aqueous solutions of two polymers.[34] The virial coefficients have been taken as purely empirical quantities adjusted to fit the phase diagrams. Such an approach leaves open the essential question of the molecular interpretation of the virial coefficients and of a mechanism by which excluded volume effects and specific monomers interactions can lead to phase separation (c.f. Ref. 35). In a classical Flory-Huggins model, Eq. (1) yields

$$G_{ij} = V_{ij} N_i N_j , \tag{4}$$

where the virial coefficient between monomers i and j, V_{ij} is a measure of the effective interaction between the different species

$$v_{ii} = v(1-2\chi_{iS}) ; \tag{5a}$$

$$v_{AB} = v(1-\chi_{AS}-\chi_{BS}+\chi_{AB}) , \tag{5b}$$

where $v \sim a^3$ (a is the monomer size) denotes the specific volume of the monomers and the solvent, respectively. Here they are supposed to be equal for the sake of simplicity.

The mean-field expression for the virial coefficients may be used only when different chains can interpenetrate freely. This is not the case for a polymer solution in a good solvent ($\chi_{AS} < 1/2$). The excluded volume interactions ($\chi_{AS} < 1/2$) and the monomer virial coefficient decay to zero as an inverse power of the molecular weight of the chains.[1-3] We discuss here the interpenetration of different macromolecules A and B in terms of simple scaling laws that are valid in the asymptotic limit where the molecular weights of both polymers are large.

This problem may be further generalized to the case where the two polymers A and B are not only chemically different but also have a different architecture. Very generally they may be considered as fractal objects characterized by their fractal dimensions D_A and D_B respectively: the radius of polymer i varies with the degree of polymerization N_i as $R_i \approx N_i^{1/D_i}$. The interaction v_{AB} between the two different types of monomers is assumed to be repulsive ($v_{AB} > 0$). The asymmetry between the two polymers is measured by the radius ratio $\sigma = R_A/R_B$.

We consider two polymers A and B with the same radius $R_A = R_B = R$, ($\sigma = R_A/R_B = 1$). When the two polymers occupy the same volume, their interaction free energy is proportional to their number of contact points. In a Flory-like model, the interpenetration free energy per chain is thus given by:

$$\frac{F_{int}}{T} \approx v_{AB} \frac{N_A N_B}{R^d} \sim R^{D_A + D_B - d} \tag{6}$$

where d is the space dimension. Whenever the interpenetration free energy F_{int} is much larger than the thermal energy kT, there is essentially no interpretation between the two polymers. They are opaque to each other and the mutual virial coefficient is proportional to the volume of the two chains.[36]

$$G_{AB} \approx R^d . \tag{7}$$

If the interpenetration free energy F_{int} is much smaller than kT, the two polymers interpenetrate almost freely. They are diaphanous to each other and a mean-field theory (Eq. 4) may be used to calculate the mutual virial coefficient G_{AB}.

The nature of the interaction between macromolecular objects reflects thus the fractal nature of the macromolecules. If the sum of the fractal dimensions $D_A + D_B$ is larger than the space dimension $d = 3$, we expect a hard sphere behavior; this will be the case for linear chains in a common good solvent ($D_A = D_B = 1/\nu \approx 5/3$), a common θ solvent ($D_A = D_B = 2$) or a selective solvent, good solvent for one polymer and θ solvent for the other. If $D_A + D_B$ is smaller than the space dimension, the polymers are diaphanous and Eq. (3) may be used. This is the case for two rodlike macromolecules ($D_A = D_B = 1$, Eq. (5a) and (5b) is then the Onsager's result up to some numerical constants which are not given by the rough scaling theory presented here) or a rodlike polymer and a linear flexible chain in a good solvent ($D_A = 1$, $D_B = 1/\nu \approx 5/3$). There is finally a marginal case when $D_A + D_B = d = 3$ which is obtained, for example for a rodlike polymer and a linear flexible chain in a θ solvent; we expect their logarithmic corrections to the hard sphere behavior.[30]

We now present briefly more explicit calculations of the mutual virial coefficients obtained with the use of des Cloizeaux' direct renormalization method[37,3] for blends of linear flexible polymers in a common good solvent, a common θ-solvent and a selective solvent and for blends of rodlike polymers and flexible polymers in a θ-solvent (marginal behavior). These calculations enable one to find (universal) prefactors relating the mutual virial coefficient to the chain volume (in Eq. 7) in the asymptotic limit. Moreover they give the corrections to the scaling behavior which explicitly depend on the interactions between unlike monomers and are actually responsible for the phase separation of flexible polymer blends in a good solvent.[22]

II.1 Flexible Polymers In A Common Good Solvent

a) Mutual virial coefficient In a continuous model, the flexible chains A and B of different chemical nature are represented by continuous curves $r_i(s)$ where s is an area ranging from zero to the Gaussian area of the chains S_i. The Gaussian area S_i is proportional to the numbers of monomers in the chain N_i and is related to the unperturbed Gaussian end-to-end distance by:[3]

$$R_i^2 = d S_i \,. \tag{8}$$

The interactions between monomers are described in terms of pseudo-potentials

$$v_{ij} = \frac{T}{2} b_{ij} \int_0^{S_i} ds \int_0^{S_j} ds' \, \delta\{r_i(s) - r_j(s')\} \,. \tag{9}$$

The b_{ij} are proportional the usual Edwards' excluded volume parameters v_{ij} (notice however that b_{ij} is not homogeneous to a volume).

In a common good solvent, we expect polymers to interact as hard spheres, we thus introduce dimensionless virial coefficients by:

$$g_{ij} = G_{ij} \left(\frac{2\pi}{d} \right)^{\frac{d}{2}} (R_i R_j)^{-d/2} \,. \tag{10}$$

In the asymptotic limit where the radii of both chains are infinite (the radius ratio $\sigma = R_A/R_B$ being kept constant) the dimensionless virial coefficients g_{AA}, g_{BB}, and g_{AB}, have a finite universal value. Starting from Eq. (9) and and applying des Cloizeaux' direct renormalization scheme, we have calculated these values in expansion of $\varepsilon = 4-d$.

When the chains have the same radius ($\sigma = 1$) one gets for infinite chains

$$g_{AA} = g_{BB} = g_{AB} = g^* = \frac{\varepsilon}{8} \{1 + \frac{\varepsilon}{8} \left(\frac{25}{4} + 4 \log 2 \right) \} \,. \tag{11}$$

The important result is that in the limit of infinite masses, different polymers are indistinguishable, the interaction between like and unlike polymers is the same. The effective repulsive interaction between different chains that provoke the phase separation between chemically different chains manifests itself only as a correction to the scaling behavior due to the finite mass of the chains. These corrections are also obtained from the renormalization calculation[22]

$$g_{AB} - g^* = a_{AB} S^{-\chi_s} \tag{12}$$

where a_{AB} is a non-universal constant roughly proportional to the Flory interaction parameter $\chi_{AB} = b_{AB} - 1/2\ (b_{AA} + b_{BB})$ and χ_S is a universal exponent which up to second order in $\varepsilon = 4-d$ reads

$$\chi_S = \varepsilon/4 - (15/128)\varepsilon^2 . \tag{13}$$

In 3-dimensions the best available approximation is $\chi_S = 0.22 \pm 0.02$[28,38]. A positive value of a_{AB} corresponds to an effective repulsion between different chains and thus a tendancy of the polymers to phase separate. A negative value of a_{AB} corresponds to compatible polymers.

The mutual virial coefficient g_{AB} has also been calculated in a polydisperse case ($\sigma \neq 1$)

$$g_{AB} = g^* \{1 - g^* \left(4\, log2 + (\frac{1}{\sigma^2} + 1)\, log\, \frac{1}{1+\sigma^2} + (1+\sigma^2)\, log\, \frac{\sigma^2}{1+\sigma^2} \right) . \tag{14}$$

In the monodisperse case, this reduces to $g_{AB} = g^*$, when σ is much smaller than one, Eq. (14) is consistent with a power law

$$\frac{g_{AB}}{g^*} = \sigma^{-\frac{\varepsilon}{4}} \{1 - \frac{\varepsilon}{8}(4 log2 - 1)\} \tag{15}$$

Notice that up to this order in ε. Eqs. (14) and (15) are consistent with a mutual virial coefficient G_{AB} varying linearly with the mass of the larger polymer B; namely, $G_{AB} \approx S_B \sim N_B$; below, we give a sample interpretation of this result in terms of a blob model.

b) Comparison with experiments Experimentally, the interactions between different polymer chains are often characterized by the so-called interpenetration functions;

$$\Psi_{ij} = \frac{G_{ij}}{4\pi^{3/2} R_{gi}^{3/2} R_{gj}^{3/2}} , \tag{16}$$

where R_{gi} is the radius of the gyration of the polymer i. The ratio of the radius of gyration R_{gi} to the end-to-end distance R_i being a universal constant, the Ψ_{ij} as well as the dimensionless virial coefficients g_{ij} are a universal constant independent of the nature of the polymer and solvent in the asymptotic limit of infinite chains. This has been demonstrated experimentally[39] for the self-interpenetration functions Ψ_{ii} which reach an asymptotic value $\Psi^* \approx 0.24$ in good agreement with the theoretical prediction of Eq. (11).

In very asymmetric systems where the radius ratio σ is either very large or very small, the mutual virial coefficient G_{AB} (or equivalently the mutual interpenetration function Ψ_{AB}) is very difficult to measure because it gives only a small contribution to the osmotic pressure or scattering intensity. In the symmetric case where the masses of the two different polymers are equal, Ψ_{AA} and Ψ_{AB} are of the same order of magnitude and their difference has been measured recently by light-scattering using the so-called optical θ-solvent method[25] which gives direct access to the monomer virial coefficient difference

$$\Delta \upsilon = \upsilon_{AB} - \frac{1}{2}(\upsilon_{AA} + \upsilon_{BB}) . \tag{17}$$

The above renormalization group theory predicts a decrease of $\Delta\upsilon$ with molecular mass

$$\Delta \upsilon = \beta_{AB} N^{-a} , \tag{18}$$

with an exponent a equal to

$$a = 2 + \chi_s - 3\nu \approx 0.45 , \tag{19}$$

where ν is the swelling exponent in a good solvent (ν = 0.588). The classical Flory theory predicts a constant value of $\Delta\upsilon$.

Both for weakly incompatible systems, PS (polystyrene)/PMMA (polymethylmetacrylate) ($a \approx 0.4 \pm 0.04$)[24,20] and strongly incompatible pairs PS/PDMS (polydimethylsiloxane) ($a = 0.42 \pm 0.02$)[20] the agreement with the renormalization theory is excellent. These experiments also give a measure of the non-universal numbers α_{AB} or β_{AB} characterizing the incompatibility of the blend

II.2 Blends in a common θ-solvent or a selective solvent: blob model

The direct renormalization method also allows the determination of the mutual virial coefficient in a common θ-solvent ($g_{AA} = g_{BB} = 0$) and a selective solvent ($g_{AA} = 0$, $g_{BB} = g^*$).[22] In both cases, for symmetric polymers, when the radius ratio is equal to unity we find a hard sphere interaction characterized by a dimensionless virial coefficient

$$g_{AB} = g_1^* = \frac{\varepsilon}{4}\{1 + \frac{\varepsilon}{4}(1 + 4\,log2)\}; (\theta - solvent)$$

$$g_{AB} = \frac{3\varepsilon}{16}\{1 + \frac{3\varepsilon}{16}\left(\frac{53}{18} + 4\,log2\right)\}. (selective\ solvent) \tag{20}$$

It is interesting to remark that in a common θ-solvent the repulsive A-B interactions lead to a much stronger chain repulsion (higher dimensionless virial coefficient) than the excluded volume interactions in a good solvent. This result is quite natural as in a θ-solvent when two chains occupy the same volume, they have many more contacts than in a good solvent. The case of a selective solvent should then lead to an intermediate value of g_{AB} which indeed is the case. It should also be noted that contrary to the case of a nonselective good solvent, even in the asymptotic limit a chain A can distinguish between a poymer A and a polymer B so that there is a tendency to demix even in the dilute regime.

The polydispersity has been studied in details only in a common θ-solvent; when the radius ratio σ is small,[22]

$$g_{AB} = g_1^* \sigma^{\frac{-\varepsilon}{2}} \{1 - \frac{\varepsilon}{4}(-1 + 4\,log2)\} . \tag{21}$$

As in a good solvent, this leads to a virial coefficient G_{AB} scaling linearly with the mass of the longer polymer.

In a selective solvent only scaling theory has been made:[22]

$$\text{If the short chain A is Gaussian, } g_{AB} \approx \sigma^{-\varepsilon/4}\,;$$

$$\text{if the long chain B is Gaussian, } g_{AB} \approx \sigma^{-\varepsilon/2}\,. \tag{22}$$

Whatever the solvent quality, we thus find a mutual virial coefficient G_{AB} proportional to the mass of the larger chain N_B. This is a signature of the fractal nature of the polymer chains. The volume excluded by the long polymer B to the short polymer A is not proportional to the volume of the B chains. There is some interpenetration between the chains.

All the above results for polydisperse blends are consistent with a blob model first suggested by de Gennes.[2] In the long chain, we group the monomers into subunits with a radius equal to the radius of the smaller chains (Fig. 1). The excluded volume G_{AB} is the total volume occupied by these blobs. Each blob having $g \approx R_A{}^{D_B}$ monomers the mutual virial coefficient is

$$G_{AB} = \frac{N_B}{g} R_A^d \approx R_B^{D_B} R_A^{d-D_B} \quad or\; g_{AB} \approx \sigma^{\frac{d}{2} - D_B}\,, \tag{23}$$

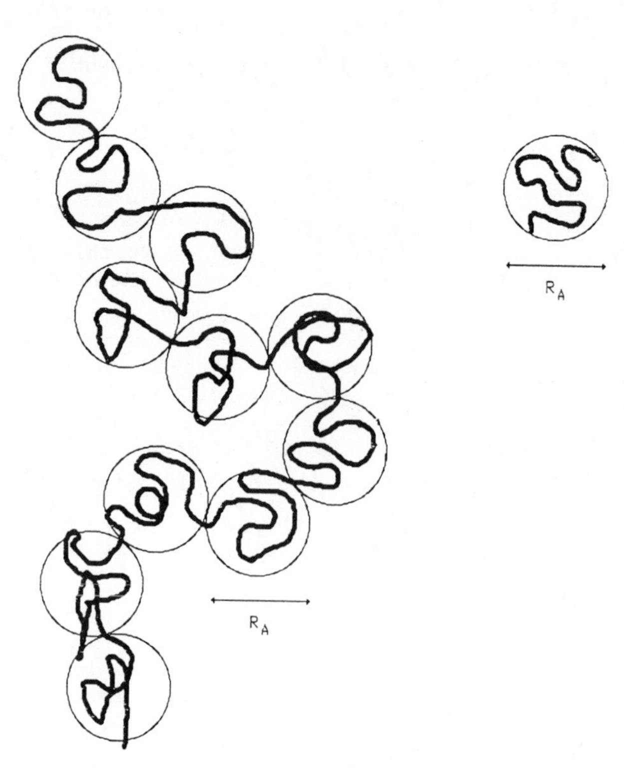

Fig. 1. Schematic representation of two interacting chains of different size. The long chain is represented as a succession of blobs of size R_A, with R_A being the radius of the smaller chain. The energy of interaction between the small chain and a blob is about T.

where D_B is the fractal dimension of polymer B: $D_B = 1/\nu_B$. In other words, G_{AB} is the fractal envelope of the longest chain with the radius of the smaller chain taken as a unit length.

II.3. Blends of Rodlike Molecules and Flexible Chains in a θ-solvent.

When both chains are flexible, they thus interact as hard spheres. This is not, however, always the case for polymers with a more general conformation. Two rods, or a rod and a flexible chain in a good solvent interpenetrate almost freely and are diaphanous. We discuss now an example of a marginal situation, a solution of rodlike molecules and flexible chains in a θ-solvent.[30]

The flexible chains A are described by the same continuous model as above. The rod has a length L proportional to the number of monomers N_B. The flexible chain being isotropic, the rod orientation is of no importance, we fix it along the z axis. The rod is represented by a continuous segment $r'(\ell) = e_z \ell$ where ℓ is a length ranging from zero to L.

The pseudopotential describing the repulsive interactions between rods and chains is written as

$$V_{AB} = Tb_{AB} \int_0^S ds \int_0^L dl\,\delta\{r(s) - r'(\ell)\}\,. \qquad (24)$$

Dimensional analysis of Eq. (24) leads to $b_{AB} \approx (\text{length})^{-3+d}$ and we recover here that the space dimension 3 is marginal for this problem.

We define the dimensionless mutual virial coefficient g_{AB} by

$$g_{AB} = \frac{3G_{AB}}{2\pi R^2 L}\,. \qquad (25)$$

The direct renormalization scheme may be applied to this problem and leads to a value of g_{AB} decreasing logarithmically with the molecular mass,

$$g_{AB} \approx \frac{1}{\log N_A}\,. \qquad (26)$$

We thus conclude that, in principle, it should be easier to dissolve two polymers in a solvent when they are rigid or say one is rigid and another is flexible than when both are flexible.

III. PHASE TRANSITIONS OF POLYMER BLENDS IN SOLUTION

The repulsive interactions between unlike polymer chains which in dilute solutions are characterized by the second virial coefficient G_{AB} monitor the polymer-polymer demixing transition which can be obtained either by increasing the total polymer concentration or decreasing the temperature. In this section we study in more details the related phase diagram, the critical behavior and the interfacial properties of polymer blends in solution. We restrict the discussion to flexible linear chains and study the role of the solvent quality and, to a lesser extent, of the molecular masses of the two polymers. We first assume that the demixing transition takes place in a dilute solution as suggestetd by the Flory-Huggins free energy, this gives sensible results for the critical demixing concentration (at constant temperature) in a common θ-solvent and in some cases in a selective solvent. In a common good solvent, however, when the chains are long and the incompatibility not too strong, the phase transition occurs always above the overlap concentration c^* in the semidilute regime: in this regime, we study in great details the situation where the two polymers have the same polymerization index.

III.1 Phase Transition in Dilute Solution

In a dilute solution quite similar to the osmotic pressure, the free energy of the solution is given by a virial expansion limited to the first order

$$\frac{F}{T} = \sum_{i=A,B} C_i \log C_i + \frac{1}{2} \sum_{ij=A,B} G_{ij} C_i C_j . \tag{27}$$

The first term is the mixing free energy of an ideal solution: the second term is the interaction free energy which has been discussed extensively in the previous section.

We define the exchange potentials by

$$\mu_i = \frac{\partial F}{\partial C_i} . \tag{28}$$

The osmotic pressure is

$$\Pi = C_A \mu_A + C_B \mu_B - F .$$

The phase diagram of a polymer blend in solution is obtained by equating the chemical potentials in the phases in equilibrium of the three components A, B and S or equivalently by equating the two exchange potentials μ_A and μ_B and the osmotic pressure.

We will limit our discussion here to the spinodal line and the critical point of the polymer-polymer phase separation at constant temperature. The spinodal line is the limit of stability of the one phase region. It is given by the change in concavity of the free energy Eq. (27):

$$\frac{\partial^2 F}{\partial C_A^2} \frac{\partial^2 F}{\partial C_B^2} - \left(\frac{\partial^2 F}{\partial C_A \partial C_B} \right)^2 = 0 . \tag{29}$$

Rather than the chain concentrations C_A and C_B, we will use in the following as variables the total monomer concentration $c = N_A C_A + N_B C_B$ and the composition x, $N_A C_A = x\, c$. With these variables, the spinodal line is written as

$$(1 + x\, c h_{AA})\{1 + (1-x) c h_{BB}\} = x\,(1-x) c^2 h_{AB}^2 , \tag{30}$$

where for convenience we have characterized the interactions by the variables h_{ij} related to the virial coefficients by

$$h_{ij} = \frac{G_{ij}}{\sqrt{N_i N_j}} . \tag{31}$$

The critical point is not in general located at the lowest concentration c_{min} of the spinodal line (this is true only in the symmetric case $N_A = N_B, h_{AA} = h_{BB}$). It is located at the point where the spinodal and the binodal line (which gives the two phases at coexistence) are tangent. Mathematically, this can be expressed as the point where a Gibbs determinant vanishes. Explicitly, writing down this condition and combining it with the spinodal condition we obtain the critical composition x_k as a solution of

$$x\,h_{AA} - (1-x)\,h_{BB} = h_{AB}\,x^{1/3}(1-x)^{1/3}\{(1-x)^{1/3}\left(\frac{N_A}{N_B}\right)^{1/6} - x^{1/3}\left(\frac{N_B}{N_A}\right)^{1/6}\}. \tag{32}$$

The critical concentration c_k is then obtained from Eq. (30) for the spinodal line.

We now apply these results for the values of the virial coefficient calculated in the previous section in a common θ-solvent, a selective solvent and a common good solvent.

a. Common θ-solvent We suppose that the chains with the larger radius are the B chains ($\sigma < 1$), we then expect the dilute solution free energy Eq. (27) to remain meaningful up to the overlap concentration of the smallest chains A

$$c_A^* \approx \frac{N_A}{R_A^3} \approx N_A^{-1/2}. \tag{33}$$

The parameter h_{ij} are given by the blob model of Eq. (21)

$$h_{AA} = h_{BB} = 0\ ; \quad h_{AB} = \frac{G_{AB}}{\sqrt{N_A N_B}} \approx N_B^{1/2}\,a^3. \tag{34}$$

Eq. (32) gives the critical composition

$$x_k = \frac{N_A}{N_A + N_B}. \tag{35}$$

The critical monomer concentration is

$$c_k = \frac{1}{h_{AB}\,\sqrt{x_k(1-x_k)}} = a^{-3}N_A^{-1/2}\left(1 + \frac{N_A}{N_B}\right). \tag{36}$$

In the symmetric case $N_A = N_B$, $x_k = 1/2$ and c_k is equal to the overlap concentration c^* as suggested by de Gennes[40,2]

In the asymmetric case $N_B >> N_A$, we get

$$c_k \approx \frac{1}{a^3 N_A^{1/2}} \approx c_A^*. \tag{37}$$

The critical concentration is thus the overlap concentration of the smaller chains. The critical concentration is different from the minimum concentration of the spinodal line

$$c_{min} \approx \frac{1}{N_B^{1/2}\,a^3} \approx c_B^*. \tag{38}$$

Notice that this result shows that the so-called quasibinary approximation where the concentration is supposed to have the same value on the two phases at equilibrium fails even when the solvent quality is the same for both polymers. The tie lines are not parallel to the A-B bases of the triangular phase diagram.

b. Selective solvent We study here only the case where A chains are in a good solvent, B chains are in a θ-solvent and where the radius ratio σ is smaller than one ($N_B > N_A^{6/5}$). We then expect the dilute solution free energy Eq.(27) to be meaningful if the concentration c is smaller than the overlap concentration of the smaller chains.

$$c_A^* = \frac{N_A}{R_A^3} \approx a^{-3} N_A^{-4/5} . \qquad (39)$$

The interaction parameters h_{ij} are given by Eqs. (31) and (23)

$$h_{BB} = 0 \, , h_{AA} = \frac{g^* R_A^3}{N_A} \approx c_A^{*-1} \, , h_{AB} \approx N_B^{1/2} \, N_A^{1/10} \, a^3 . \qquad (40)$$

The critical point is obtained from Eq. (30) and (32)

$$x_k \approx \frac{N_A}{N_B + N_A} \, ; \qquad (41)$$

$$c_k = \frac{N_B}{R_B^2 R_A} \approx \frac{1}{N_A^{3/5}} a^{-3} . \qquad (42)$$

The critical concentration lies well in the range of validity of the free energy. In the reverse limit when $\sigma > 1$, the critical concentration lies in the semidilute regime and cannot be studied from Eq. (27).

III.2 Phase Separation in a Common Good solvent

We have seen in section II that in a good solvent A-B interactions play only a marginal role in dilute solutions and excluded volume interactions dominate. In the asymptotic limit of infinite molecular weight, polymers A and B are not distinguishable, i.e., the dimensionless mutual virial coefficient g_{AB} tends to the same universal asymptotic limit g^* as g_{AA} and g_{BB} (c.f. Eq. (36)). As a result, contrary to the case of a common θ-solvent or of a selective solvent, the phase separation arises from the corrections to the scaling behavior. Let us consider the symmetric case $N_A = N_B = N$ and suppose that the solution is sufficiently dilute so that the virial expansion is valid. For such a case $x_K = 1/2$ and Eq. (30) leads to a critical concentration

$$c_k \simeq N^{1-3\nu + \chi_s} , \qquad (43)$$

where $\nu \simeq 0.588$ is the swelling exponent and $\chi_s \simeq 0.22$ is the crossover exponent given by Eq. (13).

This is an important result as it indicates that phase separation should not occur in the dilute regime but only in semidilute solutions when the overall polymer concentration is higher than the overlap concentration $c^* \simeq N^{1-3\nu}$. However, in the semidilute regime the virial expansion of the free-energy density is not valid and the prediction for the critical concentration of demixing Eq. (43) must be modified by a more refined study of monomer concentration correlations.

Before, we would like to comment on the limiting case of short chains and strong repulsive interactions χ_{AB} between unlike monomers. In this limit, the correction to the scaling behavior $g^*_{AB}-g^*$ in Eq. (12) may be comparable with the value g^* itself and varies only weakly with N. The critical concentration in this case should scale as

$$c_K \simeq N^{1-3\nu}/(g_{AB} - g^*) \simeq c^* , \tag{44}$$

and we recover the result proposed by de Gennes[2]; namely, that for short chains and strong degree of incompatibility the phase separation appears when the chains start to overlap.

a. Semidilute solutions: critical concentration of demixing In a semidilute solution, chains are forced to interpenetrate to some extent and the excluded volume effects are screened out. For distances smaller than the correlation length ξ the chains are swollen as in the dilute regime and excluded volume effects prevent contacts between like and unlike monomers. At larger distances, contacts between unlike chains cannot be avoided. When the monomer concentration c increases, the screening of excluded volume interactions becomes more efficient, the correlation length decreases as $\xi \simeq c^{-\nu/(3\nu-1)}$ and the tendency towards phase separation increases. Again as for dilute solutions the dominant term in the osmotic pressure comes from excluded volume interactions, i.e., $\Pi_0/T \simeq \xi^{-3}$ and the A-B interactions manifest themselves only through the corrections to this scaling behavior. This correction term Π_i can be postulated simply from a continuity requirement: at the crossover concentration c^* the contribution due to interactions between unlike monomers in the semidilute regime

$$\Pi_i/T \simeq x(1-x)c^m \simeq x(1-x)\xi^{-3}\, c^{\chi_{SD}} , \tag{45}$$

should scale with the polymerization index N in the same way as that obtained in dilute regime

$$\Pi_i \simeq x(1-x)(g_{AB} - g_{AA})(R^3/N^2)c^2 . \tag{46}$$

This simple approach yields the crossover exponent,

$$\chi_{SD} \simeq \chi_S/(3\nu-1) , \tag{47}$$

in agreement with a more elaborate renormalization theory of Joanny, Leibler and Ball.[22] A simple interpretation of these results can be given in terms of correlation effects and contact probability. Due to the excluded volume interactions, which are still present at the short length scale, the probability of a contact between two monomers is not proportional to c^2 or to $c_A c_B$ for A and B monomers, but it is reduced by a factor proportional to $c^{(2\nu-1)/(3\nu-1)} \simeq c^{1/4}$. The probability of contacts between unlike monomers is even more reduced; this is reflected by a correction term Π_i.

To complete the expression of the osmotic pressure, the corrections due to chains translational entropy Tc_i/N_i (i = A,B) must be added. In a semidilute regime, these contributions are small compared to the dominant excluded volume term but not to the correction to the A-B interactions term Π_i. The competition between these translational entropies and the A-B interactions determines the onset of phase separation. For the symmetric cases considered here, demixing occurs when these two contributions are comparable; namely, at a critical concentration of demixing[28]

$$c_k \simeq N^{-(3\nu-1)/(1+\chi_s)} \simeq B_{AB} N^{-b}, \tag{48}$$

where the exponent b is about 0.62. The prefactor B_{AB} characterizes the degree of incompatibility of polymers A and B and is roughly inversely proportional to the Flory interaction parameter χ_{AB}. The expression Eq. (48), first proposed by Schäfer and Kappeler,[24] is a mean-field prediction in a sense that the fluctuations of the monomer composition x are neglected. Within such an approximation the whole phase diagram, i.e., the coexistence curve and the spinodal line can be found starting from the complete expression for the free energy density of a semidilute solution[28]

$$F/T \Big|_{cm^3} = \frac{cx}{M_A} \ln x + \frac{c(1-x)}{M_B} \ln(1-x) + K\xi^{-3}[1 + K_i x(1-x) c^{\chi_{SD}}]. \tag{49}$$

where K is a universal constant and $K_i(T)$ characterizes the monomer interactions and should depend only very weakly on the nature of the solvent. It should be stressed that in the asymmetric case when the molecular weights M_A and M_B are different, the analysis must take into account the very fact that the overall polymer concentration in the coexisting phases and thus ξ are different. Moreover, we will show below that except for extremely long chains, the composition fluctuations are important near the critical point of demixing and modify the shape of the coexistence curve.[28] Still, as often occurs for phase transitions a mean-field theory may predict quite well the position of the critical point.

Indeed, systematic experiments for both the strongly incompatible case poly(styrene)-poly(dimethylsiloxane) and the weakly incompatible case poly(styrene)-poly(methyl methacrylate) in various solvents seem to confirm well the scaling prediction Eq. (48).[20] The prefactor reflects well the difference of the degree of incompatibility. Hence, because of the excluded volume correlation effects, for long chains, the demixing transition occurs well in the semidilute regime contrary to the predictions of the Flory-Huggins model.

Finally, for experiments at a fixed concentration c, Eq. (49) yields the critical temperature of demixing[28]

$$K_i(T_K) \simeq c^{-1/b} M^{-1}. \tag{50}$$

Thus, for a typical case of van der Waals interactions, when the interaction parameter $K_i(T)$ varies as $1/T$, the critical temperature should be roughly proportional to the molecular weight. However, for water-soluble polymers, such as the dextran-poly(ethylene glycol)-water system[18,41] currently used in purification of biological macromolecules, the monomer-monomer interaction parameter K_i depends on T in a more complex way[42,43] and T_K should not be simply proportional to M and Eq. (50) may be then used to find the mass dependence of T_K.

b. critical properties of demixing When the critical point of demixing is approached, composition fluctuations become more and more pronounced. As a result, the second derivatives of the free energy (e.g., osmotic compressibility) diverge. The singular critical behavior of thermodynamic quantities near the critical point may be simply described in terms of power laws. For example, the shape of the coexistence curve near the critical point (for the symmetric cases $x = 1/2$) behaves as[17]

$$\eta = x_+ - x_- \approx \eta_o \left| \varepsilon \right|^{\beta}, \tag{51}$$

where β and η_o denote the respective critical exponent and amplitude, x_+ and x_- denote the composition of A-rich and A-poor phase, respectively. Here, ε is the relative distance from the critical point, i.e., $\varepsilon = (T - T_K)/T_K$ for fixed composition experiments. The correlation length ξ_t which characterizes the spatial range of composition fluctuation diverges as

$$\xi_t \approx \xi_o \varepsilon^{-\nu_t}, \tag{52}$$

and the osmotic compressibility as

$$\kappa \simeq \kappa_o \, \varepsilon^{-\gamma}. \tag{53}$$

An important point here is that according to the theory of critical phenomena, the critical exponents take universal values essentially independent of the microscopic details of the system. The natural question then is whether the exponents characterizing the curve, the radiation scattering intensity, the correlation length, and the interfacial tension behavior in polymer-polymer-solvent systems are the same as for ternary mixtures of small molecules. It is also essential to study how the critical amplitudes depend on the molecular weight of chains.

In polymer blend solutions in a good solvent, two types of correlation effects are to be considered: the overall monomer concentration fluctuations due to the excluded volume effects and the A-B composition fluctuations.[28] For asymmetrical systems, these fluctuations are coupled. We have seen that excluded volume effects cannot be neglected even in the semidilute regime and must be incorporated in the theory. The composition fluctuations may be very important in the vicinity of the critical point. Yet, in calculating the shape of the coexistence curve, of the divergence of the osmotic compressibility from the mean-field free energy density Eq. (49) the contributions to the free energy due to the composition inhomogeneities are neglected. The critical exponents thus obtained have the classical Landau values, e.g., $\beta = 1/2$, $\gamma = 1$, $\nu_t = 1/2$.[17] A so-called Ginzburg criterion[44] permits the estimation of the temperature (concentration) range ε^* near the critical point in which the mean-field approach fails.[2,45-47] The importance of fluctuations is estimated by the ratio

$$\Sigma(\varepsilon) = \langle \delta\eta^2 \rangle / \eta^2 , \tag{54}$$

which compares the mean square fluctuations of composition $\langle \delta\eta^2 \rangle$ averaged over the correlation volume ξ_t^3 and the characteristic value of η of the composition difference in the two coexisting phases. The fluctuation spectrum and the correlation length ξ_t are calculated using a mean-field random phase approximation.[2] For $\varepsilon > \varepsilon^*$, $\Sigma(\varepsilon) < 1$ and mean-field theory should work. For polymer solutions in a good solvent, the extension of the critical regime is given by[28]

$$\varepsilon^* \approx (c_K / c^*)^{-1/(3\nu - 1)} \approx M^{-\chi_s/(1+\chi_s)}, \tag{55}$$

with a prefactor of the order of 10. Hence, for very long chains the critical regime is very narrow and the classical critical exponents should well characterize the critical behavior. In practice, however, for most cases the ratio c_K/c^* is not high enough (e.g., for PS/PMMA system with $M \approx 2.2 \times 10^6$ Fukuda et al.[26] found that c_K/c^* 10) and the non-classical region will be quite broad. Thus, solutions of polymer blends are interesting systems in which a crossover form nonclassical to classical behavior could be observed.

It is interesting to note that c_K/c^* is proportional to the number of chains interpenetrating and thus interacting with a given chain at the critical concentration. In general, as pointed out by de Gennes,[2] when this number is high, fluctuations are not important and the mean-field approximation is valid except for a very narrow region. Hence, for solutions of blends in a θ-solvent, $c_K \approx c^*$ and by a similar reasoning we could expect a nonclassical behavior to occur.

The expansion of the mean-field expression for the free energy in powers of the order parameter (composition fluctuations x -1/2 for the symmetrical case) suggests that the phase separation in polymer blends solution in a good solvent belongs to the same universality class as the Ising model. For experiments at a fixed exchange chemical potential $\mu = \partial F/\partial c$, the properties in the vicinity of the critical point $T_K(\mu)$ are described by Ising exponents.[28] Thus, one expects: $\beta \approx 0.325$ for the coexistence curve, $\nu_t \approx 0.63$ for the correlation length, $\gamma \approx 1.24$ for the osmotic compressibility and $\alpha \approx 0.11$ for the specific heat divergence.[28] The same exponents should occur for the temperature T fixed and the exchanged potential approaching the critical value $\mu_k(T)$.

The situation may be more complex in experiments which are performed at fixed concentration c rather than at fixed chemical potential. The local fluctuations of concentration can, in principle, modify the critical behavior and when T tends to $T_K(c)$ the critical singularities would be characterized by exponents renormalized by a factor $1/(1-\alpha)$, $\beta = \beta/(1-\alpha)$ or $\nu_t = \nu_t/(1-\alpha)$.[28] A similar, so called Fisher renormalization[48] could be expected for experiments at a fixed temperature T and c approaching the critical concentration of demixing $c_K(T)$. Such renormalized exponents have been observed in ternary mixtures of small molecules[49] or in binodal solutions below the θ-temperature.[50] However, for solutions of polymer blends the situation may be expected to be very different as the range of the overall monomer concentration fluctuations ξ is smaller than that of composition fluctuations $R(c)$ (the radius of the coils). Indeed, Broseta et al.[28] have shown that the renormalized exponents may be observed only in an extremely narrow region near the critical points ε^{**} much narrower than the region ε^* evaluated from the Ginzburg criterion Eq. (55). Hence, for available molecular weights of polymers, unrenormalized Ising critical exponents should be observed near the critical temperature.

This original behavior of polymer systems seems to be confirmed by (rare) experimental observations. Shinozaki et al.[21] found, for instance, that the exponent β characterizing the coexistence curve for PS-PDMS-propylbenzene systems is close to the Ising value. Also, the measurements of interfacial tension singularity seem to be compatible with the theoretical prediction for the exponent.

It is more difficult to predict the critical amplitudes that depend on microscopic details of the system. The essential issue is to predict how the amplitudes depend on the molecular weight of polymer chains. In the absence of detailed renormalization group calculations some crossover scaling arguments have been evoked to suggest that the critical amplitudes vary with molecular weight according to simple power laws[28,29] e.g.,

$$\eta_o \simeq N^{\chi_s(\beta-1/2)/(1+\chi_s)} = N^{-0.03}, \tag{56}$$

for the coexistence curve singularity Eq. (51), or

$$\kappa_o \simeq N^{1+\chi_s(1-\gamma)/(1+\chi_s)} \simeq N^{0.96}, \tag{57}$$

for the osmotic compressibility divergence Eq. (53). It is interesting to note how important are excluded volume effects for the critical behavior. Had we disregarded them, the results for critical amplitudes (which may be obtained by putting $\chi_s = 0$) would be quite different,[29] e.g., $\eta_o \simeq N^{-0.18}$, $\kappa_o \simeq N^{0.76}$. Monte Carlo simulations[29] are in striking agreement with the crossover scaling predictions[28,38] and seem to confirm well a pronounced influence of excluded volume interactions on the critical behavior.

c. Interfacial properties Below the critical point, the correlation effects determine the interfacial tension and the composition and concentration profile across the interface. The interfacial energy vanishes and the interface becomes infinitely diffuse as the critical point is approached. In the critical region, the thickness of the interface L between the polymer A-rich and the polymer B-rich phases increases like the composition fluctuations correlation length ξ_t (c.f. Eq. (52))[28]

$$L \simeq N^{(\chi_s+2\nu)/2(1+\chi_s)}\, \varepsilon^{-\nu_t} \simeq N^{0.57}\, \varepsilon^{-0.63}. \tag{58}$$

The interfacial tension γ can be obtained from a simple scaling argument by remarking that, $\gamma\, L_i^2 \simeq T_K$ so that[28]

$$\gamma \simeq N^{-f}\, \varepsilon^{-2\nu_t} \simeq N^{-1.1}\, \varepsilon^{-1.26}, \tag{59}$$

with $f = (2\nu + 2\chi_s - 2\chi_s\nu_t)/(1+\chi_s)$. This result, very different from the mean-field prediction, shows that the interfacial tension in polymer solutions is much lower than that in mixtures of small molecules. As a consequence, the ternary polymer A-polymer B-solvent systems exhibit interesting emulsifying properties.[51]

The interfacial tension remains quite low even far from the critical point of demixing. In fact, the asymptotic value of the interfacial tension for solutions of very long incompatible chains can be estimated by a simple scaling argument[52] from the classical result of Helfand and co-workers[53] obtained for strongly incompatible blends (in the absence of the solvent):

$$\gamma = (kT/a^2)(\chi_{AB}/6)^{1/2}, \tag{60}$$

and

$$L = a(6\,\chi_{AB})^{-1/2}, \tag{61}$$

by replacing the monomer length a by the concentration correlation length ξ (blob length) and the monomer A - monomer B interaction parameter by the blob A-blob B interaction parameter χ_{eff} which scales with concentration like $c\chi_{SD} = c\chi_S/^{(3/\nu-1)}$ for a symmetrical case (c.f. also section IV.2). Such a substitution and the underlying blob interpretation are a direct consequence of the free-energy density in Eq. (49).[28] Hence, for polymer blends in solutions we get[52]

$$\gamma_\infty \simeq \frac{T}{\xi^2} (\chi_{eff}/6)^{1/2} \simeq c^{1.65}, \quad (62)$$

and

$$L_\infty \simeq \xi(6\chi_{eff})^{-1/2} \simeq c^{-0.9}. \quad (63)$$

For typical values of blob interaction parameter χ_{eff} (10^{-2}-10^{-3}), the interface thickness is expected to be about few tens of Å and the interfacial tension about 10^{-3}-10^{-1} dyn/cm that is small compared with those of small molecules mixtures and with those of polymer melts without solvent.

It should be noted that the so-called dilution approximation which neglects the excluded volume effects and essentially replaces χ_{AB} by $\chi_{AB}c$ (as suggested by the Flory-Huggins free energy as in Eq. (27) leads to $\gamma \simeq c^{1.5}$ (close to the prediction in Eq. (62) and to $L \simeq c^{-0.5}$ (very different from Eq. (63).[53,54]

The asymptotic results are confirmed by detailed calculations of composition profiles52 based on Cahn-Hilliard's approach.[55] There are, however, important corrections to this limit behavior due to the finite length of chains and due to the interfacial activity of solvent which locates preferentially at the interface even in the symmetric case[52]

$$L = L_\infty[1 - 4\ln 2/(\chi_{eff} n)]; \quad (64)$$

$$\gamma = \gamma_\infty[1 - \pi^2/(6\chi_{eff} n) - 1.67\chi_{eff}],$$

where $n = N/c\xi^3$ denotes the number of blobs per chain. The preferential location of the solvent at the interface is simply due to energetics, which favors A-S-B configurations with respect to A-B-S or S-A-B configurations, as the solvent located at the interface decreases the number of unfavorable A-B contacts. It is important to stress that for most practical cases the corrections to the limiting behavior Eqs. (62,63) modify substantially the apparent exponents in the concentration-dependence of both the interfacial thickness and tension.[52]

The theory enables one to calculate the interfacial thickness and interfacial tension without any adjustable parameters since the correlation length ξ (number of blobs n) can be found for instance from light-scattering studies of semidilute solutions of A (or B) and the interaction parameter can be estimated from the knowledge of the critical concentration of demixing.[52] Systematic measurements of the interfacial tension of demixed PS-PDMS-toluene systems for various molecular weights and concentrations are in a very good agreement with theoretical predictions of Broseta.[52,38]

IV. MESOPHASE FORMATION IN BLOCK COPOLYMER SOLUTIONS

The phase diagrams of solutions of diblock copolymers A-B may be quite complex and depend both on the chemical nature of two blocks and on the solvent.[5,6] In a selective solvent, good for the B-block and poor for the A-block, intermolecular aggregates are formed in the dilute regime so that the number of unfavorable contacts A-S is limited. The shape of aggregates (e.g., spherical or wormlike micelles, vesicles), their size and polydispersity depend very much on chain composition and length. In more concentrated solutions, aggregates order in space and form mesophases, i.e., ordered microdomains rich in A (in B) (e.g., lamellar, hexagonal, cylindrical, cubic). In nonselective solvent, the repulsive interactions between unlike monomers are responsible for mesophase formation. In a sense, block copolymers may be considered as a nice model of nonionic surfactants which should exhibit some very universal features and can be studied by statistical methods developed in polymer theory.[56] Industrially, ordering and aggregation phenomena are fundamental in many applications of copolymers, e.g., thermoplastic elastomers,[5] adhesives,[57] viscosity stabilizers in oils.[58]

We discuss here two examples of aggregation of diblock copolymers in solution, the formation of spherical micelles in a highly selective solvent, good for one block and poor for the other block, and the formation of ordered mesophases in a common good solvent for both blocks.

IV.1. Micelle Formation in a Highly Selective Solvent

We consider the extreme limit where block A forms locally molten regions where neither the solvent nor the B block penetrates. In particular, an isolated chain has a swollen B- part with a radius $R_{FB} \simeq N_B^{3/5}$ and a collapsed part with a radius $R_A \simeq N_A^{1/3}a$. If the concentration of free chains is Φ_1, in a dilute solution, the chemical potential of the free chains is[32]

$$\mu_1 = T\log\Phi_1 + 4\pi\gamma_{AS}a^2N_A^{2/3}. \tag{65}$$

For the B chains, we have taken as a reference-state isolated chains in a good solvent; for the A chains the molten state. The dominant contribution to the free energy of collapsed chain is the surface energy between the solvent and the molten A globule (γ_{AS} being the relevant surface tension).

We first assume that the aggregates formed by the copolymer chains have a bimodal distribution: single chains and monodisperse spherical micelles. In this simplified model, the critical micelle concentration is a sharp transition if we neglect micelles translational entropy.[60] We then discuss the micelle size distribution.

a. Critical micelle concentration (*c.m.c.*) At extremely low concentration, only individual chains are found in the solution. As the concentration is increased, it reaches the *c.m.c.* where micelles form. We consider one individual micelle and study its stability in equilibrium with a reservoir composed of the free chains in solution which impose a chemical potential μ_{ex} and an osmotic pressure Π_{ex}.

The micelle contains p chains. Its core is a molten A region of radius R_M. The concentration of A monomers being equal to 1 in the core region, we get

$$\frac{4}{3}\pi\left(\frac{R_M}{a}\right)^3 = pN_A. \tag{66}$$

The shell is formed by the B blocks in a grafted configuration (Fig. 2). We will only consider here the case where the size L of the shell is larger than the core radius R_M.

The stability of the micelle is governed by the grand canonical free energy[32]

$$\Omega(p) = F_M(p) - \mu_{ex}p + \frac{4}{3}\pi[(R_M + L)^3 - R_M^3]\Pi_{ex}, \tag{67}$$

where $F_M(p)$ is the free energy of a micelle of p chains. The two leading terms in this free energy are the surface energy between the molten core and the solvent and the stretching energy of the B chains in the shell[61,31,32]

$$F_M(p) = 4\pi\gamma_{AS}R_M^2 + F_B\,. \tag{68}$$

In a good solvent, the energy F_B has been calculated by Daoud and Cotton[59] using blob-like arguments[31,32]

$$F_B = T p^{3/2} \log\,(N_B^{3/5} / N_A^{1/3}\; p^{2/15})\,. \tag{69}$$

The number of chains in the micelles minimizes the energy $\Omega(p)$ and the critical micelle concentration is reached when the value of Ω at the minimum is negative: the micelle is then stable. We find

$$p = N_A^{4/5}\left[\frac{\gamma_{AS}a^2}{T\log\,(N_B^{3/5} / p^{2/15} N_A^{1/3})}\right]^3, \tag{70}$$

for the optimum aggregation number and

$$\Phi_{cmc} \simeq exp\left\{-\,4\pi(\gamma_{AS}a^2/T)N_A^{2/3} + \mu_{cmc}\right\}, \tag{71}$$

for the critical micelle concentration. The critical value of the chemical potential μ_{cmc} is given by the condition $\Omega = 0$. We note that the *c.m.c.*

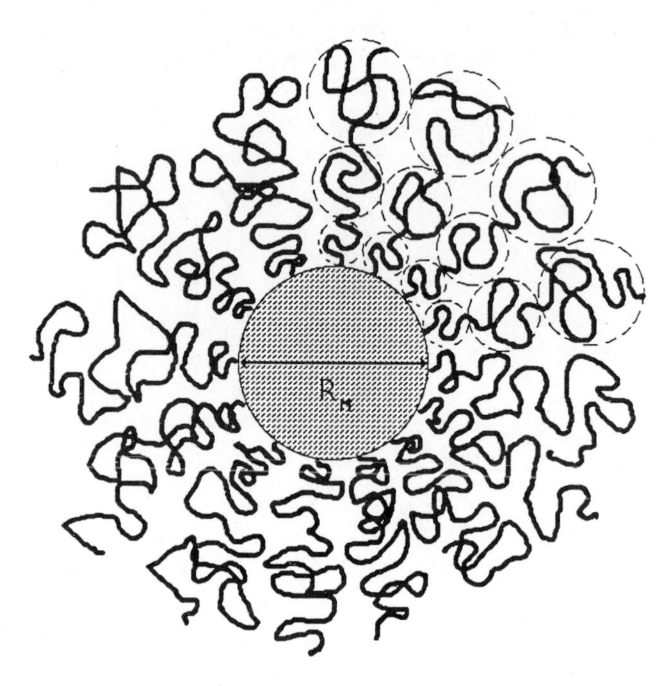

Fig. 2. Schematic representation of a diblock copolymer micelle in a selective solvent.

depends strongly on the polymerization index of A-block and is essentially independent of N_B. Also, the aggregation number p increases rapidly with increasing polymerization index N_A and is thus much larger for block copolymer micelles than for classical surfactant molecules.

Ignoring the logarithmic corrections, we find for the core and corona dimensions[31,32]

$$R_M \simeq N_A^{3/5} a (\gamma_{AS} a^2 / T)^{2/5} ; \tag{72}$$

$$L \simeq N_B^{3/5} N_A^{4/25} a (\gamma_{AS} a^2 / T)^{6/25} .$$

Hence, the A blocks are weakly stretched, their average size is larger than the Gaussian radius $N_A^{1/2}a$, whereas the shell size scales with the mass of the B-block as the Flory radius R_{FB}.

Spherical micelles have indeed been observed in various systems.[62,63] Experimental values of L of poly(styrene)-poly(butadiene) diblock copolymer micelles in n-heptane[63] seem to conform to scaling law (Eq. (42).[31]

b. Micelle size distribution The preceding model ignores the polydispersity in micelle size, we now briefly discuss the polydispersity (c.f. Ref. 64). The solution has a chain chemical potential μ_{ex} and an osmotic pressure Π_{ex}. The grand canonical free energy per unit volume is

$$G = F_{mix} + \sum_n F_M(n) \phi_n - \sum_n n \phi_n \mu_{ex} + \Pi_{ex} , \tag{73}$$

where F_M is the free energy of a micelle of n copolymer chains given by Eq. (68), ϕ_n is the number of micelles containing n chains (per unit volume) and F_{mix} is the free energy of mixing of the micelles

$$F_{mix} = T \sum_n \phi_n \log \phi_n a^3 . \tag{74}$$

The free energy Eq. (73) ignores all interactions between micelles in particular, it allows interpenetration between different micelles. This should, however, be a reasonable approximation in the vicinity of the *c.m.c* where the solution is extremely dilute.

The micelle size distribution is obtained by minimization of the free energy Eq. (73) with respect to the concentrations ϕ_n.

$$\phi_n a^3 = exp \left\{ - \frac{F_M(n) - \mu_{ex} n}{T} \right\} . \tag{75}$$

The chemical potential μ_{ex} is then expressed as a function of the concentration ϕ by writing the conservation of the block copolymer chain

$$a^3 \phi = \sum_n n \, exp \left\{ - \frac{F_M(n) - \mu_{ex} n}{T} \right\} . \tag{76}$$

The osmotic pressure is obtained by minimization of the total grand canonical free energy GV with respect to the volume v:

$$\frac{\Pi_{ex}}{T} = \sum_n \phi_n = \sum_n exp\left\{ - \frac{F_M(n) - \mu_{ex} n}{T} \right\} . \tag{77}$$

When the surfactant is a block copolymer, we expect a rather peaked distribution in micelle size around a large value $n \approx p$ (c.f. Ref. 60). We may thus approximate the distribution by a Gaussian function expanding the argument of the exponential up to second order around $n = p$

$$\phi_n a^3 = exp\left\{ \frac{-\Omega(p)}{T} \right\} exp\left\{ - \frac{(n-p)^2}{2(\delta p)^2} \right\} , \tag{78}$$

where Ω is the grand canonical potential of one micelle given by Eq. (67); the average aggregation number p minimizes this potential

$$p \simeq \left(\frac{\mu_{ex}}{T} \right)^{1/2} . \tag{79}$$

In Eq. (78), δp denotes the root mean square average of the distribution obtained from the second derivative of $\Omega(n)$ for $n = p$

$$\delta p \simeq p^{1/4} . \tag{80}$$

At the *c.m.c.* given by Eq. (71) the aggregation number scales as $N_A^{4/5}$ and the micelles are very monodisperse, since

$$\frac{\delta p}{p} = p^{-3/4} \simeq N_A^{-3/5} . \tag{81}$$

For block copolymers the bimodal model gives thus a rather good description of micelle formation.[60] Equation (77) leads then to a sharp break at the *c.m.c.* in the osmotic pressure which could be a good experimental way of determining the *c.m.c.* if it lies in an accessible range in concentration.

At higher copolymer concentrations, interactions between micelles start to play an important role and may lead to the formation of ordered structures: macrocrystals. We considered here very asymmetrical chains with $N_A << N_B$. For more symmetrical chains, one can expect that concentrating the system would favor the formation of wormlike aggregates or vesicles before the onset of an ordered mesophase.

IV.2. Ordered Phases in a Nonselective Solvent: a Blob Model

In a dilute solution in a common good solvent for both blocks, the interactions between different copolymers may be studied using the same direct renormalization procedures as the interactions between two homopolymers A and B equivalent to the two blocks.[22] As for blends, in the asymptotic limit of infinite molecular masses, the chemical difference between the two blocks is irrelevant and the dimensionless virial coefficient g_C between block copolymers defined by Eq. (10) is equal to the same value g^* as for homopolymers. The interactions which may provoke the formation of mesophases are here again due to the corrections to the scaling behavior[22]

$$g_c = g^* + \gamma_{AB} N^{-\chi_s} . \tag{82}$$

The exponent χ_S is given by Eq. (13) and the non-universal constant γ_{AB} depends both on the strength of the interaction parameter (the Flory parameter χ_{AB}) and on the architecture of the copolymer.

In a dilute solution, blends of homopolymers A and B equivalent to the two blocks of the copolymers are compatible in a common good solvent; contrary to what happens in a highly selective solvent, we do not expect mesophases to form in a dilute solution but only in semidilute and concentrated solutions. Mesophases formed by block copolymers in nonselective solvents have been studied by many experimental groups, perhaps most recently by Hashimoto et al.[55] and Williams et al.[65] The usual interpretation[65,66] of experiments is based on the "dilution approximation" in which the phase diagram of a solution can be obtained from the corresponding pure copolymer phase diagram by replacing χ_{AB} by $\Phi\ \chi_{AB}$, where Φ is the monomer volume fraction. We have seen in section III.2c that such an approach presents severe drawbacks even for the symmetric case ($N_A \cong N_B$). A more correct mean-field theory developed by Noolandi and Hong[67] yields a number of interesting predictions for concentrated solutions in the strong segregation limit. As in the dilution approximation, this theory neglects subtle correlation effects originating from excluded volume interactions which play a crucial role in semidilute regime in the weak segregation limit. Here we briefly discuss the onset of mesophases in semidilute regime following Ref. 33 and we emphasize the nonclassical excluded volume effects.

It is important to stress that in spite of the nonselectivity of the solvent, in ordered mesophases, inhomogeneities in A (B) composition are also accompanied by solvent concentration fluctuations. As for the interfaces in phase-separated polymer A - polymer B - solvent systems, this effect is due to the preferential location of solvent at the interfaces between microdomains: the solvent screens out unfavorable A-B contacts. Fredrickson and Leibler[33] have shown that solvent inhomogeneities lead to a negligible contribution to the free energy only for very long chains (when $N^{-0.22} << 1$). Also, contrary to the pure block copolymer melt, the microphase separation transition in diblock copolymer solutions corresponds to a very narrow region of coexistence between a copolymer-poor disordered phase and a copolymer-rich ordered phase. Still for long copolymers it is a reasonable first approximation[33] to neglect these effects in order to estimate the critical copolymer concentration at which the microphase separation appears. A standard procedure in semidilute polymer solutions is to group the monomers into blobs of size ξ equal to the correlation length; in a good solvent

$$\xi \simeq a\,(ca^3)^{3/4}\,, \tag{83}$$

where c is the total monomer concentration. Each blob contains $g = c\xi^3$ monomers which are in general of the same chemical nature. Each copolymer chain may be viewed as a copolymer chain of blobs with $n_A = N_A/g$ blobs of type A and $n_B = N_B/g$ blobs of type B. The copolymer solution is then equivalent to a melt of copolymers formed by these blobs. The interactions between A and B blobs, similarly to the case of polymer blends in solution, may be described by an effective Flory parameter between blobs given by

$$\chi_{eff} = \delta_{AB}\, c^{\chi_{SD}} \tag{84}$$

With these approximations, the phase diagram calculated for copolymer solutions can be obtained from previous theoretical calculations for pure copolymer melts by Leibler[11] in the mean-field theory (Fig. 3) or by

Fredrickson and Helfand[13] in a more sophisticated model which takes into account the effects of composition fluctuations that become coherent over large distances near the "critical point". The critical point for the onset of demixing occurs for $N_A = N_B$ and is given in the limit of very long chains by[11,13]

$$\left\{ \chi_{eff}(n_A + n_B) \right\}_c \simeq 10.5 \; . \tag{85}$$

Relation Eq. (85) yields, in particular, the critical concentration c_K above which mesophase is expected to be formed. This concentration scales with the total polymerization index $N = N_A + N_B$ the same way as the critical concentration of demixing in a blend solution given by Eq. (45)

$$c_K \simeq N^{-b} \, , \tag{86}$$

with $b = (3\nu - 1)/(1 + \chi_s) = 0.62$. This result seems to be in reasonable agreement with experiments on poly(styrene)-poly(isoprene) block copolymers in toluene and dioctyl phthalate by Hashimoto et al.,[65] who find $c_K \approx N^{-1/2}$. The scaling result seems to be in much better agreement than the classical mean-field prediction $c_K \approx N^{-1}$.[65-67]

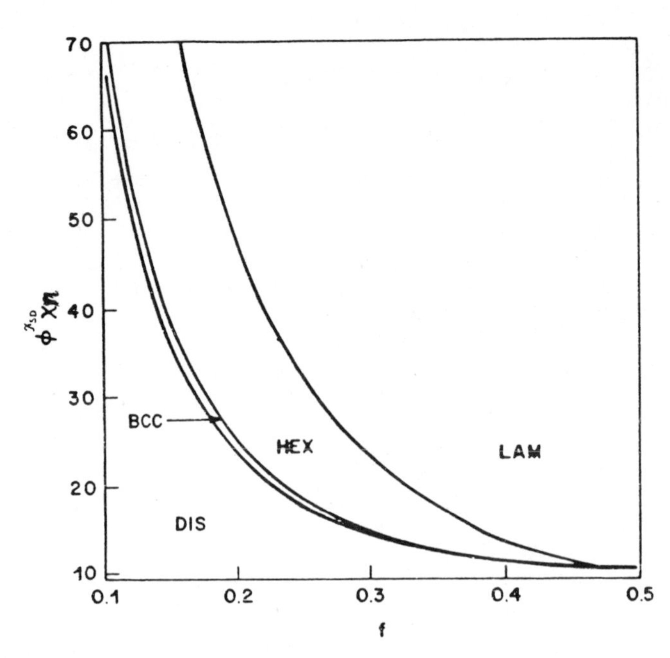

Fig. 3. Phase diagram for a semi-dilute diblock solution in a good, non selective solvent in the mean field approximation. DIS, BCC, HEX and LAM denote, respectively, disordered, body-centered cubic, hexagonal and lamellar mesophases.

The simple model discussed here predicts transitions between various ordered phases (Fig. 3). It would be interesting to see whether such transitions can indeed be observed when the solvent is evaporated from a solution.

V. CONCLUSIONS

Throughout this review we have stressed the important role of subtle excluded volume correlation effects in segregation phenomena occurring in solutions containing chemically different polymers or copolymers. Many unexplained observations have found a simple interpretation in terms of renormalization theory or scaling arguments which take these effects into account. Some predictions have been verified by recent experiments or numerical simulations. Still many questions seem to be open. Perhaps the most important seems to be that concerning the kinetics of phase separation in ternary polymer A - polymer B - solvent systems or the dynamics of interdiffusion in these systems. Another open problems concern the systems with attractive A-B interactions in which physical gels may be formed as well as calculation of complete phase diagrams of di-block or multiblock copolymers in selective or neutral solvents. It would be also interesting to extend statistical models to aqueous solutions and include specific interactions or conformation changes present in such systems.

ACKNOWLEDGEMENTS

We are deeply indebted to R. Ball, D. Broseta, G.H. Fredrickson, A. Lapp, C. Marques, L. Ould Kaddour and C. Strazielle with whom many parts of our work in the field of phase transition in polymer solutions have been carried out.

NOMENCLATURE

a	monomer size : also exponent in Eq. (41)
b	exponent (Eq. 48)
b_{ij}	monomer i monomer j interaction
c	monomer concentration
c_k	critical demixing concentration
c^*	overlap concentration
c_i	concentration of polymer i
d	space dimension
D_i	fractal dimension of chain i
F	free energy density
g_{ij}	dimensionless virial coefficient between polymers i and j
G_{ij}	second virial coefficient
h_{ij}	reduced virial coefficient
K	monomer-monomer interaction in semidilute solutions
L	size of rod; width of the interface
n	exponent
N_i	degree of polymerization of chain i
n_i	number of blob per chain
p	number of chains per micelle
R	chain end-to-end distance
R_g	radius of gyration
R_M	micelle core size
S	Gaussian area of a polymer chain
T	temperature measured in such units that the Boltzmann constant is k_b = 1
T_k	consolute temperature
v	Edwards' excluded volume parameter
x	composition of a blend
α_{AB}	renormalized interaction

β	critical exponent
β_{AB}	renormalized interaction
γ	critical exponent
γ_{ij}	interfacial tension between phases i and j
ε	shift from critical temperature or critical concentration; ε-expansion parameter $\varepsilon = 4-d$
η	shift from critical composition
θ	Flory compensation temperature
κ	osmotic compressibility
μ	chemical potential
ν	correlation length exponent
ξ	correlation length
Π	osmotic pressure
σ	radius ratio
Σ	ratio of composition fluctuations $<\delta\eta^2>$ to composition $<\eta>^2$
Φ	volume fraction
χ_{ij}	Flory interaction parameters
χ_s	chemical mismatch exponent in dilute solutions
χ_{SD}	chemical mismatch exponent in semidilute solution
Ω	grand canonical free energy

REFERENCES

1. P.J. Flory, Principles of Polymer Chemistry, Cornell University Press, Ithaca, New York (1953).
2. P.G. de Gennes, Scaling Concept in Polymer Physics, Cornell University Press, Ithaca, New York (1979).
3. J. des Cloizeaux and G. Jannink, Les Polymères en solution: leur modélisation et leur structure, les Editions de Physique, Paris (1987).
4. D.R. Paul and S. Newman Eds., Polymer Blends, Vols. 1 and 2, Academic Press, New York (1978).
5. S.L. Aggarwal Ed., Block Copolymers, Plenum Press, New York (1979).
6. I. Goodman Ed., Developments in Block Copolymers, Vols. 1-5, Applied Science Publishing, New York (1982).
7. E.L. Thomas, D.B. Alward, D.J. Kinning, D.C. Martin, D.L. Handlin and L.J. Fetters, Macromolecules, 19, 2197 (1986).
8. J.F. Joanny, Thèse de 3ème cycle, Université Pierre et Marie Curie (1978).
9. D.J. Meier, J. Polym. Sci. C, 26, 81 (1969).
10. E. Helfand, Macromolecules, 8, 552 (1975); E. Helfand and Z. Wassermann, Polym. Eng. Sci., 17, 582 (1977).
11. L. Leibler, Macromolecules, 13, 1602 (1980).
12. A.N. Semenov, Sov. Phys. JETP, 61, 733 (1985).
13. G.H. Fredrickson and E. Helfand, J. Chem. Phys, 87, 697 (1987).
14. M.D. Whitmore, J. Noolandi, Macromol. Chem., Macromol. Symp., 16, 235 (1988).
15. P.J. Flory, J. Chem. Phys., 10, 51 (1942); M.L. Huggins, J. Phys. Chem., 46, 151 (1942).
16. R.L. Scott, J. Chem. Phys., 17, 279 (1949).
17. L. Leibler in Encyclopedia of Polymer Science and Engineering, Vol. 11, 30, John Wiley, New York (1988).
18. P.Å. Albertsson, Partition of Cell Particles and Macromolecules, Almqvist and Wiksell, Stockholm (1971).
19. M. Nagata, T. Fukuda and H. Inagaki, Macromolecules, 20, 2173 (1987).
20. L. Ould Kaddour and C. Strazielle, This proceeding and references given therein (1988).
21. K. Shinozaki, Y. Saito and T. Nose, Polymer, 23, 1937 (1982).
22. J.F. Joanny, L. Leibler and R. Ball, J. Chem. Phys., 81, 4640 (1984).
23. M.K. Kosmas, J. Phys. Lett., 45, L 889 (1984).
24. L. Schäfer and Ch. Kappeler, J. Phys., (Paris) 46, 1853 (1985).

25. T. Tanaka and H. Inagaki, Macromolecules, 12, 1229 (1979).
26. T. Fukuda, M. Nagata and H. Inagaki, Macromolecules, 17, 548 (1984).
27. L. Ould Kaddour, M. Soleda Amasagasti and C. Strazielle, Makrom. Chem.Rapid Commun.
28. D. Broseta, L. Leibler and J.F. Joanny, Macromolecules, 20, 1935 (1987).
29. A. Sariban and K. Binder, J. Chem. Phys., 86, 5859 (1987); preprint (1988).
30. J.F. Joanny, submitted to Europhys. Lett.
31. A. Halperin, Macromolecules, 20, 1943 (1987).
32. C.M. Marques, J.F. Joanny and L. Leibler, Macromolecules, 21, 1051 (1988).
33. G.H. Fredrickson and L. Leibler, Macromolecules, in press.
34. E. Edmond and A.G. Ogston, Biochem. J., 109, 659 (1968); A.G. Ogston in Chemistry and Technology of Water-soluble Polymers, C.A. Finch Ed. Plenum Press, New York (1983).
35. Åke Gustafsson, H. Wennerström and F. Tjerneld, Polymer, 27, 1768 (1986).
36. T.A. Witten in Chance and Matter, J. Souletie, J. Vannimenus R. Stora Eds. North Holland, Amsterdam, 159 (1987).
37. J. des Cloizeaux, J. Phys. (Les Ulis, Fr.), 41, 749 (1980).
38. D. Broseta, Thèse de Doctorat, Université Pierre et Marie Curie (1987).
39. See for example, A. Lapp, J. Herz and C. Strazielle, Makromol. Chem., 186, 1919 (1985) and references therein.
40. P.G. de Gennes, J. Polym. Sci. Phys. Ed., 16, 1883 (1978).
41. H. Walter, D.E. Brooks and D. Fisher (Eds), Partitioning in Aqueous Two-Phase Systems: Theory, Methods Uses and Applications of Biotechnology, Academic Press, Orlando (1985).
42. G. Karlström, J. Phys. Chem., 89, 4962 (1985).
43. Å Sjöberg and G. Karlström, preprint (1988).
44. V.L. Ginsburg, Sov. Phys. Solid State, 2, 1824 (1960).
45. P.G. de Gennes, J. Phys. Lett. (Paris), 38, L441 (1977).
46. J.F. Joanny, J. Phys., A11, L117 (1978).
47. K. Binder, J. Chem. Phys., 79, 6387 (1983); Phys. Rev., A29, 341 (1984).
48. M.E. Fisher, Phys. Rev., 176, 1 (1968).
49. B. Widom, J. Chem. Phys., 46, 3324 (1967); K.K. Clark, ibid, 48, 741 (1988); G. Neece, ibid, 47, 4112 (1987); J.A. Zallweg, ibid, 55, 3 (1971); L.E. Wool, G. Pruitt and G. Morrison, ibid, 77, 1572 (1973); K. Ohbayashi and B. Chu, ibid, 68, 5066 (1978).
50. T. Dobashi, M. Nakata and M. Kaneko, J. Chem. Phys., 80, 948 (1984).
51. P. Gaillard, M. Ossenbach-Sauter and G. Reiss, Makromol. Chem. Rapid Commun., 1, 771 (1980); also in Polymer Compatibility and Incompatibility, K. Solc Ed. (MMI, Middland 1980), vol. 6.
52. D. Broseta, L. Leibler, L.Ould Kaddour and C. Strazielle, J. Chem. Phys., 87, 7248 (1987).
53. E. Helfand and Y. Tagami, J. Polym. Sci., B9, 741 (1971); J. Chem. Phys., 56, 3592 (1972); ibid, 57, 1812 (1972); E. Helfand and A.M. Sapse, J. Chem. Phys., 62, 1327 (1975).
54. K.M. Hong and J. Noolandi, Macromolecules, 13, 964 (1980) 14, 736 (1981); J. Noolandi and K.M. Hong, ibid, 15, 482 (1982), 16, 1443 (1983).
55. J.W. Cahn and J.E. Hilliard, J. Chem. Phys., 28, 258 (1958).
56. P.G. de Gennes and C. Taupin, J. Phys. Chem., 86, 2294 (1982).
57. G. Kraus and T. Hashimoto, J. Appl. Polym. Sci., 27, 1745 (1982); J.M. Widmeier, Thèse d'Etat, Université Louis-Pasteur, Strasbourg (1980).
58. G. Kraus, in Block Copolymers: Science and Technology, D.J. Meier Ed. MMI Press, Midland (1983).
59. M. Daoud and J.P. Cotton, J. Phys. (Paris), 43, 531 (1982).

60. L. Leibler, H. Orland and J.C. Wheeler, J. Chem. Phys., 79, 3550 (1983).
61. P.G. de Gennes, in Solid State Physics, Suppl. 14, L. Leibert Ed., Academic Press, New York (1978).
62. Z. Tuzar and P. Kratochvil, Adv. Colloid Interface Sci., 6, 201 (1976).
63. C. Price, Pure Appl. Chem., 55, 1563 (1983); in ref. 6.
64. R.E. Goldstein, J. Chem. Phys., 84, 3367 (1986).
65. M. Shibayama, T. Hashimoto, H. Hasegawa and H. Kawai, Macromolecules, 16, 1427 (1983); T. Hashimoto, M. Shibayama and H. Kawai, ibid, 16, 1093 (1983).
66. E.R. Pico and M.C. Williams, Polym. Eng. Sci., 17, 573 (1977); J. Appl. Polym. Sci., 22, 445 (1978).
67. J. Noolandi and K.M. Hong, Ferroelectrics, 30, 117 (1980); K.M. Hong and J. Noolandi, Macromolecules, 16, 1083 (1983).

PERCOLATION TYPE GROWTH PROCESS OF CLUSTERS NEAR THE GELATION THRESHOLD

M. Adam, M. Delsanti, J.P. Munch†

Service de Physique du Solide et de Résonance Magnétique
Institut de Recherche Fondamentale, CEA-CEN Saclay
91191 Gif-sur-Yvette, France

D. Durand

Laborat. de Chimie et Physico-Chimie Macromoléculaire
Unité assocée au CNRS
Université du Maine
Route de Laval
72017 Le Mans Cedex - France

ABSTRACT

The gelification process by polycondensation leading to polyurethane self-similar clusters is studied, near the gelation threshold. In dilute solutions, where scattering experiments are performed, the interaction between clusters is a function of the distance to the gelation threshold, ε. This is an evidence of the swelling of the clusters. The exponent values γ, which links the weight-average molecular weight to ε, and τ, which characterizes the mass distribution, are determined experimentally. Within experimental precision, they are identical with those determined by Monte Carlo simulation following percolation model.

I. INTRODUCTION

Gelation is a critical phenomenon of connectivity and as such we will use the percolation theory to describe it.[1] As percolation theory was described in detail,[2] here we recall the behaviors of measurable quantities which experimental results are given hereafter. Below the gelation threshold the system is composed of finite size polymers branched in the 3-dimensions of space. We shall call those polymers "polymers clusters" in order to distinguish them from other branched polymers as stars or combed polymers. Below the gelation threshold, the system is viscous at zero frequency. At the gelation threshold, there appears a giant polymer clus-

---

† Permanent address: Laboratoire de spectrométrie et d'imagerie ultra-sonores - Université Louis-Pasteur, 4 rue Blaise-Pascal 67070 Strasbourgh - Cedex, France.

ter which size is equal to that of the vessel in which the chemical reaction takes place. Above the gelation threshold, two phases, the infinite cluster (gel phase) and the finite size polymer clusters co-exist. For the critical phenomenon the important parameter is the relative distance to the gel point ε: if p is the degree of advancement of the chemical reaction and p_c its value at the gelation threshold then

$$\varepsilon = \left| \frac{p - p_c}{p_c} \right| .$$

Below the gel point, following the percolation model the number n_i of polymer clusters having a mass M_i is given by the mass distribution

$$n_i \sim M_i^{-\tau} f(M_i / M^*) , \tag{1}$$

where τ is the mass distribution exponent; M^* is the largest mass presents in the system; f is a cut-off function which implies that a mass $M \gg M^*$ has a zero probability to exist. The largest mass M^* increases as the gel point is approached. As we deal with a critical phenomenon the variation of M^* is a power law of the relative distance to the gel point ε

$$M^* \sim \varepsilon^{-1/\sigma} . \tag{2}$$

The correlation length ξ which is the overall largest distance by which monomers are connected

$$\left(\xi^2 = \sum_i n_i M_i^2 R_i^2 / \sum_i n_i M_i^2 \right)$$

increases as the gel point is approached:

$$\xi \sim \varepsilon^{-\nu} . \tag{3}$$

Size R_i and mass M_i of the ith polymer cluster are linked by the fractal dimension D_P: $M_i \sim R_i^{D_P}$. In particular, this relation is valid for the largest cluster $M^* \sim \xi^{D_P}$.

The weight-average mass

$$M_w \sim \frac{\sum_i n_i M_i^2}{\sum_i n_i M_i}$$

diverges at the gel point as

$$M_w \sim \varepsilon^{-\gamma} \tag{4}$$

There is a relation between exponents γ, τ, and σ:

$$\gamma = \frac{3 - \tau}{\sigma} \tag{5}$$

which means that $M_w \sim M^{*(3-\tau)}$ thus M^* and M_w both increase as the gel point is approached but they are not proportional. One has to note that as we are concerned with a polydisperse sample M_w and ξ are not related through the fractal dimension but by the relation:

$$M_w \sim \xi^{D_P(3-\tau)} . \tag{6}$$

Above the gel point the gel phase G, which coexists with finite clusters, increases as ε^{β}.

Percolation exponent values are given in Table 1, they are calculated from Monte Carlo simulations performed on a lattice at a space dimensionality d. Exponent values are very sensitive to d (the mean-field values are reached for $d > 6$) but insensitive to the exact geometry of the lattice (number of first neighbour) for a given dimensionality.

$$D_P = \frac{d}{\tau - 1} , \tag{7}$$

Table 1. Theoretical and experimental exponent values

	Theoretical exponent values		Experimental exponent values
	Percolation	Mean-field	
τ	2.20	2.5	2.2 ∓ 0.04 a
γ	1.74	1	1.71 ∓ 0.06 a
σ	0.46	1/2	0.47 ∓ 0.04 b
β	0.45	1	0.43 ∓ 0.12 c
ν	0.88	1/2	0.86 ∓ 0.1 d 0.80 ∓ 0.08 e
D_p	2.5	4	2.50 ∓ 0.08 f

Comparison between theoretical exponent values and values determined either experimentally (a) or deduced by inserting the experimental values into the following relations:
b) $\sigma = (3-\tau)/\sigma$, c) $\beta = \gamma(\tau-2)/(3-\tau)$, d) $\nu = \gamma(\tau-1)/(g-3\tau)$, e) $\nu = \nu' \mathrm{x}\, D/D_p$, f) $D_p = 3/(\tau-1)$.

The hyperscaling relation which links exponents to dimensionality of the space, means that density of the gel phase G is equal to that of the largest finite polymer cluster ξ.

The scattering experiments, here reported, are performed on systems in which crosslinks are realized through a chemical reaction. In order to be able to measure the mass and the size of the largest polymer cluster, one has to dilute the polymeric system in order to separate the polymers, to create a contrast and to minimize the interaction. Dilution has two effects:

- it swells the clusters and decreases fractal dimension $D<D_p(M_i \sim R_i^D)$;

- it reveals the interactions between the polymer clusters.

The interaction parameter, B measured by light scattering, can be calculated from the expression proposed by Yamakawa[3] taking for the co-volume V_{ij} the following form[4] $V_{ij} = (M_i/M_j)\, R_j^3$, where it is assumed that it is the smaller cluster M_j which screens the interactions in the bigger cluster M_i. It is found that:[5]

$$B \simeq \frac{\sum_{ij} n_i M_i n_j M_j V_{ij}}{M_w\left(\sum_i n_i M_i\right)^2} \simeq \sum_i n_i R_i^3 \sim \varepsilon^{-\frac{3}{\sigma}\left(\frac{1}{D}-\frac{1}{D_p}\right)} \tag{8}$$

The inverse of B is the internal concentration of monomers averaged on the polymer cluster distribution. One can see that if polymer clusters do not swell[6] under dilution ($D = D_p$), the interaction parameter B would be independent of ε. B is proportional to the intrinsic viscosity $[\eta] = (\eta-\eta_0)/\eta_0 c$ (where η and η_0 are the solution and the solvent viscosity respectively) if a hydrodynamic interaction inside polymer cluster is assumed.[1]

Performing light- or neutron-scattering experiments, we were able to measure on samples prepared below the gel point the exponents τ, γ and D.

II. EXPERIMENTAL PROCEDURE

II.1 Chemical system and location of the gel point

We chose to study a system in which the chemical reaction was a polycondensation. The chemical system was fully described in Ref. 7, here we recall the main features. Polyurethane was prepared, without solvent, by the condensation of OH groups from a trifunctional unit (poly(oxypropylene)-triol with molecular weight of 700) and NCO groups from a difunctional unit (hexamethylene diisocyanate with molecular weight of 168). One can define the stoichiometric ratio as p = [NCO]/[OH] where [NCO] and [OH] are the concentrations of NCO and OH groups present initially. p was determined by weighing, its precision was 5×10^{-5}. If $p < 1$, at the completion of the reaction, p is proportional to the number of OH groups which have reacted. The absence of the NCO groups at complete reaction was verified.[7] The sample preparation was reproducible because samples prepared in a time interval of 6 months, using the same batch of dried monomers, gave consistent results (see for example in Fig. 4). The main problem was to locate the p value corresponding to p_c, the gelation threshold. For $p \leq 0.5593$ the system could be dissolved in dioxane (a solvent), for $p \geq 0.56$ the system could not be dissolved

$$0.5593 < p_c < 0.5600 \ . \tag{9}$$

In order to determine the exponent values which link measured quantities to $\varepsilon = |(p-p_c)/p_c|$, we set $p_c = 0.5596 \mp 0.0003$.

For light-scattering experiments, the different solutions were prepared in the following way: the first dilution in stabilized 1-4-dioxane (Merck) (in which the absence of peroxide was checked regularly) was prepared at a concentration C ($0.01 < C(g/cm^3) < 0.1$) depending on the p value (lower p higher C). Then, using a Millipore filter (Millex-SR 0.5 μm), this first solution was filtered into the scattering cell in which the desired amount of filtered solvent was added. The concentration was determined by weighing and converted to g/cm^3 through $C(g/cm^3) = \rho\ C(g/g)$ where ρ, the density of dioxane, is equal to 1.034 at 20°C.

For neutron-scattering experiments, a part of the sample elaborated at a relative distance to the gel point of 2×10^{-2} was dissolved in deuterated tetrahydrofuran (TDF). The other part was fractionated by size-exclusion chromatography; the fraction investigated contained less than 10% of the total distribution and corresponded to the highest molecular weight. The characteristics of the samples (measured by light scattering) were R_z = 1000 Å, $M_w = 10^7$ and R_z = 600 Å, $M_w = 5 \times 10^5$ for the fractionated and unfractionated sample, respectively. In the fractionated sample, the radius of gyration of the smallest cluster polymer present was evaluated to be 250 Å.

II.2 Scattering experiments

Light-[8,9] and neutron-scattering[10,11] techniques were described in detail elsewhere. Here we will recall the measurable quantities. The intensity I scattered by the polymer is strongly dependent on the transfer vector $\mathbf{q} = 4\pi/\lambda \sin \theta/2$, ($\lambda$ being the wavelength in the medium and θ the scattering angle), and on the polymer concentration

$$\kappa \frac{C}{I} = \frac{1}{M_w} \times (1 + 2BC) \times \frac{1}{f(\mathbf{q}R_z)} \ , \tag{10}$$

where κ is a constant which is calculable for light and neutron scattering, M_w is the weight-average molecular weight (Eq. 4), B is the interaction parameter, and R_z is the z-average radius of gyration:

$$R_Z^2 = \frac{\sum_i n_i M_i^2 R_i^2}{\sum_i n_i M_i^2} \ ,$$

n_i, M_i, R_i are the number, the mass and the radius of gyration of the ith polymer cluster respectively.

One has to note that z-average radius of gyration is not proportional to the correlation length of connectivity ξ because polymer clusters swell under dilution (one has $M_i \sim R_i^{\ D}$ instead of $M_i \sim R_i^{\ D_p}$ with $D < D_p$).

The function $f(\mathbf{q}R_z)$ is the form factor, an unknown function, which depending on the $\mathbf{q}R_z$ value is fitted to:

$$1/f(\mathbf{q}R_z) = 1 + \frac{\mathbf{q}^2R_z^2}{3}, \quad \text{if } \mathbf{q}R_z < 1; \tag{11a}$$

$$1/f(\mathbf{q}R_z) = 1 + \frac{\mathbf{q}^2R_z^2}{3} + \frac{\mathbf{q}^4}{12}\left(\frac{4}{3}R_z^4 - L_z^4\right), \quad \text{if } \mathbf{q}R_z < 3; \tag{11b}$$

where L_z is equal to:

$$L_z = \left[\frac{\sum_i n_i M_i^2 L_i^4}{\sum_i n_i M_i^2}\right]^{1/4}. \tag{12}$$

If one supposes that the mass distribution is simply $n_i \sim M_i^{-\tau}$ one finds:

$$\frac{L_z}{R_z} = \alpha\left[\frac{(3-\tau+2/D)^2}{(3-\tau)(3-\tau+4/D)}\right]^{1/4}, \tag{13}$$

where α is a parameter which, in principle, depends on the conformation of the polymer but for a linear polymer in a theta or in a good solvent $\alpha = 1$ within 2%.[13]

If $\mathbf{q}\,R_z \gg 1$, then the form factor $f(\mathbf{q}\,R_z)$ has a power-law behavior:

$$\left.\frac{I}{C}\right|_{c\to 0} \simeq \left.\frac{I}{C}\right|_{\substack{\mathbf{q}\to 0\\ c\to 0}} \mathbf{x}(\mathbf{q}R_z)^x.$$

As we observe the polymer clusters on a scale length $\mathbf{q}^{-1} \ll R_z$, we expect that measurements will reflect the self-similarity and will yield results independent of R_z.[11,12] As $(I/C)_{\substack{\mathbf{q}\to 0\\ c\to 0}} \sim M_w \sim R_z^{D(3-\tau)}$, it follows $X = -D(3-\tau)$ and

$$\left.\frac{I}{C}\right|_{c\to 0} \sim \mathbf{q}^{-D(3-\tau)}. \tag{14a}$$

The q-dependence of the intensity scattered per monomer reflects the mass self-similarity(τ) and spatial self-similarity(D) of the clusters. In the case of a monodisperse sample ($M_w = M$, $R_z = R$) the intensity scattered at $q R \gg 1$ leads to the fractal dimension:

$$\left.\frac{I}{C}\right|_{c\to 0} \sim q^{-D} . \qquad (14b)$$

A scattering experiment allows to explore a system on a scale length q^{-1} which for a light scattering experiments lies between 280 Å and 3000 Å, and for a neutron scattering experiments lies between 9 Å and 200 Å.

Light-scattering experiments allow us to measure, by an extrapolation to zero concentration, the mean weight-average and the z-average radius of gyration both of which link to connectivity properties. From the dependence on concentration of the intensity scattered per monomer $I/C|_{q\to 0}$ the interaction parameter B is measured; B is very sensitive to the swelling of polymer clusters.

Neutron scattering allows us to explore the polymer on a scale length smaller than the size of the largest polymer cluster and thus to determine the fractal dimension D when the sample is fractionated or the product $(3 - \tau)D$ when the polydisperse sample is studied.

Now we will recall the main results obtained by light-scattering experiments ($q R_z < 1$) and by neutron-scattering experiments ($q R_z > 1$), and then we will discuss them.

III. EXPERIMENTAL RESULTS

III.1 Light-scattering experiments[9] The intensity of the light scattered by a dilute polymer solution, extrapolated to zero transfer vector q, is strongly dependent on concentration (see Fig. 1) and presents a maximum at a concentration noted as $\tilde{C}$. The experimental results used for the extrapolation to zero concentration are obtained in a concentration range smaller than $\tilde{C}$.

In Fig. 2, we plot the variation of the weight-average molecular weight M_w as a function of the distance to the gel point. The full line in the figure corresponds to the law

$$M_w = 8.1 \times 10^2 \times \varepsilon^{-\gamma} \quad with\ \gamma = 1.71 \mp 0.04 , \qquad (15)$$

a $\mp$ 3.10^{-4} variation of P_c leads to $\gamma = 1.71 \mp 0.06$.

The interaction parameter B, is dependent on the distance to the gel point ε (see Fig. 3). We find

$$B(cm^3 g^{-1}) = 10 \times \varepsilon^{-0.59 \mp 0.06} , \qquad (16)$$

a p_c variation of $\mp$ 3.10^{-4} leads to an exponent value of 0.59 $\mp$ 0.02.

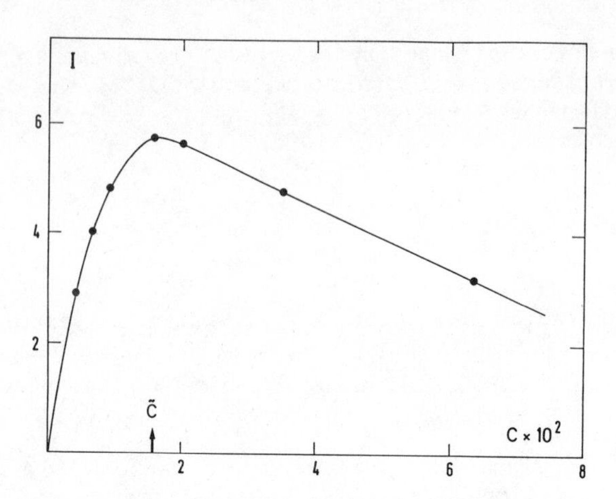

Fig. 1. Typical variation of the intensity scattered, extrapolated to zero **q** value, as a function of the concentration. The intensity presents a maximum for **C** = $\widetilde{\mathbf{C}}$.

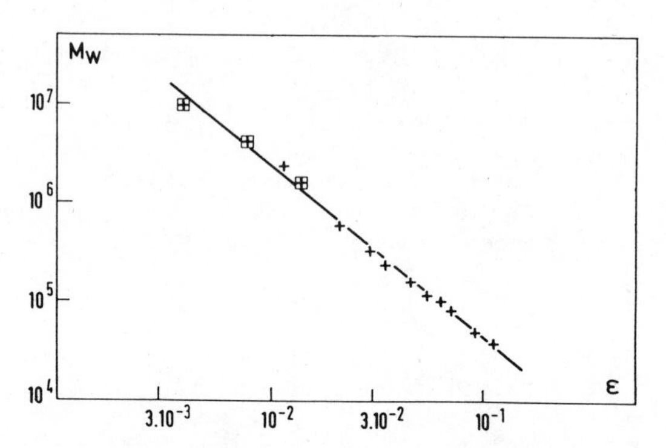

Fig. 2. Divergence of the mean weight-average molecular weight M_w as a function of the distance to the gel point: $\varepsilon = |(p-p_c)/p_c|$. The representation is log-log scale. The full line corresponds to $M_w = 8.10 \times \varepsilon^{-1.71}$. Different symbols correspond to two sets of preparation in which the same batch of dried monomers was used but the preparation was done in a 6-months time interval.

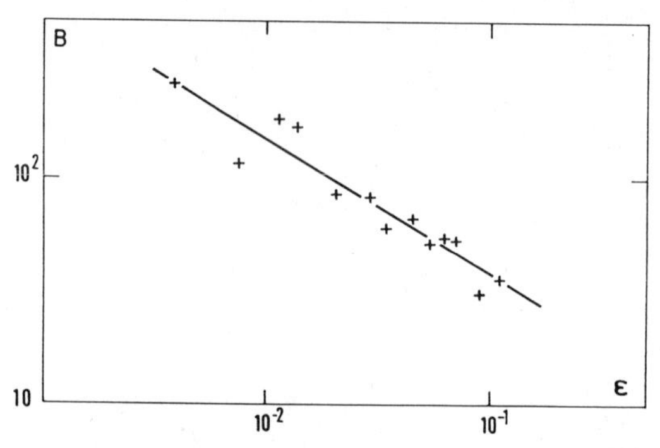

Fig. 3. Variation of the interaction parameter B (cm^3g^{-1}) as a function of the distance to the gel point ε in a log-log scale. The straight line corresponds to $B = 10 \times \varepsilon^{-0.59}$.

One has to note that intrinsic viscosity measurements, performed on the same samples prepared at different p values ($0.02 < \varepsilon < 0.1$) diluted in tetrahydrofuran (THF), lead to

$$\frac{[\eta]}{B} = 0.39 \mp 0.06 \ . \tag{17}$$

The z-average radius of gyration (see Fig. 4) extrapolated to zero concentration increases when the gel point is approached. We find:

$$R_z(\text{Å}) = 11 \times \varepsilon^{-\nu'} \quad with \ \nu' = 1.01 \mp 0.05 \ . \tag{18}$$

A p_c variation of $\mp\ 3 \times 10^{-4}$ leads to a 0.04 variation of the exponent value.

The linear behaviors, in the log-log scales (Figs. 2, 3 and 4) indicate that p_c evaluation (0.5596) is correct and its precision is adequate in the range of ε investigated (4.10^{-3}-10^{-1}).

The ratio of the z-average sizes L_z and R_z is a constant in the whole range of ε (4×10^{-3}-2×10^{-2}) where both quantities can be determined:

$$\frac{L_z}{R_z} = 1.09 \mp 0.01 \ . \tag{19}$$

In Fig. 5, we report on a log-log scale the variation of M_w as a function of R_z. The straight line corresponds to the law

$$M_w = 24 \times R_z^{1.61 \mp 0.07} \ , \tag{20}$$

this exponent corresponds to $D(3 - \tau)$.

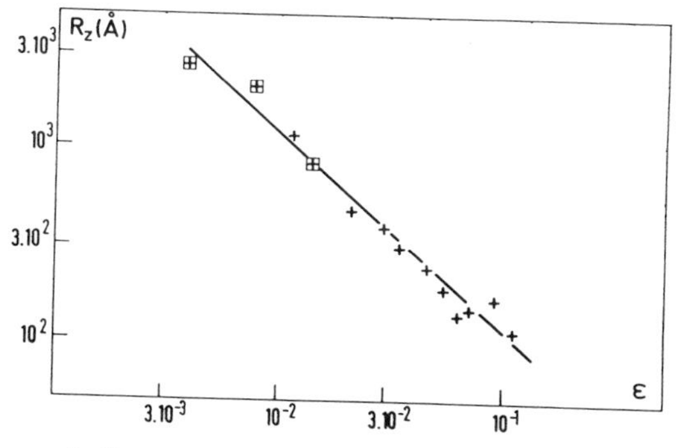

Fig. 4. Divergence of the z-average radius of gyration as a function of the distance to the gel point ε. The representation is log-log scale. The full line corresponds to $R_z(\text{Å}) = 11 \times \varepsilon^{-1.01}$.

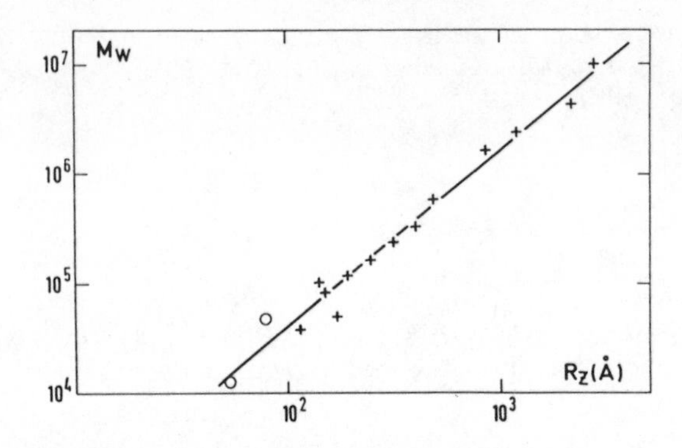

Fig. 5. Relation between z-average radius of gyration R_z and mean molecular weight M_w. The representation is a log-log scale and straight line corresponds to $M_w = 24 \times R_z^{1.61}$.

III.2 Neutron-scattering experiments[11]

Using the sample, $\varepsilon \approx 2 \times 10^{-2}$, diluted in deuterated tetrahydrofuran (TDF) at a concentration C such that $0.1 \leq BC < 0.3$, we find that for a **q**-range extending from 10^{-2} Å^{-1} to 10^{-1} Å^{-1} the scattered intensity may be described by a power law with an exponent slightly dependent on concentration: for $C = 2.8 \times 10^{-3}$ g/cm^3 and 8.3×10^{-4} g/cm^3 the exponent values are 1.51 ∓ 0.02 and 1.59 ∓ 0.07 respectively. A linear extrapolation to zero concentration leads to an exponent value of 1.62 ∓ 0.02 which, within experimental precision, corresponds to the exponent value obtained with the sample having the lowest concentration ($C = 8.3 \times 10^{-4}$g/cm^3 and $BC \approx 0.1$). The **q**-dependence of the intensity scattered by this sample is shown in Fig. 6 in a log-log scales ($5 \times 10^{-3} < \mathbf{q} < 8 \times 10^{-2}$). Following relation (14a), the slope of the straight line corresponds to the exponent

$$D(3 - \tau) = 1.59 \mp 0.05 \ . \tag{21}$$

This exponent value is identical to that measured by light-scattering experiment on different samples.

Using the same sample ($\varepsilon \approx 2 \times 10^{-2}$) fractionated and then diluted in TDF ($BC < 1$), we were able to measure the fractal dimension (see Rel. 14b) of the swollen polymer cluster because all the polymers are larger than the scale length $\mathbf{q}^{-1}$ at which the polymers are observed ($\mathbf{q}_{min} R_{min} \approx 2$ where $R_{min} \approx 250$ Å is the size of the smallest polymer cluster of the fractionated sample and $\mathbf{q}_{min} \approx 7 \times 10^{-3}$ Å^{-1} is the smallest **q** accessible with neutron scattering). The scattered intensity can be described by a power law in $\mathbf{q}(10^{-2} < \mathbf{q} < 10^{-1})$ with an exponent which is, within experimental accuracy, independent of the concentration. In Fig. 7 is plotted in log-log scale the **q**-independence of the intensity scattered by the sample having the lowest concentration ($C = 2 \times 10^{-3}$g/cm^3) the linear behaviour observed allows us to determine the fractal dimension of the swollen polymer clusters:

$$D = 1.98 \mp 0.03 \ . \tag{22}$$

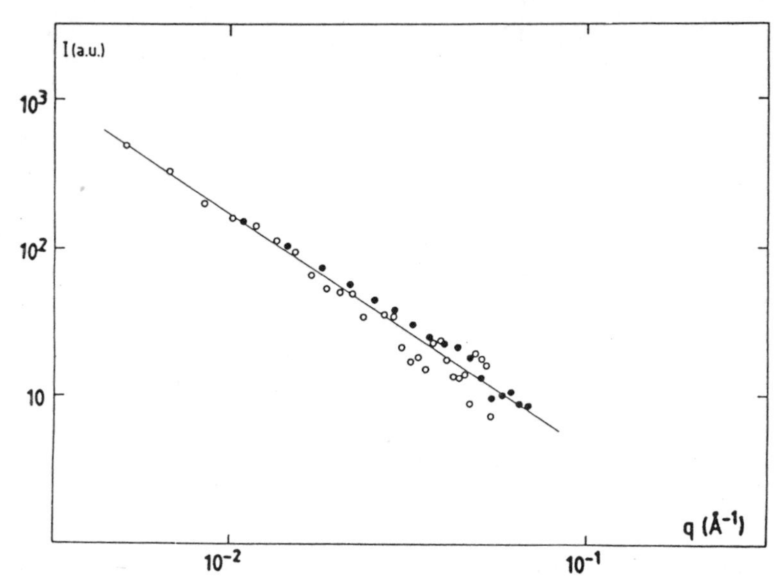

Fig. 6. Scattered intensity (arbitrary units) as a function of the scattering vector (log-log scale) measured using the polydisperse sample at a concentration of 8.3×10^{-4} g/cm^3. • and o are measurements performed with the incident wavelength of 7 Å and 15 Å, respectively. The straight line corresponds to $q^{-1.59}$ dependence.

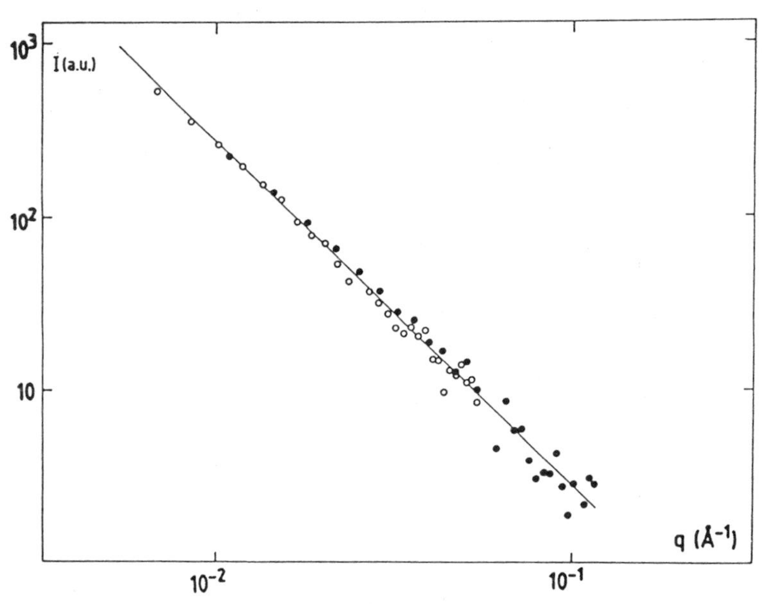

Fig. 7. Scattered intensity (arbitrary units) as a function of the scattering vector q(Å^{-1}) observed with the fractionated sample ($R_{min} \approx 250$ Å) diluted in deuterated tetrahydrofuran at a concentration of 2×10^{-3} g/cm^3. The straight line indicates a $q^{-1.98}$ dependence.

From $D = 1.98 \mp 0.03$ and $(3 - \tau)D = 1.59 \mp 0.05$ we extract

$$\tau = 2.20 \mp 0.04 \ . \tag{23}$$

IV. DISCUSSION

The exponents τ and γ which define the mass distribution and the growth process of weight-average molecular mass are found experimentally to be in good agreement with the percolation theory. The experimental determinations (see Eqs. 23 and 15) $\tau = 2.20 \mp 0.04$ and $\gamma = 1.71 \mp 0.06$ have to be compared with $\tau = 2.2$ and $\gamma = 1.74$ deduced from computer simulation following the percolation model.

The interaction parameter B, proportional to the intrinsic viscosity, is a function of the distance to the gelation threshold. This is an evidence of the swelling of the clusters. If in the expression of B(Eqs. 8 and 16) we insert the D and σ (= $(3-\tau)/\gamma$ = 0.47 ∓ 0.04) values determined experimentally, we find $D_p = 2.42 \mp 0.15$ in perfect agreement with the percolation theory ($D_p = 2.5$). This fractal dimension, which corresponds to the unswollen state is larger than that measured in the swollen state ($D = 1.98 \mp 0.03$). The ε independence of the ratio $[\eta]/B$, means that the fractal dimension D of clusters of polyurethane is identical in the two solvents used (THF and dioxane). Due to this swelling, exponent ν' linking size to the distance to the gel point cannot be directly compared with the percolation exponent value $\nu = 0.88$.

The exponent $(3 - \tau)D$ was measured using two different approaches: on different samples by determining M_w and R_z values and on one sample by determining the intensity scattered on different scale length q^{-1}. The identity of exponent values found:

$$D(3-\tau) = 1.59 \mp 0.05 \quad \textit{and} \quad D(3-\tau) = 1.61 \mp 0.07 \ .$$

with neutron- and light-scattering experiments, respectively indicate that the fractal dimensions D and D_p (via τ) do not depend on ε. One can conclude that the geometry and the growth of clusters are independent of the stoichiometric ratio p for $|p-p_c|/p_c < 0.1$.

By inserting into Eq. 13 the τ and D measured values, it is found that: $L_z/R_z = 1.1 \times \alpha$. From the measured L_z/R_z value (Eq. 19), we deduce that $\alpha \approx 1$ as found for linear polymers.

This experimental result, the ratio of z-average sizes, (L_z/R_z), is a constant independent of ε, seems to indicate that the mass distribution function, once normalized by the number of clusters having the mass M^*

$$\left(\frac{n(M)}{n(M^*)} \simeq \left(\frac{M}{M^*} \right)^{-\tau} f\left(\frac{M}{M^*} \right) \right)$$

is independent of the stoichiometric ratio p. Or in other words, the clusters obtained by polycondensation for different stoichiometric ratio p of diisocyanate and triol are mass self-similar. Experiments performed below the gel point ($p < p_c$) allow us to measure two exponents τ (2.20 $\mp$ 0.04) and γ (1.71 $\mp$ 0.06) which are in agreement with values calculated using the percolation model. Using the relations between exponents, established in the framework of critical phenomenon theory, we can deduce all of them, in particular the exponent β which links the gel fraction G to the relative distance to the gel point. For $p > p_c$ one has $G \sim \varepsilon^{\beta}$ with $\beta = \gamma(\tau-2)/(3-\tau)$ knowing $\gamma = 1.71 \mp 0.06$ and $\tau = 2.20 \mp 0.04$, β is found to be 0.43 ∓ 0.12.

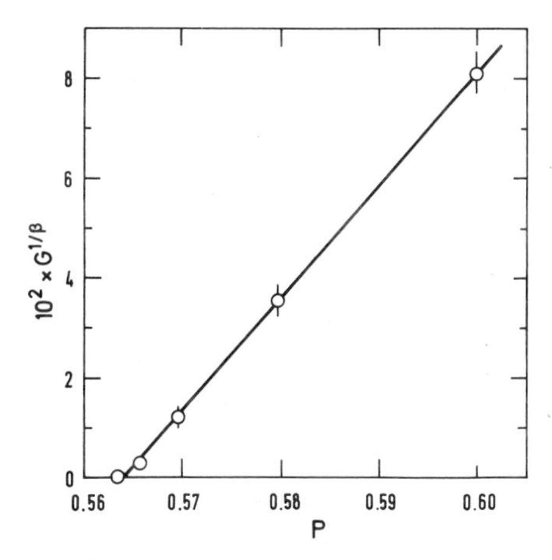

Fig. 8. Variation as a function of the stoichiometric ratio p of the gel phase to the power β^{-1}. The exponent value $\beta = 0.43$ is deduced from τ and γ exponent values measured below the gel point.

Gel phase fraction was measured on samples (see Ref. 14 for more details) formed beyond the gel point. One can see in Fig. 8 that $G^{1/\beta}$ is a linear function of the stoichiometric ratio. Concerning this figure two comments have to be made. The p_c value corresponding to the stoichiometric ratio at which the gel phase is nul (p_c = 0.5639) is very different from the ifferent batches of monomers were used for the preparation of the two series of samples. A small but finite gel phase was measured below p_c(G = 1.6 10^{-2} at p = 0.5632); this could be due to either experimental imprecision on G or the fact that this sample prepared near the gel point contains very large polymer clusters of finite size which could not pass through the membrane used for the sol extraction. This result, $G^{1/\beta}$ is a linear function of p with a β-exponent value deduced from τ and γ exponent values measured below the gel point, indicates that below and beyond the gelation threshold, the percolation theory is well adapted to describe the properties linked to connectivity of polymer clusters formed by polycondensation.

Once established the fact that the gelation threshold is described by the percolation model, we are able to picture the growth process of polymer clusters. The τ value (2.20 $\mp$ 0.04) corresponds to a fractal dimension of D_p = 2.50 ∓ 0.09 in the reaction bath. This corresponds to an internal density of one cluster on a scale $R < \xi$ of $R^{D_p - d}$; as $D_p < d$, this density is much smaller than the mean density (=1) measured on the scale of the cell. This indicates that each cluster contains clusters smaller than itself. The growth of the largest cluster is done by linkage of two large clusters, easily realized because being neighbours, these clusters do not have to diffuse towards each other. Thus, in gelation as well as in percolation, finite size clusters do not have to move, while in both cases monomers move freely.

CONCLUSIONS

Measurements performed with different techniques prove that clusters swell under dilution.

The experimental results presented here, below and beyond the gel point, show clearly that the formation of polyurethane gel by polycondensation is a critical phenomenon which cannot be described by the mean-field theory. The good agreement of exponent values found experimentally and calculated by Monte Carlo simulations shows that this type of polycondensation process can be described by the percolation model. Recent experimental results[15-20] performed on different kinds of gelation process seem to indicate that, more generally, percolation and gelation, with permanent crosslinks, belong to the same class of universality.

ACKNOWLEDGEMENTS

We would like to thank J.P. Cotton and J. Texeira for the assistance in the realization of neutron-scattering experiments; B. Bollendorff for technical assistance in the construction of the light spectrometer; S. Luzzati for the development of the spectrometer prototype and J.P. Busnel for the development of size-exclusion chromatography technique and for his cooperation in the gel-phase measurements. This work has greatly benefited by encouragements from P.G. de Gennes and continuing discussions with M. Daoud.

NOMENCLATURE

$A \simeq B$	corresponds to A = constant x B, the constant is dimensionless
$A \sim B$	A is proportional to B
$A \approx B$	means that A is of the order of B
B	interaction parameter measured by light scattering
C	monomer concentration
D	fractal dimension of the clusters at infinite dilution
D_p	fractal dimension of the clusters in the bulk state
d	space dimensionality
G	Gel phase fraction above the gel point
I	Scattered intensity by the polymer clusters in dilute solution
L_i^4	fourth moment of the average monomer-monomer separation of the ith polymer cluster
L_z	z-average of length L_i (see Eq. 2)
M_i	mass of the ith polymer cluster
M^*	mass of the largest polymer cluster
M_w	weight-average mass or weight average molecular weight in the experimental section
n_i	number of polymer clusters having a mass M_i
p	degree of advancement of the chemical reaction
p_c	degree of advancement of the chemical reaction at the gelation threshold

q	transfer vector in scattering experiment
R_i	radius of gyration of the ith polymer cluster in dilute solution
R_z	z-average radius of gyration of polymer clusters in dilute solution
V_{ij}	co-volume between the ith and j^{th} polymer
β	exponent which links G and ε
γ	exponent which links M_w and ε
ε	relative distance to the gel point $\varepsilon = \| p-p_c \|/p_c$
$[\eta]$	intrinsic viscosity
θ	scattering angle
λ	wavelength of the radiation in the medium
ν	exponent which links ξ and ε in the bulk
ν'	exponent which links R_z and ε at infinite dilution
ξ	correlation length which is proportional to the size of the largest polymer cluster in the bulk
ρ	density of solvent
σ	exponent which links M^* and ε (Eq. 2)
τ	mass distribution exponent (Eq. 1)

REFERENCES

1. D. Stauffer, A. Coniglio and M. Adam, Adv. in Polym. Sci., 44, 103 (1982).
2. See for instance; D. Stauffer, Introduction to Percolation Theory, Taylor and Francis, London (1985); D. Stauffer, Physics Reports, 54, 1 (1979).
3. H. Yamakawa, Modern Theory of Polymer Solutions, Harper and Row, New York 1971 Chap. V.
4. T.A. Witten and J.J. Prentis, J. Chem. Phys., 77, 4247 (1982).
5. M. Daoud and L. Leibler, Macromolecules, 21, 1497 (1988).
6. M. Daoud, F. Family and G. Jannink, J. Phys. Lett., 45, 199 (1984).
7. M. Adam, M. Delsanti and D. Durand, Macromolecules, 18, 2285 (1985).
8. S. Luzzati, M. Adam and M. Delsanti, Polymer, 27, 834 (1986).
9. M. Adam, M. Delsanti, J.P. Munch and D. Durand, J. Phys. (Paris), 48, 1809 (1987).
10. LLB Pace., internal report available on request from LLB 91191 Gif-sur-Yvette, Cedex, France.
11. E. Bouchaud, M. Delsanti, M. Adam, M. Daoud and D. Durand, J. Phys. (Paris), 47, 1273 (1986).
12. J.E. Martin and B.J. Ackerson, Phys. Rev., A31, 1180 (1985).
13. T.A. Witten and L. Schafer, J. Chem. Phys., 74, 2582 (1981).
14. D. Durand, F. Naveau, J.P. Busnel, M. Delsanti and M. Adam, Macromolecules, to appear in 1989.
15. S.J. Candau, M. Ankrim, J.P. Munch, P. Rempp, G. Hild and R. Okasha, Physical Optics of Dynamic Phenomena and Processes in Macromolecular Systems, Walter de Gruyter, Berlin NY (1985).
16. M. Schmidt and W. Burchard, Macromolecules, 14, 370 (1981).
17. F. Schosseler and L. Leibler, Physics of Finely Divided Matter, Springer Proceeding in Physics, 5, (1985).
18. C. Collette, F. Lafuma, R. Audebert and L. Leibler, Biological and Synthetic Polymer Networks, Ed. O. Kramer, Elsevier London 277 (1988).
19. J.E. Martin and K.D. Keefer, Phys. Rev., A34, 4988 (1986).
20. J.E. Martin, J. Wilcoxon and D.A. Dolf, Phys. Rev., A36, 1803 (1987).

THE BIASED REPTATION MODEL OF DNA GEL ELECTROPHORESIS

Gary W. Slater and **Jaan Noolandi***

Xerox Research Centre of Canada
2660 Speakman Drive
Mississauga, Ontario, Canada L5K 2L1

ABSTRACT

The electrophoresis of DNA in agarose gels is one of the most powerful tools of molecular genetics research. With the present techniques using pulsed fields, electrophoresis can now be used to separate chromosome-size molecules. The biased reptation model, which adds a field-induced bias to the original reptation model developed by S. F. Edwards and P.-G. de Gennes to deal with polymer melts, provides a good framework to understand the physical mechanisms involved in these experimental techniques. We show that forced reptation is the leading mechanism for continuous fields or low frequency pulsed field electrophoresis with parallel or crossed fields.

I. INTRODUCTION

The concept of polymer reptation as developed by de Gennes[1] and Doi and Edwards[2-4] has provided a good starting point to understand many of the transport and mechanical properties of entangled polymers both in the melt and in solution. The reptating polymer chain is assumed to evolve in a tube formed by the surrounding chains; if the entanglements are long-lived, the tube can be assumed to represent a realistic average environment seen by the chain. However, it has been shown that other mechanisms such as the tube-renewal process[5] can effectively compete with reptation for finite length chains in a melt. This means that reptation is not the only mode of migration allowed to the chains in this situation and the diffusive motion of polymer chains in a melt or in a concentrated solution is more complicated than originally proposed.

It can be argued that the motion of a polymer chain in a gel is the ideal problem for the original reptation theory as the entanglements, which are provided by the gel, are then fixed in space.[6] Although of

*To whom all correspondence should be addressed.

little interest to most engineers and physicists, this seemingly idealistic system has assumed great importance for biologists in the last few years. This is due to the development[7-12] of electrophoresis systems capable of separating mega-base nucleic acids (DNA). With the recent proposal to map or even sequence the entire human genome, these techniques are now at the forefront of the new instruments of biotechnology.

DNA gel electrophoresis is one of the most widely used tools of molecular genetics research. The biased reptation model has proven to be useful to understand most continuous field and low-frequency pulsed field effects. In this paper, we review how we have generalized the reptation model to include the electric forces acting on the DNA polymer, and we give an analysis of the main results obtained from analytical and numerical calculations.

II. DNA GEL ELECTROPHORESIS

In gel electrophoresis, charged polymeric molecules, usually DNAs or proteins, are forced to migrate through the pores of a dense gel under the action of an external electric field. In the case of DNA electrophoresis, the gel is necessary as electrophoresis in free solution does not separate DNA fragments of different sizes[13], i.e, the mobility μ_0 of DNA in free-solution is molecular-size independent and given by

$$\mu_0 = \frac{Q}{\xi} \approx 2 \times 10^{-4} \ \mathrm{cm}^2/\mathrm{Vs}, \tag{1}$$

where Q and ξ are the total electric charge and friction coefficient of the DNA molecule. Since both Q and ξ are proportional to the molecular size K (usually given in kilo-base pairs, or kbp), the ratio Q/ξ is a constant; the value quoted here is typical of what is found in the literature.[13-16]

In an agarose gel (the usual choice), the obstacles provided by the agarose fibers can be expected to lead to a molecular size-dependent electrophoretic mobility $\mu(K)$. This method has been used for more than two decades to separate nucleic acids of various sizes. However, it soon became evident that this approach was limited: for instance, it was observed experimentally that large nucleic acids (>50 kbp) possess a molecular size-independent mobility even in a dense gel; moreover, the mobility was observed to be strongly field-dependent.

Agarose gels have been studied by many authors using a variety of techniques[15-18]. Of particular interest is the "average pore size" a, which gives an idea of the porosity of the gel and the retarding effect of the gel on the migrating DNA molecules. With DNA as a probe, we found[16]:

$$a \approx 89 / A^{2/3} \ \ \mathrm{nm}\ , \tag{2}$$

where A is the agarose concentration (g/ml) expressed in %. Concentrations of 0.5-1.5% are typically used in electrophoresis experiments: this corresponds to average pore sizes of 70 to 140 nm.

Nucleic acids are "worm-like" molecules with a persistence length p of order 50 nm.[19,20] In a solution or a dilute gel, a DNA molecule has a random-walk conformation with a radius of gyration given by[21]:

$$R_g^2 = \frac{1}{3} p L \left(1 - \frac{p}{L} + \frac{p}{L} \exp(-L/p)\right), \tag{3}$$

where L is the contour length of the molecule (with 340 nm per kbp).

Figure 1 shows a schematic $L(K)$ vs $a(A)$ "phase diagram". For $L \approx p \ll a$, the molecules are rigid rods too small to be influenced by the presence of the gel; therefore their mobility is given by the size-independent value μ_0. This is the "free-solution" limit. To separate such small fragments (smaller than 1 kbp), one has to use gels with much smaller pore sizes, e.g. polyacrylamide gels[22] for which $a \sim 35$ nm or less.

For $L \gg p$ and $R_g \leq a$, the DNA molecule forms a globule with radius smaller than the average pore size of the gel. Rodbard and Chrambach[23] have proposed a theory of this "Ogston" regime where the gel is assumed to act as a molecular sieve; the mobility of a DNA molecule is then given by the total pore volume available to a globular object of radius R_g. Assuming that the gel is a random array of linear fibers[24], these authors suggested that $\mu/\mu_0 \sim \exp[-(R_g/a)^2]$. Of course, it is then expected that the mobility of the DNA molecules, or of charged spherical particles in general, will decrease very quickly when the radius of gyration R_g becomes larger than the average pore size a because of a paucity of pores large enough to admit the molecule. This has been observed for rigid spherical latex particles or viruses.[25] For DNA, with $a \approx 100$ nm and $p \approx 50$ nm, one expects the mobility of sieved globular fragments to be negligible for $R_g > 100$ nm, which corresponds approximately to $K > 2$ kbp. However, DNA molecules of 700 kbp or more have been observed to migrate with very large mobilities in agarose gels.[26]

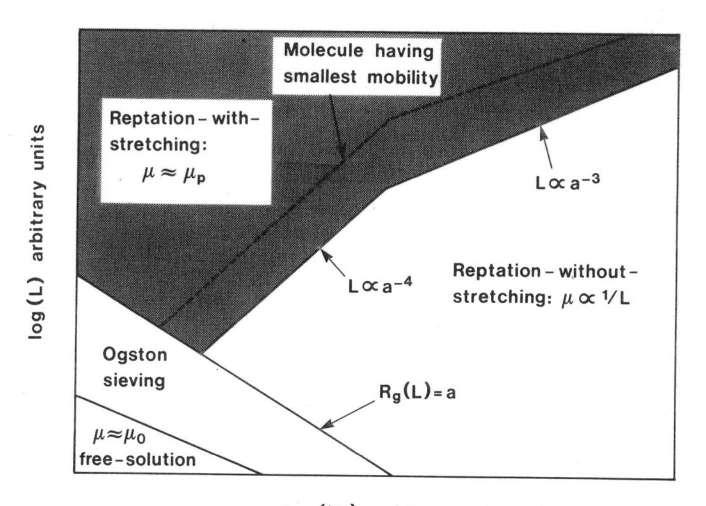

Fig.1 Schematic "phase diagram" for the electrophoretic migration of DNA in a gel; the logarithm of the molecular size L is plotted as a function of the logarithm of the inverse pore size $1/a$. The four areas of the diagram are described in the text. The dashed line shows the molecular size of the fragment having the minimum mobility.

Experiments done with large molecules (for which $R_g > a$) and low electric fields have shown that, for a certain range of molecular sizes (typically 5-50 kbp), the mobility of the fragments that should not migrate, according to the gel sieving model, follows the relation[16,27,28]:

$$\frac{\mu}{\mu_0} \propto \frac{1}{L}. \tag{4}$$

This clear departure from the sieving relation has been interpreted as a signature of reptative motion by Lerman and Frisch.[29] This relation also implies that very large molecules should have negligible mobilities, although the decrease is predicted to be slower than in the Rodbard and Chrambach theory. In fact, molecules with $R_g \gg a$ have a finite mobility according to this equation, which means that the molecules have to deform enormously in order to move through the pores of the gel.

However, for higher fields and larger molecular sizes, it is observed that the mobility ceases to decrease with the molecular size as $\mu \propto 1/L$: instead, μ then becomes molecular size-independent but strongly field-dependent.[15,16,30] This means that although $R_g \gg a$, the mobility is not influenced anymore by the relative sizes of the molecule and of the gel pores; we call this asymptotic mobility μ_p the "plateau mobility". This surprising result is unfortunate for biologists as it means that electrophoretic separation is limited to molecular sizes below approximately 50 kbp. As we will show later, this effect can be explained if the molecular stretching induced by the reptative motion of the DNA molecules is taken into account; this part of the $L(K)$ vs $a(A)$ diagram is thus called the "reptation-with-stretching regime", while the part of the phase diagram where Eq.(4) holds is called the "reptation-without-stretching regime".[16]

Finally, the biased reptation model (BRM) predicts that under some conditions of field strengths and gel concentrations, the electrophoretic mobility becomes a non-monotonic function of the

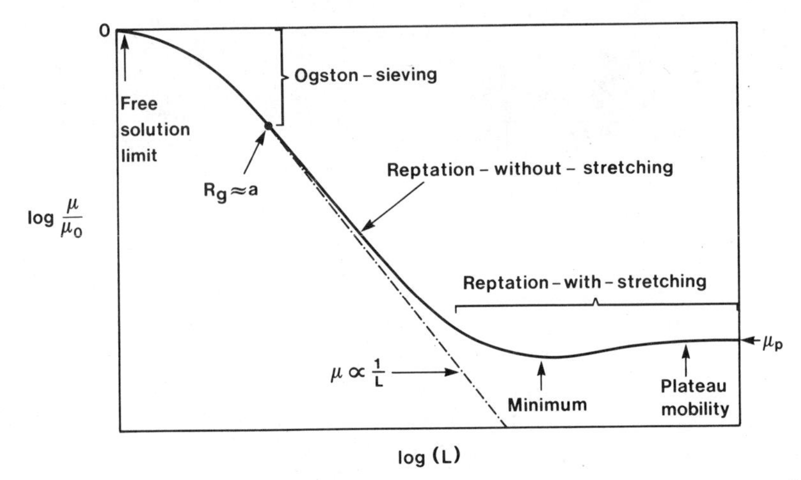

Fig.2 Schematic log-log plot of reduced mobility μ/μ_0 vs molecular size L. The various regimes refer to Fig.1.

molecular size.[31] This means that the molecule showing the minimum mobility is not necessarily the largest molecule in the system. Careful experiments have shown that this is indeed the case, and that the conditions under which this effect occurs may have been used in the past.[31,32] It is not clear at the present time if experimental results have been misinterpreted due to this effect. As we will see later, this band-inversion effect is due to dynamic self-trapping of the migrating molecules, and it can be avoided by choosing experimental conditions which minimize the self-trapping. The phase diagram on Fig. 1 also shows the band-inversion line, i.e. the molecular size showing the minimum mobility.

Figure 2 shows a schematic mobility vs molecular size diagram, with the four regimes of the phase diagram, and including the band-inversion effect. This diagram shows clearly the limitation of the technique for the separation of the increasingly large DNA molecules that biologists now study. It is a goal of this article to show how the BRM has helped us to explain the schematic results presented in Figs. 1 and 2.

In the last four years, a number of pulsed field techniques have been designed to increase the range of molecular sizes where the mobility is molecular-size dependent, i.e. to push the plateau mobility region towards larger molecular sizes. It is now possible[33] to separate fragments of up to 9000 kbp, an improvement by a factor of ≈ 200 as only molecules of <50 kbp can be separated with continuous field gel electrophoresis. Intact human chromosomes are in the size range 50-200 Mbp; we still need an improvement by an order of magnitude in order to be able to study the intact human genome with gel electrophoresis.

The BRM has been used to study the physics and the limitations of the various pulsed field techniques. In the last sections of this article, we discuss intermittent field, crossed fields and field inversion techniques in the framework of the BRM, comparing analytical as well as numerical work with some experimental results.

III. THE BIASED REPTATION MODEL

III.1 <u>The Reptation Approach</u> The large ($R_g \gg a$) nucleic acids which are of interest in this article have to deform considerably in order to enter the gel, let alone to migrate under the action of an electric field. Two possibilities can be imagined. The molecules can collapse or compress to a point where they can fit into the largest pores of the gel. However the cost in entropy and in repulsive energy (DNA is negatively charged with 2 charges per bp) is too high for this to happen regularly, especially for the Mbp fragments.

A simple calculation shows that many fibers must interfere with the molecular conformation of a large DNA molecule if the free-solution radius of gyration is retained when the molecule is in the gel; in other words, the chain gets entangled with the agarose fibers. For example, a 50 kbp DNA molecule has a radius of gyration of approximately $R_g \approx 500$ nm; with $a \approx 100$ nm, at least 5 fibers enter the domain of the DNA molecule. Figure 3a shows a schematic view of an entangled DNA chain in an agarose gel.

Figure 3a represents a typical DNA conformation in an agarose gel in the absence of an electric field. The question is therefore how the entangled molecule will move in response to the electric field; in the

BRM, it is assumed that the molecule will reptate in a biased direction. As in the original reptation model,[2] the gel obstacles (the surrounding agarose fibers) are replaced by a rigid tube of length $L_t = Na \leq L$, where N is the average number of pores (of average size a) occupied by the entangled DNA molecule (Fig. 3b). The molecule inside the tube is itself replaced by a primitive chain made of N freely-jointed and rigid

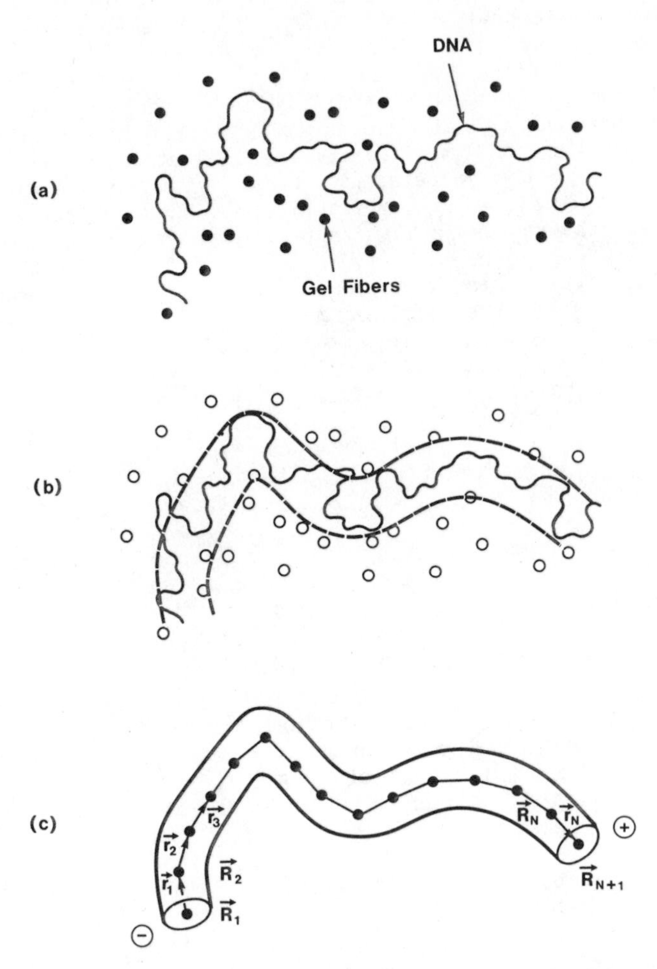

Fig.3 (a) Schematic picture of a large DNA molecule in an agarose gel. The circles represent agarose fiber cross-sections. The molecular conformation is crossed by many agarose fibers. (b) In the reptation model, the gel obstacles limiting the motion of the polymer chain are replaced by a tube whose contour follows the path described by the chain through the obstacles. The motion of the chain between the obstacles is assumed to be similar to the motion of the chain inside this fictitious tube. (c) The transverse short wavelentgh motions of the chain in the tube are neglected in the model, and the chain is replaced by a primitive chain that moves along the tube axis. The tube is defined by the $N+1$ points $\mathbf{R}_i$, and the primitive chain by the N vector-segments $\mathbf{r}_i$ joining these points.

primitive segments of length a, sitting on the tube axis (Fig. 3c) and replacing each DNA chain sections of average contour length $\ell(a)=L/N \geq a$. We thus neglect here the effects of the tube length fluctuations on the time scale of interest; similarly, it is assumed that the conformational changes within a pore are negligible in the problem. This means that the tube length L_t is unaffected by the field in this simplified picture.

III.2 Zero-Field Reptation In free solution (no gel), DNA diffuses in response to the thermal Brownian motion. If we define by ξ_0 the friction coefficient per base pair (bp) of DNA due to the buffer, stressed molecular conformations relax to random-walk like conformations on a time scale given by the Rouse time[6,34]:

$$\tau_R = \frac{\xi_0 L^2}{3\pi^2 k_B T}, \tag{5}$$

where L is the contour length of the molecule.

With a gel, the reptation model assumes that thermal diffusion occurs only along the tube axis as transverse motion is quickly stopped by the fibers defining the limits of the tube; the ends of the chain can move freely, and they define new tube sections as the thermal motion along the tube axis forces them to enter new pores.

Molecular relaxation thus takes two steps in a gel. First, the chain relaxes in its tube, i.e., it retracts or expands in the tube in order for the length of the tube L_t and the amount of DNA in each pore $\ell(a)$ to assume steady-state values; this relaxation takes a maximum time τ_R. Second, Brownian motion makes it disengage from its original tube to enter a totally new and relaxed one; this process is called reptation.[1,2,6,34]

The curvilinear diffusion constant of the chain in its tube is given by $D=k_BT/\xi$, where $\xi=(L/b)\xi_0$ is the friction coefficient of the whole chain and $b=3.4$ Å is the contour length of one bp. The time it takes to disengage completely from a tube of length Na by this diffusion process is[2]:

$$\tau_D = \frac{(Na)^2}{\pi^2 D} = 3N\tau_R \propto L^3, \tag{6}$$

where we have used the fact that the average square distance between the ends of the molecule is, for a random-walk conformation, $L_t a = Lb$. The reptation time $\tau_D\ (\propto L^3)$ is more dependent on the molecular size than the Rouse time $\tau_R\ (\propto L^2)$: this means that for a reptating molecule $(L_t/a>1)$, intramolecular relaxation is faster than tube-renewal.

It is also useful at this stage to define the Brownian time[2]:

$$\tau_{Brown} = \frac{a^2}{2D} = \frac{\xi a^2}{2k_B T} \tag{7}$$

which is the time necessary for the chain to diffuse (or reptate) over an average pore length a in the tube (the factor 2, which would be a 6 in a free solution, indicates that the diffusion is restricted to a one-dimensional space, i.e., the tube). This is the time used by Doi and Edwards[2] to replace the continuous diffusion of the molecule in its tube by a one-dimensional random-walk where the chain jumps over a distance $\pm a$ (the $\pm$ means the jumps can occur randomly forward or backward in the tube) along the tube axis every unit of time τ_{Brown}. During a jump of length $\pm a$, the end of the chain leaves the original tube and enters a new pore, thus creating a new tube section of length a. This end-segment chooses the new pore at random, which preserves the three-dimensional random-walk conformation of the tube when it evolves.

It is useful for this discussion to denote by $\mathbf{R}_i$ (with $1 \leq i \leq N+1$) the three dimensional positions of the $N+1$ entanglement points defining the reptation tube (Fig. 3c): these represent the ends of the N vector segments $\mathbf{r}_i \equiv \mathbf{R}_{i+1} - \mathbf{R}_i$ (with $1 \leq i \leq N$) of the primitive chain. In term of the $\mathbf{R}_i$'s, the stochastic (Langevin) equations defining the reptation process are[2]:

$$\mathbf{R}_i(t+\tau) = \tfrac{1}{2}[1+\eta(t)]\,\mathbf{R}_{i+1}(t) + \tfrac{1}{2}[1-\eta(t)]\,\mathbf{R}_{i-1}(t) \quad , \tag{8}$$

where the jump time is $\tau = \tau_{Brown}$ in absence of a field. In this equation, $\eta(t)$ is a stochastic variable taking the value $+1$ if the next tube section is created by the + end of the tube, and -1 if it is created by the - end of the tube; the + and - ends of the tube are defined arbitrarily (Fig. 3c). The probabilities $p_{\pm}$ of having $\eta = \pm 1$ are equal ($p_+ = p_- = 1/2$) in absence of a field because the reptation motion is not biased towards any particular end ($\pm$) of the tube in this case. The boundary conditions are:

$$\mathbf{R}_0(t) = \mathbf{R}_1(t) + \mathbf{c}(t) \; ; \tag{9a}$$

$$\mathbf{R}_{N+2}(t) = \mathbf{R}_{N+1}(t) + \mathbf{c}(t) \quad , \tag{9b}$$

where $\mathbf{c}(t)$ is a random vector (i.e. $\langle \mathbf{c}(t) \rangle = 0$) of length $|\mathbf{c}(t)| = a$ which creates a new randomly oriented tube section (or primitive segment).

The tube constraints of the reptation model are implicit in these equations of motion since the positions $\mathbf{R}_i$ have a single index i: the segments follow each other perfectly along the tube axis, and only the two end-segments can make the tube evolve.

These equations were used by Doi and Edwards[2-4] to derive the predictions of the reptation model concerning the rheological properties of polymer melts. In the next section, we show how we have modified these equations and concepts to include the effects on an electric field.

III.3 The Biased Reptation Model

III.3.1 Basic Equations Simply stated, the electric forces can modify one or both degrees of freedom left to a reptating charged polymer chain. First, it can alter the randomness of the orientations $\mathbf{c}(t)$ of the new tube sections if the end-segments $\mathbf{r}_1$ and $\mathbf{r}_N$ of the primitive chain are charged. Second, it can also alter the motion inside the tube such that the motion may become biased towards one end ($\pm$) of the tube. In other words, the field can lead to a reptative motion where most of the new tube sections are created by the same end-segment, and where these new tube sections are no longer randomly oriented in space.

The net charge per base pair (bp) is given by $q_0=2\sigma e$, where σ is the screening factor due to the counter-ions and $e=1.6\times10^{-19}$ Coulomb. With an electric field $\mathbf{E}=E\hat{\mathbf{x}}$ pointing in the x-direction, the electric force per bp is $\mathbf{f}_0=q_0\mathbf{E}=2\sigma eE\hat{\mathbf{x}}$. In presence of a gel, all electric forces transverse to the tube axis do not contribute directly to the electrophoretic migration as they only tend to bring the chain closer to the fibers defining the limits of the tube. Only those forces that are parallel to the tube axis will contribute to the net motion of the chain, which is itself along the tube axis. Therefore, a gel reduces the mobility of DNA by diminishing the effective force acting on it and forcing it to go through a series of pores which are not necessarily aligned along any given axis.

The relevant total longitudinal electric force on the molecule is thus[35]

$$F_\ell = \sum_{i=1}^{N} q_i\,\mathbf{E}\cdot\hat{\mathbf{u}}_i = \left(\frac{2\sigma eE}{b}\right)\sum_{i=1}^{N} \ell_i\,u_{ix}\,, \tag{10}$$

where q_i is the charge of DNA in the i-th pore of the tube (or, equivalently, of the i-th primitive segment, of length a_i, which goes from $\mathbf{R}_i(t)$ to $\mathbf{R}_{i+1}(t)$), and $\hat{\mathbf{u}}_i=[\mathbf{R}_{i+1}(t)-\mathbf{R}_i(t)]/a_i=\mathbf{r}_i(t)/a_i$ is the unit vector giving the orientation of the tube axis in the i-th pore. In the second part of the equation, we used the variables $\ell_i(a_i)=(q_i/2\sigma e)b$, the contour length of DNA in the i-th pore, and $u_{ix}=\hat{\mathbf{u}}_i\cdot\hat{\mathbf{x}}$, the projection of the unit vector $\hat{\mathbf{u}}_i$ on the field axis.

This equation cannot be reduced further unless we assume some specific gel architecture and some rule on how DNA distributes in the pores. For instance, with a distribution of pore sizes, one has $u_{ix}\propto\ell_i$ since the field-induced orientation u_{ix} of a tube section of length a_i is expected to be proportional to q_iE, the electric force on the end-segment that created this tube section. Therefore, Eq. (10) cannot be treated easily in general.

For simplicity, we use a single average pore size $a=\langle a_i\rangle=L_t/N$ for all the pores. In this case, each pore contains the same contour length of DNA $\ell=\langle\ell_i\rangle=L/N$ and hence the same DNA charge $q=\langle q_i\rangle$; Eq.(6) then reduces to:

$$F_\ell = qE\,h_x/a\,, \tag{11}$$

where $h_x=a\sum u_{ix}$ is the end-to-end distance of the molecule in the field direction. For a given chain, this force is time-dependent as the end-to-end distance h_x fluctuates during the electrophoretic migration.

With the total friction coefficient $\xi=(L/b)\,\xi_0$, the instantaneous velocity of the chain along the tube axis, or longitudinal velocity, is:

$$v_\ell = (F_\ell/\xi) = (qE/\xi)\cdot(h_x/a). \tag{12}$$

The mobility of the center-of-mass, or electrophoretic mobility, is then

$$\mu = \frac{\langle v_\ell \cdot (h_x/Na)\rangle}{E} = \frac{q_0}{\xi_0}\cdot\frac{\langle h_x^2\rangle}{(Na)^2} = \mu_0 \cdot \frac{\langle h_x^2\rangle}{L_t^2} \tag{13}$$

where (h_x/Na) is a geometrical factor due to the fact that the vector $\mathbf{h}=a\sum\hat{\mathbf{u}}_i$ is not in general parallel to the field axis. The brackets $\langle\ldots\rangle$ indicate an ensemble average over a large number of identical chains at a given time.[36]

Equation (13) is the central relation of the reptation theory of gel electrophoresis. It says that DNA mobility is proportional to the average end-to-end distance (square) of the chain in the field direction. Changes in DNA mobility are thus predicted to occur, in the reptation model, through changes in the extension of the molecule in the field direction brought about by the experimental conditions used.

It is easy to explain qualitatively both the $\mu(L)\propto 1/L$ relation that holds for small DNA molecules, and the size-independent mobility of large DNA molecules, with Eq.(13). If the free-solution globular conformation is retained during reptation[37], even in presence of an electric field (which is to be expected if the total longitudinal force F_ℓ is small, i.e. if we have small fields and/or small molecular sizes), we have $\langle(h_x)^2\rangle\propto L_t$ (the random-walk result), and Eq. (13) then gives directly $\mu\propto 1/L_t\propto 1/L$. This means that in this regime (Figs. 1 & 2) where one measures $\mu(L)\propto 1/L$, DNA molecules reptate in a biased direction, which accounts for their non-zero mobility, but retain their random-walk like globular conformations with $\langle(h_x)^2\rangle\propto L_t$, hence the name "reptation-without-stretching" regime.[16]

On the other hand, if reptation is accompanied by an important stretching (or orientation; stretching refers in this article to the stretching of the conformation of the tube and not of the tube length, which is assumed constant) of the tube in the field direction, such that $\langle(h_x)^2\rangle\propto(L_t)^2$, Eq. (13) gives $\mu\propto(L_t)^0$. Therefore, the "plateau" mobility μ_p can be explained by reptation if this mechanism induces stretching of the tube during the migration,[35,38] hence the name "reptation-with-stretching" regime.[16]

The model provides a coherent theory of the interplay between the field-induced tube stretching and the electrophoretic migration. There are two approaches here. One can calculate the average stretching $\langle(h_x)^2\rangle$ of the tube, and then use Eq.(13) to obtain the electrophoretic mobility $\mu(E,L)$; this approach has proven to be difficult, although some approximate solutions have been found.[39] Instead, one can derive biased stochastic equations similar to those described in section III.1, and use a computer program to simulate the motion of charged molecules according to these rules; $\mu(E,L)$ as well as $\langle(h_x)^2\rangle$ and a number of other statistical properties of DNA can be obtained from the simulation. Most of the theoretical results presented in this article have been obtained with this latter approach.[31,36]

To generalize the model to include a pore size distribution, which would be a better approximation to a real gel architecture, one has to be careful since, as mentioned before, $u_{ix} \propto \ell_i$ in Eq.(10). In general, Eq.(13) is not valid in presence of a pore size distribution (one cannot obtain Eq.(11) from Eq.(10) if $u_{ix} \propto \ell_i$), and cannot therefore be the starting point of a reptation model that would include pores of different sizes. Instead, the full summation of Eq.(10) has to be carried out. Assuming that $\ell_i(a_i) = \Lambda a_i$ with Λ a numerical factor giving the (constant) linear density of DNA contour length per unit length of tube, it is easy to generalize the model to include any pore size distribution while keeping Eq.(13). This assumption, which is the only one that allows one to obtain Eq.(11) from Eq.(10), means that the density ℓ/a of DNA contour length per unit length of tube is assumed to be strictly constant along the tube axis, which would be the case if the tube diameter is fairly constant and if the amount of DNA per unit length of tube can relax (become uniform) on the time scale of the phenomena under study. The results are qualitatively similar to those presented here[40]: a pore size distribution simply helps to fit experimental data better by adding new parameters to the theory. We will therefore discuss only the model with a uniform pore size a in this article since it reproduces all the effects of the generalized case while keeping the mathematics simple enough to allow a number of analytical results to be obtained.

Lumpkin, Déjardin and Zimm[38] used Eq.(13) to study the effect of a pore size distribution, but did not assume that $\ell_i \propto a_i$ for each pore. Instead, they calculated $\ell_i(a_i)$ for each pore size a_i by solving the Kratky-Porod equation[41]. However, as explained above, one cannot obtain Eq.(13) from the very general Eq.(10) in this case; therefore, their results, which showed that the pore size distribution has a strong effect on the electrophoretic mobility, are not reliable for quantitative comparisons.

III.3.2 Biasing the Reptation Model The reptation process is fully defined by the four elements present in the basic stochastic equations (8) and (9). These are the probabilities $p_\pm$ that the next jump will be made towards the $\pm$ end of the tube, the time duration $\tau_\pm = \tau_{Brown}$ and the length $a_\pm = a$ of this jump, and finally the vector $\mathbf{c}(t)$ giving the properties of the new tube section created by this jump. The fact that $p_\pm = \frac{1}{2}$, $\tau_\pm = \tau_{Brown}$ and $\langle \mathbf{c}(t) \rangle = 0$ in absence of a field reflects the fact that we then have Brownian motion. The tube concept and the assumption of a constant pore size a are implicit in Eq.(8).

Equation (12) gives the instantaneous field-induced longitudinal velocity $v_\ell(t)$ of a DNA chain trapped inside a "tube"; this velocity is proportional to the instantaneous end-to-end distance $h_x(t)$ of the primitive chain, but does not include the effect of the thermal agitation.

In the BRM, three of the four elements of the reptation equations (8)-(9) are modified in such a way that the effects of the field and of the Brownian motion are taken into account in a single set of equations. Only the jump length a is not changed: the pore size a still defines the unit jump of the chain since the vector $\mathbf{c}(t)$ can only be defined if its length is a (the direction $\mathbf{c}(t)$ is fixed only when the end of the chain meets a new entanglement, which is at a distance a from the previous one).

First, the electric force qE acting on the end vector-segments ($\mathbf{r}_1$ and $\mathbf{r}_N$) tends to align the vectors $\mathbf{c}(t)$ along the field direction when an end of the chain leaves the tube, because the tube-creating charged end-segment does not encounter gel fibers before it reaches an average distance a outside the tube, i.e., before it completes a full one-pore-jump. If the vector $\mathbf{c}(t)$ is assumed to have zero potential energy when it is perpendicular to the field direction, the potential energy of the segment is simply:

$$U(\theta) = -\tfrac{1}{2} q\ \mathbf{E}\cdot\mathbf{c}(t) = -\tfrac{1}{2} qEa\ cos\theta(t)\ , \tag{14}$$

where $\theta(t)$ is the angle between the vector $\mathbf{c}(t)$ and the field axis x. The probability that the vector $\mathbf{c}(t)$ has an orientation $\theta(t)$ between θ and $\theta+d\theta$ after a jump of length a is completed is thus,[35] in spherical coordinates:

$$P(\theta)\,d\theta \propto sin\theta\, d\theta\, \exp\,(qEa\, cos\theta\, /\, 2k_B T)\ , \tag{15}$$

where $0 \le \theta \le \pi$. This result leads to the following average orientations:

$$\langle cos^n\theta\rangle = \frac{\int_0^\pi P(\theta)\ cos^n\theta\ d\theta}{\int_0^\pi P(\theta)\ d\theta}\quad ; \tag{16a}$$

$$\langle cos\theta\rangle \equiv \langle \mathbf{c}(t)\cdot\hat{\mathbf{u}}_x\rangle/a = coth\,\Theta - 1/\Theta \approx \frac{1}{3}\Theta - \frac{1}{45}\Theta^3 + \ldots\quad ; \tag{16b}$$

$$\langle cos^2\theta\rangle \equiv \langle(\mathbf{c}(t)\cdot\hat{\mathbf{u}}_x)^2\rangle/a^2 = 1+2\Theta^{-2} - 2\Theta^{-1} coth\,\Theta \approx \frac{1}{3}\left[1+\frac{2}{15}\Theta^2+\ldots\right]\ , \tag{16c}$$

where $\Theta \equiv \frac{1}{2}qEa/k_B T$ is called the scaled electric field; the series approximations hold for small scaled fields, i.e., for $\Theta<1$. Note that the averages are made over the distribution function $P(\theta)$ here, i.e., over an ensemble of identical vectors $\mathbf{c}(t)$; the details of the molecular conformation of the rest of the chain, and in particular the end-to-end distance $h_x(t)$, does not play a role in this calculation.

The jump probabilities $p_\pm$ and time durations $\tau_\pm$ are obviously modified by the electric forces. We expect that for $E\to 0$, one will have $p_\pm \approx 1/2$ and $\tau_\pm \approx \tau_{Brown}$, as for the zero-field case. On the other hand, for large fields, we expect instead that $p_\pm \approx 1$, $p_\mp \approx 0$ and $\tau_\pm \approx a/|v_\ell|$. To calculate the values of $p_\pm(E)$ and $\tau_\pm(E)$ that would include both the Brownian motion and the field-induced electrophoretic drift, we use the analogy shown on Fig. 4. The completion of a jump of length $\pm a$ in the tube, in response to both the Brownian motion (with diffusion constant D) and the drift velocity v_ℓ (assumed to be constant during the jump), is similar to the problem of a particle at $s=a$ between two absorbing walls at $s=0$ and $s=2a$, with the particle having a diffusion constant D and a drift velocity v_ℓ. When the particle is absorbed by one wall, it has moved over a distance $\pm a$, which is the definition of a reptation jump in Eqs. (8)-(9).

It can be shown[42] that the probability that the particle on Fig. 4 reaches the $s=0$ wall first, and that this event occurs in a time interval $t_1<t<t_2$, is given by the integral of the distribution function $u(t,a)$ over the given time interval, with:

$$u(t,a) = \frac{\pi D}{2a^2} \exp\left[-\tfrac{1}{4}(v_\ell t+2a)v_\ell /D\right] \sum_{i=0}^{\infty} i \; sin(i\pi/2) \; \exp[-i^2\pi^2 Dt/a^2] \quad . \tag{17}$$

From this equation, it is easy to obtain the average first passage time $\tau_\pm$ at a distance $\pm a$:

$$\tau_\pm = \left[\frac{tanh\,\delta}{\delta}\right] \tau_{Brown} \tag{18a}$$

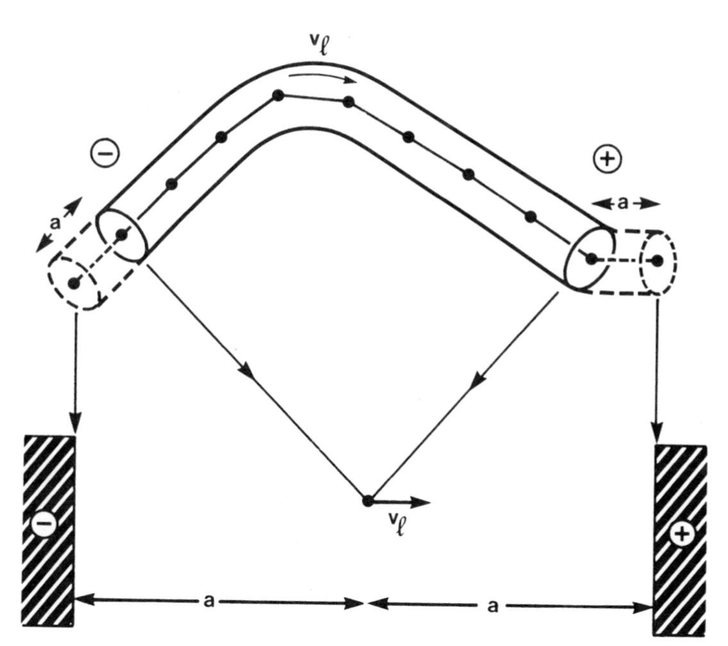

Fig.4 In the biased reptation model, the biased walk of the chain in its tube, which creates new tube sections, is similar to the motion of a point-like particle between two absorbing walls. A biased jump ends when the molecule has migrated over a distance $\pm a$ along the tube axis, i.e., when it has reached the next point defining the end of the next pore. This process is similar to the absorbtion of the particle by one of the walls, each at distance a from the starting position. The particle and the chain both have a one-dimensional velocity v_ℓ and diffusion constant D.

$$\approx \left[1 - \frac{1}{3}\delta^2 + \ldots\right] \tau_{\text{Brown}} \quad ; \qquad \text{for } |\delta| < 1 \tag{18b}$$

$$\approx |\delta|^{-1} \tau_{\text{Brown}} = a/|v_\ell| \quad , \qquad \text{for } |\delta| \gg 1 \tag{18c}$$

where

$$\delta(h_x) \equiv \Theta \frac{h_x}{a} = \frac{v_\ell \, \tau_{\text{Brown}}}{a} \quad , \tag{19}$$

is called the bias of the reptative motion as it is given by the distance moved during a time τ_{Brown} due to the electric drift, $v_\ell \tau_{\text{Brown}}$, divided by the distance moved during the same time due to the Brownian motion, a; the larger the value of $|\delta|$, the more biased the motion is inside the tube. Similarly, the probabilities $p_\pm$ calculated from Eq. (17) are

$$p_\pm = [1 + \exp(\mp 2\delta)]^{-1} \tag{20a}$$

$$\approx \frac{1}{2}\left[1 \pm \delta + \ldots\right] \quad ; \qquad \text{for } |\delta| < 1 \tag{20b}$$

$$\approx \frac{1}{2}\left[1 \pm \delta/|\delta|\right] \quad . \qquad \text{for } |\delta| \gg 1 \tag{20c}$$

These equations give the right limits for small and large biases $|\delta|$. Note that it is the sign of δ that determines the direction in which the motion in the tube is biased.

Equations (8) and (9), together with Eqs. (16), which define the field-induced bias of the vector $\mathbf{c}(t)$, and Eqs. (18)-(20), which give the modified jump durations $\tau_\pm$ and probabilities $p_\pm$, define the BRM completely.

It is interesting to note that $\tau_+ = \tau_-$ in these equations: the time duration of a jump is the same whether the jump is made in the field direction or against it. However, the probabilities $p_\pm(\delta)$ are strongly biased for $|\delta| = |\Theta h_x|/a > 1$, a condition easily met for large enough molecules since $|h_x|$ is proportional to the molecular size. From Eqs. (18a) and (20a), the net longitudinal velocity of the biased reptation motion is given by

$$v_\ell(t) = \left(\frac{p_+ - p_-}{\tau_\pm}\right) a = \frac{a\,\delta}{\tau_{\text{Brown}}} = \frac{qEh_x(t)}{a\,\xi} \quad , \tag{21}$$

in agreement with Eq. (12).

The BRM, as defined in this section, uses only two dimensionless variables: the molecular size N and the scaled electric field Θ. Equation (13) can be rewritten as:

$$\frac{\mu}{\mu_0} = \frac{\langle h_x^2 / a^2 \rangle}{N^2} = \frac{d(t)/a}{\Theta N [t/\tau_{\text{Brown}}]} \quad , \tag{22}$$

where $d(t)$ is the distance migrated in a time t. As we can see from Eqs. (19) and (22), only the variables Θ and N are relevant in this problem if distances are measured in units of a, times in units of τ_{Brown}, and mobilities in units of μ_0. We will therefore use these natural dimensionless variables in the following.

For typical experimental conditions such as $a \approx 150$ nm, $E \approx 1$ V/cm, $\ell \approx 1$ kbp/pore, $\sigma \approx 0.5$, $T \approx 300K$ and $\mu_0 \approx 2 \times 10^{-4}$ cm²/V/s, the BRM parameters correspond to:

$$\Theta \approx 0.3; \qquad \tau_{Brown} \approx 0.02\,K \ seconds; \qquad N \approx K \tag{23}$$

where K is the molecular size in kbp.

IV. CONTINUOUS-FIELD DNA GEL ELECTROPHORESIS

IV.1 The Steady-State It has not been possible yet to find analytically the exact distribution function $p(h_x)$ for the end-to-end distance h_x of the primitive chain in the steady-state, or to find the exact value of $\langle (h_x)^2 \rangle$ as a function of both N and Θ. Lumpkin, Déjardin and Zimm[38] suggested the following calculation:

Writing $\langle (h_x)^2 \rangle$ as (with $\hat{\mathbf{u}}_x$ the unit vector in the field direction)

$$\langle h_x^2 \rangle \equiv \langle \Big(\sum_{i=1}^{N} \mathbf{r}_i \cdot \hat{\mathbf{u}}_x \Big)^2 \rangle = \langle \sum_{i=1}^{N} (\mathbf{r}_i \cdot \hat{\mathbf{u}}_x)^2 \rangle + \langle \sum_{i=1}^{N} \ \sum_{j(\neq i)=1}^{N} (\mathbf{r}_i \cdot \hat{\mathbf{u}}_x)(\mathbf{r}_j \cdot \hat{\mathbf{u}}_x) \rangle , \tag{24}$$

and assuming that the direction of the segment vectors $\mathbf{r}_i$ are not correlated, they obtained, using $\langle \mathbf{r}_i \cdot \hat{\mathbf{u}}_x \rangle = a \langle cos\theta \rangle$ and $\langle (\mathbf{r}_i \cdot \hat{\mathbf{u}}_x)^2 \rangle = a^2 \langle cos^2\theta \rangle$ from Eqs.(16):

$$\langle h_x^2 \rangle / a^2 = N \langle cos^2\theta \rangle + N(N-1) \langle cos\theta \rangle^2 , \tag{25a}$$

from which Eq.(22) gives

$$\frac{\mu}{\mu_0} = \frac{\langle cos^2\theta \rangle}{N} + \frac{(N-1) < cos\theta >^2}{N} \approx \frac{1}{3N} + \frac{\Theta^2}{9} + \ldots , \tag{25b}$$

where the series expansion holds for $\Theta \leq 1$ and $N \gg 1$. This equation predicts $\mu \propto 1/N$ for $N\Theta^2 < 1$, and that the mobility becomes increasingly molecular size independent for $N\Theta^2 \gg 1$ (the second term in the series expansion becomes the leading term, but it does not include N). Therefore, the $\mu \propto 1/L$ reptation-without-stretching regime and the plateau mobility μ_p, characteristic of the reptation-with-stretching regime, are both reproduced by this equation, and the two regimes are separated by the condition $N\Theta^2 \approx 1$. The plateau mobility in this calculation is given by:

$$\frac{\mu_p}{\mu_0} = \lim_{N \to \infty} \frac{\mu}{\mu_0} = \langle cos\Theta \rangle^2 \approx \tfrac{1}{9} \Theta^2 + \ldots \tag{26}$$

The distribution function corresponding to these results is:

$$p_0(h_x) \propto exp\left(-\frac{[h_x/a - N\langle cos\theta\rangle]^2}{2N[\langle cos^2\theta\rangle - \langle cos\theta\rangle^2]}\right) \quad , \tag{27a}$$

as one can verify by calculating $\langle (h_x)^2\rangle = \int (h_x)^2 dh_x\, p_0(h_x) \;/\; \int dh_x\, p_0(h_x)$.

However, as we have shown,[36] the averages in Eq.(24) are chain ensemble averages for which the lifetime $\tau_\pm(h_x)$ of each state h_x must be taken into account, while averages in Eqs.(16) are ensemble average over the orientations of identical end-segments $\mathbf{c}(t)$ only. This means that the vectors $\mathbf{r}_i$ are not really independent if averages (16) are used in Eq.(24): for example, the $\mathbf{r}_i$'s that lead to a $h_x=0$ state are highly favoured because the lifetime $\tau_\pm(h_x=0)$ is then maximum. In fact, this calculation is only valid for $N\Theta^2\to\infty$ since fluctuations of the end-to-end distance h_x are then negligible and all states have roughly the same lifetime $\tau_\pm(h_x\sim Na\langle\cos\theta\rangle)$. Therefore, the plateau mobility μ_p as calculated in Eq.(26) is the only reliable quantitative result given by this approach.

A better choice than the distribution function $p_0(h_x)$ is obtained if $p_o(h_x)$ is weighted by the lifetimes $\tau_\pm(h_x)$ of the given h_x states:

$$p(h_x) \propto p_0(h_x)\tau_\pm(h_x) \propto \frac{tanh[\Theta h_x/a]}{[\Theta h_x/a]}\, exp\left(-\frac{[h_x/a - N\langle cos\theta\rangle]^2}{2N[\langle cos^2\theta\rangle - \langle cos\theta\rangle^2]}\right) \quad , \tag{27b}$$

where Eqs.(18a) & (19) have been used for $\tau_\pm(h_x)$. The pre-factor favours $h_x\approx 0$ conformations; therefore, the mobility given by Eq.(25b) is too large, except for the $N\to\infty$ case. We will show in the next section that the phenomenological distribution function (27b) gives a better result than Eqs.(25a,b).

IV.2 Time Scales and Transient Effects Figure 5 shows the increase of $\langle (h_x)^2\rangle/a^2$, thus of μ/μ_0, just after the field is turned on at $t=0$. The scaled field is $\Theta=1.0$ and the time is in unit of $\tau_{Brown}(N)$ for this DNA chain with $N=100$. The time taken to reach the steady-state mobility can be estimated in the following way. The steady-state is reached when the DNA chain has left is original random tube to enter a new oriented tube under the action of the electric field. For a particular chain with an initial end-to-end distance $h_x(0)$, this tube disengagement time $\tau_E(h_x(0))$ is given by the solution of (neglecting the Brownian motion)

$$Na = \int_0^{\tau_E(h_x(0))} dt\;\, v_\ell(t) \quad . \tag{28}$$

For $t>\langle\tau_E\rangle$, where the average is over an ensemble of identical chains having different initial end-to-end distances $h_x(0)$, the steady-state values of $\langle (h_x)^2\rangle$ and μ are reached. From Eq. (12), we have $v_\ell(t)=\Theta h_x(t)/\tau_{Brown}$. Assuming a homogeneous distribution of segments in the original conformation, we write $h_x(t)\simeq h_x(0)+S(t)\,[\langle\cos\theta\rangle - h_x(0)/Na]$, where $S(t)=\int_o^t v_\ell(t)\,dt$ is the net distance migrated in the tube; (28) then becomes

$$S(\tau_E) = Na = \frac{\Theta}{\tau_{Brown}} \int_0^{\tau_E} dt \left\{ h_x(0) + \left[\langle cos\theta\rangle - \frac{h_x(0)}{Na} \right] S(t) \right\} . \tag{29}$$

This integral equation has the following solution

$$\frac{S(\tau_E)}{Na} = 1 = \frac{h_x(0)}{Na\langle cos\theta\rangle - h_x(0)} \left[\exp\left(\frac{\Theta\ \tau_E}{\tau_{Brown}} \left[\langle cos\theta\rangle - h_x(0)/Na \right] \right) - 1 \right] , \tag{30}$$

from which we get

$$\frac{\langle\tau_E\rangle}{\tau_{Brown}} = \frac{1}{\Theta} \langle\ [\langle cos\theta\rangle - h_x(0)/Na]^{-1}\ \ ln[Na\langle cos\theta\rangle/h_x(0)]\ \rangle , \tag{31}$$

where the averages are made over the initial values of $h_x(0)$.

For $N\Theta^2 \gg 1$, we can drop the $h_x(0)$ in the denominator and calculate the average using the Gaussian distribution function $p_G(h_x(0)) \propto \exp[-3h_x^2(0)/2Na^2]$ to obtain

$$\frac{\langle\tau_E\rangle}{\tau_{Brown}} \approx \frac{\beta + ln[6N\langle cos\theta\rangle^2]}{2\,\Theta\,\langle cos\theta\rangle} , \qquad for\ N\Theta^2 \gg 1 \tag{32a}$$

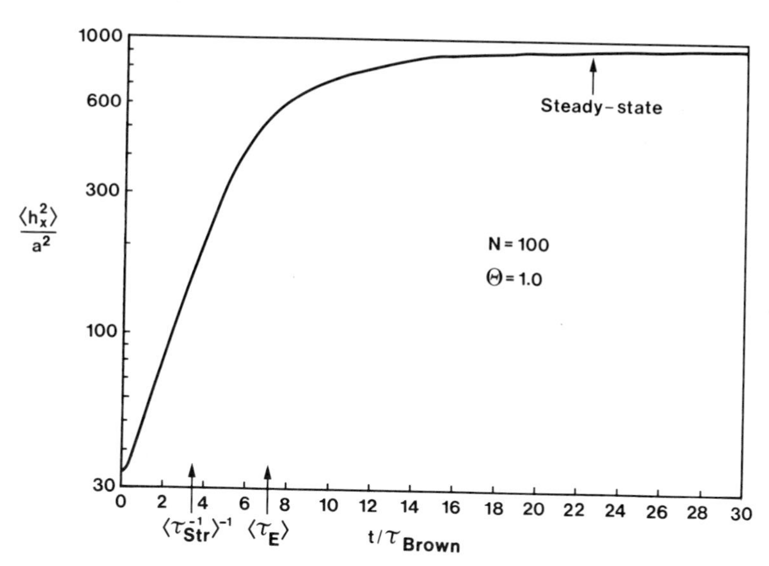

Fig.5 Result of a computer simulation of the BRM: Average end-to-end distance $\langle (h_x)^2\rangle/a^2$ as a function of time for an ensemble of 3000 chains of size $N=100$ segments. The scaled electric field $\Theta=1.0$ is turned on at $t=0$. The stretching time $\langle(\tau_{str})^{-1}\rangle^{-1}$ and the tube-renewal time $\langle\tau_E\rangle$, calculated from Eqs.(37a) and (32a) respectively, are shown. All times are measured in units of τ_{Brown}, the Brownian time for this molecule.

where $\beta=0.577...$ is the Euler number. For $N\Theta^2<1$, this equation is not valid as the pole $h_x(0)=Na\langle cos\theta\rangle$ in Eq.(31) does not have an exponentially small weight in $p_G(h_x(0))$. Instead, one can assume then that $h_x(t)\approx h_x(0)$ since biased reptation does not induce stretching for $N\Theta^2<1$. The disengagement time is then simply

$$\frac{\langle\tau_E\rangle}{\tau_{Brown}} \approx \frac{Na}{\langle v_\ell(0)\rangle\tau_{Brown}} = \frac{Na}{\Theta\langle h_x(0)\rangle} = \frac{(3\pi N/2)^{1/2}}{\Theta} \quad . \qquad for\ N\Theta^2<1 \tag{32b}$$

As we can see from Eqs. (32), $\tau_E\propto N^{3/2}\Theta^{-1}$ in the reptation-without-stretching regime $N\Theta^2<1$, while $\tau_E\propto N\Theta^{-2}ln(N\Theta^2)$ for the very large "plateau" molecules ($N\Theta^2\gg1$). This important change in behaviour indicates that the field-induced disengagement time τ_E is less molecular-size dependent but more field dependent for larger molecules.

Using Eq. (32a), we get $\tau_E\simeq7.2\,\tau_{Brown}$ for $\Theta=1.00$ and $N=100$, in good agreement with Fig.5. Another time of interest is the "stretching time", which gives the rate of stretching of the initial tube. By definition

$$\frac{\partial h_x(t)}{\partial t} = \frac{\partial[\mathbf{R}_{N+1}(t)\cdot\hat{\mathbf{u}}_x]}{\partial t} - \frac{\partial[\mathbf{R}_1(t)\cdot\hat{\mathbf{u}}_x]}{\partial t} \quad , \tag{33}$$

where it is assumed that the electric drift is towards the $\mathbf{R}_{N+1}$ end of the primitive chain. With the chain having the instantaneous longitudinal velocity $v_\ell(t)$, Eq.(33) becomes

$$\frac{\partial h_x(t)}{\partial t} \simeq v_\ell(t)\langle cos\theta\rangle - v_\ell(t)\,\frac{h_x(0)}{Na} \quad , \tag{34}$$

where the last term accounts for the end-to-end distance lost when the chain leaves its original tube. Using $v_\ell(t)=\Theta\,h_x(t)/\tau_{Brown}$, we then obtain

$$\frac{\partial h_x(t)}{\partial t} \simeq \frac{h_x(t)}{\tau_{str}(h_x(0))} \quad , \tag{35}$$

where the stretching rate $(\tau_{str})^{-1}$ is given by

$$\frac{\tau_{str}^{-1}(h_x(0))}{\tau_{Brown}^{-1}} = \Theta\left[\langle cos\theta\rangle - \frac{h_x(0)}{Na}\right] . \tag{36}$$

Equation (35) indicates that initially, the stretching grows exponentially with time, which can be seen in Fig. 5. Since $\langle h_x(0)\rangle = (2Na^2/3\pi)^{1/2}$ for a random-walk conformation, the average stretching rate is given by

$$\frac{\langle\tau_{str}^{-1}\rangle}{\tau_{Brown}^{-1}} \approx \Theta\left|\langle cos\theta\rangle - \sqrt{2/3\pi N}\right| \quad ; \tag{37a}$$

$$\approx \Theta \langle cos\theta \rangle \simeq \frac{\Theta^2}{3} + \dots \quad , \tag{37b}$$

where the last line holds for the relevant case $N\Theta^2 \gg 1$ since, obviously, stretching is assumed here. We see that $(\tau_{str})^{-1} \propto N^{-1}\Theta^2$, which means that different molecular sizes do not only take different times $\tau_E(N,\Theta)$ to stretch completely, but they also stretch at a different rate $[\tau_{str}(N,\Theta)]^{-1}$. This is an important remark as these times will play a critical role in pulsed field electrophoresis techniques where full tube disengagement is not always attained during each pulse.

Using the numbers given in Eqs. (23), we get the following numerical estimates, typical of electrophoresis experiments:

$$\tau_E\,(N\Theta^2 < 1) \approx 0.15\ K^{3/2}\ \textit{seconds} \quad ; \tag{38a}$$

$$\tau_E\,(N\Theta^2 \gg 1) \approx 0.34\,K\,[\,ln\,(K) - 2.24\,]\ \textit{seconds} \quad ; \tag{38b}$$

$$\tau_{str}\,(N\Theta^2 \gg 1) \approx 0.7\ K \quad \textit{seconds} \quad ; \tag{38c}$$

$$\tau_D \approx 4 \times 10^{-3}\,K^3 \quad \textit{seconds} \quad ; \qquad \tau_R \approx 10^{-2}\,K^2 \quad \textit{seconds} \quad , \tag{38d}$$

with K the molecular size in kbp. The first three times are of order 1-10^4 sec. for $50<K<1000$, which is the time scale of the pulses used in the new pulsed field techniques.[7–12] We pointed out previously that this is no surprise as these techniques are taking advantage of the fact that the mobility of the large ("plateau") molecules is in fact molecular-size dependent in the transient regime $0 \le t \le \tau_E(K)$. Finally, it should be noted that τ_E is usually smaller than the duration of a typical experiment ($\approx$1-3 days); therefore, the transient effects are not important in continuous field experiments where $K<50$ kbp DNA molecules are separated.

IV.3 <u>The Steady-State Mobility</u> The value of $\mu/\mu_0 = \langle (h_x)^2 \rangle / N^2$ as obtained from computer simulations is shown in Fig. 6 as a function of $1/N$ for two different values of the scaled electric field Θ. For comparison, the values obtained from the distribution function (27b) and from the LDZ Eq.(25a) are also shown. The mobility predicted by the BRM is not a monotonic function of N: instead, the mobility first decreases as $1/N$ for $N\Theta^2<1$ (this is the reptation-without-stretching regime), and is minimum for $N=N_{min}$ given by

$$N_{min} \simeq (14.43 \pm 0.03)\,\Theta^{-2} \quad , \qquad \textit{for} \quad \Theta \le 1.25 \tag{39}$$

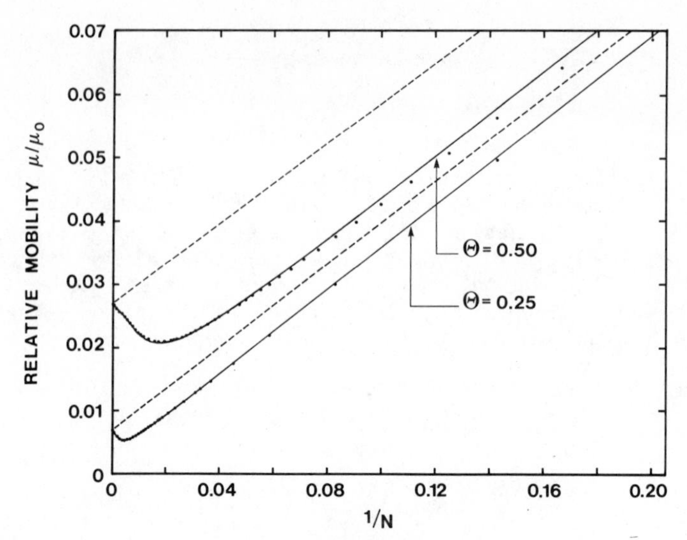

Fig.6 Scaled mobility $\mu/\mu_0 = \langle (h_x)^2 \rangle / N^2$ as a function of the inverse molecular size $1/N$ for two different values of the scaled electric field Θ. The points are from computer simulations of the BRM; the solid lines were obtained by using the distribution function given by Eq.(27b) to calculate $\langle (h_x)^2 \rangle$; the dashed lines are from Eq.(25a), and correspond to the LDZ theory.

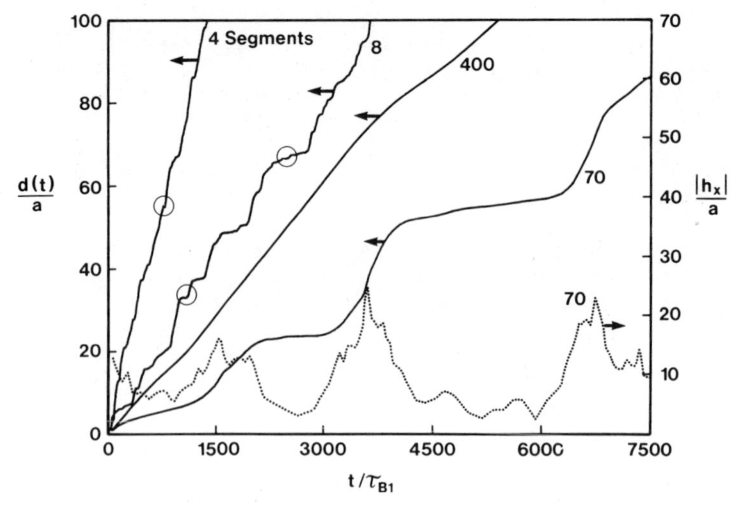

Fig.7 Result of a computer simulation of the BRM. Solid lines: Scaled migrated distance $d(t)/a$ as a function of scaled time t/τ_{B1}, where τ_{B1} is the Brownian time τ_{Brown} for a one-segment primitive chain, for four different molecular sizes; the scaled electric field is $\Theta = 0.50$. Dotted line: Scaled end-to-end distance $|h_x|/a$ as a function of scaled time t/τ_{B1} for the $N = 70$ molecule.

before it increases again to reach its plateau value given by Eq. (26). We can see that Eq. (25a) does not give the right qualitative behaviour, and that the mobilities it predicts are consistently larger than those obtained using the BRM. The phenomenological distribution function $p(h_x)$ given by Eq. (27b) is seen to give a fairly good answer.

The BRM does reproduce both the reptation-without-stretching $(\mu \propto 1/L)$ and reptation-without-stretching $(\mu \sim \mu_p)$ regimes observed experimentally and described qualitatively in section IV.1. The existence of a minimum mobility for a finite value of N_{min} is a surprising prediction of the BRM. However, this effect has been observed recently[31] for pre-separated DNA fragments in the size range 5-50 kbp with experimental conditions $E \simeq 1$-2 V/cm and $A \simeq 1$-2%. Since these conditions have been used in the past, it is possible that some successful electrophoretic separations of DNA fragments have been mis-interpreted using the non-universal rule that $\mu(K_1) \leq \mu(K_2)$ if $K_1 < K_2$ (the hypothesis of a monotonic mobility-size relationship).

To understand why the BRM predicts a non-monotonic mobility-size relationship, Fig. 7 shows a $d(t)/a$ vs t/τ_{B1} plot, where $d(t)/a$ is the distance migrated in units of a, and t/τ_{B1} is the time in units of $\tau_{B1} = \tau_{Brown}/N$ (since $\tau_{Brown} \propto N$, one has to use $\tau_{B1} = \tau_{Brown}/N$, the Brownian time for a one-segment molecule, in order to compare different molecular sizes). For $N=4$, the velocity fluctuates rapidly and the average velocity is large (the slope is large). The eight-segment molecule has a fairly small velocity at some times, and even migrates backward in at least two occasions (see circled areas). The large $N=400$ molecule has a steady-state velocity which does not show noticeable fluctuations. The $N=70$ intermediate size molecule shows remarkable effects as its velocity is almost zero for long periods of time. These zero-velocity conformations, which correspond to $h_x \simeq 0$, as can be seen from the dotted line in Fig.7, are responsible for the fact that the average velocity of the 70 segment molecule is smaller than the velocity of the larger 400 segment molecule. This will lead to a minimum in the μ vs N curve; for this value of scaled field $(\Theta = 0.5)$, the minimum mobility is found at $N_{min}=56$ (see Fig. 6).

From Eqs. (12) & (13) we see that, in the BRM, the instantaneous velocity of the center-of-mass $V_x(t)$ is proportional to $h_x^2(t)$; it is therefore for "compact" molecular conformations, for which $h_x \simeq 0$, that $V_x(t) \simeq 0$ (see Fig. 7 for $N=70$). Both ends of the molecule being close to each other (in the field direction, forming a **U** shape) in these conformations, the total longitudinal electric force $F_\ell(h_x)$ is zero. In this case, $\tau_\pm(h_x) \simeq \tau_{Brown}$ takes its maximum value: these states are long-lived because only the Brownian motion then acts on the molecule. For all these reasons, we called this phenomenon "DNA self-trapping" : DNA fragments are trapped in low-velocity long-lived compact $(h_x \simeq 0)$ conformations which reduce their average velocity.

The reason why this leads to a non-monotonic function $\mu(N)$ can be seen from the distribution function (27b): the relative probability $p(h_x \simeq 0)/p(h_x \simeq Na\langle\cos\theta\rangle)$, where $Na\langle\cos\theta\rangle$ is the expected average end-to-end distance $\langle h_x \rangle$ for a stretched molecule in the reptation model, is (for $N\Theta^2 \gg 1$)

$$\frac{p(h_x \simeq 0)}{p(h_x \simeq Na\langle\cos\theta\rangle)} \simeq \Theta N \langle\cos\theta\rangle \exp - \left[\frac{N\langle\cos\theta\rangle^2/2}{\langle\cos^2\theta\rangle - \langle\cos\theta\rangle^2} \right] \tag{40}$$

At fixed Θ, this ratio has a maximum for $N = 2[\langle\cos^2\theta\rangle/\langle\cos\theta\rangle^2 - 1] \simeq 6/\Theta^2 - 2/5$. Small molecules get trapped more easily than larger ones because $p_0(h_x \simeq 0) \propto \exp(-N)$; however, since $\tau_\pm(h_x \simeq 0) \simeq \tau_{Brown} \propto N$, the lifetime of the trapping states give a larger weight to the compact states of the larger molecules. Therefore, intermediate size molecules for which $N \approx N_{min}$ are those for which the compact states are the most critical (the ratio (40) is then large).

The mobility vs molecular size curves of Fig. 6 have some other interesting properties. From computer simulations we found that the depth of the minimum is given by

$$\frac{\mu_p}{\mu_{min}} = 1.27 \pm 0.02 \quad . \tag{41a}$$

Also, at the inflection point, i.e. immediately after the minimum, we found

$$\left.\frac{\mu(N)}{\mu_0}\right|_{inflection\ point} \propto N^{1/\nu} \quad . \qquad with \ \ \nu \approx 5.8 \pm 0.3 \tag{41b}$$

These relations, which hold for $\Theta \leq 1.25$, i.e. for the experimentally relevant range of scaled fields Θ, mean that the mobility-size curves can be superimposed if the x-axis on Fig. 5 is N/N_{min}, and the y-axis μ/μ_0 (not shown here). In fact, using the series expansions (16b-c), the distribution function (27b), Eqs. (13) for the mobility and (39) for N_{min}, we can write

$$\frac{\mu}{\mu_p} = \frac{\mu_0 \langle h_x^2 \rangle}{\mu_p N^2 a^2} = \frac{1}{(Na\langle\cos\theta\rangle)^2} \; \frac{\int h_x^2 \, dh_x \, p(h_x)}{\int dh_x \, p(h_x)}$$

$$= \frac{1}{g^2 n^2} \; \frac{\int y \, dy \, \tanh(y/3) \, e^{-g(y-n)^2/6n}}{\int \frac{dy}{y} \tanh(y/3) \, e^{-g(y-n)^2/6n}} \; , \tag{42}$$

where $n = N/N_{min}$ and $g = N_{min}\Theta^2 \simeq 14.43$ (from computer simulations). This equation shows that for $\Theta \leq 1$, μ/μ_0 is only a function of $n = N/N_{min}(\Theta)$.

Equations (41) and (39) indicate that a minimum is predicted even for very small values of Θ, which would correspond to small fields E and/or small pore sizes a (large agarose concentration A). On the other hand, if Θ is too large, a minimum will eventually be predicted for molecules too small to reptate (for example $N_{min} < 2$). Instead of reptating and getting self-trapped, these small molecules migrate according to the Ogston-sieving mechanism, although some reptation can also take place in gel areas where the pore size is locally very small. We thus expect the reptation-induced minimum mobility to disappear for large Θs. If we write $N_{min} \propto \Theta^{-2} \propto (qEa)^{-2}$, we can see easily that the minimum can be eliminated by increasing the field or/and the average pore size. Also, since both $\mu_p/\mu_0 = \langle\cos\theta\rangle^2$ and $N_{min} \approx 14.43\ \Theta^{-2}$ are proportional to the scaled electric field Θ, the BRM predicts that the relative plateau mobility μ_p/μ_0 takes a universal value when one uses the critical experimental conditions that just eliminate the minimum (i.e. that gives $N_{min} \approx 1$), regardless of the exact value of the electric field intensity E and the agarose concentration A needed to do so.

Finally, it should be noted that if one separates fragments with less than N^* segments, no minimum (or band-inversion) will exist if $N^* < N_{min}$, i.e. if $\Theta^2 < 14/N^*$. For example, with $\Theta = 0.50$, the minimum is at $N = N_{min} = 56$, and one separates $N < 56$ molecules without band-inversion (see Fig. 6). This corresponds to large agarose concentrations and/or small fields, and is also a way to avoid band-inversion in continuous field experiments.[32]

IV.4 Comparison with Experimental Results We will not attempt to fit experimental results here: we refer the reader to previous articles.[12,16,31,32] Instead, it is of interest to discuss some of the remarkable features of the theoretical μ_p/μ_0 vs N diagram, and to compare them qualitatively with experimental results.

A series of experiments have been performed[32] to study the effect of the field intensity E and the agarose concentration A on band inversion. Analysis of the results revealed that

$$\frac{\mu_p}{\mu_{min}} \approx 1.31 \pm 0.20 \quad ; \tag{43a}$$

$$\left.\frac{\mu(K)}{\mu_0}\right|_{inflection\ point} \propto K^{1/(5.6\pm 2.6)} \quad , \tag{43b}$$

where K is the molecular size in kbp. This is in fairly good agreement with Eqs (41). The large error in (43b) was due to the small number of fragments in the vicinity of the inflection point (usually 3 to 5). The error of ±0.20 in (43a) was due to the fact that the values obtained were distributed more or less uniformly between 1.10 and 1.50. In fact, the depth of the minimum was observed to increase slightly for larger A or smaller E: this latter effect is probably due to the non-uniformity of the gel.[40]

Another test of the BRM concerns the conditions under which the minimum is predicted to disappear. We showed that large fields and/or low agarose concentrations do indeed eliminate the minimum, i.e., the band-inversion phenomenon.[31,32] For example, a minimum is rarely seen for $A < 0.8\%$, which is often a working condition in the laboratory. The plateau mobility was found[32] to be $\mu_p \approx (2.3 \pm 0.4) \times 10^{-5}$ cm^2/V/s for those critical experimental conditions that eliminated the minimum. The small error on this value indicates a very good consistency between the experimentally determined values of μ_p for these critical conditions of A and E, in agreement with the model. We therefore conclude that the BRM mobility-size relationship provides a good framework to understand the results of continuous-field electrophoresis experiments. Better quantitative agreement would necessitate the use of a pore size distribution.[40]

IV.5 Effect of the Gel Concentration The gel concentration A enters in the BRM through both scaled parameters N and Θ. For large concentrations, the average pore size a is smaller than the DNA persistence length $p(\approx 50$ nm); one then expects that $\ell \propto q \propto a$, where ℓ and q are the average contour length and charge of DNA per pore (or primitive segment). Mobilities should then scale like (with $a \propto A^{-y}$):

$$\frac{\mu}{\mu_0} \approx \frac{1}{3N} = \frac{\ell}{3L} \propto A^{-y} K^{-1} \; ; \qquad \textit{for } N\Theta^2 < 1 \tag{44a}$$

$$\frac{\mu}{\mu_0} \approx \frac{\mu_p}{\mu_0} \approx \frac{\Theta^2}{9} \propto \ell^2 a^2 E^2 \propto A^{-4y} E^2 \, , \qquad \textit{for } N\Theta^2 \gg 1 \tag{44b}$$

where, from Eq. (2), $y \approx 2/3$.

In diluted gels, one has $a \gg p$, and DNA is expected to fill pores with a local random-walk conformation, giving $\ell \propto q \propto a^2$. In this case, the mobilities scale as follows:

$$\frac{\mu}{\mu_0} \approx \frac{1}{3N} = \frac{\ell}{3L} \propto A^{-2y} K^{-1} \quad ; \qquad \textit{for } N\Theta^2 < 1 \tag{45a}$$

$$\frac{\mu}{\mu_0} \approx \frac{\mu_p}{\mu_0} \approx \frac{\Theta^2}{9} \propto \ell^2 a^2 E^2 \propto A^{-6y} E^2 \quad . \qquad \textit{for } N\Theta^2 \gg 1 \tag{45b}$$

Only the reptation-without-stretching regime in a diluted gel, Eq.(45a), has been reported experimentally.[16] Larger fields will be necessary in order to observe the A^{-4y} and A^{-6y} scaling laws predicted by Eqs.(44b) and (45b), while extremely low fields and large agarose concentrations will be needed to see the scaling law A^{-y} predicted by Eq.(44a). With $p \approx 50$ nm, Eq.(2) indicates that $a \approx 2p$ at $A \approx 0.9\%$; this should thus correspond to the crossover point between diluted and concentrated gels. The experiments reported above were all done for $N\Theta^2 < 1$ and $A < 0.9\%$.

From Eq.(39), we can write that (with $\ell \propto q \propto a^2$, i.e., in diluted gels)

$$L_{min} = \ell N_{min} \propto \frac{\ell}{\Theta^2} \propto A^{4y} E^{-2} \; . \tag{46}$$

The size of the molecule having the smallest velocity should thus decrease when the field E increases or when the gel gets more diluted (A decreases). However, we already saw that the minimum is expected to disappear if $N_{min} \approx 1$, i.e. if $a \approx R_g$. With Eq.(3), this corresponds to $a^2 \approx (R_g)^2 \approx \frac{1}{3} pL$ (we can assume that $L \gg p$ here since we discuss diluted gels only). Eq. (46) then gives the following relationship between the electric field intensity and the gel concentration that will eliminate the minimum:

$$E \propto A^{3y} \, . \tag{47}$$

Smaller fields are needed in diluted gels, as observed experimentally. More experiments are needed to check the validity of this scaling law.

Figure 1 gives a schematic "phase" diagram of the migration mechanisms of DNA in continuous field gel electrophoresis (for a fixed value of the electric field E) that includes some scaling laws. The separation between the sieving and the reptation-without-stretching regimes occurs for $a \approx R_g$; for diluted gels, where $(R_g)^2 \approx \frac{1}{3}pL$, this leads to the power law $L \propto a^2$. The separation between the reptation-without-stretching and the reptation-with-stretching regimes is found at $N\Theta^2 \approx 1$, as was discussed before. With $N=L/\ell$, we can rewrite this condition as

$$L \approx \frac{\ell}{\Theta^2} \propto \frac{1}{\ell a^2 E^2} \tag{48a}$$

$$\propto a^{-4} E^{-2} \propto A^{4y} E^{-2} \quad ; \qquad for \quad a \gg p \tag{48b}$$

$$\propto a^{-3} E^{-2} \propto A^{3y} E^{-2} \quad . \qquad for \quad a < p \tag{48c}$$

Since the critical line giving the size of the slowest molecule is given by Eq. (46), the scaling laws (48) apply for this line as well.

A still incomplete "phase" diagram has been obtained[16] from experiments for $A<0.8\%$ and $K<50$ kbp. No minimum was reported, but the diagram was in agreement with the idealized "phase" diagram of Fig. 1. More experiments will be needed to obtain a realistic experimental "phase" diagram of the migration mechanisms involved in the electrophoretic separation of DNA, and to check the predictions of the BRM. This is an important step towards a complete understanding of this experimental technique. Such a diagram would be helpful to choose experimental conditions that would not only lead to an optimal separation, but also make sure that band-inversion will not occur.

IV.6 Linear Dichroism A number a linear dichroism (LD) experiments have recently been performed in order to measure the average molecular stretching (or orientation) predicted by the reptation models of DNA gel electrophoresis.[43-46] These LD measurements give the orientation factor

$$S \equiv \frac{1}{2} \langle 3 \cos^2\phi - 1 \rangle \quad , \tag{49}$$

where ϕ is the angle between the DNA chain axis and the electric field direction (we refer the reader to ref. 44 for a more complete discussion of the relation between the orientation factor S and the LD measurements). Following ref. 44, we can rewrite Eq.(49) as:

$$S(a,N,\Theta) = \Gamma(a)\, S_t(N,\Theta) \quad , \tag{50}$$

with

$$S_t(N,\Theta) = \frac{1}{2} \langle 3 \cos^2\theta - 1 \rangle \quad . \tag{51}$$

In these equations, $\Gamma(a)$ is the orientation factor of a molecule section of contour length ℓ forming one primitive segment, and $S_t(N,\Theta)$ is the orientation factor of the tube. The angle θ still refers to the angle between the local tube axis and the field direction; note however that the averages must be made here over all the primitive segments of an ensemble of identical chains, and not only over the vectors $\mathbf{c}(t)$, as in Eqs. (16).

The tube orientation factor $S_t(N,\Theta)$ can be obtained easily from computer simulations; it depends only on the two scaled variables N and Θ, and measures the average orientation of the tube in the field direction. The primitive segment orientation factor $\Gamma(a)$ measures the average orientation of the base pairs of a single segment with respect to the tube direction. Since it is assumed here that the molecular conformation of DNA inside its tube is not affected by the field, this latter factor does not depend upon E; it does depend on the pore size, however, since more DNA will pack (randomly) into larger pores for which $a \gg p$. We will not discuss the effect of the concentration A on the orientation factors S and $\Gamma(a)$ in the following, and we will assume that $\Gamma(a)$ is simply a constant.

LD measurements have shown that DNA molecules do indeed orient in the field direction during electrophoretic migration,[43-46] in agreement with the main prediction of the BRM. In the following subsections, we will give the predictions of the BRM concerning $S_t(N,\Theta)$, and we will discuss the relation between tube orientation and electrophoretic mobility.

IV.6.1 Steady-State Orientation Using Eq. (16c) for $\langle\cos^2\theta\rangle$, the steady-state tube orientation factor is found to be $S_t \approx \Theta^2/15$. This simple result predicts that the degree of orientation of the molecule increases as the second power of E, but is molecular-size independent. Experimentally,[44] LD showed that $S \propto E^2$, but the orientation factor was also observed to be molecular-size dependent[45]: for $E=10$ V/cm, $A=1.0\%$ and $K=40$-200 kbp, it was reported that $S \propto L^{0.6}$.

However, as we have mentioned, the averages in Eqs.(16c) and (51) are not of the same sort; self-trapping is expected to be important here, as it was for the mobility. Therefore, $S_t \approx \Theta^2/15$ cannot be the exact value predicted by the model; instead, it plays the role of the plateau mobility before, i.e., it is the assymptotic value for $N\Theta^2 \gg 1$. Figures 8-10 show the result of a computer simulation based on the BRM.

The steady-state orientation factor S_t increases sigmoidally with the electric field on Fig. 8 for $N=100$, and the inset shows that $S_t \propto \Theta^2 \propto E^2$ at low field, i.e., far from saturation, in agreement with experiments.[44]

Figure 9 shows the molecular size dependence of S_t for three different scaled field intensities. At high scaled electric fields ($\Theta=2.0$), the curve is almost flat with the asymptotic value $S_t \approx \Theta^2/15$, except for extremely small values of N (<5): the model therefore predicts that the high field value of S_t is essentially molecular-size independent, which seems to be at odds with experimental results.[45] At lower fields however, the curves show a shallow minimum for small molecular sizes, near the position of the self-trapping induced minimum in the mobility-size relationship. This minimum is followed by a range of molecular sizes where S_t is seen to increase approximately as $S_t \propto N^{1/3}$, ending with saturation $S_t \approx \Theta^2/15$ at still larger molecular sizes (saturation has to happen since $S_t \leq 1$ by definition). It is noticeable that the exponent found here ($\frac{1}{3}$) is exactly twice the exponent ν found in Eq. (41). It would be of great interest to measure the complete S vs L curve at low field intensities and with small molecular sizes to check the predictions of Fig. 9 as this could allow us to observe directly the decrease in tube orientation and the slowing down of tube dynamics accompanying DNA self-trapping in the BRM.

IV.6.2 Dynamics of Chain Orientation The solid line in Fig. 10 shows how the tube orientation factor S_t changes in time when a field is turned on at $t=0$, and latter turned off at $t=T_0$.

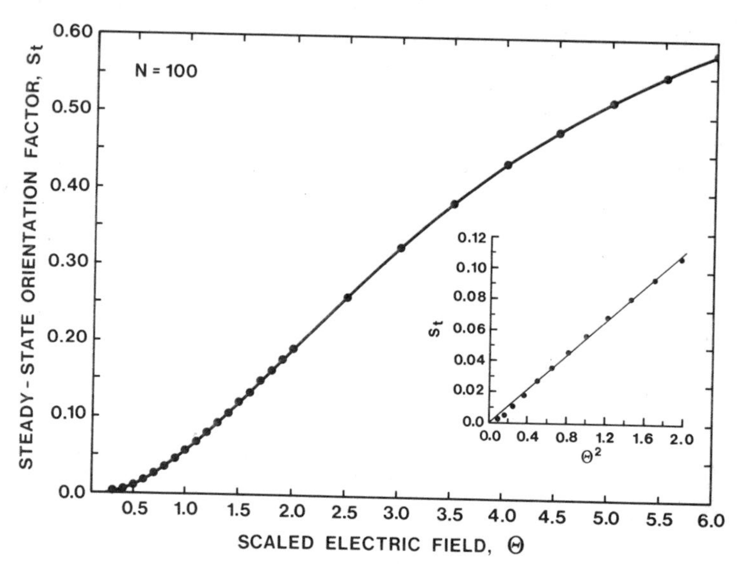

Fig.8 Result of a computer simulation of the BRM: Steady-state tube orientation factor S_t as a function of the scaled electric field intensity Θ for a molecule of $N=100$ segments. The insert shows that $S_t \propto \Theta^2$ for small fields.

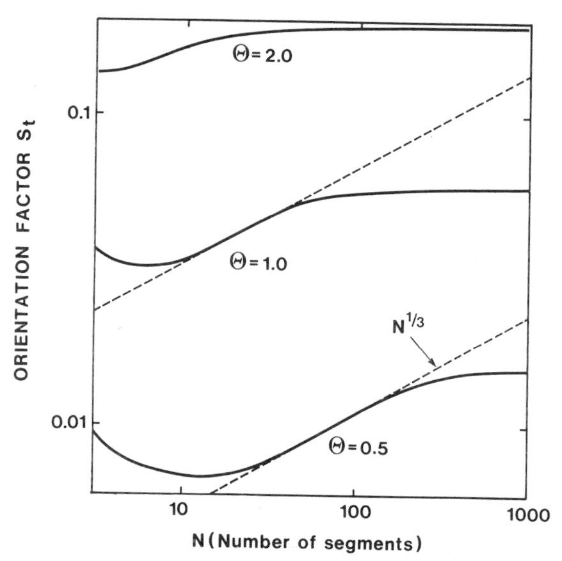

Fig.9 Result of a computer simulation of the BRM: Steady-state tube orientation factor S_t as a function of the molecular size N (in segments) for three different field intensities $\Theta=0.5$, 1.0, 2.0. The dashed lines have a slope 1/3.

For $t < \tau_E \approx 4\tau_{Brown}$, the orientation factor S_t increases sigmoidally with time, again in agreement with experiments.[45,46] The rate of chain orientation, as measured by the slope $\partial S_t(t)/\partial t$ at the inflection point, was found to increase as Θ^z for small electric fields, with $z \approx 2.7$-3.3 for $30 \le N \le 1000$, consistent with available data.[45] This suggests that reptation plays a role in the initial build-up of the molecular oriention measured by LD. We also found that $\partial S_t(t)/\partial t$ scales like N^{-1} for large molecular sizes at the inflection point, a prediction that is still to be checked experimentally.

The dashed line on Fig. 10 shows a typical experimental[45,46] curve (only schematic here). The orientation factor S goes first through an overshoot at time τ_{ov}, with $S(\tau_{ov}) > S(\tau_{ss})$, where τ_{ss} is the time needed to reach the steady-state orientation, and then through an undershoot at time τ_{un}, with $S(\tau_{ov}) > S(\tau_{ss}) > S(\tau_{un})$, before it reaches the steady-state orientation discussed in the previous subsection for $T_0 > t > \tau_{ss}$. Since the over- and under-shoots occur only for $0 < t < \tau_{ss}$, where τ_{ss} is only a few seconds, these transient effects are not likely to change the mobility of DNA fragments in continuous field experiments. However, it has been reported that the overshoot time τ_{ov} may be related to the critical pulse duration used in certain pulse-field techniques.[45] This effect may thus be very important in this case.

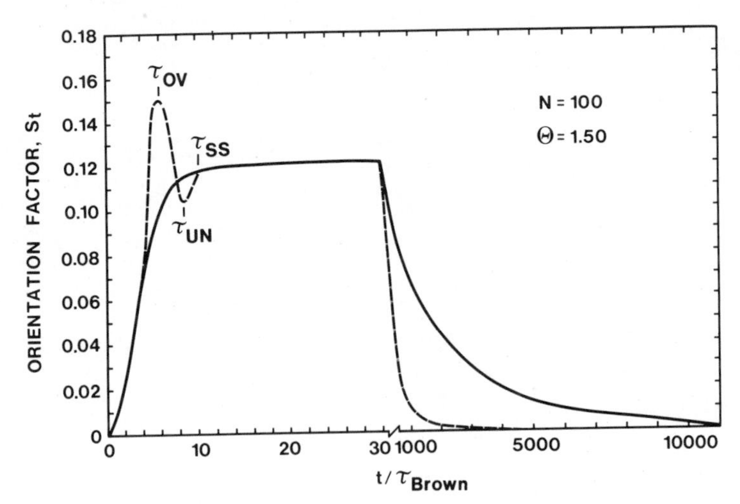

Fig.10 Result of a computer simulation of the BRM (solid line): Steady-state tube orientation factor S_t as a function of the scaled time t/τ_{Brown} for a molecular size $N = 100$. A field of scaled intensity $\Theta = 1.50$ is applied at $t = 0$, and is turned off at $t = 30\tau_{Brown}$. The time scale is larger for $t \ge 30\tau_{Brown}$ since the relaxation process is very slow. The dashed line shows schematically the results of LD experiments: there is an overshoot at time τ_{ov} and an undershoot at time τ_{un} before the steady-state is achieved, and relaxation is faster than expected from the BRM.

The BRM presented in this article does not provide a theory for τ_{ov}; however, the steady-state time τ_{ss} can be compared with the field-induced disengagement time τ_E predicted by the BRM. From Eq. (32a), which we use here because most LD experiments have been done at high fields and with fairly large molecules, we get that (dropping the logarithmic term):

$$\tau_E \propto N^1 \Theta^{-2} \propto K^1 E^{-2} \; ; \qquad \textit{low fields } (\Theta \leq 1) \tag{52a}$$

$$\tau_E \propto N^1 \Theta^{-1} \propto K^1 E^{-1} \; , \qquad \textit{high fields } (\Theta \gg 1) \tag{52b}$$

where the large field limit applies when $\langle\cos\theta\rangle \approx 1$ (saturation). The BRM predicts that the duration of the transient effects should increase linearly with the molecular size; this is in agreement with recent LD experiments.[45] Moreover, the BRM predicts that this time is $\propto E^{-2}$ for small fields and $\propto E^{-1}$ at larger fields due to saturation of the tube orientation, which is consistent with available data.[46] Finally, it should be noted that the values of τ_{ss} as measured by LD are fully consistent with the values of τ_E for typical experimental conditions. For example, with $N=100$ and $\Theta=3.0$, which would correspond fairly well with the experimental conditions used in ref. 45, Eq. (32a) gives $\tau_E \approx 1.5\,\tau_{Brown}$, where, from Eq. (23), $\tau_{Brown} \approx 2 \times 10^{-2} N$ seconds; for a 100 kbp molecule, we thus have $\tau_E \approx 3$ seconds, consistent with experiments. We thus conclude that the initial tube stretching is due to biased reptation, but that intra-tube modes also play a role on a time scale $t < \tau_E$.

IV.6.3 <u>Tube relaxation</u> In the tube model, relaxation occurs solely from random Brownian motion of the chain, i.e., unbiased reptation. This motion slowly brings the primitive chain out of its oriented tube in a time $\approx \tau_D \propto N^3$, as can be seen schematically in Fig. 10; Eq. (38d) indicates that $\tau_D \approx 1$ h for $K \approx 100$ kbp.

Experimentally,[45,46] the relaxation of the tube was found to be much faster than what is predicted by τ_D. The LD relaxation time was found to be of order 1-5 seconds even for fairly large molecules ($K>100$ kbp). Moreover, this relaxation time was found to be somewhat molecular-size independent,[45] although more experiments are clearly needed here. Finally,[46] it has been observed that the overshoot was recovered at the beginning of a second on-pulse only if the off-pulse duration T_0 separating the two on-pulses was very large; preliminary calculations indicates that the necessary value was $T_0 \approx \tau_D$.

These results indicate that some fast intra-tube modes tend to disorient the molecular conformation on a very short time scale, but that only complete tube disengagement, due to the slow unbiased reptation process, can restore the true molecular randomness.

IV.7 <u>Conclusions</u> The BRM gives a good account of continuous-field DNA gel electrophoresis. It predicts band-inversion, and the mobility-size diagram predicted by the model is in good agreement with experiments: the shape of the minimum, the $\mu(K) \propto 1/L$ and $\mu(K) \approx \mu_p$ regimes are all well accounted for in the BRM. Recently, the scaling law (45a) has also been reported. LD experiments confirmed that there is a relation between mobility and tube stretching, as predicted by the model; in particular, it was found that the average segment orientation increases as the second power of the electric field, in excellent qualitative and quantitative agreement with the model.

However, three remarks concerning the effects of short wavelength modes should be made here. First, the over- and under-shoot stretching phenomena clearly point out that some intra-tube DNA modes, neglected in the BRM, are excited if the field is applied on a truly random coil DNA. This transient effect does not seem to affect the overall reptative behaviour of DNA, but may be very important in pulsed field techniques where separation is increased by coupling the external pulsed field(s) to transient effects that may affect the molecular electrophoretic mobility.

Second, the fact that the LD orientation factor relaxes very quickly, ruling out unbiased reptation as a mechanism, indicates that the chain is not in equilibrium in its tube during continuous-field electrophoresis. The intra-tube tension, which may be due to gel inhomogeneities and tube leakages for example, is released very quickly when the field is turned off probably because this involves mainly local motion. However, the overall tube conformation cannot be renewed by these intra-tube modes.

Third, it was found that the LD orientation factor was molecular size dependent even for those large molecules whose mobility is size-independent. The strong dependence ($S \propto L^{0.6}$) reported experimentally suggests that reptation-induced self-trapping (with its lower dependence $S_t \propto L^{1/3}$) might not be a major effect here. However, since S_t and μ should be closely related in any reasonable model, it is possible that the intra-tube tensions referred to above modify the quantitative aspects of the reptative motion without changing its qualitative aspects. For example, these modes could couple to self-trapping and change the various power laws as well as the range of molecular sizes affected by this effect. More quantitative experiments are thus needed to examine the relation between the orientation factor S and the mobility μ of very large DNA molecules.

V. INTERMITTENT ELECTRIC FIELDS

One of the first techniques that has been suggested in an attempt to improve the separation of larger (K>50 kbp) DNA fragments relied on intermittent electric fields[35,47-49] where the field E is on for pulses of duration t_{ON} separated by zero-field pulses of duration t_{OFF}. The large molecules ($N\Theta^2>1$) can be expected to stretch and migrate during the on-pulses, and to relax to random-walk conformations during the off-pulses.

If complete randomnization is achieved during the off-pulses and if the on pulses are short enough to avoid complete stretching of these large molecules, one might expect that the reptation-without-stretching regime will be nearly preserved for all molecules, with $\mu \propto 1/N$ instead of $\mu \approx \mu_p$.

In practice, however, the situation is far more complicated. The condition for incomplete stretching is simply

$$t_{ON} \leq \tau_E \propto N\Theta^{-2} \; ln(N\Theta^2) \quad , \tag{53a}$$

while the condition for complete randomnization is

$$t_{OFF} \geq \tau_D \propto N^3 \; . \tag{53b}$$

The times τ_E and τ_D being proportional to the molecular size N, there is no choice of t_{ON} and t_{OFF} that will satisfy both conditions for all molecular sizes. For a given time t_{OFF}, large molecules for which $\tau_D > t_{OFF}$ will not relax completely during the off-pulses, and may therefore keep a fairly large average end-to-end distance $\langle (h_x)^2 \rangle$ during the whole experiment; intermittent fields will not permit separation of these large molecules if $N\Theta^2 > 1$ as well, unless t_{OFF} is increased.

On the other hand, for a given time t_{ON}, those large molecules for which $t_E < t_{ON}$ and $N\Theta^2 > 1$ will stretch completely during the on-pulses, which will also not permit their separation either.

Therefore, in order to be successful, an intermittent field gel electrophoresis has to use the conditions (with N_{max} the size of the larger molecule to be separated):

$$t_{OFF} \geq \tau_D (N_{max}) \quad ; \tag{54a}$$

$$t_{ON} \leq \tau_E (N \approx 1/\Theta^2) \quad , \tag{54b}$$

The first condition means that all molecules will relax completely during the off-pulses, while the second means that none of the large molecules (for which $N\Theta^2 > 1$) will stretch completely during the on-pulses. These two conditions reduce molecular stretching to a minimum and, because the molecules migrate according to an average reptation-without-stretching mechanism, they allow separation of all molecules with $N < N_{max}$ with approximately $\mu \propto 1/N$.

Using Eqs. (6) and (32b), these conditions can be rewritten

$$t_{OFF} \geq \frac{2}{\pi^2} N_{max}^3 \tau_{B1} \tag{55a}$$

$$t_{ON} \leq \sqrt{3\pi/2} \quad \Theta^{-4} \quad \tau_{B1} \tag{55b}$$

where $\tau_{B1} = \tau_{Brown} / N$ is the Brownian time per segment. For optimal results, one will want to choose t_{ON} as large as possible and t_{OFF} as small as possible. The optimal ratio is thus

$$\left(\frac{t_{OFF}}{t_{ON}} \right)_{optimal} \approx \left(\frac{8}{3\pi^5} \right)^{\frac{1}{2}} N_{max}^3 \Theta^4 \approx \frac{1}{11} N_{max}^3 \Theta^4 \tag{56}$$

This calculation points out the theoretical limitation of this approach. For $\Theta = 0.50$, it is impossible to separate DNA molecules with $N > 20$ segments in continuous electric fields; to extend this limit to $N > 40$, one would need a ratio $t_{OFF}/t_{ON} \geq 360$, which is not practical.

The calculation presented here does not take into account the fact that if $t_{ON} < \tau_E$, all molecules will not need the maximum time τ_D to relax completely during the off-pulses. In fact, one can use a time t_{ON} slightly larger than the one given in Eq.(55b) because stretching develops slowly between $N \approx 1/\Theta^2$ and $N \approx 14/\Theta^2 = N_{min}$. Similarly one can use relaxation times slightly smaller than the one given in Eq.(55a) because relaxation of an incompletely stretched conformation does not take this maximum time. However, the ratio remains proportional to a high power of N_{max}, which still limits the practical usefulness of this approach.

Figure 11 shows the result of a computer simulation of intermittent-field gel electrophoresis with $\Theta=1.0$. If only molecules with $2 \leq N \leq 14$ segments were present, separation would be possible in continuous field since $N_{min}=14$ in this case. However, if larger molecules are present, only $N<9$ molecules can be separated because $\mu_p/\mu_0 \approx \Theta^2/9 \approx 1/9 \approx \mu(N=9)$. In order to improve on this, one can use for example intermittent fields with $t_{OFF}=\tau_D(N=20) \approx 1600\tau_{B1}$ and $t_{ON}=(1/8)t_{OFF} \approx \tau_E(N=40)$. The figure shows that this makes the minimum drifts from $N_{min}=14$ to $N_{min} \approx 20$, which improves separation but increases the duration of the experiment by a factor of 9. This shows the serious limitations of this technique.

VI. CROSSED ELECTRIC FIELDS

The development of pulsed field techniques to separate larger DNA molecules by gel electrophoresis started with Schwartz and Cantor[7] pulsed field electrophoresis method. In this method, as in many other variations,[8,10] the field direction is changed alternatively between two approximately orthogonal directions. The molecules move along the diagonal formed by the two field directions, and due to the fact that the molecules have to reorient before they can start moving in a new direction each time the field is switched, one achieved separation of $\gg 50$ kbp DNA molecules.

In this section, we will discuss the ideal case where two homogeneous and equal fields are at exactly 90° from one another. In practice, the fields are often inhomogeneous, and they make an angle slightly larger than 90°. The field gradients help to sharpen the electrophoretic bands

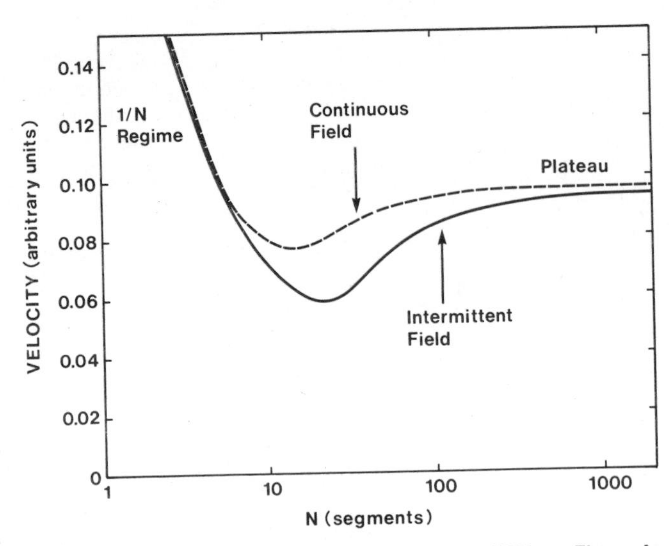

Fig.11 Result of a computer simulation of the BRM: The electrophoretic velocity is plotted as a function of the molecular size N (in segments) for a scaled electric field of $\Theta=1.0$. The dashed line is for a continuous field and has a minimum at $N \approx 14$. The solid curve is for intermittent fields with $t_{OFF}=1600\tau_{B1}$ and $t_{ON}=(1/8)t_{OFF}=200\tau_{B1}$, and has a minimum for $N \approx 20$.

in the gel by a process that we have called "self-sharpening".[50] The angle between the two fields is often a function of the position in the Cantor-Schwartz apparatus; this is sometimes equivalent to using a variable pulse duration.[51] When the angle is much larger than 90°, the effects that show up in field inversion electrophoresis (see section VII) may be expected to play a role. We focus in this section on the predictions of the BRM concerning orthogonal, uniform and equal electric fields.

VI.1 Theoretical Arguments We thus assume here that the field is given by (with i=0,1,2,3,...):

$$\mathbf{E} = E\,\hat{\mathbf{u}}_x \; ; \qquad for \quad 2i\,t_\perp \le t \le (2i+1)\,t_\perp \tag{57a}$$

$$\mathbf{E} = E\,\hat{\mathbf{u}}_y \; , \qquad for \quad (2i+1)\,t_\perp < t < 2(i+1)\,t_\perp \tag{57b}$$

where $t_\perp$ is the pulse duration. Since small DNA molecules can be separated by continuous-field gel electrophoresis, we will consider only "plateau" molecules for which $\mu \approx \mu_p$ in continuous fields; therefore, the condition $N\Theta^2 \gg 14$ is assumed to apply for all molecular sizes.

If the pulse duration is very long, a plateau molecule has enough time to reorient in the new field direction at the beginning of each new pulse before it reaches its plateau velocity. For example, with the field $\mathbf{E}=E\hat{\mathbf{u}}_x$, the molecule first stretches out in the x-direction, with $\langle (h_x)^2\rangle \approx (L_t\langle\cos\theta\rangle)^2$, and then reaches its steady-state center-of-mass velocity $V_{/\!/}=V_x=E\mu_p$, which it keeps for the rest of the pulse. The end-to-end distance $\langle (h_y)^2\rangle$ is then very small. Similarly, when the field changes direction to $\mathbf{E}=E\hat{\mathbf{u}}_y$, the molecule has to stretch out in the new field direction, with $\langle (h_y)^2\rangle \approx (L_t\langle\cos\theta\rangle)^2$, before it reaches its steady-state center-of-mass velocity $V_{/\!/}=V_y=E\mu_p$. This cycle is repeated for each pulse. We will note by $\tau_\perp$ the time necessary to reorient in the new field direction, by $(h_{/\!/})^2$ and $(h_\perp)^2$ the end-to-end distances parallel and perpendicular to the instantaneous field direction, and by $V_E=E\mu_p$ the ("plateau") velocity of a molecule that has had time to reorient during the pulse.

When $t_\perp \gg \tau_\perp$, the reorientation is only a very short transient effect, and the molecules are oriented for almost all the pulse duration; therefore,

$$\langle h_{/\!/}^2\rangle \approx L_t^2\langle\cos\theta\rangle^2 \approx (L_t\,\Theta/3)^2 \; ; \tag{58a}$$

$$\langle h_\perp^2\rangle \approx L_t\,a\,\langle\sin^2\theta\rangle \approx \tfrac{2}{3}L_t\,a\left[1-\tfrac{1}{15}\Theta^2\right] , \tag{58b}$$

$$V_{/\!/} \approx V_E = E\,\mu_p \quad , \tag{58c}$$

for most of the pulse duration. In that case, the molecule undergoes a stairwise motion in the gel (Fig. 12a) which follows roughly the diagonal of the gel. During a time $2j\,t_\perp$, i.e., after j complete cycles, the net distance travelled is

$$d = (d_x^2 + d_y^2)^{1/2} = \sqrt{2}\ d_x \approx \sqrt{2}\ j t_\perp V_E \quad . \tag{59}$$

The net mobility is thus given by:

$$\mu(t_\perp \gg \tau_\perp) \equiv \mu_\infty = \frac{d}{(2jt_\perp)E} = 2^{-1/2}\mu_p \approx \frac{\Theta^2}{9\sqrt{2}}\mu_0 \quad . \tag{60}$$

The mobility of all "plateau" DNA fragments that have plenty of time to reorient during each pulse is thus reduced by a common factor $\sqrt{2}$.

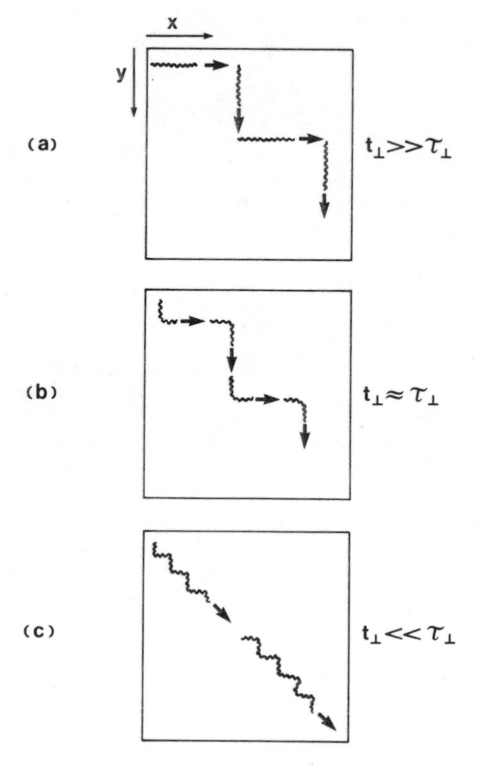

Fig.12 Schematic of the motion of a large DNA molecule during crossed field electrophoresis. (a) For long switching times $t_\perp$, the molecule reorients rapidly during the pulse, and the motion is like a series of continuous field electrophoresis experiments carried out at right angle to each others. (b) For intermediate switching times such that $t_\perp \approx \tau_\perp$, the molecule just has time to reorient before the field changes direction. (c) For short pulse durations, the molecule shown here orients and moves along the diagonal of the gel since it cannot respond to the field changes.

At the other limit, if $t_\perp \ll \tau_\perp$, the molecule does not have time to reorient in the new field direction between each pulse. The molecular conformation may then be seen as being aligned along the diagonal direction (Fig. 12c), with half its primitive segments oriented along the x-axis and the other half along the y-axis. The average end-to-end distances are then given by $\langle h_x \rangle = \langle h_y \rangle \approx \frac{1}{2} L_t \langle \cos\theta \rangle$, which lead to the average curvilinear velocity

$$\langle v_\ell \rangle \approx qE \langle h_{x,y} \rangle / \xi a \approx qEN \langle \cos\theta \rangle / 2\xi \quad . \tag{61}$$

Using Eq.(13), where h_x is now replaced by the end-to-end distance along the diagonal direction, $\langle h_d \rangle = (\langle h_x \rangle^2 + \langle h_y \rangle^2)^{1/2}$, since it is the direction of the motion of the center-of-mass, we obtain:

$$\mu(t_\perp \ll \tau_\perp) \equiv \mu_d \approx \frac{\langle v_\ell \rangle \langle h_d \rangle}{Na\ E} = 8^{-1/2}\ \mu_p \approx \frac{\Theta^2}{9\sqrt{8}}\ \mu_0 \ . \tag{62}$$

The mobility of all "plateau" DNA fragments that do not have time to reorient during the pulses is thus reduced by a common factor $\sqrt{8}$.

The increase in mobility when one goes from short to long pulses is thus:

$$\frac{\mu(t_\perp \ll \tau_\perp)}{\mu(t_\perp \gg \tau_\perp)} = \frac{\mu_d}{\mu_\infty} = \frac{1}{2} \ . \tag{63}$$

This ratio is in agreement with available experimental results.[53]

The crossed field techniques take advantage of the fact that the mobility is a function of the ratio $t_\perp/\tau_\perp$, and that the reorientation time $\tau_\perp$ can be expected to increase with the molecular size K. If $t_\perp$ is fixed, all "plateau" molecules for which $t_\perp/\tau_\perp \gg 1$ have a mobility $\mu_\infty \approx \mu_p/\sqrt{2}$, while even larger "plateau" molecules for which $t_\perp/\tau_\perp \ll 1$ have a lower mobility $\mu_d \approx \mu_p/\sqrt{8}$. The reorientation effect thus introduces a separation between these two groups of "plateau" molecules, and good electrophoretic separation can be expected for those molecular sizes where $t_\perp/\tau_\perp \approx 1$ since it is for these molecules that the molecular-size dependence of the mobility will be the most important.

If we come from the $t_\perp/\tau_\perp \gg 1$ side, we can see that the reorientation time $\tau_\perp$ should be of order τ_E: when the pulse direction is changed by 90°, the new end-to-end distance $\langle (h_\perp)^2 \rangle$ follows a quasi-random-walk distribution (see Eq.(58b)), and the molecule stretches slowly as it leaves its tube according to the description given in section 4.3.2 (the new oriented tube is obtained when the chain has disengaged completely from its original tube, which was oriented perpendicularly to the new field direction). Therefore, for a given molecular size $N_\perp$, the mobility will decrease from μ_∞ to $\mu_d = \frac{1}{2}\mu_\infty$ when the pulse duration becomes smaller than

$$\tau_\perp \approx \tau_E = \frac{\beta + ln[6N_\perp \langle cos\theta \rangle^2]}{2\,\Theta\,\langle cos\theta \rangle}\ \tau_{Brown} \propto N_\perp \Theta^{-2}\ ln[N_\perp \Theta^2] \quad . \tag{64}$$

Thus, to separate "plateau" molecules with sizes $\approx N_\perp$, Eq. (64) gives the approximate pulse duration one should use. Much smaller and much larger molecules will form two bands at $\mu=\mu_\infty$ and $\mu=\mu_d$ respectively, while molecules of size $\approx N_\perp$ will spread between these two bands.

VI.2 Computer Simulations of Crossed Field Electrophoresis We have studied crossed field electrophoresis using a computer simulation of the BRM; Fig. 13 shows a diagram of the mobility μ/μ_0 vs pulse duration $t_\perp$ for 6 different molecular sizes, 4 of them having a size-independent plateau mobility $\mu/\mu_0 \approx \mu_p/\mu_0 \approx \Theta^2/9$ in continuous fields.

The mobility of the $N=5$ molecule increases slightly for $t_\perp \geq 10\tau_{B1}$ (the times are measured in units of $\tau_{B1}=N^{-1}\tau_{Brown}$, which is independent of the molecular size); it is easy to separate this molecular size from the others for $t_\perp \leq 10\tau_{B1}$. The $N=14$ molecule has the lowest mobility of all molecules in continuous fields (as can be seen for $t_\perp \rightarrow \infty$); in crossed fields, short pulse times $t_\perp < 100\,\tau_{B1}$ make the mobility of all larger molecules slightly lower than the mobility of this molecule, but this hardly allows separation.

The crossed field technique is in fact useful to separate "plateau" molecules, e.g. for $N \geq 50$ molecules in Fig. 13. We have $\mu_p/\mu_0 \approx 1/9$ for these simulations and, as can be seen from this Figure, the large molecules all show $\mu_\infty/\mu_0 \approx 0.070 \approx \mu_p/\sqrt{2}$, and $\mu_d/\mu_0 \approx 0.035 = (1/2)\,\mu_\infty/\mu_0$, as predicted by Eqs.(60), (62) and (63). The remarkable feature in this diagram is the fact that the mobility of the various "plateau" fragments increases suddenly for $t_\perp$ larger than a critical time $\tau_\perp$ which is strongly size-dependent. Obviously, it is for this range of pulse durations $t_\perp$ where the mobility of the "plateau" molecules increases that electrophoretic separation can be achieved. For example, the $N=500$, 1000 and 2000 molecules are well separated if a pulse duration of $t_\perp \simeq 25000\,\tau_{B1}$ is used. If the $N=50$ molecule is also included, one could use $t_\perp = 10000\,\tau_{B1}$ pulses for half of the experiment, which would separate the $N=50$ molecule from the others, and then $t_\perp \simeq 25000\,\tau_{B1}$ pulses for the other half.

In Fig. 14, we plotted the reorientation time $\tau_\perp$ as a function of the molecules size N. We defined $\tau_\perp(N)$ as the pulse duration $t_\perp$ for which $\mu(N)=\frac{1}{2}(\mu_\infty+\mu_d)=3\mu_p/\sqrt{32}$ here. Clearly, $\tau_\perp \propto N^{1.0}$ for large fragments and $\tau_\perp \propto N^{1.5}$ for small ones, as expected from Eqs.(32). This has to be compared with the equation $\tau_{sep} \propto N^{4/3}$ suggested by Smith et al.[52] on the basis of experimental results, where τ_{sep} the pulse duration $t_\perp$ that allows maximum separation of the molecules of size $\approx N$; a best fit with $\tau_\perp \propto N^{4/3}$ has also been added to Fig. 14. We can see a very good correspondence between the suggested power-law of Smith et al. and our theory over this range of molecular sizes.

Schwartz[53] published a plot of mobility-vs-pulse duration showing results that are very comparable to those of Fig. 13. In particular, he found that $\mu_\infty/\mu_d \approx 2$, in agreement with Eq. (63). He also found that the mobility of large fragments often decreases slightly for intermediate pulse durations, before it increases by a factor of 2 at $t_\perp \approx \tau_\perp(N)$. We also found this behaviour with our simulations, as can been seen in Fig.13. Southern et al.[54] proposed a theory of crossed-field electrophoresis according to which the mobility should be very small for $t_\perp < \tau_\perp(N)$ due

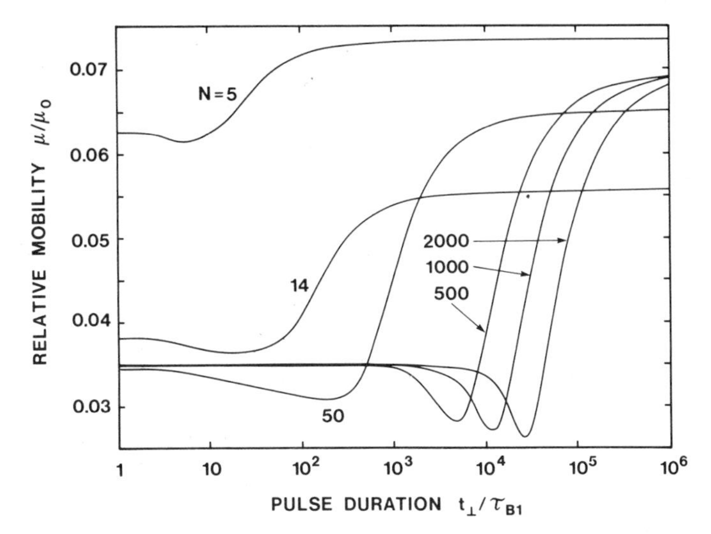

Fig.13 Result of a computer simulation of the BRM: Relative mobility μ/μ_0 as a function of the switching time $t_\perp/\tau_{B1}$ in crossed fields for six different molecular sizes, where $\Theta=1.0$ and $\tau_{B1}=N^{-1}\tau_{Brown}$ is the Brownian time per segment.

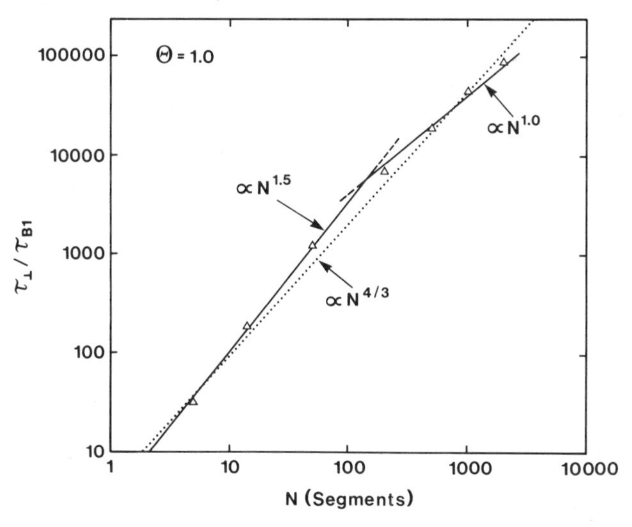

Fig.14 Result of a computer simulation of the BRM (solid line): Scaled reorientation time $\tau_\perp/\tau_{B1}$ as a function of the molecular size N, where $\tau_{B1}=N^{-1}\tau_{Brown}$ is the Brownian time per segment (this time is molecular size independent). The scaled electric field is $\Theta=1.0$. The fitted lines $\tau_\perp \propto N^{1.5}$ and $\tau_\perp \propto N^{1.0}$ are in agreement with Eqs.(32); the line $\tau_\perp \propto N^{4/3}$ is obtained using the power-law suggested by Smith et al.[52]

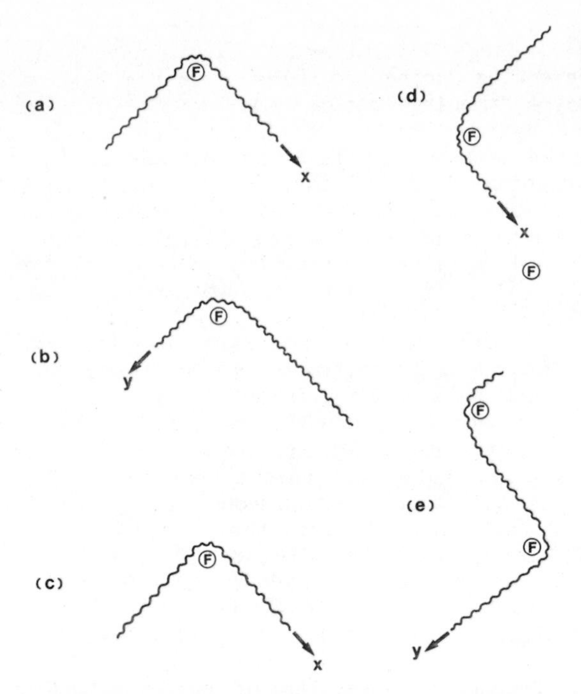

Fig.15 (a,b,c) The "hanging" self-trapping in crossed fields. The V-shaped molecular conformation cannot pass a fiber **F** during a pulse if the pulse duration $t_\perp$ is not long enough. The molecule will move back and forth around the obstacle, until fluctuations help one of its ends pass the point **F**. (c,d) In the normal zig-zag conformation, the molecule can migrate along the gel diagonal by going around obstacles easily.

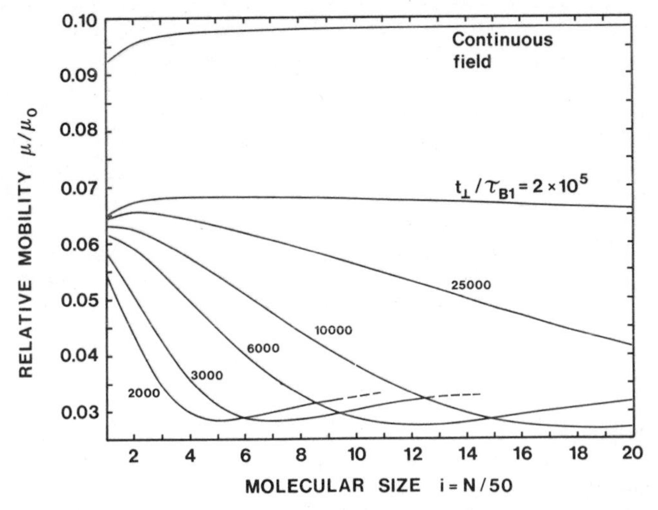

Fig.16 Result of a computer simulation of the BRM: Relative mobility μ/μ_0 as a function of the molecular size $i=N/50$ for various switching times $t_\perp/\tau_{B1}$ and $\Theta=1.0$. The mobility is constant at $\mu/\mu_0 \approx \Theta^2/9 \approx 1/9$ for very large molecules in constant fields.

to the fact that large molecules can be "hanging" in a given position instead of reorienting during the pulses. In Fig.15a, we show a molecular conformation where "hanging" occurs: if $t_\perp < \tau_\perp$, the end of the molecule cannot pass the agarose fiber **F** before the end of the E_x pulse, (Fig.15b), and the next pulse will bring the molecule back to its original V-shaped conformation (Fig.15c) that hangs around point **F**. Of course, fluctuations will destabilize this V-shaped conformation and will give a finite velocity to an otherwise trapped molecule. These fluctuations were not part of Southern's model, but they play a role in the simulation. Not surprisingly, it is for $t_\perp$ slightly smaller than $\tau_\perp(N)$ that a molecule of size N has its lowest mobility: in this case, the molecule undergoes large motions around the hanging point **F**, and only large fluctuations can make it leave this conformation. Note that this gel-induced trapping becomes more important for larger molecules because the fluctuations get weaker. When the molecules do not have the "hanging" conformations, they migrate along the diagonal in a stair-wise fashion as shown in Figs. 12c and 15d-e, and have a normal mobility. This situation is thus similar to migration in continuous fields: the molecules normally migrate with conformations having their larger end-to-end distance parallel to the direction of migration, but sometimes get trapped in long-lived, low-velocity conformations where the end-to-end distance is minimum in the field direction. Only fluctuations can reduce the effect of these "trapping" or "hanging" events.

In order to compare the results of our simulations with those of Southern et al. (which used configurations where the fields where not exactly at 90° however), we simulated the electrophoretic separation of a ladder of 20 oligomers of $N=50i$ segments, with $i=1,2,...,20$. In Figs. 16 and 17, we show the mobility μ_i vs i and the difference $(\mu_i-\mu_{i-1})$ vs i, respectively; these Figs. can be compared with Figs. 5b,c of Southern et al.

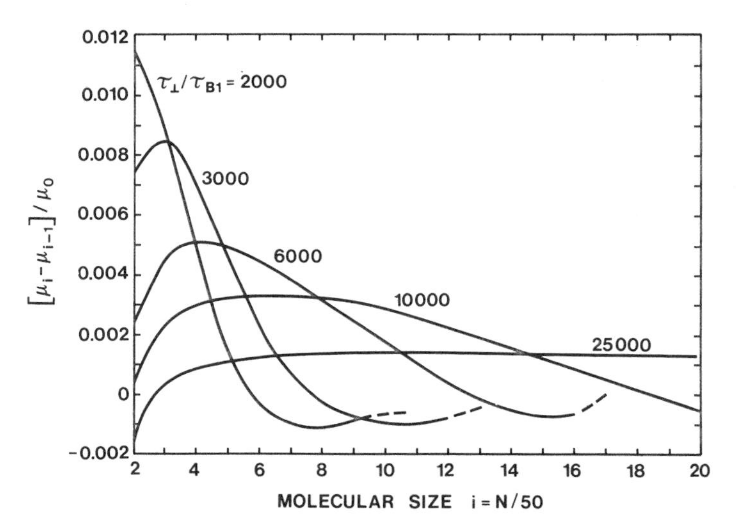

Fig.17 Result of a computer simulation of the BRM: Differential mobility $[\mu_i-\mu_{i-1}]/\mu_0$ as a function of the molecular size $i=N/50$ for various switching times $t_\perp/\tau_{B1}$ and $\Theta=1.0$. DNA molecules can be separated only if $[\mu_i-\mu_{i-1}]>0$.

For very long pulses $t_\perp$, the mobility of all molecules (Fig. 16) is $\mu_\infty \approx 0.070\mu_0 \approx \mu_p/\sqrt{2}$, as predicted by Eq.(60); this does not allow electrophoretic separation. For very short pulses $t_\perp$, the mobility of most molecules is instead $\mu_d \approx 0.035\mu_0 \approx (1/2)\mu_\infty$ as predicted by Eq.(62), which does not allow electrophoretic separation either. For intermediate pulse durations such as $t_\perp \approx 25000\tau_{B1} \approx \tau_\perp(i=20)$ however, the mobility decreases linearly with i over a large range of molecular sizes; this is in general agreement with the results of Southern et al. The $N=50$ molecule has a lower mobility because being close to the continuous field minimum $N_{min} \approx 14$, it is not a "plateau" molecule as such. The mobility goes through another minimum for a certain molecular size $N > N_{min}$, for example $\mu < \mu_d \approx 0.035\mu_0$ for $i=7$ and $t_\perp = 3000$, because of the "hanging" effect discussed before. Therefore, crossed fields greatly extend the range of molecule sizes that can be separated by electrophoresis, and lead to a second minimum in the mobility-size diagram, this one due to "hanging" or crossed-field-induced self-trapping.

Figure 17 shows that the ladder is fairly linear for $4 \leq i \leq 20$ when a $t_\perp = 25000\tau_{B1}$ pulse is used. For shorter pulses, a maximum is observed in the band-band separation for small molecules, and negative separations (i.e. band inversion) ooccurs for larger molecules. This is also in qualitative agreement with the results of Southern et al.

VI.3 Conclusions Our model presents a good qualitative description of (approximately) orthorgonal field electrophoresis. Most observed effects are reproduced, including the "hanging" effect, the right ratio $\mu_\infty/\mu_d \approx 2$, the right power law $\tau_\perp \propto N^{1\text{-}1.50}$, and the linear decrease of the mobility over that range of "plateau" molecules that are separated in crossed fields.

Using Eqs. (23) and (64), we get $\tau_\perp \approx i[4+\ln(i)]$ sec. for the ladder used in our Figures, in good quantitative agreement with the experimental conditions used by Southern et al. A good test of the model would be to check the scaling laws (52) relating the reorientation time $\tau_\perp$ and the electric field intensity.

As we will see in the next section, when field-inversion electrophoresis is used to separate nucleic acids in certain size ranges, intra-tube relaxation effects take place. Of course, as the angle between the two fields increases from 90° to 180°, we expect that the crossed field effects discussed earlier will be modified by more and more of the intra-tube effects that sometimes dominate in field-inversion electrophoresis. It is not the goal of this article to discuss the angles between 90° to 180°. The next section presents a discussion of field-inversion electrophoresis as well as the ways to optimize the reorientation-induced separation obtained with this experimental arrangement.

VII. FIELD-INVERSION ELECTROPHORESIS

A few years after the development of pulsed-field gel electrophoresis (where two approximately orthogonal fields are used), Carle, Frank and Olson[9] observed that if the angle between the two fields was 180°, i.e. if the two fields were anti-parallel, electrophoretic separation is still enhanced by pulsed fields (of course, the two pulse durations have to be different in this case). However, they also observed strong band-inversion under these conditions because the mobility of certain intermediate-sized molecules was almost zero. This latter effect increases the absolute separation between the bands because the latter spread over the whole gel, but makes it very difficult to analyze the results as the bands are not in order of molecular size. Since the size of the molecule showing zero-mobility was a function of the pulse duration, these authors used a ramp where the pulse durations increased steadily during the experiment: in this way, each molecule was the slowest at one point during the experiment and the effect of band-inversion was somewhat reduced. However, this phenomenological method for obtaining a monotonic mobility-size relationship has been criticized as it is not clear how one can choose a ramp that will work for all distributions of DNA molecules.[55]

By using two unequal fields and two unequal pulse durations, Lalande et al.[12] showed that it is indeed possible to separate yeast chromosomes (200-2200 kbp) in order of molecular size by using just a few pulse durations instead of a full ramp. A quantitative analyzis of the experimental results helped to understand the electrophoretic processes involved as well as to optimize the separation. In this section, we discuss field-inversion electrophoresis, and compare the predictions of our model with available experimental results.

VII.1 Theory In the most general case of field-inversion electrophoresis, the field is given by (with $i=0,1,2,3,\ldots$ the pulse index)

$$\mathbf{E} = E_+ \hat{\mathbf{u}}_x ; \qquad for \quad i[t_+ + t_-] \le t \le i[t_+ + t_-] + t_+ \tag{65a}$$

$$\mathbf{E} = -E_- \hat{\mathbf{u}}_x . \qquad for \quad i[t_+ + t_-] + t_+ < t < (i+1)[t_+ + t_-] \tag{65b}$$

During a complete cycle of duration $[t_+ + t_-]$, the field $\mathbf{E}=E_+\hat{\mathbf{u}}_x$ (with $E_+ \ge 0$) is applied in the forward direction (+) for a time t_+, and then the field $\mathbf{E}=-E_-\hat{\mathbf{u}}_x$ (with $E_- \ge 0$) is applied in the reverse direction (-) for a time t_-. We assume here that $E_+ \ge E_-$ and $t_+ \ge t_-$, and we consider only "plateau" molecules for which $\mu \approx \mu_p \approx \Theta^2/9$ in continuous fields. It is thus assumed that $N\Theta^2 \gg 14$ for all molecules under study.

The forward (or +) direction is the direction of the net electrophoretic migration. If the times t_+ and t_- are very long, the molecule has time to renew its tube conformation during each pulse; the velocities during the two parts of the cycle are then:

$$V_+ \approx E_+ \mu_p(E_+) = \mu_0 E_+ \langle cos\theta_+ \rangle^2 \approx \frac{1}{9} \mu_0 E_+ \Theta_+^2 \quad ; \tag{66a}$$

$$V_- \approx E_- \mu_p(E_-) = \mu_0 E_- \langle cos\theta_- \rangle^2 \approx \frac{1}{9} \mu_0 E_- \Theta_-^2 \quad , \tag{66b}$$

where the + (or -) subscripts mean that the parameters take the values they would have in a continuous field of intensity E_+ (or E_-). In this limit, the pulses have the effect of independent continuous field experiments. The net electrophoretic velocity is then:

$$V_\infty \approx \frac{V_+ t_+ - V_- t_-}{t_+ + t_-} = \mu_0 E_- \langle cos\theta_- \rangle^2 \left[\frac{R_E^3 R_t - 1}{R_t + 1} \right] \quad , \tag{67}$$

where

$$R_t \equiv \frac{t_+}{t_-} \; ; \qquad R_E^3 \equiv \frac{E_+ \langle\cos\theta_+\rangle^2}{E_- \langle\cos\theta_-\rangle^2} \approx \left(\frac{\Theta_+}{\Theta_-}\right)^3 , \tag{68}$$

are the time- and field-biases respectively. Note that if $\Theta_\pm \gg 1$, the average cosine functions saturate and one has $R_E \approx (\Theta_+/\Theta_-)^{1/3}$; here, we study only low scaled fields $\Theta_\pm$ for which $R_E \approx \Theta_+/\Theta_-$.

Since we assumed that $\Theta_+ t_+ \geq \Theta_- t_-$, the distance L_+ migrated along the tube axis during one forward (Θ_+) pulse is larger than the distance L_- migrated along the tube axis during one reverse (Θ_-) pulse. This means that in the steady-state, i.e., long after the transient effects that take place at the beginning of an experiment, one has, even for short pulses,

$$\langle h_x(t = i[t_+ + t_-] + t_+)\rangle \approx L_t \langle\cos\theta_+\rangle . \tag{69}$$

In other words (Fig. 18a), after the completion of a Θ_+ pulse, the tube conformation is oriented as if it were in a continuous field of intensity E_+. If the pulse duration t_- is short enough (Fig. 18b), we have $L_- < L_t$, and the conformation at the end of a reverse Θ_- pulse is characterized by a less oriented tube section of length L_- pointing backward. If $L_- \geq L_t$, the whole tube has a new orientation since the chain has disengaged completely from its original tube during the reverse pulse (fig. 18c). Therefore:

$$\langle h_x(t = i[t_+ + t_-])\rangle \approx L_- \langle\cos\theta_-\rangle + (L_t - L_-)\langle\cos\theta_+\rangle \; ; \qquad \text{for } L_- \leq L_t \tag{70a}$$

$$\approx L_t \langle\cos\theta_-\rangle \qquad . \qquad \text{for } L_- \geq L_t \tag{70b}$$

If $t_- \to 0$, one has $L_- \ll L_t$ and $\langle h_x\rangle \approx L_t \langle\cos\theta_+\rangle$ during the whole experiment; in this case, the velocities during the two parts of a cycle are:

$$V_+ \approx E_+ \mu_p(E_+) = \mu_0 E_+ \langle\cos\theta_+\rangle^2 \approx \frac{1}{9} \mu_0 E_+ \Theta_+^2 \quad ; \tag{71a}$$

$$V_- \approx E_- \mu_p(E_+) = \mu_0 E_- \langle\cos\theta_+\rangle^2 \approx \frac{1}{9} \mu_0 E_- \Theta_+^2 \quad , \tag{71b}$$

and the net electrophoretic velocity is given by:

$$V_0 \approx \frac{V_+ t_+ - V_- t_-}{t_+ + t_-} = \mu_0 E_- \langle\cos\theta_+\rangle^2 \left[\frac{R_E R_t - 1}{R_t + 1}\right] . \tag{72}$$

The ratio between the velocity V_∞, corrresponding to long pulses, and the velocity V_0, corresponding to short pulses, is:

$$\frac{V_\infty}{V_0} \approx \frac{\langle cos\theta_-\rangle^2}{\langle cos\theta_+\rangle^2}\left(\frac{R_E^3 R_t - 1}{R_E R_t - 1}\right) \approx \frac{R_E^3 R_t - 1}{R_E^3 R_t - R_E^2} \tag{73}$$

For realistic values such as $R_t = 2.5$ and $R_E = 2.0$, as was used in ref. 12, we get $V_\infty/V_0 = 19/16 \approx 1.19$. This indicates that for constant biases $R_t = 2.5$ and $R_E = 2.0$, an increase of the reverse time t_- (the forward time $t_+ = R_t t_-$ then increases as well) leads to an increase of the net electrophoretic velocity by 19% (in this particular case).

This situation is similar to the one we found for crossed fields. We will note by τ_- the critical reverse pulse duration t_- for which the velocity of a molecule will be $V(t_-=\tau_-) = \frac{1}{2}(V_0 + V_\infty)$; this is similar to the time $\tau_\perp$ used in section VI. This time should be approximately equal to the value of t_- for which $L_- \approx L_t$, i.e., approximately equal to the time τ_{Back} it takes to renew the tube conformation during a reverse pulse as it is only for $t_- \gg \tau_{Back}$ that Eqs. (66) apply, and only for $t_- \ll \tau_{Back}$ that Eq.(70a) applies.

As usual, this critical time τ_- is expected to be molecular-size dependent. Therefore, for a given set of experimental conditions (i.e. a fixed value of t_-), all "plateau" molecules for which $\tau_-(N) \gg t_-$ will have a velocity V_0, while all those "plateau" molecules for which $\tau_-(N) \ll t_-$ will have a larger velocity V_∞. The field-induced reorientation of the reptation tube occurring during the reverse pulse is therefore responsible for a separation of the "plateau" molecules in two groups, and good electrophoretic separation can thus be expected in the critical size range where $\tau_-(N) \approx t_-$.

Note however that $V_\infty = V_0$ when $R_E = 1.0$; this means that the BRM predicts no enhanced separation of the "plateau" molecules if the two fields are of equal intensities ($E_+ = E_-$). This is because the end-to-end distance is then always given by $\langle h_x\rangle \approx L_t \langle \cos\theta\rangle$, giving a time-independent velocity; moreover, since the probability of jumping in the tube against the field is, from Eq.(20a), $p \approx \exp(-2\delta)$, with $\delta = \Theta h_x/a \approx N\Theta^2/3 \gg 1$, the tube conformation has an I-shape that is perfectly symmetrical as regard to field inversion. This result is in contradiction with the experimental results of Carle et al.[9]; we will discuss this matter in section VII.4.

VII.2 <u>The Critical Time</u> To obtain the critical time $\tau_-(N)$ and the tube-renewal time τ_{Back}, we will calculate the velocity $V(N,t_-)$ of a molecule of size N. As explained before, it is assumed here that, at the beginning of a reverse (E_-) pulse, the end-to-end distance of the tube is given by (since we neglect fluctuations in this calculations, we drop the $\langle ...\rangle$ averages for the h_x's):

$$h_x(0) \approx L_t \langle cos\theta_+\rangle . \tag{74}$$

Therefore, the equation of motion for $h_x(t)$ during this pulse is:

$$h_x(t) = h_x(0) - \int_0^t dt\, v_\ell(t) \left[\langle cos\theta_+\rangle - \langle cos\theta_-\rangle\right] \quad ; \qquad t \le \tau_{Back} \tag{75a}$$

$$= L_t \langle cos\theta_- \rangle \quad , \qquad t \geq \tau_{Back} \tag{75b}$$

where τ_{Back} is as before the time needed to leave the original tube (Fig. 18a), with an end-to-end distance given by (74), and enter a less oriented tube (Fig.18c), with an end-to-end distance given by (70b), during a reverse pulse of intensity E_-. Using Eq.(12), rewritten as $v_\ell(t)=\Theta_- h_x(t)/\tau_{Brown}$, we obtain the following integral equation from (75a):

$$h_x(t) = h_x(0) - \tau_{Brown}^{-1}\,\Theta_-\,\Delta \int_0^t dt\, h_x(t) \quad , \qquad t \leq \tau_{Back} \tag{76}$$

where $\Delta \equiv [\langle\cos\theta_+\rangle - \langle\cos\theta_-\rangle] \approx \frac{1}{3}\Theta_-[R_E-1]$. The solution of this equation is:

$$h_x(t) = L_t \langle cos\theta_+ \rangle \, exp\Big(-\Theta_-\,\Delta\,[t/\tau_{Brown}]\Big) \; . \qquad t \leq \tau_{Back} \tag{77}$$

The end-to-end distance thus decreases exponentially with time during a reverse pulse to eventually take the minimum value given by (75b) for $t \geq \tau_{Back}$; from (77), the tube renewal time τ_{Back} is given by:

$$\frac{\tau_{Back}}{\tau_{Brown}} = (\Theta_-\,\Delta)^{-1}\, ln\left(\frac{\langle cos\theta_+\rangle}{\langle cos\theta_-\rangle}\right) \approx 3\,(\Theta_-)^{-2}\,\frac{ln\,(R_E)}{(R_E-1)} \; . \tag{78}$$

The center-of-mass velocity during the reverse pulse being given by $V(t) = \mu_0 E_- \langle (h_x(t)/Na)^2 \rangle$, the distance migrated backward during this pulse is:

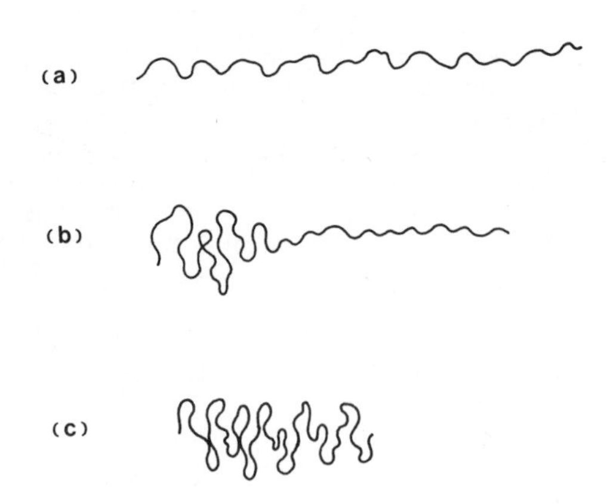

Fig.18 Schematic of the motion of a large DNA molecule during field-inversion electrophoresis. (a) After a forward pulse of duration t_+, the reptation tube is fully oriented in the field direction as if it were in a continuous field experiment with a field intensity E_+. (b) After a short reverse pulse of duration t_-, the new tube section of length L_- is less oriented than the rest because it has been created by the reverse field $E_- < E_+$. (c) For long reverse pulse of durations $t_- \geq \tau_{Back}$, the new tube section is of length $L_- = L_t$, i.e. the whole tube has the orientation it would have during a continuous field experiment with a field intensity E_-. Note that this orientation is less than in the case (a) above since $E_- < E_+$.

$$d_{\text{Back}}(t_-) = \int_0^{t_-} dt\, E_-\, \mu_0\, \langle cos\theta_+\rangle^2\, exp\left(-2\Theta_-\, \Delta\,[\,t/\,\tau_{\text{Brown}}\,]\right)$$

$$= \frac{L_t \langle cos\theta_+\rangle^2}{2\,\Delta\theta} \left[1 - exp\left(-2\Theta_-\,\Delta\,[\,t_-\,/\,\tau_{\text{Brown}}\,]\right)\right] \;; \qquad if\;\; t_- \leq \tau_{\text{Back}}$$

$$d_{Back}(t_-) = \tfrac{1}{2} L_t \left[\langle cos\theta_+\rangle + \langle cos\theta_-\rangle\right] + (t_- - \tau_{\text{Back}})\,\mu_0 E_-\,\langle cos\theta_-\rangle^2 \;. \quad if\;\; t_- \geq \tau_{\text{Back}} \tag{79b}$$

At the beginning of a forward (E_+) pulse, the tube has the conformation shown in Fig.18b if $t_-<\tau_{\text{Back}}$, and the conformation shown in Fig.18c if $t_-\geq\tau_{\text{Back}}$. During the first stage of the forward pulse, the molecule has to disengage from the tube in order to recover a fully extended tube (Fig. 18a) for which the end-to-end distance is given by Eq.(69). This process lasts for a time τ_{For}; the equation of motion for the end-to-end distance $h_x(t)$ is:

$$h_x(t) = h_x(0) + \Delta \int_0^{t} dt\; v_\ell(t) \quad ; \qquad t\leq\tau_{\text{For}} \tag{80a}$$

$$= L_t \langle cos\theta_+\rangle \qquad\qquad , \qquad t\geq\tau_{\text{For}} \tag{80b}$$

where $h_x(0)$ is now given by Eq.(77) for $t_-<\tau_{\text{Back}}$ and by Eq.(75b) if $t_-\geq\tau_{\text{Back}}$. Using again Eq.(12), now rewritten as $v_\ell(t)=\Theta_+ h_x(t)/\tau_{\text{Brown}}$, we find that the solution of the integral equation (80a) is:

$$h_x(t) = L_t \langle cos\theta_+\rangle\; exp\left(\frac{\Theta_-\,\Delta}{\tau_{\text{Brown}}}\left[R_E\, t - min\,(t_-, \tau_{Back})\right]\right) \;. \qquad t\leq\tau_{For} \tag{81}$$

Since τ_{For} is defined by $h_x(t\geq\tau_{\text{For}})=L_t\langle cos\theta_+\rangle$, Eq. (81) gives:

$$\frac{\tau_{\text{For}}}{\tau_{\text{Brown}}} = R_E^{-1}\; min\,(t_-, \tau_{\text{Back}}) \;. \tag{82}$$

The distance migrated forward during this pulse is thus:

$$d_{\text{For}}(t_-) = \int_0^{\tau_{\text{For}}} dt\, \mu_0\, E_+ \langle cos\theta_+\rangle^2\, exp\left[2\Theta_+\,\Delta\,\frac{(t-\tau_{\text{For}})}{\tau_{\text{Brown}}}\right] + \int_{\tau_{\text{For}}}^{t_+} dt\; \mu_0\, E_+ \langle cos\theta_+\rangle^2$$

$$= \frac{L_t \langle cos\theta_+\rangle^2}{2\,\Delta} \left[1 - exp\left(-2\Theta_+\,\Delta\,\frac{\tau_{\text{For}}}{\tau_{\text{Brown}}}\right)\right] + \mu_0\, E_+ \langle cos\theta_+\rangle^2\,(t_+ - \tau_{\text{For}}) \;, \tag{83}$$

which applies for all t_+ as it is assumed in this calculation that $t_+\geq\tau_{\text{For}}$.

Using Eqs.(79a) and (83), we find that during a complete cycle of duration (t_++t_-), the net electrophoretic velocity is, for $t_-\leq\tau_{\text{Back}}$:

$$V(t_- \leq \tau_{Back}) = \frac{d_{For} - d_{Back}}{t_+ + t_-} = \mu_0 E_- \langle cos\,\theta_+ \rangle^2 \left[\frac{R_E R_t - 1}{R_t + 1} \right] = V_0 \quad . \tag{84}$$

This result indicates that the net electrophoretic velocity is, in fact, V_0 for <u>all</u> pulse durations t_- smaller than the tube-renewal time τ_{Back} under these conditions. Using Eqs. (79b) and (83), we find that the net electrophoretic velocity is, for $t_- \geq \tau_{Back}$:

$$V(t_- \geq \tau_{Back}) = V_\infty \left[1 - \frac{\tau_{Back}}{t_-} \left(\frac{R_E^2 - 1}{R_E^3 R_t - 1} \right) \right] \quad , \tag{85}$$

where we have used V_∞ as defined in Eq. (67). It can be checked from this equation that $V(t_- = \tau_{Back}) = V_0$, and that $V(R_E=1) = V_0(R_E=1) = V_\infty(R_E=1)$, as it should be. This equation indicates, as expected, that the velocity decreases when t_- decreases from $t_- \gg \tau_{Back}$ to $t_- = \tau_{Back}$. The critical reverse pulse duration $t_- = \tau_-$ for which the velocity of a molecule is $V(t_- = \tau_-) = \frac{1}{2}(V_0 + V_\infty)$ is, from Eqs.(85) and (73):

$$\tau_- = 2\,\tau_{Back} \approx 6\,(\Theta_-)^{-2} \left[\frac{ln\,R_E}{R_E - 1} \right] \tau_{Brown} \propto \frac{N\;ln\,[E_+ / E_-]}{E_+ E_-} \quad , \tag{86}$$

where the last approximation holds only for large biases R_E.

The critical time τ_- is thus a function of the electric fields E_- and E_+, but not of the ratio $R_t = t_+/t_-$; this is because the critical process, when $E_+ t_+ > E_- t_-$, is the disengagement of the chain from its tube during the reverse pulse (E_-), which takes a time $\tau_{Back} = \frac{1}{2}\tau_-$. Those large "plateau" molecules for which $\tau_{Back}(N) \gg t_-$ do not change their tube conformation appreciably during a reverse pulse as they do not have time to disengage from their tube; therefore, they have a velocity V_0. Smaller "plateau" molecules for which $\tau_{Back}(N) \ll t_-$ do leave their tube very quickly during a reverse pulse, and their velocity V_∞ indicates that they migrate as if in a series of continuous-field electrophoresis runs with field intensities E_- and E_+. For intermediate molecular sizes such that $\tau_{Back}(N) \approx t_-$, the velocity is intermediate between V_0 and V_∞ as the molecules just have time to renew their reptation tube during the low-field reverse pulses.

The ratio R_t enters in the ratio V_∞/V_0 however, and thus determines, with the field bias $R_E = E_+/E_-$, the actual distance between the large and small fragments on the gel. Therefore, the fields $E_\pm$ determine the range of pulse durations to be used to allow the separation of molecules of size N, while the ratios R_E and R_t determine the relative band spreading on the gel. We will show results of computer simulations in the next section, and we will discuss the relation with experimental results.

VII.3 <u>Computer Simulation of Field Inversion Electrophoresis</u> Figure 19 shows a diagram of relative velocity V/V_0 as a function of the reverse pulse duration t_- for constant field and time ratios $R_E = 2.0$ and $R_t = 1.0$, with $\Theta_+ = 1.0$. The arrows indicate where the pulse duration t_- is equal to the critical time $\tau_{Back} = \frac{1}{2}\tau_-$ for the three molecular sizes shown here. We see that the relative velocity V/V_0 increases from 1.0 to $V/V_0 \approx V_\infty/V_0 = 1.75$ (from Eq.(73)) when the reverse pulse duration t_- increases above $\tau_{Back}(N)$, as expected. Since the time $\tau_{Back}(N) \propto N$, electrophoretic separation of these three fragments can be achieved if one chooses $t_- \approx \tau_-(N=100) \approx 1664\,\tau_{B1}$.

This effect has been observed by Lalande et al.[12] who measured an increase of velocity from $\approx$5.1 cm/65 hrs to $\approx$6.5 cm/65 hrs for yeast chromosomes when the reverse pulse duration t_- was increased from 1 sec to thousands of seconds, with $R_E = 2.0$ and $R_t = 2.5$. With these biases, the expected spreading ratio was $V_\infty/V_0 = 19/16 \approx 1.19$, while a spreading of $V_\infty/V_0 = 6.5/5.1 \approx 1.27$ was observed, in good agreement with theory.

Good electrophoretic separation is predicted by the BRM using this field-induced tube re-orientation if three conditions are satisfied: (1) the ratio V_∞/V_0 has to be as large as possible; (2) the velocity V_∞ has to be large enough for experimental times to be reasonable; (3) the critical times $\tau_-(N,\Theta_+,\Theta_-)$ have to be small since the width of the transition region where the velocity increases from V_0 to V_∞ in Fig.19 is of order τ_- (a narrower region increases the selectivity of the technique). One can use the equations of the BRM to choose experimental conditions that will optimize these three conditions.

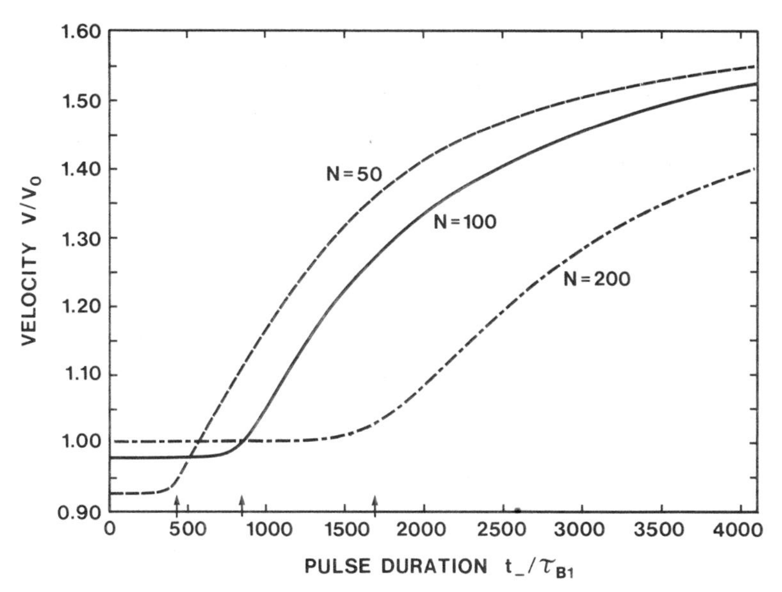

Fig.19 Result of a computer simulation of the BRM: Electrophoretic velocity V/V_0 as a function of reverse pulse duration t_-/τ_{B1} (where V_0 is the velocity for $t_- \to 0$ and $\tau_{B1} = N^{-1}\tau_{Brown}$ is the Brownian time per segment) for three different molecular sizes. The ratios are $R_E = 2.0$ and $R_t = 1.0$, and the forward scaled field is $\Theta_+ = 1.0$. The arrows are at $t_- = \tau_{Back}(N=50) = 416\tau_{B1}$, $t_- = \tau_{Back}(N=100) = 832\tau_{B1}$ and $t_- = \tau_{Back}(N=200) = 1664\tau_{B1}$, and indicate where the velocity should start to increase according to our analytical calculations. Good separation can be obtained for $t_- \approx \tau_-(N=100) = 2\tau_{Back}(N=100) = 1664\tau_{B1}$.

However, the systematic experimental study of Lalande et al.,[12] and the results of Carle et al.,[9] indicate that, together with the increase in mobility found at $t_-=\tau_-(N)$, which would in fact be $V_\infty/V_0=1.0$ for $R_E=1.0$ as in Carle et al., a very large decrease in mobility occurs for $t_-=\tau^*(N)<\tau_-(N)$. This "resonance-like" decrease in velocity was also observed to be broader and more complete for larger molecules, and the characteristic time $\tau^*(N)$ was found to be proportional to the molecular size. In Lalande et al., velocities as low as ≈1.0 cm/65 hrs have been reported for 1500-2200 kbp yeast chromosomes. This effect leads to strong band-inversion. More experiments are clearly needed here to fully characterize this phenomenon. This resonance-like behaviour suggests greater separation power since some fragments show negligible velocities when others are not affected by this effect at a given pulse duration t_-.

VII.4 Conclusion In conclusion, the BRM predicts that field-inversion gel electrophoresis should improve the separation of "plateau" molecules if two unequal fields are used. In this case, the reverse field, which is of lesser intensity, serves to reduce the orientation of the reptation tube; since this is field-induced, this is much faster than the tube relaxation process which is the basis of the intermittent-field technique. Experimentally, this reptation induced effect has been observed, but a potentially even more powerful "resonance-like" effect has been observed to exist as well. This latter effect leads to band-inversion however, and since it occurs for pulse durations $t_-=\tau^*(N)<\tau_-(N)$, it involves chain movements inside its tube and/or movements of only parts of the tube, which have not been discussed in this article.

VIII. DISCUSSION

The BRM is a natural generalization of the original reptation model where the motion of the primitive chain in its "tube" is considered to become biased when an electric field is applied. The electrophoretic properties of the DNA molecules are related in the BRM to the effect of the field on the conformation of the reptation tube: since this conformation tends to orient in the field direction when the primitive chain creates new tube sections, the electrophoretic velocity becomes a nonlinear function of the electric field. This tube alignment also reduces the effectiveness of the entanglements in opposing the electrophoretic drift, with the consequence that, except for transient effects or very small molecules, the mobility becomes molecular-size independent in continuous fields.

The BRM has been successful in explaining continuous-field gel electrophoresis data. In particular, band-inversion, the "plateau" mobility and the $\mu\propto 1/L$ law are all accounted for by the BRM. Better quantitative agreement would be expected with a BRM which would include a pore size distribution, corresponding to a more realistic model of a gel, but a simple uniform gel model already describes most observed effects. A number of power laws predicted by the BRM still need to be checked experimentally, however.

In this paper, it has been assumed that the chain is in equilibrium in its tube, i.e., that the conformation of the chain inside its tube is no different when the field is present; this assumption allows one to replace the polymer by a primitive chain, which greatly simplifies the approach.

In fact, one does expect that the intra-tube molecular conformations will be influenced by the electric field, at least on short time- and length- scales. For example, electric forces may induce tube leakages.

Likewise, these forces may generate inhomogeneous distributions of DNA inside the tube, especially in areas of the gel where the agarose fibers are closer to each other (smaller pores). Finally, it should be noted that since the forces $f_i=q\ \mathbf{E}\cdot\hat{\mathbf{u}}_i$ (see Eq. (10)) acting on the primitive segments $1\leq i\leq N$ are not equal (the unit vectors $\hat{\mathbf{u}}_i$ can point in all directions), there are tensions in the tube between the different parts of the chain, with possible redistributions of the DNA charge along the tube axis.

In continuous fields, these intra-tube effects change the effective values of the parameters τ_{Brown}, N and Θ one needs to fit experimental results, but leave the main predictions of the model intact.

One exception to that is apparently the steady-state orientation factor measured by LD. Our BRM predicts that this orientation factor should be slightly molecular-size dependent in the vicinity of the band-inversion phenomenon, but otherwise size-independent. Experiments showed a fairly strong size dependence up to large molecular sizes, even if these large molecules had a size-independent mobility. This is possibly due to an intra-tube effect which can induce local fluctuations in the average tube orientation but, since the tube stays oriented, retain the molecular-size independence of the mobility.

Although intra-tube effects may be expected to play a larger role in pulsed field techniques and transient effects, the form of the model which averages over these effects offers a good framework to discuss the experimental results even in these cases. The orientation overshoot observed with linear dichroism is a transient effect that occurs at a time $t=\tau_{\mathrm{ov}}$ which is smaller than the tube-renewal time τ_{E}; we thus conclude that the phenomenon involves short length- (i.e. smaller than the tube length L_{t}) and time- (i.e. shorter than the field-induced tube-renewal time τ_{E}) scale intra-tube molecular motions. Similarly, the fast relaxation of the molecular orientation clearly suggests that some fast intra-tube relaxation modes play a role; note however that reptation-induced (tube) relaxation has also been observed.

Intermittent electric fields have been suggested in the past as a way to reduce the average molecular orientation during an experiment, with the predicted effect that this will increase the size dependence of the mobility of the large DNA molecules. Our work indicates that although this is in principle a correct idea, it is limited in practice by the very slow tube relaxation process. For molecules of more than 100 kbp, waiting times of hours will be needed between electric field pulses of a few seconds, which is clearly unpractical. Experimentally, it was observed that the velocity of the large molecules was in fact decreasing very quickly for short waiting periods, but this did not allow electrophoretic separation.[48] This is an agreement with the LD measurements quoted above: intra-tube effects seem to disorient the chain rather quickly, but the tube relaxation still occurs by the very slow field-free reptation of the molecules. It is thus the average tube orientations that changes the power of separation of electrophoresis in this case, and not the intra-tube molecular conformation.

Crossed-field electrophoresis provides an interesting alternative to intermittent-field electrophoresis. The idea is simply to change by 90° the direction of the field each time the tube is fully oriented in the field direction because we know that this orientation destroys the molecular-size dependence of the electrophoretic velocity. Although some intra-tube effects can be expected to play a role here, especially because crossed fields may induce the creation of a large number of transverse tube leakages, the BRM reproduces most experimentally observed effects up to the present time. In particular, the critical time $\tau_{\perp}(N,\Theta)$ (Fig. 14), the ratio $\mu_{\infty}/\mu_{d}=2$ between the mobilities for long (μ_{∞}) and short pulses (μ_{d}), the mobility-vs-pulse duration ($t_{\perp}$) diagram (Fig. 13), and the mobility-vs-molecular size diagram (Fig. 16), correlate well with available data. The BRM rationalizes the model of Southern et al., and takes into account the effect of the fluctuations. The problem of molecular "hanging", which is similar to self-trapping in continuous fields, seems to limit the molecular size one can separate with crossed fields. An increase of the pulse time $t_{\perp}$ increases the range of molecular sizes that can be separated by crossed fields, but since the ratio μ_{∞}/μ_{d} is always equal to two, the spreading of the bands on gel does not increase when more bands are allowed to separate: therefore, the crossed field technique is intrinsically limited according to the biased reptation model.

Field-inversion electrophoresis is predicted to give a much better spreading of the bands. The idea here is to use a secondary, weaker negative field to speed up the relaxation of the reptation tube between the primary (forward) electrophoretic pulses. This is therefore similar to intermittent field electrophoresis, where a secondary field instead of Brownian motion is used to decrease the tube orientation. By choosing the field and time ratios R_{E} and R_{t} properly, one can optimize the band spreading ratio V_{∞}/V_{0} while keeping the critical times $\tau_{-}(N)$ small enough to be useful. Experimentally however, the predicted effect has been observed to be accompanied (but not replaced) by an unexpected resonance-like decrease of the electrophoretic velocity in a limited range of the pulse frequencies. Comparison with theory and LD measurements indicate that this is due to intra-tube effects not discussed in this article. Although this new effect looks promising, especially because the velocity can be made arbitrarily small by choosing the pulse duration properly, it also leads to strong band-inversion, an undesirable phenomenon in practical applications. It is therefore of great interest to study the intra-tube modes in order to understand, quantify and eventually exploit this new effect. In particular, since this effect tends to decrease the velocity of the affected molecules, while reptation tends to increase the velocity of the other molecules when the pulse duration increases, the combination of both effects increases the spreading of the bands on the gel.

IX. CONCLUSION

The biased reptation model provides a good framework to discuss the experimental results of the various gel electrophoresis techniques used to separate nucleic acids. Although more experiments are needed to fully characterize these techniques, available results indicate that the simplified version of the model discussed in this paper is satisfactory when low-frequency pulsed fields are used, or when transient intra-tube effects are not dominant. This is the case in continuous fields, for small molecules in intermittent fields, and possibly also for crossed fields. However, intra-tube effects are observed to play a role in field-inversion electrophoresis, for long molecules in intermittent fields, and during the first stages of an experiment (where an orientation overshoot is observed).

The large number of cases where the biased reptation model is reliable indicates that the intra-tube effects do not rule out reptation as the basic migration mechanism. Further theoretical advances will include the effects of both intra-tube molecular orientation and tube orientation on the electrophoretic properties of large nucleic acids.

NOMENCLATURE

a	average pore size of the gel
a_i	size of i-th pore forming the tube
A	concentration of the agarose gel in %
b	contour length of one base pair = 3.4Å
$\mathbf{c}(t)$	orientation of the new tube section created at time t
$d(t)$	distance migrated by the center-of-mass at time t
$d_{x,y}$	distance migrated by the center-of-mass in the x,y direction
$d_{For}; d_{Back}$	distance migrated during the forward (backward) motion in field-inversion electrophoresi
D	curvilinear diffusion coefficient of the DNA molecule
e	electron charge = 1.6×10^{-19} Coulomb
$E; \mathbf{E}$	electric field
$E_+; E_-$	electric field intensities in field-inversion electrophoresis
f_0	electric force applied on one base pair
$\mathbf{F}$	fiber position
F_ℓ	total longitudinal force acting on the chain in its tube
g	numerical constant equal to $N_{min}\Theta^2$
h	end to end distance of the chain conformation
$h_{x,y,\perp,//}$	component of h in the $x,y,\perp,//$ directions
h_d	component of h along the diagonal of the gel
j	number of pulses
k_B	Boltzmann constant
$K; K_1; K_2$	molecular size in kbp
ℓ	contour length of DNA in an average pore
ℓ_i	contour length of DNA in the i-th pore
L	contour length of the molecule
L_t	contour length of the reptation tube
L_-	distance migrated in the tube during the backward pulse in field-inversion electrophoresis
n	scaled molecular size $=N/N_{min}$
N	molecular size in number of primitive segments
N^*	molecular size of the largest DNA fragment
N_{min}	molecular size of the slowest DNA fragment
N_{max}	molecular size of the largest DNA fragment to be separated
$N_\perp$	molecular size of the DNA fragment which has the critical size in crossed fields when the pulse duration is $t_\perp = \tau_\perp$
p	persistence length of DNA
$p_\pm$	probability of jumping towards the ± end of the tube
$p_0(h_x)$	probability distribution function for the end-to-end distance h_x corresponding to the Lumpkin, Déjardin and Zimm result
$p(h_x)$	probability distribution function for the end-to-end distance h_x
$P(\theta)$	probability distribution for the angle θ
q	net electric charge of one primitive segment
q_i	net electric charge of the i-th primitive segment
q_0	net electric charge of one base pair
Q	net electric charge of the DNA molecule
$\mathbf{r}_i$	i-th vector segment forming the primitive chain
R_E	ratio of field intensities in field-inversion electrophoresis
R_g	radius of gyration of the molecule
$\mathbf{R}_i$	position of the i-th entanglement point defining the tube
R_t	ratio of pulse durations in field-inversion electrophoresis
$S(t)$	distance migrated along the tube axis after time t
S	orientation factor of the DNA molecule
S_t	orientation factor of the reptation tube

t	time
$t_\perp$	time duration of the pulses in crossed-field electrophoresis
$t_+;t_-$	time duration of the forward (backward) pulses in field-inversion electrophoresis
t_{ON}	duration of the on-pulses in intermittent fields
t_{OFF}	duration of the off-pulses in intermittent fields
T	temperature in Kelvin
T_0	duration of the pulse
$\hat{\mathbf{u}}_i$	unit vector along the i-th primitive segment vector $\mathbf{r}_i$
u_{ix}	component of the unit vector $\hat{\mathbf{u}}_i$ on the x-axis
$\hat{\mathbf{u}}_{x,y}$	unit vectors in the x- or y-direction
$u(t,a)$	probability distribution function for a Brownian particle between two absorbing walls
$U(\theta)$	energy of a segment making an angle θ with the field
v_ℓ	total longitudinal velocity of the chain in its tube
V	velocity of the center-of-mass
V_E	velocity of a DNA "plateau" molecule in continuous fields
V_0	velocity of large DNAs in field-inversion electrophoresis with short pulses
V_∞	velocity of large DNAs in field-inversion electrophoresis with long pulses
$V_{/\!/}$	velocity of a large DNA molecule in the instantaneous field direction for long pulses in crossed-field electrophoresis
$V_{x,y}$	velocity of the center-of-mass in the field direction
$V_\pm$	velocity of the center-of-mass during a long ± pulse in field-inversion electrophoresis
y	exponent in the power-law $a \propto A^{-y}$
$\hat{\mathbf{x}}$	unit vector in the x-direction
β	Euler's number
$\Gamma(a)$	orientation factor of a DNA segment of length a
δ	bias of the reptative motion in the tube
$\eta(t)$	stochastic variable taking the value ±1 at time t
θ	angle between the primitive segment and the field direction
Θ	scaled electric field
$\Theta_+;\ \Theta_-$	scaled electric fields (field-inversion electrophoresis)
Λ	linear density of DNA contour length per unit length of tube
μ	mobility of a DNA fragment
μ_{min}	mobility of the slowest DNA fragment
μ_0	mobility of DNA in free solution (no gel)
μ_p	mobility of a large ("plateau") DNA
μ_d	net mobility of large DNAs in crossed-field electrophoresis with short pulses
μ_∞	net mobility of large DNAs in crossed-field electrophoresis with long pulses
ν	exponent in the power-law relating mobility to molecular size at the inflexion point in continuous fields
ξ	curvilinear friction coefficient of the DNA molecule in its tube
ξ_0	friction coefficient of one base pair
σ	screening factor due to counterions
τ	time duration of one Brownian jump in the tube
τ_{Back}	backward tube-renewal time in field-inversion electrophoresis
τ_{Brown}	time duration of one Brownian jump (no electric field)
τ_{B1}	time duration of one Brownian jump (no electric field) for a one segment DNA molecule
τ_D	reptation time
τ_E	tube-renewal time in presence of an electric field

τ_{For} forward tube-renewal time in field-inversion electrophoresis

τ_{ov} time where the "over-stretching" is observed after the field is turned on

τ_R longest Rouse relaxation time

$\tau_{sep.}$ optimal pulse times found experimentally by Smith et al.[52]

τ_{ss} time where the steady-state orientation is observed after the field is turned on

$\tau_{str.}$ stretching time of the DNA chain in an electric field

τ_{un} time where the "under-stretching" is observed after the field is turned on

$\tau_{\pm}$ time duration of one Brownian jump in the ± direction in the tube (with an electric field)

τ_{-} critical pulse duration in field-inversion electrophoresis

$\tau_{\perp}$ tube-renewal time in presence of crossed electric fields

τ^* critical pulse duration for the "resonance-like" effect observed in field-inversion electrophoresis

REFERENCES

1. P.-G. de Gennes, J. Chem. Phys. 55, 572-579 (1971).
2. M. Doi and S. F. Edwards, J. Chem. Soc. Faraday Trans. II 74, 1789-1801 (1978).
3. M. Doi and S. F. Edwards, J. Chem. Soc. Faraday Trans. II 74, 1802-1817 (1978).
4. M. Doi and S. F. Edwards, J. Chem. Soc. Faraday Trans. II 74, 1818-1832 (1978).
5. W. W. Graessley, Adv. Polym. Sci. 47, 67-117 (1982).
6. P.-G. de Gennes, Scaling Concepts in Polymer Physics, Cornell University Press, Ithaca (1979).
7. D. C. Schwartz and C. R. Cantor, Cell 37, 77-84 (1984).
8. G. F. Carle and M. V. Olson, Nucleic Acids Res. 12, 5647-5664 (1984).
9. G. F. Carle, M. Frank and M. V. Olson, Science 232, 65-68 (1986).
10. K. Gardiner, W. Laas and D. Patterson, Somatic Cell and Mol. Gen. 12, 185-195 (1986).
11. G. Chu, D. Vollrath and R. W. Davis, Science 234, 1582-1585 (1986).
12. M. Lalande, J. Noolandi, C. Turmel, J. Rousseau and G.W. Slater, Proc. Natl. Acad. Sci. USA 84, 8011-8015 (1987).
13. B. M. Olivera, P. Baine and N. Davidson, Biopolymers 2, 245-257 (1964).
14. P. Serwer and J. L. Allen, Biochemistry 23, 922-927 (1984).
15. N. C. Stellwagen, Biopolymers 24, 2243-2255 (1985).
16. G. W. Slater, J. Rousseau, J. Noolandi, C. Turmel and M. Lalande, Biopolymers 27, 509-524 (1988).
17. P. G. Righetti, in: Electrophoresis '81 (R. C. Allen and P. Arnaud, eds), pp. 3-16, Walter de Gruyter & Co., New York (1981).
18. P. Serwer and J. L. Allen, Electrophoresis 4, 273-276 (1983).
19. G. S. Manning, Acc. Chem. Res. 12, 443-449 (1979).
20. N. Borochov, H. Eisenberg and Z. Kam, Biopolymers 20, 231-235 (1981).
21. H. Benoit and P. Doty, J. Phys. Chem. 57, 958-963 (1953).
22. N. C. Stellwagen, Biochemistry 22, 6186-6193 (1983).
23. D. Rodbard and A. Chrambach, Proc. Natl. Acad. Sci. USA 65, 970-977 (1970).
24. A. G. Ogston, Trans. Far. Soc. 54, 1754-1757 (1958).
25. P. Serwer and S. J. Hayes, in: Electrophoresis '81 (R. C. Allen and P. Arnaud, eds), pp. 237-243, Walter de Gruyter & Co., New York (1981).
26. W. L. Fangman, Nucleic Acids Res. 5, 653-665 (1978).

27. E. M. Southern, Anal. Biochem. 100, 319-323 (1979).
28. P. Edmonson and D. M. Gray, Biopolymers 23, 2725-2742 (1986).
29. L. S. Lerman and H. L. Frisch, Biopolymers 21, 995-997 (1982).
30. M. W. McDonnell, M. N. Simon and F. W. Studier, J. Mol. Biol. 110, 119-146.
31. J. Noolandi, J. Rousseau, G. W. Slater, C. Turmel, and M. Lalande, Phys. Rev. Lett. 58, 2428-2431 (1987).
32. G. W. Slater, C. Turmel, M. Lalande and J. Noolandi, submitted for publication.
33. C. Turmel and M. Lalande, Nucl. Acid. Res., 16, 4727 (1988).
34. M. Doi and S. F. Edwards, The theory of polymer dynamics, Clarendon Press, Oxford (1986).
35. G. W. Slater and J. Noolandi, Biopolymers 25, 431-454 (1986).
36. G. W. Slater, J. Rousseau and J. Noolandi, Biopolymers 26, 863-872 (1987).
37. O. J. Lumpkin and B. H. Zimm, Biopolymers 21, 2315-2316 (1982).
38. O. J. Lumpkin, P. Déjardin and B. H. Zimm, Biopolymers 24, 1573-1593 (1985).
39. M. Doi, T. Kobayashi, Y. Makino, M. Ogawa, G. W. Slater and J. Noolandi, submitted for publication (1988).
40. J. Rousseau, A Generalized Reptation Model of DNA Gel Electrophoresis, M. Sc. Thesis, Waterloo University, Ontario, Canada (1988).
41. O. Kratky and G. Porod, Rec. Trav. Chim. Pays-Bas 68, 1106-1122 (1949).
42. W. Feller, An Introduction to Probability Theory and Its Application, John Wiley & Sons, New York (1968).
43. B. Åkerman, M. Jonsson and B. Nordén, J. Chem. Soc., Chem Commun. 422-423 (1985).
44. M. Jonsson, B. Åkerman and B. Nordén, Biopolymers 27, 381-414 (1988).
45. G. Holzwarth, C. B. McKee, S. Steiger and G. Crater, Nucleic Acids Res. 15, 10031-10044 (1987).
46. B. Åkerman, M. Jonsson and B. Nordén, submitted for publication.
47. T. Jamil and L.S. Lerman, J. Biomol. Struct. Dyn. 2, 963-966 (1985).
48. M. Lalande, J. Noolandi, C. Turmel, R. Brousseau, J. Rousseau and G. W. Slater, Nucleic Acids Res., 16, 5427-5437 (1988).
49. J. L. Viovy, Biopolymers 26, 1929-1940 (1987).
50. G. W. Slater and J. Noolandi, Electrophoresis 1988, in press (1988).
51. C. L. Smith, P. E. Warburton, A. Gaal and C. R. Cantor in: Genetic Engineering, Principles and Methods (J. K. Setlow and Al. Hollaender, eds.), Vol. 8, pp. 45-70, Plenum Press, New York (1986).
52. C.L. Smith, T. Matsumoto, O. Niwa, S. Klco, J.-B. Fan, M. Yanagida and C. R. Cantor, Nucleic Acids Res. 15, 4481-4489 (1987).
53. D. C. Schwartz, Giga-Dalton Sized DNA Molecules, Ph. D. Thesis, Columbia University, New York (1985).
54. E. M. Southern, R. Anand, W. R. A. Brown and D. S. Fletcher, Nucleic Acids Res. 15, 5925-5943 (1987).
55. T. H. N. Ellis, W. G. Cleary, K. W. G. Burcham and B. A. Bowen, Nucleic Acids Res. 15, 5489 (1987).

FINAL REMARKS

P.G. de Gennes

Collége de France
11, Place Marcelin-Berthelot
75231 Paris CEDEX 05 France

First, I would like to thank all the participants to this symposium, for their willingness to come - sometimes from very distant lands, and in a time where we are submerged with too many conferences. I must also express my gratitude to the sponsors, who have made the whole thing possible, and in particular Dr. Woo Young Lee, from the Mobil Chemical Company who sponsored this ACS award in Polymer Chemistry. You also know that chemical reactions always require a catalyst: here the active element was Dr. Lieng Huang Lee from Xerox: without his everlasting energy this symposium would not have existed, and I want to thank him very warmly.

During the last 4 days, I have heard many kind words from many colleagues! Compliments are always pleasant, and I enjoy them - but I would not like our younger participants to get from this symposium a distorted image of reality. The truth is that I have made many mistakes; even if I restrict myself to polymer science, the list of my mistakes is quite long. Here are a few examples:

a) On the **self-contacts** of a single chain in good solvents, I had a completely naive view: Des Cloiseaux showed that the reality is much more complex.

b) On the ternary systems: polymer A + polymer B + good solvent, I had a naive approach, valid only when A and B are chemically very similar. Strazielle and Leibler have shown to us this week that common situations are different, and even involve a new exponent.

c) On the interdiffusion of two polymers (A + B) in a melt I produced an incorrect theory (the so-called "slow theory") nearly 10 years ago. It was corrected by ad hoc arguments by the Cornell group - but only recently did we get a completely clear discussion (by F. Brochard) allowing the "tubes" to drift but fixing their velocity by a strict balance of forces!

d) Long ago I started some of our experimentalists on a rather stupid program: the idea was to dissolve conventional polymers in nematic solvents, and to measure the many viscosities, or friction coefficients, in the nematic - to obtain some new information on dilute coils. In fact, we know now that, for very general reasons, conventional polymers usually precipitate in a nematic solvent - but it took a long time to understand this. . .

I could quote more mistakes; and there must be many others which have not been detected yet! But the way to survive, in situations like this, is to have active coworkers, who do not take your ideas for granted, and who come with new ideas and new experiments. I have had the luck to interact with many of these, plus some friends abroad, such as S. Alexander and P. Pincus: let me thank them all, from all my heart, for all these years of trial and error.

Letter to the Editor

CONFORMATION OF A POLYMER IN LIQUID CRYSTAL POLYMER*

L. Noirez, J.P. Cotton, P. Keller, M. Lambert, F. Moussa and G. Pépy

Laboratoire Léon Brillouin
(CEA-CNRS) - CEN-Saclay
91191 Gif-sur-Yvette
Cedex, France

F. Hardouin

Centre de Recherche Paul Pascal
Bordeaux I, 33405 Talence
Cedex, France

A side-chain liquid crystal polymer is a polymer where on each monomer is grafted a mesogenic molecule. A good example is given by the liquid crystal derived from poly(methyl methacrylate):

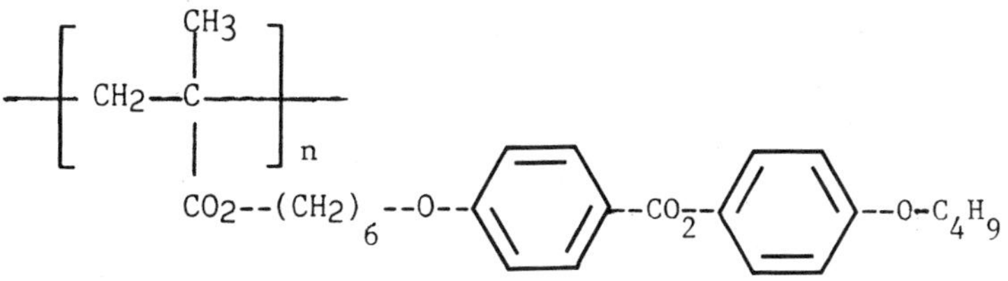

$$\left[CH_2 - \overset{\displaystyle CH_3}{\underset{\displaystyle CO_2-(CH_2)_6-O-C_6H_4-CO_2-C_6H_4-O-C_4H_9}{C}} \right]_n$$

The bulk of such polymers shows the characteristic phases of liquid crystals (isotropic, nematic and smectic). It raises the following problem: how is the polymer backbone inserted inside the mesomorphic molecules?

Following Kirste and Ohm [1], the answer can be obtained from Small Angle Neutron Scattering with samples obtained from having deuterated backbone, mixed with their hydrogenated homologues:

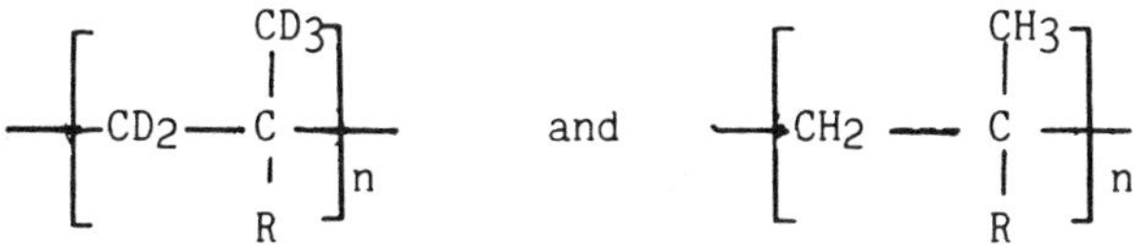

$$\left[CD_2 - \overset{\displaystyle CD_3}{\underset{\displaystyle R}{C}} \right]_n \quad \text{and} \quad \left[CH_2 - \overset{\displaystyle CH_3}{\underset{\displaystyle R}{C}} \right]_n$$

*This was presented by Prof. G. Jannink as a "quickie" to the Symposium.

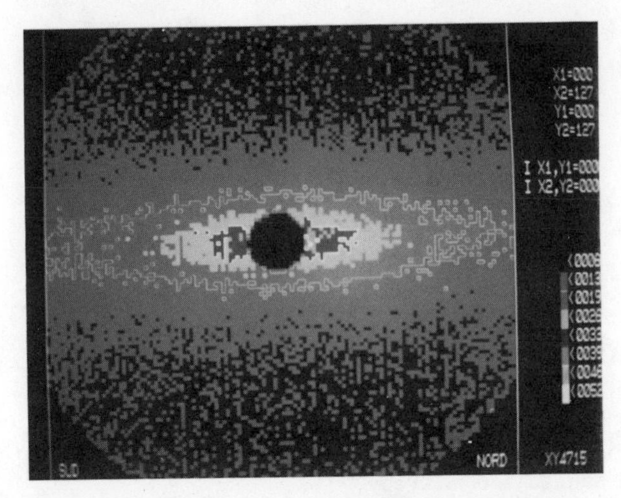

Fig. 1. Intensity scattered on a XY-multidetector by a sample containing 50% chains with deuterated backbones. The scattering pattern is elongated along the direction of the magnetic field which orients the mesogenic groups.

The experiments [2] made at Saclay (spectrometer PAXY) show for the first time the anisotropy of the backbone in the smectic phase (see Fig. 1). In fact, the data analysis reveals that the backbone of chain is confined two smectic layers only (the larger dimension is perpendicular to the magnetic field).

REFERENCES

1. R.G. Kirste and H.G. Ohm, Makromol. Chem. Rapid Commun., 6, 179 (1985).
2. P. Keller, B. Carvalho, J.P. Cotton, M. Lambert, F. Moussa and G. Pépy, J. Physique Lett. (Paris), 46, 1065 (1985).
3. F. Moussa, J.P. Cotton, F. Hardouin, P. Keller, M. Lambert and G. Pépy, J. Physique, (Paris) 48, 1079 (1987).
4. F. Hardouin, L. Noirez, P. Keller, M. Lambert, F. Moussa and G. Pépy, Mol. Cryst. and Liq. Cryst., 155, 389 (1988).
5. G. Pépy, J.P. Cotton, F. Hardouin, P. Keller, M. Lambert, F. Moussa and L. Noirez, Makromol. Chem. Makromol. Sympo., 15, 251 (1988).
6. L. Noirez, J.P. Cotton, F. Hardouin, P. Keller, F. Moussa, G. Pépy and C. Strazielle, Macromolecules, 1988, (in press).

Discussion

On the Paper by M. Adam, M. Desanti, J.P. Munch and D. Durand

Augstin E. González (National University of Mexico): I am puzzled by one thing: In the percolation model one assumes that the monomers are fixed in space while the bonds are being formed, while here the clusters formed should migrate, more like in a cluster-cluster aggregation model. Why should we expect percolation exponents instead of cluster-cluster aggregation exponents?

P.G. de Gennes (College de France): I think that what happens is that the polymerizing system is not very dilute, thus preventing the formed clusters to have a strong movement and migration. In this case we can neglect the kinetics of the clusters. Perhaps, if we had started with a more dilute polymerizing system, we could see a crossover from percolation to cluster-cluster aggregation.

On the Paper by G. Slater and J. Noolandi

R.P. Wool (University of Illinois, Urbana): A few years ago, we determined changes in the Hermans molecular orientation function vs molecular weight, electric field strength, by computer simulation on reptation chains. Have you looked at experimental DNA systems and how do they compare with your result?

Gary Slater (Xerox Research Center of Canada): The changes in the orientation function have been measured by linear dichroism; our theory of DNA chains which include stretching describes this transient orientation when the field is applied, as well as the overdamped oscillatory behavior that follows the initial stretching.

APPENDIX

List of Publications

(1956 - 1988)

by

P.G. de Gennes

LIST OF PUBLICATIONS OF P.G. DE GENNES

I. BOOKS

1) *Superconducting Metals and Alloys*,

Benjamin (1964).

Russian Edition (Mir 1968).

Japanese Edition (1975).

2) *The Physics of Liquid Crystals*,

Oxford University Press - 1st edition (1974); 2nd edition (1976).

Russian Edition (1976)

3) *Scaling Concepts in Polymer Physics*,

Cornell University Press of Ithaca, New York (1979).

Russian edition (1984)

Japanese edition (1986)

II. SCIENTIFIC ARTICLES*

1. "Excitations collectives de spin dans les métaux ferromagnétiques," Compt. rend. 222, 1730 (1956).
2. "Séparation isotropique par passage du courant dans un métal fondu", J. phys rad., 17, 343 (1956).
3. "Diffusion critique des neutrons dans les ferromagnétiques" (avec A. Herpin), Compt. rend., 243, 1611 (1956).
4. "Calcul des corrélations de spin dans une substance magnétique", Compte-Rendu du Colloque National du Magnétisme, Strasbourg (1957) (édition du C.N.R.S.) p. 287.
5. "Effet du désordre de spin sur la résistivité" (avec J. Friedel), C.R. du Colloque National du Magnétisme, Strasbourg (édition du C.N.R.S.) p. 287.
6. "Inelastic magnetic scattering of neutrons at high temperatures", C.R. de la Conference de Stockholm sur l'utilisation des neutrons lents (Stockholm 1957).
7. "Anomalies de résistivité dans les métaux ferromagnétiques" (avec J. Friedel) J. Phys. Chem. Solids, 4, 223 (1958).
8. "Neutron scattering by antiferromagnets above Néel point", J. Phys. Chem. Solids, 6, 411 (1958).
9. "Etudes de magnétisme réalisées avec les neutrons" (avec M. Ericson, A. Herpin, B. Jacrot, et P. Mériel), J. phys rad., 19, 647 (1958).
10. "Sur la relaxation nucléaire dans les cristaux ioniques", J. Phys. Chem. Solids, 7, 345 (1958).
11. "Note sur les corrélations dans les gaz comprimés", Supp. Nuovo Cimento, Vol. IX n°1, 240 (1958).
12. "Sur un phénomène de propagation dans un milieu désordonné" (avec P. Lafore et J.P. Millot), J. phys. rad., 20, 624 (1959).
13. "Amas accidentels dans les solutions solides désordonées" (avec P. Lafore et J.P. Millot) J. Phys. Chem. Solids, 11, 105 (1959).
14. "Sur le magnétisme des métaux de terres rares", Compt rend, 247, 1836 (1959).
15. "Liquid dynamics and inelastic scattering of neutrons", Physica, 25, 825 (1959).
16. "Fluctuations d'aimantation et opalescence critique", (avec J. Villain), J. Phys. Chem. Solids, 13, 10 (1960).
17. "Theory of ferromagnetic resonance in rare earth garnets II. (Lines with (with C. Kittel and A.M. Portis), Phys. Rev., 116, 323 (1959).
18. "Canted spin arrangements", Phys. Rev. Lett. 3, 209 (1959).
19. "Effect of double exchange in magnetic crystals", Phys. Rev., 118, 141 (1960).
20. "Sur l'origine de la largeur de raie dans la résonance nucléaire des métaux ferromagnétiques" (avec J. Friedel), Comp. rend., 251, 1283 (1960).

* Les titres soulignés corrrespondent à quelques contributions qui ont été citées de facon durable dans la littérature.

21. "Sur une méthode de détermination du noyau composé dans la diffusion résonante des neutrons par des noyaux polarisés", Compt. rend., 252, 3571 (1961).

22. "Relaxation nucléaire de 57 Fe dans le grenat de fer et d'ytrium" (avec F. Hartmann), Compt. rend., 253, 2922 (1961).

23. "Modes de déformation d'une paroi de Bloch" (avec F. Hartmann), Compt. rend., 253, 1662 (1961).

24. "Nuclear magnetic resonance modes in magnetic materials" (with P. Pincus, J.M. Winter and F. Hartmann), Phys. Rev., 129, 1105 (1963).

25. "Polarisation de charge ou de spin au voisinage d'une impureté dans un alliage", J. phys. rad., 23, 630 (1962).

26. "Theory of neutron scattering by magnetic material" "Magnetism" Tome III, (édité par Shul et Rado).

27. "Interactions indirectes entre couches 4f dans les métaux de terres rares", J. phys.rad., 23, 510 (1962).

28. "Effet de libre parcours moyen sur le pas des hélices dans les métaux de terres rares" (with D. Saint-James), Solid State Commun., 1, (6), 132 (1963).

29. "Relations between magnetic and superconducting properties" (with G. Sarma), J. Appl. Phys., 34, (2) 1380(1963).

30. "Sur certaines propriéts des alliages supraconducteurs non magnétiques" (avec C. Caroli, et J. Matricon). J. phys. rad., 23, 707 (1962).

31. "Nature of the driving force on flux creep phenomena" (with J. Friedel and J. Matricon), Appl. Phys. Lett., 2, 199 (1963).

32. "Coherence length and penetration depth of dirty superconductors" (with C. Carol and, J. Matricon), Phys. Cond. Matter, 1, 176 (1963).

33. "Elementary excitations in the vicinity of a normal metal superconducting metal contact" (with D. Saint-James), Phys. Lett., 4, 151 (1963).

34. "Collective modes of hydrogen bonds", Solid State Commun., 1, 132 (1963).

35. "Onset of superconductivity in decreasing fields" (with D. Saint-James) Phys. Lett., 7, 306 (1963).

36. "Collective modes of vortex lines in type II superconductors" (with J. Matricon), Rev. Mod. Phys, 36, (1964).

37. "Microscopic derivation of the Josephson currents", Phys. Lett., 5, 22 (1963).

38. "Boundary effects in superconductors", Rev. Mod. Phys., 36, 225 (1964).

39. "Surface superconductivity", Rapport à la Conférence de Cleveland sur les superconducteurs de type II (1964).

40. "Vortex lines in superconductors" Rapport à la 9ème Conférence de Physique des Basses Températures (Columbus, Ohio 1964).

41. "Magnetic behavior of very small superconducting particles" (with M. Tinkham) Physics, 1, 107-126 (1964).

42. "Bound fermion state on a vortex line in a type II superconductor" (with C. Caroli and J. Matricon), Phys. Lett., 9, 307 (1965).

43. "Behavior of dirty superconductors in high magnetic fields", Phys. Cond. Matter, 3, 79 (1964).

44. "Superconducting metals and alloys" Benjamin, New York (1964).

45. "Contribution of crystals field to the line width in rare-Earth Yttrium iron garnet" (with F. Hartmann and D. Saint-James) Phys. Lett., 1, 273 (1962).

46. "Processional modes of flux tubes in superconductors of type I" (with P. Nozières), Phys. Lett., 15, 216 (1965).

47. "Anomalous penetration depths for superconductors coated with normal metals" (with J. Matricon), Solid State Commun., 3, 151-153 (1965).

48. "Excitations électroniques dans les supraconducteurs de 2ème espèce. II Excitations de basse énergie" (avec C. Caroli et J. Matricon), Phys. Kond. Mat., 3, 380-401 (1965).

49. "Vortex nucleation in type II superconductors", Solid Stat. Commun., 3, 127 (1965).

50. "The magnetic behavior of dirty superconductors" (with C. Caroli and M. Cyrot), Solid Stat. Commun., 4, 17-19 (1966).

51. "Proximity effects under magnetic fields, II interpretation of 'Breakdown'" (with J.P. Hurault), Phys. Lett, 17, 181-182 (1965).

52. "Propriétés magnétiques de supraconducteurs de deuxième espèce", L'onde électrique, n° 11, 460 (1965).

53. "Strong field effects at the surface of a superconductor" (Orsay groups on superconductivity), Quantum fluids, North Holland (1966).

54. "The Landau - Ginsburg equations and the properties of type II superconductors", Tokyo Summer Lectures in Theoretical Physics, p. 117 (1966).

55. "Excitation spectrum of superimposed normal and superconducting films" (with S. Mauro), Solid State Commun., 3, 381 (1965).

56. "L'observation des corrélations de vitesse et de pression dans un écoulement turbulent", Compt. rend., 262, S.A. 74 (1966).

57. "Note sur la résonance paramagnétique des supraconducteurs de 2ème espèce", Solid State Commun., 4, 95 (1966).

58. "Ultrasons et phonons dans les supraconducteurs", J. Phys., 28, CI - 169 (1967).

59. "Quasi-elastic scattering of neutrons by dilute polymer solutions : I. free draining limit", Phys. (N.Y.) 3, 37 (1967).

60. "Some properties of high field superconductors with magnetic impurities" (with G. Sarma), Solid State Commun., 4, 449 (1966).

61. "Coupling between ferromagnets through a superconducting layer", Phys. Lett., 23, n° 1, 10 (1966).

62. "Proximity effects" (with G. Deutscher), in "Superconductivity", R. Park ed., M. Dekker, New York, (1969).

63. "Loi d'aimantation d'un eutectique supraconducteur" (avec L. Dobrosavlyevic), Solid State Commun., 5, 177 (1967).

64. "Quasi-elastic scattering by dilute, ideal, polymer solutions : II effects of hydrodynamic interactions" (with E. Dubois-Violette). Phys. (N.Y.) 3, n° 4, 181-198 (1967).

65. "Dénaturation des acides déoxyribonucléiques", Ed. C.N.R.S. p. 15 (1967) Coll. Phys. Théorique de Marseille.

66. "Régimes transitoires dans une transition complète hélice pilote statistique", J. Chem. Phys., 65, n° 5, 962 (1968).

67. "A soluble model for fibrous structures with steric constraints", J. Chem. Phys., 48, 2257 (1968).

68. "Vibration spectra of hydrogen-bonds", Comments Solid State Phys., 1, 65 (1968).

69. "Liquids Crystals", Comments Solid State Phys., 1, 213 (1969).

70. "Fluctuations d'orientation et diffusion Rayleigh dans un cristal nématique", C.R.A.S. 266, 15-17 (1968).

71. "Quasi-Elastic scattering by semi-dilute polymer solutions" (with G. Jannink), J. Chem. Phys., 48, 2260 (1968).

72. "Statistics of branching and Hairpin helices for dAT copolymer", Biopolymers, 6, 715 (1968).

73. "Calcul de la distorsion d'une structure cholestérique par un champ magnétique", Solid Stat. Commun., 6, 163 (1968).

74. "Critical opalescence of macromolecular solutions", Phys. Lett., 26A, 313 (1968).

75. "Structure des cloisons de Grand-Jean-Cano", C.R.A.S., 266, 571 .s.B. (1968).

76. "Upper critical field of weakly couples linear chains", (with S. Barisic), Solid State Commun., 6, 281 (1968).

77. "Isotropic ferromagnets", (with P. Pincus) Solid State Commun., 7, 339 (1969).

78. "Possibiliéts offertes par la réticulation de polymères en présence d'un cristal liquide", Phys. Lett., 28A, 725 (1969).

79. "Boucles de disclination dans les cristaux liquides" (avec J. Friedel), C.R.A.S., 268, 257 (1969).

80. "Vibrations de basse fréquence dans certaines structures biologigues" (avec M. Papoular), volume jubilaire en l'honneur de A. Kastler (1969).

81. "Theory of spin echoes in turbulent fluids", Phys. Lett., 29A, 20 (1969).

82. "Modes "Péristaltiques" d'un film de savon", C.R.A.S., 268, 1207 (1969).

83. "Conjectures sur l'état smectique" J. Phys., (Paris) C4, 30, n° 11, C4-65 (1969).

84. "Diffusion de la lumière dans une transition hélice-pelote", C.R.A.S., 269, 560 (1969).

85. "Dépolarisation de la lumière diffusée lors d'une transition hélice-pelote statistique", C.R.A.S., 269, 705 (1969).

86. "Phenomenology of short range order effects in the isotropic phase of nematic materials", Phys. Lett., 30A, 454 (1969).

87. "Some conformation problems for long macromolecules", Rep. progr. Phys., 32, 187 (1969).

88. "Some applications of path integrals and diagrammatic methods to chemical physics", C.R. de la Conférence de Physique Théorique, Trieste (1968), (publié par IAEA (Vienne, 1969)).

89. "Long range order and thermal fluctuations in liquid crystals", Mol. Cryst. Liq. Cryst., 7, 325 (1969).

90. "Nucleic acid denaturation", Comments Solid State Phys., 2, 49 (1969).

91. "Dynamics of fluctuations in nematic-liquid crystals" (Groupe d'études des Cristaux Liquides), J. Chem. Phys., 51, 816 (1969).

92. "Structures en domaines dans un nématique sous champ magnétique", Solid State Commun., 8, 213 (1970).

93. "Sur l'impossibilité de certaines synthèses asymmétriques", C.R.A.S., 270, 891 (1970).

94. "Theory of X-ray scattering by liquid macromolecules with heavy atom labels", J. Phys., (Paris) 31, 235 (1970).

95. "Theory of light scattering by liquid nematics" (Orsay Liquid Cryst. Group), Liquid Crystals and Ordered Fluids, p. 195 (1970).

96. "Electrohydrodynamique des nématiques" (avec P. Parodi et E. Dubois-Violette), J. Phys., (Paris) 32, 305 (1970).

97. "Theory of magnetic suspensions in liquid crystals" (with F. Brochard) J. Phys., (Paris) 31, 691 (1970).

98. "Pair correlations in a ferromagnetic colloid" (with P. Pincus), Phys. Kond. Mat., 11, S 189 (1970).

99. "Microstructure of liquid crystals" (with M. Kléman), in "Liquid Crystalline Systems", Editors, Gray and Winsor, Academic Press, New York (1971).

100. "Short range order in the isotropic phase of nematic", Mol. Cryst., 12, 193 (1971).

101. "Concept de reptation pour une chaine polymérique", J. Chem. Phys., 55, 572 (1971).

102. "Mouvements de parois dans un nématique sous champ tournant", J. Phys., (Paris) 32, 789 (1971).

103. "Electrohydrodynamic effects in nematics II", Comments Solid State Phys., 3, 148 (1971).

104. "Possible experiments on 2-dimensional nematics" Symp. Faraday Society, 5, 16 (1971).

105. "Remarques sur la diffusion des rayons X par les fluides nématiques", C.R.A.S., 274, 142-144 (1972).

106. "Tentative model for the smectic B phase" (with G. Sarma), Phys. Lett., 38A, 219 (1972).

107. "Exponents for the excluded-volume problem as derived by the Wilson method", Phys. Lett., 38A, 339 (1972).

108. "Sur la structure du coeur des coniques focales dans les smectiques A", C.R.A.S., 275, Série 549 (1972).

109. "Sur la transition smectique A ↔ smectique C", C.R.A.S., 274, 758 (1972).

110. "Analogie entre smectique A et supraconducteurs", Solid State Commun., 10, 753 (1972).

111. "Types de singularités permises dans une phase ordonnée", C.R.A.S., 275, 319 (1972).

112. "Dislocations coin dans un smectique A", C.R.A.S., 275, 939 (1972).

113. "Mecanochromatic effect in nematics", Phys. Lett., 41A, 479 (1972).

114. "Some remarks on the polymorphism of smectics", Mol. Cryst. Liq. Cryst., 21, 49-76 (1973).

115. "Long range distortions in an anisotropic superfluid", Phys. Lett., 44A, 271 (1973).

116. "Statistical properties of focal conic textures in smectic liquid crystals" (with R. Bidaux, N. Boccara, G. Sarma, L. de Sèze and O. Parodi), J. Phys., (Paris) 34, 661 (1973).

117. "Remarks on anisotropic superfluids" Nobel, 24, 112 (1973).

118. "Eclatement des membranes d'érythrocyte sous tension mécanique", (unpublished) (1973).

119. "Nematodynamics" Les Houches-lectures Notes, Gordon and Breach, London and New York (1973).

120. "Recent results in the physics of Liquid Crystals", Nobel, 24, 228 (1973).

121. "Alignment of anisotropic by flow" (with D. Rainer), Phys. Lett., 46A, 429 (1974).

122. "General features of lipid organization", Phys. Lett., 47A, 123 (1974).

123. "Viscous flow in smectic A liquid crystals", Phys Fluids, 17, 1645 (1974).

124. "Coil-stretch transition of dilute flexible polymer under ultra-high velocity gradients", J. Chem. Phys., 60, 5030 (1974).

125. "Remarks on entanglements and rubber elasticity", J. Phys., (Paris) 35, L-133 (1974).

126. "Fluctuations géantes et phénomènes critiques", La Recherche, 5, 1022 (1974).

127. "Gravitational instabilities of liquid crystals" in Cooperative Phenomena, H. Haken, éd. North Holland New York, Amsterdam 1974.

128. "Landau-Ginsburg equations for an anisotropic superfluid", (with V. Ambegaokar and D. Rainer), Phys. Rev., A9, 2676 (1974).

129. "Light scattering from random disclinations in a nematic fluid", J. Phys. Lett., (Orsay, Fr.) 35, L-217 (1974).

130. "Molecular films of block copolymers: theoretical equation of state" (unpublished).

131. "Effet d'un écoulement sur les domaines de Williams", C.R.A.S., 280, 9 (1975).

132. "Brownian motion of a classical partical through potential barriers. Application to the helix-coil transitions of hetero-polymers", J. Statis. Phys., 12, 463 (1975).

133. "Collapse of a polymer chain in poor solvents" J. Phys Lett. (Orsay, Fr.), 36, L-55 (1975).

134. "Large scale organization of flexible polymers" Congrès de I.U.P.A.C. Jérusalem Israël, Isr. J. Chem., 14, 154 (1975).

135. "Sur une éventuelle application à l'étude des interfaces fluides et des smectiques" J. Phys. (Paris) 36, 603 (1975).

136. "Réflexions sur un type de polymères nématiques" C.R.A.S., 281, 101 (1975).

137. "Local Frederiks transitions near a solid/nematic interface" (avec E. Dubois-Violette), J. Phys. Lett., 36, L-255 (1975).

138. "Critical dimensionality for a special percolation problem", J. Phys., 36, 1049 (1975).

139. "Solutions of flexible polymers : neutron experiments and interpretation" (with M. Daoud, J.P. Cotton, B. Farnoux, G. Jannink, G. Sarma and H. Benoit) Macromolecules, 8, 804 (1975).

140. "Reptation of stars", J. Phys., (Paris) 36, 1199 (1975).

141. "Hydrodynamic properties of fluid lamellar phases of lipid/water", (with F. Brochard), Pranama, Supp. n° 1, p. 1-21 (1975).

142. "Phase transition and turbulence : an introduction" in "Fluctuations, instabilities and phase transitions"", T. Riste ed., Plenum Press, New York. (1975).

143. "Effects of long range van der Waals forces on the anchoring of a nematic fluid at an interface" J. Colloid Interface Sci., 57, 403 (1976).

144. "On a relation between percolation theory and the elasticity of gels", J. Phys. Lett., 37, L-1 (1976).

145. "Conformation of a polymer chain in certain mixed solvents" J. Phys. Lett., (Orsay, Fr.) 37, L-59 (1976).

146. "Dynamics of entangled polymer solutions I. Inclusion of hydrodynamic interactions, II. The rouse model", Macromolecules, 9, 587-593, 594-598 (1976).

147. "Effect of shear flows on critical fluctuations in fluids", Liq. Cryst. Lett, 34, 91 (1976).

148. "Instabilities under mechanical tension in a smectic cylinder", J. Phys., (Paris) 37, 1359 (1976).

149. "Remarks on polyelectrolyte conformation" (with P. Pincus, R.M. Velasco and F. Brochard) J. Phys. (Paris) 37, 1461 (1976).

150. "Convection and permeation in cholesteric fluid", J. Colloid Interface Sci., 57, n° 403 (1976).

151. "Scaling theory of polymer adsorption", J. Phys., (Paris) 37, 1445 (1976).

152. "Surface tension and deformations of membrane structures : relation to two-dimensional phase transitions", J. Phys., (Paris) 37, 1099 (1976).

153. "La notion de percolation", La Recherche n° 72, p. 919 (1976).

154. "Conformation properties of one isolated polyelectrolyte chain in D dimensions" (with P. Pfeuty and R.M. Velasco) J. Phys., (Paris) 38, L-5 (1977).

155. "Principe de nouvelles mesures sur les écoulements par échauffements optiques localisés", J. Phys., (Paris) 38, L-1 (1977).

156. "Origin of internal viscosities in dilute polymers solutions", J. Chem. Phys., 66, 12 (1977).

157. "Dynamics of confined polymer chains", J. Chem. Phys., 67, 52 (1977).

158. "One long chain among shorter chains", J. Appl. Polym. Sci. Polym. Symp., 61, 313-315 (1977).

159. "Theoretical methods of polymer statistics", Rivista del Nuovo Cimento, 7, 363 (1977).

160. "Dynamical scaling for polymer in θ solvents", Macromolecules, 10, 1157 (1977).

161. "Hazard et nécessité dans les phénomènes coopératifs", Diogéne n°, 100, Casalini libri ed., Firenzo, Italy (1977).

162. "Transitions de démixition dans des pores" (avec F. Brochard) C.R.A.S., 284, 251 (1977).

163. "Qualitative features of polymer demixtion", J. Phys., (Paris) 38, L-441 (1977).

164. "Statistics of macromolecular solutions trapped into small pores" (with M. Daoud), J. Phys., (Paris) 38, 85 (1977).

165. "Nematic fluids; some easy demonstrations experiments" (Collectif Orsay), Contemp. Phys., 18, n° 3, p. 247-264 (1977).

166. "One long chain among shorter chains", J. Appl. Polym. Sci. Polym. Symp., 61, 313 (1977).

167. "Polymeric liquid crystal: Frank elasticity and light scattering", Mol. Cryst. Lett., 34, n° 8, 177 (1977).

168. "Les surprises du désordre", La Recherche, n° 8, 1082 (1977).

169. "Critical behavior for vulcanization processes", J. Phys., (Paris) 38, L-355 (1977).

170. "Introduction to the physics of transitions" J. Microsc. Spectrosc. Electr., 2, 331, n° 4 (1977). (Transitions phases Colloq. Annu. Nice 1977).

171. "Régimes transitoires de pression dans un matériau poreux", C.R.A.S., 284, 357 (1977).

172. "Transverse acoustic waves in semi-dilute polymer solutions", (with P. Pincus) J. Chim. Phys., (Paris) 74, n° 5 (1977).

173. "Macromolecules and liquid crystals : reflexions on certain lines of research" in Liquid Crystals (Seitz and Turnbull, ed.) Supplément 14 L. Liébert (1978). (with F. Brochard) J. Chem. Phys., 67, 52 (1977).

174. "Global molecular shapes in polyelectrolytes solutions", Colston paper, p. 69-81, vol. 29; communication lors d'une réunion de la Faraday Society à Bristol (ions Macromol. Biol. Syst.) (1978).

175. "Lois générales pour l'injection d'un fluide dans un milieu poreux aléatoire" (avec E. Guyon), J. Mecan., 17, n° 3 (1978).

176. "Scaling law for incompatible polymer solutions", J. Polym. Sci., 16, 1883 (1978).

177. "Instabilités de suspensions anisotropes par passage du courant" (avec M. Adam et P. Pieranski), J. Phys. Lett., (Orsay, Fr.) 39, L-47 (1978).

178. "Cross-over in polymer solutions", (with B. Farnoux, F. Boué, J.P. Cotton, M. Daoud, G. Jannink and N. Nierlich), J. Phys., (Paris) 39, 77 (1978).

179. "Viscosité près d'une transition sol-gel", C.R.A.S., 286, 131 (1978).

180. "Statistics of self-avoiding walks confined to strips and capillaries", (with F.T. Wall, W.A. Seitz and J.C. Chin) Proc. Natl. Aca. Sci., (U.S.A.) 75, 2069 (1978).

181. "Collapse of a flexible polymer chain II", J. Phys., (Paris) 39, L-299 (1978).

182. "Liquid crystals", Encyclopedia of Physics (à paraitre).

183. "Liquid-state Research. Why?" in Microscopic Structure and Dynamics of Liquids, Ed. J. Dupuy and A.J. Dianoux, Plenum, New York and London (1977), p.507.

184. "Phénomènes aux parois dans un mélange binaire critique: physique des colloides" (avec M. Fisher) C.R.A.S., 287, 207 (1978).

185. "Sur le pouvoir torsadant des molécules chirales", (1978).

186. "Some features of phase transition in polymer systems", Ann. Isr. Phys. Soc., 2, n° 1, 75-82 (1978).

187. "Remarques sur les contributions des physiciens en biologie", (Coll. d'intérêt général en Suisse Romande, 1978).

188. "Nematic polymers" (with P. Pincus) J. Polym. Sci., 65, 85-90 (1978).

189. "A short guide to polymer physics", Course 7 (summary), Les Houches (1978).

190. "Quelques réflexions sur la dynamique des chaines flexibles" (exposé au 30ème anniversaire du Centre de Recherche sur les Macromolécules) J. Chim. Phys., (Paris) 76, n° 9 (1979).

191. "Ecoulements "bouchon" en physique des suspensions, modèle de percolation" J. Phys., (Paris) 40, 783 (1979).

192. "Conjectures on the transition from Poiseuille to plug flow in suspensions", J. Phys., (Paris) 40, 783 (1979).

193. "Effect of cross-links on a mixture of polymers" J. Phys. Lett., (Orsay, Fr.) 40, L-69 (1979).

194. "Effects of polymer solutions on colloidal stability" (with J.F. Joanny and L. Leibler), J. Polym. Sci., 17, 1073-1084 (1979).

195. "Incoherent scattering near a sol gel transition", J. Phys., (Paris) 40, L-197 (1979).

196. "Some remarks on the dynamics of polymer melts" (with M. Daoud), J. Polym. Sci., : PPE 17, 1971 (1979).

197. "La matière et ses artisans", La Recherche n° 100, vol. 10, 515-516 (1979).

198. "Ecoulement viscométriques de polymères enchevêtrés", C.R.A.S., 288, 219 (1979).

199. "Conformations de polymères fondus dans des pores très petits" (avec F. Brochard), J. Phys., (Paris) 40, L-399 (1979).

200. "Les gels polymériques : présent et avenir", (unpublished) (1979).

201. "Suspensions colloïdales dans une solution de polymères", C.R.A.S., 288, 359 (1979).

202. "Brownian motions of flexible polymer chains", Nature 282, 367-370 (1979).

203. "Stabilisation des mousses par des solutions de polymères", C.R.A.S., 289, 103 (1979).

204. "Comment on papers 8 (J. Klein) and 9 (J.D. Hoffmann, E. di Marzio, C.M. Guttmann), Faraday Society Discussions, Roy. Soc. Chem., n° 68, (1979).

205. "Theory of long-range correlations in polymer melts", Faraday Society Discussions, Roy. Soc. Chem., n° 68, (1979).

206. "Small angle neutron scattering by semi-dilute solutions of polyelectrolyte", J. Phys., 40, 701 (1979).

207. "Quelques réflexions sur les erreurs des sciences "exactes"", "Le. Debat" Paris (1979).

208. "Erosions et sédimentation : un régime simple", C.R.A.S., 289, 265 (1979).

209. "Some physical properties of polymer solutions and melts", Makromol. Chem. Suppl., 3, 195 (1979).

210. "Mouvement d'une suspension à l'intérieur d'un matériau poreux", C.R.A.S., 289, 329 (1979).

211. "Percolation : quelques systèmes nouveaux" J. Phys., (Paris) Colloque C3, supplément au n° 4, 41, C 3-17 (1980).

212. "Ilots en apesanteur dans une couche de Langmuir" C.R.A.S., 290, 119 (1980).

213. "Weak nematic gels", in Springer Ser. Chem. Phys., Liquid Crystal One- and Two-Dimension Order, p. 231 (1980).

214. "Chimie Physique : quelques frontières nouvelles", Actual. Chim., No. 8, 7 (1980).

215. "Conformations of polymers attached to an interface", Macromolecules, 13, 1069 (1980).

216. "A Collapse of one polymer coil in poor solvents B. Collapse in a mixture AB" Ferroelectrics, 30, pp. 33-47 (1980).

217. "Sur une règle de somme pour des chaines polymériques semi-diluées près d'une paroi", C.R.A.S., 290, 509 (1980).

218. "Correlations and dynamics of polyelectrolyte solutions" (with J. Hayter, G. Jannink and F. Brochard), J. Phys. Lett., (Orsay, Fr.) 41, 451 (1980).

219. "Amas sur des réseaux : sur la différence entre la percolation et la statistique des "animaux"" C.R.A.S., 291, 17 (1980).

220. "Dynamics of fluctuations and spinodal decomposition in polymer blends", J. Chem. Phys., 72, 4756 (1980).

221. "Pierre Curie et le rôle de la symétrie dans les lois physiques", Compres Rendus du Colloque Pune Curie (Editions IDSET Paris) (1980).

222. "Sur la soudure des polymères amorphes" C.R.A.S., 291, 219 (1980).

223. "Dynamics of concentrated dispersions : a list of problems", Physico-Chemical Hydrodynamics, 2, n° 1, pp. 31-44 (1981).

224. "Motions of one stiff molecule in an entangled polymer melt", J. Phys., (Paris) 42, 473 (1981).

225. "Properties of macromolecules in liquid crystal solvents" (with M. Veyssié), International Union of Pure Applied Chemistry Structural Order in Polymers (1981).

226. "Coherent scattering by one reptating chain", J. Phys., (Paris) 42, 735 (1981).

227. "Diamagnétisme de grains supraconducteurs près d'un seuil de percolation", C.R.A.S., 292, 9 (1981).

228. "Champ critique d'une boucle supraconductrice ramifiée", C.R.A.S., 292, 279 (1981).

229. "Polymer solutions near an interface. 1. Adsorption and depletion layers", Macromolecules, 14, 1637 (1981).

230. "Suspensions colloïdales dans un mélange binaire critique", C.R.A.S., 292, 701 (1981).

231. Médaille d'or du C.N.R.S. - Discours de Monsieur Pierre-Gilles de Gennes, (unpublished) (1981).

232. "Les gels", (with M. Rinaudo) (unpublished).

233. "Some effects of long-range forces on interfacial phenomena", J. Phys. Lett., (Orsay, Fr.) 42, n° 16, L-377 (1981).

234. "Couples de polymères compatibles : propriétés spéciales en diffusion et en adhésion", C.R.A.S., 292, 1505 (1981).

235. "Les polymères enchevêtrés : nouvelles études en mécanique, en physique et en chimie", Institut de France, Académie des Sciences, exposé sur la physique (1981).

236. "Effet Soret des macromolécules flexibles" (avec F. Brochard), C.R.A.S., 293, 1025 (1981).

237. "Polymers at an interface. 2. Interaction between two plates carrying adsorbed polymer layers" Macromolecules, 15, 492 (1982).

238. "Mechanical properties of nematic polymers" Polymer Liquid Crystals, A. Ciferri, W.C. Krigbaum and R.B. Meyer, Editors, pp. 115-131 Academic, New York (1982).

239. "Kinetics of diffusion-controlled processes in dense polymer systems. I. Nonentangled regimes", J. Chem. Phys., 76, 3316 (1982).

240. "Kinetics of diffusion-controlled processes in dense polymer systems. II. Effects of entanglements", J. Chem. Phys., 76, 3322 (1982).

241. "The formation of polymer/polymer junctions" Microscopic aspects of adhesion and lubrication, pp. 355-367 Tribology Serles, Elsevier, London (1982).

242. "Microemulsions and the flexibility of oil/water interfaces" (with C. Taupin) J. Phys. Chem., 86, 2294 (1982).

243. "Dynamics of entangled polymer chains" (with L. Léger), Ann. Rev. Phys. Chem., 33, pp. 49-61 (1982).

244. "Mobilité électrophorétique de polyions rigides dans un gel", C.R.A.S., 294, 827 (1982).

245. "Phase equilibria involving microemulsions (Remarks on the Talmon-Prager model)' (with J. Jouffroy, P. Levinson) J. Phys. (Paris) 43, pp. 1241-1248 (1982).

246. "Dynamics of compatible polymer mixtures" (with F. Brochard) Physica, 118A, 289 (1983).

247. "Some neutron experiments on flexible polymers", Physica, 120B, 407 (1983).

248. "Distribution en masse moléculaire des "boucles" dans une couche diffuse de polymères adsorbs", C.R.A.S., 294, 1317 (1982).

249. "Theory of slow biphasic flows in porous media", Physico-Chemical Hydrodynamics, 4, n°2, pp. 175-185 (1983).

250. "Effects of shear flows on polymer-polymer reaction rates", (unpublished) (1982).

251. "Kinetics of polymer dissolution", Physico-Chemical Hydrodynamics, 4, n°4, pp. 313-322 (1983).

252. "Continu et discontinu : l'example de la percolation", Encyclopedia Universalie, p.313 (1982).

253. "Entangled polymers", Phys. Today, 36, 33-39 (1983).

254. "Effet Soret intrinsèque d'un poreux imprégné", C.R.A.S., 295, 959 (1982).

255. "Transfert d'excitation dans un milieu aléatoire", C.R.A.S., 295, 1061 (1982).

256. "Scaling theory of polymer adsorption : proximal exponent", (with P. Pincus) J. Phys. Lett., 44, L-241 (1983).

257. "Reptation d'une chaine hétérogène", J. Phys., (Paris) 44, L-225 (1983).

258. "Capture d'une "fourmi" par des pièges sur un amas de percolation", C.R.A.S., 296, 881 (1983).

259. "Statistics of "Starburst" polymers" (with H. Hervet), J. Phys. Lett., 44, L-351 (1983).

260. "Hydrodynamic dispersion in unsaturated porous media", J. Fluid Mech., 136, pp. 189-200 (1983).

261. "Diffusion-controlled reactions in polymer melts", Rad. Phys. Chem., 22, 193 (1983).

262. "Ellipsometric formulas for an inhomogeneous layer with arbitrary refractive index profile" (with J.C. Charmet), J. Opt. Soc. Am., 73, 1777-1784 (1983).

263. ""Pincements" de Skoulios et structures incommensurables", J. Phys. Lett., 44, L-657 (1983).

263. "Phase transitions of binary mixtures in random media" (avec F. Brochard) J. Phys. Lett., 44, L-785 (1983).

265. "Transitions de phase en présence de perturbations spatialement périodiques" (avec F. Brochard), C.R.A.S., 297, 223 (1983).

266. "Effet des forces à longue portée sur les transitions de mouillage", C.R.A.S., 297, 9 (1983).

267. "Partially connected systems" ou : "Les transitions de connectivité", Portgal. Phys., 15, fasc. 1-2, pp. 1-7 (1984).

268. "Tight knots", Macromolecules, 17, 703 (1984).

269. "Adhésion : une liste de questions" Proceedings of the PISA summer school on interfaces (Nato), Plenum Press, New York (1983).

270. "Quelques états de la matière", Helv. Phys. Acta, 57, 157-164 (1984).

271. "Mobilité électrophorétique de grains avec polymères adsorbés", C.R.A.S., 297, 883 (1983).

272. "Dynamique d'étalement d'une goutte", C.R.A.S., 298, 111 (1984).

273. "Flexible polymers : an introduction", (unpublished) (1983).

274. "A model for contact angle hysteresis" (with J.F. Joanny), J. Chem. Phys., 81, 552 (1984).

275. "Comment s'étale une goutte?" Pour la Science, 79, 88 (1984).

276. "Film mouillant en ascension verticale", C.R.A.S., 298, 439 (1984).

277. "Lois d'étalement pour des gouttes microscopiques", C.R.A.S., 298, 475 (1984).

278. "Capillary rise between closely spaced plates : effect of van der Waals forces" (with B. Legait) J. Phys. Lett., 45, L-647 - L-652 (1984).

279. "Spreading laws for liquid polymer droplets : interpretation of the 'foot'" (with F. Brochard), J. Phys. Lett., 45, L-597 (1984).

280. "'Flexible' nematic polymers : stiffening near the clearing point", Mol. Cryst. Liq. Cryst. (Lett.) 102, pp. 95-104 (1984).

281. "Liquid-liquid demixtion inside a rigid network : qualitative features", J. Phys. Chem., 88, n°26 (1984).

282. "Structure statique des films de mouillage et des lignes de contact" (with J.F. Joanny) C.R.A.S., 299, 279 (1984).

283. "Dynamique du mouillage : films précurseurs sur solide "sec" (avec H. Hervet) C.R.A.S., 299, 499 (1984).

284. "Compétition entre mouillage et forces macroscopiques adverses" C.R.A.S., 299, 605 (1984).

285. "Wetting : statics and dynamics", Rev. Mod. Phys., 57, n°3, 827 (1985).

286. "Adsorption de polymères linéaires flexibles sur une surface fractale", C.R.A.S., 299, 913 (1984).

287. "The dynamics of wetting/general features and open questions for low temperature experimentation", (1984).

288. "Etalement "sec" d'une goutte sur une surface aléatoire", C.R.A.S., 300 129 (1985).

289. "Partial filling of a fractal structure by a wetting fluid", A Paraitre dans le livre en l'honneur du Prof. N. Mott Physics of Disordered Materials, D. Adler, H. Fritzsche and S.R. Oshinsky, p. 227, Plenum, New York (1985).

290. "Dry spreading of liquids on solids", Physico-Chemical Hydrodynamics, 6, n°5/6, pp. 579-583 (1985).

291. "Stabilité de films polymère/solvant", C.R.A.S., 300, 839 (1985).

292. "Role of long range forces in heterogeneous nucleation", (with J.F. Joanny), J. Colloid Interface Sci 111, 94, (1986).

293. "Kinetics of collapse for a flexible coil", J. Phys.Lett., 46, L-639 - L-642 (1985).

294. "Pattern recognition by flexible coils", (unpublished) (1985).

295. "Transitions de monocouches à molécules polaires" (avec D. Andelman, F. Brochard et J.F. Joanny), C.R.A.S., 301, 675 (1985).

296. "Conjectures on the transport of a melt through a gel" Macromolecules, 19, 1245 (1986).

297. "Upward creep of a wetting fluid : a scaling analysis" (with J.F. Joanny) J. Phys., 47, 121-127 (1986).

298. "Dynamique d'une couche de polymères adsorbés", C.R.A.S., 302, 765 (1986).

299. "Solutions de polymères conducteurs". (avec Lois d'échelles), C.R.A.S., 302, 1 (1986).

300. "Pénétration d'une chaine dans une couche adsorbée: échanges solution/adsorbat et pontage de grains colloïdaux", C.R.A.S., 301, 1399 (1985).

301. "Polymer-polymer interdiffusion" (with F. Brochard), Europhys. Lett., 1, pp. 221-224 (1986).

302. "Conducting polymers in solution : scaling laws for charge transport", Physica, 138A, 206-219 (1986).

303. "Decomposition of Langmuir-Blodgett layers", Colloid Polym. Sci., 264, 463-465 (1986).

304. "Wetting of polymer covered surfaces" (with A. Halperin) J. Phys., (Paris) (1986).

305. "Dynamique d'une ligne triple", C.R.A.S., 302, 731 (1986).

306. "Imperfect hele - Shaw cells" J. Phys., (Paris) 47, 1541-1546 (1986).

307. "Neutron scattering by adsorbed polymer layers" (with L. Auvray), Europhys. Lett., 2, pp. 647-650 (1986).

308. "A cascade theory of drag reduction" (with M. Tabor), Europhys. Lett., 2, (7), pp. 519-522 (1986).

309. "Model polymers at interfaces" in "Physical Basis of Cell/Cell Adhesion", P. Bongrand ed., CRC press (1987).

310. "Nucléation en régime de mouillage total" (avec J.F. Joanny), C.R.A.S., 303, 337 (1986).

311. "L'arête produite par un coin liquide près de la ligne triple de contact solide/liquide/fluide" (avec Martin et E.R. Shanahan), C.R.A.S., 302, 517 (1986).

312. "Time effects in viscoelastic fingering", Europhys. Lett., 3, (2), pp. 195-197 (1987).

313. "Towards a scaling theory of drag reduction", Physica, 140A, 9 (1986).

314. "Flexible polymers at solid/liquid interfaces" Annali di Chimica, 77, 389-410, Società Chimica Italiana, (1987).

315. "Mécanismes de transformation d'un film de savon horizontal", C.R.A.S., 303, 1275 (1986).

316. "'Flambage électrostatique' de bicouches chirales", C.R.A.S., 304, 259 (1987).

317. "Introduction à la dynamique des polymères flexibles" Lezione Lincei, Milan (décembre 1986).

318. "Rupture fragile en présence d'un liquide passif", C.R.A.S., 304, 547 (1987).

319. "Dynamics of drying and film-thinning", in Proceedings of the International Winter School on the Physics of Amphiphilic Layers held at les Houghes, France, J. Meunier, D. Langevin and N. Boccara, Eds. Springer-Verlag, Berlin (1987).

320. "Stabilité des films de savons 'jeunes', C.R.A.S., 305, 9 (1987).

321. "Bursting of a soap film in a viscous environment" (with J.F. Joanny), Physica 147A, 238-255 (1987).

322. "Bubbles", Phys. Today, 40, 17 (1987).

323. "Rôle du double échange dans les oxydes de cuivre à valence mixte" C.R.A.S, 305, 345 (1987).

324. "Transport of polymer solutions through non-adsorbing porous media", Proceeding of the Cargise Symposium on Physical Hydrodynamics (1987).

325. "Antiferromagnétisme et champs cristallins dans les oxydes de cuivre supraconducteurs", (avec V. Ambegaokar et A.M. Séguin Tremblay) C.R.A.S., 305, 757 (1987).

326. "Polymers at an interface; a simplified view" Adv. Colloid Interface Science, 27, 189-209 (1987).

327. "Interactions entre deux plaques via un polymère fondu" C.R.A.S., 305, 1181 (1987).

328. "Some dynamical features of adsorbed polymers", in Molecular Conformation and Dynamics of Macromolecules in Condensed Systems, M. Nagasawa (Ed.) Studies in Polymer Science, Volume 2, pages 315-331, Elsevier Science Publishers B.V., Amsterdam (1988).

329. "Effondrement d'une structure fragile", C.R.A.S., 306, 5 (1988).

330. "Mobilité d'une chaîne adsorbdée", C.R.A.S., 306, 183 (1988). + Erratum C.R.A.S., 306, 739 (1988).

331. "Electrophordse d'un obvjet fractal", (avec. F. Brochard) C.R.A.S., (1987).

332. "Dynamic capillary pressure in porous media" Europhys. Lett., 5, (8). pp. 689-691 (1988).

333. "Ségrégation par traction dans un homopolymère" (avec F. Brochard), C.R.A.S., 306, 699 (1988).

334. "Discrimination chirale dans une monocouche de Langmuir" (avec. D. Andelman) C.R.A.S, 307, 233 (1988).

335. "Instabilités d'un tas de sable en vibration", (unpublished) (1988).

336. "Dynamics of adsorbed polymers", in New Trends in Physics and Physical Chemistry of Polymers, Ed. L.H. Lee, to be published by Plenum, New York, 1989.

337. "The Dynamics of Wetting," in Fundamentals of Adhesion, Ed. L.H. Lee, to be published by Plenum, New York, 1989.

ABOUT THE CONTRIBUTORS

Professor P.G. de Gennes has been one of the most inventive theoretical physicists in the world studying organized molecular systems (liquid crystals, polymers) over the last 30 years (See his list of publications in the Appendix).

His various researches in physics of condensed matter - magnetism of rare earth and, more particularly, on superconductivity - had already made him one of the most famous theoretical solid state physicists.

His book on second kind superconductors remains the basic theoretical reference on the field. At the same time, de Gennes had developed powerful methods (correlation functions, scaling approaches) which he naturally applied to his further research. The ten years of activity on liquid crystals corresponds to the period of major development of the field. His book, The Physics of Liquid Crystals, is the basic original reference of the subject matter.

In parallel with this activity, de Gennes had started considering the scaling properties of polymers, following the original statistical approach of Flory. Again here, he brought his original view of the field; his understanding of polymers in solution with the various dilute, semi-dilute and concentrated regimes is a remarkable mixture of deep understanding of scaling methods (polymer statistics described as an analytical continuation of magnetic problems), detailed knowledge of the chemistry of polymers and intuition in the description of the systems in simple model terms (such as "blobs" for the semidilute case and "reptation" for the dynamics of individual chains in the presence of entangled other ones).

Dr. de Gennes activated strong experimental activities (Collège de France, Strasbourg, Saclay) using, in particular, tools such as neutron- and light-scattering techniques which he masters particularly well and where his correlation function type approaches can be applied. This theoretical and experimental activities have spread again over the world in particular in the U.S.A. The book, Scaling Concepts in Polymers is another milestone in de Gennes activity.

P. Pincus, Ph.D., University of California at Berkeley. 1962 - 1982, Physics Department, UCLA; 1982 - 1984, Exxon Research and Engineering Company. Since 1984, Materials Department, University of California at Santa Barbara.

Frank P. Buff, Ph.D., California Institute of Technology, 1949. Professor of Chemistry; University of Rochester 1950-. Research interests: statistical mechanical theories of fluids, interfaces, phase transitions, capillarity and chemical kinetics.

Mark A. Robert, Ph.D., Swiss Federal Institute of Technology, Lausanne, 1980. Associate Professor of Chemical Engineering; Rice University, 1984-. thermodynamics and statistical mechanics of fluid phase transitions, critical phenomena, surfaces and interfaces.

Matthew Tirrell, B.S. Ch.E., Northwestern University, 1973. Ph.D., Polymer Science and Engineering, University of Massachusetts, 1977. Shell Distinguished Chair in Chemical Engineering, University of Minnesota. Published over 100 papers.

Sanjay Patel, obtained Bachelor of Technology from Indian Institute of Technology, Kampur, India (1982). Ph.D., under Professor M. Tirrell, University of Minnesota, 1988. Set up the surface forces laboratory at the University, studied homopolymer and block copolymer adsorption. Currently he is at the AT&T Bell Laboratories, Murray Hill, N.J.

Jacob Klein, Ph.D., Cambridge University, England, 1977. 1984 - present, Professor of Polymer Physics, Weizmann-Institute, Israel, Research interests: physics of polymers and of interfacial phenomena.

D. Langevin, Ph.D. in physics, Ecole Normale Supérièure, France, 1974. Post-doctoral work at Collège de France. Research work at Ecole Normale - experimental studies with optical techniques of surfactant systems: monolayer, micelles, microemulsions. Currently, Directeur de Recherche at the CNRS.

Francis Rondelez, Ph.D., Universite de Orsay, Paris, 1973. Post-doc at M.I.T. (73-75). 1975 - 85, he was a Group Leader at Collège de France. Research work: polymer solutions, reptation, monolayers, Forced Rayleigh work on polymers in solutions and interfaces. Since 1987, he is also Director of Research in the Lab of Structure and Reactivity at Interfaces, University Paris VI.

Maya Dvolaitzky, Ph.D., Sorbonne, 1964. Director de Recherche au C.N.R.S. She is an organic chemical researcher and has been collaborating with Professor P.G. de Gennes since 1974 in physico-chemistry of surfaces (microemulsions, liquid crystal lyotropic phases).

Lieng-Huang Lee is a Senior Scientist of Xerox Corporation. He received his B.S. degree in Chemistry from Xiamen (Amoy) University, Xiamen, China, and his M.S. and Ph.D. degrees in Chemistry from Case Institute of Technology, Cleveland, Ohio. He is the editor of ten books on adhesion, friction and wear of polymers. Currently, he is editing two new books on Fundamentals of Adhesion and Fundamentals of Adhesive Bonding to be published by Plenum, New York. In addition, he has published over sixty technical papers and holds twenty-seven U.S. patents. He is a member of the Board of Editors for the Journal of Adhesion. In 1976, he was the Chairman of the Division of Organic Coatings and Plastics Chemistry of the American Chemical Society. In 1983, he was chosen as one of the distinguished scholars by the National Academy of Sciences and the Chinese Academy of Sciences. In 1986, he was awarded a visiting professorship in chemistry by Xiamen University. Dr. Lee is listed in several biographical directories, including World's Who's Who in Science. Currently Dr. Lee is an Honorary Professor of the Chinese Academy of Sciences.

Richard P. Wool joined the faculty of the University of Illinois at Urbana-Champaign as an Assistant Professor in the Department of Metallurgy and Mining in 1977 and has attained the rank of Professor in the Department of Materials Science and Engineering. He received a Ph.D. in Materials Science and Engineering from the University of Utah in 1974 after having done graduate research in mechanical engineering at the University of Colorado and in materials science and engineering at the University of Utah. Professor Wool's research interests include biodegradable plastics, polymer interfaces, molecular deformation, composites, mechanical properties, vibration spectroscopy and simulation.

F. Brochard-Wyart, Ph.D. Orsay, France, 1974. 1974 - 1986 - research in dynamics of polymer at Collège de France. 1988 - present, Professor at University P.& M. Curie. Current research interests: wetting/dewetting/polymer fibers.--. She is also an animator of a group carrying out research in structure, reactivity at interface.

Lilane Léger, Ph.D. Orsay, France, 1976. Joined the Laboratoire de Physique de la Matière Condensée in the Collège de France in 1977. Since 1985, started a new activity on polymers at interfaces and spreading and is currently the Group Leader for that area of research.

Anne-Marie Cazabat graduated from the Ecole Normale Supérieure in Paris. She worked first at the E.N.S. laboratory, applying light-scattering techniques to various fields: relaxation processes in gases and liquids (Rayleigh Brillouin Scattering), or dynamic processes in microemulsions. She is currently Professor of Physics at the Université Pierre et Marie (Paris VI), while studying wetting phenomena in the laboratory of Professor de Gennes in the Collège de France.

Benjamin Chu received his Ph.D. degree from Cornell University in 1959. After a four-year post-doctoral training with the late Professor Peter J.W. Debye, he went to the University of Kansas as an Assistant Professor in 1962. He has been Professor of Chemistry at the State University of New York at Stony Brook since 1968 and Professor of Materials Science and Engineering since 1982. His research interests are in light scattering and transient grating techniques, small-angle X-ray scattering, polymer physics and colloid science.

CL. Strazièlle Docteur ès-Sciences, University of Strasbourg, France (1966). Laboratory: Institut C. Sadron (CRM), CNRS, Strasbourg (since 1961). Field of interest: Physicochemical properties of polymer solutions (dilute and semidilute solutions = Characterization of polymers, thermodynamic of solutions and conformation of polymers - Ternary systems: Polymer - polymer - solvent and polymer - solvent - solvent.

Renyuan Qian, (Jen-Yuan Chien), B.S., Zhejian University, China, 1939. Research in Raman Spectroscopy at the University of Wisconsin, at Madison around 1956. Professor, Academia Sinica since 1951. Joined Institute of Chemistry, Beijing, China, 1956. Director, Institute of Chemistry, Academia Sinica, 1980-84. President, Chinese Chemical Society, 1982-86; Executive President, 1984-85.

G. Jannink, Ph.D. Currently at Laboratoire Léon Brilloun, CEN-Saclay, France. The following is an excerpt from his own words about his co-workers, P.G. de Gennes et al.: "I remember playing tennis with P.G. de Gennes in Grammar School; he was excellent. I met him again in 1956, on my return from a long period at the University of Michigan where I studied polymer dynamics. In Paris, with M. Daoud, J.P. Cotton, B. Farnoux and H. Benoît we shared the aspiration to develop new Polymer Physics as proposed by P.G. de Gennes and S.F. Edwards; I became interested in experimental tests using small-angle scattering at the Saclay Nuclear Center. I coauthored with J. des Cloizeaux a book on polymer structure in solutions."

Andrzej R. Altenberger, b. 1942, M.S. (1966) University of Warsaw, Department of Chemistry, Ph.D. (1972) Polish Academy of Sciences, Institute of Physical Chemistry. Visiting Assistant Professor, University of Minnesota since 1982. Member of: American Chemical Society, New York Academy of Sciences. Research Interests: Statistical mechanics, polymer rheology, light scattering, electrolytes.

John S. Dahler, b. 1930, B.S. (1951) and M.S. (1952) University of Wichita; Ph.D. (1955), University of Wisconsin at Madison; University of Minnesota, Departments of Chemistry and of Chemical Engineering and Materials Science, Asst. Prof. (1959-61), Assoc. Prof. (1961-64), Prof. (1964-). Research Interests: Non-equilibrium statistical mechanics and quantum theory of collisions.

M. Desanti, Ph.D. in Polymer Science, 1978. Post-doc, University of Wisconsin, 1979. Currently, physicist at Commissariat à lénergie atomique, France.

Isaac Sanchez, Ph.D. in physical chemistry from University of Delaware, 1963. From 1969 - 1971 he was a NRC/NAS Postdoctoral Assoc. at the National Bureau of Standards. After spending a year at the Xerox research laboratories in Webster, New York, he accepted a position in 1972 as Assist. Professor in the Polymer Science and Engineering Dept. at the U. of Massachusetts. In 1977 he returned to NBS as a research and supervisory chemist. He still retains academic ties with UMASS through an adjunct professorship. In 1986 he left NBS for his position as Fellow at the Alcoa Laboratories located just outside Pittsburgh. Currently he is a Professor at the University of Texas at Austin.

Peter F. Green, Member of the technical staff at Sandia National Laboratories in Albuquerque, New Mexico. He received a BA and a MA in Physics from Hunter College, New York, in 1981. In 1985 he received a Ph.D. in Materials Science from Cornell University. His research interests include polymer surfaces and ion beam analysis of diffusion in polymers.

Russell J. Composto, B.A., Gettysburg College, 1962. M.S., 1984 and Ph.D., 1987, Cornell University. He is currently a post-doctoral fellow in the Department of Polymer Science and Engineering at the University of Massachusetts. Working with Prof. Richard S. Stein, they are studying the diffusion of unentangled molecules and the kinetics of phase separation in polymer blends.

Leoncio Garrido From 1979 to 1982, preparation of the work for the doctoral dissertation of the Instituto de Plasticon Y Caucho, Spanish Research Council, Madrid (Spain). In 1982, Ph.D. in Chemistry, the University of Vallaclolid (Spain). From 1982 to 1987, research fellow at Harvard Medical School and Massachusetts General Hospital, Boston and from 1987 to the present, physicist at MGH.

B. Deloche, Ph.D., 1978. Thesis "NMR Studies of Molecular Dynamics in Liquid Crystals". 1980, Postdoctoral fellow at the University of Connecticut in the laboratory of Professor E.T. Samulski. 1987, Winter-Klein prize of the Academie des Sciences (Paris).

Christos Tsenoglou. He received a Dipolma in Chemical Engineering at the National Tech. University of Athens, and M.S. and Ph.D. (1985), in the same discipline at Northwestern University (Evanston, Illinois). He is currently an assistant professor in the Dept. of Chemistry and Chemical Engineering at Stevens Institute of Technology. His research interests are in the areas of experimental and theoretical rheology of polymer melts, solutions, blends and suspensions; molecular aspects of viscoelasticity and diffusion of macromolecular systems, rubber elasticity, kinetic theories for the viscoplasticity and thixotropy of suspensions, and non-Newtonian fluid mechanics in polymer processing and mixing.

A. Silberberg B.Sc. (Eng.) in Chemical Engineering, University of Witwatersrand, Johannesburg. Entered the University of Basel, Switzerland, to study under Prof. W. Kuhn. Received the Ph.D. degree in 1952, "Summa cum laude", for a thesis in Physical Chemistry entitled "Interfacial Tension and Phase Separation in Two Polymer-Solvent Systems". Appointed to the Weizmann Institute of Science, Rehovot, Israel, in 1953, promoted to Research Associate in 1954, to Senior Scientist in 1959, to Associate Professor in 1963, to Professor in 1970.

Recipient of the Poiseuille Gold Medal Award of the International Society of Biorheology, 1981. First incumbent of the Joseph and Marian Robbins Chair of Biorheology at the Weizmann Institute of Science.

Tom A. Kavassalis Ph.D., is a member of Research Staff at the Xerox Research Centre of Canada. His current interests include the rheology of polymeric fluids, polymer blends and nonlinear irreversible phenomena in fluids.

Jaan Noolandi Ph.D., is a senior research fellow at the Xerox Research Center of Canada. He is active in the areas of polymer blends, polymer dynamics and the application of reptation concepts to gel electrophoresis of DNA macromolecules.

Michael Rubinstein, received B.S. with honors from California Institute of Technology in 1979, majoring in physics. Won the Carnation Prize at Caltech in 1978. Received M.A. in 1980 and Ph.D. in 1983 in theoretical physics from Harvard University under the supervision of Prof. David R. Nelson. For the next two years worked as a postdoctoral fellow at AT&T Bell Laboratories. In 1985 joined the Research Laboratories of Eastman Kodak Company, where in 1987 won C.E.K. Mees Award for "Excellence in Scientific Research and Reporting". An adjunct professor at the Department of Physics and Astronomy of the University of Rochester since 1987.

Henri Benoît, Thèse d'Etat Strasbourg, 1950. Assistant Professor then Professor at Strasbourg University Louis Pasteur from 1949 till now. Assistant Director 1954-1967 then Director 1967-1978 of the Centre de Recherches sur les Macromolécules (C.R.M.). Ford Prize Award of the High Polymer Physics Section of the American Physical Society 1978. Witco Award of the Polymer Division, ACS, 1979. Dr. Honoris Causa Uppsal (1971), Aberdeen (1973), Lodz (Poland) (1977).

Sonja Krause, obtained her Ph.D. in Physical Chemistry at the University of California at Berkeley in 1957. She worked as a Senior Physical Chemist at the Rohm and Haas Company until 1964 when she joined the U.S. Peace Corps, teaching for two years in Universities in Nigeria and Ethiopia. She then taught for one year at the University of Southern California in Los Angeles and has been at Rensselaer Polytechnic Institute, presently as Professor of Physical Chemistry, since 1967. Her research interests include polymer-polymer miscibility, phase-separated polymer systems, small-angle scattering, and transient electric birefringence of macromolecules in solution.

Doctor ès-Sciences Strasbourg 1969 - Directeur de Recherches CNRS at the ICS(CRM) Strasbourg - Specialist Light and Neutron Scattering and on Solution and Bulk Polymers.

J.F. Joanny, Ph.D., 1985. Former student of Professor P.G. de Gennes. Worked on polymer solutions and wetting problems, both static and dynamic.

L. Leibler, Ph.D. in solid state physics, 1976. Post-doc at Professor P.G. de Gennes Lab., 1977-1978. Worked at Centre de Recherches sur les Macromolecules, Strasbourg, France, 1979-84, and at E.S.P.C.I. since 1984.

G.W. Slater, Ph.D. Université de Sherbrooke, Sherbrooke (Quebec), 1984. His supervisor for his Master degree was Dr. Alain Caillé, who post-docted with de Gennes in the 70's. He has been working on reptation problems with Dr. Jaan Noolandi for the last 3 years. This research is done in collaboration with the Biotechnology Research Institute, a National Research Council institute located in Montreal.

AUTHOR INDEX

C

D

E

F

G

H

I

J

K

L

Q

R

S

SUBJECT INDEX

D

N

O

P

S